Mathematical Techniques

An Introduction for the
Engineering, Physical, and
Mathematical Sciences

D. W. JORDAN
and
P. SMITH

Department of Mathematics
Keele University

Oxford   Toronto   Melbourne
OXFORD UNIVERSITY PRESS

Oxford University Press, Walton Street, Oxford OX2 6DP

Oxford   New York
Athens   Auckland   Bangkok   Bombay
Calcutta   Cape Town   Dar es Salaam   Delhi
Florence   Hong Kong   Istanbul   Karachi
Kuala Lumpur   Madras   Madrid   Melbourne
Mexico City   Nairobi   Paris   Singapore
Taipei   Tokyo   Toronto
and associated companies in
Berlin   Ibadan

Oxford is a trade mark of Oxford University Press

Published in the United States
by Oxford University Press Inc., New York

© D. W. Jordan and P. Smith, 1994

Reprinted with corrections 1994
Reprinted 1995

A catalogue record for this book is available from the British Library

Library of Congress Cataloging in Publication Data
Jordan, D. W. (Dominic William)
Mathematical techniques : an introduction for the engineering,
physical, and mathematical sciences / D. W. Jordan and P. Smith.
Includes index.
1. Mathematical analysis. I. Smith, Peter, 1935– . II. Title.
QA300.J82   1994   515—dc20   94-15961
ISBN 0 19 856268 3 (Hbk)
ISBN 0 19 856267 5 (Pbk)

Printed in Great Britain by Butler & Tanner Ltd, Frome

Preface

This book is a student text covering the mathematical techniques usually taught in the early stages of science and engineering degree courses. It also provides the groundwork of 'methods' needed by first-year mathematics specialists. The requirements of such students have influenced the content and presentation of the book in many ways suggested by the authors' long and continuous experience of teaching such courses to a variety of joint degree students at Keele University, including those who have only minimal entry-level background in mathematics. We have also tried to take account of the likely background of future students, who will probably have a more broadly based pre-university education.

There are two main strands in the subjects treated. The calculus sequence covers the theory and applications of differentiation, integration, differential equations, the phase plane, Laplace transforms, Fourier series, functions of more than one variable, multiple integrals, and line integrals. The algebra-like strand contains complex numbers, matrices, vectors, linear systems of equations, eigenvalues, sets, Boolean algebra, graph (network) theory, and difference equations. The later chapters are largely independent and partial choices of material can be made.

We have organized the book so as to enable students to use it with the minimum of guidance. The same features will help teachers to make selections from the book according to the length, emphasis, and prerequisities of different courses.

From Chapter 2 onwards, every subject treated starts from scratch. For example, it is not assumed that the reader necessarily knows any calculus, although most science and engineering students are likely to have had some previous encounter with the subject; if so, the text can be used for revision as well as extension of various topics. Most of the chapters are short (the average length is about 18 pages), and the sections within the chapters include not more than one or two new ideas. All the principal results are displayed in summary form in numbered and shaded 'boxes'. For revision purposes, or in desperate cases, progress could be made by attending only to the boxes.

The reader is encouraged to look out for some kind of geometrical or numerical reality to illustrate symbolic statements. Where possible we make use of graphical justifications; science students in particular should benefit by cultivating geometrical reasoning. Numerical methods for equation-solving, integration, and the solution of differential equations and systems are introduced at the points where they can effectively illustrate the main text, rather than being collected together in a separate chapter on numerical analysis.

Attempts to generate interest by using specialized examples from physics, chemistry, engineering, etc., are liable to misfire with many students, who have enough to do at first in grasping the underlying mathematical processes, and are confused by layers of scientific vocabulary and unfamiliar notations. We have therefore taken out into separate chapters certain technical applications such as the harmonic oscillator, phasors, and circuit analysis, and given these topics a fuller treatment. Certain other applications are confined to separate sections, so that they can be avoided if they do not suit a particular class. Most of the applications in the main text are drawn from common knowledge, or are such as can easily be understood.

The last chapter contains a collection of projects to introduce the use of symbolic computation and graphics manipulation, following the text chapter by chapter. A short first-year course at Keele University introducing mathematics students to *Mathematica*† software has proved successful with students who often have no previous experience of computing. Symbolic computation is a useful interactive facility for graph plotting and routine manipulations in, for example, linear algebra, differentiation, and integration, although it is no substitute for understanding methods and principles. Almost all the graphs of curves and surfaces in this book have their origins in *Mathematica* graphical outputs. Specific *Mathematica* programs which provide solutions for these projects can be obtained on disk at nominal cost by writing to the authors (see p. 601). Mostly they incorporate standard *Mathematica* commands, and they are readily adaptable to a variety of alternative inputs for specific functions, matrices, vectors, determinants, Fourier series, Laplace transforms, etc.

There are nearly 450 fully-worked examples in the book, a large number of exercises and problems including simple programming applications, over 100 projects for symbolic computation, and several Appendices consisting of tables of standard results for reference, as well as answers and hints to selected problems and projects.

---

† *Mathematica* is a registered trade mark of Wolfram Research Inc.

Confidence is half the battle in mathematics: it is very encouraging to learn how to do something so as to be able to get it nearly right most times. Continual practice, even at the point of repetitive drill, is the way to achieve this. The beneficial effects of practice can be obtained without using very complicated and difficult exercises, so on the whole we have avoided such problems.

We should like to thank John Bentin for numerous corrections, criticisms and suggestions, and to record our appreciation of the helpfulness of the staff of Oxford University Press throughout the writing and production of this book.

Keele   D.W.J.
January 1994   P.S.

Contents

1

Standard functions and techniques

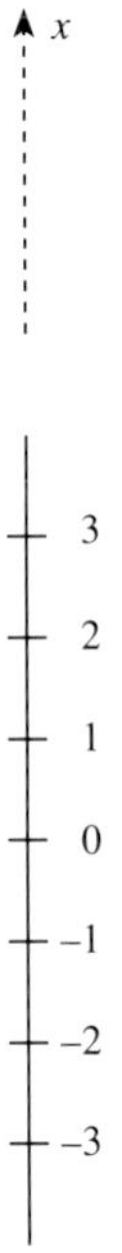

Fig. 1.1
The number line, or x axis.

1.1 Number line, intervals

Draw a straight line containing a point O, called the **origin**, and indicate a scale starting at O as in Fig. 1.1, with positive scale markings above O and negative markings below. Imagine the line to be infinitely long in both directions. This is called a **number line**, and every real number, positive or negative, has a place on it. We shall use x to denote a general number.

The **signs of inequality** $<$, $\leqslant$, $>$, $\geqslant$ have the following meanings:

$<$ 'is less than', $\leqslant$ 'is less than or is equal to';

$>$ 'is greater than', $\geqslant$ 'is greater than or is equal to'.

If we are given two numbers, then the one which is higher on the number line is the **greater** one. Therefore $-2 > -3$, $-3 < -2$, $3 > 0$, $-3 < 0$, and so on.

Obviously $2 < 3$. But it is also true that $2 \leqslant 3$, because 2 is certainly *either* less than *or* equal to 3. For similar reasons, all these are true: $1 = 1$, $1 \leqslant 1$, $1 \geqslant 1$.

A single piece, or a segment, of the number line is called an **interval**. The piece of the line between $x = 2$ and $x = 3$ which includes both the **end-points** $x = 2$ and 3 can be specified by the expression

'the interval $2 \leqslant x \leqslant 3$'

which means 'all the values of x between, and including, 2 and 3'. The interval $2 < x < 3$ means all values of x between 2 and 3, but excluding the end values. **Infinite intervals** can be expressed in two ways; for example, the interval

$$x \geqslant 2 \quad \text{or} \quad 2 \leqslant x < \infty$$

contains all the numbers x which are greater than or equal to 2.

The 'size' of a number is denoted by the symbol

$$|x| = \begin{cases} x & \text{if } x \geqslant 0, \\ -x & \text{if } x < 0, \end{cases} \tag{1.1}$$

which is called the **modulus** of x or mod x. Thus $|3| = 3$, $|-4| = 4$. We can use the modulus notation to define intervals. The inequality $|x| \leqslant 2$ defines the same interval as $-2 \leqslant x \leqslant 2$; $|x - 1| \leqslant 3$ is the same as $-3 \leqslant x - 1 \leqslant 3$ or $-2 \leqslant x \leqslant 4$.

If a is a negative number, then it has no real square root: $\sqrt{a}$ or $a^{\frac{1}{2}}$ means the number whose square is equal to a, and the square of a number must be positive. The same applies to $a^{\frac{1}{4}}$ and other indices which have an even denominator. Suppose that $a > 0$; then the signs $\sqrt{a}$ and $a^{\frac{1}{2}}$ always stand for the **positive number** whose square is a. If we want the negative square root, we must indicate it by a '$-$': the two solutions of $x^2 = a$ are $\pm\sqrt{a}$.

1.2 Coordinates in the plane

The location of a point in a plane can be specified in terms of **right-handed cartesian axes**, illustrated in Fig. 1.2. These are effectively two number lines at right angles, meeting at the common origin O. The position of a point is determined by two **coordinates** (x, y), as illustrated in Fig. 1.2a. They represent, in order, its **signed distances** from the y and x axes respectively, as read off on the axis scales. (Axes are called **right-handed** as is customary if, when we walk along the x axis in the direction of increasing x, the positive y axis is on the left. If you look at Fig. 1.2 in a mirror, you will see **left-handed axes**.)

Thus, the point A has **coordinates** $(1.2, 1)$, B has coordinates $(2.3, -2)$, etc. We refer to particular points using the notation $A : (1.2, 1)$ and $B : (2.3, -2)$. For a general point P represented by the coordinates (x, y), x is known as the **abscissa** and y the **ordinate** of P.

The distance of $P : (x, y)$ from the origin is OP; by Pythagoras' theorem,

$$OP = \sqrt{(OU^2 + UP^2)},$$

where U is the base of the perpendicular from P onto the x axis (Fig. 1.2b). If we put $OP = r$, then

$$r = \sqrt{(x^2 + y^2)}. \tag{1.2}$$

Note that **distances**, such as OP and r, are always non-negative numbers.

Similarly, for any two points $P_1 : (x_1, y_1)$ and $P_2 : (x_2, y_2)$ in the plane, the distance $P_1 P_2$ between them – see Fig. 1.2b – is given by

$$P_1 P_2 = \sqrt{[(x_1 - x_2)^2 + (y_1 - y_2)^2]}. \tag{1.3}$$

1.3 Straight lines and curves

If x and y are related by an equation, then this relation can be represented by a curve or curves in the (x, y)-plane known as the **graph** of the equation.

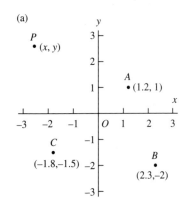

(a)

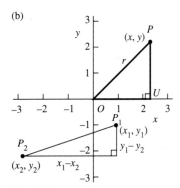

(b)

Fig. 1.2
Right-handed cartesian axes, x, y.

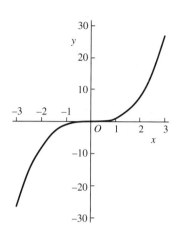

Fig. 1.3
Graph of $y = x^3$.

Example 1.1. Sketch the graph of $y = x^3$.

We decide over what interval of values of (say) x we wish to sketch the graph. Let $-3 \leqslant x \leqslant 3$. Construct a table of (x, y) values as shown below:

x	-3	-2	-1	0	1	2	3
y	-27	-8	-1	0	1	8	27

We then plot the points corresponding to this set of coordinates and draw a smooth curve through them, as shown in Fig. 1.3. The greater the number of values of x in the interval, the greater is the reliability of the graph. It is assumed that the curve has smooth or regular behaviour between consecutive plotted points.

Example 1.2. Find the equation of the straight line through the points $A:(1, 2)$ and $B:(-1, -1)$.

The line is shown in Fig. 1.4. Let $P:(x, y)$ be any point on the line. The triangles ABR and PBQ are similar, so that

$$\frac{PQ}{QB} = \frac{y + 1}{x + 1} = \frac{AR}{RB} = \frac{3}{2},$$

Therefore

$$2(y + 1) = 3(x + 1),$$

or

$$y = \tfrac{3}{2}x + \tfrac{1}{2}.$$

This represents the equation of the straight line through the points $(1, 2)$ and $(-1, -1)$.

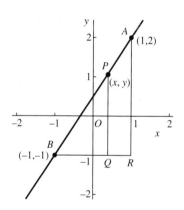

Fig. 1.4

Any equation of the form

$$y = mx + c$$

has a straight line graph, and any straight line can be expressed in this form unless it is parallel to the y axis, when its equation is

$$x = b.$$

The method of Example 1.2 can be used to find the equation of a line passing through any two given points $A:(x_1, y_1)$ and $B:(x_2, y_2)$ (see Fig. 1.5). Let $P:(x, y)$ be any point on the line. Then

$$\frac{PQ}{AQ} = \frac{BR}{AR},$$

or

$$\frac{y - y_1}{x - x_1} = \frac{y_2 - y_1}{x_2 - x_1}. \tag{1.4}$$

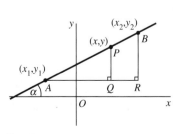

Fig. 1.5

This equation can be rearranged in the form $y = mx + c$, where

$$m = \frac{y_2 - y_1}{x_2 - x_1}, \qquad c = \frac{x_2 y_1 - x_1 y_2}{x_2 - x_1}. \qquad (1.5)$$

In the right-angled triangle ABR of Fig. 1.5, the angle α is given by

$$\tan \alpha = \frac{y_2 - y_1}{x_2 - x_1}, \qquad (1.6)$$

and this gives the standard measure of the **slope** or **gradient** of the straight line. The line slopes upwards or downwards from left to right according as $\tan \alpha$ is positive or negative respectively; the slope is zero when α or $\tan \alpha$ is zero; and the larger the size (or modulus – see (1.1)) of $\tan \alpha$, the steeper is the line. Also, according to (1.5), if the equation is in the form $y = mx + c$, then

$$\text{slope} = \tan \alpha = m. \qquad (1.7)$$

Finally, if we require the line through $A : (x_1, y_1)$ with given slope m, its equation is

$$\frac{y - y_1}{x - x_1} = m. \qquad (1.8)$$

The following result is often needed:

Perpendicular straight lines

The condition for the straight lines $y = m_1 x + c$ and $y = m_2 x + d$ to be perpendicular is $\qquad (1.9)$

$$m_1 m_2 = -1.$$

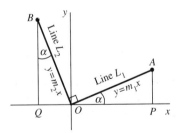

Fig. 1.6

This is proved as follows. First suppose that $m_1 m_2 = -1$; we have to deduce that the lines are perpendicular. Translate the two lines, so that their directions are unchanged, to meet at the origin O as in Fig. 1.6. The *angle of intersection is unaffected* by translation, and the equations become

$$\text{line } L_1 : \quad y = m_1 x; \qquad \text{line } L_2 : \quad y = m_2 x;$$

and $m_1 m_2 = -1$. Let $A : (a, b)$ be any point on L_1. Then $b = m_1 a$ so that $m_1 = b/a$. Since $m_1 m_2 = -1$, we have $m_2 = -a/b$; therefore the point $B : (-b, a)$ lies on L_2, as shown. Since the lines containing the right angles at P and Q have length a and b in each case, the triangles OPA and BQO are congruent, and the angles $\widehat{POA}$ and $\widehat{QBO}$ are corresponding angles. Put $\widehat{POA} = \alpha$; then

$$\widehat{POA} = \widehat{QBO} = \alpha.$$

Therefore $\widehat{QOB} = 90° - \alpha$, and finally

$$\widehat{AOB} = 180° - \alpha - (90° - \alpha) = 90°,$$

as required. The converse result, that if L_1 and L_2 are at right angles, then $m_1 m_2 = -1$, can be proved in a similar way.

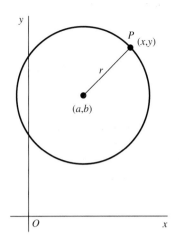

Fig. 1.7

A **circle** consists of all the points which are a constant distance from a given point. In Fig. 1.7, a circle has radius r, and its centre is at (a, b). The point $P : (x, y)$ represents any point on the circle. Equation (1.3) for the distance between two points gives

$$\sqrt{[(x - a)^2 + (y - b^2)]} = r.$$

Square this expression to get rid of the square root, and we have the **equation of a circle in its standard form**:

> **Equation of a circle, centre (a, b) and radius r**
>
> $$(x - a)^2 + (y - b)^2 = r^2 \qquad\qquad \textbf{(1.10)}$$

Example 1.3. Find the centre and radius of the circle

$$4x^2 + 4y^2 - 4x + 8y - 11 = 0. \qquad \textbf{(i)}$$

To convert (i) to the form (1.10), rewrite it in the form

$$x^2 - x + y^2 + 2y = \tfrac{11}{4}, \qquad \textbf{(ii)}$$

Take the terms involving x and reorganize them:

$$x^2 - x = (x - \tfrac{1}{2})^2 - \tfrac{1}{4}$$

(this process is used in many different contexts and is called **completing the square**). Treat the terms in y similarly:

$$y^2 + 2y = (y + 1)^2 - 1.$$

Replace the terms in (ii) by the new forms; we get

$$(x - \tfrac{1}{2})^2 - \tfrac{1}{4} + (y + 1)^2 - 1 = \tfrac{11}{4},$$

or

$$(x - \tfrac{1}{2})^2 + (y + 1)^2 = 4.$$

Therefore the centre is at $(\tfrac{1}{2}, -1)$, and the radius is 2.

Notice that (1.10) implies that, if we are given an equation $ax^2 + by^2 + cx + dy + e = 0$, it can only represent a circle if $a = b$. (The equation might not represent anything, as with $x^2 + y^2 + 1 = 0$, but if it does, it will be a circle.)

Figure 1.8 shows other important types of second-degree curve.

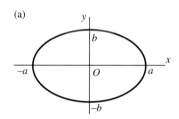

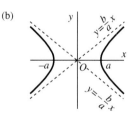

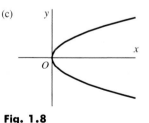

Fig. 1.8
(a) An ellipse
$x^2/a^2 + y^2/b^2 = 1$.
(b) A hyperbola
$x^2/a^2 - y^2/b^2 = 1$.
(c) A parabola $y^2 = x$.

1.4 Functions

The area A of a circle depends on its radius r, and the dependence is expressed in the **formula** $A = \pi r^2$. In general, suppose that the values of a certain **independent variable** x, say, determine values of a **dependent variable** y in such a way that if a numerical value of x is given, a single value of y is determined. Then we say that y is a **function** of x, and write for example,

$$y = f(x), \qquad y = g(x),$$

and so on, where the letter f, g, etc. can be used to distinguish different forms of dependence, which can be thought of pictorially in terms of different graphs. The letter f, g, and so on, standing alone, need not be associated with a *formula* in the usual sense. They can stand for any rule, programme, or calculation process which produces a definite single value for y when we offer a number x to it. A function can be thought of as an input-output device as in Fig. 1.9.

Functions can also be defined **implicitly** by means of formulae. For example

$$x^2 + y^2 = 1$$

represents a circle, centre the origin and radius 1. But if we sort out y from the relation we obtain $y = \pm(1 - x^2)^{\frac{1}{2}}$, which is not a single function, but two separate, single-valued, functions:

$$y = (1 - x^2)^{\frac{1}{2}} \quad \text{and} \quad y = -(1 - x^2)^{\frac{1}{2}},$$

representing the upper and lower semicircles respectively.

The following result is required several times in the rest of the book. Suppose that c is a positive constant, and that we are given a function f, with graph $y = f(x)$. The graph

$$y = f(x - c)$$

is exactly the same as that of $f(x)$, except that it is moved, or **translated a distance c to the right** along the x axis. There is a similar result for $f(x + c)$, the movement being to the left. Therefore

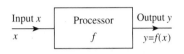

Input x Processor   Output y
x f $y=f(x)$

Fig. 1.9

Translation of $y = f(x)$ along the x axis

Let c be a positive number, and f any function. Then
$$y = f(x - c) \quad \text{and} \quad y = f(x + c) \qquad (1.11)$$
represent translation of $y = f(x)$ a distance c along the x axis to the right and left respectively.

It is useful to have terms in which symmetry of a graph can be described. For example, the graphs of $y = x^2$ and $y = \cos x$ are symmetrical about the y axis; the two halves are reflections of each other. Such functions are called **even functions**. On the other hand $y = x^3$ and $y = \sin x$ are antisymmetrical about the y axis; they are called **odd functions** (see also Section 13.9). The corresponding algebraic properties are

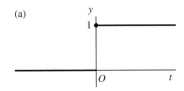

(a)

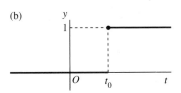

(b)

Fig. 1.10
(a) Graph of H(t). (b) Graph of H($t - t_0$).

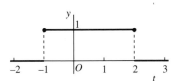

Fig. 1.11

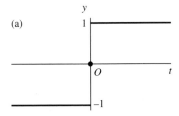

(a)

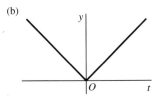

(b)

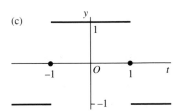

(c)

Fig. 1.12
(a) $y = \text{sgn}(t)$. (b) $y = t\,\text{sgn}(t)$.
(c) $y = \text{sgn}(1 - t^2)$.

Even and odd functions

(a) $f(x)$ is **even** if $f(-x) = f(x)$, **(1.12)**

(b) $f(x)$ is **odd** if $f(-x) = -f(x)$.

For example, in plotting $y = f(x) = x^3$ in Example 1.1, we did not really have to calculate x^3 for negative values of x. All that was necessary was to notice that x^3 is an odd function; $(-x)^3 = -(x)^3$; and this gives the table for negative x by changing the sign of the entries for x positive.

Some functions of practical significance have graphs that are not entirely smooth. For example, we may wish to model a device that is turned on at a given time, being quiescent before that time but active afterwards. A sudden change in the state of the device can be represented by a function that has a **jump** or **discontinuity** in its graph at the critical moment. The basic building block for functions with a jump is the **unit step function** H(t) (also known as the Heaviside function after its inventor, and sometimes denoted by U(t)) which we shall define by

$$H(t) = \begin{cases} 0 & \text{when } t < 0, \\ 1 & \text{when } t \geqslant 0; \end{cases} \quad \textbf{(1.13)}$$

(see Fig. 1.10a). If switch-on is required at $t = t_0$ then we can use

$$H(t - t_0) = \begin{cases} 0 & \text{when } t < t_0, \\ 1 & \text{when } t \geqslant t_0, \end{cases}$$

shown in Fig. 1.10b: it is the same graph translated to the right a distance t_0, by (1.11).

Example 1.4. Sketch the graph of $f(t) = H(2 - t) + H(t + 1) - 1$.
For $t < -1$, $f(t) = 1 + 0 - 1 = 0$. For $-1 \leqslant t < 2$, $f(t) = 1 + 1 - 1 = 1$. For $t > 2$, $f(t) = 0 + 1 - 1 = 0$. The graph is shown in Fig. 1.11.

The function denoted by sgn, given by

$$\text{sgn } t = H(t) - H(-t) = \begin{cases} -1 & \text{when } t < 0, \\ 0 & \text{when } t = 0, \\ 1 & \text{when } t > 0; \end{cases}$$

is called the **signum function** ('signum' means 'sign'). H(t) and sgn t can be used along with other functions to produce a variety of functions having discontinuities in either value or direction at assigned points. Figure 1.12 shows some examples.

1.5 Radian measure of angles

For everyday purposes angles are measured in degrees, so we are still following the Babylonian practice of dividing the circle into 360 sectors. For mathematical purposes, a less arbitrary measure is desirable. The absolute unit is the **radian**, which represents about 57°. The special properties which makes the unit valuable is its connection with length. Figure 1.13 shows a circle of radius R with a sector AOB containing an angle θ. The length of the arc $\overset{\frown}{AB}$ is obviously *proportional* to R, and it is proportional to θ whatever the angular units, so it is *proportional* to the product $R\theta$. We shall look for a unit of angle such that $\overset{\frown}{AB}$ **is numerically equal to** $R\theta$.

This requirement must include

$$\overset{\frown}{AB} = R\theta$$

for the *whole circle*. But, for the whole circle,

$$\overset{\frown}{AB} = 2\pi R.$$

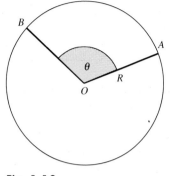

Fig. 1.13

Therefore the radian measure of angle for the *whole circle* must be

$$\theta = 2\pi.$$

Since the whole circle measures 360 in degrees, it must be that

$$360° = 2\pi \text{ radians,}$$

or   1 radian = $180/\pi$ degrees (about 57°).

The following table summarizes some useful information:

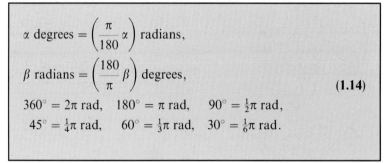

$$\alpha \text{ degrees} = \left(\frac{\pi}{180}\alpha\right) \text{ radians,}$$

$$\beta \text{ radians} = \left(\frac{180}{\pi}\beta\right) \text{ degrees,}$$

$$360° = 2\pi \text{ rad,} \quad 180° = \pi \text{ rad,} \quad 90° = \tfrac{1}{2}\pi \text{ rad,}$$
$$45° = \tfrac{1}{4}\pi \text{ rad,} \quad 60° = \tfrac{1}{3}\pi \text{ rad,} \quad 30° = \tfrac{1}{6}\pi \text{ rad.}$$

(1.14)

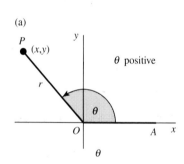

(a)

θ positive

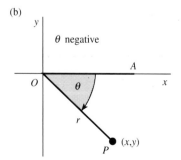

(b)

θ negative

Fig. 1.14

1.6 Trigonometric functions

We assume that the reader knows the meaning of *sine*, *cosine*, and *tangent* in ordinary trigonometry. We shall extend the meaning to incorporate angles larger than 90°, and negative angles.

In Fig. 1.14, consider any point $P : (x, y)$, with

$$\widehat{AOP} = \theta, \qquad OP = r > 0.$$

The angle θ is to be measured **from the positive x axis,** in the **anticlockwise direction if θ is positive**, and in a clockwise direction if θ is negative. The angle θ may take any value, however large: positive, negative, or zero. The quantity r **is always positive or zero**. Then the trigonometric functions are defined as follows for positive r:

(a)

(b)

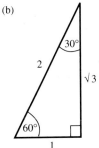

Fig. 1.15
(a) $\sin 45° = \cos 45° = 1/\sqrt{2}$.
(b) $\sin 60° = \cos 30° = \sqrt{3}/2$.
$\cos 60° = \sin 30° = 1/2$.

Trigonometric functions for any angle

$$\cos\theta = \frac{x}{r}, \quad \sin\theta = \frac{y}{r}, \quad \tan\theta = \frac{\sin\theta}{\cos\theta} = \frac{y}{x}, \qquad (1.15)$$

$$\sec\theta = 1/\cos\theta, \quad \csc\theta = 1/\sin\theta, \quad \cot\theta = 1/\tan\theta.$$

If r and θ are specified, they locate a point P in the plane just as well as x and y do. They therefore constitute an alternative pair of coordinates, called **polar coordinates**.

If on the other hand a point P is specified, as in Fig. 1.14, there is a definite value of r associated with it, but there are many values of θ that will do. In Fig. 1.14a, for example, the value of θ shown is the one we are likely to adopt. But we are brought to the same point P if we take the angular coordinate to be (in radians)

$$\theta \pm 2\pi, \quad \theta \pm 4\pi, \quad \text{etc.},$$

because every increase or decrease of 2π merely represents a complete revolution. This ambiguity will be dealt with in the book as it is encountered. To sum up: from Fig. 1.14:

Polar and cartesian coordinates.

$$x = r\cos\theta, \qquad y = r\sin\theta; \qquad (1.16)$$

$$r = (x^2 + y^2)^{\frac{1}{2}}, \qquad \tan\theta = y/x.$$

(a)

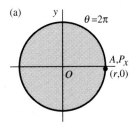

(b)

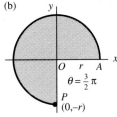

(c)

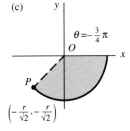

Fig. 1.16

Example 1.5. Obtain (a) $\sin\frac{1}{3}\pi$, (b) $\tan\frac{1}{6}\pi$, (c) $\cos\frac{1}{4}\pi$. (The angles are in radians.)

From (1.14) we obtain the values of the angles in degrees. Use the familiar triangles in Fig. 1.15.

(a) $\sin\frac{1}{3}\pi = \sin 60° = \sqrt{3}/2$. (b) $\tan\frac{1}{6}\pi = \tan 30° = 1/\sqrt{3}$. (c) $\cos\frac{1}{4}\pi = \cos 45° = 1/\sqrt{2}$.

Example 1.6. Obtain (a) $\cos 2\pi$, (b) $\sin\frac{3}{2}\pi$, (c) $\sin(-\frac{3}{4}\pi)$.

From Fig. 1.16, and eqn (1.15):

(a) P is $(r, 0)$, so $\cos 2\pi = x/r = r/r = 1$;
(b) P is $(0, -r)$, so $\sin\frac{3}{2}\pi = y/r = -1$;
(c) P is $(-r/\sqrt{2}, -r/\sqrt{2})$, so $\sin(-\frac{3}{4}\pi) = y/r = -1/\sqrt{2}$.

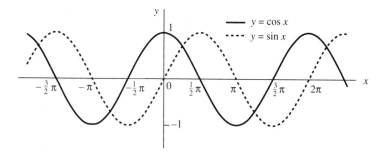

Fig. 1.17

(a)
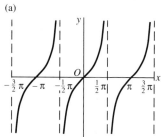

(b)
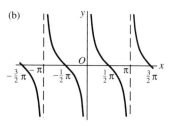

(c)
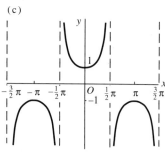

(d)
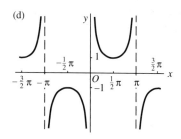

Fig. 1.18
(a) $y = \tan x$. (b) $y = \cot x$.
(c) $y = \sec x$. (d) $y = \csc x$.

The graphs $y = \cos x$ and $y = \sin x$ are shown in Fig. 1.17. The following points should be noticed.

1. $y = \cos x$ and $y = \sin x$ have identical shape, but are displaced a distance $\frac{1}{2}\pi$ (radians) from each other: thus (see (1.11))

$$\sin x = \cos(x - \tfrac{1}{2}\pi), \qquad \cos x = \sin(x + \tfrac{1}{2}\pi). \tag{1.17}$$

2. The curves repeat themselves after every interval of length 2π. This is evident from the definition (1.15), because 2π radians is a complete revolution in Fig. 1.14, and therefore delivers the same point $P : (x, y)$. The function $\cos x$ and $\sin x$ are therefore called **periodic functions** with **period** (or **wavelength**) equal to 2π (see also Section 18.1).

3. $y = \cos x$ is **symmetrical** (or **even**) about the y axis; $y = \sin x$ is **antisymmetrical** (or **odd**) about the y axis (see also Section 13.9).

4. They swing between ± 1 in every complete period of length 2π.

There are many trigonometric identities in common use; the following are the most important (a more extensive list can be found in Appendix B).

Trigonometric identities

(a) Sums of angles:

$$\sin(A \pm B) = \sin A \cos B \pm \cos A \sin B,$$
$$\cos(A \pm B) = \cos A \cos B \mp \sin A \sin B,$$
$$\tan(A \pm B) = \frac{\tan A \pm \tan B}{1 \pm \tan A \tan B}.$$

(b) Products as sums:

$$\cos A \cos B = \tfrac{1}{2}[\cos(A + B) + \cos(A - B)];$$
$$\cos A \sin B = \tfrac{1}{2}[\sin(A + B) - \sin(A - B)];$$
$$\sin A \sin B = \tfrac{1}{2}[-\cos(A + B) + \cos(A - B)].$$

(c) $\cos^2 A = \tfrac{1}{2}(1 + \cos 2A);$
$\sin^2 A = \tfrac{1}{2}(1 - \cos 2A).$

$$\tag{1.18}$$

Graphs of tan x, cot x, sec x, and csc x are shown in Fig. 1.18.
A function of the form

$$c \cos(\omega x + \phi),$$

where c, ω (Greek omega), and ϕ (Greek phi) are constants, is called a **harmonic function** or a sinusoidal function. It includes sine functions, by virtue of (1.17). Any function of the form

$$a \cos \omega x + b \sin \omega x$$

can be put into this form. Numbers c and ϕ can be found graphically by plotting the point $P : (a, b)$ in cartesian coordinates as in Fig. 1.19. Let $c = (a^2 + b^2)^{\frac{1}{2}} > 0$ be its distance from O, and choose ϕ as **minus the polar angle** θ: $\phi = -\theta$. Then

$$a \cos \omega x + b \sin \omega x = c \cos(-\phi) \cos \omega x + c \sin(-\phi) \sin \omega x$$
$$= c \cos \phi \cos \omega x - c \sin \phi \sin \omega x$$
$$= c \cos(\omega x + \phi),$$

after using (1.18a). Therefore we have the following result.

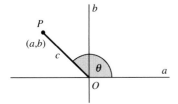

Fig. 1.19

An important identity.

$$a \cos \omega x + b \sin \omega x = c \cos(\omega x + \phi),$$

where the point (a, b) has polar coordinates

$$r = c \;(>0) \quad \text{and} \quad \theta = -\phi.$$

(1.19)

A typical graph of $y = c \cos(\omega x + \phi)$ is shown in Fig. 1.20. The **wavelength**, or the **period** if the variable stands for time, is $2\pi/\omega$. The positive number c is called the **amplitude**, ω is the **angular frequency** in radians per second. The **frequency** in cycles per second is $\omega/2\pi$, and ϕ is called the **phase** angle. More detail is given in Chapters 18 and 19.

A general function f is said to be **periodic with period** p if

$$f(x + p) = f(x)$$

for every value of x. It then repeats itself in any interval of length p. If p is a period, so obviously are $2p$, $3p$, and so on. The period usually meant when we say a function has period p is the smallest period. Thus in Fig. 1.18, tan x and cot x have period π.

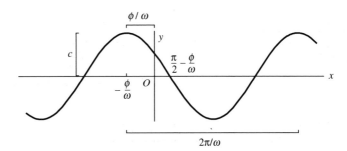

Fig. 1.20

1.7 Inverse functions

Figure 1.21a shows the graph of $y = x^3$, and x and y have the *same scales*. Choose a number a. To find a^3, locate $x = a$ at A and follow the track ABC. Naturally, the point C on the y axis represents a^3.

Now suppose we have a number b and want to find $b^{\frac{1}{3}}$. Read the graph backwards: find $y = b$ at U on the y axis and follow the track UVW. Then W represents $b^{\frac{1}{3}}$. Therefore exactly the same curve can be used to find both cubes and cube roots of numbers:

$$x = y^{\frac{1}{3}} \quad \text{is the same curve as} \quad y = x^3.$$

It order to find cube roots, we might prefer to have a separate graph which can be read off in the ordinary way: from the x axis to the y axis. We can do this by changing round the labels x and y as in Fig. 1.21b. To turn it into a standard graph, with x horizontal and y vertical, simple flip the figure over, around a diagonal axis as shown in Fig. 1.21c: the appropriate axis is the $45°$ line $y = x$, since the x and y scales are the same. The resulting graph is the graph of the **inverse function** of $y = x^3$, namely $y = x^{\frac{1}{3}}$. In the end we have simply **reflected the graph in the line** $y = x$ to obtain the graph of the inverse, as shown in Fig. 1.21b. (By the same token, the inverse function for $y = x^{\frac{1}{3}}$ is $y = x^3$).

Back to Fig. 1.21a, follow the sequence of operations indicated by the path $ABCBA$ from the point $A : (a, 0)$. In following ABC we find a^3 at C, then on CBA we find its cube root, and are back to a again; algebraically

$$(a^3)^{\frac{1}{3}} = a;$$

and similarly

$$(b^{\frac{1}{3}})^3 = b$$

by considering the path $UVWVU$.

These two properties are perfectly general for functions and their inverses: when applied successively, a function f and its inverse f_I neutralize each other and return the original value. This is the **inverse function property**:

$$f_I\big(f(x)\big) = x, \qquad f\big(f_I(x)\big) = x. \tag{1.20}$$

We can consider other powers and their inverses, such as x^2 and $x^{\frac{1}{2}}$ However, no negative number has a real square root, and it is only exceptionally true that negative numbers permit fractional powers. In Fig. 1.22 we show the general character of positive whole-number powers x^n and their inverses $x^{1/n}$ for $0 \leqslant x \leqslant 1$. Notice the symmetry of the inverse pairs $(x^n, x^{1/n})$ about the $45°$ line $y = x^1 = x$. (The graphs of fractional powers such as $x^{\frac{5}{2}}$ lie between the whole-number graphs in a regular way.)

(a)

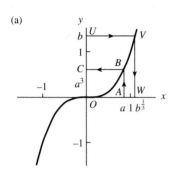

(b)

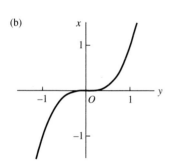

(c)

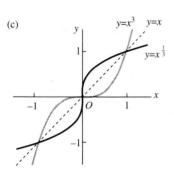

Fig. 1.21

Now consider the inverse problem for the function

$$y = \sin x.$$

The inverse function would answer the question: what angle has its sine equal to x? Evidently $-1 \leqslant x \leqslant 1$, or there is no such angle. Moreover if $-1 \leqslant x \leqslant 1$, then there is an infinite number of angles, not just one (for example, $\sin \frac{1}{6}\pi = \frac{1}{2}$, but $\sin \frac{13}{6}\pi$ and $\sin \frac{5}{6}\pi$ also equal $\frac{1}{2}$). The universal convention in use for tables, computers and calculators is to offer only the **smallest angle in magnitude**, positive or negative. This is done by **restricting** the angle to the range $-\frac{1}{2}\pi$ to $\frac{1}{2}\pi$ (in radians). The resulting function inverse to $y = \sin x$ is denoted by arcsin (the notation $\sin^{-1}$ is also widely used):

$$y = \arcsin x, \qquad y \text{ restricted to } -\tfrac{1}{2}\pi \leqslant y \leqslant \tfrac{1}{2}\pi.$$

Its graph is the reflection of $y = \sin x$ in the line $y = x$ as shown in Fig. 1.23a (the x and y scales are the same). Also displayed are the inverse functions for the cosine and tangent, denoted by arccos x and arctan x respectively. The angular range is restricted in a similar way (with a different range for cos x) so as to make the inverse functions single-valued.

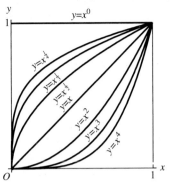

Fig. 1.22

Fig. 1.23
(a) $y = \arcsin x$, $-1 \leqslant x \leqslant 1$, $-\frac{1}{2}\pi \leqslant y \leqslant \frac{1}{2}\pi$. (b) $y = \arccos x$, $-1 \leqslant x \leqslant 1$, $0 \leqslant y \leqslant \pi$. (c) $y = \arctan x$, any x, $-\frac{1}{2}\pi < y < \frac{1}{2}\pi$.

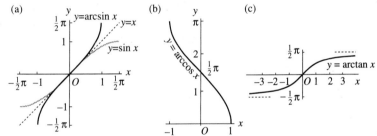

The various inverse functions are connected, as in the following example.

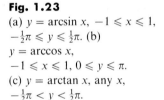

(a)

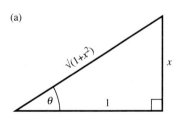

(b)

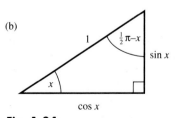

Fig. 1.24
$\cos(\arctan x) = \cos \theta$
$= 1/\sqrt{(1 + x^2)}$.

Example 1.7. Simplify (a) cos(arctan x), (b) arcsin(cos x)

(a) In the triangle of Fig. 1.24a, $\tan \theta = x$. Therefore $\theta = \arctan x$. The hypotenuse has length $\sqrt{(1 + x^2)}$, so that $\cos \theta = 1/\sqrt{(1 + x^2)}$
(b) In the triangle in Fig. 1.24b, the angles are as indicated, and cos x is correctly given. But

$$\sin(\tfrac{1}{2}\pi - x) = \cos x,$$

so that

$$\arcsin(\cos x) = \tfrac{1}{2}\pi - x.$$

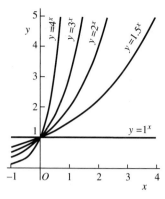

Fig. 1.25

1.8 Exponential functions

A class of functions not so far considered are of the form

$$y = a^x, \qquad a > 0$$

(we must have $a > 0$ since, if a is negative, the power a^x only exists for occasional values of x: consider the case $a^{\frac{1}{2}}$). Several instances are shown in Fig. 1.25. If $a > 1$, then $a^x \to \infty$ as $x \to \infty$, and $a^x \to 0$ as $x \to -\infty$. All pass through the point $(0, 1)$. (Cases where $a < 1$ can be considered by putting, for example, $y = (\frac{1}{5})^x = 1/5^x$.)

The number a is called the **base** for the **exponential function** a^x. Functions with different bases are all closely related. For example, $4^x = 2^{2x}$; and it can be confirmed by using a calculator that, to four decimals, $2^x = 3.5^{0.5533x}$, and so on. In Example 1.8, we show that all the curves $y = a^x$ amount to **the same curve plotted with a different x scale**, so we really need only one base to describe them all. This special base is denoted by the letter e, which occurs in mathematics as often as does π.

The numerical value of e can be obtained by requiring that **the rate of growth of e^x at any value of x is equal to** e^x. Choose any value of x (see Fig. 1.26), and another value $x + h$ very close to it. The rate of growth of e^x at P is nearly equal to NQ/PN and becomes more accurate the smaller the value of h. Thus

$$\frac{NQ}{PN} = \frac{e^{x+h} - e^x}{h} \approx e^x.$$

Since $e^{x+h} = e^x e^h$, we can cancel e^x and obtain the condition

$$\frac{e^h - 1}{h} \approx 1 \quad \text{or} \quad e^h \approx 1 + h,$$

which is equivalent to

$$e \approx (1 + h)^{1/h},$$

the approximation becoming more accurate, the smaller h is. The following table, which was calculated and can be extended on a hand calculator, shows how the value of e is approached as h is taken smaller and smaller

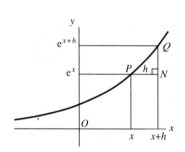

Fig. 1.26

h	0.1	0.01	0.001	0.0001	...
$(1 + h)^{1/h}$ ($\approx$ e)	2.5937...	2.7048...	2.7169...	2.7181...	...

To seven decimal places, the value of e is given by

$$e = 2.7182818 \ldots.$$

The function has a growth rate at $x = 0$ of $e^0 = 1$; that is to say the **graph of $y = e^x$ cuts the y axis at 45°**, provided that x and y have the same scale.

The function

$$y = e^x \quad \text{or} \quad y = \exp(x)$$

is the basic **exponential function**. Its rate of increase is very rapid, as indicated by the following table (to 2 significant figures).

x	-9	-6	-3	0	3	6	9
e^x	1.25×10^{-4}	2.5×10^{-3}	5.0×10^{-2}	1	2.0×10^{1}	4.0×10^{2}	8.1×10^{3}

When x increases by a step of 3, e^x is **multiplied** by a factor of about 20.

1.9 The logarithmic function

The inverse function corresponding to the exponential function $y = e^x$ is called the **logarithm** of x (historically the 'natural' logarithm):

$$y = \ln x \quad \text{(or sometimes } \log_e x \text{)},$$

read as 'log of x'. It answers the typical question $e^? = x$, or alternatively it solves the equation for y:

$$e^y = x.$$

For example, the equation

$$e^y = 3$$

has the solution

$$y = \ln 3.$$

This can be confirmed by using a scientific calculator, on which e^x and $\ln x$ are always present: $\ln 3 = 1.098$ and $e^{1.098} = 3$. The graphs $y = e^x$ and $y = \ln x$ are shown in Fig. 1.27, but with different x and y scales. Note that there exists **no logarithm of a negative number**. The logarithm has the following properties.

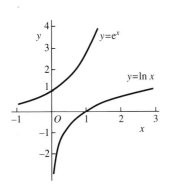

Fig. 1.27

$$e^{\ln x} = x \quad \text{if } x > 0, \qquad \ln e^x = x \quad \text{for any } x \qquad \textbf{(1.21)}$$

(since $\ln x$ and e^x are inverse functions).

$$\ln 1 = 0, \qquad \ln e = 1; \qquad \textbf{(1.22)}$$

(since $e^0 = 1$ and $e^1 = e$).

From (1.21), $e^{\ln a} = a$. $e^{\ln b} = b$, and $e^{\ln ab} = ab$. Therefore

$$ab = e^{\ln ab} = e^{\ln a} e^{\ln b} = e^{\ln a + \ln b},$$

from which it follows that

$$\ln ab = \ln a + \ln b, \qquad \ln \frac{a}{b} = \ln a - \ln b. \qquad \textbf{(1.23)}$$

(To obtain the second result, use the result $(a/b)b = a$.)

From (1.21), $e^{\ln a} = a$; so, if m is any number, then
$$a^m = e^{\ln(a^m)}; \text{ and } a^m = (e^{\ln a})^m = e^{m \ln a},$$
Therefore $e^{\ln a^m} = e^{m \ln a}$, or

$$\ln a^m = m \ln a. \tag{1.24}$$

Example 1.8. Prove that if a is any positive number, then $a^x = e^{x \ln a}$.
(This result shows that the curves $y = 2^x$, $y = 3^x$, etc. become identical if the scale of the x axis is changed by a factor $\ln 2$, $\ln 3$, etc.)
$$a^x = (e^{\ln a})^x = e^{x \ln a},$$
by the ordinary rule for powers.

Example 1.9. Find x in the equation $2^{3x} = 2(3^x)$.
If x is the solution then it is also true that
$$\ln 2^{3x} = \ln 2(3^x),$$
or
$$3x \ln 2 = \ln 2 + x \ln 3$$
(using (1.24) and (1.23)), or
$$x = \ln 2/(3 \ln 2 - \ln 3).$$

Example 1.10. Find y in terms of x, if $\ln y = 3 \ln x + 2$.
If y is the right function of x, then it is also true that
$$y = e^{\ln y} = e^{3 \ln x + 2}$$
or
$$y = e^{3 \ln x} e^2 = (e^{\ln x})^3 e^2,$$
or
$$y = e^2 x^3.$$

There exist logarithms using bases other than e, which have occasional use. The properties are analogous to those involving e.

1.10 Exponential growth and decay

Here we shall use t (for time) in place of x, and consider the function
$$y = A e^{ct},$$
where A and c are constants. These include functions such as 2^t. By Example 1.8, these can be expressed in the form e^{ct} since
$$2^t = e^{t \ln 2} = e^{0.69 \ldots t}.$$

If $c > 0$, then y is said to have **exponential growth**. To obtain an idea of what this entails we consider the **doubling period** of $y = A\, e^{ct}\ (c > 0)$. Choose *any* moment of time t_0. At some later time $t_0 + T$, y will have doubled, that is to say,

$$A\, e^{c(t_0 + T)} = 2A\, e^{ct_0}, \quad \text{or} \quad A\, e^{ct_0}\, e^{cT} = 2A\, e^{ct_0}.$$

After cancelling e^{ct_0}, we have an equation for T:

$$e^{cT} = 2.$$

Therefore $cT = \ln 2$, or

$$T = \frac{1}{c}\ln 2.$$

Exponential doubling principle

$y = A\, e^{ct}$ doubles in every interval of length $T = \dfrac{1}{c}\ln 2$ $\qquad$ **(1.25)**

Example 1.11. The number N of scientists and engineers in the U.S.A. doubled every 10 years between 1900 and 1935, and in 1935 they numbered about 1.5×10^5. This suggests exponential growth $N = A\, e^{ct}$. Find c, and predict the number N for 1990 on the assumption that the trend continued.

Suppose that we count 1900 as $t = 0$. The doubling period is 10 years; so $N = A\, e^{ct}$, where, by (1.25),

$$c = \tfrac{1}{10}\ln 2 = 0.0693.$$

Thus

$$N = A\, e^{0.0693t}.$$

In 1935, where $t = 35$ (years), $N = 1.5 \times 10^5$, so that

$$1.5 \times 10^5 = A\, e^{0.0693 \times 35},$$

or $\quad A = 13265.$

Therefore

$$N = 13265\, e^{0.0693t}.$$

In 1990 $t = 90$, from which it follows that

$$N = 6.8 \times 10^6.$$

Exponential growth occurs when a quantity increases at a rate proportional to the amount already accumulated. In the short term, animal populations, epidemics, and explosions have this characteristic. **Exponential decay** may also occur. If c is a positive number and

$$y = A\, e^{-ct} = A/e^{ct},$$

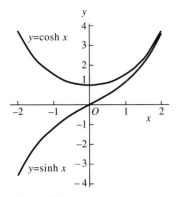

Fig. 1.28
Graphs of the hyperbolic
functions cosh x and sinh x.

then y *halves* itself in *every* interval of length $(1/c)\ln 2$. This occurs in radioactive decay, the period being called the half-life period of a radioactive substance. The half-life period provides a convenient, memorable measure of the time it would take for the substance to become less harmful.

1.11 Hyperbolic functions

It is often convenient to represent certain combinations of exponential functions by separate functions. The **hyperbolic cosine** and **hyperbolic sine** functions, denoted by cosh and sinh respectively, are defined by the following formulae.

> **Hyperbolic functions**
> $$\cosh x = \tfrac{1}{2}(e^x + e^{-x}), \qquad \sinh x = \tfrac{1}{2}(e^x - e^{-x}). \tag{1.26}$$

Since

$$\cosh(-x) = \tfrac{1}{2}(e^{-x} + e^{-(-x)}) = \tfrac{1}{2}(e^{-x} + e^x) = \cosh x,$$

it follows that the graph of cosh x is symmetrical about the y axis. By a similar argument, it can be shown that sinh x is antisymmetrical about the y axis. Graphs of the two functions are shown in Fig. 1.28.
From the definitions (1.26)

$$\cosh x + \sinh x = e^x, \qquad \cosh x - \sinh x = e^{-x}.$$

The remaining hyperbolic functions are defined in a similar manner to their trigonometric counterparts. Thus

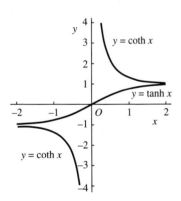

> $$\tanh x = \frac{\sinh x}{\cosh x}, \qquad \coth x = \frac{\cosh x}{\sinh x},$$
> $$\operatorname{sech} x = \frac{1}{\cosh x}, \qquad \operatorname{cosech} x = \frac{1}{\sinh x}. \tag{1.27}$$

Graphs of tanh x, coth x, sech x, and cosech x are shown in Fig. 1.29.
From the definitions, a number of identities follow which parallel those for trigonometric functions but with important sign differences. Some are derived below.

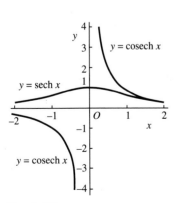

Fig. 1.29.

> (a) $\cosh^2 x + \sinh^2 x = \cosh 2x,$
> (b) $\cosh^2 x - \sinh^2 x = 1.$ $\tag{1.28}$

For (a):

$$\cosh^2 x + \sinh^2 x = \tfrac{1}{4}(e^{2x} + 2 + e^{-2x}) + \tfrac{1}{4}(e^{2x} - 2 + e^{-2x})$$
$$= \tfrac{1}{2}(e^{2x} + e^{-2x}) = \cosh 2x.$$

For (b):

$$\cosh^2 x - \sinh^2 x = \tfrac{1}{4}(e^{2x} + 2 + e^{-2x}) - \tfrac{1}{4}(e^{2x} - 2 + e^{-2x}) = 1.$$

To obtain the identity

$$\sinh(x_1 + x_2) = \sinh x_1 \cosh x_2 + \cosh x_1 \sinh x_2,$$

start with the right-hand side:

$$\sinh x_1 \cosh x_2 + \cosh x_1 \sinh x_2$$
$$= \tfrac{1}{2}(e^{x_1} - e^{-x_1})\tfrac{1}{2}(e^{x_2} + e^{-x_2}) + \tfrac{1}{2}(e^{x_1} + e^{-x_1})\tfrac{1}{2}(e^{x_2} - e^{-x_2})$$
$$= \tfrac{1}{4}(e^{x_1 + x_2} - e^{-x_1 + x_2} + e^{x_1 - x_2} - e^{-x_1 - x_2})$$
$$+ \tfrac{1}{4}(e^{x_1 + x_2} + e^{-x_1 + x_2} - e^{x_1 - x_2} - e^{-x_1 - x_2})$$
$$= \tfrac{1}{2}(e^{x_1 + x_2} - e^{-x_1 - x_2}) = \sinh(x_1 + x_2).$$

To sum up similar identities:

$$\sinh(x_1 \pm x_2) = \sinh x_1 \cosh x_2 \pm \cosh x_1 \sinh x_2,$$
$$\cosh(x_1 \pm x_2) = \cosh x_1 \cosh x_2 \pm \sinh x_1 \sinh x_2, \qquad \textbf{(1.29)}$$
$$\tanh(x_1 \pm x_2) = \frac{\tanh x_1 \pm \tanh x_2}{1 \pm \tanh x_1 \tanh x_2}.$$

The inverse hyperbolic functions are defined in a fairly obvious way. Thus

$$y = \sinh^{-1} x \quad \text{(for all } x),$$
$$y = \cosh^{-1} x \quad \text{(for } x \geqslant 1), \qquad \textbf{(1.30)}$$
$$y = \tanh^{-1} x \quad \text{(for } -1 < x < 1).$$

Traditionally the index -1 is used to symbolize the inverses in (1.30). Do not mistake it as meaning a reciprocal: $\sinh^{-1} x$ **does not mean** $1/\sinh x$.

Inverse hyperbolic functions have alternative representations in terms of logarithms. Since

$$x = \cosh y = \tfrac{1}{2}(e^y + e^{-y}),$$

it follows that y satisfies

$$e^{2y} - 2x\, e^y + 1 = 0,$$

which is a quadratic equation in e^y. Its solutions are

$$e^y = \tfrac{1}{2}[2x \pm \sqrt{(4x^2 - 4)}] = x \pm \sqrt{(x^2 - 1)}.$$

Since, by definition, $y \geqslant 0$, we must discount the negative sign. Hence

$$y = \cosh^{-1} x = \ln[x + \sqrt{(x^2 - 1)}].$$

In a similar way, it can be shown that

$$\sinh^{-1} x = \ln[x + \sqrt{(x^2 + 1)}].$$

1.12 Partial fractions

We shall first make a distinction between an **equation** and an **identity**. The word 'equation' has many uses, but for the present we shall think of an *equation* as something like

$$x^2 + 2 = -3x,$$

since it is true only for certain particular values of x, namely -1 and -2. On the other hand,

$$x^2 + 3x + 2 \equiv (x + 1)(x + 2)$$

is an *identity*, meaning that it is true automatically, or for all values of x. Just for the purposes of this section we shall write $\equiv$ instead of $=$ when we want to draw attention to an identity.

It is easy to test the truth of the following identities by adding up the fractions on the right.

(i) $\qquad \dfrac{1}{x^2 - 1} \equiv \tfrac{1}{2}\dfrac{1}{x - 1} - \tfrac{1}{2}\dfrac{1}{x + 1},$

(ii) $\qquad \dfrac{x}{4x^2 - 1} \equiv \tfrac{1}{4}\dfrac{1}{2x - 1} + \tfrac{1}{4}\dfrac{1}{2x + 1},$

(iii) $\qquad \dfrac{3x + 2}{x^2(x + 1)} \equiv \dfrac{1}{x} + \dfrac{2}{x^2} - \dfrac{1}{x + 1},$

The terms on the right are individually simpler than the function on the left. This break-up into simpler constituents is useful for many purposes; in this section, we show how to break up a complicated function into simpler terms of the type above.

A **rational function** is a function which takes the form

$$f(x) = P(x)/Q(x),$$

where $P(x)$ and $Q(x)$ are **polynomials**. For example, $1/x^2(3x - 2)$ and $(2x^3 + 1)/(x - 1)^2$ are rational functions, but $x^{\frac{1}{2}}/(x + 1)$ and $(\cos x)/(x + 1)$ are not rational functions. We shall be concerned only with rational functions, and initially we shall suppose that

$$\text{degree of } P(x) < \text{degree of } Q(x),$$

something like proper fractions in arithmetic. Such functions can be broken up into **partial fractions**, like the examples at the beginning. No proofs will be given here, but the reader should learn the techniques.

It is the denominator of $Q(x)$ which determines what the form of the constituent partial fractions will take. Suppose that the denominator is broken up into factors as far as possible. For example,

$$2x^4 + x^3 - 4x^2 + x - 6 = (2x - 3)(x + 2)(x^2 + 1),$$

and it cannot be factorized any further. We shall consider only the cases where the factors are of the type

$ax + b$ (a **simple factor**), $(cx + d)^n$ (a **repeated factor**, order n), $px^2 + qx + r$ (an **irreducible quadratic**, with $q^2 < 4pr$)

The rules affecting these are as follows

Partial fractions for rational functions $P(x)/Q(x)$ (degree of $P(x)$) < (degree of $Q(x)$)

Each factor of $Q(x)$ gives rise to a partial fraction (or partial fractions) as below. Capitals denote constants: their values are unique.

(a) *Simple factors.* To each factor $ax + b$ of $Q(x)$, a term $K/(ax + b)$. (1.31)

(b) *Repeated simple factors.* To each factor $(cx + d)^n$ of $Q(x)$, there are n terms:

$$L_1/(cx+d) + L_2/(cx+d)^2 + \cdots + L_n/(cs+d)^n.$$

(c) *Irreducible quadratic.* To each factor $px^2 + qx + r$ of $Q(x)$, a term $(Mx + N)/(px^2 + qx + r)$.

$P(x)$ is involved in these rules only in that it will affect the values of the **coefficients** K etc. The following examples show how to determine the values of the coefficients.

Example 1.12. Express $x/(x - 1)(x + 2)$ in partial fractions.

We can use any convenient letters for the unknown coefficients of the terms. The denominator has two simple factors, $x - 1$ and $x + 2$, so (1.30a) says that the partial fractions must have the form

$$\frac{x}{(x - 1)(x + 2)} = \frac{A}{x - 1} + \frac{B}{x + 2} \tag{i}$$

Multiply through by $(x - 1)(x + 2)$:

$$x = A(x + 2) + B(x - 1). \tag{ii}$$

The constants must be chosen so that this becomes an **identity**. An identity has to be true for any x, so if we put *any two* values of x into (ii), the results must be correct. But the two substitutions of numbers for x form *two simultaneous equations* for the two unknown constants A and B. For example, if we put $x = -10$ and $x = 100$ we obtain

$$-10 = -8A - 11B, \qquad 100 = 102A + 99B.$$

The numbers chosen are inconvenient, but according to (1.31) we get the same A and B whatever values of x we use. Therefore, choose values that make the equations as simple as possible:

$$x = -2 \quad \text{gives} \quad -2 = 0 - 3B, \quad \text{so} \quad B = \tfrac{2}{3},$$
$$x = 1 \quad \text{gives} \quad 1 = 3A + 0, \quad \text{so} \quad A = \tfrac{1}{3}.$$

Therefore, from (i),

$$\frac{x}{(x-1)(x+2)} \equiv \frac{\tfrac{1}{3}}{x-1} + \frac{\tfrac{2}{3}}{x+2}.$$

Example 1.13. Express $(3x - 1)/(2x + 1)(x - 1)^2$ in partial fractions.

According to (1.30),

$$\frac{3x - 1}{(2x + 1)(x - 1)^2} = \frac{A}{2x + 1} + \frac{B}{x - 1} + \frac{C}{(x - 1)^2}. \tag{i}$$

Multiply by $(2x + 1)(x - 1)^2$ to give

$$3x - 1 = A(x - 1)^2 + B(2x + 1)(x - 1) + C(2x + 1). \tag{ii}$$

We need three values of x to obtain the three equations for A, B, C. Obvious choices are $x = 1$ and $x = -\tfrac{1}{2}$. For the third, choose, say, $x = 0$. From (ii):

$$x = 1 \quad \text{gives} \quad 2 = 0 + 0 + 3C, \quad \text{so} \quad C = \tfrac{2}{3},$$
$$x = -\tfrac{1}{2} \quad \text{gives} \quad -\tfrac{5}{2} = \tfrac{9}{4}A + 0 + 0, \quad \text{so} \quad A = -\tfrac{10}{9},$$
$$x = 0 \quad \text{gives} \quad -1 = A - B + C, \quad \text{so} \quad B = 1 + A + C = \tfrac{5}{9}.$$

Finally,

$$\frac{3x - 1}{(2x + 1)(x - 1)^2} \equiv -\tfrac{10}{9}\frac{1}{2x + 1} + \tfrac{5}{9}\frac{1}{x - 1} + \tfrac{2}{3}\frac{1}{(x - 1)^2}.$$

Example 1.14. Express $1/x(x^2 + 1)$ in partial fractions.

Here, $x^2 + 1$ is an irreducible quadratic; so, by (1.30),

$$\frac{1}{x(x^2 + 1)} = \frac{A}{x} + \frac{Bx + C}{x^2 + 1}.$$

Multiply by $x(x^2 + 1)$:

$$1 = A(x^2 + 1) + (Bx + C)x \tag{i}$$

Then $x = 0$ gives $1 = A + 0$, so

$$A = 1.$$

There are no other very easy values of x to choose. Put the value of A just found into (i) and rearrange, and cancel an x: we get

$$-x = Bx + C. \tag{ii}$$

It is easiest just to notice that (ii) is satisfied only if

$$B = -1 \quad \text{and} \quad C = 0.$$

Therefore

$$\frac{1}{x(x^2 + 1)} \equiv \frac{1}{x} - \frac{x}{x^2 + 1}.$$

If **the degree of the numerator is greater than or equal to the degree of the denominator**, the case is not covered by (1.31), but we can treat it as follows.

Example 1.15. Put $(x^3 + 1)/x(x - 1)$ into the form of a polynomial plus partial fractions.

Carry out polynomial division, until the remainder is of lower degree than the divisor:

$$
\begin{array}{r}
x + 1 \\
x^2 - x \,\overline{\big)\, x^3 + 1} \\
\text{subtract } x^3 - x^2 \\
\hline
x^2 + 1 \\
\text{subtract } x^2 - x \\
\hline
\text{remainder } x + 1
\end{array}
$$

Therefore

$$\frac{x^3 + 1}{x(x - 1)} \equiv x + 1 + \frac{x + 1}{x(x - 1)}.$$

The last term is of the right type for partial fractions, and finally

$$\frac{x^3 + 1}{x(x - 1)} \equiv x + 1 - \frac{1}{x} + \frac{2}{x - 1}.$$

1.13 Summation sign; geometric series

The sign $\sum$ is a large Greek capital S, standing for 'the sum of …'. It is used in the following way. Suppose, for example, we are provided with a string of six quantities indexed in order, say

$$u_1, u_2, u_3, \ldots, u_6.$$

This is called a **sequence** consisting of six **terms**. We can denote the **general term** by (say) u_n, where n can take any value within the range 1 to 6. Suppose we want to add them all up. Then

$$u_1 + u_2 + u_3 + u_4 + u_5 + u_6$$

is denoted by

$$\sum_{n=1}^{6} u_n,$$

which is read 'the sum of all the u_n between $n = 1$ and $n = 6$'. Similarly

$$u_2 + u_3 + u_4 + u_5 \quad \text{is written} \quad \sum_{n=2}^{5} u_n.$$

Any letter can be used as the **counting index** instead of n, provided that there is no conflict; so we could also write, for instance,

$$u_3 + u_4 + u_5 + u_6 = \sum_{i=3}^{6} u_i.$$

The letters m, n, r, i, j, k are often used.

We index a sequence according to convenience. The first index does not have to be 1. For example consider the important sequence

$$1, \quad x, \quad x^2, \quad x^3, \ldots,$$

which is the same as

$$x^0, \quad x^1, \quad x^2, \quad x^3, \ldots.$$

This is called, for historical reasons, a **geometric sequence**. Each term in turn is got from its predecessor by multiplying by the **common ratio** x. The natural way to index such a sequence is to run the counter starting with $n = 0$ instead of $n = 1$. Suppose then we want the sum of the first 6 terms. It can be expressed as

$$1 + x + x^2 + x^3 + x^4 + x^5 = \sum_{n=0}^{5} x^n,$$

though we could put

$$\sum_{n=1}^{6} x^{n-1}$$

instead. Such a sum, whether or not it starts with the x^0 term, is called a **geometric series**, and it appears frequently, though often the shape is disguised somewhat.

We will obtain an expression for the sum S of a geometric series having any value of the common ratio x, and which runs from the term in x^0 to the term in x^N. Thus $S = \sum_{n=0}^{N} x^n$. Note that it **contains** $N + 1$ **terms** (i.e. not N terms). Written at length:

$$S = 1 + x + x^2 + \cdots + x^{N-1} + x^N$$

Then

$$xS = x + x^2 + x^3 + \cdots + x^N + x^{N+1}.$$

Subtract the second line from the first. All the terms cancel except for two; we obtain

$$S(1 - x) = 1 - x^{N+1}, \quad \text{so} \quad S = (1 - x^{N+1})/(1 - x).$$

Sum of a geometric series

$$\sum_{n=0}^{N} x^n = 1 + x + x^2 \cdots + x^N$$

$$= \frac{1 - x^{N+1}}{1 - x}.$$

(1.32)

Example 1.16. Find the following sums. (a) $\sum_{n=0}^{4} (0.1)^n$, (b) $\sum_{n=0}^{6} \frac{1}{2^n}$,

(c) $\sum_{n=0}^{N} e^{nx}$, (d) $\sum_{n=0}^{N} (-1)^n$, (e) $\sum_{n=0}^{5} 2^n$.

(a) $x = 0.1$ and $N = 4$, so

$$S = \frac{1 - (0.1)^5}{1 - (0.1)} = \frac{0.99999}{0.9} = 1.1111,$$

(as becomes obvious if you write out the terms individually).

(b) $1/2^n = (\frac{1}{2})^n$, so $x = \frac{1}{2}$, and $N = 6$. By (1.31),

$$S = \frac{1 - (\frac{1}{2})^7}{1 - \frac{1}{2}} = 2(1 - \frac{1}{128}) = \frac{127}{64}.$$

(c) $e^{nx} = (e^x)^n$, so the common ratio is e^x, in place of x in (1.31):

$$S = \frac{1 - (e^x)^{N+1}}{1 - e^x} = \frac{1 - e^{(N+1)x}}{1 - e^x}.$$

(d) Here $x = -1$, so

$$S = \frac{1 - (-1)^{N+1}}{1 - (-1)} = \tfrac{1}{2}[1 - (-1)^{N+1}].$$

The sums of $N = 1, 2, 3, 4, \ldots$ terms of the sequence are successively $1, 0, 1, 0, 1, \ldots$.

(e) $x = 2$ and $N = 5$. Therefore

$$1 + 2 + 2^2 + 2^3 + 2^4 + 2^5 = \frac{1 - 2^6}{1 - 2} = 63.$$

Example 1.17. Find an expression for the sum

$$ar^2 + ar^5 + ar^8 + ar^{11} + ar^{14}.$$

This can be written

$$ar^2(1 + r^3 + r^6 + r^9 + r^{12}).$$

The bracket contains a series of type (1.31), with common ratio r^3 and with the number of terms $N + 1$ equal to 5: therefore $N = 4$. We have

$$S = ar^2 \frac{1 - (r^3)^5}{1 - (r^3)} = ar^2 \frac{1 - r^{15}}{1 - r^3}.$$

(In terms of $\sum$, we wanted $S = \sum_{n=0}^{4} ar^{2+3n}$. It is perhaps easier to see what to do when the series is written out fully.)

Problems, Chapter 1

1.1. (Section 1.3). Sketch graphs of the following equations over the intervals stated:
(a) $y = x^4$, $-1.5 \leqslant x \leqslant 1$;
(b) $y = x(1 - x)$, $-1 \leqslant x \leqslant 2$;
(c) $y = 1 + x + x^2$, $|x - 1| \leqslant 2$;
(d) $y = |x - 1|$, $-3 \leqslant x \leqslant 3$;
(e) $y = |x| + |x - 3| + |x + 2|$, $-3 \leqslant x \leqslant 4$;
(f) $y = ||x| - 1|$, $-2 \leqslant x \leqslant 2$;
(g) $y = \sqrt{(x^2 + 1)}$, $|x| \leqslant 2$.

1.2. (straight lines, Section 1.3). Find the straight lines through the following pairs of points:
(a) $(1, 1), (-1, 2)$; (b) $(0, 1), (2, 1)$; (c) $(2, 1), (-1, -1)$.
Sketch the triangle formed by these lines. Find the lengths of each side of the triangle.

1.3. (straight line, Section 1.3). What are the slopes of the following straight lines, and where do they cut the coordinate axes?
(a) $y = x - 1$; (b) $3y = x - 2$; (c) $2x + 5y = 4$.

1.4. (straight line, Section 1.3). Find the equations of the following straight lines:
(a) passing through $(1, 2)$ inclined at $45°$ to the x-axis;
(b) passing through $(-1, -2)$ with slope -2;
(c) with slope 0.5 and x axis intercept $x = 1$;
(d) through $(1, 2)$ parallel to the line $y = 3x - 4$;
(e) through $(-1, 3)$ perpendicular to the line $y = 4x - 1$.

1.5. Show that the following pairs of lines are mutually perpendicular:
(a) $2x + 3y - 2 = 0$ and $-3x + 2y - 3 = 0$;
(b) $y = 2x + 1$ and $y = \frac{1}{2}(3 - x)$;
(c) $y = 2x$ and $y = -\frac{1}{2}x$; (d) $y = \pm x$.

1.6. (straight line, Section 1.3). Show that the equation
$$(x + y + 1) + \alpha(2x - 3y - 2) = 0,$$
where α is a constant, represents a straight line through the intersection of $x + y + 1 = 0$ and $2x - 3y - 2 = 0$. Find the line joining this point with the point $(1, 1)$.

1.7. (circles, Section 1.3). Find the centre and radius for each of the following circles:
(a) $x^2 + y^2 = 9$; (b) $(x - 1)^2 + y^2 = 4$;
(c) $x^2 + y^2 - 2x - 2y - 21 = 0$;
(d) $4x^2 - 4x + 4y^2 + 4y = 9$.

1.8. (circles, Section 1.3). Find the equations of the circle centred at $(1, -2)$ with radius 3.

1.9. Find the points of intersection of the following circles and lines:
(a) $x^2 + y^2 = 8$ with $x = 2$;
(b) $x^2 + y^2 - 2x + 2y - 4 = 0$ and $y = 2x + 1$;
(c) $x^2 + y^2 = 1$ and $x + y = \sqrt{2}$.

1.10. The following table contains experimental data:

x	1.06	0.84	0.72	0.44	0.23
y	0	0.53	0.71	0.78	1.1

The hypothesis is that the points should lie on a circle centre at the origin in the (x, y) plane. Find the distance of each point from the origin. Calculate the average of these values, and write down the equation of the circle approximation.

1.11. (functions, Section 1.4). Draw sketches of the following functions in the (x, t) plane over the intervals indicated:
(a) $x = H(t + 1) - H(t - 1)$ for $-2 \leqslant t \leqslant 2$;
(b) $x = \text{sgn}(1 + t) + \text{sgn}(1 - t)$ for $-2 \leqslant t \leqslant 2$;
(c) $x = tH(t - 1)$ for $0 \leqslant t \leqslant 2$;
(d) $x = (t^2 - 1)[\text{sgn}(t + 1) + \text{sgn}(1 - t)]$ for $-2 \leqslant t \leqslant 2$.

1.12. (functions, Section 1.4). Using Heaviside and signum functions, construct a single formula in each case for $f(t)$ where

(a) $f(t) = \begin{cases} 0 & (t < -1), \\ 1 & (-1 < t < 2), \\ 0 & (t > 2); \end{cases}$

(b) $f(t) = \begin{cases} 0 & (t < 0), \\ 2t & (t > 0); \end{cases}$

(c) $f(t) = \begin{cases} 0 & (t < 0), \\ t & (0 < t < 1), \\ 1 & (1 < t < 2), \\ -t + 3 & (2 < t < 3), \\ 0 & (t > 3). \end{cases}$

1.13. (Section 1.5). What are the radian measures of the following angles: (a) $30°$, (b) $100°$?

1.14. (trigonometric functions, Section 1.6). Using the methods of Examples 1.4 and 1.5, obtain
(a) $\sin \frac{1}{4}\pi$; (b) $\sin \frac{1}{2}\pi$; (c) $\sin \pi$;
(d) $\sin(-\frac{3}{4}\pi)$; (e) $\cos \frac{1}{6}\pi$; (f) $\cos \frac{5}{6}\pi$.

1.15. (trigonometric functions, Section 1.6). Use (1.18c) to show that
(a) $\cos^4 A = \frac{1}{8}(3 + 4\cos 2A + \cos 4A)$;
(b) $\sin^4 A = \frac{1}{8}(3 - 4\cos 2A + \cos 4A)$.

1.16. (trigonometric functions, Section 1.6). Use (1.18) to express the following in terms of $\sin x$ and $\cos x$:
(a) $\cos(x + \frac{1}{2}\pi)$; (b) $\sin(x + \frac{1}{2}\pi)$; (c) $\sin(x - \frac{1}{2}\pi)$;
(d) $\cos(x \pm \pi)$; (e) $\sin(x \pm \pi)$.

1.17. (trigonometric functions, Section 1.6). Use (1.18) to express the following in terms of the cos and sin of $\frac{1}{2}(x + y)$ and $\frac{1}{2}(x - y)$:
(a) $\cos x + \cos y$; (b) $\sin x - \sin y$; (c) $\cos x - \cos y$.

1.18. State where the graphs of the following functions cross the x-axis:
(a) $\sin x$; (b) $\cos x$; (c) $\sin \frac{1}{2}x$; (d) $\cos 3\pi x$;
(e) $\cos(2x - \frac{1}{2}\pi)$; (f) $e^{-x} \sin \frac{1}{2}\pi x$.

1.19. State the amplitude, angular frequency, period, and phase of the following harmonic outputs:
(a) $2\cos(0.2t + 3.2)$; (b) $1.5\sin[0.2(t - 2.4)]$;
(c) $2\cos(0.2t + 0.12) + 2\cos(0.2t - 0.39)$.

1.20. (Section 1.7). State the inverse $f_1(x)$ for each of the following functions $f(x)$ in for the values of x stated:
(a) $f(x) = 4x^2$, $x \leqslant 0$;
(b) $f(x) = 2x + 3$, $-\infty < x < \infty$;
(c) $\sin 2x$, $0 \leqslant x \leqslant \frac{1}{4}\pi$; (d) $2\sin x$, $0 \leqslant x \leqslant \frac{1}{2}\pi$;
(e) $\cos x^2$, $0 \leqslant x \leqslant \sqrt{\pi}$; (f) $\sin(\frac{1}{2}\pi \cos x)$, $0 \leqslant x \leqslant \frac{1}{2}\pi$;
(g) $f(x) = x^{-\frac{1}{4}}$, $x > 0$; (h) $x^2 + x$, $x > -\frac{1}{2}$.

1.21. Sketch a graph of the function inverse to the function $x^3 - x + 1$. (There is no need to try to solve the equation $x^3 - x + 1 = y$ for x.)

1.22. (Sections 1.8, 1.9). Solve the following equations for x:
(a) $e^{2x} = 3$; (b) $\ln 3x = 2$;
(c) $\ln x^{-\frac{1}{3}} = 1$; (d) $3e^{3x} = 1$;
(e) $e^x + e^{-x} = 2$ (hint: multiply through by e^x first);
(f) $e^{\ln 2x} = 4$; (g) $\ln e^{2x} + 3\ln e^{5x} = 2$;
(h) $\ln(x + 1) + \ln(x - 1) = 0$;
(i) $\ln(x + 1) + \ln(x - 1) = e$;
(j) $2^x = 3$; (k) $3^{2x} = \frac{1}{2}$;
(l) $\sinh 2x = 4$; (m) $2\sinh x = 2\cosh x + 3$.

1.23. Express 2^x as a power of e.

1.24. (Section 1.10). Prove that 10^x doubles its value in any interval of length equal to
$$\frac{\ln 2}{\ln 10}.$$

1.25. Sketch regions in the (x, y) plane defined by the following inequalities:
(a) $(x - 1)^2 + y^2 \leqslant 9$;
(b) $x \geqslant 0$, $y \geqslant 0$ and $x + y \leqslant 1$;
(c) $\dfrac{x^2}{4} + \dfrac{y^2}{9} \leqslant 1$;
(d) $x^2 + y^2 \leqslant 1$ and $x \geqslant 0$; (e) $|x| + |y| \leqslant 1$

1.26. Prove that $\sinh^{-1} x = \ln[x + \sqrt{(x^2 + 1)}]$.

1.27. Figure 1.30 shows a cross-section of a simple model of a piston and crankshaft. The crankshaft rotates at 4000 rpm (revolutions per minute). If $AB = 2.5$ and $BC = 5$ (in cm), show that the displacement AC is given by
$$AC = 2.5[\sin \omega t + \sqrt{(4 - \cos^2 \omega t)}]$$
(in cm), where t is measured from $\theta = 0$, and state ω in radians per second.

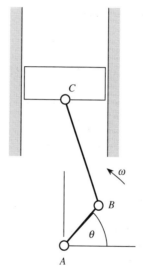

Fig. 1.30

1.28. An oscillation takes the form
$$x = 3\cos \omega t + 4\sin \omega t.$$
By finding numbers c and ϕ such that
$$c \cos \phi = 3, \quad c \sin \phi = 4$$
express x as a single sinusoidal term. What are the amplitude and phase of the oscillation?

1.29. The exponential function $f(t) = C e^{-\alpha t}$ satisfies the conditions $f(0) = 2$ and $f(1) = 0.5$. Find the constants C and α. What is the value $f(2)$?

1.30. A yacht, which has a draught of 2 metres, is anchored in a tidal estuary, in which the depth of water around the yacht is

$$5 + 4.5 \sin 0.5t$$

(in metres), where the time t is measured in hours. What is the tidal period in hours? Over how many hours in one period can the yacht float free of the estuary?

1.31. Draw sketches of the graphs of the following curves given in polar coordinates, by constructing a table of values of r for equally spaced angles (say $15°$ intervals):
(a) the cardioid $r = 0.5(1 + \cos \theta)$;
(b) the folium $r = (4 \sin^2 \theta - 1) \cos \theta$;
(c) the four-leaved rose $r = \sin 2\theta$;
(d) the Archimidean spiral $r = 0.04\theta$ (extend the interval in θ to $[0, 6\pi]$);
(e) the equiangular spiral $r = 0.1e^{0.1\theta}$ (extend the interval in θ to $[0, 6\pi]$).

1.32. Sketch the graphs of the following functions:
(a) sgn sin x;   (b) sgn cos $2x$;   (c) H(x) sin x;
(d) $\sin^2 x$;   (e) $|\sin x|$;   (f) $\sin|x|$;   (g) H($x - \pi$) sin x.

1.33. The coordinates of three vertices of a rectangle are given by

$$(-7, 3),   (1, -3),   (4, 1).$$

Find the coordinates of the fourth vertex. Determine also the area of the rectangle.

1.34. State which of the following functions are periodic, and, if so, find the (minimum) period:
(a) $\sin 4x$;   (b) $\cos(\pi + t)$;   (c) $\sin t + \cos 2t$;
(d) $\sin x^2$;   (e) $e^{-\sin x}$;   (f) $\cos^2 x$;   (g) $x \sin x$;
(h) $|\sin x|$;   (i) $1/(4 + \sin^2 t)$;   (j) $\sin \frac{1}{4}t$.

1.35. Decide which of the following functions are even, odd, or neither even nor odd:
(a) $x^2 + x^3$;   (b) $x^2 + 2x^4$;   (c) $x + \sin x$;
(d) $\sin x \cos x$;   (e) e^{-x^2};   (f) $\ln(1 + x^2)$;   (g) $e^{\sin x}$.

1.36. (partial fractions, Section 1.11). Express the following in partial fractions:

(a) $\dfrac{1}{(x - 2)(x + 3)}$;   (b) $\dfrac{x}{(x + 1)(x + 2)}$;

(c) $\dfrac{2x - 1}{x(x - 1)}$;   (d) $\dfrac{1}{x(x + 1)(x + 2)}$;

(e) $\dfrac{1}{x(x^2 - 1)}$;   (f) $\dfrac{1}{x(x + 2)^2}$;

(g) $\dfrac{x^2}{(x + 1)(x + 2)^2}$;   (h) $\dfrac{x - 1}{x^2 - 2x - 3}$;

(i) $\dfrac{1}{(x - 1)^2}$;   (j) $\dfrac{1}{x^3 + x^2}$.

1.37. (partial fractions, Section 1.11). In the following problems with irreducible factors, express the functions in partial fractions:

(a) $\dfrac{1}{x(x^2 + x + 1)}$;   (b) $\dfrac{x}{(x - 1)(x^2 + 1)}$;

(c) $\dfrac{x}{(x + 1)(x^2 + 2x + 6)}$.

1.38. Express the following in partial fractions:

(a) $\dfrac{1}{x^2(x^2 + 1)}$;   (b) $\dfrac{x^3}{(x + 1)(x + 2)}$;

(c) $\dfrac{x^3}{(x^2 - 9)(x + 1)}$.

1.39. Write down in full all the terms in each of the series:

(a) $\displaystyle\sum_{j=2}^{5} 2^j$;   (b) $\displaystyle\sum_{n=0}^{4} \dfrac{1}{1 + n^2}$;   (c) $\displaystyle\sum_{n=1}^{4} nx^n$.

1.40. (geometric series, Section 1.12). Using the geometric series find the sums of the following series:

(a) $\displaystyle\sum_{n=1}^{7} (0.5)^n$;   (b) $\displaystyle\sum_{n=2}^{6} \left(\dfrac{1}{3}\right)^n$;   (c) $\displaystyle\sum_{n=0}^{5} e^{-2n}$;

(d) $\displaystyle\sum_{n=1}^{6} n2^n$;   (e) $\displaystyle\sum_{n=1}^{10} (-\tfrac{1}{2})^n$;

(f) $\displaystyle\sum_{n=0}^{6} [2(0.5)^n + 3(0.6)^n]$.

1.41. Find a formula for the sum of

$$x + x^5 + x^9 + \cdots + x^{1+4n} + \cdots + x^{41}.$$

1.42. ABC is a triangle with sides a, b, c opposite to the corresponding angles. Prove the **cosine rule**; that

$$c^2 = a^2 + b^2 - 2ab \cos C.$$

(Hint: drop a perpendicular from B on to b. Look for $a \cos C$, and then for an opportunity to use Pythagoras' theorem.)

Differentiation

2

2.1 The slope of a graph

Figure 2.1 shows the graph of a straight line. The x and y coordinates are assumed to have the **same scale**. Choose any two points $P : (x_1, y_1)$ and $Q : (x_2, y_2)$ which lie on the line. If we measure the angle α **from the positive x direction** then

$$\tan \alpha = \frac{y_2 - y_1}{x_2 - x_1} \tag{2.1}$$

(see (1.6)). The value of $\tan \alpha$ remains the same whether Q is to the right or left of P, since the value of the fraction on the right is unchanged. The angle α itself will differ in the two cases by an amount equal to π (or $180°$), but this does not affect the value of $\tan \alpha$. (If we refer to α itself to indicate the steepness of a line, we choose the value that lies between $\pm 90°$, but normally we only need $\tan \alpha$). Notice that if the x and y scales differed, the angle α as depicted would not satisfy (2.1); it would be too great or too small.

The **slope** or **gradient** of a straight line is defined to be the quantity $\tan \alpha$. If the line is horizontal, $\tan \alpha$ is zero. It is positive or negative according as the line slopes upwards or downwards as we go **from left to right**. It increases or decreases as the inclination increases or decreases, becoming $\pm \infty$ when $\alpha = \pm 90°$.

Consider now the **slope** or **gradient of a curve** at a point. Figure 2.2a shows a typical curve. By the **slope of the curve** at the point P we mean the **slope of the tangent line to the curve** at P. We can think of the tangent line as the line joining two points on the curve which are 'infinitely close together'; but it is no use making P and Q coincide, since we simply get $\tan \alpha = 0/0$, which has no definite meaning. It is necessary to carry out an indirect process.

Let P be the fixed point (see Fig. 2.2b). Take any other point Q on the curve and join PQ by a straight line, called the **chord** PQ. If Q is some distance from P, then the slope of PQ will not be close to that of PT; but if we take a succession of points Q closer

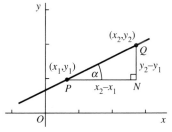

Fig. 2.1

(a)

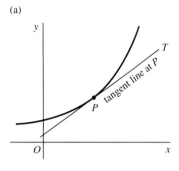

(b)

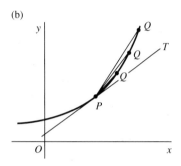

Fig. 2.2

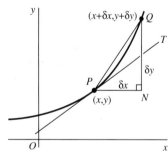

Fig. 2.3

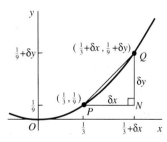

Fig. 2.4.

and closer to P, then the slope of the chord PQ can be made as close as we wish to that of PT. The points Q that we consider are said to **approach** P. The corresponding value of the slope of PQ then **approaches a limit** or a **limiting value**, and this is equal to the slope of the curve at P. We use the sign $\rightarrow$ to signify 'approaches'; so we can write:

$$\text{as } Q \rightarrow P, \text{ slope of } PQ \rightarrow \text{slope of the curve at } P. \tag{2.2}$$

We shall be able to obtain the **exact value** of the slope of PT, which is the same as the slope of the curve at P, by carrying out the approach of Q to P in algebraic terms. To do this, we introduce a new symbol

$$\delta x$$

pronounced 'delta-x'. This is a **single symbol**: the Greek letter δ stands for the words '**the increment in**' or '**the change in**' – in this case, the increment in the value of x as we move from P to Q. For the two points $P : (x_1, y_1)$ and $Q : (x_2, y_2)$, the change in x on moving from P to Q is

$$\delta x = x_2 - x_1. \tag{2.3}$$

(The change in moving from Q to P is given by $\delta x = x_1 - x_2$, which has the opposite sign.) Similarly we use the symbol δy to indicate the change in y as we move from P to Q.

In Fig. 2.3, $P : (x, y)$ is the point on a graph at which we want to find the slope (i.e. we want the slope of the tangent line PT). Take another point Q on the curve, and suppose that it approaches P. The separation between P and Q is indicated by the sides of length δx and δy of the triangle PNQ. Then, by (2.1),

$$\text{slope of } PQ = \frac{\delta y}{\delta x}. \tag{2.4}$$

Now let $\delta x \rightarrow 0$ so that $Q \rightarrow P$: the ratio $\delta y / \delta x$ approaches a number which is equal to the value of the slope at P. We first show what happens **numerically** in a particular case.

Example 2.1. Find the slope of the curve $y = x^2$ at the point $P : (\frac{1}{3}, \frac{1}{9})$ on the curve (Fig. 2.4).

At P, we have $x = \frac{1}{3}$ and $y = \frac{1}{9}$. We shall make a table of values of $\delta y / \delta x$ for diminishing values of δx; that is, for points Q which are approaching P (from either side). We first need to express δy in terms of δx:

$$\delta y = (y \text{ at } Q) - (y \text{ at } P) = (\tfrac{1}{3} + \delta x)^2 - (\tfrac{1}{3})^2$$
$$= (\tfrac{1}{9} + \tfrac{2}{3} \delta x + \delta x^2) - \tfrac{1}{9} = \tfrac{2}{3} \delta x + \delta x^2. \tag{2.5}$$

For Q to the left of P:

δx	-0.1	-0.001	-0.0001	$\cdots$
δy	$-0.05\dot{6}$	$-0.0065\dot{6}$	$-0.000665\dot{6}$	$\cdots$
$\dfrac{\delta y}{\delta x}$	$0.5\dot{6}$	$0.65\dot{6}$	$0.665\dot{6}$	$\cdots$

For Q to the right of P:

δx	0.1	0.001	0.0001	$\cdots$
δy	$0.07\dot{6}$	$0.0067\dot{6}$	$0.000667\dot{6}$	$\cdots$
$\dfrac{\delta y}{\delta x}$	$0.7\dot{6}$	$0.67\dot{6}$	$0.667\dot{6}$	$\cdots$

($\dot{6}$ means that the number 6 recurs: e.g. $0.\dot{6} = 0.66666\cdots$.)

First notice that if we put $\delta x = 0$ we obtain $\delta y/\delta x = 0/0$, which gives no information at all, so we have to look at the sequence of values for $\delta y/\delta x$ as $\delta x \to 0$. Inspection suggests that each term is formed from its predecessor in a regular way; so we can predict that:

as $\delta x \to 0$, slope of $PQ \to 0.66666\cdots = \frac{2}{3}$.

The **exact value of the slope** at P is therefore $\frac{2}{3}$.

In fact, we could have worked out the slope at P in this example without doing any calculation. From (2.5),

$$\frac{\delta y}{\delta x} = \frac{\frac{2}{3}\delta x + \delta x^2}{\delta x} = \frac{2}{3} + \delta x$$

provided that $\delta x \neq 0$. Therefore, when $\delta x \to 0$, $\delta y/\delta x \to \frac{2}{3}$, as before.

Finally, in exactly the same way, we can find a general formula giving the slope at any point $P : (x, y)$ on the graph of $y = x^2$. The value of δy corresponding to a value of δx is given by

$$\delta y = (x + \delta x)^2 - x^2 = x^2 + 2x\,\delta x + \delta x^2 - x^2$$
$$= 2x\,\delta x + \delta x^2.$$

Therefore

$$\frac{\delta y}{\delta x} = \frac{2x\,\delta x + \delta x^2}{\delta x} = 2x + \delta x.$$

Now let $\delta x \to 0$; we obtain:

the slope of $y = x^2$ at (x, y) is $2x$.   **(2.6)**

This process is a model for treating other functions: for example, you could now show in the same way that the slope of the graph of $y = x^3$ at any point is equal to $3x^2$.

2.2 The derivative: notation and definition
We shall need to find the value approached by $\delta y/\delta x$ as $\delta x \to 0$ in many different situations. There is a special notation used to signify

the process:

Limit notation

Let $y = f(x)$, so that $\delta y = f(x + \delta x) - f(x)$. Then the value approached by $\delta y/\delta x$ when $\delta x \to 0$ is denoted by

$$\lim_{\delta x \to 0} \frac{\delta y}{\delta x}.$$

(2.7)

Read this as 'The **limit**, or the **limiting value**, of $\delta y/\delta x$ as $\delta x \to 0$'. (The lim sign is used in many other contexts too.)

The result of the process $\lim_{\delta x \to 0} \delta y/\delta x$, where $\delta y = f(x + \delta x) - f(x)$, is called the **derivative of y with respect to** x, or **the derivative of** $f(x)$. The process is called **differentiation**. We worked our earlier that, if $y = f(x) = x^2$, then the derivative is equal to $2x$. The following notations are **standard short ways of indicating a derivative**:

$$\frac{dy}{dx}, \quad \frac{df(x)}{dx}, \quad \text{or} \quad \frac{d}{dx} f(x) \quad \text{signify} \quad \lim_{\delta x \to 0} \frac{\delta y}{\delta x}.$$

The symbol dy/dx is usually pronounced 'dee-y by dee-x'. Notice that the letter used is an ordinary d, not δ.

Differentiation

Let $y = f(x)$ and $\delta y = f(x + \delta x) - f(x)$. Then the derivative of y with respect to x, signified by

$$\frac{dy}{dx}, \quad \frac{df(x)}{dx} \quad \text{or} \quad \frac{d}{dx} f(x),$$

(2.8)

means the result of taking the limit of $\delta y/\delta x$ as $\delta x \to 0$:

$$\frac{dy}{dx} = \lim_{\delta x \to 0} \frac{\delta y}{\delta x}.$$

Thus our earlier result for $y = x^2$ can be written in several ways:

$$\frac{dy}{dx} \quad \text{or} \quad \frac{d(x^2)}{dx} \quad \text{or} \quad \frac{d}{dx} x^2 = 2x.$$

Strictly speaking, dy/dx should be regarded as a **single shorthand symbol** representing the longer expression $\lim_{\delta x \to 0} \delta y/\delta x$, and not as a ratio which can be taken to pieces. However, its great usefulness is that it often behaves just like an ordinary ratio of nonzero quantities, and we shall later see cases where this property guides us to true results and makes them easy to remember.

It is sometimes useful to think of the symbol

$$\frac{d}{dx}$$

standing alone as meaning: 'differentiate what follows'. It is also called an **operator**, meaning that we operate on one function (x^2 say) to produce another (i.e. $2x$). Sometimes the symbol D is used to stand for the operator d/dx. We would then write

$$Dx^2 = 2x.$$

2.3 **Rates of change**

The quantity $\lim_{\delta x \to 0} \delta y/\delta x$ is usually needed to solve problems which have no immediate connection with the slope of graphs: this was just introduced to give the reader a picture to hold on to. Moreover, it is not always appropriate to call the variables x and y if other letters arise more naturally.

For example, suppose that a car is moving along a straight road, represented by an x axis, and that at time t its **displacement** from the origin is given by

$$x = f(t).$$

We can deduce its **velocity** from moment to moment from this information.

Choose any moment t, and suppose that, between times t and $t + \delta t$, the car moves from x to $x + \delta x$. Then δx must be given by

$$\delta x = f(t + \delta t) - f(t).$$

The quantities δt and δx could be imagined as being recorded with a stopwatch and distance meter, and the average velocity over the interval δt would be

$$\frac{\text{distance travelled}}{\text{time taken}} = \frac{f(t + \delta t) - f(t)}{\delta t} = \frac{\delta x}{\delta t}.$$

The smaller that δt is, the more nearly will this ratio approximate to the instantaneous velocity v at time t. Therefore, let $\delta t \to 0$; using the notation (2.7), we obtain

$$v = \lim_{\delta t \to 0} \frac{\delta x}{\delta t},$$

or alternatively, by (2.8),

$$v = \frac{dx}{dt}.$$

We can borrow the result (2.6) to complete the calculation in one case. Suppose that

$$x = t^2.$$

Equation (2.6) says in effect that

$$\text{if} \quad y = x^2 \quad \text{then} \quad \frac{dy}{dx} = 2x,$$

and just by changing the letters we obtain:

$$\text{if} \quad x = t^2 \quad \text{then} \quad \frac{dx}{dt} = 2t.$$

Therefore the velocity is

$$v = 2t.$$

Another way of expressing the meaning of velocity is that **velocity is the rate of change of displacement with time**. Similarly, **acceleration a is the rate of change of velocity with time**:

$$a = \frac{dv}{dt}.$$

For the case when $x = t^2$ we have

$$\delta v = 2(t + \delta t) - 2t = 2\,\delta t,$$

so

$$a = \frac{dv}{dt} = \lim_{\delta t \to 0} \frac{\delta v}{\delta t} = 2.$$

As seen in the next example, the idea of rate of change is quite general and need not involve time.

Example 2.2. Find the rate of change of the area of a circle with respect to its radius.

Call the radius r and the area A. The rate of change of A with respect to r is

$$\frac{dA}{dr} \quad \text{or} \quad \lim_{\delta r \to 0} \frac{\delta A}{\delta r}.$$

Since $A = \pi r^2$, we have

$$\delta A = \pi(r + \delta r)^2 - \pi r^2 = \pi(2r\,\delta r + \delta r^2).$$

Therefore

$$\frac{\delta A}{\delta r} = \pi[2r + \delta r].$$

Now let $\delta r \to 0$; we obtain

$$\frac{dA}{dr} = \lim_{\delta r \to 0} \frac{\delta A}{\delta r} = 2\pi r.$$

This result could have been obtained by using our previous result

$$\frac{d(t^2)}{dt} = 2t$$

and multiplying it by π. (Notice also that $2\pi r$ is the circumference: the result can be interpreted as meaning that if we increase r by a small amount δr, then the area increase is nearly equal to that of a narrow strip of length $2\pi r$ and breadth δr.)

2.4 Derivative of x^n $(n = 0, 1, 2, 3, \ldots)$

The following is our first general result:

(a) if $y = c$, where c is a constant, then

$$\frac{dy}{dx} = 0.$$

(b) If $y = x^n$, where n $= 1, 2, 3, \ldots$, then

$$\frac{dy}{dx} = nx^{n-1}.$$

$$(2.9)$$

To prove (a): the graph of $y = c$ is a horizontal straight line; therefore its slope is zero, so $dy/dx = (d/dx)c = 0$.

To prove (b) in the most elementary way, we shall use an identity: if n is a positive whole number and a, b are any numbers,

$$a^n - b^n = (a - b)(a^{n-1} + a^{n-2}b + a^{n-3}b^2 + \cdots + b^{n-1}).$$

This can be verified by multiplying out the two brackets on the right; everything cancels except for the two terms on the left.

Follow (2.8), with $f(x) = x^n$, so that $\delta y = (x + \delta x)^n - x^n$:

$$\frac{dy}{dx} = \lim_{\delta x \to 0} \frac{\delta y}{\delta x} = \lim_{\delta x \to 0} \frac{1}{\delta x} \{(x + \delta x)^n - x^n\}.$$

Put $x + \delta x = a$ and $x = b$ into the identity, noticing that

$$a - b = (x + \delta x) - x = \delta x.$$

$$\frac{\delta y}{\delta x} = \frac{1}{\delta x} (\delta x)[(x + \delta x)^{n-1} + (x + \delta x)^{n-2}x + \cdots + x^{n-1}]$$

$$= (x + \delta x)^{n-1} + (x + \delta x)^{n-2}x + \cdots + x^{n-1}$$

when $\delta x \neq 0$. Now let $\delta x \to 0$; we obtain

$$\frac{dy}{dx} = \lim_{\delta x \to 0} \frac{\delta y}{\delta x} = x^{n-1} + x^{n-1} + \cdots + x^{n-1}.$$

There are n terms on the right, each equal to x^{n-1}, so finally,

$$\frac{dy}{dx} = nx^{n-1}.$$

(In Section 3.4 we show that (2.9b) is in fact true for *all* values of n.)

Example 2.3. Obtain (a) the general expression for dy/dx when $y = x^3$; (b) the slope of the curve $y = x^3$ at $P : (2, 8)$; (c) the angle of inclination of the tangent line to $y = x^3$ at the point P; (d) the equation of the tangent line through P; (e) the velocity v and acceleration a of a point with coordinate x, when $x = t^3$.

(a) From (2.9) with $n = 3$, we have

$$\frac{dy}{dx} = 3x^{3-1} = 3x^2.$$

(b) The slope of the curve at P is equal to the value of dy/dx at $x = 2$, which is 12.

(c) The slope is equal to $\tan \alpha$, where α is the angle of inclination, so $\alpha = 85.2°$.

(d) Let (x, y) now represent **any point on the tangent line** at $(2, 8)$. The slope of the tangent is equal to 12, so from (2.1)

$$\frac{y - 8}{x - 2} = 12.$$

Therefore the equation of the tangent line is $y = 12x - 16$.

(e) From Section 2.3, $v = dx/dt = (d/dt)t^3 = 3t^2$. Also

$$a = \frac{dv}{dt} = \frac{d}{dt} 3t^2$$

$$= 3 \frac{d}{dt} t^2 = 6t.$$

A little thought about the process of finding $\lim_{\delta t \to 0} \delta v/\delta t$ will persuade the reader that it is right to take the constant 3 from under the differentiation sign in the last line; see also the next section.

2.5 Sums of terms: multiplication by constants

The following are general rules which become obvious when the definition (2.8) is applied to them.

Linear combinations of functions

(a) If C is a constant, then

$$\frac{d}{dx}[Cf(x)] = C\frac{d}{dx}f(x).$$

(b) $\dfrac{d}{dx}[f(x) + g(x)] = \dfrac{d}{dx}f(x) + \dfrac{d}{dx}g(x).$ **(2.10)**

(c) If $A, B, C, \ldots$ are constants, then

$$\frac{d}{dx}[Af(x) + Bg(x) + Ch(x) + \cdots]$$

$$= A\frac{d}{dx}f(x) + B\frac{d}{dx}g(x) + C\frac{d}{dx}h(x) + \cdots.$$

The result (2.10c) follows easily by repeated use of (a) and (b). We can use this rule together with (2.9) to obtain the derivatives of polynomials, as in the following example.

Example 2.4. Obtain dy/dx when (a) $y = 3x^3 - \frac{1}{2}x^2 + 5$; (b) $y = 3x(x^2 - 2)$.

(a) From (2.10c),

$$\frac{dy}{dx} = \frac{d}{dx}(3x^3 - \tfrac{1}{2}x^2 + 5) = 3\frac{d(x^3)}{dx} - \frac{1}{2}\frac{d(x^2)}{dx} + \frac{d(5)}{dx}$$

$$= 3(3x^2) - \tfrac{1}{2}(2x) + 0 = 9x^2 - x.$$

(b) It is necessary in this case to express y without brackets, i.e. as a polynomial:

$$\frac{dy}{dx} = \frac{d}{dx}[3x(x^2 - 2)] = \frac{d}{dx}(3x^3 - 6x)$$

$$= 3\frac{d(x^3)}{dx} - 6\frac{dx}{dx} \quad \text{(from (2.10c))}$$

$$= 9x^2 - 6.$$

These derivatives, of course, represent the slopes of the corresponding graphs.

Example 2.5. A car travels along a straight road with varying velocity v for one hour. At time t hours, its displacement from the starting point O is given by $x = 60t^2(3 - 2t)$ kilometres. Find expressions for (a) the velocity v; (b) the acceleration a.

(a) The velocity is the rate of change of displacement with time:

$$v = \frac{dx}{dt} = \frac{d}{dt}[60t^2(3 - 2t)]$$

$$= 60\frac{d}{dt}(3t^2 - 2t^3) = 60\left(3\frac{d(t^2)}{dt} - 2\frac{d(t^3)}{dt}\right)$$

$$= 60[3(2t) - 2(3t^2)] = 360(t - t^2)) \text{ in km hr}^{-1}.$$

(b) Acceleration is the rate of change of velocity with time:

$$a = \frac{dv}{dt}.$$

Therefore

$$a = 360(1 - 2t) \quad (\text{in km hr}^{-2}).$$

Example 2.6. The potential energy V of a pendulum of length l with a bob of mass m is given approximately by $V = mgl(\theta - \frac{1}{6}\theta^3)$ when the angle of inclination θ (radians) is small. Find the rate of change of V with respect to θ. (This quantity is associated with the moment exerted by gravity.)

Using the letters suggested by the question, we require

$$\frac{dV}{d\theta} = \frac{d}{d\theta}[mgl(\theta - \frac{1}{6}\theta^3)] = mgl\left(\frac{d\theta}{d\theta} - \frac{1}{6}\frac{d(\theta^3)}{d\theta}\right)$$

$$= mgl(1 - \frac{1}{2}\theta^2).$$

2.6 Three important limits

In order to increase the repertory of functions that we can differentiate, three important limits are needed. The reader might not need to learn the proofs, but the results (2.11), (2.13), and (2.14) are essential, and you should try to acquire a feeling for what is happening by examining the numerical tables given; or better by working out tables of your own.

Instead of δx, the letter ε (greek epsilon) will be used to represent the quantity that tends to zero. (The limit is not affected by the letter we use.)

First consider

$$\lim_{\varepsilon \to 0} \frac{e^\varepsilon - 1}{\varepsilon}.$$

If we put $\varepsilon = 0$ we get $0/0$, which is meaningless, but the approach to a limit can be seen in the following table:

ε	0.1	0.01	0.001	$\cdots$
$\dfrac{e^\varepsilon - 1}{\varepsilon}$	1.0517	1.0050	1.0005	$\cdots$

(An approach to zero through negative ε values is similar.) It looks as if the limit is equal to 1.

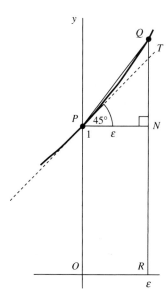

Fig. 2.5

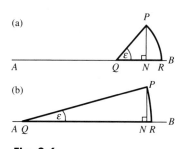

Fig. 2.6

To prove this, recall that in Section 1.8 it was shown that the graph $y = e^x$ intersects the y axis at $45°$; that is to say, its slope there is equal to 1. (This is the characteristic property of the base $e = 2.7128 \cdots$.) The same thing is true if we plot $y = e^\varepsilon$ against ε, as in Fig. 2.5. Referring to this Figure:

$$\frac{e^\varepsilon - 1}{\varepsilon} = \frac{RQ - OP}{PN} = \frac{NQ}{PN},$$

which represents the slope of the chord PQ. When $\varepsilon \to 0$, the slope of the chord PQ approaches the slope of the tangent PT, which is equal to 1. Therefore we have proved that

$$\lim_{\varepsilon \to 0} \frac{e^\varepsilon - 1}{\varepsilon} = 1. \tag{2.11}$$

The second limit to be considered is

$$\lim_{\varepsilon \to 0} \frac{\sin \varepsilon}{\varepsilon},$$

ε being measure in radians. (Degrees would give a different result: the numerator would be unaltered but the denominator would be larger.) The approach to the limit is shown in the following table, which includes negative values of ε:

ε	± 0.1	± 0.08	± 0.06	± 0.04	± 0.02
$\dfrac{\sin \varepsilon}{\varepsilon}$	± 0.99833	± 0.99893	± 0.99940	± 0.99973	± 0.99993

The limit looks as if it might equal 1.

To prove this, consider Fig. 2.6a. PN is any line segment perpendicular to the base line AB, with Q any point to the left of N, and we allow ε to represent the angle PQN (in radians). The arc PR is a circular arc with centre Q and radius PQ. Then

$$\frac{PN}{PQ} = \sin \varepsilon, \quad \text{so} \quad PN = PQ \sin \varepsilon.$$

Also (radian property, Section 1.5)

$$\text{arc } PR = PQ \times \widehat{RPQ} = PQ \, \varepsilon.$$

Therefore

$$\frac{\sin \varepsilon}{\varepsilon} = \frac{PN}{\text{arc } PR}. \tag{2.12}$$

Now let Q recede some distance towards the left, as illustrated in Fig. 2.6b. The angle ε decreases, and $\varepsilon \to 0$ as Q recedes to infinity. At the same time the arc PR approaches the straight line PN, tending ultimately to coincide with it. Therefore, when $\varepsilon \to 0$, the length of the arc PR approaches the length of PN; so, from (2.12),

$$\lim_{\varepsilon \to 0} \frac{\sin \varepsilon}{\varepsilon} = 1. \qquad (2.13)$$

Finally we consider

$$\lim_{\varepsilon \to 0} \frac{\ln(1 + \varepsilon)}{\varepsilon}.$$

Figure 2.7 shows the graphs of $y = \ln \varepsilon$, and $y = \ln(1 + \varepsilon)$. The graph $y = \ln \varepsilon$ (see Fig. 1.27) passes through the point $(1, 0)$ at $45°$ to the ε axis. The graph $y = \ln(1 + \varepsilon)$ is the same graph moved over to the left by a distance 1, so it passes through the origin O at $45°$: that is to say, it has slope equal to 1 at the origin. An argument similar to that which led to (2.11) then shows that

$$\lim_{\varepsilon \to 0} \frac{\ln(1 + \varepsilon)}{\varepsilon} = 1. \qquad (2.14)$$

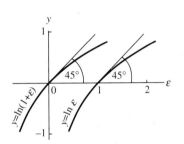

Fig. 2.7

The reader may be glad to know that there are no more complicated limits to be evaluated.

2.7 Derivatives of e^x, $\sin x$, $\cos x$, $\ln x$

These follow from the definition (2.8) and the limits obtained in the previous section.

First let

$$y = e^x.$$

Then according to the definition (2.8),

$$\frac{dy}{dx} = \lim_{\delta x \to 0} \frac{e^{x+\delta x} - e^x}{\delta x}$$

$$= \lim_{\delta x \to 0} \frac{e^x e^{\delta x} - e^x}{\delta x} = \lim_{\delta x \to 0} e^x \frac{e^{\delta x} - 1}{\delta x}.$$

Now put

$$\delta x = \varepsilon.$$

The previous expression becomes

$$e^x \lim_{\varepsilon \to 0} \frac{e^\varepsilon - 1}{\varepsilon} = e^x,$$

by (2.11). Therefore

$$\frac{d}{dx} e^x = e^x. \tag{2.15}$$

The rate of increase of e^x is therefore numerically equal to e^x itself. The simplicity of (2.15) is the reason why the number e is a desirable base for exponential functions. The result is not so simple for any other base.

Next, consider

$$y = \sin x.$$

Then

$$\frac{dy}{dx} = \lim_{\delta x \to 0} \frac{\sin(x + \delta x) - \sin x}{\delta x}.$$

From the formula in Appendix A(d)

$$\sin C - \sin D = 2 \sin \tfrac{1}{2}(C - D) \cos \tfrac{1}{2}(C + D)$$

for any C and D. Put $C = x + \delta x$ and $D = x$ into the identity. Then we have

$$\frac{dy}{dx} = \lim_{\delta x \to 0} \frac{2 \sin \tfrac{1}{2}\delta x \cos(x + \tfrac{1}{2}\delta x)}{\delta x}$$

$$= \lim_{\delta x \to 0} \frac{\sin \tfrac{1}{2}\delta x}{\tfrac{1}{2}\delta x} \cos(x + \tfrac{1}{2}\delta x).$$

Putting $\tfrac{1}{2}\delta x = \varepsilon$, we have from (2.13),

$$\frac{dy}{dx} = \lim_{\varepsilon \to 0} \frac{\sin \varepsilon}{\varepsilon} \cos(x + \varepsilon)$$

$$= \cos x.$$

Therefore

$$\frac{d}{dx} \sin x = \cos x. \tag{2.16}$$

By a closely similar argument, it can be shown that

$$\frac{d}{dx} \cos x = -\sin x. \tag{2.17}$$

(Notice the minus sign which occurs here.)

Finally, suppose that

$$y = \ln x.$$

Then from (2.8),

$$\frac{dy}{dx} = \lim_{\delta x \to 0} \frac{\ln(x + \delta x) - \ln x}{\delta x}$$

$$= \lim_{\delta x \to 0} \frac{1}{\delta x} \ln \frac{x + \delta x}{x} = \lim_{\delta x \to 0} \frac{1}{\delta x} \ln\left(1 + \frac{\delta x}{x}\right).$$

By putting

$$\frac{\delta x}{x} = \varepsilon, \quad \text{or} \quad \delta x = x\varepsilon,$$

the previous equation becomes

$$\frac{dy}{dx} = \lim_{\varepsilon \to 0} \frac{1}{x} \frac{\ln(1 + \varepsilon)}{\varepsilon}.$$

By equation (2.14) the limit of the part containing ε is 1, so we have

$$\frac{d}{dx} \ln x = \frac{1}{x}. \tag{2.18}$$

(Remember that x must be positive for $\ln x$ to have a meaning.)

2.8 A basic table of derivatives

We assemble the results (2.15) to (2.18) from Section 2.7, and (2.9) for powers of x, in a short table of derivatives.

Derivatives of the elementary functions

Function	*Derivative*	
$y = f(x)$	dy/dx or $df(x)/dx$	
c (c = constant)	zero	
x^n ($n = 1, 2, \ldots$)	nx^{n-1}	(2.19)
e^x	e^x	
$\sin x$	$\cos x$	
$\cos x$	$-\sin x$	
$\ln x$ (x positive)	$1/x$ (or x^{-1})	

The derivatives of more complicated functions can be obtained from these by using the rules described in the next chapter. A more extensive table is given in Appendix D. Remember the rule (2.10) for addition of functions and multiplication by constants.

Example 2.8. Obtain the equation of the tangent line at the point $(\frac{1}{2}\pi, \pi)$ on the graph of $y = 2x - 3\cos x$.

At a general point on the curve,

$$\frac{dy}{dx} = \frac{d}{dx}(2x - 3\cos x) = 2\frac{dx}{dx} - 3\frac{d}{dx}\cos x \quad \text{(by (2.10))}$$

$$= 2 - 3(-\sin x) = 2 + 3\sin x$$

(from the table). At $(\frac{1}{2}\pi, \pi)$ this becomes equal to

$$2 + 3\sin \tfrac{1}{2}\pi = 5,$$

and this is the slope of the tangent line at the point. The equation of the tangent line is therefore

$$\frac{y - \pi}{x - \frac{1}{2}\pi} = 5,$$

or $y = 5x - \frac{3}{2}\pi.$

2.9 Higher-order derivatives

We may differentiate a function, and then differentiate the result. For example:

if $y = x^4,$

then $\dfrac{dy}{dx} = 4x^3,$

which we shall sometimes call the **first derivative** of x^4. By differentiating again, we obtain

$$\frac{d}{dx}\frac{dy}{dx} = 12x^2,$$

which we call the **second derivative** of x^4; and so on.

In general, if $y = f(x)$, we use the notation

$$\frac{d}{dx}\frac{dy}{dx} = \frac{d^2y}{dx^2} \quad \text{or} \quad \frac{d^2f(x)}{dx^2}.$$

(Notice where the indices 2 are placed: the locations are different above and below.) If we differentiate again, we get

$$\frac{d}{dx}\frac{d}{dx}\frac{dy}{dx} = \frac{d}{dx}\frac{d^2y}{dx^2} = \frac{d^3y}{dx^3} \quad \text{or} \quad \frac{d^3f(x)}{dx^3},$$

and so on.

Example 2.9. Show that $d^4y/dx^4 = 0$ when $y = 2x^3 + 3x^2 - 1$.

Differentiating four times, we have

$$\frac{dy}{dx} = 6x^2 + 6x, \qquad \frac{d^2y}{dx^2} = 12x + 6,$$

$$\frac{d^3y}{dx^3} = 12, \qquad \frac{d^4y}{dx^4} = 0.$$

Example 2.10. Write down the sequence y, dy/dx, $d^2y/dx^2, \ldots$, d^7y/dx^7 when $y = \sin x$.

The sequence is $\sin x$, $\cos x$, $-\sin x$, $-\cos x$, $\sin x$, $\cos x$, $-\sin x$, $-\cos x$ (and it continues in this regular way).

2.10 An interpretation of the second derivative

The second derivative has a simple interpretation. Suppose that $y = f(x)$. From Section 2.9,

$$\frac{d^2y}{dx^2} = \frac{d}{dx}\frac{dy}{dx}.$$

dy/dx represents the slope of the graph, so d^2y/dx^2 gives the **rate of change of the slope** with respect to x as we move from left to right on the graph. Where d^2y/dx^2 is positive, the slope is increasing; where it is negative, the slope is decreasing.

If d^2y/dx^2 is consistently positive, then the slope dy/dx steadily increases; it might even increase from negative values (downward slope) through a zero value (tangent horizontal) to positive values (upward slope). If d^2y/dx^2 is consistently negative, then the slope steadily decreases. Figure 2.8 shows two curves upon which, respectively, the second derivative is positive and negative all the way along.

Example 2.11. Sketch one period 2π of the graph of $\sin x$ and indicate the signs of dy/dx and d^2y/dx^2.

We have

$$y = \sin x, \qquad \frac{dy}{dx} = \cos x, \qquad \frac{d^2y}{dx^2} = -\sin x.$$

The signs of the derivatives are shown in Fig. 2.9a; dy/dx is zero at the points marked Z. In Fig. 2.9b, dy/dx is sketched to show explicitly how it varies.

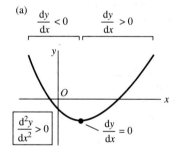

(a) $\frac{dy}{dx} < 0$ $\frac{dy}{dx} > 0$

$\frac{d^2y}{dx^2} > 0$ $\frac{dy}{dx} = 0$

(b) $\frac{dy}{dx} > 0$ $\frac{dy}{dx} < 0$

$\frac{dy}{dx} = 0$ $\frac{d^2y}{dx^2} < 0$

Fig. 2.8.

(a)

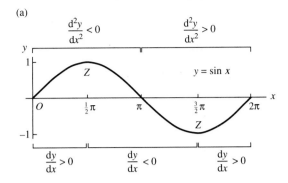

(b)

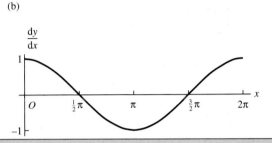

Fig. 2.9

Problems, Chapter 2

2.1. (computational). A point P is given on each of the following curves. Choose a sequence of points Q which lie closer and closer to P on the curve, and make a table giving the slopes of the chords PQ. From this table, estimate the slope of the curve at P. (Consider points on both sides of P.)

(a) $y = x^3$ at $P : (1, 1)$; (b) $y = x^{\frac{1}{2}}$ at $P : (1, 1)$;

(c) $y = \cos x$ at $P : (\frac{1}{4}\pi, 2^{-\frac{1}{2}})$;

(d) $y = e^x$ at $P : (0, 1)$; (e) $y = e^{2x}$ at $P : (0, 1)$;

(f) $y = x^3 + x^{\frac{1}{2}}$ at $P : (1, 2)$ (compare (a) and (b));

(g) $y = \ln x$ at $P : (1, 0)$.

2.2. (Sections 2.1, 2.2). Obtain dy/dx in each of the following cases at the given point P. Do this from **first principles**; that is, find δy in terms of δx, simplify $\delta y/\delta x$, and let $\delta x \to 0$ to obtain $\lim_{\delta x \to 0} \delta y/\delta x$, or dy/dx.

(a) $y = 3x$ at $P : (2, 6)$; (b) $y = 3 - 2x$ at $P : (1, 1)$;

(c) $y = 3x^2$ at $P : (1, 3)$; (d) $y = x^3$ at $P : (1, 1)$;

(e) $y = 1/x$ at $P : (2, \frac{1}{2})$; (f) $y = 3x + 2x^2$ at $P : (1, 5)$;

(g) $y = (1 + 2x)^2$ at $P : (-1, 1)$.

2.3. (Secctions 2.1, 2.2). Obtain dy/dx from first principles (see Problem 2.2) at a general point $P : (x, y)$ on the given curves.

(a) $y = 3x^2$; (b) $y = x^3$; (c) $y = 1/x$;

(d) $y = x + \frac{1}{2}$; (e) $y = x + 1/x$; (f) $y = 2x^2 - 3$.

2.4. (Section 2.3). Let x be the displacement of a point moving on a straight line, and let t represent the time elapsed. Form a table by taking the given value of t and calculating the average velocity between t and $t + \delta t$ for diminishing values of δt. Use the table to estimate the velocity at time t.

(a) $x = 3t$ at $t = 1$; (b) $x = 5t^2$ at $t = 3$;

(c) $x = 2t - 5t^2$ at $t = 1$; (d) $x = 2t - 5t^2$ at $t = 0.2$.

2.5. Use the formula (2.9) to find dy/dx at the given points in the following cases.

(a) $y = x$ at any point; (b) $y = x^3$ at $x = 3$;

(c) $y = x^4$ at $x = 2$ and at $x = -2$.

2.6. From (2.9), write down the derivatives, dy/dx or $(d/dx)f(x)$, for the given functions $f(x)$. Use this information to sketch rough graphs of $f(x)$ (notice the sign and the magnitude of the slope of $y = f(x)$).

(a) $y = x$; (b) $y = x^2$; (c) $y = x^3$;

(d) $y = x^4$; (e) $y = x^5$.

2.7. Sketch a velocity–time graph and an acceleration–time graph for a point moving on a straight line with displacement $x = t^3$. Use these to sketch a graph of acceleration against distance. (See Example 2.5.)

2.8. In the following, different letters for the variables are used in place of the usual x and y. Write down the derivatives in the appropriate form. (For example,

if $w = r^3$, then $dw/dr = 3r^2$.)
(a) $V = \frac{4}{3}\pi r^3$;   (b) $S = \pi d^2$;
(c) $E = kT^4$ (k is a constant);
(d) $I = V/R$ (R is a constant);
(e) $H = RI^2$ (R is a constant).
(f) $V = RT/P$ (R and P are constant).

2.9. Differentiate the following functions by using (2.10).
(a) $3x^2 - 2x + 1$;   (b) $x^7 - 3x^6 + x + 1$;
(c) $x + C$ (where C is a constant);
(d) $x(x - 1)$;   (e) $x^2(x^2 + 1) - 1$;
(f) $ax^2 + bx + c$ (where a, b, c, are constants);
(g) $(x - 1)^2$.

2.10. Prove that the following pairs of curves intersect in a right angle at the points given. (Hint: find dy/dx at the point for each curve.)
(a) $y = 1 + x - x^2$ and $y = 1 - x + x^2$ at (1, 1);
(b) $y = \frac{1}{2}(1 - x^2)$ and $y = x - 1$ at (1, 0);
(c) $y = 1 - \frac{1}{3}x^3$ and $y = \frac{1}{6} + \frac{1}{2}x^2$ at $(1, \frac{2}{3})$.

2.11. Find the angle between the following curves at their points of intersection. (Hint: the angle of intersection is the angle between the tangents to the curves at the point; then consider (1.7) and the tangent formula of (1.18a) for the difference of angles.)
(a) $y = x^2$ and $y = 1 - x^2$;
(b) $y = \frac{1}{3}x^3$ and $y = x^2 - 2x + \frac{4}{3}$.

2.12. (See Section 2.6). Find the limits of the following functions when $\varepsilon \to 0$. (Remember: 0/0 has no definite meaning.)

(a) $\dfrac{\varepsilon}{\varepsilon}$;   (b) $\dfrac{\varepsilon}{2\varepsilon}$;   (c) $\dfrac{\varepsilon^2}{\varepsilon}$;

(d) $\dfrac{e^{2\varepsilon} - 1}{2\varepsilon}$;   (e) $\dfrac{e^{2\varepsilon} - 1}{\varepsilon}$;   (f) $\dfrac{\sin 2\varepsilon}{2\varepsilon}$;

(g) $\dfrac{\sin 2\varepsilon}{\varepsilon}$;   (h) $\dfrac{\ln(1 + \varepsilon^2)}{\varepsilon^2}$;

(i) $\dfrac{\sin \varepsilon}{\varepsilon}$ when ε is an angle measured in *degrees*;

(j) $\dfrac{\tan \varepsilon}{\varepsilon}$;   (k) $\dfrac{\sinh \varepsilon}{\varepsilon}$;   (l) $\dfrac{e^{-\varepsilon} - 1}{\varepsilon}$.

2.13. (See Section 2.7). Obtain $d(\cos x)/dx$ in the same way that (2.16), for $\sin x$, was obtained.

2.14. (See Section 2.7). (a) Differentiate e^{2x} by following the method leading to (2.15).
(b) Differentiate $\sin 2x$ by following the method leading to (2.16).
(c) Prove that $(d/dx) e^{-x} = -e^{-x}$ by following part-way the method leading to (2.15). (Hint:

$$\lim_{\varepsilon \to 0}[(e^{-\varepsilon} - 1)/(-\varepsilon)] = 1.)$$

Use this result to differentiate $\sinh x$ and $\cosh x$ (see (1.26) for the definitions).

2.15. Differentiate the following functions.
(a) $2 \sin x - 3 \cos x$;
(b) $\ln 3x$ (see (1.9) for the properties of the logarithm);
(c) $\ln x^3$ (see (1.9));   (d) $\sin x - x$;
(e) $e^x - 1 - x - \frac{1}{2}x^2$.

2.16. Find the equations of the tangent lines in the following cases.
(a) $y = x^3$ at (1, 1);   (b) $y = x^4 - 2x^2 + 1$ at (2, 9);
(c) $y = \cos x$ at $(\frac{1}{2}\pi, 0)$;   (d) $y = \ln x$ at (e, 1);

(e) $y = \dfrac{1}{\sqrt{2}} \sin x + \dfrac{1}{\sqrt{2}} \cos x$ at $(\frac{1}{4}\pi, 1)$;

(f) $y = 3e^x - 4x$ at (0, 3).

2.17. Obtain dy/dx, d^2y/dx^2, d^3y/dx^3 in the following cases.
(a) $y = x^6$;   (b) $y = 3x^2 - 2x + 2$;
(c) $y = x^6 - x^2$;   (d) $y = 2 \sin x - 3 \cos x$;
(e) $e^x - 1 - x - \frac{1}{2}x^2$.

2.18. Show that, if N is a positive whole number, then $(d^N/dx^N)x^N = N!$

2.19. For the curve $y = x^2(x^2 - 3)$, find the ranges in x for which (a) dy/dx is positive (so that y is increasing); (b) dy/dx is negative (so that y is decreasing); (c) d^2y/dx^2 is positive (so that the slope is increasing); (d) d^2y/dx^2 is negative (so that the slope is decreasing). Deduce the general shape of the curve from these facts. (Hint: if dy/dx changes sign at some point, then dy/dx must be zero at the point. But dy/dx does not *necessarily* change sign where $dy/dx = 0$.)

3

Further techniques for differentiation

The table of elementary derivatives (2.9) is not sufficient to satisfy basic needs; for example, it does not even tell us the derivative of $\sin 2x$. But fortunately we need not start afresh every time we meet a new function. By using the rules of combination given in this chapter it is possible to differentiate functions that are developments of those given in (2.9), no matter how complicated they are.

3.1 The product rule

The derivatives of a product of several functions can be obtained when the derivatives of its individual components are known. Examples of such products are

$$x^2 \, e^x, \quad e^x \sin x, \quad x \, e^x \cos x.$$

Suppose firstly that y takes the form of a product of two functions $u(x)$ and $v(x)$:

$$y(x) = u(x)v(x),$$

where $y(x)$ is written to display the dependence of y on x. We require $\mathrm{d}y/\mathrm{d}x$ in terms of u and v. Fix a value for x, and change it by an amount δx so that

$$x \text{ becomes } x + \delta x.$$

Then u, v, and y all change:

$$u \text{ becomes } u + \delta u, \quad v \text{ becomes } v + \delta v, \quad \text{and } y \text{ becomes } y + \delta y,$$

where u, v, y represent the values at x. Since

$$\delta y = (u + \delta u)(v + \delta v) - uv,$$

we obtain

$$\frac{\delta y}{\delta x} = \frac{(u + \delta u)(v + \delta v) - uv}{\delta x} = \frac{uv + u \, \delta v + v \, \delta u + \delta u \, \delta v - uv}{\delta x}$$

$$= u \frac{\delta v}{\delta x} + v \frac{\delta u}{\delta x} + \delta u \frac{\delta v}{\delta x}.$$

Now let $\delta x \to 0$, so that $\delta y/\delta x$, $\delta u/\delta x$, $\delta v/\delta x$ become $\mathrm{d}y/\mathrm{d}x$, $\mathrm{d}u/\mathrm{d}y$, $\mathrm{d}v/\mathrm{d}x$ respectively. Also, since $\delta u \to 0$ when $\delta x \to 0$, the final term becomes zero, and we obtain the product rule:

Product rule

If $y(x) = u(x)v(x)$, then

$$\frac{dy}{dx} = \frac{d}{dx}(uv) = u\frac{dv}{dx} + v\frac{du}{dx}.$$

(3.1)

Example 3.1. Find dy/dx when $y = x^2 e^x$.

Put $u = x^2$, $v = e^x$, $y = x^2 e^x = uv$. Then

$$\frac{du}{dx} = 2x, \qquad \frac{dv}{dx} = e^x.$$

Therefore, by (3.1),

$$\frac{dy}{dx} = u\frac{dv}{dx} + v\frac{du}{dx}$$

$$= x^2(e^x) + e^x(2x) = (x^2 + 2x)\,e^x.$$

Example 3.2. Find dx/dt when $x = e^t \cos t$.

We have to interpret (3.1) in terms of the new symbols. Put

$$u = e^t, \qquad v = \cos t, \qquad x = e^t \cos t = uv.$$

Then (refer if necessary to the table (2.19) with the appropriate changes of letters)

$$\frac{du}{dt} = e^t, \qquad \frac{dv}{dt} = -\sin t.$$

Changing the symbols in (3.1) we have

$$\frac{dx}{dt} = u\frac{dv}{dt} + v\frac{du}{dt}$$

$$= e^t(-\sin t) + (\cos t)\,e^t = e(\cos t - \sin t)$$

$$= e^t(\cos t - \sin t).$$

Example 3.3. Find dy/dx when $y = x e^x \sin x$.

This product has three terms, and we have to carry out the differentiation in two stages. Write

$$y = (x e^x)\sin x$$

and put $u = x e^x$ and $v = \sin x$. By (3.1),

$$\frac{dy}{dx} = x\,e^x \frac{d}{dx}\sin x + \sin x \frac{d}{dx}(x\,e^x)$$

$$= x\,e^x \cos x + \sin x \frac{d}{dx}(x\,e^x). \tag{i}$$

To evaluate $(d/dx)(x\,e^x)$, use the product rule again, putting $u = x$ and $v = e^x$. Then

$$\frac{d}{dx}(x\,e^x) = x\frac{d}{dx}e^x + e^x\frac{dx}{dt} = x\,e^x + e^x. \tag{ii}$$

Replace (ii) into (i):

$$\frac{dy}{dx} = x\,e^x \cos x + (\sin x)(x\,e^x + e^x)$$

$$= e^x(x\cos x + x\sin x + \sin x).$$

Another method of dealing with the product of several terms, which is usually more convenient, is given in Section 3.7. The reader is strongly recommended to write out all the steps completely at first, otherwise mistakes are likely to occur.

3.2 Quotients and reciprocals
Suppose that

$$y(x) = \frac{u(x)}{v(x)}.$$

Proceed as for the product rule: let x change to $x + \delta x$, so that u becomes $u + \delta u$, v becomes $v + \delta v$, and y becomes $y + \delta y$. Then

$$\frac{\delta y}{\delta x} = \left(\frac{u + \delta u}{v + \delta v} - \frac{u}{v}\right)\frac{1}{\delta x}$$

$$= \frac{uv + v\,\delta u - uv - u\,\delta v}{v(v + \delta v)\,\delta x} = \frac{v\,\delta u - u\,\delta v}{v(v + \delta v)\,\delta x}$$

$$= \frac{1}{v(v + \delta v)}\left(v\frac{\delta u}{\delta x} - u\frac{\delta v}{\delta x}\right).$$

Let $\delta x \to 0$; then $\delta y/\delta x$, $\delta u/\delta x$, $\delta v/\delta x$ become dy/dx, du/dx, dv/dx, and $\delta v \to 0$. Therefore

$$\frac{dy}{dx} = \frac{d}{dx}\frac{u}{v} = \frac{1}{v^2}\left(v\frac{du}{dx} - u\frac{dv}{dx}\right).$$

It is worth noting the special case of the reciprocal of a function. In that case, $u(x) = 1$, so $du/dx = 0$. Finally we have

Quotient and reciprocal rules

(a) If $y(x) = \dfrac{u(x)}{v(x)}$, then

$$\frac{dy}{dx} = \frac{d}{dx}\frac{u}{v} = \frac{1}{v^2}\left(v\frac{du}{dx} - u\frac{dv}{dx}\right). \tag{3.2}$$

(b) If $y(x) = \dfrac{1}{v(x)}$ (i.e. if $u(x) = 1$), then

$$\frac{dy}{dx} = \frac{d}{dx}\frac{1}{v} = -\frac{1}{v^2}\frac{dv}{dx}.$$

Example 3.4. Obtain dy/dx when $y = \tan x$.

Express y in the form

$$y = \tan x = \frac{\sin x}{\cos x}.$$

Put $u = \sin x$, $v = \cos x$, $y = u/v$. Then

$$\frac{du}{dx} = \cos x, \qquad \frac{dv}{dx} = -\sin x.$$

From (3.2),

$$\frac{dy}{dx} = \frac{1}{v^2}\left(v\frac{du}{dx} - u\frac{dv}{dx}\right)$$

$$= \frac{1}{\cos^2 x}[\cos x \cos x - \sin x(-\sin x)]$$

$$= \frac{1}{\cos^2 x}(\cos^2 x + \sin^2 x) = \frac{1}{\cos^2 x}.$$

(Remember that $\cos^2 A + \sin^2 A = 1$.)

Example 3.5. Find dy/dx when $y = (x + 3)/(2x^3 + 1)$.

Put $u = x + 3$, $v = 2x^3 + 1$, $y = u/v$. Then

$$\frac{du}{dx} = 1, \qquad \frac{dv}{dx} = 6x^2.$$

By (3.2),

$$\frac{dy}{dx} = \frac{1}{(2x^3 + 1)^2}[(2x^3 + 1)(1) - (x + 3)(6x^2)]$$

$$= \frac{1 - 18x^2 - 4x^3}{(2x^3 + 1)^2}.$$

Example 3.6. Obtain dy/dx when (a) $y = 1/x$; (b) $y = 1/x^2$.

(a) Put $v = x$ into the reciprocal rule (3.2b) (or $u = 1$ and $v = x$ into (3.2a)):

$$\frac{dy}{dx} = -\frac{1}{v^2}\frac{dv}{dx} = -\frac{1}{x^2}.$$

(b) Put $v = x^2$ into (3.2b):

$$\frac{dy}{dx} = -\frac{1}{x^4}2x = -\frac{2}{x^3}.$$

If we had put $n = -1$ and -2 respectively into the formula (2.9),

$$\frac{d}{dx}(x^n) = nx^{n-1},$$

proved only for *positive* integer n, the correct result is obtained. The formula is in fact correct for all values of n, as will be shown in Section 3.4.

3.3 The chain rule

The **chain rule** will be used continually in future chapters. It is also called the **function-of-a-function rule**, for the following reason. Suppose we require dy/dx in the special case where y can be expressed as a function of a variable u, where u is a function of x. We shall express this by the notation

$$y = y(u), \quad \text{where} \quad u = u(x).$$

An example of this is

$$y = \cos(x^3),$$

which we can rewrite in the form

$$y = y(u) = \cos u, \quad \text{where} \quad u = u(x) = x^3.$$

Another example is when $y = \cos^3 x$. Write it in the form $y = (\cos x)^3$, so that

$$y = u^3, \quad \text{where} \quad u = \cos x.$$

The rule for such cases is the following.

The chain rule

If $y = y(u)$ where $u = u(x)$, then

$$\frac{dy}{dx} = \frac{dy}{du}\frac{du}{dx}.$$

(3.3)

The form of this result is *easy to remember* if you first write

$$\frac{dy}{dx} = \frac{dy}{\bullet} \frac{\bullet}{dx},$$

then put du in place of the dots. Sometimes it is inconvenient to use u; any letter not already in use can be used in place of u.

To prove (3.3), fix on any value of x. Consider a nearby value $x + \delta x$, and denote the corresponding small changes in u and y by δu and δy. When x becomes $x + \delta x$, then u becomes $u + \delta u$ and y becomes $y + \delta y$. Evidently

$$\frac{\delta y}{\delta x} = \frac{\delta y}{\delta u} \frac{\delta u}{\delta x}$$

since the terms δu cancel. Now let $\delta x \to 0$. Then $\delta u \to 0$, and consequently $\delta y/\delta x$, $\delta y/\delta u$, $\delta u/\delta x$ approach dy/dx, dy/du, du/dx respectively. Thus we obtain

$$\frac{dy}{dx} = \frac{dy}{du} \frac{du}{dx}.$$

The following examples show how to recognize when it is appropriate to use the chain rule. You should lay out every application in the systematic way shown until you are used to it.

Example 3.7. (a) You are given that $(d/dx)\, e^x = e^x$ (see the table, (2.19)). Deduce that $(d/dx)\, e^{ax} = a\, e^{ax}$, where a is any constant. (b) Find the derivative of e^{-x}. (c) Use this result to obtain the derivatives of $\sinh x$ and $\cosh x$ (see (1.26) for the definitions of these functions).

(a) Rewrite $y = e^{ax}$ in the form

$$y = e^u, \quad \text{where} \quad u = ax.$$

To use the chain rule (3.3), we need dy/du and du/dx:

$$\frac{dy}{du} = e^u \quad \text{and} \quad \frac{du}{dx} = a.$$

The chain rule gives

$$\frac{dy}{dx} = \frac{dy}{du} \frac{du}{dx} = e^u\, a = a\, e^{ax},$$

after restoring the variable x.

(b) For e^{-x}, the constant a is -1, so

$$\frac{d}{dx} e^{-x} = -e^{-x}.$$

(c) $\sinh x = \frac{1}{2}(e^x - e^{-x})$ and $\cosh x = \frac{1}{2}(e^x + e^{-x})$, so the result (b) gives

$$\frac{d}{dx}(\sinh x) = \frac{d}{dx}[\tfrac{1}{2}(e^x - e^{-x})] = \tfrac{1}{2}(e^x + e^{-x})$$

$$= \cosh x,$$

$$\frac{d}{dx}(\cosh x) = \frac{d}{dx}[\tfrac{1}{2}(e^x + e^{-x})] = \tfrac{1}{2}(e^x - e^{-x})$$

$$= \sinh x.$$

Example 3.8. Find dy/dx when $y = (x^2 + 1)^{10}$.

We could expand $(x^2 + 1)^{10}$ as a polynomial by means of the binomial theorem, but the chain rule is far simpler, Put

$$y = u^{10}, \quad \text{where} \quad u = x^2 + 1.$$

Then

$$\frac{dy}{du} = 10u^9 \quad \text{and} \quad \frac{du}{dx} = 2x.$$

By the chain rule,

$$\frac{dy}{dx} = \frac{dy}{du}\frac{du}{dx} = 10u^9 2x = 20x(x^2 + 1)^9.$$

Example 3.9. Find dy/dx when (a) $y = \sin(x^3)$; (b) $y = \sin^3 x$.

(a) Put $y = \sin u$, where $u = x^3$. Then

$$\frac{dy}{du} = \cos u \quad \text{and} \quad \frac{du}{dx} = 3x^2,$$

By the chain rule,

$$\frac{dy}{dx} = \frac{dy}{du}\frac{du}{dx} = (\cos u)3x^2 = 3x^2 \cos(x^3).$$

(b) Put $y = u^3$, where $u = \sin x$. Then

$$\frac{dy}{du} = 3u^2 \quad \text{and} \quad \frac{du}{dx} = \cos x.$$

By the chain rule,

$$\frac{dy}{dx} = \frac{dy}{du}\frac{du}{dx} = 3u^2 \cos x = 3 \sin^2 x \cos x.$$

Example 3.10. Find dy/dx when $y = 1/(x^2 + 1)$.

Put $y = 1/u$, where $u = x^2 + 1$. Then

$$\frac{dy}{du} = -\frac{1}{u^2} \quad \text{and} \quad \frac{du}{dx} = 2x,$$

(where the reciprocal rule (3.2b) was used for differentiating $1/u$.)

By the chain rule,

$$\frac{dy}{dx} = -\frac{1}{u^2} 2x = -\frac{2x}{(x^2 + 1)^2}.$$

Example 3.11. Find du/dt when $u = a \cos k(x - ct)$ where a, k, c, and x are constant, t and u being the only variables.

We should not use u for the intermediate variable in the chain rule (3.3), because it is already in use (as the name of the dependent variable). Instead of u, use an uncommitted letter such as w as the intermediate variable, putting

$$u = a \cos w, \quad \text{where} \quad w = kx - kct.$$

The chain rule takes the form

$$\frac{du}{dt} = \frac{du}{dw}\frac{dw}{dt},$$

in which

$$\frac{du}{dw} = -a \sin w \quad \text{and} \quad \frac{dw}{dt} = -kc.$$

Therefore

$$\frac{du}{dt} = (-a \sin w)(-kc) = akc \sin k(x - ct).$$

3.4 Derivative of x^n for any value of n

Consider the derivative of $y = x^n$, where n may have any value, an integer or not, positive or negative. The rule turns out to take the same form as (2.9), in which n was limited to positive integers:

> **Derivative of x^n**
>
> If $y = x^n$, where n may take any value whatever, then   (3.4)
>
> $$\frac{dy}{dx} = nx^{n-1}.$$

To prove (3.4), we use the chain rule (3.3). Note that $x = e^{\ln x}$ (see (1.21)), so that

$$y = x^n = (e^{\ln x})^n = e^{n \ln x}.$$

To use the chain rule, we put this in the form

$$y = e^u, \quad \text{where} \quad u = n \ln x,$$

so that

$$\frac{dy}{du} = e^u \quad \text{and} \quad \frac{du}{dx} = \frac{n}{x}.$$

Then

$$\frac{dy}{dx} = \frac{dy}{du}\frac{du}{dx} = e^u \frac{n}{x} = x^n \frac{n}{x} = nx^{n-1}$$

(where we used $e^u = y = x^n$ again).

Example 3.12. Find dy/dx when (a) $y = x^{\frac{3}{2}}$, (b) $y = 1/x^{\frac{3}{2}}$, (c) $y = 1/\sqrt{x}$, (d) $y = 1/(2x^{\frac{1}{3}} + x)$.

(a) Here $n = \frac{3}{2}$ in (3.4), so $\dfrac{d}{dx}(x^{\frac{3}{2}}) = \frac{3}{2}x^{\frac{1}{2}}$.

(b) This may be written $y = x^{-\frac{3}{2}}$, so $n = -\frac{3}{2}$ in (3.4), and

$$\frac{d}{dx}(x^{-\frac{3}{2}}) = -\frac{3}{2}x^{-\frac{5}{2}}.$$

(c) $y = x^{-\frac{1}{2}}$, so $\dfrac{dy}{dx} = -\frac{1}{2}x^{-\frac{3}{2}}$.

(d) Write $y = (2x^{\frac{1}{3}} + x)^{-1}$. We can use the chain rule: put $y = u^{-1}$, where $u = 2x^{\frac{1}{3}} + x$.

Then

$$\frac{dy}{dx} = -u^{-2} \quad \text{(by (3.4))}, \qquad \frac{du}{dx} = \frac{2}{3}x^{-\frac{2}{3}} + 1 \quad \text{(by (3.4))}.$$

Therefore, by the chain rule (3.3),

$$\frac{dy}{dx} = \frac{dy}{du}\frac{du}{dx} = (-u^{-2})(\frac{2}{3}x^{-\frac{2}{3}} + 1) = -\frac{\frac{2}{3}x^{-\frac{2}{3}} + 1}{(2x^{\frac{1}{3}} + x)^2}.$$

3.5 Functions of $ax + b$

A frequently occurring application of the chain rule (3.3) is in connection with functions like e^{ax+b}, $\sin(ax + b)$, $(ax + b)^n$, and in general $f(ax + b)$. The spirit of the chain rule is to say: 'If the functions were e^x, $\sin x$, x^n, $f(x)$, then they would be easy. Therefore, try the chain rule with $u = ax + b$.'

Suppose that, in general, we want to differentiate y when

$$y = f(ax + b),$$

and that we know how to differentiate $f(x)$. Write

$$u = ax + b, \qquad y = f(u).$$

Then the chain rule gives

$$\frac{dy}{dx} = \frac{dy}{du}\frac{du}{dx} = a\frac{df(u)}{du},$$

in which the derivative occurring on the right is already known.

The following special cases, in which $b = 0$, should be noticed;

they consitute an extension of the table (2.19).

Function	Derivative	
e^{ax}	$a\,e^{ax}$	(3.5)
$\sin ax$	$a \cos ax$	
$\cos ax$	$-a \sin ax$	

3.6 An extension of the chain rule

In Section 3.3, the chain rule (3.3) was looked upon as a way of differentiating a function of a function, say $y\big(u(x)\big)$. Sometimes we need to consider 'a function of a function of a function of ...'. These can always be worked through by repeated applications of (3.3), but it may be less complicated to proceed as in the following example.

Example 3.13. Obtain dy/dx when $y = e^{\sin(x^2 + 1)}$.

Instead of using one intermediate variable u, introduce two variables, u and v. Put

$$y = e^v, \qquad v = \sin u, \qquad u = x^2 + 1.$$

Then by exactly the same type of arguments as led to (3.3),

$$\frac{dy}{dx} = \frac{dy}{dv}\frac{dv}{du}\frac{du}{dx}.$$

We have

$$\frac{du}{dx} = 2x, \qquad \frac{dv}{du} = \cos u, \qquad \frac{dy}{dv} = e^v,$$

so

$$\frac{dy}{dx} = (e^v \cos u)2x = 2x\, e^{\sin(x^2 + 1)} \cos(x^2 + 1).$$

The result can be extended in an obvious way to any number of intermediate variables, but it is seldom that more than two would be needed:

Extended chain rule

Suppose that

$$y = y(v), \quad v = v(u), \quad \text{and} \quad u = u(x).$$

Then

$$\frac{dy}{dx} = \frac{dy}{dv}\frac{dv}{du}\frac{du}{dx}.$$

(3.6)

3.7 Logarithmic differentiation

To differentiate a product

$$y = u(x)v(x)w(x)$$

consisting of three terms, the product rule (3.1) can be applied twice, as in Example 3.3. An alternative procedure, which is often simpler, is the following.

Since $y = uvw$,

$$\ln y = \ln(uvw) = \ln u + \ln v + \ln w.$$

By the chain rule (3.3), if $u(x)$ is any function of x,

$$\frac{d}{dx}\ln u = \frac{1}{u}\frac{du}{dx},$$

and we may have y, v, or w in place of u. Therefore

$$\frac{1}{y}\frac{dy}{dx} = \frac{1}{u}\frac{du}{dx} + \frac{1}{v}\frac{dv}{dx} + \frac{1}{w}\frac{dw}{dx},$$

By multiplying through by $y = uvw$, we obtain:

Logarithmic differentiation

If $y = uvw$, then

$$\frac{dy}{dx} = uvw\left(\frac{1}{u}\frac{du}{dx} + \frac{1}{v}\frac{dv}{dx} + \frac{1}{w}\frac{dw}{dx}\right) \qquad (3.7)$$

(and so for any number of terms).

Example 3.14. Find dy/dx when $y = (x^{\frac{1}{2}}\sin^2 x)/(x^2 + 1)$.

Put $y = uvw$, where

$$u = x^{\frac{1}{2}}, \qquad v = \sin^2 x, \qquad w = (x^2 + 1)^{-1}.$$

Then

$$\ln y = \ln(x^{\frac{1}{2}}) + \ln(\sin^2 x) + \ln(x^2 + 1)^{-1}.$$
$$= \tfrac{1}{2}\ln(x) + 2\ln(\sin x) - \ln(x^2 + 1).$$

Notice that we did not just copy the formula (3.7) rigidly: the logarithm is useful for getting rid of awkward powers and we might have missed this. Differentiate this expression:

$$\frac{1}{y}\frac{dy}{dx} = \frac{1}{2x} + \frac{2}{\sin x}\frac{d}{dx}(\sin x) - \frac{1}{x^2 + 1}\frac{d}{dx}(x^2 + 1)$$

$$= \frac{1}{2x} + \frac{2\cos x}{\sin x} - \frac{2x}{x^2 + 1}.$$

Multiply through by $y = (x^{\frac{1}{2}}\sin^2 x)/(x^2 + 1)$ to give dy/dx:

$$\frac{dy}{dx} = \frac{x^{\frac{1}{2}}\sin^2 x}{x^2 + 1}\left(\frac{1}{2x} + \frac{2\cos x}{\sin x} - \frac{2x}{x^2 + 1}\right).$$

3.8 Implicit differentiation

An equation of the form

$$f(x, y) = c \quad \text{(a constant)}$$

represents a curve or curves; for example $x^2 + y^2 = 1$ represents a circle, otherwise expressed by $y = y(x) = \pm(1 - x^2)^{\frac{1}{2}}$. This latter relation is **implicit** in the first form, which is called an **implicit equation**.

Suppose that the implicit equation for y is $f(x, y) = c$ and its explicit equation is $y = y(x)$; that is to say, both equations specify the same curve. Then $f(x, y(x)) = c$ **for all values of** x: it is an **identity**. Since the value of $f(x, y(x))$ remains constant, its derivative is zero:

$$\frac{\mathrm{d}}{\mathrm{d}x} f(x, y(x)) = 0,$$

for every relevant value of x.

Notice further that if y is a function of x, then the chain rule (3.3), using y as the intermediate variable instead of u, gives results such as

$$\frac{\mathrm{d}}{\mathrm{d}x} y^2 = 2y \frac{\mathrm{d}y}{\mathrm{d}x} \quad \text{and} \quad \frac{\mathrm{d}}{\mathrm{d}x} \cos y = -\sin y \frac{\mathrm{d}y}{\mathrm{d}x}.$$

This fact can be used in the following way to obtain an expression for $\mathrm{d}y/\mathrm{d}x$, even in cases when we cannot solve the implicit equation to obtain y as a function of x explicitly.

Example 3.15. Find a general expression for $\mathrm{d}y/\mathrm{d}x$ at any point on the curve given by $f(x, y) = x + y + \sin x + \cos y = 1$.

So long as we stay on the curve, $f(x, y)$ does not change when x changes, so $\mathrm{d}f(x, y)/\mathrm{d}x = 0$. Therefore

$$1 + \frac{\mathrm{d}y}{\mathrm{d}x} + \cos x - \sin y \frac{\mathrm{d}y}{\mathrm{d}x} = 0,$$

so finally

$$\frac{\mathrm{d}y}{\mathrm{d}x} = \frac{\cos x + 1}{\sin y - 1}.$$

Such a result is not quite so good as its neatness suggests, because we would still find it hard to say what values of y are to be associated with a particular value of x in the new formula: this would in effect involve solving the original equation for y in terms of x.

3.9 Derivatives of inverse functions

The derivatives of functions such as $\ln x$, $\arctan x$, $\arccos x$, and $\arcsin x$, which are the respective inverses of e^x, $\tan x$, $\cos x$, and $\sin x$ can be obtained by a standard procedure. We need the general result:

$$\frac{dy}{dx} = 1 \bigg/ \frac{dx}{dy} \qquad\qquad (3.8)$$

To illustrate what (3.8) means, take as an example the case when $y = \ln x$. As described in Section 1.9, two statements

$$y = \ln x \text{ for } x > 0 \quad \text{and} \quad x = e^y \text{ for all } y$$

are different ways of saying the same thing; the two graphs depicting the relation between x and y are the same graph. For small corresponding increments δx and δy on this graph,

$$\frac{\delta y}{\delta x} = 1 \bigg/ \frac{\delta x}{\delta y}.$$

Since δx and δy approach zero together, we obtain $dy/dx = 1/(dx/dy)$, which is (3.8).

To find dy/dx when $y = \ln x$, write equivalently

$$x = e^y.$$

Then

$$\frac{dx}{dy} = e^y;$$

so, by (3.7),

$$\frac{dy}{dx} = \frac{1}{e^y} = \frac{1}{x}.$$

This, of course, agrees with (2.18), where a more direct method was used.

Example 3.16. Find dy/dx when $y = \arctan x$.

If $y = \arctan x$, then $x = \tan y$. From Example 3.4, interchanging x and y,

$$\frac{dx}{dy} = \frac{1}{\cos^2 y};$$

so, by (3.7),

$$\frac{dy}{dx} = \cos^2 y.$$

To express $\cos^2 y$ in terms of $\tan y$ (i.e. in terms of x) draw the triangle in Fig. 3.1, in which y is represented by the angle A. This is a right-angled triangle because the sides conform with Pythagoras' theorem. Evidently $\cos^2 y = 1/(1 + \tan^2 y)$, as can be checked by

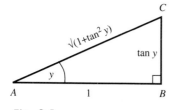

Fig. 3.1

putting $\tan^2 y = (\sin^2 y)/\cos^2 y$. Therefore

$$\frac{dy}{dx} = \frac{1}{1 + \tan^2 y} = \frac{1}{1 + x^2}.$$

3.10 Derivative as a function of a parameter

If x and y are functions of a **parameter**, or supplementary variable, t, so that

$$x = x(t), \qquad y = y(t),$$

then the point $(x(t), y(t))$ follows a curve as t varies. Suppose that t changes from t to $t + \delta t$; then x changes to $x + \delta x$ and y to $y + \delta y$. Obviously

$$\frac{\delta y}{\delta x} = \frac{\delta y}{\delta t} \bigg/ \frac{\delta x}{\delta t},$$

since δt cancels on the right-hand side. Let $\delta t \to 0$; then $\delta x \to 0$, and we have

---

Differentiation in terms of a parameter

If $x = x(t)$ and $y = y(t)$, then

$$\frac{dy}{dx} = \frac{dy}{dt} \bigg/ \frac{dx}{dt}.$$

(3.9)

Example 3.17. A curve is given in polar coordinates by $r = \sin \theta$. Find dy/dx at the point where $\theta = \frac{1}{8} \pi$.

We can use θ as the parameter in the following way. The universal relation between polar and cartesian coordinates is

$$x = r \cos \theta, \qquad y = r \sin \theta.$$

On the special curve described by $r = \sin \theta$, these equations become

$$x = \sin \theta \cos \theta, \qquad y = \sin^2 \theta.$$

Then

$$\frac{dx}{d\theta} = -\sin^2 \theta + \cos^2 \theta = \cos 2\theta$$

(by the product rule and the identity in Appendix B) and

$$\frac{dy}{d\theta} = 2 \sin \theta \cos \theta = \sin 2\theta$$

(by the chain or product rule and the identity in Appendix B). Therefore

$$\frac{dy}{dx} = \frac{dy}{d\theta} \bigg/ \frac{dx}{d\theta} = \tan 2\theta.$$

At the point where $\theta = \frac{1}{8}\pi$,

$$\frac{dy}{dx} = \tan \frac{1}{4}\pi = 1.$$

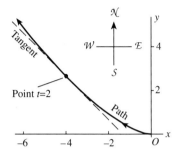

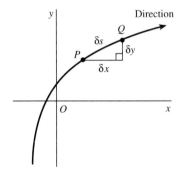

Fig. 3.2

Example 3.18. The map coordinates of a moving vehicle are given by $x = -t^2$, $y = \frac{1}{3}t^3$, where t is time and $t > 0$. Find the direction the vehicle is facing when $t = 2$.

From (3.8),

$$\frac{dy}{dx} = \frac{dy}{dt} \bigg/ \frac{dx}{dt} = \frac{t^2}{-2t} = -\frac{1}{2}t.$$

This equals -1 when $t = 2$. The slope of the curve is negative at this point, so the tangent to the path slopes downwards from left to right as shown in Fig. 3.2. The actual direction in which the vehicle is moving is, however, from right to left. It is facing north west as shown.

From information such as that given in the previous example, the speed of a moving point can be calculated. Suppose that a point moves so that

$$x = x(t), \qquad y = y(t),$$

where t represents **time**. Figure 3.3 shows the effect of changing t to $t + \delta t$, where δt is small: the point moves from P to Q, a short distance δs say along the curve (δs is called an element of arc length). Then the **average speed** over this short time is given by

$$\frac{\text{arc length } PQ}{\delta t} = \frac{\delta s}{\delta t}$$

$$\approx \frac{\text{straight distance } PQ}{\delta t}$$

$$= \frac{(\delta x^2 + \delta y^2)^{\frac{1}{2}}}{\delta t}$$

$$= \left[\left(\frac{\delta x}{\delta t} \right)^2 + \left(\frac{\delta y}{\delta t} \right)^2 \right]^{\frac{1}{2}}$$

Now let $\delta t \to 0$. Then $\delta x/\delta t$ and $\delta y/\delta t$ become dx/dt and dy/dt, and finally we have the result:

Fig. 3.3

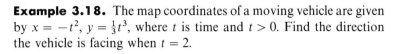

> **Speed of a moving point**
>
> Let $x = x(t)$ and $y = y(t)$, where t is time.
> The speed of the point is given by
>
> $$\text{speed} = \frac{ds}{dt} = \left[\left(\frac{dx}{dt} \right)^2 + \left(\frac{dy}{dt} \right)^2 \right]^{\frac{1}{2}},$$
>
> (3.10)
>
> where ds stands for an element of arc length.

Example 3.19. Find the speed of the vehicle in Example 3.18 when $t = 2$.

$$\text{In general } \frac{ds}{dt} = \left[\left(\frac{dx}{dt} \right)^2 + \left(\frac{dy}{dt} \right)^2 \right]^{\frac{1}{2}},$$

$$= (4t^2 + t^4)^{\frac{1}{2}}.$$

The speed is therefore $4\sqrt{2}$ when $t = 2$. (Speed is always counted as a non-negative number: when we want to make a distinction as to direction, the word 'velocity' is used.)

Problems, Chapter 3

3.1. (Product rule, (3.1)). Obtain $df(x)/dx$ for the following $f(x)$.

(a) $x\,e^x$; (b) $x \sin x$; (c) $x \cos x$; (d) $e^x \sin x$;
(e) $x \ln x$; (f) $x^2 \ln x$; (g) $e^x \ln x$; (h) $x^2 e^x$;
(i) $\sin x \cos x$; (j) $x^2 x^3$ (this is the same as x^5: show that the result is the same for both forms).

3.2. (Quotient and reciprocal rule, (3.2)). Obtain $df(x)/dx$ for the following $f(x)$.

(a) $\cot x$; (b) $x/(x + 1)$; (e) $(\sin x)/x$; (d) e^x/x;
(e) $(x^2 - 1)/(x^2 + 1)$; (f) $(\tan x)/x^2$;
(g) $(\sin x + \cos x)/(\sin x - \cos x)$;
(h) $\sec x \ (= 1/\cos x)$; (i) $\operatorname{cosec} x \ (= 1/\sin x)$;
(j) $x/(3x^2 - 2)$; (k) $1/x(x^3 + 1)$; (l) $1/\ln x$;
(m) x^n where n is a negative whole number
$(x^n = 1/x^{-n})$; (n) $1/(x + 1)$; (o) $e^{-x} \ (= 1/e^x)$;
(p) $1/\tan x$; (q) $x^{-2} \ln x$.

3.3. Find the first, second, and third derivatives of
(a) $1/(1 - x)$; (b) $x \sin x$; (c) $x/(x - 1)$;
(d) $f(x)g(x)$, where f and g are any functions.

3.4. (Chain rule, (3.3). Obtain $df(x)/dx$ for the following $f(x)$. (Set out the calculation systematically, as in the examples in Section 3.3.)

(a) $\sin^2 x$; (b) $\cos^2 x$; (c) $\sin x^2$; (d) $\cos x^2$;
(e) $\tan^2 x$; (f) $\tan x^2$; (g) $\cos(1/x)$;

(h) e^{-x} (compare Problem 3.2o); (i) $(x + 1)^5$;
(j) $(x^3 + 1)^4$; (k) $\sin 3x$; (l) $\cos \frac{1}{2}x$;
(m) $\tan \frac{1}{2}x$; (n) e^{-3x}; (o) $\sin(2x + 1)$;
(p) $\cos(3x - 2)$; (q) $\tan(1 - 2x)$; (r) $e^{1/x}$;
(s) a^x (write as a power of e).

3.5. (General powers of x, Section 3.4). Differentiate the following.

(a) x^{-2}; (b) x^{-1}; (c) $x^{\frac{1}{3}}$; (d) $x^{-\frac{1}{3}}$; (e) $x^{\frac{3}{2}}$;
(f) $\sqrt{x}$; (g) $\sqrt{(x^3)}$; (h) $1/x$; (i) $1/\sqrt{x}$.

3.6. Differentiate the following (the independent variable is not always x, and more than one rule is needed).

(a) $x^{\frac{1}{2}} \sin x$; (b) $\sin^{\frac{1}{3}} x$; (c) $(x^2 + 1)^{-\frac{1}{2}}$;
(d) $\sin^2(3t + 1)$; (e) $e^{-t} \cos t$; (f) $e^{-t} \sin t$;
(g) $e^{-2t} \cos 3t$; (h) $e^{-3t} \cos 2t$;

(i) $\sin x \cos^2 x$; (j) $\sin^2 x \cos x$; (k) $\left(\dfrac{\sin x}{x} \right)^2$;

(l) $x \sin^3 x$; (m) $x \cos^3 x$.

3.7. Differentiate $\cos^2 x$ and $\sin^2 x$, (a) by using the identities $\cos^2 A = \frac{1}{2}(1 + \cos 2A)$ and $\sin^2 A = \frac{1}{2}(1 - \cos 2A)$, (b) by using the product rule, (c) by using the chain rule.

3.8. Confirm the correctness of the following statements. The letters A, B, C, D, and n stand for any constants.

(a) If $x = A \cos 2t + B \sin 2t$, then $\dfrac{d^2 x}{dt^2} + 4x = 0$;

(b) If $x = A \cos nt + B \sin nt$, then $\dfrac{d^2 x}{dt^2} + n^2 x = 0$;

(c) If $x = A e^{3t} + B e^{-3t}$, then $\dfrac{d^2 x}{dt^2} - 9x = 0$;

(d) If $x = A e^{nt} + B e^{-nt}$, then $\dfrac{d^2 x}{dt^2} - n^2 x = 0$;

(e) If $x = A e^{-t} \cos t + B e^{-t} \sin t$, then
$$\frac{d^2 x}{dt^2} + 2\frac{dx}{dt} + 2x = 0;$$

(f) If $y = A e^x + B e^{-x} + C \cos x + D \sin x$, then
$$\frac{dy^4}{dx^4} - y = 0.$$

3.9. (Chain rule (3.3); or, more easily, the extension (3.5)). Differentiate the following functions.
(a) $e^{\cos^2 x}$; (b) $e^{-\cos x^2}$; (c) $\ln(\cos x^2)$; (d) $(e^{x^2} - 1)^4$.

3.10. (Logarithmic differentiation, Section 3.7, is easiest). Differentiate the following.
(a) $x e^x \sin x$; (b) $t e^t \cos t$; (c) $x^{\frac{1}{2}} e^{2x} \sin^{\frac{1}{2}} 3x$.

3.11. (Implicit differentiation, Section 3.8). Proceed as in Example 3.8 to obtain expressions for dy/dx in the following.
(a) Show that if $x^2 + y^2 = 4$, then $dy/dx = -x/y$. Check the correctness of the expression by testing it with $y = \pm(4 - x^2)^{\frac{1}{2}}$. Interpret the result geometrically by sketching the circle $x^2 + y^2 = 4$ and considering the meaning of dy/dx in terms of slope. (b) $x^{\frac{1}{2}} + y^{\frac{1}{2}} = 1$; (c) $x^3 + xy - y^3 = 0$; (d) $x \sin y - y \sin x = 1$.

3.12. The same expression for dy/dx in Problem 3.11a is obtained when the radius is changed; for example, if $x^2 + y^2 = 9$, we still get $dy/dx = x/y$. Is this paradoxical? (Notice that even in the general case of $f(x, y) = c$, a constant, the expression for dy/dx will not depend on c: think of the difference between the expression and the values it takes.)

3.13. Find expressions for dy/dx and then $d^2 y/dx^2$ if $xy^2 - x^2 y = 1$.

3.14. Differentiate the following inverse functions, using the method of Section 3.9. The results are quite important, and are included in the table of derivatives, Appendix D.
(a) $\arcsin x$; (b) $\arccos x$; (c) $\arctan x$;
(d) $\operatorname{arcsinh} x$; (e) $\operatorname{arccosh} x$; (f) $\operatorname{arctanh} x$.

3.15. (Parametric differentiation, Section 3.10). The curves in the following are in polar coordinates. Find dy/dx at the point specified.
(a) $r = \sin \frac{1}{2}\theta$ at $\theta = \frac{1}{2}\pi$; (b) $r = 1 + \sin^2 \theta$ at $\theta = \frac{1}{4}\pi$.

3.16. Obtain dy/dx in terms of t, then re-express it in terms of x, when the path of a point is given parametrically by the following.
(a) $x = t^3$, $y = t^2$; (b) $x = 2 \cos t$, $y = 2 \sin t$.

3.17. The path of a point is given parametrically by
$$x = a \cos t, \quad y = b \sin t.$$

Show that the point travels around the ellipse

$$\frac{x^2}{a^2} + \frac{y^2}{b^2} = 1.$$

Express dy/dx in terms of t. Suppose that t represents time. Express the speed as a function of t.

Applications of differentiation

4

Reminder. A basic table of derivatives, and a summary of the various rules for combining functions which were obtained in Chapters 2 and 3, are given in Appendix D at the end of the book.

4.1 Function notation for derivatives

So far, we have used the dy/dx or $(d/dx)f(x)$ notation for derivatives. The usefulness of the dy/dx notation is illustrated by the chain rule (3.3) and by (3.7)–(3.8): it strongly suggests the truth of certain results and makes them easy to remember. However, it is sometimes desirable to use another notation, $f'(x)$, which means exactly the same thing:

$$f'(x) \text{ means the same as } \frac{d}{dx} f(x).$$

By itself the symbol f' stands for the **derivative function**, because it is 'derived' from the original function f. Think of f' in the following way. Choose a 'neutral' letter for the independent variable, u say, which is not being used for anything else at the moment, and specify f in terms of u. For example, suppose that

$$f(u) = u^2 - 3u.$$

Then f' stands for the function specified by

$$f'(u) = \frac{d}{du} f(u) = 2u - 3.$$

Knowing now the form of the function f' (i.e. its formula), we can put anything we like in place of u, so that

$$f'(x) = 2x - 3, \qquad f'(t) = 2t - 3, \qquad f'(5) = 2.5 - 3 = 7,$$

$$f'(x^3) = 2x^3 - 3, \qquad f'(x - ct) = 2(x - ct) - 3,$$

$$f'(g(x)) = 2g(x) - 3 \quad \text{(where } g \text{ is any function), and so on.}$$

The following examples show how this notation can be used.

Example 4.1. The function f is defined by $f(u) = \sin u$. Obtain
(a) $f(x^2)$; (b) $\dfrac{d}{dx} f(x^2)$; (c) $f'(x^2)$.

(a) $f(x^2) = \sin x^2$.

(b) $\dfrac{d}{dx} f(x^2) = \dfrac{d}{dx} \sin x^2 = 2x \cos x^2$

(by using the chain rule (3.3) with $u = x^2$).

(c) The **first thing to do is to obtain the function** f':

$$f'(u) = \frac{d}{du} f(u) = \frac{d}{du} (\sin u) = \cos u.$$

Now put $u = x^2$; then

$$f'(x^2) = \cos x^2.$$

Notice that **the result** (c) **is different from the result** (b): $f'(x^2)$ **is not the same as** $(d/dx)f(x^2)$. In (b) we **first find** $f(x^2)$ and then differentiate with respect to x; in (c) we **first find** $f'(u)$ and then put $u = x^2$.

Example 4.2. Express (a) the product rule (3.1); (b) the quotient rule (3.2a); (c) the chain rule (3.3); in terms of the 'dash' notation.

(a) **Product rule**

$$\frac{d}{dx} [u(x)v(x)] = u(x)v'(x) + v(x)u'(x).$$

or simply

$$(uv)' = uv' + vu'.$$

(b) **Quotient rule**

$$\left(\frac{u}{v}\right)' = \frac{1}{v^2} (vu' - uv').$$

(c) **Chain rule**

$$\frac{d}{dx} f(u(x)) = f'(u(x))u'(x).$$

Example 4.3. (a) Suppose that f represents any function. Express $(d/dx)f(5x - 3)$ in any terms available. (b) Verify the correctness of (a) in the special case when $f(5x - 3) = \sin(5x - 3)$.

(a) Since the particular function f is not specified, the only thing to be done is to express $(d/dx)f(5x - 3)$ in terms of f', which is also unspecified. Then, from the chain rule (c) in Example 4.2, with $u = 5x - 3$,

$$\frac{d}{dx} f(5x - 3) = 5f'(5x - 3).$$

It is awkward to express the right-hand side without using the dash notation. One alternative is to write it as

$$\left[\frac{d}{du} f(u)\right]_{u=5x-3}$$

(b) In this case $f(u) = \sin u$, so

$$f'(u) = \cos u.$$

The result in (a) predicts that

$$\frac{d}{dx} f(5x - 3) = 5 \cos(5x - 3).$$

This is the same as the result obtained by working out

$$(d/dx) \sin(5x - 3)$$

directly by using the chain rule with $u = 5x - 3$.

The dash notation extends to higher derivatives: we put

The dash notation

$$f'(x) = \frac{d}{dx} f(x), \quad f''(x) = \frac{d^2}{dx^2} f(x),$$

$$f'''(x) = \frac{d^3}{dx^3} f(x), \quad \dots \ . \tag{4.1}$$

If $y = f(x)$, then the notation $y', y'', y''', \dots$, is also used.

(a)

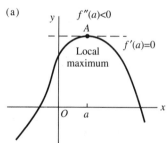

(b)

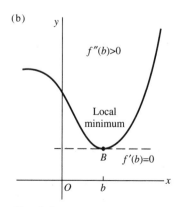

Fig. 4.1

4.2 Maxima and minima

A prominent feature in the graph of any function

$$y = f(x)$$

is any point at which the graph 'turns over'. For example, in Fig. 2.9, the graph of $y = \sin x$ turns over at $x = \frac{1}{2}\pi$ and $x = \frac{3}{2}\pi$. These are points where **the slope changes sign** from positive to negative or negative to positive. The derivative of $f(x)$ is zero (the tangent is horizontal) at such turnover points: for example it is easy to verify that, for $y = \sin x$,

$$f'(\tfrac{1}{2}\pi) = 0 \quad \text{and} \quad f'(\tfrac{3}{2}\pi) = 0.$$

Therefore

$$f'(x) = 0$$

can be looked on as an **equation whose solution gives all the possible points at which the graph turns over**.

However, **graphs do not necessarily turn over at points where** $f'(x) = 0$. For example, if $y = x^3$, then $f'(x) = 3x^2$. This is zero at $x = 0$, but the graph does not turn over at $x = 0$ (see e.g. Fig. 1.3): it just flattens instantaneously and then continues upward.

Figure 4.1 sketches two typical cases in which the graph does turn over, at A $(x = a)$ in Fig. 4.1a and at B $(x = b)$ in Fig. 4.1b. Then $f'(a) = 0$ and $f'(b) = 0$.

If $f'(x) = 0$ at a point $x = c$: that is, if

$$f'(c) = 0,$$

then $x = c$ is called a **stationary point** of $f(x)$. A stationary point such as A in Fig. 4.1a is called a **maximum** of the function $f(x)$. More precisely, the function is said to have a **local maximum** at $x = c$, because the value of $f(x)$ at $x = c$ is greater than its value at any point in the immediate neighbourhood. (There may be local maxima elsewhere that are either greater or smaller than this one.) Similarly a point such as B in Fig. 4.1b is called a **local minimum** of $f(x)$.

To distinguish between types of stationary point algebraically, consider also the second derivative of $f''(x)$ at $x = c$. Suppose that

$$f'(c) = 0 \quad \text{and} \quad f''(c) < 0.$$

Now (see also Section 2.10)

$$f''(x) = \frac{d^2 y}{dx^2} = \frac{d}{dx} \frac{dy}{dx},$$

and this is **negative** at $x = c$. Therefore dy/dx, or $f'(x)$, is **decreasing** across $x = c$, and since $f'(x) = 0$ at $x = c$, $f'(x)$ must be positive on the left of c and negative on the right. Thus the graph is of the type shown in Fig. 4.1a, and the point is a **local maximum**.

If $x = c$ is a point where

$$f'(c) = 0 \quad \text{and} \quad f''(c) > 0,$$

then $f'(x)$ is **increasing**, and therefore goes from negative to positive, across $x = c$. The point is therefore a **minimum**, like the point B in Fig. 4.1b.

In the special case when

$$f'(c) = 0 \quad \text{but} \quad f''(c) = 0$$

there might occur a maximum (as with $y = -x^4$ at $x = 0$), or a minimum (as with $y = x^4$ at $x = 0$), or another feature called a **stationary point of inflection** (as with $y = \pm x^3$ at $x = 0$). These cases are illustrated in Fig. 4.2. One way to classify such a point is to examine directly the sign of dy/dx on both sides of the point.

To summarize:

(a)

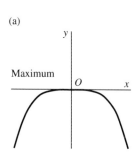

Maximum

(b)

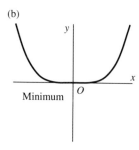

Minimum

(c)

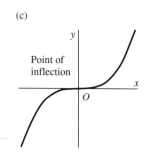

Point of inflection

(d)

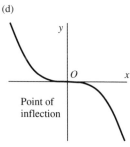

Point of inflection

Fig. 4.2
Cases for which $f'(0) = 0$ and $f''(0) = 0$. (a) $y = -x^4$.
(b) $y = x^4$. (c) $y = x^3$.
(d) $y = -x^3$.

> **Stationary points of** $f(x)$
>
> Let $f'(c) = 0$; i.e. $x = c$ is a stationary point of $f(x)$.
> Then
> (a) If $f''(c) < 0$, $f(x)$ has a local maximum at $x = c$.   **(4.2)**
> (b) If $f''(c) > 0$, $f(x)$ has a local minimum at $x = c$.
> (c) If $f''(c) = 0$, the stationary point might be a maximum, a minimum, or a point of inflection. Examine the sign of $f'(x)$ on both sides of $x = c$.

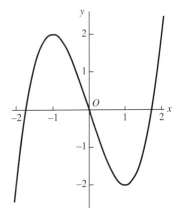

Fig. 4.3

Example 4.4. Classify the stationary points of $f(x) = x^3 - 3x$.

The stationary points are where $f'(x) = 0$; that is, where

$$3x^2 - 3 = 0, \quad \text{or} \quad x = \pm 1.$$

We need the signs of $f''(\pm 1)$, where

$$f''(x) = 6x.$$

Then

$$f''(1) = 6,$$

which is positive, so there is a minimum at $x = 1$. Also

$$f''(-1) = -6$$

which is negative, so there is a maximum at $x = -1$.
   The values of $f(x)$ at these points are

$$f(1) = -2, \quad f(-1) = 2;$$

so the graph has the shape shown in Fig. 4.3. Alternatively, we could simply have checked the signs of $f'(x) = 3x - 3$ on both sides of the stationary points directly, instead of using the test (4.1).

Example 4.5. In the circuit shown in Fig. 4.4, V is a constant voltage and R and x represent two resistances: R is fixed and x is variable. The rate of heat generation y in resistance x is equal to I^2x where I is the current. Show that y is a maximum when $x = R$.
   Current equals voltage divided by total resistance, so

$$I = \frac{V}{R + x}.$$

Therefore the rate of heat generation is

$$y = \frac{V^2 x}{(R + x)^2} = f(x),$$

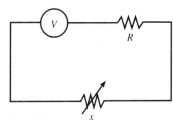

Fig. 4.4

say. If there is a maximum, it will occur when $f'(x) = 0$. From the quotient rule (3.2)

$$f'(x) = \frac{V^2}{(R+x)^4}\left[(R+x)^2 - x\cdot 2(R+x)\right]$$

$$= V^2\frac{R-x}{(R+x)^3}. \qquad \text{(i)}$$

This is zero when $x = R$.

To show that $f(x)$ has a maximum when $x = R$ we may work out the sign of $f''(R)$. From (i),

$$f''(x) = \frac{V^2}{(R+x)^6}\left[(R+x)^3(-1) - (R-x)\cdot 3(R+x)^2\right]$$

$$= \frac{V^2}{(R+x)^4}(-4R + 2x).$$

Therefore

$$f''(R) = -V^2/8R^3,$$

which is negative, so $x = R$ corresponds to a maximum of y.

However, it is easier to look instead at the expression (i) for $f'(x)$. When $x < R$, we have $f'(x) > 0$, so $f(x)$ is increasing. When $x > R$, we have $f'(x) < 0$, so $f(x)$ is decreasing. This ensures that a maximum has been obtained without the need to differentiate again.

Example 4.6. x and y are two numbers subject to the restriction that $x + y = 1$. Find the maximum possible value of xy.

There are two variables, x and y, but we can reduce the problem to one involving only x by using the fact that $x + y = 1$, so that

$$y = 1 - x. \qquad \text{(i)}$$

In that case,

$$xy = x(1-x) = x - x^2 = f(x),$$

say. Now $f(x)$ has a stationary point (a maximum, minimum, or point of inflection) where $f'(x) = 0$, that is to say, where

$$1 - 2x = 0, \quad \text{or} \quad x = \tfrac{1}{2}.$$

By (4.2), this value of x delivers a maximum, because $f''(x) = -2$ (for any value of x) which is negative. From (i), $y = \tfrac{1}{2}$ when $x = \tfrac{1}{2}$, so the maximum value of xy is $\tfrac{1}{4}$.

4.3 Exceptional cases of maxima and minima

The method of finding local maxima and local minima by solving $f'(x) = 0$ reveals only points where the slope of the graph of $y = f(x)$ is horizontal. Sometimes there is a maximum or minimum at an **end-point of an interval**, even if the graph is not horizontal there.

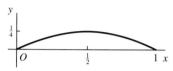

Fig. 4.5
$y = x - x^2, 0 \leqslant x \leqslant 1.$

Example 4.7. Suppose that the values of x to be considered are restricted to lying between 0 and 1 inclusive: that is, $0 \leqslant x \leqslant 1$. Find the points **on this interval** at which $x - x^2$ takes maximum and minimum values.

The graph of $y = f(x) = x - x^2$ between $x = 0$ and $x = 1$ is shown in Fig. 4.5. The maximum of $f(x) = x - x^2$ which we found in Example 4.6 at $x = \frac{1}{2}$ can be seen. But, understood in a common-sense way, there are minimum values at $x = 0$ and $x = 1$, the end points of the restricted interval. These cannot be detected by the method of differentiation. Whether we are interested in them would depend on the demands of any practical problem from which the question originated.

In problems of the type illustrated in Example 4.6, this situation can arise naturally, as in the following example.

Example 4.8. Find the maximum and minimum values of $x^2 - y^2$ on the circle $x^2 + y^2 = 1$.

It is evident that the point (x, y) can only be on the circle if x and y both have values between -1 and 1 inclusive, that is if

$$-1 \leqslant x \leqslant 1 \quad \text{and} \quad -1 \leqslant y \leqslant 1. \tag{i}$$

A restricted interval therefore arises naturally in the problem. On the circle $x^2 + y^2 = 1$, we have

$$y^2 = 1 - x^2, \tag{ii}$$

so
$$x^2 - y^2 = 2x^2 - 1 = f(x), \tag{iii}$$

say. To find the stationary points of $f(x)$ we see that $f'(x) = 4x$, which is zero when $x = 0$. Also $f''(0) = 4 > 0$, so $x = 0$ is a local minimum of $f(x)$, whose value is $f(0) = -1$.

However, we have overlooked something. In Fig. 4.6, we show the graph of $f(x) = 2x^2 - 1$, within the permitted interval $-1 \leqslant x \leqslant 1$. The local minimum at $x = 0$ can be seen, but there are also maxima at the end points $x = -1$ and $x = 1$, where $f(x)$ takes the values $+1$.

Alternatively, the maxima at $x = \pm 1$ can be found by substituting for x instead of y at the first stage. Put $x^2 = 1 - y^2$, so that

$$x^2 - y^2 = 1 - 2y^2 = g(y),$$

say, and solve $g'(y) = 0$: we then find a local maximum at $y = 0$, where $x = \pm 1$. However, we also lose sight of the minima we found before. The subject is discussed again in Section 27.2.

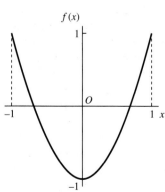

Fig. 4.6
$f(x) = 2x^2 - 1, -1 \leqslant x \leqslant 1.$

Another possibility is that there may be points at which the graph of $y = f(x)$ does not have a definite tangent. Then $f'(x)$ or $\mathrm{d}y/\mathrm{d}x$

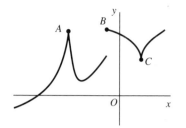

Fig. 4.7

has no meaning at such points. For example, in Fig. 4.7, there is no tangent at the points A, B, and C. The points A and B could qualify as local maxima, and C as a local minimum, but at A and C the graph suddenly changes direction, and at B there is a jump in the value of $f(x)$. These points cannot be located by solving $f'(x) = 0$, because $f'(x)$ does not exist at A, B, and C.

4.4 Sketching graphs of functions

To **sketch a graph** is to indicate its general shape so as to draw attention to its most important features without being concerned with accurate plotting. To do this it is necessary for the reader to have a clear idea of the shape of the graphs of the basic functions

$$x^a, \quad e^{ax}, \quad \sin ax, \quad \cos ax, \quad \ln x.$$

Example 4.9. Sketch the graph $y = 1 - 1/(1 + x)^2$.

This can be done in stages, as shown in Fig. 4.8. Figure 4.8c is obtained from 4.8b by using the rule (1.11) with $c = 1$; it simply involves sliding the graph $y = -1/x$ one unit to the left. To get from 4.8c to 4.8d, we add 1, which moves the graph up the y axis by one unit.

(a)

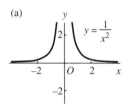

(b)

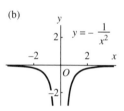

(c)

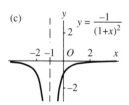

(d)

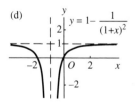

Fig. 4.8

In Fig. 4.8d of Example 4.9 we see that, as x increases, becoming large and positive, the value of

$$y = 1 - \frac{1}{(1 + x)^2}$$

gets closer and closer to 1. The same is true when x becomes large and negative. This is obviously an important feature of the graph. It can be seen to be true by thinking what happens to y when we put a **large value** of x into the formula for y (think of a *very* large number: $x = 1,000,000$ rather than $x = 10$). Then obviously $1/(1 + x)^2$ is very small, so y gets very close to 1, and the larger x becomes, the nearer y is to 1. The same is true when x is large and negative.

We say that, as x increases, the graph **approaches the line** $y = 1$, which in general terms is called an **asymptote** of the graph. When x approaches -1, the graph approaches the vertical line $x = -1$; this is also called an asymptote. The two continuous halves of the graph to the left and right of $x = -1$ are called **branches**.

Suppose that $y = f(x)$ is to be sketched. A general question to be asked is 'What happens to y when x increases towards infinity (or decreases towards minus infinity)?' We normally say 'as x **approaches** $\pm \infty$', and as usual indicate the approach by '$\rightarrow$':

$$x \rightarrow \pm \infty.$$

For example, $1/x \to 0$ when $x \to -\infty$. Also

$$\frac{x-1}{3x+2} \to \tfrac{1}{3} \quad \text{when} \quad x \to \infty \ (\text{or } x \to -\infty).$$

To see this, think of the effect of giving x an immense value. Only the terms x and $3x$ are significant; they are said to **dominate** the expression; so

$$\frac{x-1}{3x+2} \to \frac{x}{3x} = \tfrac{1}{3}.$$

The **limit notation** can be used in this context (see Section 2.2). We can write, for example

$$\lim_{x \to \infty} \frac{1-2x}{1+x} = -2.$$

The reasoning is the same as in the earlier case: think of a very large value of x.

Very often the function has no definite limit as $x \to \infty$. For example $\lim_{x \to \infty} \sin x$ **does not exist**; no definite single number is approached, since $\sin x$ simply goes up and down between ± 1 for ever. However, it is quite usual to write, say,

$$\lim_{x \to \infty} x^2 = \infty,$$

even though ∞ is not a number.

Notice the following result.

$$\lim_{x \to \infty} a x^n \, e^{-cx} = 0,$$

where a and n are any constants, and c is a *positive* constant.   **(4.3)**

We shall not prove (4.3); but, to convey the feel of it, a table of values is given for the special case of $x^3\, e^{-x}$:

x	0	1	2	3	4	8	10
$x^3\,e^{-x}$	0	0.36	1.08	1.34	1.17	0.18	0.05

Fairly large values are needed before the function settles down to approach zero, because x^3 is increasing, and therefore competes with e^{-x} in the early stage. However, e^{-x} will beat any power of x down to zero eventually. In the following example, we sketch the graph of the function in the table without using the calculated values above.

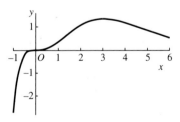

Fig. 4.9

Example 4.10. Sketch the graph $y = x^3\, e^{-x}$.

Do it in stages, using any easily obtained facts you can think of.

(a)

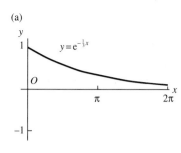

(b)

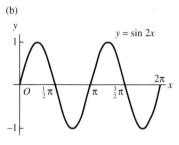

(c)

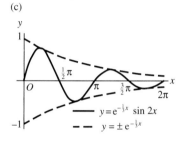

Fig. 4.10

(a)

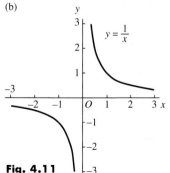

(b)

Fig. 4.11

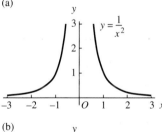

1. *Are there any points where it is easy to obtain values?* At $x = 0$, $y = 0$.

2. *Are there any definite points where $x^3 e^{-x}$ is infinite?* There are no such points.

3. *Are there any points where the graph crosses the x axis?* Only the point found in (1).

4. *Are there any maxima/minima?*

$$\frac{dy}{dx} = 3x^2 e^{-x} - x^3 e^{-x} = x^2(3 - x) e^{-x}.$$

This is zero when $x = 0$ and $x = 3$, so these are stationary points. dy/dx is positive when $x < 3$, and negative when $x > 3$, so $x = 3$ is a maximum. Since $e^3 \approx 20$, at this point $y \approx 1\frac{1}{3}$.

Near the other stationary point $x = 0$, dy/dx is positive on both sides, so $x = 0$ is a point of inflexion.

5. *Behaviour as $x \to \infty$.* According to (4.3), $y \to 0$ as $x \to \infty$.

6. *Behaviour as $x \to -\infty$.* As $x \to -\infty$, we have $x^3 \to -\infty$ and $e^{-x} \to \infty$ (think, for example, of $x = -1000$). Therefore, $x^3 e^{-x} \to -\infty$ (very rapidly).

The sketch is shown in Fig. 4.9.

Example 4.11. Sketch the graph of $e^{-\frac{1}{3}x} \sin 2x$ for $0 \leqslant x \leqslant 2\pi$.

(x is assumed to be in radians.) Split the expression into its two factors, $e^{-\frac{1}{3}x}$ and $\sin 2x$. These are shown in Fig. 4.10a,b. The value of $e^{-\frac{1}{3}x}$ drops to about $\frac{1}{8}$ at $x = 2\pi$. Also, $\sin 2x$ is zero when

$$2x = 0, \pi, 2\pi, \ldots, 6\pi,$$

or when

$$x = 0, \tfrac{1}{2}\pi, \pi, \tfrac{3}{2}\pi, 2\pi.$$

The product of the two is shown in Fig. 4.10c. The graph crosses the x axis (i.e. $y = 0$) where $\sin 2x = 0$, and nowhere else. The height of the peaks and troughs of $e^{-\frac{1}{3}x} \sin 2x$ are estimated by the size of the factor $e^{-\frac{1}{3}x}$, shown as a broken line, which multiplies the maxima and minima of $\sin 2x$. The new maxima and minima do not occur at exactly the same points: it is left to the reader to show that the new maxima and minima occur at values of x which satisfy the equation $\tan 2x = 6$.

It is useful to be able to distinguish between the behaviour of functions such as those shown in Fig. 4.11a,b. The function $1/x$ of Fig. 4.11b is infinite at $x = 0$, but the sign changes across $x = 0$. The terminology used to describe $y = 1/x$ near $x = 0$ is

$$y \to -\infty \text{ as } x \to 0 \text{ from the left};$$

$$y \to \infty \quad \text{ as } x \to 0 \text{ from the right}.$$

Example 4.12. Sketch the graph of $y = 1/(x - 2)(x + 1)$.

Look out for the obvious things first. At $x = 0$, we have $y = -\frac{1}{2}$. The function is infinite at $x = 2$ and $x = -1$. It does not cross the x axis anywhere.

 Now consider the sign of $1/(x - 2)(x + 1)$. It is positive when $x < -1$ (try e.g. $x = -3$). It is positive when $x > 2$ (try e.g. $x = 3$). It is negative when $-1 < x < 2$, which is linked with the facts that the graph does not cross the x axis and, as we already know, that y is negative when $x = 0$.

 We know now that

$$y \to \infty \quad \text{as } x \to -1 \text{ from the left;}$$
$$y \to -\infty \text{ as } x \to -1 \text{ from the right;}$$
$$y \to -\infty \text{ as } x \to 2 \quad \text{from the left;}$$
$$y \to \infty \quad \text{as } x \to 2 \quad \text{from the right;}$$

so Fig. 4.12 is emerging.

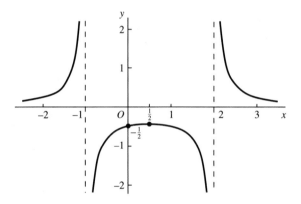

Fig. 4.12
$y = 1/(x - 2)(x + 1)$.

 We now locate precisely the obvious maximum between $x = -1$ and 2, and make sure there are no other stationary points. By the reciprocal rule (3.2b), we have

$$\frac{dy}{dx} = \frac{1}{(x - 2)^2(x + 1)^2} \frac{d}{dx}[(x - 2)(x + 1)] = \frac{2x - 1}{(x - 2)^2(x + 1)^2}.$$

This is zero at $x = \frac{1}{2}$ and nowhere else. There is no need to use the test (4.1); the point can only be a maximum since there are no other stationary points. The value of y there is $-\frac{4}{9}$.

 We now return to asymptotes and show that there can be asymptotes that slope. Consider the function

$$y = \frac{x^2 - 1}{2x + 1},$$

when x is large, positive or negative. The term $x^2 - 1$ is dominated by x^2; meaning that the part -1 is negligible compared with x^2 when x is large. Likewise the **dominant term** in $2x + 1$ is $2x$. It is therefore obvious that

$$y \to \pm \infty \quad \text{when} \quad x \to \pm \infty.$$

However, we can do much better than this, because

$$y = \frac{x^2 - 1}{2x + 1} = \tfrac{1}{2}x - \tfrac{1}{4} - \frac{\tfrac{3}{4}}{2x + 1}.$$

Therefore the graph will approach the straight line

$$y = \tfrac{1}{2}x - \tfrac{1}{4}$$

when x is large. As in the earlier instances we have seen, the line $y = \tfrac{1}{2}x - \tfrac{1}{4}$ is said to be an **asymptote** of the original graph. The notation

$$y \sim \tfrac{1}{2}x - \tfrac{1}{4} \quad \text{when} \quad x \to \pm \infty$$

is sometimes used, meaning that the curve approaches the line $y = \tfrac{1}{2}x - \tfrac{1}{4}$ when x is large. The curve is sketched in Example 4.13.

In the same way, a function may be an **asymptotic to a curve** as $x \to \pm \infty$. For example, if

$$y = \frac{1}{x} - \frac{1}{x^3} \sin x,$$

then

$$y \sim \frac{1}{x} \quad \text{when} \quad x \to \pm \infty.$$

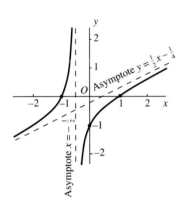

Fig. 4.13

Example 4.13. Sketch the graph of $y = (x^2 - 1)/(2x + 1)$.

The curve cuts the x axis ($y = 0$) at $x = \pm 1$. Also $y = -1$ when $x = 0$. The function is infinite at $x = -\tfrac{1}{2}$.

Also, as shown above, the straight line $y = \tfrac{1}{2}x - \tfrac{1}{4}$ is an asymptote for large values of x. This is shown as the broken line in Fig. 4.13.

From the quotient rule (3.2a),

$$\frac{dy}{dx} = \frac{(2x + 1)(2x) - (x^2 - 1)2}{(2x + 1)^2} = 2\frac{x^2 + x + 1}{(2x + 1)^2}.$$

This is never zero, because the equation $x^2 + x + 1 = 0$ has no real solutions. Therefore, there are no stationary points.

4.5 Estimating small changes

Let

$$y = f(x).$$

Suppose that the value of x changes by a small amount δx. Then y will change by a small amount δy. It will be shown that there is a simple approximate relation between δy and δx which is important for practice and theory.

Fix a particular value of x: say $x = a$; the small deviation δx will be made from this value of x. The derivative at $x = a$ is $f'(a)$. According to (2.8), to obtain $f'(a)$ we take a nearby point $x = a + \delta x$ and form the ratio

$$\frac{\delta y}{\delta x} = \frac{f(a + \delta x) - f(a)}{\delta x},$$

and

$$\frac{\delta y}{\delta x} \to f'(a) \quad \text{as } \delta x \to 0.$$

If δx is small enough, $\delta y/\delta x$ will become close in value to $f'(a)$:

$$\frac{\delta y}{\delta x} \approx f'(a),$$

so that

$$\delta y \approx f'(a) \, \delta x.$$

This is how to obtain an **approximation to the change δy in y due to a small change from** $x = a$ **to** $x = a + \delta x$. It is easier to **remember** the result in the form

$$\delta y \approx \frac{dy}{dx} \, \delta x,$$

near a general point x (which again shows the usefulness of the dy/dx notation in suggesting true results). We call this the **incremental approximation** for functions of a single variable.

Incremental approximation

For a small increment δx from $x = a$:

(a) $\delta y \approx f'(a) \, \delta x$ (4.4)

(b) (Mnemonic form)

$$\delta y \approx \frac{dy}{dx} \, \delta x.$$

Example 4.14. Let $y = x + 1/x$. Estimate the change δy in y when x changes from $x = 2$ to $x = 1.8$. Compare the estimate with the exact value of δy.

Put

$$y = x + \frac{1}{x} = f(x).$$

Then

$$\frac{dy}{dx} \text{ or } f'(x) = 1 - \frac{1}{x^2},$$

so that $f'(2) = 0.75$. Here $\delta x = 1.8 - 2.0 = -0.2$; so, by (4.4a),

$$\delta y \approx 0.75 \times (-0.2) = -0.15.$$

(The exact value is given by $\delta y = (1.8 + 1/1.8) - (2.0 + 1/2.0) = -0.1444\cdots$.)

Example 4.15. The volume V of a sphere of radius r is given by $V = \frac{4}{3}\pi r^3$. Estimate the change in volume if the radius increases from 2.0 to 2.1 metres.

We shall use the letters that the question offers, considering δV and δr. Put

$$V = \tfrac{4}{3}\pi r^3 = f(r).$$

Then

$$f'(r) = \frac{\mathrm{d}V}{\mathrm{d}r} = 4\pi r^2,$$

so, by (4.4b),

$$\delta V \approx 4\pi r^2 \, \delta r. \tag{i}$$

(Notice that $4\pi r^2$ is the formula for the *surface area* of a sphere: the change in volume is nearly equal to the surface area times the thickness δr.) Now put $r = 2$:

$$f'(2.0) = 16\pi \quad \text{and} \quad \delta r = 2.1 - 2.0 = 0.1.$$

Then, by (i),

$$\delta V \approx 16\pi \times 0.1 = 5.02.$$

The exact value is $\delta V = 5.282 \cdots$ (cubic metres), implying an error of 5% in our estimate.

The number $\delta V \approx 5.02$ in Example 4.15 might not seem to qualify as a *small* change. Furthermore, if we express the identical problem in different units, say in centimetres rather than metres, the numbers are even larger; then δr becomes 10 (cm) and δV is about 5×10^6 (cm^3). But nothing at all is changed except the units of measurement. We still get only a 5% error in the estimate. On the other hand, if the units had been kilometres then δr and δV would have looked very small indeed. The reason for this is that the ratio

$$\text{Estimated } \delta V / \text{Exact } \delta V = [f'(r) \, \delta r] / [f(r + \delta r) - f(r)]$$

is dimensionless; that is to say it is unaffected by the choice of units. There is no easy way to predict when the method will work well: geometrically speaking, we are content to guess that the graph sticks sufficiently closely to its tangent line at a within the interval $a \pm \delta x$.

Example 4.16. The cosine rule for a triangle ABC is

$$c^2 = a^2 + b^2 - 2ab \cos C.$$

In a triangle for which $a = 3$ and $b = 4$, estimate the change in c when C increases from 60° to 65°.

Put in the fixed numbers, $a = 3$ and $b = 4$; then

$$c^2 = 25 - 24 \cos C$$

or

$$c = (25 - 24 \cos C)^{\frac{1}{2}} = f(C),$$

say. By the chain rule (3.3) with $u = 25 - 24 \cos C$,

$$f'(C) = \frac{dc}{dC} = -(25 - 24 \cos C)^{-\frac{1}{2}}(24 \sin C).$$

The quantity δC must be measured in radians, because radian measure was assumed in obtaining the derivatives of the sin and cos functions. So we put

$$C = 60° = \tfrac{1}{3}\pi \text{ radians}, \qquad \delta C = \tfrac{5}{180}\pi = 0.087 \text{ radians}.$$

We know that $\cos C = \tfrac{1}{2}$ and $\sin C = \tfrac{1}{2}\sqrt{3}$, so

$$f'(\tfrac{1}{3}\pi) = 6\sqrt{3}/\sqrt{13}.$$

Therefore, by the incremental approximation (4.4),

$$\delta c \approx (6\sqrt{3}/\sqrt{13}) \times 0.087 = 0.25.$$

(The exact change is $\delta c = 0.2489 \cdots$.)

4.6 Numerical solution of equations: Newton's method

It is often necessary to solve equations for which there is no standard method of solution. (In fact, this is true for practically all equations.) Simple examples are the equations

$$x^4 + x^3 - 1 = 0 \quad \text{and} \quad e^{-x} - x = 0.$$

For such cases, there are many methods for obtaining **numerical solutions**, which are applicable no matter how complicated the equation is. We describe one of them here.

To apply the method, it is necessary first of all to obtain at least a rough idea of the location of the solution we are seeking. There are various ways of doing this: for example we can plot a rough graph. Taking the first example above, the graph

$$y = x^4 + x^3 - 1$$

is sketched in Fig. 4.14 using only five values of x; namely -1.5, $-1, 0, 0.5$, and 1. The solutions occur where it crosses the axis; there seems to be one not far from -1.3 and one not far from 0.8.

Suppose now that we have a general equation to solve:

$$f(x) = 0;$$

and that, by drawing its graph, or by some other method, we have established that one of its solution is not far from the value

$$x = x_0,$$

say. We show how to locate this solution accurately.

Figure 4.15 shows one possibility for the shape of the graph of $y = f(x)$ close to its (unknown) solution $x = c$, say, corresponding to the point C. (If the graph is different from this, the discussion is much the same.) The initial estimate $x = x_0$ corresponds to A_0. The point A_0 could be on the left of the solution C as shown, or on the right; we are not likely to be sure: again the argument is much the same (see Problem 4.14).

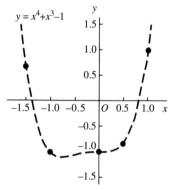

Fig. 4.14
$y = x^4 + x^3 - 1$.

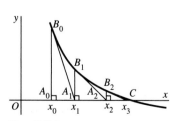

Fig. 4.15

Perform, in imagination, the following steps.

1. Starting at the point A_0, where $x = x_0$. Draw the perpendicular A_0B_0, intersecting the curve at B_0. Construct the tangent at B_0, and continue it to intersect the x axis at A_1, where $x = x_1$. Then A_1 is nearer to the solution C than A_0.

2. Repeat the process, starting with the improved estimate A_1. We arrive at A_2, where $x = x_2$, which is a still better approximation.

3. Using the new estimates as starting values as they arise, keep repeating the process to produce a sequence of approximations

$$A_0 \ (x = x_0), \quad A_1 \ (x = x_1), \quad A_2 \ (x = x_2), \quad A_3 \ (x = x_3), \ldots,$$

and stop when the accuracy attained is satisfactory.

The steps 1–3 can be carried out algebraically:

1. Starting with $A_0 \ (x = x_0)$, the equation of the tangent line at B_0 is given by

$$\frac{y - f(x_0)}{x - x_0} = f'(x_0).$$

At $A_1 \ (x = x_1)$, we have $y = 0$, so

$$\frac{-f(x_0)}{x_1 - x_0} = f'(x_0).$$

Therefore, the new approximation $A_1 \ (x = x_1)$ is given by

$$x_1 = x_0 - \frac{f(x_0)}{f'(x_0)}.$$

2. x_1 takes the place of x_0, and x_2 takes the place of x_1 so

$$x_2 = x_1 - \frac{f(x_1)}{f'(x_1)}.$$

3. Once the nth approximation x_n is available, the $(n + 1)$th value x_{n+1} is given by

$$x_{n+1} = x_n - \frac{f(x_n)}{f'(x_n)}.$$

This process, in which essentially we do exactly the same thing over and over again, using for each step the information obtained from the previous step, is called a **step-by-step process** or an **iterative process**. It is summarized in the following **algorithm** (or recipe), known as **Newton's method**.

> **Newton's method for the numerical solution of**
> $f(x) = 0$
> Find a value $x = x_0$ sufficiently close to the solution
> required. Then carry out the following step-by-step
> process until the desired accuracy is obtained: **(4.5)**
>
> $$x_{n+1} = x_n - \frac{f(x_n)}{f'(x_n)},$$
>
> for $n = 0, 1, 2, 3, \ldots,$ successively.

The following example works through the equation with which
we opened the section.

Example 4.17. The equation $x^4 + x^3 - 1 = 0$ has a solution near
$x = 0.8$. Find it to five-decimal accuracy.

We have

$$x_0 = 0.8, \qquad f(x) = x^4 + x^3 - 1, \qquad f'(x) = 4x^3 + 3x^2.$$

Then, in (4.4),

$$x_{n+1} = x_n - \frac{x_n^4 + x_n^3 - 1}{4x_n^3 + 3x_n^2}.$$

Starting with $x_0 = 0.8$, we obtain the following table:

x_0	x_1	x_2	x_3
0.80000	0.81975	0.81917	0.81917

Evidently we do not have to pursue the sequence any further.

Example 4.18. The equation $e^{-x} = x$ has a solution near to
$x = 0.5$. Find the solution accurately to five decimal places.

We have

$$x_0 = 0.5, \qquad f(x) = e^{-x} - x, \qquad f'(x) = -e^{-x} - 1.$$

From (4.4),

$$x_{n+1} = x_n - \frac{e^{-x_n} - x_n}{-e^{-x_n} - 1} = \frac{x_n + 1}{e^{x_n} + 1}$$

(the last step for simplicity of calculation). We obtain the following
sequence:

x_0	x_1	x_2	x_3
0.50000	0.56631	0.56714	0.56714.

This repetitive process described is easy to program for a computer for individual cases as they arise, and then the complexity of the equation is of no importance. The same program can be adapted to scan a range of x in order to get a provisional idea of where the solutions are to be found. A simple program combined for safety's sake with inspection of the whole sequence of values output would satisfy most requirements.

However, to write a program which will *automatically*, without intervention, find the solutions for *any* function $f(x)$ that might be presented to it is a very different matter. For example, we would have to find means to be absolutely sure that none of the possible tangents would by chance carry us an irrecoverable distance away from the solution we are seeking (see e.g. Problem 4.15) by designing a way of automatically recognizing and rectifying the situation if it occurs.

Problems, Chapter 4

4.1. (see Section 4.1 on the 'dash' notation). The function f is defined by $f(u) = u^2$. Obtain the following.

(a) $f'(t)$; (b) $f'(t^2)$; (c) $\dfrac{d}{dt} f(t^2)$;

(d) $f'(t^{\frac{1}{2}})$; (e) $\dfrac{d}{dt} f(t^{\frac{1}{2}})$; (f) $f''(t^{\frac{1}{2}})$.

4.2. (see Section 4.11). Find the stationary points of the following functions and classify them as maxima, minima, or points of inflection.
(a) $x^2 - x$; (b) $x^2 - 2x - 3$; (c) $x \ln x$ $(x > 0)$;
(d) $x e^{-x}$; (e) $1/(x^2 + 1)$; (f) $x^2 - 3x + 2$;
(g) $e^x + e^{-x}$; (h) $x^2 + 4x + 2$; (i) $x - x^3$;
(j) $x^2(x - 1)$; (k) $\sin x - \cos x$ (in $0 < x < 2\pi$).
(l) $\sin x \cos x$ $(-\pi < x < \pi)$; (m) $e^{-x} \sin x$;
(n) $e^{-\frac{1}{3}x} \sin 2x$ (see Example 4.11);
(o) $x(1 - \cos x)$; (p) $2 e^x - \frac{1}{2} e^{2x}$;
(q) $x^2 e^{-x}$; (r) $(\ln x)/x$ $(x > 0)$; (s) $(1 - x)^3$;
(t) $\sin^3 x$; (u) e^{-x^2}; (v) $e^{x^2 - x}$;
(w) $x + x^{-1}$; (x) $x^3 e^{-x}$.

4.3. Let $y = f(u(x))$. Use two successive applications of the chain rule in the form of Example 4.2c to show that

$$\frac{d^2 y}{dx^2} = f''(u(x))[u'(x)]^2 + f'(u(x))u''(x).$$

Show that if $f'(u)$ is always greater than zero, or always less than zero, then $f(u(x))$ and $u(x)$ have the same stationary points. Consider, e.g. Problem 4.2v

in this connection, with $f(u) = e^u$ and $u(x) = x^2 - x$: it becomes rather obvious.

4.4. A rectangular piece of ground is to be marked out, which must have a given area A. Find the dimensions of the plot which requires the minimum length of perimeter fence. (This is a 'restricted' problem, like Example 4.6. Call the sides x and y.)

4.5. A tunnel cross-section is to have the shape of a rectangle surmounted by a semicircular roof. The total cross-sectional area must be A, but the perimeter minimized to save building costs. Find its dimensions.

4.6. A circular-cylindrical oil drum is required to have a given surface area (including its lid and base). Find the proportions of the design which contain the greatest volume.

4.7. Solve Problem 4.6 for the case when the lid is not included in the restriction.

4.8. Sketch the graphs of the following functions.
(a) $1/(x^2 + 1)$ (this is an even function: see (1.12)).
(b) e^{-x^2}. (c) $x/(x - 1)$. (d) $x e^{-x}$.
(e) $x^2 e^{-x}$. (f) $x^3 e^{-x}$. (g) $e^{2x} - 4 e^x$.
(h) $(\ln x)/x$ for $x > 0$ $((\ln x)/x \to 0$ when $x \to \infty$; this can be proved by putting $x = e^u$ and letting $u \to \infty)$.
(i) $[\ln(-x)]/x$ for $x < 0$ (compare (h)).
(j) $x \ln x - x$ for $x > 0$ $(x \ln x \to 0$ when $x \to 0$; this can be seen by writing $x = e^{-u}$ and letting $u \to \infty)$.

(k) $\sin 1/x$ (Start by finding where it crosses the axis, using the fact that $\sin u = 0$ when $u = 0, \pm\pi, \pm2\pi, \ldots$.)

(l) $(x^2 - 1)^2$ (this is an even function: see (1.12)).

(m) $x(x^2 - 1)^2$ (this is an odd function: see (1.12)).

(n) $(\sin x)/x$ (You will not be able to find the exact positions of the maxima and minima; be content to indicate the trend. It is an even function: see (1.12). For the value approached at $x = 0$, see (2.13).)

4.9. Sketch the graphs of the following functions.

(a) $1/(x^2 - 1)$ (hint: write $x^2 - 1 = (x + 1)(x - 1)$, and then follow Example 4.12; alternatively, sketch $y = x^2 - 1$ and imagine taking its reciprocal).

(b) $x/(x^2 - 1)$. (c) $1/x(x - 2)$.

(d) $x^3/(1 - x)$ (hint: see the note on curved asymptotes following Example 4.12)

(e) $(x + 2)/(x - 1)$ (see the hint in (d)).

(f) $1/(x + 1) + 1/(x + 2)$.

4.10. (see Section 4.5). Find the approximate value of the change δy in y due to a small change δx in x using the incremental approximation (4.4) in the following cases. Compare the approximate and exact values of δy.

(a) $y = x^3$ when $x = 2$ and $\delta x = 0.1$;

(b) $y = x \sin x$ when $x = \frac{1}{2}\pi$ and $\delta x = -0.2$;

(c) $y = \cos x$ when $x = \frac{1}{4}\pi$ and $\delta x = 0.1$;

(d) $y = (1 + x)/(1 - x)$ when $x = 2$ and $\delta x = -0.2$;

(e) $y = \tan x$ when $x = \frac{1}{4}\pi$ and $\delta x = 0.1$;

(f) $y = 1/(1 - x^2)$ when $x = 0.5$ and $\delta x = \pm0.1$.

4.11. (a) If the focal length of a lens is f, and a viewed object is at distance u, then the image is at distance v where $v = uf/(u - f)$. Let $f = 0.75$ (m). Find approximately the change in v if u changes from 1.25 to 1.30 (m).

(b) In a Wheatstone bridge circuit, the out-of-balance voltage v is given by

$$v = E(R_1R_4 - R_2R_3)/(R_1 + R_2)(R_3 + R_4),$$

where E is the applied voltage and R_1, R_2, R_3, R_4 represent the resistances in the branches. Suppose that $E = 5$, $R_1 = 4$, $R_2 = 2$, $R_3 = 6$, and $R_4 = 3$, so that the circuit is initially balanced. Obtain an approximate expression for δv in terms of a small change δR_1 in R_1.

(c) In a triangle ABC with corresponding sides a, b, c, the formula $a = b \sin A/\sin B$ applies. Show that $\delta a \approx -a \cot B \, \delta B$.

(d) In a triangle ABC with corresponding sides a, b, c, the area A is given by

$$A = [s(s - a)(s - b)(s - c)]^{\frac{1}{2}},$$

where $s = \frac{1}{2}(a + b + c)$. Find an approximate expression for δA in terms of δa. (Hint: use logarithmic differentiation to shorten the working.) Estimate δA when $a = 2$, $b = 4$, $c = 5$, and $\delta c = 0.1$, with a and b remaining constant.

4.12. (Computational). Growth on a deposit by compound interest is given by the formula $C = P(1 + r)^n$, where P is the amount deposited, r is the compound-interest annual growth rate, n is the time of deposit in years or fractions of a year, and C is the accumulated balance. Obtain approximating expressions for δC when (a) r changes by a small amount δr; (b) n changes by a small amount δn. (c) Consider plausible values of P, r, and n, and experiment with the accuracy of the formulae for various values of δr and δn.

4.13. (Newton's method, Section 4.6). It is not too difficult to make these calculations on a hand-held calculator. Find the solutions of the following equations within the broad ranges indicated, which contain exactly one solution.

(a) $x^4 + 2x^2 - x - 1 = 0$ (range $0.5 < x < 1$);

(b) $x^4 + x^{\frac{1}{3}} - 1 = 0$ ($0.5 < x < 0.75$);

(c) $x \ln x = -0.3$ ($0.1 < x < 0.2$);

(d) $e^x = 4x^3$ ($0 < x < 1$);

(e) $\tan x = 2x$ ($0 < x < \frac{1}{2}\pi$);

(f) $(e^x \sin x)/(1 + x) = 2$ ($1.5 < x < 1.9$).

4.14. The equation $f(x) = x e^{-x} + 1 = 0$ is known to have exactly one solution (not far from $x = -0.6$). Demonstrate numerically that it is of no use to start off Newton's method for this equation with a value of x greater than 1. Sketch the graph of the function, and make the construction based on Fig. 4.15 to explain why this is so.

4.15. (a) Supposing $y = f(x)$ to have a continuous graph, illustrate graphically that the following principle is true:

If $f(a)$ and $f(b)$ have opposite signs, then there is at least one solution of the equation $f(x) = 0$ in the range $a < x < b$.

(b) (Computational) The equation $e^x - 3x = 0$ has exactly two solutions, and they are in the range $0 < x < 2.5$. Use the principle in (a) to narrow the ranges in which they are known to lie, so as to produce starting values x_0 for Newton's method. One systematic technique is to start with the given end points $x = 0$ and 2.5, then to halve the interval repeatedly, considering the signs at the ends of the subdivisions.

4.16. (Computational). (a) Suppose that an equation

$f(x) = 0$ is known to have exactly one solution in a particular finite interval $a < x < b$. Write a program, using the principle described in Problem 4.15, to obtain a closer starting value x_0 for Newton's method. (Since there is only one solution, any subdivision you find across which the sign of $f(x)$ does not change can be ignored. Arrange for the process to stop when the solution is located within a small preset interval of length E.)

(b) Try this with, say, the equation $x(e^x - 1) = 1$, whose single solution lies between 0 and 1.

(c) By choosing E to be very small, the process can by itself locate the solution to any degree of accuracy if E is small enough. (This is called the **bisection method** for solving equations.) Obtain the number of iterations required to locate the single solution of $f(x) = 0$ in $0 < x < 1$ to 2, 4, and 6 decimal accuracy. (The number is the same for any such equation.)

(d) Solve the equation in (b) by Newton's method to 2, 4, and 6 decimal accuracy, starting with $x = 0.5$, and compare the number of iterations required with the number required by the bisection method.

5

Taylor series and approximations

5.1 The index notation for derivatives of any order

We shall use yet another standard notation for derivatives in this chapter. Since we shall have to keep track of derivatives of high orders we modify the 'dash' notation of (4.1) as follows to provide a brief form:

Index notation for derivatives

For the first, second, third, ... derivatives respectively of $f(x)$, write

$$f'(x) = f^{(1)}(x), \quad f''(x) = f^{(2)}(x), \tag{5.1}$$
$$f'''(x) = f^{(3)}(x), \dots .$$

If $y = f(x)$, the notation $y^{(1)}, y^{(2)}, y^{(3)}, \dots$, is also used.

Thus if $f(x) = x^3$, then $f^{(1)}(x) = 3x^2$, $f^{(2)}(x) = 6x$, and $f^{(3)}(x) = 6$. As with the dash notation, we encounter such forms as $f^{(2)}(u) = 6u$, $f^{(2)}(0) = 0$ and $f^{(2)}(x - c) = 6(x - c)$.

5.2 Taylor polynomials

Firstly we shall show how to obtain **approximations to a given** $f(x)$ **for use when x is a small number**. Suppose, for example, that

$$f(x) = \frac{1}{1 - x}.$$

Since $f(0) = 1$, we can be sure that

$$\frac{1}{1 - x} \approx 1$$

so long as x is small enough. This is shown in Fig. 5.1a. It is, of course, a poor approximation, acceptable only very close to $x = 0$.

A better approximation near $x = 0$ is given by the equation of the tangent line at $x = 0$ (Fig. 5.1b). Since $f^{(1)}(x) = 1/(1 - x)^2$, the

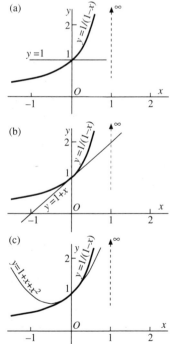

(a)

(b)

(c)

Fig. 5.1

slope at $x = 0$ is

$$f^{(1)}(0) = 1.$$

The equation of the tangent line at $x = 0$ is therefore $y = 1 + x$, so

$$\frac{1}{1 - x} \approx 1 + x,$$

when x is small enough.

We need a way to continue improving the approximation, $P(x)$ say, to a further stage and beyond. At present we have reached the tangent approximation $P(x) = 1 + x$, which was chosen so that $P(0) = f(0)$ and $P^{(1)}(0) = f^{(1)}(0)$. To obtain the next approximation choose a $P(x)$ which also **matches the second derivative at** $x = 0$:

$$P(0) = f(0), \qquad P^{(1)}(0) = f^{(1)}(0), \qquad P^{(2)}(0) = f^{(2)}(0).$$

This involves adding a term in x^2, and we can choose its coefficient so that the extra condition is satisfied **without disturbing the two terms we have already found**. Continuing with the example,

$$f^{(2)}(x) = \frac{2}{(1 - x)^3}, \quad \text{so} \quad f^{(2)}(0) = 2.$$

It is easy to check that $P(x) = 1 + x + x^2$ satisfies the three conditions. Therefore

$$\frac{1}{1 - x} \approx 1 + x + x^2$$

is an improved approximation. This represents the parabolic curve shown in Fig. 5.1c.

We can carry out this process for any function $f(x)$, and take it to any level of approximation we wish. Successive approximations will consist of **polynomials** of increasing degree. However, we must not expect too much of it: we cannot go too far from the origin and still expect a good approximation.

To deal with the general case, we need the following simple result.

Derivatives of a polynomial in x at $x = 0$

$$P(x) = a_0 + a_1 x + a_2 x^2 + \cdots + a_N x^N$$

is a polynomial of degree N. Then

$$P(0) = a_0, \quad P^{(1)}(0) = a_1, \quad P^{(2)}(0) = 2! a_2, \qquad \text{(5.2)}$$

and in general

$$P^{(n)}(0) = n! \, a_n$$

for $n = 1, 2, 3, \ldots, N$.

It is easy to verify (5.2) by working out the first few derivatives.

Now suppose that we wish to approximate to a general function $f(x)$ (near $x = 0$) by means of a polynomial

$$P(x) = a_0 + a_1 x + a_2 x^2 + \cdots + a_N x^N.$$

We require that

$$P(0) = f(0), \quad P^{(1)}(0) = f^{(1)}(0), \quad P^{(2)}(0) = f^{(2)}(0), \quad \cdots \quad .$$

According to (5.2), the coefficients are given by

$$a_0 = P(0) = f(0),$$

$$a_1 = \frac{1}{1!} P^{(1)}(0) = \frac{1}{1!} f^{(1)}(0),$$

$$a_2 = \frac{1}{2!} P^{(2)}(0) = \frac{1}{2!} f^{(2)}(0),$$

$$a_3 = \frac{1}{3!} P^{(3)}(0) = \frac{1}{3!} f^{(3)}(0),$$

and so on. By writing the coefficients a_n in terms of the known values $f^{(n)}(0)$ we obtain the **Taylor polynomial approximation**:

Taylor polynomial $P(x)$ of degree N near $x = 0$

Let $P(x)$ be the Nth degree polynomial

$$f(0) + \frac{1}{1!} f^{(1)}(0)x + \frac{1}{2!} f^{(2)}(0)x^2 + \cdots + \frac{1}{N!} f^{(N)} x^N. \qquad (5.3)$$

Then for x sufficiently close to zero,

$$f(x) \approx P(x).$$

Example 5.1. Obtain a fifth-degree polynomial which approximates to e^x for values of x which are not too large.

Use (5.3), putting $f(x) = e^x$. This case is simple:

$$f(x) = f^{(1)}(x) = f^{(2)}(x) = \cdots = f^{(5)}(x) = e^x,$$
so $f(0) = f^{(1)}(0) = f^{(2)}(0) = \cdots = f^{(5)}(0) = 1$. Therefore

$$e^x \approx 1 + \frac{1}{1!} x + \frac{1}{2!} x^2 + \frac{1}{3!} x^3 + \frac{1}{4!} x^4 + \frac{1}{5!} x^5 = P(x).$$

(If we take higher degree approximations, the terms continue according to the same rule.) We show e^x and its approximation $P(x)$

in the following table for a few values of x.

x	-4	-3	-2	-1	-0.5
e^x	0.0183	0.0498	0.1353	0.3679	0.6065
$P(x)$	-3.533	-0.6500	0.0667	0.3666	0.6065

x	0	0.5	1	2	3	4
e^x	1	1.6487	2.7183	7.3891	20.086	54.598
$P(x)$	1	1.6487	2.7167	7.2667	18.400	42.867

The approximating polynomial $P(x)$ clings to the true values for a considerable range around the origin.

Example 5.2. (a) Obtain the Taylor polynomial approximation of any order N for the function $1/(1 - x)$ near $x = 0$. (b) Obtain an expression for the error in the approximation.

(a) Putting $1/(1 - x) = f(x)$, the sequence of derivatives of $f(x)$ is

$$f^{(1)}(x) = \frac{1}{(1 - x)^2}, \quad f^{(2)}(x) = \frac{2 \cdot 1}{(1 - x)^3}, \quad f^{(3)}(x) = \frac{3 \cdot 2 \cdot 1}{(1 - x)^4}$$

and in general

$$f^{(n)}(x) = \frac{n!}{(1 - x)^{n+1}}.$$

Therefore, referring to (5.3), the Taylor polynomial of degree N is

$$1 + x + x^2 + x^3 + \cdots + x^N.$$

(b) The error in an estimation using this approximation is equal to

$$P(x) - f(x) = 1 + x + \cdots + x^N - \frac{1}{1 - x}$$

$$= \frac{(1 - x)(1 + x + x^2 + \cdots + x^N) - 1}{1 - x}$$

$$= \frac{-x^{N+1}}{1 - x}.$$

The reader should experiment with this expression using various values of N and x. If x is very small, the error involved is very small even if N is only 2 or 3. If we take any *fixed* value of x in the range $-1 < x < 1$, the error will approach zero when we take approximations of higher and higher degree (because, when $-1 < x < 1$, x^{N+1} approaches zero as N increases: try this numerically with, say, $x = 0.9$). The approximation fails altogether if $x > 1$ or $x < -1$. The error will be large, and to increase N will make it still larger because $|x^{N+1}|$ increases when N increases.

5.3 A note on infinite series

In the previous section we did not put any limit on the degree of the approximating polynomial, and there seems to be no reason why we should not let the terms run on for ever: in fact, let the degree N approach infinity. If we extend the polynomial approximation of Example 5.1 for $f(x) = e^x$, we obtain an example of a so-called **infinite series**:

$$1 + \frac{1}{1!}x + \frac{1}{2!}x^2 + \frac{1}{3!}x^3 + \cdots \quad \text{or} \quad \sum_{n=0}^{\infty} \frac{1}{n!}x^n.$$

It might be that by extending the approximating polynomials in this way, *approximation* will become *equality*, so that the sum of the series will be *equal to* the original function instead of being just an *approximation* to it, but this is only true with reservations.

There are many types of infinite series (see, e.g. Chapter 24 on Fourier series). Consider first what is meant by the **sum of an infinite series**. When x is given any particular value, the terms to be added become simply numbers. We cannot in practice add an infinite number of numbers: no matter how many operations we carry out we never reach the end. However, this does not mean that the infinite series does not add up to a definite number, only that we cannot reach it exactly by simply piling on more and more terms.

Consider the simpler infinite series that we get from putting $x = 0.1$ into the Taylor polynomial for $1/(1 - x)$ (see Example (5.2), and letting the degree increase to infinity. It is

$$1 + 0.1 + 0.1^2 + 0.1^3 + 0.1^4 + \cdots.$$

This is the same as

$$1 + 0.1 + 0.01 + 0.001 + 0.0001 + \cdots.$$

If we record the sum of $1, 2, 3, 4, \ldots,$ terms successively, we obtain what is called a **sequence of partial sums** ('partial' because we only take a finite number of terms into account). The sequence is

$$1, \quad 1.1, \quad 1.11, \quad 1.111, \quad 1.1111, \quad \ldots \ .$$

The number that is being approached is obviously $1.11111, \ldots,$ which is equal to $10/9$. This number is equal to the value of $1/(1 - x)$ when $x = 0.1$, so in this case the *infinite* series has delivered the value required. Similarly, if we put $x = \frac{1}{2}$, the infinite series is

$$1 + \tfrac{1}{2} + (\tfrac{1}{2})^2 + (\tfrac{1}{2})^3 + (\tfrac{1}{2})^4 + \cdots = 1 + \tfrac{1}{2} + \tfrac{1}{4} + \tfrac{1}{8} + \tfrac{1}{16} + \cdots.$$

For the sum of $1, 2, 3, 4, \ldots$ terms, we obtain the sequence of partial sums

$$1, \quad 1\tfrac{1}{2}, \quad 1\tfrac{3}{4}, \quad 1\tfrac{7}{8}, \quad 1\tfrac{15}{16}, \quad \ldots \ ,$$

which is obviously approaching the value 2. (This is indeed the value of $1/(1 - x)$ when $x = \frac{1}{2}$.) Infinite series whose **partial sums**

approach a definite value as we take more and more terms are said to **converge** to this value, which is called the **sum of the infinite series**.

However, not all infinite series converge. For example, if we form successively the sum of 1, 2, 3, ... terms of the infinite series

$$1 + 1 + 1 + \cdots,$$

then we obtain

$$1, 2, 3, 4, \quad \ldots \quad ,$$

which is obviously going to infinity. The infinite series

$$1 - 1 + 1 - 1 + \cdots$$

has the successive partial sums

$$1, 0, 1, 0, 1, \quad \ldots \quad ,$$

which is not going anywhere. Such series are said to **diverge**. The reader might be surprised to know that the infinite series

$$1 + \tfrac{1}{2} + \tfrac{1}{3} + \tfrac{1}{4} + \cdots$$

diverges: the partial sums go to infinity. (It is worth experimenting with this series: even using a computer you might take a while to convince yourself that it really does diverge.)

5.4 Infinite Taylor expansions

We return to the subject of general Taylor polynomials of the type (5.3) when we extend the polynomial to an infinite number of terms, so that we have an infinite series instead of a polynomial expression. This is called a **Taylor series** or an infinite **Taylor expansion about the origin** $x = 0$ for the function $f(x)$.

The mathematical theory of infinite series, and in particular of Taylor series, cannot be discussed in this book. In the previous section it is indicated that pitfalls might arise when the polynomials are extended into infinite series. Moreover, it seems obvious, for example, that the values of a function and its derivatives *at the origin only* cannot possibly predict values elsewhere if we allow functions to be *completely arbitrary* at other points.

However, the ordinary functions do follow the simple pattern illustrated by the case of $f(x) = 1/(1 - x)$ in Example (5.2). Each function has an individual range of values of x, called its **interval of validity**, in which the Taylor series **converges to the exact value of** $f(x)$. Elsewhere, the series must not be used for approximation. In Problem 5.2, the reader is invited to verify the coefficients in the following series:

Standard Taylor expansions about $x = 0$

| | Interval |
| Function and expansion | of validity |

(a) $e^x = 1 + \dfrac{1}{1!} x + \dfrac{1}{2!} x^2 + \cdots$ any x

(b) $\sin x = x - \dfrac{1}{3!} x^3 + \dfrac{1}{5!} x^5 - \cdots$ any x

(c) $\cos x = 1 - \dfrac{1}{2!} x^2 + \dfrac{1}{4!} x^4 - \cdots$ any x

(d) $(1 + x)^\alpha = 1 + \alpha x + \dfrac{\alpha(\alpha - 1)}{2!} x^2$ (5.4)

$\quad + \dfrac{\alpha(\alpha - 1)(\alpha - 2)}{3!} x^3 + \cdots$ $-1 < x < 1$

(e) $\ln(1 + x) = x - \frac{1}{2}x^2 + \frac{1}{3}x^3 - \cdots$ $-1 < x \leqslant 1$

(If a series is cut short after the term in x^N, the result is an Nth-degree Taylor polynomial approximation.)

Equality is achieved over the intervals stated: the *infinite* series *exactly* represents the original function. Notice (5.4d): it is the binomial theorem (see Appendix A(c)) extended to arbitrary values of α, positive or negative (if α is a positive integer N, it ends at the term in x^N).

When the series is used to provide approximations by taking only a finite number of terms, it is necessary to estimate how many terms to take so as to obtain a desired degree of accuracy. It is usually sufficient to observe the size of the terms involved, as in the following example.

Example 5.3. Find how many terms of the Taylor series for $\sin x$ are needed to obtain three-decimal accuracy over the range $-1 \leqslant x \leqslant 1$ (in radians: 1 radian is about $57°$).

The intuitive requirement is that we should stop at the point where we can see that taking further terms is not likely to affect the third decimal place. The magnitude (modulus) of the terms in (5.4b) increases when the magnitude of x increases, so it should be sufficient to provide an approximation good for the largest value, $x = 1$. The magnitudes of successive terms when $x = 1$ are equal to

$\quad 1, \quad 0.1\dot{6}, \quad 0.08\dot{3}, \quad 0.0002, \quad 2 \cdot 10^{-6}, \ldots$.

It is therefore enough to retain three terms of the series; that is to say we should retain powers of x up to x^5. To three decimals, then,

$$\sin x \approx x - \frac{1}{3!}x^3 + \frac{1}{5!}x^5 \quad \text{for} \quad -1 \leqslant x \leqslant 1.$$

5.5 Manipulation of Taylor series

We can obtain new Taylor series from the standard ones in (5.4).

Example 5.4. Find the Taylor expansion about $x = 0$ for the function $(2 - x)^{\frac{1}{2}}$, and state its range of validity.

Write $(2 - x)^{\frac{1}{2}} = 2^{\frac{1}{2}}(1 - \frac{1}{2}x)^{\frac{1}{2}} = 2^{\frac{1}{2}}[1 + (-\frac{1}{2}x)]^{\frac{1}{2}}$.

We can use the binomial expansion (5.4d), with $\alpha = \frac{1}{2}$, and with $-\frac{1}{2}x$ in place of x. The expansion will be valid, provided that $-1 < -\frac{1}{2}x < 1$ i.e. when $-2 < x < 2$. Therefore

$$(2 - x)^{\frac{1}{2}} = 2^{\frac{1}{2}}[1 + (-\tfrac{1}{2}x)]^{\frac{1}{2}}$$

$$= 2^{\frac{1}{2}}\left(1 + \tfrac{1}{2}(-\tfrac{1}{2}x) + \frac{\frac{1}{2}(\frac{1}{2} - 1)}{2!}(-\tfrac{1}{2}x)^2 \right.$$

$$\left. + \frac{\frac{1}{2}(\frac{1}{2} - 1)(\frac{1}{2} - 2)}{3!}(-\tfrac{1}{2}x)^3 + \cdots \right)$$

$$= 2^{\frac{1}{2}}(1 - \tfrac{1}{4}x - \tfrac{1}{32}x^2 - \tfrac{1}{128}x^3 + \cdots)$$

when $-2 < x < 2$.

To find the first few terms in the Taylor series for a composite function $f(x)$ such as

$$f(x) = \frac{e^{-x}}{(1 + x)^{\frac{1}{2}}},$$

it is usually best *not* to start from first principles by calculating $f(0)$, $f^{(1)}(0)$, $f^{(2)}(0)$ and so on, which can lead to great complication, but to manipulate standard expansions as in the following examples.

Example 5.5. Approximate to $(\sin x/x)^2$ by a polynomial of degree 4, and compare the approximate and exact values when $x = 0$, $\frac{1}{4}$, $\frac{1}{2}$, 1, 2.

From (5.4b),

$$\left(\frac{\sin x}{x}\right)^2 = \left(\frac{x - \frac{1}{3!}x^3 + \frac{1}{5!}x^5 - \cdots}{x}\right)^2$$

$$= \left(1 - \frac{1}{3!}x^2 + \frac{1}{5!}x^4 - \cdots\right)^2$$

$$= 1 - \frac{2}{3!}x^2 + \left(\frac{2}{5!} + \frac{1}{3!^2}\right)x^4 + \cdots,$$

where only terms up to x^4 are retained. Write the approximating polynomial $P(x)$ as

$$P(x) = 1 - 0.3333x^2 + 0.0444x^4$$

to obtain the table

x	0	0.25	0.5	1.0	2.0
$[\sin x/x]^2$	1	0.9793	0.9179	0.6861	0.2067
$P(x)$	1	0.9793	0.9194	0.7111	0.3772

Example 5.6. Approximate to $e^{-x}/(1 + x)^{\frac{1}{2}}$ near $x = 0$ by a polynomial of degree 2.

Write

$$f(x) = e^{-x}(1 + x)^{-\frac{1}{2}}.$$

Use (5.4a) with $-x$ in place of x, and carry it to degree 2:

$$e^{-x} \approx 1 - x + \frac{1}{2!}x^2.$$

Also, by (5.4d) (the binomial theorem) with $\alpha = -\frac{1}{2}$:

$$(1 + x)^{-\frac{1}{2}} \approx 1 + (-\tfrac{1}{2})x + \frac{(-\frac{1}{2})(-\frac{1}{2} - 1)}{2!}x^2.$$

Then by multiplying the two polynomials we obtain

$$f(x) \approx 1 - \tfrac{3}{2}x + \tfrac{11}{8}x^2,$$

when x is small. (Reject powers higher than 2 in the final product – they would not be correct since we neglected such terms in the original approximations.)

Example 5.7. Obtain the first three nonzero terms of the Taylor expansion for $1/\cos x$.

There are several ways of doing this problem.

Working from (5.3). The reader might try this, but it is very arduous.

Using the power series for cos x. Write

$$\frac{1}{\cos x} = \frac{1}{1 - \dfrac{1}{2!}x^2 + \dfrac{1}{4!}x^4 - \cdots}.$$

The problem is to find the first three terms in the *reciprocal* of the infinite series; we then have a Taylor polynomial. Anticipate that only the even powers of x will occur, as in the expansion of cos x.

Then we expect

$$\frac{1}{\cos x} = \frac{1}{1 - \frac{1}{2!}x^2 + \frac{1}{4!}x^4 - \cdots} = b_0 + b_2 x^2 + b_4 x^4 + \cdots.$$

We have to find b_0, b_2, b_4. To do this, cross-multiply:

$$1 = \left(1 - \frac{1}{2!}x^2 + \frac{1}{4!}x^4 - \cdots\right)(b_0 + b_2 x^2 + b_4 x_4 + \cdots)$$

$$= b_0 + (b_2 - \tfrac{1}{2}b_0)x^2 + (b_4 - \tfrac{1}{2}b_2 + \tfrac{1}{24}b_0)x^4 + \cdots$$

(retaining only powers up to x^4). Match the coefficients of powers of x on both sides, starting with the constant term; we obtain

$$b_0 = 1,$$

and, since the coefficients of x^2 and x^4 on the left are zero,

$$b_2 - \tfrac{1}{2}b_0 = 0 \quad \text{and} \quad b_4 - \tfrac{1}{2}b_2 + \tfrac{1}{24}b_0 = 0.$$

The last two equations can be solved successively to give

$$b_2 = \tfrac{1}{2} \quad \text{and} \quad b_4 = \tfrac{5}{24}.$$

Finally

$$1/\cos x \approx 1 + \tfrac{1}{2}x^2 + \tfrac{5}{24}x^4.$$

Polynomial division. We can evaluate $1/(1 - \frac{1}{2}x^2 + \frac{1}{24}x^4 - \cdots)$ by long division, setting it out like this, ignoring powers higher than x^4:

$$
\begin{array}{r}
1 + \tfrac{1}{2}x^2 + \tfrac{5}{24}x^4 \\
1 - \tfrac{1}{2}x^2 + \tfrac{1}{24}x^4 \,\overline{\big)\, 1 } \\
1 - \tfrac{1}{2}x^2 + \tfrac{1}{24}x^4 \\
\text{subtract:} = \overline{\tfrac{1}{2}x^2 - \tfrac{1}{24}x^4} \\
\tfrac{1}{2}x^2 - \tfrac{1}{4}x^4 \\
\text{subtract:} = \overline{\tfrac{5}{24}x^4} \\
\tfrac{5}{24}x^4
\end{array}
$$

5.6 Approximations for large values of x

When x is large, $1/x$ is small. This fact can sometimes be used to obtain approximations valid when x is large, as in the following example.

Example 5.8. Obtain a three-term approximation to $(1 + 1/x)^{\frac{1}{2}}$ valid when x is **large** enough.

Translate the binomial theorem, (5.4d), with $\alpha = \tfrac{1}{2}$, in terms of a neutral variable, say u:

$$(1 + u)^{\frac{1}{2}} = 1 + \tfrac{1}{2}u + \frac{\tfrac{1}{2}(\tfrac{1}{2} - 1)}{2!} u^2 + \cdots \quad \text{when } -1 < u < 1,$$

so $(1 + u)^{\frac{1}{2}} \approx 1 + \tfrac{1}{2}u - \tfrac{1}{8}u^2$ when u is small enough, the approximation improving as u gets smaller. Now put $u = 1/x$; we obtain

$$\left(1 + \frac{1}{x}\right)^{\frac{1}{2}} \approx 1 + \frac{1}{2x} - \frac{1}{8x^2},$$

when x is *large* enough (positively or negatively), the approximation improving as x gets larger.

5.7 Taylor series at other points

The Taylor series at $x = 0$ for

$$f(x) = \frac{1}{1 - x} = 1 + x + x^2 + x^3 + \cdots$$

does not work when $x = 2$: we get $1 + 2 + 2^2 + \cdots$, which is infinite. However, we can obtain a *different* Taylor-type series which represents $1/(1 - x)$ near $x = 2$ by a process which amounts to changing the origin, as in the following Example.

Example 5.9. Find a Taylor-type series which represents $1/(1 - x)$ for values of x near $x = 2$.

Look for a series of this type:

$$\frac{1}{1 - x} = b_0 + b_1(x - 2) + b_2(x - 2)^2 + \cdots,$$

because we want a series that works when x is close to 2; which is to say when $x - 2$ is small, rather than when x is small as before. Therefore we need a series consisting of powers of $x - 2$. We can bring the element $x - 2$ into view by writing

$$\frac{1}{1 - x} = \frac{1}{1 - (x - 2 + 2)} = -\frac{1}{1 + (x - 2)}.$$

Now expand the final term by using (5.4d) (the binomial theorem) with $\alpha = -1$, and $x - 2$ in place of x, obtaining

$$\frac{1}{1 - x} = -1 + (x - 2) - (x - 2)^2 + (x - 2)^3 + \cdots,$$

valid if $-1 < x - 2 < 1$, that is, if

$$1 < x < 3.$$

Example 5.10. Obtain a Taylor series about the point $x = \pi$ for the function $\cos x$.

There exists already the series (5.4c) which is valid at $x = \pi$. However, if we are interested in approximating to $\cos x$ near $x = \pi$, an expansion in powers of $x - \pi$ should be more economical and expressive than one consisting of powers of x. We show two ways of finding the series.

(a) *On the lines of Example 5.9.* Write

$$\cos x = \cos[\pi + (x - \pi)]$$
$$= \cos \pi \cos(x - \pi) - \sin \pi \sin(x - \pi) = -\cos(x - \pi).$$

We can use (5.4c) to expand this, by putting $x - \pi$ in place of x. We obtain

$$\cos x = -\cos(x - \pi) = -1 + \frac{1}{2!}(x - \pi)^2 - \frac{1}{4!}(x - \pi)^4 + \cdots.$$

This is valid for all values of x. A two-term approximation shows that $\cos x$ has a parabolic shape near $x - \pi = 0$ or $x = \pi$, where $\cos x$ has a local minimum.

(b) *Matching the value and the derivatives at $x = \pi$.* The derivatives of $f(x) = \cos x$ at $x = \pi$ are given by

$$f(\pi) = \cos \pi = -1, \qquad f^{(1)}(\pi) = -\sin \pi = 0,$$
$$f^{(2)}(\pi) = -\cos \pi = 1,$$

and so on. The same relations hold good between the coefficients of a polynomial in powers of $x - \pi$ and the values of its derivatives at $x = \pi$, as was stated in (5.3) for polynomials in x at $x = 0$. We simply put $x - \pi$ in place of x in (5.3). The required Taylor series is

$$f(x) = f(\pi) + \frac{1}{1!} f^{(1)}(\pi)(x - \pi) + \frac{1}{2!} f^{(2)}(\pi)(x - \pi)^2 + \cdots,$$

which is the same as the result obtained in (a).

The general result is the following:

Taylor series about a point $x = c$

$$f(x) = f(c) + \frac{1}{1!} f^{(1)}(c)(x - c)$$

$$+ \frac{1}{2!} f^{(2)}(c)(x - c)^2 + \cdots.$$

(5.5)

(The range of validity depends upon $f(x)$.)

Problems, Chapter 5

5.1. Obtain a four-term Taylor polynomial approximation valid near $x = 0$ for each of the following. Estimate the ranges of x over which three-term polynomials will give two-decimal accuracy (you cannot usually tell until you have seen the next-higher term).
(a) $e^{\frac{1}{2}x}$;   (b) $(1 + x)^{\frac{1}{2}}$;   (c) $(1 + x)^{-\frac{1}{3}}$;
(d) $\sin 2x$;   (e) $\cos \frac{1}{2}x$;   (f) $\ln(1 + x)$;
(g) $(1 + x^2)^{\frac{1}{2}}$ (hint: consider $(1 + u)^{\frac{1}{2}}$; then put $u = x^2$);
(h) $\ln(1 + 3x)$ (see the hint in (g)).

5.2. Verify the coefficients of each of the infinite Taylor series shown in (5.4) (taking the ranges of validity for granted); namely
(a) e^x;   (b) $\sin x$;   (e) $\cos x$;
(d) $(1 + x)^x$;   (e) $\ln(1 + x)$.

5.3. For each of the following series give the Taylor polynomial having the lowest degree which you think will safely give four-decimal accuracy over the ranges given.
(a) e^x over $-2 \leqslant x \leqslant 2$;
(b) $\sin x$ over $-2 \leqslant x \leqslant 2$;
(c) $\cos x$ over $-2 \leqslant x \leqslant 2$;
(d) $(1 + x)^{\frac{1}{2}}$ over $-0.5 \leqslant x \leqslant 0.5$;
(e) $\ln(1 + x)$ over $-0.5 \leqslant x \leqslant 0.5$.

5.4. Obtain the first two nonzero terms in the Taylor series at the origin for the following.
(a) $\arcsin x$;   (b) $\arccos x$;   (c) $\arctan x$;
(d) $e^{-x} \sin x$;   (e) $e^{-x} \cos x$.

5.5. (See Section 5.5). Find three nonzero terms in the Taylor series at $x = 0$ for the following functions and state the ranges of validity.
(a) $1/(1 + 3x)$;   (b) $1/(2 - x)$;   (c) $(3 - x)^{\frac{1}{3}}$;
(d) $(x - 3)^{\frac{1}{3}}$;   (e) $\ln(9 - x)$;   (f) $\cos \frac{1}{2}x$;
(g) $\sin x^{\frac{1}{2}}$ (Consider the series for $\sin u$; then put $u = x^{\frac{1}{2}}$). This is not strictly a Taylor series – it is spoken of as 'a Taylor series in $x^{\frac{1}{2}}$' – but it still is useful for approximations.
(h) $\cos x^{\frac{1}{2}}$.

5.6. (See Section 5.5). Find the first three nonzero terms in the Taylor series at $x = 0$ for the following.
(a) $e^{-x}/(1 + x)$;   (b) $(1 - x)^{\frac{1}{2}} e^x$;
(c) $[\ln(1 - x)]^2/x^2$.

5.7. (See Example 5.7). Find the first three nonzero terms in the Taylor expansions at the origin for the following.
(a) $1/(1 + \ln[1 + x])$;   (b) $\tan x$;   (c) $1/(1 + e^x)$;
(d) $\tanh x$, or $(e^x - e^{-x})/(e^x + e^{-x})$. (It is less complicated if you firstly reduce this to a more manageable form);
(e) $x/\sin x$.

5.8. (See Section 5.6). Find a three-term approximation, valid for *large* enough values of x, in each of the following cases.
(a) $\left(1 - \dfrac{1}{x}\right)^{\frac{1}{2}}$;   (b) $\ln\left(1 + \dfrac{1}{x^{\frac{1}{2}}}\right)$;
(c) $x^{\frac{1}{2}}/(1 + x)^{\frac{1}{2}}$;   (d) $\ln(1 + x + x^2)$
(e) $(1 + x)^{\frac{1}{2}}$ $(= x^{\frac{1}{2}}(1 + x^{-1})^{\frac{1}{2}})$;
(f) $1/\sin x^{-1}$.

5.9. (a) Show that $1/\sin x \approx 1/x + \frac{1}{6}x^2$ when x is nonzero but small enough.
(b) Show that, when x is *large* enough, $(1 + x)^{\frac{1}{2}} \approx x^{\frac{1}{2}} + 1/2x^{\frac{1}{2}}$.
(c) Show that, when x is *large* enough, $(2 + x)^{\frac{1}{2}} - (1 + x)^{\frac{1}{2}} \approx 1/2x^{\frac{1}{2}}$.
(d) Show that, when x is nonzero but small enough, then $1/(1 - \cos x)^{\frac{1}{2}} \approx 2/x^2 + \frac{1}{6}$.

5.10. (a) (See Section 5.6). Write

$$\ln x = \ln(1 + [x - 1]),$$

and so obtain the Taylor series for $\ln x$ at $x = 1$. State the range of validity.
(b) Obtain the Taylor series at $x = \frac{1}{2}\pi$ for $\cos x$, and state the range of validity.
(c) Obtain the Taylor series at $x = 1$ for the function $(1 + x)^{\frac{1}{2}}$, and state the range of validity.

5.11. Suppose that $f(x)$ has a stationary point at $x = c$. Write down the form of its Taylor series at $x = c$, taking this into account.
(a) By considering the first three terms, rediscover the conditions on $f''(c)$ which determine the type of stationary point (see (4.2)).
(b) By considering further terms of the Taylor series, extend the criteria to obtain a general rule which covers the case $f''(c) = 0$.

6 Complex numbers

6.1 Definitions and rules

The quadratic equation

$$x^2 - 2x + 2 = 0$$

can be written in the form

$$(x - 1)^2 = -2 + 1 = -1$$

using the method known as 'completing the square'. If we solve this equation by taking the square root, then

$$x - 1 = \pm\sqrt{-1} \quad \text{or} \quad x = 1 \pm \sqrt{-1}.$$

However, there is no real number whose square is -1. In order to solve it we must extend the numbers into the set of **complex numbers**. We use the symbol j to represent $\sqrt{-1}$ (mathematicians normally use the symbol i). For $-\sqrt{-1}$ we write $-j$. The **complex roots** of the quadratic equation can be expressed as

$$x = 1 + j, \quad \text{and} \quad x = 1 - j, \quad \text{or} \quad x = 1 \pm j.$$

For the general quadratic equation

$$ax^2 + bx + c = 0, \tag{6.1}$$

assume that $a > 0$. The left-hand side of this equation can be written as

$$ax^2 + bx + c = a\left(x^2 + \frac{b}{a}x\right) + c = a\left(x + \frac{b}{2a}\right)^2 + c - \frac{b^2}{4a}.$$

The quadratic equation (6.1) becomes

$$\left(x + \frac{b}{2a}\right)^2 = \frac{b^2 - 4ac}{4a^2}.$$

Taking the square root

$$x + \frac{b}{2a} = \pm\frac{\sqrt{(b^2 - 4ac)}}{2a}.$$

Hence the roots of the quadratic equation (6.1) are

$$x = -\frac{b}{2a} \pm \frac{1}{2a}\sqrt{(b^2 - 4ac)}.$$

The roots are distinct real numbers if $b^2 > 4ac$, equal if $b^2 = 4ac$, and complex if $b^2 < 4ac$. In the last case they can then be written in terms of j as

$$x = -\frac{b}{2a} \pm j\frac{1}{2a}\sqrt{(4ac - b^2)},$$

where $\sqrt{(4ac - b^2)}$ is a real number.

A complex number is any number of the form

$$z = x + jy,$$

where x and y are real numbers. This is known as the **standard form** of the complex number. In this expression x is the **real part** of z, written as $x = \mathrm{Re}\,z$, and y is the **imaginary part**, written as $y = \mathrm{Im}\,z$ (note that it is y, not jy, which is called the imaginary part.) If $y = 0$, then z is a **real number**, and if $x = 0$, then z is an **imaginary number**.

We need to put together rules for manipulating complex numbers: the rules are natural consequences of operations of addition, multiplication, etc. on numbers containing j. The only exception to normal algebra is that, whenever j^2 appears, we can substitute -1.

Example 6.1. Express j^2, j^3, j^4, j^5, and j^6 in standard form.

The standard forms are

$$j^2 = -1, \qquad j^3 = j^2 j = -1 \times j = -j.$$

Since $-j$ can be written $0 + (-1)j$, it follows that $\mathrm{Re}\,j^3 = 0$ and $\mathrm{Im}\,j^3 = -1$.

$$j^4 = j^2 j^2 = (-1)(-1) = 1, \qquad j^5 = jj^4 = j, \qquad j^6 = jj^5 = -j.$$

Example 6.2. Express the following in standard form, and state the real and imaginary parts in each case:
(a) $(2 + j) - (3 + 3j)$;   (b) $j(j + 2)$;
(c) $(1 - j)(1 + 2j)$;   (d) $(1 - 3j)(2 + 3j)$.

(a) $(2 + j) - (3 + 3j) = 2 + j - 3 - 3j = -1 - 2j$.
Real part $= -1$; imaginary part $= -2$.
(b) $j(j + 2) = j^2 + 2j = -1 + 2j$.
Real part $= -1$; imaginary part $= 2$.
(c) $(1 - j)(1 + 2j) = 1 + 2j - j - 2j^2 = 1 + 2j - j + 2 = 3 + j$.
Real part $= 3$; imaginary part $= 1$.
(d) $(1 - 3j)(2 + 3j) = 2^2 - (3j)^2 = 4 - 9j^2 = 4 + 9 = 13$.
Real part $= 13$; imaginary part $= 0$.

Let $z_1 = x_1 + jy_1$ and $z_2 = x_2 + jy_2$. In formal terms, the principal rules are as follows.

1. Two complex numbers z_1 and z_2 are said to be **equal** if $x_1 = x_2$ and $y_1 = y_2$: we write $z_1 = z_2$.

2. The **sum** of two complex numbers z_1 and z_2 is given by

$$z_1 + z_2 = (x_1 + jy_1) + (x_2 + jy_2) = (x_1 + x_2) + j(y_1 + y_2),$$

its real part being the sum of the real parts, and its imaginary part the sum of the imaginary parts of z_1 and z_2.

3. Similarly the **difference** is

$$z_1 - z_2 = (x_1 + jy_1) - (x_2 + jy_2) = (x_1 - x_2) + j(y_1 - y_2).$$

4. The **product** of z_1 and z_2 is

$$\begin{aligned}
z_1 z_2 &= (x_1 + jy_1)(x_2 + jy_2) \\
&= x_1 x_2 + jy_1 x_2 + x_1 jy_2 + jy_1 jy_2 \\
&= x_1 x_2 + jy_1 x_2 + jx_1 y_2 + j^2 y_1 y_2 \\
&= x_1 x_2 + jy_1 x_2 + jx_1 y_2 - y_1 y_2 \quad (\text{since } j^2 = -1) \\
&= (x_1 x_2 - y_1 y_2) + j(y_1 x_2 + x_1 y_2).
\end{aligned}$$

In order to carry out **division**, a special result is needed. Suppose that $z = x + jy$, where x and y are real. Then the number $x - jy$, where we have changed j to $-j$, is called the **complex conjugate** of z, and will be written $\bar{z}$. The product $z\bar{z}$ is given by

$$z\bar{z} = (x + jy)(x - jy) = x^2 - (jy)^2 = x^2 + y^2,$$

which is a *real positive number*. (This was illustrated in Example 6.2d.)

Property of the conjugate

Let $z = x + jy$, with x, y real. Then $\bar{z} = x - jy$ and   **(6.2)**

$$z\bar{z} = x^2 + y^2.$$

---

5. The **reciprocal** of a complex number in standard form. Let $z = x + jy$, and consider

$$\frac{1}{z} = \frac{1}{x + jy}.$$

This is not in standard form $a + jb$: to reduce it to standard form, multiply it by the factor

$$\frac{x - jy}{x - jy},$$

$x - jy$ being the conjugate of $x + jy$. This factor is equal to 1, and it will not affect the value of $1/z$. Hence

$$\frac{1}{z} = \frac{1}{x + jy} = \frac{x - jy}{x - jy} = \frac{x - jy}{x^2 + y^2}$$

$$= \frac{x}{x^2 + y^2} - j\frac{y}{x^2 + y^2} \quad \text{(from (6.2))}$$

in standard form. This process also enables us to reduce **quotients** to standard form, as in Example 6.3(c) below.

Example 6.3. Reduce to standard form

(a) $\dfrac{1}{2 + 3j}$; (b) $\dfrac{1}{2 - 3j}$; (c) $\dfrac{1 - j}{1 + j}$; (d) $\dfrac{1}{j}$.

The standard forms are:

(a) $\dfrac{1}{2 + 3j} = \dfrac{1}{2 + 3j}\dfrac{2 - 3j}{2 - 3j} = \dfrac{2 - 3j}{2^2 + 3^2} = \dfrac{2}{13} - \dfrac{3}{13}j$;

(b) $\dfrac{1}{2 - 3j} = \dfrac{1}{2 - 3j}\dfrac{2 + 3j}{2 + 3j} = \dfrac{2 + 3j}{2^2 + 3^2} = \dfrac{2}{13} + \dfrac{3}{13}j$;

(c) $\dfrac{1 - j}{1 + j} = \dfrac{1 - j}{1 + j}\dfrac{1 - j}{1 - j} = \dfrac{1 - 2j + j^2}{1^2 + 1^2} = -j$;

(d) $\dfrac{1}{j} = \dfrac{1}{j}\dfrac{-j}{-j} = -j$.

The general quotient rule is

$$\frac{z_1}{z_2} = \frac{(x_1 + jy_1)(x_2 - jy_2)}{(x_2 + jy_2)(x_2 - jy_2)},$$

$$= \frac{(x_1x_2 + y_1y_2) + j(x_2y_1 - x_1y_2)}{x_2^2 + jy_2x_2 - jx_2y_2 + y_2^2},$$

$$= \frac{(x_1x_2 + y_1y_2) + j(x_2y_1 - x_1y_2)}{x_2^2 + y_2^2}.$$

Example 6.4. Find the standard form of the complex numbers (a) $z_1 + z_2$, (b) $2z_1 - 3z_2$, (c) z_1z_2, (d) z_1^2/z_2, where $z_1 = -1 + 2j$ and $z_2 = 2 - 3j$.

(a) $z_1 + z_2 = (-1 + 2j) + (2 - 3j) = 1 - j$.

(b) $2z_1 - 3z_2 = 2(-1 + 2j) - 3(2 - 3j)$
$$= -2 + 4j - 6 + 9j = -8 + 13j.$$

(c) $z_1z_2 = (-1 + 2j)(2 - 3j)$
$$= -2 + 4j + 3j - 6j^2 = -2 + 4j + 3j + 6 = 4 + 7j.$$

(d) First
$$z_1^2 = (-1 + 2j)(-1 + 2j) = 1 - 4j + 4j^2 = -3 - 4j.$$

Then

$$\frac{z_1^2}{z_2} = \frac{-3 - 4j}{2 - 3j} = \frac{(-3 - 4j)(2 + 3j)}{(2 - 3j)(2 + 3j)},$$

$$= \frac{-6 - 8j - 9j - 12j^2}{4 - 6j + 6j - 9j^2},$$

$$= \frac{6 - 17j}{4 + 9} = \frac{6}{13} - \frac{17}{13}j.$$

The conjugate $\bar{z}$ of z has the following properties, which are simply applications of the rule:

'to obtain the conjugate, change j to $-$j wherever it appears'.

Properties of the conjugate

(a) $\overline{z_1 + z_2} = \bar{z}_1 + \bar{z}_2$.

(b) $\overline{z_1 z_2} = \bar{z}_1 \bar{z}_2$.

(c) $\overline{\left(\dfrac{z_1}{z_2}\right)} = \dfrac{\bar{z}_1}{\bar{z}_2}$.   (6.3)

(d) $x = \operatorname{Re} z = \dfrac{1}{2}(z + \bar{z}), \quad y = \operatorname{Im} z = \dfrac{1}{2j}(z - \bar{z}).$

Property 6.3(c) was illustrated by Example 6.2(a)(b). Similarly, for example,

the conjugate of $\dfrac{(2 + 3j)(4 + j)(3 - 2j)}{(1 - 4j)}$ is $\dfrac{(2 - 3j)(4 - j)(3 + 2j)}{(1 + 4j)}$;

we do not have to work the whole thing out first and then find the conjugate. These rules become important later.

6.2 The Argand diagram and complex numbers in polar form

A complex number $z = x + jy$ can be regarded as a pair of real numbers (x, y) known as an **ordered pair**. This means that the pair of numbers can be interpreted as the cartesian coordinates of a point in the plane in the usual way. The complex number $z = x + jy$ has an abscissa x and an ordinate y. In Fig. 6.1, the x-axis is known as the **real axis** and the y axis is the **imaginary axis**. The number $z = x + jy$ is represented by the point $P : (x, y)$. A figure showing complex numbers is known as the **Argand diagram** of the complex numbers.

The length $OP = r = \sqrt{(x^2 + y^2)}$ is called the **modulus** of z (or simply 'mod z') and written $|z|$.

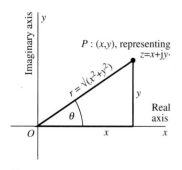

Fig. 6.1
The Argand diagram.

Example 6.5. Obtain $|z|$ where (a) $z = 2 + 3j$; (b) $z = 2 - 3j$.

(a) $|z| = |2 + 3j| = (2^2 + 3^2)^{\frac{1}{2}} = \sqrt{13}$.

(b) $|z| = |2 + (-3)j| = [2^2 + (-3)^2]^{\frac{1}{2}} = \sqrt{13}$.

Thus $|2 + 3j| = |2 - 3j|$.

Example 6.6. Let $z_1 = 1 - 3j$ and $z_2 = 3 - 2j$. Find

(a) $|z_1 + z_2|$;   (b) $|z_1| + |z_2|$;   (c) $|z_1 z_2|$;

(d) $|z_1||z_2|$;   (e) $\left|\dfrac{z_1}{z_2}\right|$;   (f) $\dfrac{|z_1|}{|z_2|}$.

(a) $z_1 + z_2 = (1 - 3j) + (3 - 2j) = 4 - 5j$

(this *must* be worked out in standard form first). Hence

$$|z_1 + z_2| = |4 - 5j| = [4^2 + (-5)^2]^{\frac{1}{2}} = \sqrt{41}.$$

(b) $|z_1| + |z_2| = [1^2 + (-3)^2]^{\frac{1}{2}} + [3^2 + (-2)^2]^{\frac{1}{2}} = \sqrt{10} + \sqrt{13}$.

(c) $|z_1 z_2| = |(1 - 3j)(3 - 2j)| = |-3 - 11j|$

$$= [(-3)^2 + (-11)^2]^{\frac{1}{2}} = \sqrt{130}.$$

(d) $|z_1||z_2| = |1 - 3j||3 - 2j| = \sqrt{10}\sqrt{13} = \sqrt{130}$

(that is, the same as (c))

(e) $\dfrac{z_1}{z_2} = \dfrac{1 - 3j}{3 - 2j}\,\dfrac{3 + 2j}{3 + 2j} = \dfrac{1}{13}(9 - 7j).$

Therefore

$$\left|\frac{z_1}{z_2}\right| = \frac{1}{13}|9 - 7j| = \frac{1}{13}\sqrt{130} = \frac{\sqrt{10}}{\sqrt{13}}.$$

(f) $\dfrac{|z_1|}{|z_2|} = \dfrac{|1 - 3j|}{|3 - 2j|} = \dfrac{\sqrt{10}}{\sqrt{13}}$ (that is, the same as (e)).

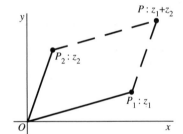

Fig. 6.2
Parallelogram law of addition

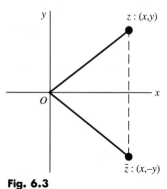

Fig. 6.3
Argand diagram showing z
and its conjugate $\bar{z}$.

The following results hold good for the modulus; they are illustrated in Example 6.6 above.

> **Properties of the modulus**
> If $z = x + jy$ (x and y real), then $|z| = (x^2 + y^2)^{\frac{1}{2}}$, and   (6.4)
> (a) $|\bar{z}| = |z|$,   (b) $z\bar{z} = |z|^2$,
> (c) $|z_1 z_2| = |z_1||z_2|$,   (d) $|z_1/z_2| = |z_1|/|z_2|$.

The identity (6.4b) follows directly from (6.2). We shall defer the proof of (c) and (d), but the truth is illustrated in Example 6.6. A sum or difference *cannot* be split in this way: contrast the results of Examples 6.6a and 6.6b.

The sum of two complex numbers can be interpreted by the **parallelogram law of addition** in the Argand diagram, as in Fig. 6.2.

Construct a parallelogram on OP_1 and OP_2, where P_1 and P_2 correspond to the complex numbers z_1 and z_2. The corner P of the parallelogram represents the sum $z_1 + z_2$. This follows from the addition rule for complex numbers. If you know anything about vectors, you will recognize that complex numbers add like vectors.

The conjugate $\bar{z} = x - jy$ is the reflection of z in the x axis, shown in Fig. 6.3.

6.3 Complex numbers in polar coordinates

In Fig. 6.1, r and θ obviously serve as **polar coordinates**, as defined in (1.16). In the context of complex numbers, θ is called the **argument** of z and denoted by arg z. The same point P can be described by the angles $\theta + 2\pi n$, where $n = \pm 1, \pm 2, \ldots$. Usually we use the **principal value of the argument**, denoted by Arg z (capital A), which is the smallest numerically; its limits are given by $-\pi < \theta \leqslant \pi$, i.e.

$$-\pi < \text{Arg } z \leqslant \pi.$$

The pair of equations

$$\cos \theta = \frac{x}{r}, \qquad \sin \theta = \frac{y}{r},$$

has exactly one solution within this range.

Example 6.7. Find the moduli and principal values of the arguments of the following complex numbers: (a) $z_1 = 2j$; (b) $z_2 = -1 - j$; (c) $z_3 = -2$; (d) $z_4 = \frac{1}{2} + \frac{1}{2}j\sqrt{3}$.

The moduli are given by:
(a) $|z_1| = |2j| = 2$;
(b) $|z_2| = |-1 - j| = \sqrt{(1 + 1)} = \sqrt{2}$;
(c) $|z_3| = |-2| = 2$;
(d) $|z_4| = |\frac{1}{2} + \frac{1}{2}j\sqrt{3}| = \sqrt{(\frac{1}{4} + \frac{3}{4})} = 1$.

A sketch of the Argand diagram for the complex numbers helps to decide their arguments. Figure 6.4 shows their locations. Thus
(a) (i) Arg $z_1 = \theta_1 = \frac{1}{2}\pi$; (b) Arg $z_2 = \theta_2 = -\frac{3}{4}\pi$; (c) Arg $z_3 = \theta_3 = \pi$;
(d) Arg $z_4 = \theta_4 = \frac{1}{3}\pi$.

In Fig. 6.1, the coordinates (x, y) and the polar coordinates (r, θ) are related by

$$x = r \cos \theta, \qquad y = r \sin \theta.$$

Hence the complex number $z = x + jy$ can be written

$$z = r \cos \theta + j \sin \theta = r(\cos \theta + j \sin \theta), \tag{6.5}$$

which is the **polar form** of the complex number z. Note that $r \geqslant 0$ and that $-\pi < \theta \leqslant \pi$ is assumed to take its principal value.

(a)

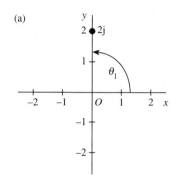

(b)

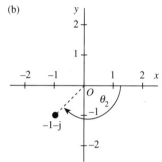

(c)

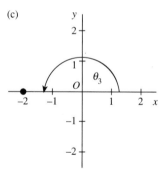

(d)

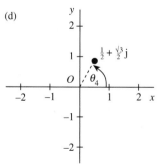

Fig. 6.4

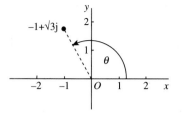

Fig. 6.5

Example 6.8. Express $-1 + \sqrt{3}j$ in polar form.

Here

$$r = \sqrt{[(-1)^2 + (\sqrt{3})^2]} = \sqrt{(1 + 3)} = 2$$

and θ is given by (see Fig. 6.5)

$$\cos \theta = -\tfrac{1}{2}, \qquad \sin \theta = \tfrac{1}{2}\sqrt{3}.$$

Hence from Fig. 6.5, $\theta = \tfrac{2}{3}\pi$, and

$$-1 + \sqrt{3}j = 2(\cos \tfrac{2}{3}\pi + j \sin \tfrac{2}{3}\pi).$$

Example 6.9. Obtain (a) $|\cos \theta + j \sin \theta|$;   (b) $|1/(\cos \theta + j \sin \theta)|$.

(a) $|\cos \theta + j \sin \theta| = (\cos^2 \theta + \sin^2 \theta)^{\frac{1}{2}} = 1.$

(b) $|1/(\cos \theta + j \sin \theta)| = 1/|\cos \theta + j \sin \theta|$ (by (6.4d))

$$= 1 \quad \text{(from (a))}$$

6.4 Complex numbers in exponential form

Consider the function

$$f(\theta) = \cos \theta + j \sin \theta,$$

where θ can take any value. Its derivative with respect to θ is

$$\frac{df(\theta)}{d\theta} = -\sin \theta + j \cos \theta = j(\cos \theta + j \sin \theta) = jf(\theta).$$

Hence $f(\theta)$ satisfies a relation in which the derivative is 'proportional' to itself notwithstanding that the constant of proportionality is j. As we saw in Chapter 1, a function with this property is the exponential function $k\,e^{j\theta}$, where k is a constant. We conclude that

$$\cos \theta + j \sin \theta = k\,e^{j\theta}$$

for some value of k. In particular this must be true for the value $\theta = 0$, from which $k = 1$. Hence, we obtain the important result

$$e^{j\theta} = \cos \theta + j \sin \theta. \tag{6.6}$$

The conjugate formula is

$$e^{-j\theta} = \cos \theta - j \sin \theta. \tag{6.7}$$

applying the rule of replacing j by $-$j. Hence any complex number can be written in the **exponential form** ('Euler's formula')

$$z = r(\cos \theta + j \sin \theta) = r\,e^{j\theta},$$

with its conjugate

$$\bar{z} = r(\cos \theta - j \sin \theta) = r\,e^{-j\theta}.$$

Properties of $e^{j\theta}$

(a) $e^{j\theta} = \cos\theta + j\sin\theta,$ (b) $|e^{j\theta}| = 1,$ (6.8)

(c) conjugate of $e^{j\theta}$ is $e^{-j\theta}$.

Exponential form for a complex number

$$z = r\,e^{j\theta} = r\cos\theta + jr\sin\theta,$$ (6.9)

where $r = |z|$ and θ is any value of arg z.

Example 6.10. Express the following in standard form (a) $2\,e^{\frac{1}{2}\pi j}$; (b) $3\,e^{-\pi j}$; (c) $e^{3\pi j}$; (d) $2\,e^{-\frac{1}{2}\pi j}$; (e) $3\,e^{\frac{1}{4}\pi j}$.

Remember that, in $r\,e^{j\theta}$, the numbers r and θ are polar coordinates. For these simple cases, we can therefore put the points straight on an Argand diagram and read off the coordinates, without needing to work out $\cos\theta$ and $\sin\theta$.

(a) $r = 2$, $\theta = \frac{1}{2}\pi$ (90°). Hence $2\,e^{\frac{1}{2}\pi j} = 2j$.

(b) $r = 3$, $\theta = -\pi$ ($-180°$). Hence $3\,e^{-j\pi} = -3$.

(c) $r = 1$, $\theta = 3\pi$. Hence $e^{3\pi j} = -1$.

(d) $r = 2$, $\theta = -\frac{1}{2}\pi$ ($-90°$). Hence $2\,e^{-\frac{1}{2}\pi j} = -2j$.

(e) $r = 3$, $\theta = \frac{1}{4}\pi$ (45°). Hence $3\,e^{\frac{1}{4}\pi j} = \frac{3}{2}\sqrt{2} + j\frac{3}{2}\sqrt{2}$.

It follows by treating (6.6) and (6.7) as two simultaneous equations for $\sin\theta$ and $\cos\theta$, that

$$\cos\theta = \frac{1}{2}(e^{j\theta} + e^{-j\theta}), \qquad \sin\theta = \frac{1}{2j}(e^{j\theta} - e^{-j\theta}).$$ (6.10)

Equation (6.6) will still be true if we replace the angle θ by $n\theta$, where n is an integer. Hence we obtain De Moivre's theorem:

$$\cos n\theta + j\sin n\theta = e^{nj\theta}$$
$$= (e^{j\theta})^{n}$$
$$= (\cos\theta + j\sin\theta)^{n}.$$

De Moivre's theorem

If n is any integer, then   (6.11)

$$(\cos\theta + j\sin\theta)^{n} = \cos n\theta + j\sin n\theta$$

Example 6.11. Express the following complex numbers in exponential form with principal arguments: (a) j; (b) $-5j$; (c) -3; (d) $4 - 4j$; (e) $3 - 4j$.

In each example, we put $r \cos \theta$ equal to the real part and $r \sin \theta$ equal to the imaginary part of the given complex number. In each case find θ by plotting the point on an Argand diagram.

(a) $r \cos \theta = 0$, $r \sin \theta = 1$. Hence $r = 1$ and, in the interval $-\pi < \theta \leqslant \pi$, we obtain $\theta = \frac{1}{2}\pi$. The polar form is

$$j = e^{\frac{1}{2}j\pi}.$$

(b) $r \cos \theta = 0$, $r \sin \theta = -5$. Hence $r = 5$ and $\theta = -\frac{1}{2}\pi$. The polar form is

$$-5j = 5 e^{-\frac{1}{2}j\pi}.$$

(c) $r \cos \theta = -3$, $r \sin \theta = 0$. Hence $r = 3$ and $\theta = \pi$. The polar form is

$$-3 = 3 e^{j\pi}.$$

(d) $r \cos \theta = 4$, $r \sin \theta = -4$. Hence $r = \sqrt{(16 + 16)} = 4\sqrt{2}$ while $\theta = -\frac{1}{4}\pi$. Hence

$$4 - 4j = 4\sqrt{2} e^{-\frac{1}{4}j\pi}.$$

(e) $r \cos \theta = 3$, $r \sin \theta = -4$. Hence $r = \sqrt{(9 + 16)} = 5$ while the angle α is the principal argument such that

$$\cos \alpha = \tfrac{3}{5}, \qquad \sin \alpha = -\tfrac{4}{5}.$$

The polar form is

$$3 - 4j = 5 e^{j\alpha}.$$

where $\alpha = -53.1°$ or -0.927 radians.

Example 6.12. By expressing $-1 + j$ in the form $r e^{j\theta}$, find $(-1 + j)^{-8}$ as a complex number in standard form

First $r = |-1 + j| = \sqrt{2}$. From its position on an Argand diagram $\theta = 3 \times 45°$, or $\frac{3}{4}\pi$ in radians. Therefore

$$-1 + j = \sqrt{2} e^{\frac{3}{4}\pi j}.$$

Then

$$(-1 + j)^{-8} = (\sqrt{2} e^{\frac{3}{4}\pi j})^{-8} = (\sqrt{2})^{-8} e^{-6\pi j} = \tfrac{1}{16} e^{-6\pi j}.$$

On an Argand diagram the polar coordinates are $r = \frac{1}{16}$ and $\theta = -6\pi = -3(2\pi)$. This value of θ, equivalent to 3 complete revolutions, puts us on the positive real axis again, so that

$$(-1 + j)^{-8} = \tfrac{1}{16}.$$

6.5 The general exponential form

The advantage of the exponential form of a complex number is that it is particularly easy to differentiate, integrate, and to combine with other exponentials, including ordinary ones. We are not tied to the Argand diagram when we manipulate the exponentials, so we shall often use letters other than r and θ.

Example 6.13. Prove that

$$\cos(A + B) = \cos A \cos B - \sin A \sin B,$$
$$\sin(A + B) = \sin A \cos B + \cos A \sin B.$$

We have

$$e^{j(A+B)} = e^{jA} e^{jB}.$$

In terms of the definition (6.8), this becomes

$$\cos(A+B) + j \sin(A+B) = (\cos A + j \sin A)(\cos B + j \sin B)$$
$$= (\cos A \cos B - \sin A \sin B)$$
$$+ j(\sin A \cos B + \cos A \sin B).$$

The real and imaginary parts of the two sides of the equation must both be equal, and so we have the result immediately.

Example 6.14. Express $\cos 5\theta$ in terms of power of $\cos \theta$.

Since

$$e^{5j\theta} = (e^{j\theta})^5,$$

it follows that

$$\cos 5\theta + j \sin 5\theta = (\cos \theta + j \sin \theta)^5.$$

Expand the right-hand side by the binomial theorem. Thus

$$\cos 5\theta + j \sin 5\theta = \cos^5 \theta + 5 \cos^4 \theta \cdot j \sin \theta$$
$$+ 10 \cos^3 \theta (j \sin \theta)^2 + 10 \cos^2 \theta (j \sin \theta)^3$$
$$+ 5 \cos \theta (j \sin \theta)^4 + (j \sin \theta)^5$$
$$= \cos^5 \theta + 5j \cos^4 \theta \sin \theta$$
$$- 10 \cos^3 \theta \sin^2 \theta - 10j \cos^2 \theta \sin^3 \theta$$
$$+ 5 \cos \theta \sin^4 \theta + j \sin^5 \theta.$$

Equate real parts on both sides of the equation:

$$\cos 5\theta = \cos^5 \theta - 10 \cos^3 \theta \sin^2 \theta + 5 \cos \theta \sin^4 \theta.$$

Finally replace $\sin^2 \theta$ by $1 - \cos^2 \theta$ and simplify:

$$\cos 5\theta = \cos^5 \theta - 10 \cos^3 \theta (1 - \cos^2 \theta) + 5 \cos \theta (1 - \cos^2 \theta)^2,$$
$$= 16 \cos^5 \theta - 20 \cos^3 \theta + 5 \cos \theta.$$

(Incidentally, $\sin 5\theta$ can also be found in terms of powers of $\sin \theta$.)

Example 6.15. Prove the result in (6.4d), that

$$\left|\frac{z_1}{z_2}\right| = \frac{|z_1|}{|z_2|}.$$

Put $z_1 = r_1\,e^{j\theta_1}$ and $z_2 = r_2\,e^{j\theta_2}$. Then

$$\frac{z_1}{z_2} = \frac{r_1\,e^{j\theta_1}}{r_2\,e^{j\theta_2}} = \frac{r_1}{r_2}\,e^{j(\theta_1-\theta_2)}.$$

Therefore

$$\left|\frac{z_1}{z_2}\right| = \frac{r_1}{r_2} = \frac{|z_1|}{|z_2|},$$

as required.

Consider the number z where

$$z = r\,e^{p+jq},$$

and p and q are any real numbers whatever, and $r > 0$. We have

$$z = r\,e^{p+jq} = r\,e^{p}\,e^{jq} = r\,e^{p}(\cos q + j\sin q),$$

(q is assumed to be in radians). Therefore we have

The form $z = c\,e^{p+jq}$ $(p, q, c$ **real, with** $c > 0)$

(a) $|z| = c\,e^{p}$,   (b) $\arg z = q$,   (6.12)

(c) $\operatorname{Re} z = c\,e^{p}\cos q$, $\operatorname{Im} z = c\,e^{p}\sin q$

In science and engineering, complex exponentials of the type in (6.12) are often used to describe oscillations of various kinds. Instead of $r\,e^{\alpha+j\beta}$, the kind of symbols that occur may look like

$$A\,e^{(-k+j\omega)t}, \quad \text{or} \quad c\,e^{(\alpha+j\beta)t},$$

in which t represents time. Recast $e^{(\alpha+j\beta)t}$ by writing it as $c\,e^{\alpha t+j\beta t}$, and we have

---

The form $c\,e^{(\alpha+j\beta)t}$ $(\alpha, \beta, t$ **real**)

(a) $c\,e^{\alpha t}\cos\beta t = \operatorname{Re} c\,e^{(\alpha+j\beta)t}$,   (6.13)

(b) $c\,e^{\alpha t}\sin\beta t = \operatorname{Im} c\,e^{(\alpha+j\beta)t}$,

---

Example 6.16. The damped vibration of a piece of machinery is described by $x = 0.01\,e^{-0.02t}\cos 15t$. Write this in the form $x = \operatorname{Re} c\,e^{(\alpha+j\beta)t}$.

We have

$$x = 0.01\,e^{-0.02t}\cos 15t = 0.01\,e^{-0.02t}\operatorname{Re} e^{j15t}$$
$$= \operatorname{Re}(0.01\,e^{-0.02t}\,e^{j15t}) = \operatorname{Re}(0.01\,e^{(-0.02+15j)t}).$$

Example 6.17. The current $i(t)$ in a branch of a circuit is given by

$$i(t) = c\, e^{-kt} \sin(\omega t + \phi).$$

Write this in the form of (a) the imaginary part of a complex function; (b) the real part of a complex function.

(a) $i(t) = \mathrm{Im}(c\, e^{-kt} e^{j(\omega t + \phi)}) = \mathrm{Im}(c\, e^{(-k + j\omega)t + j\phi}).$

(b) Note that, if $z = x + jy$, then

$$y = \mathrm{Im}\, z = \mathrm{Re}(-jz).$$

Therefore

$$i(t) = \mathrm{Re}(-j\, e^{(-k + j\omega)t + j\phi}) = \mathrm{Re}(c\, e^{-\frac{1}{2}\pi j}\, e^{(-k + j\omega)t + j\phi}).$$
$$= \mathrm{Re}(c\, e^{(-k + j\omega)t + j(\phi - \frac{1}{2}\pi)}).$$

6.6 Hyperbolic functions

The hyperbolic functions cosh and sinh are related to the trigonometric functions cos and sin. The hyperbolic functions were defined in Section 1.11 by

$$\cosh x = \tfrac{1}{2}(e^x + e^{-x}), \qquad \sinh x = \tfrac{1}{2}(e^x - e^{-x}).$$

It follows that

$$\cosh jx = \tfrac{1}{2}(e^{jx} + e^{-jx}) = \cos x,$$
$$\sinh jx = \tfrac{1}{2}(e^{jx} - e^{-jx}) = j \sin x,$$

by (6.10). Similarly

$$\cos jx = \tfrac{1}{2}(e^{j^2 x} + e^{-j^2 x}) = \tfrac{1}{2}(e^{-x} + e^x) = \cosh x,$$

$$\sin jx = \frac{1}{2j}(e^{j^2 x} - e^{-j^2 x}) = -\tfrac{1}{2}j(e^{-x} - e^x) = j \sinh x.$$

Example 6.18. Solve the equation $\cosh z = -1$

Since, for real z, we have $\cosh z \geqslant 1$, we expect the equation to have complex roots. In exponential form,

$$\tfrac{1}{2}(e^z + e^{-z}) = -1,$$

or

$$e^{2z} + 2e^z + 2 = 0,$$

or

$$(e^z + 1)^2 = 0.$$

Hence

$$e^z = -1 = e^{(2n+1)\pi j} \quad (n = 0, \pm 1, \pm 2, \ldots).$$

Here we have considered **all the representations** of the number -1, in the form $e^{(2n+1)\pi j}$. It is important when finding *all* the roots of an equation to include **all possible arguments**, not just the principal one. By matching the exponents, the roots are

$$z = (2n + 1)\pi j \quad (n = 0, \pm 1, \pm 2, \ldots).$$

6.7 Miscellaneous applications

The polar form of complex numbers can be used to solve polynomial equations as in the following example.

Example 6.19. Find all roots of $z^5 = 4 - 4j$.

We first express $4 - 4j$ in polar form $\rho\, e^{j\alpha}$. Thus

$$\rho \cos \alpha = 4, \qquad \rho \sin \alpha = -4,$$

from which it follows that $\rho = \sqrt{32}$ and $\alpha = -\frac{1}{4}j\pi + 2n\pi j$, using an Argand diagram. Hence, in polar form,

$$4 - 4j = 2\sqrt{2}e^{-\frac{1}{4}j\pi + 2n\pi j} = 2^{\frac{5}{2}} e^{j\pi(-\frac{1}{4} + 2n)} \quad (n = 0, \pm 1, \pm 2, \ldots).$$

Let $z = r\, e^{j\theta}$. Then

$$r^5\, e^{5j\theta} = 2^{\frac{5}{2}} e^{j\pi(-\frac{1}{4} + 2n)},$$

or

$$r\, e^{j\theta} = \sqrt{2}\, e^{\frac{1}{5}j\pi(-\frac{1}{4} + 2n)} \quad (n = 0, \pm 1, \pm 2, \ldots).$$

Comparing the two sides gives

$$r = \sqrt{2}, \qquad \theta = \tfrac{1}{20}(-1 + 8n)\pi.$$

Five successive values of n give distinct roots: other values of n merely duplicate existing roots. In full, the roots are

$$\sqrt{2}e^{\frac{1}{20}j\pi}, \quad \sqrt{2}e^{\frac{7}{20}j\pi}, \quad \sqrt{2}e^{\frac{3}{4}j\pi}, \quad \sqrt{2}e^{\frac{23}{20}j\pi}, \quad \sqrt{2}e^{\frac{31}{20}j\pi}.$$

The exponential form (6.8a) can be used to express $\cos n\theta$ and $\sin n\theta$ in terms of powers of $\cos\theta$ and $\sin\theta$ respectively, and conversely, to express $\cos^n\theta$ and $\sin^n\theta$ in terms of cosines and sines of multiple angles.

Example 6.20. Expand $\cos^6\theta$ in terms of multiple angles.

Let $z = \cos\theta + j\sin\theta$. By de Moivre's theorem, with n an integer,

$$z^n = \cos n\theta + j\sin n\theta, \qquad \frac{1}{z^n} = \cos n\theta - j\sin n\theta.$$

By adding these two results, it follows that

$$\cos n\theta = \frac{1}{2}\left(z^n + \frac{1}{z^n}\right). \tag{6.14}$$

Hence

$$(2\cos\theta)^6 = \left(z + \frac{1}{z}\right)^6,$$

$$= z^6 + 6z^4 + 15z^2 + 20 + \frac{15}{z^2} + \frac{6}{z^4} + \frac{1}{z^6}$$

$$= \left(z^6 + \frac{1}{z^6}\right) + 6\left(z^4 + \frac{1}{z^4}\right) + 15\left(z^2 + \frac{1}{z^2}\right) + 20$$

$$= 2\cos 6\theta + 12\cos 4\theta + 30\cos 2\theta + 20,$$

by repeatedly use of (6.5). Finally

$$\cos^6 \theta = \tfrac{1}{32} \cos 6\theta + \tfrac{3}{16} \cos 4\theta + \tfrac{15}{32} \cos 2\theta + \tfrac{5}{16}.$$

We can also use the polar form to sum certain series as in the following example.

Example 6.21. Find the sum of the series

$$f(\theta) = 1 + \cos \theta + \frac{\cos 2\theta}{2!} + \frac{\cos 3\theta}{3!} + \cdots.$$

In summation notation, the series can be written

$$f(\theta) = \sum_{n=0}^{\infty} \frac{\cos n\theta}{n!}.$$

Since $\cos n\theta = \mathrm{Re}\; e^{nj\theta}$, consider the series

$$S(\theta) = 1 + e^{j\theta} + \frac{e^{2j\theta}}{2!} + \frac{e^{3j\theta}}{3!} + \cdots,$$

the real part of whose sum is the required sum $f(\theta)$. Thus

$$S(\theta) = 1 + e^{j\theta} + \frac{(e^{j\theta})^2}{2!} + \frac{(e^{j\theta})^3}{3!} + \cdots$$

$$= \exp e^{j\theta} = e^{\cos \theta + j \sin \theta}$$

$$= e^{\cos \theta}[\cos(\sin \theta) + j \sin (\sin \theta)],$$

using the formula (5.4a) for the power series of the exponential function. Thus

$$f(\theta) = \mathrm{Re}\; S(\theta) = e^{\cos \theta} \cos(\sin \theta).$$

Problems, Chapter 6

6.1. (Section 6.1). Find the roots of the following quadratic equations:
(a) $x^2 + 2x + 5 = 0$; (b) $x^2 - 6x + 10 = 0$;
(c) $x^2 + 2jx + 3 = 0$.

6.2. (Section 6.1). Find all the complex roots of $x^4 + 3x^2 - 4 = 0$.

6.3. (Section 6.1). Express the following complex numbers in standard form:
(a) $(1 - j) + (3 + 4j)$; (b) $2(3 - j) + 3(-1 - j)$;
(c) $3(-1 + j) - 4(2 - 3j)$; (d) $3(1 + j)(2 - j)$;
(e) $\dfrac{2 + j}{3 - j}$; (f) $\dfrac{(2 + j)(7 + 5j)}{3 - j}$; (g) $(-1 + 2j)^2$;
(h) $(-1 + 2j)^2 + \dfrac{1}{(-1 + 2j)^2}$; (i) $(1 + j)^5$.

6.4. (Section 6.1). Find the boundary curve in the (p, q)-plane which separates the (p, q) values giving real roots from those for the complex roots in the quadratic equation

$$x^2 + px + q = 0,$$

where p and q are real parameters. Of the real roots, where in the (p, q) plane do the roots which are both negative lie?

6.5. (Section 6.1). Let $z_1 = 3 - j$ and $z_2 = 1 + 2j$. Find, in standard form, the complex numbers
(a) $z_1 + z_2$; (b) $z_1 z_2$; (c) z_1/z_2; (d) z_1/z_2^2.

6.6. (Section 6.1). Let $z_1 = 2 + 3j$ and $z_2 = -2 + j$. Find the following complex conjugates:
(a) $\overline{z_1 + z_2}$; (b) $\overline{z_1 z_2}$; (c) $\overline{z_1/z_2}$; (d) $\overline{z_1}/\overline{z_2}$.

6.7. Let $z = 1 + j$. Find the following complex numbers in standard form and plot their corresponding points in the Argand diagram:
(a) $\bar{z}$; (b) z^2; (c) $\bar{z}^2$; (d) $1/\bar{z}$; (e) $z/\bar{z}$.

6.8. (Section 6.5). Three complex numbers are given by $z_1 = 2e^{1+j}$, $z_2 = 3e^{-j}$, and $z_3 = \frac{1}{2}e^{-1+2j}$. Express the following complex numbers in standard form:
(a) $z_1 + z_2 + z_3$;   (b) $z_1 z_2 + \bar{z}_3$;   (c) $z_1 z_2 z_3$;
(d) $z_1 \bar{z}_2 / z_3$;   (e) $z_1^2 z_2^2 - 2z_3^2$.

6.9. (Section 6.3). Find the modulus and principal argument of each of the following complex numbers:
(a) $z_1 = -2 + 2j$;   (b) $z_2 = 4 - 4\sqrt{3}j$;   (c) $z_3 = -5j$;
(d) $z_4 = -3$;   (e) $z_5 = 3 + 4j$.

6.10. (Section 6.2). Let $z = x + jy$. Express each of the following equations in the complex variable z in real form in terms of x and y. Sketch, and identify in each case, the corresponding curve in the Argand diagram:
(a) $z\bar{z} = 1$;   (b) Im $z = 2$;
(c) $|z - a| = 1$, where a is a complex number;
(d) $(z - \bar{z})^2 = -8(z + \bar{z})$;   (e) $|z - 1| + |z + 1| = 4$;
(f) Arg $z = \frac{1}{4}\pi$ (see Section 6.3);   (g) $|z| = \arg z$.

6.11. (Section 6.4). Express the following complex numbers in exponential form with principal arguments:
(a) $-1 + j$;   (b) -2;   (c) $-3j$;   (d) $7 - 7j\sqrt{3}$;
(e) $(1 - j)(1 + j\sqrt{3})$;   (f) $\dfrac{1 - j}{1 + \sqrt{3}}$;   (g) e^{2+j};
(h) $(1 + j)e^{2j}$;   (i) $(1 - j\sqrt{3})^9$;   (j) $\dfrac{(1 + j)^4}{2 - 2j}$.

6.12. (Section 6.3). Using Euler's formula (6.8) for $e^{\pm j\theta}$, obtain the trigonometric identities for $\cos(\theta_1 \pm \theta_2)$ and $\sin(\theta_1 \pm \theta_2)$.

6.13. (Section 6.3). Using the parallelogram law, sketch the locations in the Argand diagram, for general complex numbers z_1 and z_2, the following points:
$z_1 + z_2$, $\bar{z}_1 + \bar{z}_2$, $z_1 - z_2$, $\bar{z}_1 + z_2$, $z_1 - \bar{z}_2$.

6.14. Let $f(\theta) = \cos \theta + j \sin \theta$. Verify that

$$\frac{d^2 f(\theta)}{d\theta^2} = -f(\theta).$$

Show that it is still true if

$$f(\theta) = a \cos \theta + b \sin \theta,$$

where a and b are arbitrary complex numbers.

6.15. Prove that $\tan ja = j \tanh a$, where a is a real number.

6.16. Find all the complex roots of the following equations:
(a) $\cosh z = 1$;   (b) $\sinh z = 1$;
(c) $e^z = -1$;   (d) $\cos z = \sqrt{2}$.

6.17. The **logarithm** of a complex number $z = r\,e^{j\theta}$ is defined by

$$\log z = \ln r + j\theta,$$

which will be a *multivalued* function because of the term $j\theta$. The **principal value** of the logarithm is denoted by Log z (note the capital letter L), and defined by

$$\text{Log } z = \ln r + j\theta,$$

where $\pi < \theta \leqslant \pi$.
(a) Find the principal value Log$(1 + j\sqrt{3})$, and indicate its location on the Argand diagram.
(b) Find all roots of the equation $\log z = \pi j$.
(c) Express Log(ej) in standard form.

6.18. If $z \neq 0$ and c are complex numbers, then z^c is defined by

$$z^c = e^{c \ln z}.$$

(a) Express 2^j in standard form.
(b) Find the principal value of j^j.
(c) Find all complex roots of $z^j = -1$.

6.19. (Section 6.4). Find all complex roots of $z^5 = -1$, and sketch their locations on the Argand diagram.

6.20. (Section 6.5). Find the modulus, argument, real and imaginary parts of each of the following complex numbers:
(a) $2e^{3+j2}$;   (b) $4e^j$;   (c) $5e^{\cos\frac{1}{4}\pi + j\sin\frac{1}{4}\pi}$;   (d) e^{1+j}.

6.21. (Section 6.5). An oscillation in a system is given by $x = 0.04e^{-0.01t} \sin 12t$. Write this in the form

$$x = \text{Re}(c\,e^{\alpha + j\beta}).$$

6.22 (Section 6.5). The current in a branch of a circuit is given by

$$i(t) = c\,e^{-0.05t} \sin(0.4t + 0.5).$$

Write this in the form of the real part of a complex function.

6.23. A function $f(z)$, where $z = x + jy$, is known as a **function of complex variable** z. Find the real and imaginary parts of the following functions in terms of x and y:
(a) z^2;   (b) $z + 2z^2 + 3z^3$;   (c) $\sin z$;
(d) $\cos z$;   (e) $e^z \cos z$;   (f) e^{z^2}.

6.24. Let $w = f(z)$, where $z = x + jy$ and $w = u + jv$ are complex variables. If $f(z) = z^2$, find u and v in terms of x and y. The relation represents a **mapping** between two Argand diagrams. What curves do the hyperbolas $x^2 - y^2 = 1$ and $xy = 1$ map into the (u, v) plane?

6.25. Show that the mapping

$$w = z + \frac{c}{z},$$

where $z = x + jy$ and $w = u + jv$ and c is a real number, maps the circle $|z| = 1$ in the z plane into an ellipse in the w plane; and find its equation.

6.26. (Section 6.4). Show that

$$\cos^6 \theta = \tfrac{1}{32}(\cos 6\theta + 6 \cos 4\theta + 15 \cos 2\theta + 10),$$

and find $\sin^6 \theta$.

6.27. The damped oscillation of a vibrating block is given by

$$x = \text{Re } z, \qquad z = e^{(-0.2 + 0.5j)t},$$

in terms of the time t. Find x, and determine the values of t where x is zero. Find the velocity of the block
(a) as dx/dt;
(b) as $\text{Re } dz/dt$;
and confirm that the answers are the same.

6.28. Given that $2 + j$ is a solution of the equation

$$z^4 - 2z^3 - z^2 + 2z + 10 = 0,$$

find the other solutions.

6.29. Find the sum of the series

(a) $1 - \sin \theta + \dfrac{\sin 2\theta}{2!} - \dfrac{\sin 3\theta}{3!} + \cdots;$

(b) $1 + 2 \cos \theta + \dfrac{2^2 \cos 2\theta}{2!} + \dfrac{2^3 \cos 3\theta}{3!} + \cdots.$

7 Matrix algebra

7.1 Matrix definition and notation

In many applications in physics and engineering it is useful to be able to represent and manipulate data in tabular or array form. An array which obeys certain algebraic rules of operation is known as a **matrix**. Capital letters are usually used to denote matrices. Thus

$$A = \begin{bmatrix} 1 & 2 & -1 \\ 0 & 3 & -4 \end{bmatrix}$$

is a matrix with 2 *rows* and 3 *columns*. The individual terms are known as *elements*: the element in the second row and third column is -4. This matrix is said to be of *order* 2×3, or a 2×3 matrix. A general $m \times n$ matrix, one with m rows and n columns, can be represented by the notation

$$A = \begin{bmatrix} a_{11} & a_{12} & \cdots & a_{1n} \\ a_{21} & a_{22} & \cdots & a_{2n} \\ \vdots & \vdots & \ddots & \vdots \\ a_{m1} & a_{m2} & \cdots & a_{mn} \end{bmatrix} = [a_{ij}], \quad (1 \le i \le m, 1 \le j \le n),$$

where a_{ij} is the element in the ith row and jth column of A; or by

$$A = [a_{ij} : i = 1, \ldots, m; j = 1, \ldots, n],$$

or simply

$$A = [a_{ij}],$$

for brevity, if it is clear in context that the matrix is $m \times n$. A 1×1 matrix is simply a number: for example, $[-5] = -5$.

Matrices which have either one row or column are known as **vectors**. Thus

$$[1.3 \quad -1.1 \quad 2.9 \quad 4.6] \quad \text{and} \quad \begin{bmatrix} -1.1 \\ 6.5 \\ -2.0 \end{bmatrix}$$

are respectively **row** and **column vectors**.

A matrix in which the number of rows equals the number of columns is called a **square matrix**: if $m \neq n$ then the matrix is said to be **rectangular**.

7.2 Rules of matrix algebra

We need to define consistent algebraic rules for manipulating matrices, such concepts as addition, multiplication, etc. As we shall see, these rules have their origins in the representation of linear equations and linear transformations, but for the present we simply state them as a list of rules.

1. *Equality.* Two matrices can only be equated if they are of the same *order*, that is, if they each have the same number of rows and the same number of columns. They are then said to be **equal** if the corresponding elements are equal. Thus if

$$A = \begin{bmatrix} a & b \\ c & d \end{bmatrix} \quad \text{and} \quad B = \begin{bmatrix} e & f \\ g & h \end{bmatrix},$$

then $A = B$ if and only if $a = e$, $b = f$, $c = g$, and $d = h$. In general, if $A = [a_{ij}]$ and $B = [b_{ij}]$ are both $m \times n$ matrices, then $A = B$ if and only if $a_{ij} = b_{ij}$ for $i = 1, 2, \ldots, m$ and $j = 1, 2, \ldots, n$.

Example 7.1. Solve the equation $A = B$ when

$$A = \begin{bmatrix} x & 1 & 2 \\ 0 & x^2 - y & 3 \end{bmatrix}, \quad B = \begin{bmatrix} 1 & 1 & 2 \\ 0 & 2 & 3 \end{bmatrix}.$$

Since A and B must have the same elements, if follows that $x = 1$ and $x^2 - y = 2$. Hence, $y = x^2 - 2 = -1$. The solution is $x = 1$, $y = -1$.

---

2. *Multiplication by a constant.* Let k be a constant or **scalar**. By the product kA we mean the matrix in which **every element of A is multiplied by** k. Thus, if

$$A = \begin{bmatrix} 2.0 & 1.5 & 3.1 \\ -1.2 & 3.0 & -4.6 \end{bmatrix},$$

and $k = 10$, then

$$kA = \begin{bmatrix} 10 \times 2.0 & 10 \times 1.5 & 10 \times 3.1 \\ 10 \times -1.2 & 10 \times 3.0 & 10 \times -4.6 \end{bmatrix}$$

$$= \begin{bmatrix} 20 & 15 & 31 \\ -12 & 30 & -46 \end{bmatrix}.$$

Equally, we can 'factorize' a matrix. Thus

$$A = \begin{bmatrix} 5 & 25 & -30 \\ 10 & 15 & -5 \end{bmatrix} = 5 \begin{bmatrix} 1 & 5 & -6 \\ 2 & 3 & -1 \end{bmatrix}.$$

3. *Zero matrix.* Any matrix in which every element is zero is called a **zero matrix**. If A is a zero matrix, we can simply write $A = 0$. (A better notation is $O_{m \times n}$ to denote the zero matrix with m rows and n rows, but it is not widely used.)

4. *Matrix sums and differences.* The sum of two matrices A and B is defined if A and B are of the same order, in which case $A + B$ is defined as the matrix C whose elements are the sums of the corresponding elements in A and B. We write $C = A + B$. Thus, if $A = [a_{ij}]$ and $B = [b_{ij}]$ are both $m \times n$ matrices, then

$$C = A + B = [a_{ij} + b_{ij}].$$

Example 7.2. If

$$A = \begin{bmatrix} 1 & 3 \\ 2 & 2 \\ 3 & 1 \end{bmatrix} \quad \text{and} \quad B = \begin{bmatrix} -4 & -6 \\ -5 & -5 \\ -6 & -4 \end{bmatrix},$$

then find $A + B$, $B + A$, and $A + 2B$.

Thus

$$A + B = \begin{bmatrix} 1-4 & 3-6 \\ 2-5 & 2-5 \\ 3-6 & 1-4 \end{bmatrix} = \begin{bmatrix} -3 & -3 \\ -3 & -3 \\ -3 & -3 \end{bmatrix} = -3\begin{bmatrix} 1 & 1 \\ 1 & 1 \\ 1 & 1 \end{bmatrix}$$

(by Rule 2)

Also

$$B + A = \begin{bmatrix} -4+1 & -6+3 \\ -5+2 & -5+2 \\ -6+3 & -4+1 \end{bmatrix} = \begin{bmatrix} -3 & -3 \\ -3 & -3 \\ -3 & -3 \end{bmatrix} = A + B.$$

Further

$$A + 2B = \begin{bmatrix} 1 & 3 \\ 2 & 2 \\ 3 & 1 \end{bmatrix} + 2\begin{bmatrix} -4 & -6 \\ -5 & -5 \\ -6 & -4 \end{bmatrix}$$

$$= \begin{bmatrix} 1 & 3 \\ 2 & 2 \\ 3 & 1 \end{bmatrix} + \begin{bmatrix} -8 & -12 \\ -10 & -10 \\ -12 & -8 \end{bmatrix} \quad \text{(by Rule 2)}$$

$$= \begin{bmatrix} -7 & -9 \\ -8 & -8 \\ -9 & -7 \end{bmatrix}.$$

As the second sum suggests, the commutative property of the real numbers, namely $a_{ij} + b_{ij} = b_{ij} + a_{ij}$, implies the **commutative property of matrix addition**, that is,

$$A + B = B + A.$$

The difference of two matrices is written as $A - B$ which is interpreted as $A + (-1)B$, using Rule 2 for the multiplication of B by the number -1, and then Rule 4 for the sum of A and $(-1)B$. In practice, we simply take the difference of corresponding elements.

Example 7.3. Find $A - B$ and $2A - 3B$, if

$$A = \begin{bmatrix} 1 & -1 & 2 \\ 0 & 2 & -3 \end{bmatrix}, \quad B = \begin{bmatrix} 2 & -2 & -3 \\ 1 & 0 & -1 \end{bmatrix}.$$

We have

$$A - B = \begin{bmatrix} 1 & -1 & 2 \\ 0 & 2 & -3 \end{bmatrix} - \begin{bmatrix} 2 & -2 & -3 \\ 1 & 0 & -1 \end{bmatrix}$$

$$= \begin{bmatrix} 1-2 & -1+2 & 2+3 \\ 0-1 & 2-0 & -3+1 \end{bmatrix} = \begin{bmatrix} -1 & 1 & 5 \\ -1 & 2 & -2 \end{bmatrix}.$$

Also

$$2A - 3B = 2\begin{bmatrix} 1 & -1 & 2 \\ 0 & 2 & -3 \end{bmatrix} - 3\begin{bmatrix} 2 & -2 & -3 \\ 1 & 0 & -1 \end{bmatrix}$$

$$= \begin{bmatrix} 2 & -2 & 4 \\ 0 & 4 & -6 \end{bmatrix} - \begin{bmatrix} 6 & -6 & -9 \\ 3 & 0 & -3 \end{bmatrix}$$

$$= \begin{bmatrix} -4 & 4 & 13 \\ -3 & 4 & -3 \end{bmatrix}.$$

The rules of arithmetic as applied to the elements of matrices lead to the following results for matrices for which addition can be defined:

(a) $A + (B + C) = (A + B) + C$ (**associative law of addition**). In other words, the order of addition of matrices is immaterial.

(b) $k(A + B) = kA + kB,$ $(k + l)A = kA + lA.$

5. *Matrix multiplication.* We now need to define the concept of the product of two matrices. Not all matrices can be multiplied: they must have the right shape, or be **conformable** for multiplication to be defined. The product of A and B, in this order, is written as AB (no product sign is used), but it is only defined if the **number of columns in A equals the number of rows in B**. Let us look at the case where A is a 1×3 matrix, that is, a row vector, and B is a 3×1 matrix, that is, a column vector, given by

$$A = [a_{11} \quad a_{12} \quad a_{13}], \qquad B = \begin{bmatrix} b_{11} \\ b_{21} \\ b_{31} \end{bmatrix}.$$

The product AB is defined as the 1×1 matrix C given by

$$AB = [a_{11} \quad a_{12} \quad a_{13}] \begin{bmatrix} b_{11} \\ b_{21} \\ b_{21} \end{bmatrix}$$

$$= [a_{11}b_{11} + a_{12}b_{21} + a_{13}b_{31}] = C. \tag{7.1}$$

Here, the single remaining element is the sum of the products of corresponding elements from the row in A and the column in B. Thus the product of a 1×3 matrix and a 3×1 matrix is a 1×1 matrix (or number). This is known as a **row-on-column** operation.

Suppose now that A is a 2×3 matrix and that B a 3×2 matrix which are given by

$$A = \begin{bmatrix} a_{11} & a_{12} & a_{13} \\ a_{21} & a_{22} & a_{23} \end{bmatrix}, \qquad B = \begin{bmatrix} b_{11} & b_{12} \\ b_{21} & b_{22} \\ b_{31} & b_{32} \end{bmatrix}.$$

The product AB is now a 2×2 matrix C given by

$$AB = \begin{bmatrix} a_{11} & a_{12} & a_{13} \\ a_{21} & a_{22} & a_{23} \end{bmatrix} \begin{bmatrix} b_{11} & b_{12} \\ b_{21} & b_{22} \\ b_{31} & b_{32} \end{bmatrix}$$

$$= \begin{bmatrix} a_{11}b_{11} + a_{12}b_{21} + a_{13}b_{31} & a_{11}b_{12} + a_{12}b_{22} + a_{13}b_{32} \\ a_{21}b_{11} + a_{22}b_{21} + a_{23}b_{31} & a_{21}b_{12} + a_{22}b_{22} + a_{23}b_{32} \end{bmatrix}$$

$$= C. \tag{7.2}$$

Note that each row in A 'operates' on each column in B giving four elements in the 2×2 matrix C.

Example 7.4. Find AB if

$$A = \begin{bmatrix} 1 & -1 & 0 \\ 2 & 1 & -3 \end{bmatrix}, \qquad B = \begin{bmatrix} 0 & 3 \\ 1 & -1 \\ -2 & 4 \end{bmatrix}.$$

Thus

$$AB = \begin{bmatrix} 1 & -1 & 0 \\ 2 & 1 & -3 \end{bmatrix} \begin{bmatrix} 0 & 3 \\ 1 & -1 \\ -2 & 4 \end{bmatrix}$$

$$= \begin{bmatrix} 1 \times 0 + (-1) \times 1 + 0 \times (-2) & 1 \times 3 + (-1) \times (-1) + 0 \times 4 \\ 2 \times 0 + 1 \times 1 + (-3) \times (-2) & 2 \times 3 + 1 \times (-1) + (-3) \times 4 \end{bmatrix}$$

$$= \begin{bmatrix} -1 & 4 \\ 7 & -7 \end{bmatrix}.$$

We can use a *summation* (Section 1.13) notation to condense the expanded sums of products which occur in matrices. The sum of a string of numbers, say, $c_1 + c_2 + c_3$ can be expressed as

$$\sum_{i=1}^{3} c_i,$$

where i runs through all the integers between the lower limit on i under the $\sum$ symbol, and the upper limit above. Thus, for example,

$$\sum_{i=3}^{8} h_i = h_3 + h_4 + h_5 + h_6 + h_7 + h_8.$$

We can also use the summation notation with the double-suffix notation, as in

$$\sum_{i=1}^{4} h_{i6} = h_{16} + h_{26} + h_{36} + h_{46}.$$

The product given by (7.1) can be written

$$AB = [a_{11}b_{11} + a_{12}b_{21} + a_{13}b_{31}] = \left[\sum_{j=1}^{3} a_{1j}b_{j1} \right] = \sum_{j=1}^{3} a_{1j}b_{j1}.$$

Similarly the elements in (7.2) can be expressed as

$$AB = \left[\sum_{k=1}^{3} a_{ik}b_{kj} : i,j = 1, 2 \right].$$

The summation formulae give a clue to the general expression for the product of an $m \times n$ matrix A and an $n \times p$ matrix B. Remember that the number of columns in A must always equal the number of rows in B for the product to be defined. Thus, the row-on-column definition of the product is the $m \times p$ matrix

$$AB = \left[\sum_{k=1}^{n} a_{ik}b_{kj} : i = 1, \ldots, m; j = 1, \ldots, p \right].$$

Multiplication rule

The element in the ith row and jth column of the product consists of the row-on-column product of the ith row in A and the jth column in B. (7.3)

Example 7.5. If A is a 5×4 matrix, B is a 4×5 matrix, and C is a 6×4 matrix, which of the following products are defined: AB, BA, AC, CB, $(AB)C$, $(CB)A$?

AB is a 5×5 matrix.
BA is a 4×4 matrix.
AC is not defined since A has 4 columns and C has 6 rows.
CB is a 6×5 matrix.
$(AB)C$ is not defined since AB is a 5×5 and C is a 6×4 matrix.
CB is a 6×5 matrix; hence $(CB)A$ is a 6×4 matrix.

One conclusion which can be inferred from the previous example is that **matrix multiplication does not commute**, that is, in general, $AB \neq BA$. As the previous example indicates, one or both products may not be defined; when both are defined, AB and BA may be of different order; and, even when both are defined and of the same order, AB is generally not equal to BA. So we must be careful about the order of multiplication. In the product AB, we say that A is **multiplied on the right** by B, or that B is **multiplied on the left** by A. The expressions 'A **postmultiplied** by B' and 'B **premultiplied** by A' are also used. Statements such as 'A is multiplied by B' can be ambiguous without carefully stating how the product occurs.

Example 7.6. If

$$A = \begin{bmatrix} 1 & -1 & 0 \\ 3 & -2 & -1 \end{bmatrix}, \qquad B = \begin{bmatrix} 1 & 2 \\ 1 & 2 \\ 1 & 2 \end{bmatrix},$$

calculate AB and BA.

Thus

$$AB = \begin{bmatrix} 1 & -1 & 0 \\ 3 & -2 & -1 \end{bmatrix} \begin{bmatrix} 1 & 2 \\ 1 & 2 \\ 1 & 2 \end{bmatrix} = \begin{bmatrix} 0 & 0 \\ 0 & 0 \end{bmatrix} = 0,$$

and

$$BA = \begin{bmatrix} 1 & 2 \\ 1 & 2 \\ 1 & 2 \end{bmatrix} \begin{bmatrix} 1 & -1 & 0 \\ 3 & -2 & -1 \end{bmatrix} = \begin{bmatrix} 7 & -5 & -2 \\ 7 & -5 & -2 \\ 7 & -5 & -2 \end{bmatrix}.$$

This example illustrates the point that AB can be a zero matrix without either A or B or BA being zero. Also, as a consequence, $A(B - C) = 0$ does not necessarily imply $B = C$.

We state the following results concerning sums and products, but proofs are omitted:

(a) $A(B + C) = AB + AC$ (distributive law of addition),
(b) $A(BC) = (AB)C$ (associative law of multiplication),

provided that the products are defined.

7.3 Special matrices

Many matrices have special properties. We list, define, and illustrate a number of them here. Some properties apply to rectangular matrices: others are specific to square matrices.

1. *Transpose.* The **transpose** of any matrix is one in which the rows and columns are interchanged. Thus the first row becomes the first column, the second row the second column, and so on. We denote the transpose of A by A^T. Hence,

$$\text{if} \quad A = \begin{bmatrix} a_{11} & a_{12} \\ a_{21} & a_{22} \\ a_{31} & a_{32} \end{bmatrix} \quad \text{then} \quad A^T = \begin{bmatrix} a_{11} & a_{21} & a_{31} \\ a_{12} & a_{22} & a_{32} \end{bmatrix}. \tag{7.4}$$

The 3×2 matrix A becomes the 2×3 *matrix* A^T.

Example 7.7. Find the transposes of A, B, $A + B^T$, AB, and $B^T A^T$, where

$$A = \begin{bmatrix} 1 & 2 \\ 0 & 1 \\ -1 & 1 \end{bmatrix}, \quad B = \begin{bmatrix} 3 & -1 & 0 \\ 1 & 2 & -2 \end{bmatrix}$$

Thus

$$A^T = \begin{bmatrix} 1 & 0 & -1 \\ 2 & 1 & 1 \end{bmatrix}, \quad B^T = \begin{bmatrix} 3 & 1 \\ -1 & 2 \\ 0 & -2 \end{bmatrix},$$

$$(A + B^T)^T = \left(\begin{bmatrix} 1 & 2 \\ 0 & 1 \\ -1 & 1 \end{bmatrix} + \begin{bmatrix} 3 & 1 \\ -1 & 2 \\ 0 & -2 \end{bmatrix} \right)^T = \begin{bmatrix} 4 & 3 \\ -1 & 3 \\ -1 & -1 \end{bmatrix}^T$$

$$= \begin{bmatrix} 4 & -1 & -1 \\ 3 & 3 & -1 \end{bmatrix} = A^T + (B^T)^T,$$

$$(AB)^T = \left(\begin{bmatrix} 1 & 2 \\ 0 & 1 \\ -1 & 1 \end{bmatrix} \begin{bmatrix} 3 & -1 & 0 \\ 1 & 2 & -2 \end{bmatrix} \right)^T = \begin{bmatrix} 5 & 3 & -4 \\ 1 & 2 & -2 \\ -2 & 3 & -2 \end{bmatrix}^T$$

$$= \begin{bmatrix} 5 & 1 & -2 \\ 3 & 2 & 3 \\ -4 & -2 & -2 \end{bmatrix},$$

$$B^T A^T = \begin{bmatrix} 3 & 1 \\ -1 & 2 \\ 0 & -2 \end{bmatrix} \begin{bmatrix} 1 & 0 & -1 \\ 2 & 1 & 1 \end{bmatrix} = \begin{bmatrix} 5 & 1 & -2 \\ 3 & 2 & 3 \\ -4 & -2 & -2 \end{bmatrix}.$$

Note that $(AB)^T = B^T A^T$.

Provided that the sum $A + B$ and product AB are defined for two matrices A and B, the last example points to the following two results concerning transposes:

(a) $(A + B)^{\mathrm{T}} = A^{\mathrm{T}} + B^{\mathrm{T}}$;

(b) $(AB)^{\mathrm{T}} = B^{\mathrm{T}}A^{\mathrm{T}}$.

2. *Symmetric matrices.* A *square* matrix is said to be **symmetric** if $A = A^{\mathrm{T}}$. Since rows and columns are interchanged in the transpose, this is equivalent to $a_{ij} = a_{ji}$ for all elements if $A = [a_{ij}]$. Symmetric matrices are easy to recognize since their elements are reflected in the **leading diagonal**, the diagonal string of elements from the top left to the bottom right of the matrix. Thus

$$A = \begin{bmatrix} 1 & 3 & -2 \\ 3 & 2 & 4 \\ -2 & 4 & -1 \end{bmatrix}$$

is a 3×3 *symmetric matrix.*

A square matrix A for which $A = -A^{\mathrm{T}}$ is said to be **skew-symmetric**. Note that, if A is any square matrix, then $A + A^{\mathrm{T}}$ is symmetric and $A - A^{\mathrm{T}}$ is skew-symmetric. The elements along the leading diagonal of a skew-symmetric matrix must all be zero. Thus

$$\begin{bmatrix} 0 & 1 & 2 \\ -1 & 0 & -3 \\ -2 & 3 & 0 \end{bmatrix}$$

is skew-symmetric.

3. *Row and column vectors.* As we defined them in Section 4.1, a row vector is a matrix with one row, and a column vector is one with one column. For vectors, we usually use bold-faced small letters and write, for example,

$$a = \begin{bmatrix} a_1 \\ a_2 \\ \vdots \\ a_n \end{bmatrix}, \qquad b^{\mathrm{T}} = [b_1 \quad b_2 \quad \dots \quad b_n]$$

The transpose of a row vector is a column vector and vice versa. If A is an $m \times n$ matrix, then Aa is a column vector with m rows.

Example 7.8. If

$$A = \begin{bmatrix} 1 & -1 & 2 \\ 3 & 1 & -4 \\ -1 & 2 & 1 \end{bmatrix}, \qquad x = \begin{bmatrix} x \\ y \\ z \end{bmatrix}, \qquad d = \begin{bmatrix} 2 \\ 1 \\ -1 \end{bmatrix},$$

find the set of equations for x, y, z represented by $Ax = d$.

The matrix equation in full is

$$\begin{bmatrix} 1 & -1 & 2 \\ 3 & 1 & -4 \\ -1 & 2 & 1 \end{bmatrix} \begin{bmatrix} x \\ y \\ z \end{bmatrix} = \begin{bmatrix} 2 \\ 1 \\ -1 \end{bmatrix},$$

or

$$\begin{bmatrix} x - y - 2z \\ 3x + y - 4z \\ -x + 2y + z \end{bmatrix} = \begin{bmatrix} 2 \\ 1 \\ -1 \end{bmatrix}.$$

The set of **linear equations** for x, y, z is

$$\begin{aligned} x - \ y + 2z &= \ \ 2, \\ 3x + \ y - 4z &= \ \ 1, \\ -x + 2y + \ z &= -1. \end{aligned}$$

We shall say more about the solutions of linear equations in Chapter 8.

4. *Diagonal matrices.* A square matrix all of whose elements off the leading diagonal are zero is called a **diagonal matrix**. Thus, if $A = [a_{ij}]$ is an $n \times n$ matrix, then A is diagonal if $a_{ij} = 0$ for all $i \neq j$. Hence

$$A = \begin{bmatrix} 1 & 0 & 0 \\ 0 & -2 & 0 \\ 0 & 0 & 3 \end{bmatrix}$$

is an example of a 3×3 diagonal matrix.

A diagonal matrix is obviously symmetric. If A and B are diagonal matrices of the same order then $A + B$ and AB are also both diagonal.

5. *Identity matrix.* The diagonal matrix with all diagonal elements 1 is called the **identity** or **unit matrix** I_n. Hence, the 3×3 identity is

$$I_3 = \begin{bmatrix} 1 & 0 & 0 \\ 0 & 1 & 0 \\ 0 & 0 & 1 \end{bmatrix}.$$

(If there is no confusion likely to arise, I_n or I_3 are simply replaced by the universal symbol I.) The reason for the definition becomes clear if we multiply a 3×3 matrix by I_3. If A is a general 3×3 matrix, then

$$AI_3 = \begin{bmatrix} a_{11} & a_{12} & a_{13} \\ a_{21} & a_{22} & a_{23} \\ a_{31} & a_{32} & a_{33} \end{bmatrix} \begin{bmatrix} 1 & 0 & 0 \\ 0 & 1 & 0 \\ 0 & 0 & 1 \end{bmatrix} = \begin{bmatrix} a_{11} & a_{12} & a_{13} \\ a_{21} & a_{22} & a_{23} \\ a_{31} & a_{32} & a_{33} \end{bmatrix} = A.$$

Similarly $I_3 A = A$.

A need not be square: provided that the products are defined, $AI = A$ and $IA = A$ for the appropriate identity in each case.

6. *Powers of matrices.* If A is a square matrix of order n, then we write AA as A^2, AA^2 as A^3 and so on.

If A is diagonal, as in

$$A = \begin{bmatrix} d_1 & 0 & 0 \\ 0 & d_2 & 0 \\ 0 & 0 & d_3 \end{bmatrix},$$

then

$$A^2 = \begin{bmatrix} d_1^2 & 0 & 0 \\ 0 & d_2^2 & 0 \\ 0 & 0 & d_3^2 \end{bmatrix}, \qquad A^3 = \begin{bmatrix} d_1^3 & 0 & 0 \\ 0 & d_2^3 & 0 \\ 0 & 0 & d_3^3 \end{bmatrix}, \qquad \text{etc.}$$

In particular $I_m^n = I_m$ for all positive integers n.

7.4 The inverse matrix

If A and B are square matrices, each of order n, which satisfy the equations

$$AB = BA = I_n,$$

then B is called the **inverse** of A. We say *the* inverse because, if a matrix B exists with this property, then it is uniquely determined by A (although we shall not prove this here). We write $B = A^{-1}$ (*not* $B = I/A$). Since the definition is 'symmetric', it follows that A is the inverse of B, that is, $A = B^{-1}$. The inverse matrix defines 'division' for matrices, but analogies with numbers must not be taken too far. It is a particularly useful operation since it enables us to manipulate matrix equations. Thus, if $AB = C$, and the inverse of B exists, then we can solve the equation and find A as $A = CB^{-1}$.

How do we find the inverse? Does it always exist? Let us look first at the case in which A is a 2×2 matrix, and consider the equation

$$Ax = d,$$

where

$$A = \begin{bmatrix} a_{11} & a_{12} \\ a_{21} & a_{22} \end{bmatrix}, \qquad x = \begin{bmatrix} x_1 \\ x_2 \end{bmatrix}, \qquad d = \begin{bmatrix} d_1 \\ d_2 \end{bmatrix}.$$

Thus

$$\begin{bmatrix} a_{11} & a_{12} \\ a_{21} & a_{22} \end{bmatrix} \begin{bmatrix} x_1 \\ x_2 \end{bmatrix} = \begin{bmatrix} d_1 \\ d_2 \end{bmatrix},$$

or

$$a_{11}x_1 + a_{12}x_2 = d_1, \tag{7.5}$$

$$a_{21}x_1 + a_{22}x_2 = d_2. \tag{7.6}$$

Eliminate x_2 by multiplying (7.5) by a_{22}, (7.6) by a_{12} and by subtracting the two equations so that

$$(a_{11}a_{22} - a_{21}a_{12})x_1 = a_{22}d_1 - a_{12}d_2.$$

Similarly, elimination of x_1 leads to

$$-(a_{11}a_{22} - a_{21}a_{12})x_2 = a_{21}d_1 - a_{11}d_2.$$

Provided that $a_{11}a_{22} - a_{21}a_{12} \neq 0$, it follows that

$$x_1 = \frac{a_{22}d_1 - a_{12}d_2}{a_{11}a_{22} - a_{21}a_{12}}, \qquad x_2 = \frac{-a_{21}d_1 + a_{11}d_2}{a_{11}a_{22} - a_{21}a_{12}}.$$

We can now express this pair of equations in matrix form:

$$x = \begin{bmatrix} x_1 \\ x_2 \end{bmatrix} = \frac{1}{a_{11}a_{22} - a_{21}a_{12}} \begin{bmatrix} a_{22}d_1 - a_{12}d_2 \\ -a_{21}d_1 + a_{11}d_2 \end{bmatrix} = Cd,$$

where

$$C = \frac{1}{\det A} \begin{bmatrix} a_{22} & -a_{12} \\ -a_{21} & a_{11} \end{bmatrix}, \quad \text{with} \quad \det A = a_{11}a_{22} - a_{21}a_{22}.$$

If $Ax = d$ is multiplied on the left by the inverse A^{-1}, then

$$A^{-1}Ax = I_2 x = x = A^{-1}d.$$

Hence A^{-1} can be identified with the matrix C. In other words,

$$A^{-1} = C = \frac{1}{\det A} \begin{bmatrix} a_{22} & -a_{12} \\ -a_{21} & a_{11} \end{bmatrix}. \tag{7.7}$$

It is worth remembering the rule for 2×2 matrices by which A^{-1} can be constructed from A.

Rule for 2 × 2 inverse

the diagonal elements are interchanged, the signs are changed for the other two elements, and the matrix is divided by det A. (7.8)

The number $\det A = a_{11}a_{22} - a_{21}a_{12}$ is known as the **determinant** of A. It may also be written directly in terms of the corresponding matrix as

$$\det A = \det \begin{bmatrix} a_{11} & a_{12} \\ a_{21} & a_{22} \end{bmatrix},$$

or, more briefly, as

$$\det A = \begin{vmatrix} a_{11} & a_{12} \\ a_{21} & a_{22} \end{vmatrix}.$$

The determinant is a function of the corresponding matrix, but is a number—*not* an array. If $\det A = 0$, then we cannot find x_1 and x_2, and the matrix has no inverse. It is then said to be **singular**: if A has an inverse, then A is said to be **nonsingular**. We shall say more about determinants in the next Chapter.

Example 7.9. Decide whether A, where

$$A = \begin{bmatrix} 1 & 3 \\ -1 & 4 \end{bmatrix}$$

is singular or not. If it is nonsingular, find its inverse.

Here $a_{11} = 1$, $a_{12} = 3$, $a_{21} = -1$, and $a_{22} = 4$. Hence

$$\det A = \begin{vmatrix} 1 & 3 \\ -1 & 4 \end{vmatrix} = 1 \times 4 - (-1) \times 3 = 4 + 3 = 7.$$

Since $\det A \neq 0$, then A is nonsingular. Its inverse is, by the rule above,

$$A^{-1} = \tfrac{1}{7} \begin{vmatrix} 4 & -3 \\ 1 & 1 \end{vmatrix}.$$

Example 7.10. If

$$A = \begin{bmatrix} 1 & 3 \\ -1 & 4 \end{bmatrix}, \qquad B = \begin{bmatrix} 1 & 2 \\ 1 & -1 \end{bmatrix},$$

find A^{-1}, B^{-1}, and $(AB)^{-1}$.

Always check the determinants first. Here

$$\det A = 4 - (-3) = 7, \qquad \det B = -1 - 2 = -3.$$

Hence

$$A^{-1} = \tfrac{1}{7} \begin{bmatrix} 4 & -3 \\ 1 & 1 \end{bmatrix}, \qquad B^{-1} = \tfrac{1}{3} \begin{bmatrix} -1 & -2 \\ -1 & 1 \end{bmatrix}.$$

Also

$$AB = \begin{bmatrix} 1 & 3 \\ -1 & 4 \end{bmatrix}\begin{bmatrix} 1 & 2 \\ 1 & -1 \end{bmatrix} = \begin{bmatrix} 4 & -1 \\ 3 & -6 \end{bmatrix}.$$

Thus, $\det(AB) = -24 + 3 = -21$, and

$$(AB)^{-1} = -\tfrac{1}{21} \begin{bmatrix} -6 & 1 \\ -3 & 4 \end{bmatrix}.$$

Note that

$$B^{-1}A^{-1} = -\tfrac{1}{21} \begin{bmatrix} -1 & -2 \\ -1 & 1 \end{bmatrix}\begin{bmatrix} 4 & -3 \\ 1 & 1 \end{bmatrix}$$

$$= -\tfrac{1}{21} \begin{bmatrix} -6 & 1 \\ -3 & 4 \end{bmatrix} = (AB)^{-1}.$$

This last result suggests the following correct rule for the inverse of the product of two square matrices, namely

$$(AB)^{-1} = B^{-1}A^{-1}.$$

For the inverse of a 3×3 matrix we can adopt the same approach as for the 2×2 case by eliminating pairs from x_1, x_2, x_3 from the set of equations $A\boldsymbol{x} = \boldsymbol{d}$ or

$$
\left.
\begin{aligned}
a_{11}x_1 + a_{12}x_2 + a_{13}x_3 &= d_1, \\
a_{21}x_1 + a_{22}x_2 + a_{23}x_3 &= d_2, \\
a_{31}x_1 + a_{32}x_2 + a_{33}x_3 &= d_3.
\end{aligned}
\right\}
\tag{7.9}
$$

The result is

$$
\boldsymbol{x} = A^{-1}\boldsymbol{d},
$$

where

$$
A^{-1} = \frac{1}{\det A}
\begin{bmatrix}
a_{22}a_{33} - a_{32}a_{23} & -(a_{12}a_{33} - a_{32}a_{13}) & a_{12}a_{23} - a_{22}a_{13} \\
-(a_{21}a_{33} - a_{31}a_{23}) & a_{11}a_{33} - a_{31}a_{13} & -(a_{11}a_{23} - a_{21}a_{13}) \\
a_{21}a_{32} - a_{31}a_{22} & -(a_{11}a_{32} - a_{31}a_{12}) & a_{11}a_{22} - a_{21}a_{12}
\end{bmatrix}
\tag{7.10}
$$

with

$$
\det A = a_{11}(a_{22}a_{33} - a_{32}a_{23}) - a_{12}(a_{21}a_{33} - a_{31}a_{23})
$$
$$
+ a_{13}(a_{21}a_{32} - a_{31}a_{22}),
\tag{7.11}
$$

provided that $\det A \neq 0$. Again $\det A$ is known as the determinant of A and denoted by

$$
\det A =
\begin{vmatrix}
a_{11} & a_{12} & a_{13} \\
a_{21} & a_{22} & a_{23} \\
a_{31} & a_{32} & a_{33}
\end{vmatrix}.
$$

Equation (7.10) gives the inverse matrix, as can be verified by calculation of the products AA^{-1} and $A^{-1}A$. Even for 3×3 matrices; the formula for the inverse is quite complicated. If $\det A = 0$, then the matrix is singular.

The determinants which have arisen in the context of inverse matrices have important properties particularly with regard to their evaluation. They will be discussed in more detail in the next chapter.

Example 7.11. Verify by direct multiplication that

$$
A =
\begin{bmatrix}
0 & 1 & 1 \\
-1 & 1 & 1 \\
1 & -1 & 1
\end{bmatrix}
$$

has the inverse

$$
B = \tfrac{1}{2}
\begin{bmatrix}
2 & -2 & 0 \\
2 & -1 & -1 \\
0 & 1 & 1
\end{bmatrix}
$$

Check the matrix product BA:

$$BA = \tfrac{1}{2} \begin{bmatrix} 2 & -2 & 0 \\ 2 & -1 & -1 \\ 0 & 1 & 1 \end{bmatrix} \begin{bmatrix} 0 & 1 & 1 \\ -1 & 1 & 1 \\ 1 & -1 & 1 \end{bmatrix} = \begin{bmatrix} 1 & 0 & 0 \\ 0 & 1 & 0 \\ 0 & 0 & 1 \end{bmatrix} = I_3.$$

Hence $B = A^{-1}$.

Note that we need only verify that either $BA = I_3$ or $AB = I_3$, not both. If $BA = I_3$, then $AB = I_3$, and vice versa.

Example 7.12. Using formula (7.10) find A^{-1}, where

$$A = \begin{bmatrix} 2 & 1 & 0 \\ 1 & -1 & 5 \\ -1 & -1 & 2 \end{bmatrix}.$$

We first find det A:

$$\det A = 2 \times [(-1) \times 2 - (-1) \times 5] - 1$$

$$\times [1 \times 2 - (-1) \times 5] + 0$$

$$\times [1 \times (-1) - (-1) \times (-1)]$$

$$= 2 \times 3 - 1 \times 7$$

$$= -1.$$

Thus, from (7.10),

$$A^{-1} = \frac{1}{-1} \begin{bmatrix} -1 \times 2 - (-1) \times 5 & -[1 \times 2 - (-1) \times 0] & 1 \times 5 - (-1) \times 0 \\ -[1 \times 2 - (-1) \times 5] & 2 \times 2 - (-1) \times 0 & -(2 \times 5 - 1 \times 0) \\ 1 \times (-1) - (-1) \times (-1) & -[2 \times (-1) - (-1) \times (-1)] & 2 \times (-1) - 1 \times (-1) \end{bmatrix}$$

$$= - \begin{bmatrix} 3 & -2 & 5 \\ -7 & 4 & -10 \\ -2 & 1 & -3 \end{bmatrix}$$

$$= \begin{bmatrix} -3 & 2 & -5 \\ 7 & -4 & 10 \\ 2 & -1 & 3 \end{bmatrix}.$$

This is the 'formula' method of finding the inverse of a 3×3

matrix, but it is not an efficient procedure numerically. There are better methods using row operations which will be explained in Chapter 10.

Problems, Chapter 7

7.1. (Section 7.1). The matrix $A = [a_{ij}]$ is given by

$$A = \begin{bmatrix} 1 & 2 & 3 \\ -1 & 0 & 1 \\ 2 & -2 & 4 \\ 1 & 5 & -3 \end{bmatrix}.$$

Identify the elements a_{13}, and a_{31}.

7.2. (Section 7.2). Solve the equation $A = B$, where

$$A = \begin{bmatrix} 1 & -2 \\ 3 & 1 \\ -1 & 2 \end{bmatrix}, \quad B = \begin{bmatrix} 1 & x \\ y - x & 1 \\ -1 & 2 \end{bmatrix},$$

for x and y.

7.3. (Section 7.2). Given that

$$A = \begin{bmatrix} 1 & 2 & -3 \\ -1 & 0 & 4 \end{bmatrix}, \quad B = \begin{bmatrix} 2 & -1 & 3 \\ 4 & 1 & 2 \end{bmatrix},$$

find the matrices $A + B$, $A - B$, and $2A - 3B$.

7.4. (Section 7.2). Given that

$$A = \begin{bmatrix} 1 & 3 & 0 \\ 2 & 1 & 1 \end{bmatrix}, \quad B = \begin{bmatrix} 1 & 0 \\ 2 & 1 \\ -1 & -1 \end{bmatrix},$$

$$C = \begin{bmatrix} 2 & 1 \\ -1 & 1 \\ 0 & 1 \end{bmatrix},$$

verify the distributive law $A(B + C) = AB + AC$ for the three matrices.

7.5. (Section 7.2). Let

$$A = \begin{bmatrix} -1 & 2 & -1 \\ 2 & 3 & 1 \end{bmatrix}, \quad B = \begin{bmatrix} -1 & 0 \\ 1 & 2 \\ 3 & -1 \end{bmatrix},$$

$$C = \begin{bmatrix} 1 & 1 \\ -1 & 2 \end{bmatrix}.$$

Verify the associative law $A(BC) = (AB)C$ for these matrices.

7.6. (Section 7.2). Let

$$A = \begin{bmatrix} 4 & 2 \\ 2 & 1 \end{bmatrix}, \quad B = \begin{bmatrix} -2 & -1 \\ 4 & 2 \end{bmatrix}.$$

Show that $AB = 0$, but that $BA \neq 0$.

7.7. (Section 7.3). Let

$$A = \begin{bmatrix} 2 & 1 & 3 \\ 1 & -1 & 2 \\ -2 & 1 & 1 \end{bmatrix}.$$

Find a matrix C such that $A + C$ is the identity matrix I_3. Deduce that $AC = CA$. Find AC, and hence the matrix $A^2 + C^2$.

7.8. (Section 7.3). A general $n \times n$ matrix is given by

$$A = [a_{ij}].$$

Show that $A + A^{\mathrm{T}}$ is a symmetric matrix, and that $A - A^{\mathrm{T}}$ is skew-symmetric.

Express the matrix

$$A = \begin{bmatrix} 2 & 1 & 3 \\ -2 & 0 & 1 \\ 3 & 1 & 2 \end{bmatrix}$$

as the sum of a symmetric matrix and a skew-symmetric matrix.

7.9. (Section 7.3). Let

$$A = \begin{bmatrix} 1 & 3 \\ -1 & 2 \\ 0 & 1 \end{bmatrix}.$$

Write down A^{T}, and find the products AA^{T} and $A^{T}A$.

7.10. (Section 7.3). If

$$A = \begin{bmatrix} 1 & -1 & 2 \\ 3 & 0 & 1 \\ -1 & 2 & -3 \end{bmatrix}, \quad x = \begin{bmatrix} x \\ y \\ z \end{bmatrix},$$

$$d = \begin{bmatrix} 2 \\ 0 \\ -1 \end{bmatrix},$$

write down the set of equations defined by $Ax = d$. Confirm that the same set of equations is given by $x^{\mathrm{T}}A^{\mathrm{T}} = d^{\mathrm{T}}$.

7.11. (Section 7.3). Let

$$A = \begin{bmatrix} 1 & 0 & 0 \\ a & -1 & 0 \\ b & c & 1 \end{bmatrix}.$$

Find A^2. For what relation between a, b, and c is $A^2 = I_3$? In this case, what is the inverse matrix of A? What is the inverse matrix of A^{2n-1} (n a positive integer)?

7.12. If

$$A = \begin{bmatrix} 2 & 0 & 1 \\ 2 & -2 & 2 \\ 0 & 4 & -4 \end{bmatrix},$$

$$B = \begin{bmatrix} 0 & \frac{1}{2} & \frac{1}{4} \\ 1 & -1 & -\frac{1}{4} \\ 1 & -1 & -\frac{1}{2} \end{bmatrix},$$

find the products AB and BA, and confirm that B is the inverse of A.

7.13. Let

$$A = \begin{bmatrix} 2 & 0 & 1 \\ 2 & -2 & 2 \\ 0 & 4 & 1 \end{bmatrix}.$$

Find the powers A^2 and A^3, and verify that

$$A^3 - A^2 - 12A = -12I_3.$$

Hence find the inverse matrix A^{-1} by multiplying the equation on both sides by A^{-1}.

7.14. (Section 7.4). Using the rule for inverses of 2×2 matrices, write down the inverses of:

(a) $\begin{bmatrix} 1 & 1 \\ 2 & -1 \end{bmatrix}$;   (b) $\begin{bmatrix} 2 & 3 \\ -7 & 11 \end{bmatrix}$;

(c) $\begin{bmatrix} 1 & 0 \\ 0 & -2 \end{bmatrix}$;   (d) $\begin{bmatrix} 10 & -7 \\ 8 & 0 \end{bmatrix}$;

(e) $\begin{bmatrix} -99 & 100 \\ 97 & 98 \end{bmatrix}$.

7.15. (Section 7.4) The **sparsely filled** matrix A is given by

$$A = \begin{bmatrix} 0 & 1 & 0 & 0 \\ 0 & 0 & 1 & 0 \\ 1 & 0 & 0 & 0 \\ 0 & 0 & 0 & 1 \end{bmatrix}.$$

Thinking about the row-on-column rule for matrix multiplication, can you guess the columns in the inverse matrix A^{-1}? How would this rule generalize to the matrix

$$A = \begin{bmatrix} 0 & a & 0 & 0 \\ 0 & 0 & b & 0 \\ c & 0 & 0 & 0 \\ 0 & 0 & 0 & d \end{bmatrix}?$$

7.16. (Section 7.4). Write down the set of equations given by $A\mathbf{x} = \mathbf{d}$, where

$$A = \begin{bmatrix} 0 & 1 & 1 \\ 1 & -2 & 2 \\ 1 & 0 & 1 \end{bmatrix}, \qquad \mathbf{x} = \begin{bmatrix} x \\ y \\ z \end{bmatrix},$$

$$\mathbf{d} = \begin{bmatrix} 6 \\ 3 \\ -9 \end{bmatrix}.$$

Find A^{-1} and calculate the product $A^{-1}\mathbf{d}$. What is the solution of the equation?

7.17. (Section 7.4). If A and B are both $n \times n$ matrices with A nonsingular, show that

$$(A^{-1}BA)^2 = A^{-1}B^2A.$$

Let $A = \begin{bmatrix} 1 & 2 \\ -1 & 1 \end{bmatrix}$ and $B = \begin{bmatrix} 1 & 2 \\ -1 & 0 \end{bmatrix}$. Calculate $A^{-1}B^4A$.

7.18. (Section 7.4). For interpolation purposes for given data, it is required that the parabola $y = a + bx + cx^2$ should pass through the three points with coordinates (x_1, y_1), (x_2, y_2), and (x_3, y_3) in the (x, y) plane. Show that the matrix equation for the constants a, b, and c can be written as

$$\begin{bmatrix} 1 & x_1 & x_1^2 \\ 1 & x_2 & x_2^2 \\ 1 & x_3 & x_3^2 \end{bmatrix} \begin{bmatrix} a \\ b \\ c \end{bmatrix} = \begin{bmatrix} y_1 \\ y_2 \\ y_3 \end{bmatrix}.$$

Verify that the inverse of the 3×3 matrix on the left is

$$\begin{bmatrix} \dfrac{x_2 x_3}{(x_2 - x_1)(x_3 - x_1)} & \dfrac{x_3 x_1}{(x_3 - x_2)(x_1 - x_2)} & \dfrac{x_1 x_2}{(x_1 - x_3)(x_2 - x_3)} \\[2ex] -\dfrac{x_2 + x_3}{(x_2 - x_1)(x_3 - x_1)} & -\dfrac{x_3 + x_1}{(x_3 - x_2)(x_1 - x_2)} & -\dfrac{x_1 + x_2}{(x_1 - x_3)(x_2 - x_3)} \\[2ex] \dfrac{1}{(x_2 - x_1)(x_3 - x_1)} & \dfrac{1}{(x_3 - x_2)(x_1 - x_2)} & \dfrac{1}{(x_1 - x_3)(x_2 - x_3)} \end{bmatrix}.$$

provided that certain conditions are met. What are they, and what implications have they for the given points in the plane? Find a, b, and c in terms of the given data. Find the equation of the parabola through the points $(-2, 0)$, $(1, -2)$, $(3, 4)$.

7.19. The elements in a 3×3 matrix $A = [a_{ij}]$ are given by the rule

$$a_{ij} = (-j)^i - ij.$$

Write down the matrix A. Calculate det A and the inverse of A.

7.20. If

$$A = \begin{bmatrix} 2 & 1 & 3 \\ 1 & -1 & 2 \\ 1 & 2 & 1 \end{bmatrix},$$

show that $A^3 - 2A^2 - 9A = 0$, but that $A^2 - 2A - 9I_3 \neq 0$. Does the inverse of A exist?

7.21. An nth-order square matrix A satisfies $A^2 = A$ and $A \neq I_n$. Show that

(a) det $A = 0$;
(b) $(I_n + A)^{-1} = I_n - \frac{1}{2}A$;
(c) $(I_n + A)^m = I_n + (2^m - 1) A$ for any positive integer m.

7.22. Let $A_1 = \begin{bmatrix} x_1 & y_1 \\ -y_1 & x_1 \end{bmatrix}$, $A_2 = \begin{bmatrix} x_2 & y_2 \\ -y_2 & x_2 \end{bmatrix}$. Calculate $A_1 + A_2$, $A_1 A_2$, $A_2 A_1$, A_1^{-1}. Compare your results with $z_1 + z_2$, $z_1 z_2$, and $1/z$, where $z_1 = x_1 + jy_1$ and $z_2 = x_2 + jy_2$, where these are complex numbers (see Chapter 6). Consider the possibility of developing further parallels, such as to $|z|$ and e^z.

8

Determinants

8.1 The determinant of a square matrix

As we saw in (7.8) (Section 7.4), certain combinations of elements from a square matrix appear as the denominator in the construction of the inverse matrix. If this number, called the **determinant** of the matrix, turns out to be zero, then the matrix is singular and no inverse exists. Here we look at the definition of the determinant of a matrix and its properties. Special emphasis will be placed on the 2×2 and 3×3 determinants which suggest generalizations to higher-order cases.

Given the matrix

$$A = \begin{bmatrix} a_{11} & a_{12} \\ a_{21} & a_{22} \end{bmatrix},$$

then the determinant of A is denoted and defined by

$$\det A = \begin{vmatrix} a_{11} & a_{12} \\ a_{21} & a_{22} \end{vmatrix} = a_{11}a_{22} - a_{21}a_{12}. \tag{8.1}$$

(The notation $|A|$ is also used extensively for the determinant.) For the 3×3 matrix

$$A = \begin{bmatrix} a_{11} & a_{12} & a_{13} \\ a_{21} & a_{22} & a_{23} \\ a_{31} & a_{32} & a_{33} \end{bmatrix},$$

its determinant is (see (7.11)) defined as

$$\det A = \begin{vmatrix} a_{11} & a_{12} & a_{13} \\ a_{21} & a_{22} & a_{23} \\ a_{31} & a_{32} & a_{33} \end{vmatrix},$$

$$= a_{11}a_{22}a_{33} - a_{11}a_{32}a_{23} - a_{12}a_{21}a_{33} + a_{12}a_{31}a_{23}$$
$$+ a_{13}a_{21}a_{32} - a_{13}a_{31}a_{22}. \tag{8.2}$$

In (8.2) there are six terms, each of which is the product of three elements. Each term contains three elements, each from a different row and column. In other words, there are never two elements in any term from the same row or column. It can be seen that there must be just $3 \times 2 \times 1 = 6$ terms of this form, because three elements can be chosen from row 1, two from the two remaining elements in row 2, and one element from row 3.

Each term is prefixed by either $+1$ or -1. This is decided according to the following rule. Write each term in the form

$$a_{1j_1} a_{2j_2} a_{3j_3},$$

in which the first suffixes are in consecutive increasing order. Examine the second suffix **permutation** $j_1 j_2 j_3$. The permutation is said to be even (odd) if it has an even (odd) number of **inversions**. An inversion occurs whenever a larger integer precedes a smaller one. Thus the permutation 132 is odd, since 3 precedes 2, but 312 is even since there are two inversions because 3 precedes 1 and 2. If the number of permutations is even, then a + sign is attached; if the number is odd, then a − sign is attached. This rule can be extended to a determinant of any order.

While this expansion of the determinant says something about the structure of the determinant, it is not really a practical rule for evaluating determinants. Returning to (8.2), we can rewrite det A as

$$\det A = a_{11}(a_{22}a_{33} - a_{32}a_{23}) - a_{12}(a_{21}a_{33} - a_{31}a_{23})$$
$$+ a_{13}(a_{21}a_{32} - a_{31}a_{22}).$$

The terms in parentheses are themselves 2×2 determinants. Thus

$$\det A = \begin{vmatrix} a_{11} & a_{12} & a_{13} \\ a_{21} & a_{22} & a_{23} \\ a_{31} & a_{32} & a_{33} \end{vmatrix}$$

$$= a_{11} \begin{vmatrix} a_{22} & a_{23} \\ a_{32} & a_{33} \end{vmatrix} - a_{12} \begin{vmatrix} a_{21} & a_{23} \\ a_{31} & a_{33} \end{vmatrix} + a_{13} \begin{vmatrix} a_{21} & a_{22} \\ a_{31} & a_{32} \end{vmatrix}. \quad (8.3)$$

This expression is called an **expansion by the top row**. The term associated with a_{11}, namely

$$C_{11} = \begin{vmatrix} a_{22} & a_{23} \\ a_{32} & a_{33} \end{vmatrix}$$

is known as the **cofactor** of a_{11}: it is obtained from A by deleting the row and column through a_{11} and writing down the determinant of the elements of the remaining 2×2 submatrix with a + or − sign attached. The cofactors of a_{12} and a_{13} are constructed similarly as

$$C_{12} = - \begin{vmatrix} a_{21} & a_{23} \\ a_{31} & a_{33} \end{vmatrix}, \qquad C_{13} = \begin{vmatrix} a_{21} & a_{22} \\ a_{31} & a_{32} \end{vmatrix},$$

where the signs attached should be noted. In the same way the cofactors of the elements in the second and third rows are defined as follows:

$$C_{21} = - \begin{vmatrix} a_{12} & a_{13} \\ a_{32} & a_{33} \end{vmatrix}, \qquad C_{22} = \begin{vmatrix} a_{11} & a_{13} \\ a_{31} & a_{33} \end{vmatrix}, \qquad C_{23} = - \begin{vmatrix} a_{11} & a_{12} \\ a_{31} & a_{32} \end{vmatrix},$$

$$C_{31} = \begin{vmatrix} a_{12} & a_{13} \\ a_{22} & a_{23} \end{vmatrix}, \qquad C_{32} = - \begin{vmatrix} a_{11} & a_{13} \\ a_{21} & a_{23} \end{vmatrix}, \qquad C_{33} = \begin{vmatrix} a_{11} & a_{12} \\ a_{21} & a_{22} \end{vmatrix}.$$

The signs associated with the cofactors alternate, starting with a $+$ at the top left as we move across or down from the top left-hand corner as shown:

$$\begin{vmatrix} + & - & + \\ - & + & - \\ + & - & + \end{vmatrix}.$$

Example 8.1. Let

$$A = \begin{bmatrix} 1 & -1 & 0 \\ 2 & 3 & -2 \\ 1 & -1 & 1 \end{bmatrix}.$$

Evaluate det A by expanding by the first row. Find the cofactors C_{13}, C_{23}, C_{33} of the elements in the third column. Calculate

$$a_{13}C_{13} + a_{23}C_{23} + a_{33}C_{33},$$

and verify that it also equals det A.

By (8.3),

$$\det A = 1 \times \begin{vmatrix} 3 & -2 \\ -1 & 1 \end{vmatrix} - (-1) \times \begin{vmatrix} 2 & -2 \\ 1 & 1 \end{vmatrix} + 0 \times \begin{vmatrix} 2 & 3 \\ 1 & -1 \end{vmatrix},$$

$$= (3 - 2) + (2 + 2) = 5.$$

The cofactors are (with due regard to the sign convention)

$$C_{13} = \begin{vmatrix} 2 & 3 \\ 1 & -1 \end{vmatrix} = -5, \qquad C_{23} = - \begin{vmatrix} 1 & -1 \\ 1 & -1 \end{vmatrix} = 0,$$

$$C_{33} = \begin{vmatrix} 1 & -1 \\ 2 & 3 \end{vmatrix} = 5.$$

Hence expansion by the third column gives

$$a_{13}C_{13} + a_{23}C_{23} + a_{33}C_{33} = 0 \times (-5) + (-2) \times 0 + 1 \times 5 = 5,$$

which is the same as det A. (See also Rule 4 below.)

Example 8.2. Evaluate the determinant

$$\det A = \begin{vmatrix} 1 & 2 & k \\ 2 & -1 & 3 \\ -1 & 4 & -2 \end{vmatrix},$$

for any k. Find the value of k for which the determinant is zero.

Expanding by the first row gives

$$\det A = \begin{vmatrix} 1 & 2 & k \\ 2 & -1 & 3 \\ -1 & 4 & -2 \end{vmatrix}$$

$$= 1 \times \begin{vmatrix} -1 & 3 \\ 4 & -2 \end{vmatrix} - 2 \times \begin{vmatrix} 2 & 3 \\ -1 & -2 \end{vmatrix} + k \times \begin{vmatrix} 2 & -1 \\ -1 & 4 \end{vmatrix}$$

$$= 1 \times (2 - 12) - 2 \times (-4 + 3) + k \times (8 - 1)$$

$$= -10 + 2 + 7k = -8 + 7k.$$

Hence $\det A = 0$ if $k = \frac{7}{8}$.

The notion of cofactors generalizes to higher-order determinants. The alternating-sign rule applies from the top left-hand corner. For example, a 4×4 determinant has 16 cofactors, each of which is a 3×3 determinant.

8.2 Properties of determinants

We list here some properties of determinants. Many of them are useful in evaluating determinants. We shall not aim for complete generality but illustrate the rules mainly in the 3×3 case. However, the rules have obvious generalizations to higher orders.

1. $\det A^{\mathrm{T}} = \det A$, *where A^{T} is the transpose of A (see Section 7.3)*.

The determinants of a square matrix and its transpose are equal since

$$\det A^{\mathrm{T}} = \begin{bmatrix} a_{11} & a_{21} & a_{31} \\ a_{12} & a_{22} & a_{32} \\ a_{13} & a_{23} & a_{33} \end{bmatrix}$$

$$= a_{11}a_{22}a_{33} - a_{11}a_{23}a_{32} - a_{21}a_{12}a_{33} + a_{21}a_{13}a_{32}$$
$$+ a_{31}a_{12}a_{23} - a_{31}a_{13}a_{22},$$

and all terms in this expansion can be identified with those in (8.2). Hence $\det A^{\mathrm{T}} = \det A$.

All results applicable to rows will be equally applicable to columns.

Example 8.3. Evaluate

$$\det A = \begin{vmatrix} 1 & 28 & -29 \\ 0 & 1 & -4 \\ 0 & -2 & 5 \end{vmatrix}.$$

Since the determinant has two zeros in the first column, it is advantageous to use Rule 1. The determinant of the transpose of A

is given by

$$\det A^{\mathrm{T}} = \begin{vmatrix} 1 & 0 & 0 \\ 28 & 1 & -2 \\ -29 & -4 & 5 \end{vmatrix},$$

which now has two zeros in the first row. Hence the expansion by the top row becomes particularly easy:

$$\det A^{\mathrm{T}} = 1 \times \begin{vmatrix} 1 & -2 \\ -4 & 5 \end{vmatrix} = 5 - 8 = -3.$$

---

2. *If every element of any single row or column of the matrix A is multiplied by a scalar k, then the determinant of this matrix is* $k \det A$.

(Note: this rule is different from Rule 2, Section 7.2, for matrices.)

This is a self-evident result, since just one element from every row and column appears in every term. Thus, by (8.2), if every element of the second row in A is multiplied by k, then

$$\begin{vmatrix} a_{11} & a_{12} & a_{13} \\ ka_{21} & ka_{22} & ka_{23} \\ a_{31} & a_{32} & a_{33} \end{vmatrix} = a_{11}ka_{22}a_{33} - a_{11}ka_{23}a_{32} - a_{12}ka_{21}a_{33}$$
$$+ a_{12}ka_{23}a_{31} + a_{13}ka_{21}a_{32} - a_{13}ka_{22}a_{31},$$
$$= k \det A.$$

By putting $k = 0$ in this result, note that any determinant must have zero value if all the elements of any row or column are zeros.

Example 8.4. Evaluate the determinant

$$\Delta = \begin{vmatrix} -1 & 99 & 1 \\ 2 & 33 & -2 \\ 3 & 55 & 1 \end{vmatrix}.$$

Since the second column obviously has a factor of 11, then we can remove this factor from the second column before expansion. Thus, by Rule 2,

$$\Delta = 11 \times \begin{vmatrix} -1 & 9 & 1 \\ 2 & 3 & -2 \\ 3 & 5 & 1 \end{vmatrix}$$

$$= 11 \times \left((-1) \times \begin{vmatrix} 3 & -2 \\ 5 & 1 \end{vmatrix} - 9 \times \begin{vmatrix} 2 & -2 \\ 3 & 1 \end{vmatrix} + 1 \times \begin{vmatrix} 2 & 3 \\ 3 & 5 \end{vmatrix} \right)$$

$$= 11 \times [(3 + 10) - 9 \times (2 + 6) + (10 - 9)]$$

$$= 11 \times (13 - 72 + 1) = -638.$$

---

3. *If B is obtained from A by interchanging two rows (or columns) then* det $B = -$det A.

Suppose, for example, that rows 1 and 3 are interchanged, so that

$$A = \begin{bmatrix} a_{11} & a_{12} & a_{13} \\ a_{21} & a_{22} & a_{23} \\ a_{31} & a_{32} & a_{33} \end{bmatrix}, \qquad B = \begin{bmatrix} a_{31} & a_{32} & a_{33} \\ a_{21} & a_{22} & a_{23} \\ a_{11} & a_{12} & a_{13} \end{bmatrix}.$$

Then, by analogy with (8.2), the expansion of B by its first row is given by

$$\det B = a_{31}\begin{vmatrix} a_{22} & a_{23} \\ a_{12} & a_{13} \end{vmatrix} - a_{32}\begin{vmatrix} a_{21} & a_{23} \\ a_{11} & a_{13} \end{vmatrix} + a_{33}\begin{vmatrix} a_{21} & a_{22} \\ a_{11} & a_{12} \end{vmatrix},$$

$$= a_{31}a_{22}a_{13} - a_{31}a_{23}a_{12} - a_{32}a_{21}a_{13} + a_{32}a_{23}a_{11}$$

$$+ a_{33}a_{21}a_{12} - a_{33}a_{22}a_{11}.$$

These are the same terms as those present in (8.2) except that the sign of every term is changed. Therefore in this case

$$\det A = -\det A.$$

The same is true whichever row or column pairs are exchanged.

The rule applies to a determinant of any order.

Example 8.5. Evaluate the determinant

$$\Delta = \begin{vmatrix} 1 & 2 & 1 & 2 \\ 0 & 2 & 0 & 0 \\ -1 & 3 & 0 & 4 \\ -1 & 2 & 0 & -1 \end{vmatrix}.$$

There are several ways of approaching the evaluation of this determinant since the second row and third column each have three zeros. It is obviously advantageous to have as many zeros as possible in the top row. With this in view, interchange rows 1 and 2 using Rule 3:

$$\Delta = -\begin{vmatrix} 0 & 2 & 0 & 0 \\ 1 & 2 & 1 & 2 \\ -1 & 3 & 0 & 4 \\ -1 & 2 & 0 & -1 \end{vmatrix}.$$

Expanding by row 1, remembering the sign rule for cofactors:

$$\Delta = 2 \times \begin{vmatrix} 1 & 1 & 2 \\ -1 & 0 & 4 \\ -1 & 0 & -1 \end{vmatrix}.$$

Now successively use Rule 1 and interchange rows with columns, and then Rule 3 and interchange the new rows 1 and 2:

$$\Delta = 2 \times \begin{vmatrix} 1 & -1 & 1 \\ 1 & 0 & 0 \\ 2 & 4 & -1 \end{vmatrix}$$

$$= (-2) \times \begin{vmatrix} 1 & 0 & 0 \\ 1 & -1 & 1 \\ 2 & 4 & -1 \end{vmatrix}$$

$$= (-2) \times \begin{vmatrix} -1 & 1 \\ 4 & -1 \end{vmatrix}$$

$$= (-2) \times (1 - 4) = 6.$$

4. *Expansion by any row or column.*

From (8.2), by grouping the terms differently, we can write, for example,

$$\det A = a_{31}(a_{12}a_{23} - a_{13}a_{22}) - a_{32}(a_{11}a_{23} - a_{13}a_{21})$$
$$+ a_{33}(a_{11}a_{22} - a_{12}a_{21})$$
$$= a_{31} \begin{vmatrix} a_{12} & a_{13} \\ a_{22} & a_{23} \end{vmatrix} - a_{32} \begin{vmatrix} a_{11} & a_{13} \\ a_{21} & a_{23} \end{vmatrix} + a_{33} \begin{vmatrix} a_{11} & a_{12} \\ a_{21} & a_{22} \end{vmatrix},$$
$$= a_{31}C_{31} + a_{32}C_{32} + a_{33}C_{33}.$$

Here the elements a_{31}, a_{32}, a_{33} constitute the third row, and we call this the expansion of det A by the third row.

It can be shown that the expansion can be written down similarly using any row or column. Thus

$$\det A = a_{12}C_{12} + a_{22}C_{22} + a_{32}C_{32}$$

is an expansion by the second column.

Example 8.6. Evaluate

$$\Delta = \begin{vmatrix} 1 & 3 & 0 & 1 \\ 1 & 0 & 0 & 2 \\ -1 & 2 & 2 & 4 \\ 2 & 1 & 0 & -1 \end{vmatrix}.$$

Since column 3 contains three zeros, expand by this column. The cofactor of the element in row 3, column 3, is the 3×3 determinant obtained from A by deleting the third row and third column in A.

It is associated with a + sign. Hence

$$A = 2 \begin{vmatrix} 1 & 3 & 1 \\ 1 & 0 & 2 \\ 2 & 1 & -1 \end{vmatrix} \quad \text{(remember the sign rule)}$$

$$= 2 \left(-1 \times \begin{vmatrix} 3 & 1 \\ 1 & -1 \end{vmatrix} - 2 \times \begin{vmatrix} 1 & 3 \\ 2 & 1 \end{vmatrix} \right) \quad \text{(expanding by row 2)}$$

$$= 2(4 + 10) = 28.$$

---

5. *If two rows (or columns) of A are identical, then* $\det A = 0$. This is a direct consequence of Rule 3. Interchange the two identical rows (columns). The determinant looks the same, but its value is now $-\det A$. Hence $\det A = -\det A$, which implies that $\det A = 0$.

Also, as a consequence of Rule 2, it follows that, if the corresponding elements of two rows (columns) are in the same ratio, then the value of the determinant is zero. Thus, for example,

$$\begin{vmatrix} 99 & 18 & 63 \\ 11 & 2 & 7 \\ -2 & 3 & 4 \end{vmatrix} = 9 \begin{vmatrix} 11 & 2 & 7 \\ 11 & 2 & 7 \\ -2 & 3 & 4 \end{vmatrix} \quad \text{(by Rule 2)}$$

$$= 0 \quad \text{(by Rule 4)}$$

6. *If the matrix B is constructed from A by adding k times one row (or column) to another row (column) then* $\det B = \det A$: *in other words, any number of such operations on rows and on columns has no effect on the value of* $\det A$.

For our standard matrix A, consider the matrix B which is obtained from A by adding k times the elements in the first row to the elements in the third row. Thus

$$\det B = \begin{vmatrix} a_{11} & a_{12} & a_{13} \\ a_{21} & a_{22} & a_{23} \\ a_{31} + ka_{11} & a_{32} + ka_{12} & a_{33} + ka_{13} \end{vmatrix}$$

$$= (a_{31} + ka_{11})C_{31} + (a_{32} + ka_{12})C_{32} + (a_{33} + ka_{13})C_{33}$$

(expanding by row 3)

$$= a_{31}C_{31} + a_{32}C_{32} + a_{33}C_{33}$$
$$\quad + k(a_{11}C_{31} + a_{12}C_{32} + a_{13}C_{33})$$

$$= \det A + k \begin{vmatrix} a_{11} & a_{12} & a_{13} \\ a_{21} & a_{22} & a_{23} \\ a_{11} & a_{12} & a_{13} \end{vmatrix},$$

$$= \det A,$$

since the second determinant vanishes by Rule 6.

Note that
$$a_{11}C_{31} + a_{12}C_{32} + a_{13}C_{33} = 0,$$
that is, in its general form, **the sum of the products of the elements of one row (or column) and the cofactors of the elements of *another* row (column) is zero**. This follows since the left-hand side must arise from a matrix with two identical rows (columns).

Rule 5 is a particularly useful rule for simplifying the elements in a determinant before expansion and evaluation. We illustrate a number of these points in the next example.

Example 8.7. Evaluate
$$\Delta = \begin{vmatrix} 2 & 99 & -99 \\ 999 & 1000 & 1001 \\ 1000 & 1001 & 998 \end{vmatrix}.$$

Usually we use the rules (particularly 6) either to introduce zeros into the matrix or to reduce the size of elements as far as possible. It is important to list the operations in order to make the sequence of operations intelligible. For this purpose we identify the current rows by $r_1, r_2, \ldots$, and the current columns by $c_1, c_2, \ldots$. Denote the new rows and columns which have been changed by $r_1', r_2', \ldots$ and $c_1', c_2', \ldots$. There are many ways of approaching the evaluation of Δ. A first step in this example could be to add column 3 (c_3) to column 2 (c_2) since this produces a zero at the top of column 2. This operation is represented by $c_2' = c_2 + c_3$, and we list the operations on the right-hand side as we proceed. The second operation is to subtract the new row 3 from the new row 2. A decision is taken at each step in the light of the new matrix. By Rule 5, these operations do not affect the value of Δ. Hence

$$\Delta = \begin{vmatrix} 2 & 99 & -99 \\ 999 & 1000 & 1001 \\ 1000 & 1001 & 998 \end{vmatrix}$$

$$= \begin{vmatrix} 2 & 0 & -99 \\ 999 & 2001 & 1001 \\ 1000 & 1999 & 998 \end{vmatrix} \quad (c_2' = c_2 + c_3)$$

$$= \begin{vmatrix} 2 & 0 & -99 \\ -1 & 2 & 3 \\ 1000 & 1999 & 998 \end{vmatrix} \quad (r_2' = r_2 - r_3)$$

$$= \begin{vmatrix} 2 & 0 & -99 \\ -2 & 2 & 2 \\ 0.5 & 1999 & -1.5 \end{vmatrix} \quad \left(\begin{matrix} c_1' = c_1 - \frac{1}{2}c_2 \\ c_3' = c_3 - \frac{1}{2}c_2 \end{matrix} \right)$$

$$= \begin{vmatrix} 2 & 0 & -93 \\ -2 & 2 & -4 \\ 0.5 & 1999 & 0 \end{vmatrix} \quad (\mathbf{c}_3' = \mathbf{c}_3 + 3\mathbf{c}_1)$$

$$= 2(4 \times 1999) - 93[(-2 \times 1999) - 1],$$

$$= 387\,899.$$

Note that while $\mathbf{r}_2' = \mathbf{r}_2 + k\mathbf{r}_3$ does not affect the value of the determinant, $\mathbf{r}_2' = k\mathbf{r}_2 + \mathbf{r}_3$ will increase its value by a factor k.

8.3 The adjoint and inverse matrices

We can now rewrite the formula for the inverse given in Section 7.4, using cofactors from the previous section. The *transposed* matrix of cofactors given by

$$\text{adj } A = \begin{bmatrix} C_{11} & C_{21} & C_{31} \\ C_{12} & C_{22} & C_{32} \\ C_{13} & C_{23} & C_{33} \end{bmatrix},$$

is known as the **adjoint** of A. Hence the inverse matrix of A given by eqn. (7.10) becomes

$$A^{-1} = \frac{\text{adj } A}{\det A},$$

in terms of the adjoint and determinant of A.

Example 8.8. Find the inverse of

$$A = \begin{bmatrix} 1 & 2 & -1 \\ 0 & 1 & -1 \\ 1 & -1 & -2 \end{bmatrix}.$$

We evaluate $\det A$ first. Thus

$$\det A = 1 \times (-2 - 1) - 2 \times (0 + 1) - 1 \times (0 - 1) = -4.$$

The cofactors are

$$C_{11} = -3, \quad C_{12} = -1, \quad C_{13} = -1,$$
$$C_{21} = 5, \quad C_{22} = -1, \quad C_{23} = 3, \quad .$$
$$C_{31} = -1, \quad C_{32} = 1, \quad C_{33} = 1$$

Hence

$$A^{-1} = -\tfrac{1}{4} \begin{bmatrix} -3 & 5 & -1 \\ -1 & -1 & 1 \\ -1 & 3 & 1 \end{bmatrix}.$$

We can now confirm just why in general

$$A^{-1} = \frac{\text{adj } A}{\det A}.$$

Thus

$$A \frac{\operatorname{adj} A}{\det A} = \begin{bmatrix} a_{11} & a_{12} & a_{13} \\ a_{21} & a_{22} & a_{23} \\ a_{31} & a_{32} & a_{33} \end{bmatrix} \begin{bmatrix} C_{11} & C_{21} & C_{31} \\ C_{12} & C_{22} & C_{32} \\ C_{13} & C_{23} & C_{33} \end{bmatrix} \frac{1}{\det A},$$

$$= \frac{1}{\det A} \begin{bmatrix} a_{11}C_{11} + a_{12}C_{12} + a_{13}C_{13} & a_{11}C_{21} + a_{12}C_{22} + a_{13}C_{23} & a_{11}C_{31} + a_{12}C_{32} + a_{13}C_{33} \\ a_{21}C_{11} + a_{22}C_{12} + a_{23}C_{13} & a_{21}C_{21} + a_{22}C_{22} + a_{23}C_{23} & a_{21}C_{31} + a_{22}C_{32} + a_{23}C_{33} \\ a_{31}C_{11} + a_{32}C_{12} + a_{33}C_{13} & a_{31}C_{21} + a_{32}C_{22} + a_{33}C_{23} & a_{31}C_{31} + a_{32}C_{32} + a_{33}C_{33} \end{bmatrix}$$

$$= \frac{1}{\det A} \begin{bmatrix} \det A & 0 & 0 \\ 0 & \det A & 0 \\ 0 & 0 & \det A \end{bmatrix} = I_3.$$

This formula for the inverse uses the results that the sum of the products of the elements of one row and their cofactors is the value of the determinant whilst the sum of the products of one row and the cofactors of another row is zero.

The definition of the adjoint generalizes to matrices of higher order. However, the adjoint of a 4×4 contains sixteen 3×3 determinants, which is about the limit of hand calculations unless the determinant is sparsely filled with nonzero elements or can be reduced to such a determinant. Such computations become a fertile source of errors. There are computer packages available which will quickly perform the arithmetic operations for determinants of reasonable size.

Summary of results for 3×3 matrices

(a) (*determinant of A*)

$$\det A = a_{11} \begin{vmatrix} a_{22} & a_{23} \\ a_{32} & a_{33} \end{vmatrix} - a_{12} \begin{vmatrix} a_{21} & a_{23} \\ a_{31} & a_{33} \end{vmatrix}$$
$$+ a_{13} \begin{vmatrix} a_{21} & a_{22} \\ a_{31} & a_{32} \end{vmatrix};$$

(b) (*adjoint of A*)

$$\operatorname{adj} A = \begin{bmatrix} C_{11} & C_{21} & C_{31} \\ C_{12} & C_{22} & C_{32} \\ C_{13} & C_{23} & C_{33} \end{bmatrix};$$

(c) (*inverse of A*)

$$A^{-1} = \frac{\operatorname{adj} A}{\det A}.$$

(8.4)

Problems, Chapter 8

8.1. Evaluate the following determinants.

(a) $\begin{vmatrix} 1 & 2 \\ -1 & 3 \end{vmatrix}$; (b) $\begin{vmatrix} 1 & 0 & 1 \\ 0 & 1 & 0 \\ 1 & 0 & 1 \end{vmatrix}$;

(c) $\begin{vmatrix} 1 & -1 & 2 \\ 3 & 1 & -1 \\ 2 & 1 & -1 \end{vmatrix}$;

(d) $\begin{vmatrix} 2 & 1 & 0 & -1 \\ 0 & 0 & 2 & 0 \\ 3 & -1 & 2 & 1 \\ 0 & 1 & -1 & 1 \end{vmatrix}$;

(e) $\begin{vmatrix} 0 & 1 & 0 & 0 & 0 \\ 1 & 0 & 0 & 0 & 0 \\ 0 & 0 & 0 & 0 & 1 \\ 0 & 0 & 1 & 0 & 0 \\ 0 & 0 & 0 & 1 & 0 \end{vmatrix}$;

(f) $\begin{vmatrix} 2 & 1 & 0 & 0 & 0 \\ 1 & 2 & 1 & 0 & 0 \\ 0 & 1 & 2 & 1 & 0 \\ 0 & 0 & 1 & 2 & 1 \\ 0 & 0 & 0 & 1 & 2 \end{vmatrix}$.

8.2. Without evaluating the following determinants, explain why they are all zero:

(a) $\begin{vmatrix} 2 & 3 & 4 \\ 4 & 6 & 8 \\ 1 & -1 & 2 \end{vmatrix}$; (b) $\begin{vmatrix} -1 & 2 & 3 \\ 3 & 1 & -2 \\ -2 & -3 & -1 \end{vmatrix}$;

(c) $\begin{vmatrix} a & b & c \\ b & c & a \\ a-b & b-c & c-a \end{vmatrix}$;

(d) $\begin{vmatrix} 1 & 1 & 1 \\ 3 & 0 & 0 \\ 5 & 0 & 0 \end{vmatrix}$.

8.3. Given that

$$\Delta = \begin{vmatrix} a & b & c \\ b & c & a \\ c & a & b \end{vmatrix},$$

what is the value of

$$\begin{vmatrix} a^3 & ab & ac^2 \\ ab & c & ac \\ ac & a & bc \end{vmatrix}$$

in terms of Δ?

8.4. Simplify first and then evaluate the following determinants:

(a) $\begin{vmatrix} 99 & 100 & 200 \\ 98 & 102 & 199 \\ -1 & 2 & 3 \end{vmatrix}$;

(b) $\begin{vmatrix} 77 & 84 & 55 \\ 75 & 87 & 57 \\ 1 & -2 & 3 \end{vmatrix}$;

(c) $\begin{vmatrix} 2 & -1 & 1 \\ 99 & 98 & 55 \\ 200 & 197 & 111 \end{vmatrix}$;

(d) $\begin{vmatrix} 87 & 84 & 83 & 81 \\ 77 & 76 & 77 & 75 \\ 54 & 53 & 52 & 54 \\ -43 & -44 & -46 & -4 \end{vmatrix}$.

8.5. Explain why the determinant

$$\Delta = \begin{vmatrix} 1 & 1 & 1 \\ a & b & c \\ a^2 & b^2 & c^2 \end{vmatrix},$$

has factors $b - c$, $c - a$, and $a - b$. Express the value of Δ as the product of factors.

8.6. Factorize the determinant

$$\Delta = \begin{vmatrix} 1 & 1 & 1 \\ a & b & c \\ a^3 & b^3 & c^3 \end{vmatrix}.$$

8.7. Explain, using one of the rules for determinants, why the equation

$$\begin{vmatrix} x & y & 1 \\ a_1 & b_1 & 1 \\ a_2 & b_2 & 1 \end{vmatrix} = 0$$

represents the equation of the straight line through the points (a_1, b_1) and (a_2, b_2) in the (x, y) plane. If X_1 and X_2 are the cofactors of x and y, what is the slope of this line in terms of the cofactors? Using this method, find the equation of the straight line through the points:
(a) $(1, -1)$ and $(2, 3)$;   (b) $(-1, 0)$ and $(4, -1)$.

8.8. Find the value of a which makes the determinant

$$\begin{vmatrix} 1 & 1 & -1 \\ 1 & a & 2 \\ -1 & 1 & 2 \end{vmatrix}$$

equal to zero.

8.9. Explain why

$$\begin{vmatrix} x & 2 & -2 \\ 2 & x & 3 \\ x & -1 & x \end{vmatrix} = 0$$

will be at most a cubic equation in x, but that

$$\begin{vmatrix} 1 & 1 & 2 \\ 3 & x & 2 \\ x & 1 & x \end{vmatrix} = 0$$

will be at most a quadratic equation in x. Solve both equations, and find all roots including any complex ones.

8.10. Show that

$$\begin{vmatrix} a_{11} + b_{11} & a_{12} + b_{12} & a_{13} + b_{13} \\ a_{21} & a_{22} & a_{23} \\ a_{31} & a_{32} & a_{33} \end{vmatrix}$$

$$= \begin{vmatrix} a_{11} & a_{12} & a_{13} \\ a_{21} & a_{22} & a_{23} \\ a_{31} & a_{32} & a_{33} \end{vmatrix} + \begin{vmatrix} b_{11} & b_{12} & b_{13} \\ a_{21} & a_{22} & a_{23} \\ a_{31} & a_{32} & a_{33} \end{vmatrix}$$

8.11. The determinant

$$\begin{vmatrix} a_{11} + b_{11} & a_{12} + b_{12} & a_{13} + b_{13} \\ a_{21} + b_{21} & a_{22} + b_{22} & a_{23} + b_{23} \\ a_{31} + b_{31} & a_{32} + b_{32} & a_{33} + b_{33} \end{vmatrix}$$

is required as the sum of determinants each of which has just as or bs in columns. How many determinants are there in the sum? If the determinant is $n \times n$, how many determinants would there be in the sum?

8.12. Show that

$$\begin{vmatrix} 1 & a_1 - b_1 & a_1 + b_1 \\ 1 & a_2 - b_2 & a_2 + b_2 \\ 1 & a_3 - b_3 & a_3 + b_3 \end{vmatrix} = 2 \begin{vmatrix} 1 & a_1 & b_1 \\ 1 & a_2 & b_2 \\ 1 & a_3 & b_3 \end{vmatrix}.$$

8.13. Let D_n be the $n \times n$ **tridiagonal** determinant defined by

$$D_n = \begin{vmatrix} 2 & 1 & 0 & \cdots & 0 \\ 1 & 2 & 1 & \ddots & 0 \\ 0 & 1 & 2 & \ddots & \vdots \\ \cdots & \ddots & \ddots & \ddots & 1 \\ 0 & 0 & \cdots & 1 & 2 \end{vmatrix}.$$

Show that

$$D_n = 2D_{n-1} - D_{n-2}.$$

If $Q_n = D_n - D_{n-1}$, deduce that

$$Q_n = Q_{n-1} = \cdots = Q_3 = 1.$$

Show that $D_n = n + 1$.

8.14. Find all values of x for which

$$\begin{vmatrix} x & a & b & c \\ a & x & b & c \\ a & b & x & c \\ a & b & c & x \end{vmatrix}$$

is zero.

8.15. Let A and B be the two 2×2 matrices

$$A = \begin{bmatrix} a_{11} & a_{12} \\ a_{21} & a_{22} \end{bmatrix}, \qquad B = \begin{bmatrix} b_{11} & b_{12} \\ b_{21} & b_{22} \end{bmatrix}.$$

Write down $\det A$ and $\det B$.
(a) Find the product AB and its determinant $\det AB$. Confirm that

$$\det AB = \det A \det B.$$

Show also that

$$\det A^2 = (\det A)^2.$$

(b) Write down A^{T} and find its determinant $\det A^{\mathrm{T}}$. Confirm that

$$\det A^{\mathrm{T}} = \det A.$$

(c) Find A^{-1} and det A^{-1}, assuming that det $A \neq 0$. Confirm that

$$\det A^{-1} = 1/\det A.$$

(d) Show that

$$\det \text{adj } A = \det A.$$

(These formulae for 2×2 matrices suggest generalizations for $n \times n$ matrices. Thus, for two $n \times n$ matrices A and B,

$$\det AB = \det A \det B,$$

$$\det A^n = (\det A)^n,$$

$$\det A^{\mathrm{T}} = \det A,$$

$$\det A^{-1} = 1/\det A,$$

$$\det \text{adj } A = (\det A)^{n-1}.$$

We shall not attempt to prove these formulae here.)

8.16. If

$$A = \begin{bmatrix} 1 & 2 & -1 \\ 0 & 1 & 2 \\ 1 & 3 & -1 \end{bmatrix}, \quad B = \begin{bmatrix} 1 & 2 & -1 \\ 0 & 3 & 1 \\ 2 & 1 & 3 \end{bmatrix},$$

calculate det A, det B, det AB, A^{T}, det A^{T}, adj A, det adj A, A^{-1} and det A^{-1}. Confirm the results conjectured at the end of the previous problem.

8.17. The elements in a 3×3 matrix $A = [a_{ij}]$ are given by the formula

$$a_{ij} = \alpha j + (-1)^i 2^j \quad (i, j = 1, 2, 3).$$

Show that det $A = 0$ for all real α.

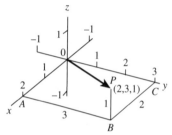

Fig. 9.1
Right-handed cartesian axes.
$CB = x = 2$, $AB = y = 3$,
$BP = z = 1$.

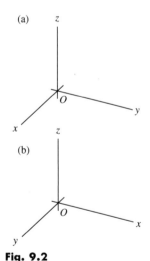

Fig. 9.2
(a) Right-handed axes.
(b) Left-handed axes.

9 Vectors in two and three dimensions

9.1 Three-dimensional cartesian coordinates

To locate points in a plane, two axes are needed. For three-dimensional space, introduce a third axis, Oz, perpendicular to the other two and drawn from the **origin** O in the direction shown in Fig. 9.1. The position of a point P is then specified by the triplet of **coordinates** (x, y, z), determined by reference to the three axes Ox, Oy, Oz. For the point P in the figure we have $x = 2$, $y = 3$, and $z = 1$.

The axes shown are *right-handed axes* (left-handed axes are seldom used). At the start we had a choice of two opposite directions for the positive z axis. The two possibilities are shown in Fig. 9.2. The two sets of axis are mirror images of each other, like a right-hand and a left-hand glove, and they cannot be superposed no matter how we turn them about.

9.2 Position vectors

In Fig. 9.1, join the origin O to P : (2, 3, 1). The line OP has **length**

$$OP = \sqrt{(OA^2 + AP^2)} = \sqrt{(OA^2 + AB^2 + BP^2)}$$
$$= \sqrt{(2^2 + 3^2 + 1^2)} = \sqrt{14}.$$

It also has a **direction indicated by the arrow**, from O to P. Any line having its **length and direction specified** is called a **directed line segment**. A directed line segment starting at the origin O and ending at a point P is called *a* the **position vector** of P. Given that **it starts at O and ends at P**, a position vector is completely specified by the coordinates (x, y, z) of P. Denote it (conventionally) by $\overline{OP}$. Then we can write for the position vector in Fig. 9.1

$$\overline{OP} = (2, 3, 1),$$

in this context, the **coordinates of P** are called the **components of the position vector $\overline{OP}$**.

The geometric notation $\overline{OP}$ is not always convenient, and is usually replaced in print by bold-faced letters for vectors (in hand-written work the underline is used to distinguish vectors). The symbol r is often used for position vectors. Thus, we can write $r = (2, 3, 1)$ to represent the point with coordinates $(2, 3, 1)$.

A position vector is always anchored at the origin and points towards the given point whose position it determines. For a general point $A : (a_1, a_2, a_3)$ (Fig. 9.3), its position vector is $r = (a_1, a_2, a_3)$. The **length** or **magnitude** of the vector is denoted by $|r|$ or r, and

$$r = |r| = \sqrt{(a_1^2 + a_2^2 + a_3^2)}.$$

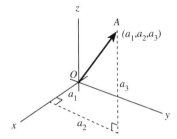

Fig. 9.3

9.3 General vectors

The position vector has three attributes: length, direction, and location at the origin. If the last of these is no longer a requirement then the quantity is simply a **vector**; it has length and direction but may be situated anywhere. These vectors represent geometrically the set of *all* line segments which have the same length and direction. If s is such a vector, then it could be represented by any of the straight line segments in Fig. 9.4.

We need this generalization for vectors so that they can represent such directed quantities as **force**, **velocity**, and **acceleration**, which are associated with objects moving in space and not necessarily situated at the origin.

Rules for manipulating vectors are now needed. Vectors have common features with row vectors in matrix algebra (Chapter 7).

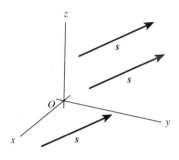

Fig. 9.4

1. *Equality.* Two vectors a and b are said to be equal if they have the same lengths and directions. We write $a = b$.

The two vectors obtained by joining $A : (-1, 2, 3)$ to $B : (4, -2, 6)$ and $C : (-3, 0, 1)$ to $D : (2, -4, 4)$ are equal, since

$$a = \overline{AB} = (4-(-1), (-2) - 2, 6 - 3) = (5, -4, 3),$$

and

$$b = \overline{CD} = (2 - (-3), -4, -0, 4 - 1) = (5, -4, 3) = a.$$

2. *Multiplication by a scalar.* If k is a real number, then ka is the vector whose components are each multiplied by k. Thus, if $a = (a_1, a_2, a_3)$, then

$$ka = (ka_1, ka_2, ka_3).$$

If k is a *negative* number then the *direction* of the vector is reversed.

3. *Addition and subtraction of vectors.* As with matrices, to add or subtract two vectors, we add or subtract the corresponding components. Thus, if $a = (a_1, a_2, a_3)$ and $b = (b_1, b_2, b_3)$, then

$$a \pm b = (a_1, a_2, a_3) \pm (b_1, b_2, b_3)$$
$$= (a_1 \pm b_1, a_2 \pm b_2, a_3 \pm b_3).$$

The addition rule is known as the **parallelogram law of addition** for the following reason. Suppose that a and b are drawn at the

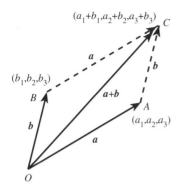

Fig. 9.5
Parallelogram law of addition.

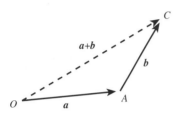

Fig. 9.6
Triangle law of addition.

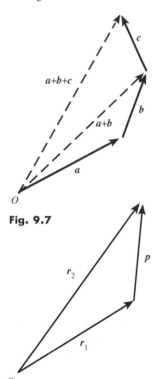

Fig. 9.7

Fig. 9.8

origin O (Fig. 9.5) so that $a = \overline{OA}$ and $b = \overline{OB}$. Construct the fourth point $C : (c_1, c_2, c_3)$ which makes $OACB$ a parallelogram.

Noticing that $\overline{AC} = b$ and $\overline{BC} = a$, it can be seen that the coordinates of C are given by

$$(c_1, c_2, c_3) = (a_1 + b_1, a_2 + b_2, a_3 + b_3).$$

Therefore the sum of two vectors drawn from the same point is the diagonal of the parallelogram drawn on the vectors. For subtraction, write $a - b = a + (-b)$.

It can also be viewed as a **triangle law**. If the vector a is drawn from O, and b is drawn from the end of a, then $a + b$ is the third side of the triangle formed by a and b (Fig. 9.6). Interpreted in geometric notation, the result is

$$\overline{OC} = \overline{OA} + \overline{AC}.$$

Any number of vectors can be added by placing them end to end. This follows from repeated use of the triangle law. Figure 9.7 shows geometrically the sum $(a + b) + c$ which, as can be seen from the figure, is the same as $a + (b + c)$. This confirms the associative law of addition, which follows also from the component definition of a vector.

If the vector p joins the two points with positive vectors $r_1 = (a_1, a_2, a_3)$ and $r_2 = (b_1, b_2, b_3)$ as shown in Fig. 9.8, then, by the triangle law,

$$r_2 = r_1 + p \quad \text{or} \quad p = r_2 - r_1 = (b_1 - a_1, b_2 - a_2, b_3 - a_3).$$

The **components** of p are the differences $(b_1 - a_1, b_2 - a_2, b_3 - a_3)$. The **magnitude**, $|p|$ or p, of p is the **distance between the points**, given by

$$p = |p| = \sqrt{[(b_1 - a_1)^2 + (b_2 - a_2)^2 + (b_3 - a_3)^2]}.$$

9.4 Unit vectors

Any vector of unit length or magnitude is known as a **unit vector**. Thus $a = (a_1, a_2, a_3)$ is a unit vector if $|a| = 1$, that is, if $\sqrt{(a_1^2 + a_2^2 + a_3^2)} = 1$. For example,

$$a = (-\tfrac{2}{7}, \tfrac{3}{7}, \tfrac{6}{7})$$

is a unit vector since

$$\sqrt{[(-\tfrac{2}{7})^2 + (\tfrac{3}{7})^2 + (\tfrac{6}{7})^2]} = \sqrt{\tfrac{49}{49}} = 1.$$

Unit vectors in the directions of the coordinate axes have special symbols. The vector $(1, 0, 0)$ is a unit vector in the direction of the x-axis, since, drawn as a position vector it would join the origin to the point 1 unit along the x-axis. We denote it by

$$\hat{\imath} = (1, 0, 0),$$

(called 'i hat'). Similarly, we define $\hat{\jmath} = (0, 1, 0)$ and $\hat{k} = (0, 0, 1)$. Figure 9.9 shows $\hat{\imath}, \hat{\jmath}, \hat{k}$ as position vectors. These particular **basis vectors** are introduced since they provide a convenient way of expressing any vector in terms of $\hat{\imath}, \hat{\jmath}, \hat{k}$. Thus we can write

$$(a_1, 0, 0) = a_1\hat{\imath}, \qquad (0, a_2, 0) = a_2\hat{\jmath}, \qquad (0, 0, a_3) = a_3\hat{k}.$$

Also

$$a = (a_1, a_2, a_3) = (a_1, 0, 0) + (0, a_2, 0) + (0, 0, a_3)$$
$$= a_1\hat{\imath} + a_2\hat{\jmath} + a_3\hat{k}.$$

The components become the coefficients of $\hat{\imath}, \hat{\jmath}, \hat{k}$. Any vector can be expressed in this form.

Position vectors are always drawn from the origin. The position vector r of the point $(1, 2, -4)$ is

$$r = \hat{\imath} + 2\hat{\jmath} - 4\hat{k}.$$

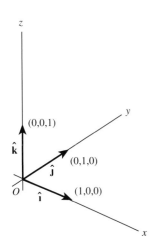

Fig. 9.9

For general vectors, a **unit vector in the direction of a given vector** can be found by simply dividing the vector by its magnitude, or length:

Unit vectors

A unit vector is a vector of length unity. $\hfill$ (9.1)
The vector $a/|a|$ is a unit vector in the direction of a.

The **notation** $\hat{a}$ **denotes the unit vector in the direction of** a. (Note that $-a/|a|$ is a unit vector in the opposite direction to a.) For example, if $a = -2\hat{\imath} + 6\hat{\jmath} + 3\hat{k}$, then a unit vector in the direction of a is

$$\hat{a} = \frac{a}{|a|} = \frac{1}{\sqrt{(2^2 + 6^2 + 3^2)}}(-2\hat{\imath} + 6\hat{\jmath} + 3\hat{k}),$$

$$= \frac{1}{\sqrt{49}}(-2\hat{\imath} + 6\hat{\jmath} + 3\hat{k}),$$

$$= -\tfrac{2}{7}\hat{\imath} + \tfrac{6}{7}\hat{\jmath} + \tfrac{3}{7}\hat{k}.$$

9.5 The scalar product

There are two types of product for vectors in three dimensions which are useful in applications. They are known as the **scalar product**, sometimes referred to as the **inner product** or **dot product**, and the **vector product**. The scalar product is treated in this section and the vector product in the next.

The definition of the scalar product, together with certain properties which follow immediately, is as follows.

> **Scalar or 'dot' product**
>
> If $a = (a_1, a_2, a_3)$ and $b = (b_1, b_2, b_3)$, the scalar product of a and b is defined by
>
> $$a \cdot b = a_1 b_1 + a_2 b_2 + a_3 b_3.$$
>
> (a) $a \cdot b = b \cdot a$ (i.e. the scalar product is commutative);
> (b) $a \cdot (b + c) = a \cdot b + a \cdot b$ (i.e. the scalar product is distributive);
> (c) $a \cdot a = a_1^2 + a_2^2 + a_3^2 = |a|^2$.
>
> (9.2)

Products involving the unit basis vectors $\hat{\imath}, \hat{\jmath}, \hat{k}$ are important. From (9.2c), we have

$$\hat{\imath} \cdot \hat{\imath} = \hat{\jmath} \cdot \hat{\jmath} = \hat{k} \cdot \hat{k} = 1,$$

since their lengths are equal to 1. But

$$\hat{\imath} \cdot \hat{\jmath} = (1, 0, 0) \cdot (0, 1, 0) = 0,$$

and similarly for the other mixed products. This has the consequence that, if we write

$$a \cdot b = (a_1 \hat{\imath} + a_2 \hat{\jmath} + a_3 \hat{k}) \cdot (b_1 \hat{\imath} + b_2 \hat{\jmath} + b_3 \hat{k}),$$

and expand it at length using (9.2a,b) (there are nine terms), only the terms

$$a_1 b_1 \hat{\imath} \cdot \hat{\imath} + a_2 b_2 \hat{\jmath} \cdot \hat{\jmath} + a_3 b_3 \hat{k} \cdot \hat{k} = a_1 b_1 + a_2 b_2 + a_3 b_3$$

survive, and so this process correctly returns the definition (9.2).

The x, y, z **components** of a vector are picked out by scalar multiplication by $\hat{\imath}, \hat{\jmath},$ or $\hat{k}$: for example, if $a = (a_1, a_2, a_3)$, then

$$a_2 = a \cdot \hat{\jmath} \quad \text{or} \quad \hat{\jmath} \cdot a,$$

so any vector can be written in the form

$$a = (a \cdot \hat{\imath}) \hat{\imath} + (a \cdot \hat{\jmath}) \hat{\jmath} + (a \cdot \hat{k}) \hat{k}.$$

Summing up these properties:

> **Scalar products in $\hat{\imath}, \hat{\jmath}, \hat{k}$**
>
> (a) $\hat{\imath} \cdot \hat{\imath} = \hat{\jmath} \cdot \hat{\jmath} = \hat{k} \cdot \hat{k} = 1;$
> $\hat{\imath} \cdot \hat{\jmath} = \hat{\jmath} \cdot \hat{k} = \hat{k} \cdot \hat{\imath} = 0.$
>
> (b) The components of any vector a are given by $\hat{\imath} \cdot a,$ $\hat{\jmath} \cdot a, \hat{k} \cdot a.$
>
> (9.3)

An alternative expression for the scalar product is

Scalar product: connection with angle
$$a \cdot b = |a||b| \cos \theta,$$ (9.4)
where θ is the angle between a and b.

This is proved by using the cosine rule (see Appendix B). In Fig. 9.10,

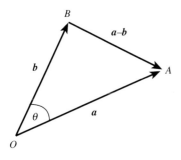

Fig. 9.10

$$\overline{OA} = \overline{OB} + \overline{BA},$$

so $\overline{BA} = a - b$. By the cosine rule,
$$AB^2 = OA^2 + OB^2 - 2OA \cdot OB \cos \theta,$$
or $\quad |a - b|^2 = |a|^2 + |b|^2 - 2|a||b| \cos \theta.$

By using (9.2c), the squared lengths can be written as scalar products:
$$(a - b) \cdot (a - b) = a \cdot a + b \cdot b - 2|a||b| \cos \theta;$$
But also
$$(a - b) \cdot (a - b) = a \cdot a - 2a \cdot b + b \cdot b.$$
After cancellation, we have $a \cdot b = 2|a||b| \cos \theta$ as required. This can be written
$$\cos \theta = a \cdot b / |a||b|,$$
which tells us the angle between a and b. It also gives a frequently used condition for perpendicularity:

Angle between two vectors
(a) The angle θ between a and b is obtained from
$$a \cdot b = |a||b| \cos \theta.$$ (9.5)
(b) The condition that (nonzero) vectors a and b be perpendicular is that $a \cdot b = 0$.

Example 9.1. Find the numbers α, β, and γ which make the vectors
$$a = \alpha\hat{\imath} + \hat{\jmath} + 2\hat{k}, \qquad b = \hat{\imath} + \beta\hat{\jmath} - \hat{k}, \qquad c = \hat{\imath} - \hat{\jmath} + \gamma\hat{k},$$
mutually perpendicular.

We require simultaneously $a \cdot b = 0$, $b \cdot c = 0$, and $c \cdot a = 0$. Thus
$$a \cdot b = (\alpha\hat{\imath} + \hat{\jmath} + 2\hat{k}) \cdot (\hat{\imath} + \beta\hat{\jmath} - \hat{k}) = \alpha + \beta - 2 = 0,$$
$$b \cdot c = (\hat{\imath} + \beta\hat{\jmath} + \hat{k}) \cdot (\hat{\imath} - \hat{\jmath} + \gamma\hat{k}) = 1 - \beta - \gamma = 0,$$
$$c \cdot a = (\hat{\imath} - \hat{\jmath} + \gamma\hat{k}) \cdot (\alpha\hat{\imath} + \hat{\jmath} + 2\hat{k}) = \alpha - 1 + 2\gamma = 0.$$
Hence α, β, γ satisfy
$$\alpha + \beta \qquad = \quad 2,$$ (i)
$$\qquad - \beta - \gamma = -1,$$ (ii)
$$\alpha \qquad + 2\gamma = \quad 1.$$ (iii)

Substitute α from (i) and γ from (ii) into (iii) to give

$$(2 - \beta) + 2(1 - \beta) = 1.$$

Hence $\beta = 1$. From (ii), $\gamma = 1 - \beta = 0$, and from (i), $\alpha = 2 - \beta = 1$. The required vectors are

$$\boldsymbol{a} = \hat{\boldsymbol{\imath}} + \hat{\boldsymbol{\jmath}} + 2\hat{\boldsymbol{k}}, \qquad \boldsymbol{b} = \hat{\boldsymbol{\imath}} + \hat{\boldsymbol{\jmath}} - \hat{\boldsymbol{k}}, \qquad \boldsymbol{c} = \hat{\boldsymbol{\imath}} - \hat{\boldsymbol{\jmath}}.$$

9.6 The vector product

The **vector product** or **cross product** of two vectors $\boldsymbol{a}$ and $\boldsymbol{b}$ is a vector $\boldsymbol{p}$ defined and denoted by

$$\boldsymbol{p} = \boldsymbol{a} \times \boldsymbol{b} = (a_2 b_3 - a_3 b_2)\hat{\boldsymbol{\imath}} + (a_3 b_1 - a_1 b_3)\hat{\boldsymbol{\jmath}} + (a_1 b_2 - a_2 b_1)\hat{\boldsymbol{k}}.$$

It finds application in problems involving moments, angular velocity, and other rotational features.

The elements in the expression $\boldsymbol{p}$ have a familiar look in the context of determinants (Chapter 8). The vector product can be written formally as

Vector product

$$\boldsymbol{a} \times \boldsymbol{b} = \begin{vmatrix} \hat{\boldsymbol{\imath}} & \hat{\boldsymbol{\jmath}} & \hat{\boldsymbol{k}} \\ a_1 & a_2 & a_3 \\ b_1 & b_2 & b_3 \end{vmatrix}.$$

(9.6)

To evaluate it, we expand by the first row, using the determinant expansion rule. The unit vectors $\hat{\boldsymbol{\imath}}, \hat{\boldsymbol{\jmath}}, \hat{\boldsymbol{k}}$ must always be in the first row, the first vector $\boldsymbol{a}$ in the second row and the second vector $\boldsymbol{b}$ in the third row. Usually this rule is the most convenient way of evaluating vector products.

Example 9.2. Find the vector products $\boldsymbol{a} \times \boldsymbol{b}$ and $\boldsymbol{b} \times \boldsymbol{a}$ where $\boldsymbol{a} = 2\hat{\boldsymbol{\imath}} - \hat{\boldsymbol{\jmath}} + 3\hat{\boldsymbol{k}}$, $\boldsymbol{b} = -\hat{\boldsymbol{\imath}} + 2\hat{\boldsymbol{\jmath}} + 4\hat{\boldsymbol{k}}$.

Thus

$$\boldsymbol{a} \times \boldsymbol{b} = \begin{vmatrix} \hat{\boldsymbol{\imath}} & \hat{\boldsymbol{\jmath}} & \hat{\boldsymbol{k}} \\ 2 & -1 & 3 \\ -1 & 2 & 4 \end{vmatrix}$$

$$= [(-1) \times 4 - 3 \times 2]\hat{\boldsymbol{\imath}} + [3 \times (-1) - 2 \times 4]\hat{\boldsymbol{\jmath}}$$

$$+ [2 \times 2 - (-1) \times (-1)]\hat{\boldsymbol{k}}$$

$$= -10\hat{\boldsymbol{\imath}} - 11\hat{\boldsymbol{\jmath}} + 3\hat{\boldsymbol{k}}.$$

Also
$$\boldsymbol{b} \times \boldsymbol{a} = \begin{vmatrix} \hat{\boldsymbol{\imath}} & \hat{\boldsymbol{\jmath}} & \hat{\boldsymbol{k}} \\ -1 & 2 & 4 \\ 2 & -1 & 3 \end{vmatrix}$$
$$= [2 \times 3 - 4 \times (1)]\hat{\boldsymbol{\imath}} + [4 \times 2 - (-1) \times 3]\hat{\boldsymbol{\jmath}}$$
$$+ [(-1) \times (-1) - 2 \times 2]\hat{\boldsymbol{k}}$$
$$= 10\hat{\boldsymbol{\imath}} + 11\hat{\boldsymbol{\jmath}} - 3\hat{\boldsymbol{k}}$$
$$= -(\boldsymbol{a} \times \boldsymbol{b}).$$

A number of properties of the vector product are now listed.

1. $\boldsymbol{a} \times \boldsymbol{b} = -\boldsymbol{b} \times \boldsymbol{a}$.

The previous example indicates that the vector product does not commute. From the definition (9.6), $\boldsymbol{b} \times \boldsymbol{a}$ involves interchanging the second and third rows in the determinant, which changes its sign.

2. $\boldsymbol{a} \times \boldsymbol{a} = \boldsymbol{0}$.

From the definition (9.6), the corresponding determinant has two identical rows, so it is zero. If, for any scalar λ, we have $\boldsymbol{b} = \lambda\boldsymbol{a}$, then a similar argument shows that $\boldsymbol{a} \times \boldsymbol{b} = \boldsymbol{0}$. In other words, **if $\boldsymbol{a}$ and $\boldsymbol{b}$ are parallel, then $\boldsymbol{a} \times \boldsymbol{b} = \boldsymbol{0}$.**

3. $\boldsymbol{a} \times \boldsymbol{b}$ *is perpendicular to both* $\boldsymbol{a}$ **and** $\boldsymbol{b}$.

Consider the scalar product of $\boldsymbol{a}$ and $\boldsymbol{a} \times \boldsymbol{b}$.
$$\boldsymbol{a} \cdot (\boldsymbol{a} \times \boldsymbol{b}) = a_1(a_2b_3 - a_3b_2) + a_2(a_3b_1 - a_1b_3)$$
$$+ a_3(a_1b_2 - a_2 - b_1)$$
$$= \begin{vmatrix} a_1 & a_2 & a_3 \\ a_1 & a_2 & a_3 \\ b_1 & b_2 & b_3 \end{vmatrix} \quad \text{(by reversing the determinant expansion rule (7.11))}$$
$$= 0 \qquad \text{(since row 1 = row 2)}.$$

Hence the scalar product of $\boldsymbol{a}$ with $\boldsymbol{a} \times \boldsymbol{b}$ vanishes, which implies that $\boldsymbol{a}$ and $\boldsymbol{a} \times \boldsymbol{b}$ are perpendicular (see (9.5b)). A similar argument proves that $\boldsymbol{b} \cdot (\boldsymbol{a} \times \boldsymbol{b}) = 0$.

Suppose now that the axes are rotated about the origin in such a way that $\boldsymbol{a} = a_1\hat{\boldsymbol{\imath}}$ and $\boldsymbol{b} = b_1\hat{\boldsymbol{\imath}} + b_2\hat{\boldsymbol{\jmath}}$ with $a_1 > 0$ and $b_2 > 0$. This can always be achieved. Then
$$\boldsymbol{a} \times \boldsymbol{b} = \begin{vmatrix} \hat{\boldsymbol{\imath}} & \hat{\boldsymbol{\jmath}} & \hat{\boldsymbol{k}} \\ a_1 & 0 & 0 \\ b_1 & b_2 & 0 \end{vmatrix} = a_1b_2\hat{\boldsymbol{k}}.$$

Since $a_1b_2 > 0$, the vector product will always be a vector in the direction of the new positive z axis. Since we started with any two vectors $\boldsymbol{a}$ and $\boldsymbol{b}$, we have proved that $\boldsymbol{a} \times \boldsymbol{b}$ has the following properties:

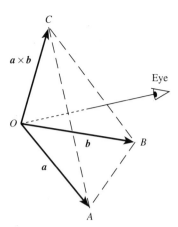

Fig. 9.11
Direction of the vector product.

> **Direction of $a \times b$** (see Fig. 9.11)
>
> (a) $a \times b$ is perpendicular to a and b.
> (b) (Direction of $a \times b$) When the origin O is viewed
> *through the triangle ABC*, where $\overline{OC} = a \times b$, an   **(9.7)**
> anticlockwise circuit of the triangle encounters the
> points A, B, C in order. (In other words, the system
> of vectors $a, b, a \times b$ is right-handed.)

The vector products of the basis vectors of right-handed cartesian
axes (taken in the order $\hat{\imath}, \hat{\jmath}, \hat{k}$) conform with this rule:

> **Vector product of unit basis vectors**
> $$\hat{\jmath} \times \hat{k} = \hat{\imath}, \qquad \hat{k} \times \hat{\imath} = \hat{\jmath}, \qquad \hat{\imath} \times \hat{\jmath} = \hat{k}. \qquad \textbf{(9.8)}$$

4. (*Scalar triple product*) $a \cdot (b \times c) = (a \times b) \cdot c = (c \times a) \cdot b$.
From the definition of the vector product,

$$a \cdot (b \times c) = a_1(b_2 c_3 - b_3 c_2) + a_2(b_3 c_1 - b_1 c_3)$$
$$+ a_3(b_1 c_2 - b_2 c_1)$$

$$= \begin{vmatrix} a_1 & a_2 & a_3 \\ b_1 & b_2 & b_3 \\ c_1 & c_2 & c_3 \end{vmatrix}$$

$$= \begin{vmatrix} c_1 & c_2 & c_3 \\ a_1 & a_2 & a_3 \\ b_1 & b_2 & b_3 \end{vmatrix} \qquad \text{(two applications of rule 3,}$$
$$\text{Section 8.2)}$$

$$= c \cdot (a \times b).$$

A similar shuffling of the rows gives the other result. Note, however,
that the **cyclic order** a, b, c **must be preserved**. If the cyclic order
of the letters is changed, a minus sign must be introduced; for
example

$$a \cdot (c \times b) = -a \cdot (b \times c).$$

$a \cdot (b \times c)$ is a **scalar quantity**, and is called a **scalar triple
product**.

5. *If a, b, c are three nonzero vectors and $a \cdot (b \times c) = 0$, then a, b, c
are coplanar.*

Vectors are said to be coplanar if, when drawn from the same
point, they lie in the same plane. See Fig. 9.12: from Rule 3, b and
c are both perpendicular to $b \times c$. Since $a \cdot (b \times c) = 0$, the vector a
is also perpendicular to $b \times c$. Therefore a, b, c are coplanar, being

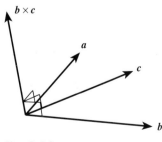

Fig. 9.12

all perpendicular to the same vector. The converse is also true: if
a, b, c are coplanar, then $a \cdot (b \times c) = 0$. This includes the case when
a scalar triple product contains two parallel vectors. We sum up
these properties as follows.

Scalar triple product

$$a \cdot (b \times c) = \begin{vmatrix} a_1 & a_2 & a_3 \\ b_1 & b_2 & b_3 \\ c_1 & c_2 & c_3 \end{vmatrix}.$$

(9.9)

(a) Cyclic permutations:
  $a \cdot (b \times c) = b \cdot (c \times a) = c \cdot (a \times b)$.
  Other permutations: $b \cdot (a \times c) = -a \cdot (b \times c)$ etc.
(b) a, b, c are coplanar if, and only if, $a \cdot (b \times c) = 0$.
(c) $a \cdot (b \times c) = 0$ if any two of the vectors are parallel.

6. (*Vector triple product*). $a \times (b \times c) = (a \cdot c)b - (a \cdot b)c$.
  Translate a, b, c so that they are all drawn from the same point.
Then $a \times (b \times c)$ is perpendicular to $b \times c$, so it lies in the plane
defined by b and c. Therefore there exist constants λ and μ such that

$$a \times (b \times c) = \lambda b + \mu c.$$

But, from (9.9c), $a \cdot [a \times (b \times c)] = 0$, so $a \cdot (\lambda b + \mu c) = 0$, or

$$\lambda a \cdot b + \mu a \cdot c = 0.$$

This equation is satisfied if we put

$$\lambda = ka \cdot c, \qquad \mu = -ka \cdot b,$$

where k is a further constant. Then

$$a \times (b \times c) = k[(a \cdot c)b - (a \cdot b)c].$$

It remains to show that $k = 1$. This result must be true for all $a, b,$
and c, and in particular for the simple choice

$$a = \hat{\imath}, \qquad b = \hat{\jmath}, \qquad c = \hat{\imath} + \hat{k}.$$

Thus

$$a \times (b \times c) = \hat{\imath} \times [\hat{\jmath} \times (\hat{\imath} + \hat{k})] = \hat{\imath} \times (-\hat{k} + \hat{\imath}) = \hat{\jmath}, \qquad (9.9)$$

and

$$(a \cdot c)b - (a \cdot b)c = [\hat{\imath} \cdot (\hat{\imath} + \hat{k})]\hat{\jmath} - (\hat{\imath} + \hat{\jmath})(\hat{\imath} + \hat{k}) = \hat{\jmath}.$$

Hence, comparing these two equations, we conclude that $k = 1$,
which established the result.
  The vector $a \times (b \times c)$ is an example of a **vector triple product**.

7. $|a \times b| = |a||b| \sin \theta$, *where θ is the smaller angle* $(0 < \theta \leqslant 180°)$
between a and b.
  Using previous results

$$|\boldsymbol{a} \times \boldsymbol{b}|^2 = (\boldsymbol{a} \times \boldsymbol{b})\cdot(\boldsymbol{a} \times \boldsymbol{b}) \quad \text{(by 9.5c)}$$

$$= [(\boldsymbol{a} \times \boldsymbol{b}) \times \boldsymbol{a}]\cdot\boldsymbol{b} \quad \text{(by Rule 5)}$$

$$= -[\boldsymbol{a} \times (\boldsymbol{a} \times \boldsymbol{b})]\cdot\boldsymbol{b} \quad \text{(by Rule 1)}$$

$$= -(\boldsymbol{a}\cdot\boldsymbol{b})(\boldsymbol{a}\cdot\boldsymbol{b}) + (\boldsymbol{a}\cdot\boldsymbol{a})(\boldsymbol{b}\cdot\boldsymbol{b}) \quad \text{(by Rule 6)}$$

$$= -|\boldsymbol{a}|^2|\boldsymbol{b}|^2 \cos^2\theta + |\boldsymbol{a}|^2|\boldsymbol{b}|^2 \quad \text{(by 9.4)}$$

$$= |\boldsymbol{a}|^2|\boldsymbol{b}|^2 \sin^2\theta.$$

Hence, since the magnitude of a vector must be nonnegative,

$$|\boldsymbol{a} \times \boldsymbol{b}| = |\boldsymbol{a}||\boldsymbol{b}| \sin\theta, \qquad 0 \leqslant \theta \leqslant \pi.$$

If $\boldsymbol{a} \times \boldsymbol{b} = 0$ and neither $\boldsymbol{a}$ nor $\boldsymbol{b}$ is zero, then the result above implies that the vectors are parallel (either pointing in this same direction or in opposite directions).

8. *The vector product also satisfies the laws*

$$(\boldsymbol{a} + \boldsymbol{b}) \times \boldsymbol{c}, \qquad \lambda(\boldsymbol{a} \times \boldsymbol{b}) = (\lambda\boldsymbol{a}) \times \boldsymbol{b} = \boldsymbol{a} \times (\lambda\boldsymbol{b}).$$

9.7 Direction cosines and ratios

Let the point P have position vector $\boldsymbol{r} = x\hat{\boldsymbol{i}} + y\hat{\boldsymbol{j}} + z\hat{\boldsymbol{k}}$. Let α, β, γ be the angles between this positive vector and the positive directions of the x, y, and z axes (see Fig. 9.13). The scalar product of $\boldsymbol{r}$ with each of the unit vectors $\hat{\boldsymbol{i}}, \hat{\boldsymbol{j}}, \hat{\boldsymbol{k}}$ leads to

$$\boldsymbol{r}\cdot\hat{\boldsymbol{i}} = x = |\boldsymbol{r}||\hat{\boldsymbol{i}}| \cos\alpha = r \cos\alpha,$$

$$\boldsymbol{r}\cdot\hat{\boldsymbol{j}} = y = r \cos\beta,$$

$$\boldsymbol{r}\cdot\hat{\boldsymbol{k}} = z = r \cos\gamma.$$

Put $\cos\alpha = l$, $\cos\beta = m$, $\cos\gamma = n$. Then

$$l = x/r, \qquad m = y/r, \qquad n = z/r,$$

where $r = |\boldsymbol{r}| > 0$. The numbers l, m, n are **the direction cosines of any straight line parallel to the line** $\overline{OP}$ and they specify its three-dimensional inclination. Treated as vector components, however, the vector (l, m, n) specifies a particular direction (in the sense of an arrow) as well: that of the vector $\boldsymbol{r}$. The vector (l, m, n) is a **unit vector**, because

$$l^2 + m^2 + n^2 = \cos^2\alpha + \cos^2\beta + \cos^2\gamma$$

$$= (x/r)^2 + (y/r)^2 + (z/r)^2 = 1. \tag{9.10}$$

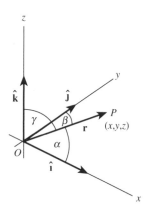

Fig. 9.13
Direction cosines of OP.

Example 9.3. Obtain the direction cosines of the vector $\boldsymbol{a} = \hat{\boldsymbol{i}} + 2\hat{\boldsymbol{j}} - 2\hat{\boldsymbol{k}}$. Find the angles between $\boldsymbol{a}$ and the coordinate axes. The vector $\boldsymbol{a}$ can be written as $(1, 2, -2)$. The length (magnitude) of a is given by $|a| = \sqrt{[1^2 + 2^2 + (-2)^2]} = 3$. Therefore

$$\cos\alpha = \tfrac{1}{3}, \qquad \cos\beta = \tfrac{2}{3}, \qquad \cos\gamma = -\tfrac{2}{3}.$$

The corresponding angles in radians are $\alpha = 1.231,\ \beta = 0.841,$ $\gamma = 2.301$.

Any triple of numbers a_1, a_2, a_3 defines a vector (a_1, a_2, a_3), so they can be used to specify a direction, namely that of any straight line that is parallel to the vector (a_1, a_2, a_3). In terms of coordinate geometry, a_1, a_2, a_3 constitute a set of **direction ratios** for a straight line. Direction ratios do not satisfy (9.10), but they can be replaced by direction cosines by dividing them by the length of the vector (a_1, a_2, a_3). Thus, if the direction ratios are $1, 2, -3$, the corresponding l, m, n are given by

$$(l, m, n) = (1, 2, -3)/\sqrt{[1^2 + 2^2 + (-3)^2]}$$
$$= (1/\sqrt{14}, 2/\sqrt{14}, -3/\sqrt{14}).$$

9.8 Rotation of axes

Frequently in applications, axes need to be rotated about the origin. For example, if an object is rotating about a fixed point, axes fixed in the body will rotate relative to axes fixed in space. Figure 9.14 shows two sets of rectangular axes which have a common origin. Denote the axes by $Oxyz$ and $OXYZ$. Let l_1, m_1, n_1 be the direction cosines of the X axis with respect to the $Oxyz$ axes. If $\alpha_1, \beta_1, \gamma_1$ are the angles between OX and Ox, Oy, Oz respectively, then

$$l_1 = \cos\alpha_1, \qquad m_1 = \cos\beta_1, \qquad n_1 = \cos\gamma_1.$$

Let $\hat{\imath}, \hat{\jmath}, \hat{k}$ be the usual unit vectors along the axes Ox, Oy, Oz, and let $\hat{I}, \hat{J}, \hat{K}$ be unit vectors along the axes OX, OY, OZ. The coordinates of a point with position vector $\hat{I}$ are (l_1, m_1, n_1) in the $Oxyz$ coordinates. Hence

$$\hat{I} = l_1\hat{\imath} + m_1\hat{\jmath} + n_1\hat{k}.$$

Let (l_2, m_2, n_2) and (l_3, m_3, n_3) be the direction cosines of the axes OX and OZ, and let $\alpha_2, \beta_2, \gamma_2$ and $\alpha_3, \beta_3, \gamma_3$ be the angles between OY and Ox, Oy, Oz, and OZ and Ox, Oy, Oz respectively. Then

$$\hat{J} = l_2\hat{\imath} + m_2\hat{\jmath} + n_2\hat{k}, \qquad \hat{K} = l_3\hat{\imath} + m_3\hat{\jmath} + n_3\hat{k}.$$

Since $\hat{J}\cdot\hat{K} = \hat{K}\cdot\hat{I} = \hat{I}\cdot\hat{J} = 0$, the direction cosines must satisfy the relations

$$l_2l_3 + m_2m_3 + n_2n_3 = 0,$$
$$l_3l_1 + m_3m_1 + n_3n_1 = 0,$$
$$l_1l_2 + m_1m_2 + n_1n_2 = 0.$$

Inverting the view of the two sets of axes, we can see that the direction cosines of Ox, Oy, Oz with respect to the $OXYZ$ frame must be

$$(l_1, l_2, l_3), \quad (m_1, m_2, m_3), \quad (n_1, n_2, n_3).$$

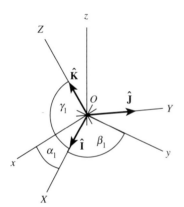

Fig. 9.14
Rotation of axes.

Hence

$$\left. \begin{aligned} \hat{\imath} &= l_1 \hat{I} + l_2 \hat{J} + l_3 \hat{K}, \\ \hat{\jmath} &= m_1 \hat{I} + m_2 \hat{J} + m_3 \hat{K}, \\ \hat{k} &= n_1 \hat{I} + n_2 \hat{J} + n_3 \hat{K}. \end{aligned} \right\} \tag{9.11}$$

Suppose that a **fixed point** has position vector

$$r = x\hat{\imath} + y\hat{\jmath} + z\hat{k}$$

in the first set of axes, or **frame of reference**, and

$$R = X\hat{I} + Y\hat{J} + Z\hat{K},$$

in the second. Then

$$\begin{aligned} r &= x\hat{\imath} + y\hat{\jmath} + z\hat{k} \\ &= x(l_1\hat{I} + l_2\hat{J} + l_3\hat{K}) + y(m_1\hat{I} + m_2\hat{J} + m_3\hat{K}) \\ &\quad + z(n_1\hat{I} + n_2\hat{J} + n_3\hat{K}) \\ &= (xl_1 + ym_1 + zn_1)\hat{I} + (xl_2 + ym_2 + zn_2)\hat{J} \\ &\quad + (xl_3 + ym_3 + zn_3)\hat{K}, \tag{9.12} \end{aligned}$$

using (9.11). Comparison of (9.11) and (9.12) implies that

$$\begin{aligned} X &= l_1 x + m_1 y + n_1 z, \\ Y &= l_2 x + m_2 y + n_2 z, \\ Z &= l_3 x + m_2 y + n_3 z, \end{aligned}$$

or, in matrix form

$$R = Pr, \tag{9.13}$$

where

$$P = \begin{bmatrix} l_1 & m_1 & n_1 \\ l_2 & m_2 & n_2 \\ l_3 & m_3 & n_3 \end{bmatrix}.$$

The matrix P of direction cosines in an example of an **orthogonal matrix**. We shall say more about orthogonal matrices in Section 11.8.

Example 9.4. Show that axes with direction cosines

$$\begin{aligned} l_1 &= \tfrac{1}{3}, & m_1 &= -\tfrac{2}{3}, & n_1 &= \tfrac{2}{3}, \\ l_2 &= -\tfrac{2}{3}, & m_2 &= \tfrac{1}{3}, & n_2 &= \tfrac{2}{3}, \\ l_3 &= \tfrac{2}{3}, & m_3 &= \tfrac{2}{3}, & n_3 &= \tfrac{1}{3}, \end{aligned}$$

form a rectangular cartesian set.

For $i = 1, 2, 3$ we need to confirm that

$$l_i^2 + m_i^2 + n_i^2 = 1,$$

and that, for $i, j = 1, 2, 3$ and $i \neq j$,

$$l_i l_j + m_i m_j + n_i n_j = 0.$$

One condition will be checked in each case: the others should be checked by the reader. Thus

$$l_1^2 + m_1^2 + n_1^2 = (\tfrac{1}{3})^2 + (-\tfrac{2}{3})^2 + (\tfrac{2}{3})^2 = 1,$$

and

$$l_2 l_3 + m_2 m_3 + n_2 n_3 = (-\tfrac{2}{3})(\tfrac{2}{3}) + (\tfrac{1}{3})(\tfrac{2}{3}) + (\tfrac{2}{3})(\tfrac{1}{3}) = 0.$$

9.9 Vector equation of the straight line

Consider the problem of finding the equation of the straight line which passes through a given point A and points in a given direction. Let the given point A have position vector $\boldsymbol{a} = a_1\hat{\boldsymbol{i}} + a_2\hat{\boldsymbol{j}} + a_3\hat{\boldsymbol{k}}$, and let the given direction be specified by the vector $\boldsymbol{p} = p_1\hat{\boldsymbol{i}} + p_2\hat{\boldsymbol{j}} + p_3\hat{\boldsymbol{k}}$ (Fig. 9.15). Consider a general point P with position vector $\boldsymbol{r} = x\hat{\boldsymbol{i}} + y\hat{\boldsymbol{j}} + z\hat{\boldsymbol{k}}$ which lies on the line. By the triangle law, $\overline{AP} = \boldsymbol{r} - \boldsymbol{a}$, and this vector must be parallel to $\boldsymbol{p}$ for every point P. Hence

$$\boldsymbol{r} - \boldsymbol{a} = \mu\boldsymbol{p}, \quad \text{or} \quad \boldsymbol{r} = \boldsymbol{a} + \mu\boldsymbol{p}.$$

As we allow μ to run between $\pm\infty$, the point P traces out the line. The varying constant μ is called a **parameter** for the equation, and the equation is said to be a **parametric vector equation** of the line.

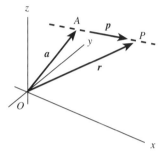

Fig. 9.15
Straight line through $\boldsymbol{a}$ in the direction $\boldsymbol{p}$.

> **Parametric vector equation for the straight line through $\boldsymbol{a}$ parallel to $\boldsymbol{p}$.**   (9.14)
>
> $$\boldsymbol{r} = \boldsymbol{a} + \mu\boldsymbol{p} \quad (-\infty < \mu < \infty).$$

In component form, this becomes

$$x = a_1 + \mu p_1, \qquad y = a_2 + \mu p_2, \qquad z = a_3 + \mu p_3.$$

Since $(x - a_1)/p_1 = \mu$, and similarly for the other two equations, we have

> **Cartesian equations for the straight line through (a_1, a_2, a_3) with direction ratios (p_1, p_2, p_3).**   (9.15)
>
> $$\frac{x - a_1}{p_1} = \frac{y - a_2}{p_2} = \frac{z - a_3}{p_3}.$$

The interpretation of the equations in (9.11) is that, in effect, the line is expressed as the intersection of two planes; for example, the plane $(x - a_1)/p_1 = (y - a_2)/p_2$ and the plane $(y - a_2)/p_2 = (z - a_3)/p_3$. (The general equation for a plane is $ax + by + cz = d$: see (9.17) in the next section.)

Example 9.5. Find the equation of the line through $(1, 2, -2)$ in the direction $\hat{\boldsymbol{i}} - \hat{\boldsymbol{j}} + 3\hat{\boldsymbol{k}}$. Where does the line cut the (x, y) plane?

In this example, $a = \hat{i} + 2\hat{j} - 2\hat{k}$ and $p = \hat{i} - \hat{j} + 3\hat{k}$. Hence the line has the equation

$$r = a + \mu p = (\hat{i} + 2\hat{j} - 3\hat{k}) + \mu(\hat{i} - \hat{j} + 3\hat{k})$$
$$= (1 + \mu)\hat{i} + (2 - \mu)\hat{j} + (-3 + 3\mu)\hat{k} \qquad \text{(i)}$$

or

$$x = 1 + \mu, \qquad y = 2 - \mu, \qquad z = -3 + 3\mu,$$

or

$$\frac{x - 1}{1} = \frac{y - 2}{-1} = \frac{z + 3}{3} \quad (=\mu).$$

The line meets the (x, y) plane where the $\hat{k}$ component in (i) is zero. Hence $-3 + 3\mu = 0$ or $\mu = 1$. The position vector of this point is $r = (1 + 1)\hat{i} + (2 - 1)\hat{j} = 2\hat{i} + \hat{j}$.

Example 9.6. Find the equations of the two lines L_1 and L_2 which pass through (a) $a_1 = 2\hat{i} - \hat{j} + \hat{k}$ and $a_2 = 5\hat{j} - 7\hat{k}$; (b) $b_1 = \hat{i} - \hat{j} + \hat{k}$ and $b_2 = \hat{i} - 4\hat{j} + 5\hat{k}$. Show that the lines intersect.

(a) A vector in the direction joining the points a_1 and a_2 is

$$p_1 = a_2 - a_1 = (0 - 2)\hat{i} + [5 - (-1)]\hat{j} + (-7 - 1)\hat{k}$$
$$= -2\hat{i} + 6\hat{j} - 8\hat{k}.$$

The line L_1 has the equation

$$r = a_1 + \mu_1 p_1$$
$$= (2\hat{i} - \hat{j} + \hat{k}) + \mu_1(-2\hat{i} + 6\hat{j} - 8\hat{k})$$
$$= (2 - 2\mu_1)\hat{i} + (-1 + 6\mu_1)\hat{j} + (1 - 8\mu_1)\hat{k},$$

or

$$x = 2 - 2\mu_1, \qquad y = -1 + 6\mu_1, \qquad z = 1 - 8\mu_1. \qquad \text{(i)}$$

(b) In this case,

$$p_2 = b_2 - b_1 = (1 - 1)\hat{i} + (-4 - (-1))\hat{j} + (5 - 1)\hat{k}$$
$$= -3\hat{j} + 4\hat{k}.$$

This line L_2 has the equation

$$r = b_1 + \mu_2 p_2 = \hat{i} + (-1 - 3\mu_2)\hat{j} + (1 + 4\mu_2)\hat{k}, \qquad \text{(ii)}$$

where μ_2 is another parameter, or

$$x = 1, \qquad y = -1 - 3\mu_2, \qquad z = 1 + 4\mu_2.$$

The two lines will intersect if (i) and (ii) have at least one point in common. This will be the case if

$$2 - 2\mu_1 = 1, \qquad -1 + 6\mu_1 = -1 - 3\mu_2, \qquad 1 - 8\mu_1 = 1 + 4\mu_2,$$

are consistent. We can check that all are satisfied if $\mu_1 = \frac{1}{2}$ and $\mu_2 = -1$. The point of intersection is $(1, 2, -3)$.

9.10 Vector equation of the plane

Let us now find the equation of the plane which passes through a given point A and is perpendicular to a given direction. Let $a = a_1\hat{i} + a_2\hat{j} + a_3\hat{k}$ be the position vector of A and $p = p_1\hat{i} + p_2\hat{j} + p_3\hat{k}$ the direction of the perpendicular, or **normal**, to the

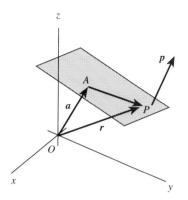

Fig. 9.16
Plane through point A with its normal in the direction p.

plane (Fig. 9.16). Let $r = x\hat{\imath} + y\hat{\jmath} + z\hat{k}$ be the position vector of a general point P. If P lies in the plane, then $\overline{AP}$ must always be perpendicular to p. Now $\overline{AP} = r - a$, by the triangle law, and $r - a$ and p are perpendicular if their scalar product is zero.

The vector equation of the plane

A general point r on the plane through the point a with normal parallel to p satisfies

$$(r - a)\cdot p = 0.$$

(9.16)

In component form, (9.16) becomes

$$[(x - a_1)\hat{\imath} + (y - a_2)\hat{\jmath} + (z - a_3)\hat{k}]\cdot[p_1\hat{\imath} + p_2\hat{\jmath} + p_3\hat{k}] = 0,$$

or $\quad (x - a_1)p_1 + (y - a_2)p_2 + (z - a_3)p_3 = 0.$

Therefore the cartesian equation of the plane is

$$p_1 x + p_2 y + p_3 z = a_1 p_1 + a_2 p_2 + a_3 p_3.$$

Note that the coefficients of x, y, z identify the direction of the normal, p. Since a and p are arbitrary and this covers all planes, we have the following.

General cartesian equation for a plane

(a) The general equation of a plane is

$$ax + by + cz = d.$$

(9.17)

(b) The line with direction ratios a, b, c is perpendicular, or normal, to the plane.

Example 9.7. Find a unit vector normal to the plane

$$x - 2y + 2z = 4.$$

Also find the parametric equation of a line through the origin which is perpendicular to the plane.

The coefficients $(1, -2, 2)$ of x, y, z define a vector perpendicular to the plane. Hence a unit vector in this direction is

$$\hat{n} = (\tfrac{1}{3}, -\tfrac{2}{3}, \tfrac{2}{3})$$

is also a unit normal but in the opposite direction.

The required line passes through $r = 0$, and lies in the direction $\hat{n}$. Hence its equation is

$$r = \mu\hat{n} = \mu(\tfrac{1}{3}\hat{\imath} - \tfrac{2}{3}\hat{\jmath} + \tfrac{2}{3}\hat{k}),$$

or $\quad x = \tfrac{1}{3}\mu, \qquad y = -\tfrac{2}{3}\mu, \qquad z = \tfrac{2}{3}\mu.$

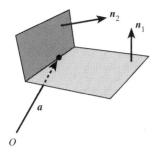

Fig. 9.17

Example 9.8. Two planes pass through the point with position vector a and have unit normals $\hat{n}_1$ and $\hat{n}_2$. Find the equation of their line of intersection (see Fig. 9.17).

Let p be a vector in the direction of the line of intersection. Then p must be perpendicular to both $\hat{n}_1$ and $\hat{n}_2$. Let $p = \hat{n}_1 \times \hat{n}_2$ (which must be a non-zero vector since the planes are distinct). Hence the vector equation of the line is (see 9.14)

$$r = a + \mu(\hat{n}_1 \times \hat{n}_2).$$

Example 9.9. Find the equation of the plane through three non-collinear points A, B, and C which have position vectors a, b, and c respectively (see Fig. 9.18).

By the triangle law, $\overline{AB} = b - a$ and $\overline{AC} = c - a$. A normal n to the plane will be in the direction of the vector product of these two vectors. Let

$$n = (b - a) \times (c - a),$$

$$= b \times c - a \times c - b \times a + a \times a,$$

$$= a \times b + b \times c + c \times a. \quad \text{(Rules 1–2 of Section 9.6)}$$

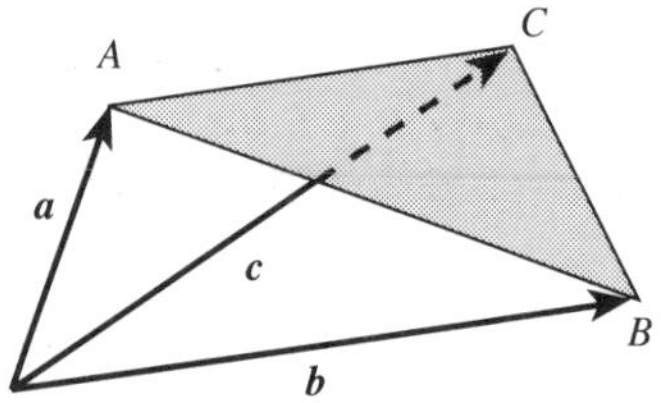

Hence, the vector equation of the plane can be written

$$(r - a) \cdot (a \times b + b \times c + c \times a) = 0.$$

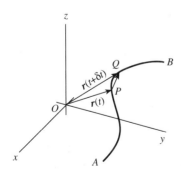

Fig. 9.18

9.11 Differentiation of vectors

Suppose that the position vector r is a function of a parameter t, namely, $r(t)$. If t lies in the interval $a \leqslant t \leqslant b$, then the corresponding position vector traces out a **curve** C in space (Fig. 9.19) from point A to point B where $t = a$ and $t = b$. Each component of r is a function of t, namely

$$r(t) = x(t)\hat{\imath} + y(t)\hat{\jmath} + z(t)\hat{k}.$$

If each component is continuous in t, then the curve will be connected (that is, in one piece). If each component also has a derivative with respect to t for $a < t < b$, then the curve will be smooth.

Denote the current location of the position vector by P on C, and consider a point Q with position vector $r(t + \delta t)$ close to P. The chord $\overline{PQ}$ may be represented as $r(t + \delta t) - r(t)$ by the triangle law; also, as $\delta t \to 0$, the direction of $\overline{PQ}$ approaches the tangent to the

Fig. 9.19

curve at P. The limit

$$\frac{\mathrm{d}r}{\mathrm{d}t} = \lim_{\delta t \to 0} \frac{r(t + \delta t) - r(t)}{\delta t}$$

$$= \lim_{\delta t \to 0} \frac{[x(t+\delta t) - x(t)]\hat{\imath} + [y(t+\delta t) - y(t)]\hat{\jmath} + [z(t+\delta t) - z(t)]\hat{k}}{\delta t}$$

defines the **derivative of the position vector**. It is itself a vector, parallel to the tangent at P in the direction of increasing t.

The derivative of $r(t)$ on the curve

$$r(t) = (x(t), y(t), z(t))$$

$$\frac{\mathrm{d}r}{\mathrm{d}t} = \frac{\mathrm{d}x}{\mathrm{d}t}\hat{\imath} + \frac{\mathrm{d}y}{\mathrm{d}t}\hat{\jmath} + \frac{\mathrm{d}z}{\mathrm{d}t}\hat{k} \qquad\qquad (9.18)$$

points in the direction of the tangent for increasing t.

If $r(t)$ represents the position vector of a particle moving in space and the parameter t is **time**, then $v = \dot{r} = \mathrm{d}r/\mathrm{d}t$ is the **velocity** of the particle. Its direction, being tangential to the curve, defines instantaneously the direction in which the particle is moving.

The second derivative $f = \ddot{r} = \mathrm{d}^2r/\mathrm{d}t^2$ of the position vector defines the **acceleration** of the particle.

Example 9.10. The position vector of a particle is given by

$$r = (a \cos \omega t)\hat{\imath} + (a \sin \omega t)\hat{\jmath}$$

where a (>0) and ω are constants. Find the velocity v and acceleration f, of the particle.

The velocity and acceleration are given by

$$v = \frac{\mathrm{d}r}{\mathrm{d}t} = -(a\omega \sin \omega t)\hat{\imath} + (a\omega \cos \omega t)\hat{\jmath},$$

$$f = \frac{\mathrm{d}v}{\mathrm{d}t} = -(a\omega^2 \cos \omega t)\hat{\imath} - (a\omega^2 \sin \omega t)\hat{\jmath} = -\omega^2 r.$$

Their directions are shown in Fig. 9.20. Finally note that the velocity v and acceleration f have constant magnitudes, and that $v \cdot f = 0$, so that v and f are perpendicular.

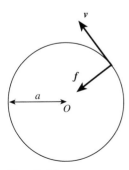

Fig. 9.20

The usual rules for differentiation still apply to addition, and to scalar and vector products of two or more vectors. Let $a(t)$, $b(t)$, and $c(t)$ be vector functions of t, and let $f(t)$ represent a **scalar function**. The important rules are as follows.

1. $\dfrac{d}{dt}[a(t) + b(t)] = \dfrac{da(t)}{dt} + \dfrac{db(t)}{dt};$

2. $\dfrac{d}{dt}[f(t)a(t)] = \dfrac{df(t)}{dt}a(t) + f(t)\dfrac{da(t)}{dt};$

3. $\dfrac{d}{dt}[a(t) \cdot b(t)] = \dfrac{da(t)}{dt} \cdot b(t) + a(t) \cdot \dfrac{db(t)}{dt};$

4. $\dfrac{d}{dt}[a(t) \times b(t)] = \dfrac{da(t)}{dt} \times b(t) + a(t) \times \dfrac{db(t)}{dt}.$

The order of the products in 4 is significant.

Suppose now that, in Fig. 9.19, the parameter t is the **arc length** s measured from some point on the curve, so that $r = r(s)$ defines C. In this case dr/ds is a **unit** tangent vector denoted by $\hat{t}$, because $|r(t + \delta t) - r(t)| = |\delta r| = \delta s$. Since $|\hat{t}| = 1$,

$$\hat{t} \cdot \hat{t} = 1,$$

so, by rule 4 above,

$$\frac{d\hat{t}}{ds} \cdot \hat{t} + \hat{t} \cdot \frac{d\hat{t}}{ds} = 0,$$

so that

$$\hat{t} \cdot \frac{d\hat{t}}{ds} = 0.$$

Hence $d\hat{t}/ds$ is perpendicular to the unit tangent $\hat{t}$ in the direction of increasing s, and consequently is normal to the curve. If $\hat{n}$ is a *unit* normal in this direction, then

$$\frac{d\hat{t}}{ds} = \kappa\hat{n}, \quad \text{where} \quad \kappa = \left|\frac{d\hat{t}}{ds}\right|;$$

κ is called the **curvature** of the curve at the point. Its reciprocal $\rho = 1/\kappa$ is called the **radius of curvature.**

9.12 Plane vectors

In many applications, the behaviour of the system is confined to a plane, as in the sliding of a ball on a table top, a projectile moving under gravity, or a satellite orbiting the sun. It can be arranged that the plane coincides with the (x, y) plane, so that any position vector has only two components, as in

$$r = x\hat{i} + y\hat{j}.$$

We can still think of these vectors as existing in a three-dimensional space. The vector product of two plane vectors $a = a_1\hat{i} + a_2\hat{j}$ and

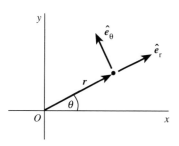

Fig. 9.21
Polar unit vectors.

$b = b_1\hat{\imath} + b_2\hat{\jmath}$ is given by

$$c = \begin{vmatrix} \hat{\imath} & \hat{\jmath} & \hat{k} \\ a_1 & a_2 & 0 \\ b_1 & b_2 & 0 \end{vmatrix} = \hat{k}(a_1 b_2 - a_2 b_1),$$

which is always a vector perpendicular to the plane.

For some applications, it is more convenient to use a different set of unit vectors instead of $\hat{\imath}$ and $\hat{\jmath}$. If polar coordinates (r, θ) are appropriate to the geometry of the application, then we can associate with these coordinates unit vectors $\hat{e}_r$ and $\hat{e}_\theta$ (Fig. 9.21). Their directions are respectively in the direction of the position vector r (such that θ is constant) and perpendicular to that direction (such that r is constant). While the vectors $\hat{e}_r$ and $\hat{e}_\theta$ have unit magnitudes, their directions vary from point to point. They can be related to the unit vectors $\hat{\imath}$ and $\hat{\jmath}$ by the triangles shown in Fig. 9.22. By the triangle law of addition,

$$\hat{e}_r = \hat{\imath} \cos\theta + \hat{\jmath} \sin\theta, \qquad \hat{e}_\theta = -\hat{\imath} \sin\theta + \hat{\jmath} \cos\theta.$$

These unit vectors are functions of the angle θ. Their rates of change with respect to θ are

$$\frac{d\hat{e}_r}{d\theta} = -\hat{\imath} \sin\theta + \hat{\jmath} \cos\theta = \hat{e}_\theta, \tag{9.16}$$

$$\frac{d\hat{e}_\theta}{d\theta} = -\hat{\imath} \cos\theta - \hat{\jmath} \sin\theta = -\hat{e}_r. \tag{9.17}$$

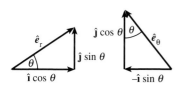

Fig. 9.22

Example 9.11. A particle moves in a plane in such a way that its polar coordinates are given by $r = r(t)$ and $\theta = \theta(t)$ in terms of the time t. Find the polar components of its velocity and acceleration.

The position vector of the particle is

$$r = r\hat{e}_r.$$

Its velocity is

$$\frac{dr}{dt} = \frac{d(r\hat{e}_r)}{dt} = \frac{dr}{dt}\hat{e}_r + r\frac{d\hat{e}_r}{dt},$$

$$= \dot{r}\hat{e}_r + r\frac{d\hat{e}_r}{d\theta}\frac{d\theta}{dt},$$

$$= \dot{r}\hat{e}_r + r\dot{\theta}\hat{e}_\theta,$$

by (9.16). The **radial** component of the velocity is $\dot{r}$ and its **transverse** component is $r\dot{\theta}$.

The acceleration of the particle is

$$\frac{d^2 r}{dt^2} = \frac{d}{dt}(\dot{r}\hat{e}_r + r\dot{\theta}\hat{e}_\theta),$$

$$= \ddot{r}\hat{e}_r + \dot{r}\frac{d\hat{e}_r}{dt} + \frac{d(r\dot{\theta})}{dt}\hat{e}_\theta + r\dot{\theta}\frac{d\hat{e}_\theta}{dt},$$

$$= \ddot{r}\hat{e}_r + \dot{r}\dot{\theta}\hat{e}_\theta + \frac{d(r\dot{\theta})}{dt}\hat{e}_\theta - r\dot{\theta}^2\hat{e}_r,$$

$$= (\ddot{r} - r\dot{\theta}^2)\hat{e}_r + \left(\dot{r}\dot{\theta} + \frac{d(r\dot{\theta})}{dt}\right)\hat{e}_\theta,$$

$$= (\ddot{r} - r\dot{\theta}^2)\hat{e}_r + (2\dot{r}\dot{\theta} + r\ddot{\theta})\hat{e}_\theta,$$

$$= (\ddot{r} - r\dot{\theta}^2)\hat{e}_r + \frac{1}{r}\frac{d(r^2\dot{\theta})}{dt}\hat{e}_\theta.$$

Hence its radial component is $\ddot{r} - r\dot{\theta}^2$ whilst its transverse component is

$$\frac{1}{r}\frac{d(r^2\dot{\theta})}{dt}.$$

Problems, Chapter 9

9.1. State the distances between the following pairs of points. (a) $(0, 0, 0)$, $(1, 2, 3)$; (b) $(1, 2, 3)$, $(3, 2, 1)$; (c) $(1, 0, -1)$, $(-1, 1, 0)$.

9.2. Find $2a$, $3b$, and $2a - 3b$ when
(a) $a = (1, 2, 1)$, $b = (2, 1, 2)$;
(b) $a = (3, 2, 3)$, $b = (1, 1, 2)$;
(c) $a = (6, 3, 1)$, $b = (4, 2, 1)$.
How do you recognize that $2a - 3b$ is parallel to the x, y plane in (b), and parallel to the z axis in (c)?

9.3. r is the position vector $(2, 3, 1)$ and $a = (1, 1, 2)$ is a general vector. R is a position vector given by $R = r + a$. What are the coordinates of the terminal point of R? Make a sketch illustrating your result.

9.4. Write down a unit vector $\hat{}$ which has the same direction as the vector a, for the following vectors a.
(a) $(1, 1, -2)$; (b) $(-1, -1, 2)$; (c) $(1, -2, 1)$; (d) $3\hat{i} - \hat{j} + \hat{k}$; (e) $\hat{i} + \hat{j} + \hat{k}$.

9.5. State the projections on to the three axes of the directed line segment $\overline{PQ}$ when P is the point $(1, 2, 1)$ and Q is $(2, 2, 3)$.

9.6. Draw a diagram to show that if A, B, and C are any given points, then $\overline{AB} + \overline{BC} + \overline{CA} = 0$. Formulate a similar result for any number of points.

9.7. Draw a diagram to show that if A, B, C, D are four given points, then $\overline{CD} = \overline{CB} + \overline{BA} + \overline{AD}$.

9.8. The points P: $(1, 1, 0)$, Q: $(1, 1, 1)$, and S: $(1, 2, 1)$ are vertices of a parallelogram with sides PQ and PS. Find the position vector of the other vertex by vector addition, and find the mid-point of the parallelogram.

9.9. Two points A and B have position vectors a and b, respectively. Find the position vector of the point P which divides AB internally in the ratio $AP/AB = \lambda$, where $0 < \lambda < 1$.

9.10. Show that the line joining the points with position vectors $\hat{i} - \hat{j} + 2\hat{k}$ and $2\hat{i} - 2\hat{j} - 3\hat{k}$ intersects the z axis.

9.11. A set of position vectors given by $r = a\hat{i} + b\hat{j} + c\hat{k}$ satisfies the inequality $|a| + |b| + |c| \leqslant 1$. Describe the shape of the region which includes all the points with these position vectors.

9.12. An aircraft flying at a constant speed V is circling horizontally at height h above an airfield which lies in the (x, y) plane with its runway occupying the strip $|x| \leqslant k$, $|y| \leqslant l$. The centre of its circling path is at $x = k$, $y = 0$; at time $t = 0$, the aircraft is at $(k + r, 0, h)$. Construct a position vector for the path of the aircraft if its direction is counterclockwise when viewed from above.

9.13. Find a unit vector (a) in the same direction as $a = 7\hat{\imath} + 8\hat{\jmath} - 5\hat{k}$, (b) in the direction opposite to a.

9.14. Determine the vectors which have start and end points given respectively by the pairs of points with position vectors given by
(a) $\hat{\imath} + \hat{\jmath} + \hat{k}$ and $-2\hat{\imath} + 3\hat{\jmath} + 5\hat{k}$;
(b) $\hat{\imath} + 2\hat{\jmath} - \hat{k}$ and $3\hat{\imath} - \hat{\jmath} - 2\hat{k}$;
and find the length of the vector in each case.

9.15. Given $a = \hat{\imath} + 2\hat{\jmath} - \hat{k}$ and $b = \hat{\imath} + 3\hat{\jmath} + \hat{k}$, find the following scalar and vector products:
(a) $a \cdot b$;   (b) $(a - b) \cdot (a + b)$;   (c) $a \times b$;
(d) $a \cdot (a \times b)$;   (e) $a \times (a \times b)$;   (f) $(a - b) \times (a + b)$.

9.16. Let $a = \hat{\imath} + \hat{\jmath} - \hat{k}$ and $b = 2\hat{\imath} - \hat{\jmath} + 2\hat{k}$. Find the angle between a and b. Find a vector which is perpendicular to both a and b.

9.17. Find the value of λ such that the vectors

$$a = \lambda\hat{\imath} + 2\hat{\jmath} - \hat{k} \quad \text{and} \quad b = \hat{\imath} + \hat{\jmath} - 3\lambda\hat{k}$$

are mutually perpendicular.

9.18. Determine numbers α, β, and γ which ensure that the vectors

$$a = \alpha\hat{\imath} + 2\hat{\jmath} - 3\hat{k},$$
$$b = -\hat{\imath} + 2\beta\hat{\jmath} + 2\hat{k},$$
$$c = 2\hat{\imath} + \hat{\jmath} - 3\gamma\hat{k}$$

are mutally perpendicular.

9.19. The vertices of the corners of a triangle are given by the position vectors a, b, and c. Show that the area of the triangle is given by

$$\tfrac{1}{2}|b \times c + c \times a + a \times b|.$$

A second triangle has vertices with position vectors $a + \lambda(b - c)$, b, and c, where λ is a scalar. Show that the areas of the two triangles are the same. What simple geometric result does the equality exhibit?
Find the area of the triangle formed by the points

$$a = \hat{\imath} - 2\hat{\jmath} - \hat{k},$$
$$b = \hat{\imath} - \hat{\jmath} + 2\hat{k},$$
$$c = \hat{\imath} + 2\hat{\jmath} - \hat{k}.$$

9.20. If $a = \hat{\imath} + 2\hat{\jmath} + \hat{k}$, $b = -\hat{\imath} + \hat{\jmath} - 2\hat{k}$, and $c = 2\hat{\imath} - \hat{\jmath} + 3\hat{k}$, find the following multiple products:
(a) $a \cdot (b \times c)$;   (b) $(a \times b) \cdot c$;   (c) $a \times (b \times c)$;
(d) $(a \times b) \times c$;   (e) $(a \times b) \cdot (b \times c)$;
(f) $(a \times b) \times (b \times c)$.

9.21. Find the direction cosines of the vector $\hat{\imath} - 2\hat{\jmath} + 2\hat{k}$.

9.22. A particle describes a plane path with position vector

$$r = (a \cos \omega t)\hat{\imath} + (b \sin \omega t)\hat{\jmath}.$$

Show that
(a) the path is the ellipse

$$\frac{x^2}{a^2} + \frac{y^2}{b^2} = 1,$$

(b) the acceleration of the particle is directed towards the origin.

9.23. The position vector of a particle is given by

$$r = (a \cos \omega t \sin vt)\hat{\imath} + (a \sin \omega t \sin vt)\hat{\jmath}$$
$$+ (a \cos vt)\hat{k},$$

where a, ω, and v are constants. Show that the particle moves on a sphere of radius a. Find the velocity of the particle, and show that its magnitude at any time is

$$a(v^2 + \omega^2 \sin^2 vt)^{\frac{1}{2}}.$$

Deduce that the minimum speed must occur at the highest and lowest points of the sphere. Where does the maximum speed occur?

9.24. A rigid body rotates about a point of itself with **angular velocity** ω (what this means is that at any instant, the body is rotating at a rate $|\omega|$ about the line in the instantaneous direction of the vector ω). The velocity of a point of the body with position vector $r = x\hat{\imath} + y\hat{\jmath} + z\hat{k}$ is given by

$$v = \omega \times r.$$

Explain why every point of the body is moving instantaneously perpendicular to ω. Find the matrix S such that $v = S\omega$. Show that the square of the speed of the point is

$$|v| = \omega^T S^T S \omega,$$

where

$$S^T S = \begin{bmatrix} y^2 + z^2 & -xy & -zx \\ -xy & x^2 + z^2 & -yz \\ -zx & -yz & y^2 + x^2 \end{bmatrix}.$$

9.25. The vector moment M of a force $F = F_x\hat{\imath} + F_y\hat{\jmath} + F_z\hat{k}$ about the origin O is defined as $M = r \times F$, where $r = x\hat{\imath} + y\hat{\jmath} + z\hat{k}$ is the position vector of the point where F acts. Find the components of M.
A force of magnitude 4 acts at the point $(1, -1, 2)$ in the direction $\hat{\imath} + 2\hat{\jmath} - 2\hat{k}$. Find the vector moment of the force about (a) the origin, (b) the point $(-2, 1, 2)$.

9.26. If two moving points A and B have position vectors $r_A(t)$ and $r_B(t)$, then the position vector of B relative to A is $r_A(t) - r_B(t)$, and the velocity of B relative to A is $\dot{r}_A(t) - \dot{r}_B(t)$. The two points are $A: (t, -t^2, t)$ and $B: (t^3, 2t^2, 1 + 3t)$. Find the velocity of B relative to A and the velocity of A relative to B. At what time is the relative speed between the points a minimum?

9.27. A cyclist rides due north along a straight road at 10 km/hr. The wind appears to come from the east. If she increases her speed to 20 km/hr, then the wind appears to blow from the north-east. Represent the wind by the vector w. Using relative velocities determine the speed and direction of the wind.

9.28. Two aircraft are flying in straight lines with flight paths given by the vectors $0.41\hat{\imath} + 148t\hat{\jmath} + 0.99\hat{k}$ and $100t\hat{\imath} + 250t\hat{\jmath} + 250t\hat{k}$, where distances are measured in kilometres and the time t is in hours. The second aircraft takes off from an airfield at O at time $t = 0$. Show that a near-miss between the aircraft will occur: determine the time at which this happens.

9.29. The centre of mass of n particles with masses m_i located at points with position vectors r_i ($i = 1, 2, \ldots, n$) is given by

$$\bar{r} = \sum_{i=1}^{n} m_i r_i.$$

Find the centre of mass of three particles of masses 1 kg, 2 kg, and 3 kg located at the points with position vectors $\hat{\imath} + \hat{\jmath} + 2\hat{k}$, $-2\hat{\imath} + 3\hat{\jmath} - 5\hat{k}$, and $3\hat{\jmath} + 2\hat{k}$.

9.30. The moment of momentum h of a particle of mass m located at the point with position vector r is given by

$$h = mr \times v,$$

where $v = \dot{r}$ is the velocity of the particle. Show that

$$\dot{h} = M$$

(see Problem 9.17). In other words, the moment of the force about the origin equals the rate of change of the moment of momentum.

9.31. The position vector of a point on a circular helix is given by

$$r = a[(\cos t)\hat{\imath} + (\sin t)\hat{\jmath} + (t \cot \alpha)\hat{k}].$$

Find the unit tangent vector to the helix at any point. Show that the curvature at any point is

$$\kappa = (1/a) \sin^2 \alpha.$$

Find the unit normal, and show that it always passes through the z axis.

9.32. The position vector of a particle is given by

$$r = \sec t, \qquad \theta = t,$$

in polar coordinates. Draw a sketch of the path of the particle for $0 \leqslant t < \frac{1}{2}\pi$. Find the radial and transverse components of velocity and acceleration.

9.33. Express the vector $2\hat{\imath} + 3\hat{\jmath} - \hat{k}$ in the form $c + d$, where c lies in the plane of $\hat{\imath} - \hat{\jmath} - \hat{k}$ and $2\hat{\imath} - \hat{k}$, and d is perpendicular to this plane.

9.34. Give the vector equation of the line through $(1, 1, 1)$ in the direction $\hat{\imath} + 2\hat{\jmath} - \hat{k}$. Find the intersection of this line with the plane $x - y + z = -2$.

9.35. Let A, B, and C have coordinates $(0, -1, 3)$, $(1, 0, 3)$, and $(0, 0, 5)$ respectively. Let P_1 be the plane through A with normal in the direction $-\hat{\jmath} + \hat{k}$, and P_1 be the plane through A, B, and C.
(a) Find equations of P_1 and P_2, and determine the angle between them.
(b) Determine the perpendicular distance from the origin O to P_1.
(c) Find an equation of the line L in which P_1 and P_2 intersect, and decide whether L intersects the line OD, where D is the point $(1, 4, -4)$; if so, determine the point of intersection.

9.36. Let A, B, and C have coordinates $(-1, -2, 1)$, $(-1, -2, 0)$, and $(-1, 0, 3)$ respectively.
(a) Find the angle at A in the triangle ABC and the area of the triangle. Find also an equation of the plane through A, B, and C.
(b) Let P_1 be the plane through B with normal vector $n_1 = \hat{\jmath} + \hat{k}$, and P_2 the plane through C with normal vector $n_2 = 2\hat{\imath} - \hat{\jmath} + 3\hat{k}$. Show that the line AC meets P_1 at right angles, and determine an equation of the line of intersection of the planes P_1 and P_2.

9.37. Prove that

$$a \times (b \times c) + b \times (c \times a) + c \times (a \times b) = 0.$$

9.38. Let A, B, C, and D be four points in space with A, B, and C not collinear (that is not in the same straight line), and let $\overline{AB} = r$, $\overline{AC} = s$, and $\overline{AD} = t$. Show that the distance of D from the plane passing through A, B, and C is

$$\frac{|t \cdot r \times s|}{|r \times s|}.$$

9.39. Show that the straight line which passes through the point with position vector b, and intersects the straight line

$$r = a + \lambda n,$$

at right angles is given by

$$r = b + \mu n \times [(a-b) \times n].$$

9.40. The position vectors of four points given by a, b, c, and d define a quadrilateral. Using a vector method, show that the lines joining the mid-points of adjacent sides of the quadrilateral form a parallelogram. Show also that the area of the parallelogram is one-half that of the quadrilateral.

9.41. An observer S with an eye at the point with position vector $s = \hat{\imath} + \hat{\jmath} + \hat{k}$ views objects on the other side of a plane screen with equation

$$r \cdot (1.1\hat{\imath} + 1.1\hat{\jmath} + 1\hat{k}) = 1, \quad \text{where} \quad r = x\hat{\imath} + y\hat{\jmath} + z\hat{k}.$$

What are the coordinates on the screen of a general point P_1 with position vector

$$r_1 = x_1\hat{\imath} + y_1\hat{\jmath} + z_1\hat{k}$$

behind the screen as seen by the observer? (Hint: find the equation of the line joining the points S and P_1 and find where it cuts the screen.)

9.42. Show that the point on the line of intersection of the planes

$$r \cdot n_1 = p_1 \quad \text{and} \quad r \cdot n_2 = p_2$$

that is closest to the origin has the position vector

$$\alpha(n_1 \times n_2) \times n_1 + \beta(n_1 \times n_2) \times n_2.$$

Find the constants α and β.

10 Linear equations

10.1 Solution of linear equations by elimination

A matrix equation

$$Ax = d, \tag{10.1}$$

defines a set of **linear equations** (we referred to them previously in connection with the inverse matrix in Section 7.4). In general, A will be an $m \times n$ matrix, while x and d are column n-vectors. Usually, but not always, we are interested in the case where the number of unknowns in the equations equals the number of equations. In other words, there is neither a surplus of unknowns nor equations. This is the normal situation in applications. For example, the set of equations defined by

$$A = \begin{bmatrix} 1 & 2 & 1 \\ -2 & 3 & -1 \\ 1 & 1 & -2 \end{bmatrix}, \qquad x = \begin{bmatrix} x \\ y \\ z \end{bmatrix}, \qquad d = \begin{bmatrix} 1 \\ -7 \\ -7 \end{bmatrix}$$

is

$$x_1 + 2x_2 + x_3 = 1, \tag{10.2a}$$

$$-2x_1 + 3x_2 - x_3 = -7, \tag{10.2b}$$

$$x_1 + 4x_2 - 2x_3 = -7. \tag{10.2c}$$

Consider now the case in which A is an arbitrary square matrix. If the inverse of A exists, then multiplication of (10.1) on the left by A^{-1} leads to the solution vector

$$x = A^{-1}d = \frac{\operatorname{adj} A}{\det A} d,$$

using the formula for the inverse given in Section 7.4. Let $n = 3$; then, for our standard matrix

$$A = \begin{bmatrix} a_{11} & a_{12} & a_{13} \\ a_{21} & a_{22} & a_{23} \\ a_{31} & a_{32} & a_{33} \end{bmatrix},$$

we have

$$x = \begin{bmatrix} x_1 \\ x_2 \\ x_3 \end{bmatrix} = \frac{\text{adj } A}{\det A} d$$

$$= \frac{\text{adj } A}{\det A} \begin{bmatrix} d_1 \\ d_2 \\ d_3 \end{bmatrix} = \frac{1}{\det A} \begin{bmatrix} C_{11}d_1 + C_{21}d_2 + C_{31}d_3 \\ C_{12}d_1 + C_{22}d_2 + C_{32}d_3 \\ C_{13}d_1 + C_{23}d_2 + C_{33}d_3 \end{bmatrix},$$

where $C_{11}, C_{12}, \ldots$ are the cofactors of $a_{11}, a_{12}, \ldots$ (see Section 8.1). Thus, comparison of elements in the vectors leads to

$$x_1 = \frac{1}{\det A}(C_{11}d_1 + C_{21}d_2 + C_{31}d_3) = \frac{1}{\det A}\begin{vmatrix} d_1 & a_{12} & a_{13} \\ d_2 & a_{22} & a_{23} \\ d_3 & a_{32} & a_{33} \end{vmatrix},$$

$$x_2 = \frac{1}{\det A}\begin{vmatrix} a_{11} & d_1 & a_{13} \\ a_{21} & d_2 & a_{23} \\ a_{31} & d_3 & a_{33} \end{vmatrix}, \qquad x_3 = \frac{1}{\det A}\begin{vmatrix} a_{11} & a_{12} & d_1 \\ a_{21} & a_{22} & d_2 \\ a_{31} & a_{32} & d_3 \end{vmatrix}.$$

This representation of the solution is known as **Cramer's rule**. It is systematic in that, for x_1, the determinant in the numerator has the first column of A replaced by d; for x_2, the second column is replaced by d; and so on. The generalization to n linear equations in n unknowns is fairly clear from this formula. It is a useful theoretical result, but not generally a recommended method of solving more than four equations in four unknown equations. High-order determinant evaluation is complicated.

   Short of using computer software, the simplest method of solving equations involves systematic elimination. Consider the three equations

$$x_1 + 2x_2 + x_3 = 1, \tag{10.3a}$$

$$-2x_1 + 3x_2 - x_3 = -7, \tag{10.3b}$$

$$x_1 + 4x_2 - 2x_3 = -7. \tag{10.3c}$$

We can perform three **elementary operations** on linear equations which do not affect the solution. They are

---

Elementary row operators
   (i) any equation can be multiplied by a nonzero constant,
   (ii) any two equations can be interchanged,   **(10.4)**
   (iii) any equation can be replaced by the sum of itself and any multiple of another equation.

Step 1. Eliminate x_1 from (10.3b,c) by subtracting multiples of (10.3a) from (10.3b) and (10.3c):

$$x_1 + 2x_2 + x_3 = 1,$$
$$7x_2 + x_3 = -5 \quad (\mathbf{r}_2' = \mathbf{r}_2 + 2\mathbf{r}_1),$$
$$2x_2 - 3x_3 = -8 \quad (\mathbf{r}_3' = \mathbf{r}_3 - \mathbf{r}_1).$$

The required operations between the equations are listed on the right.

Step 2. We now proceed to eliminate x_2 from (10.2c) using a multiple of the new row 2. Hence

$$x_1 + 2x_2 + x_3 = 1,$$
$$7x_2 + x_3 = -5,$$
$$-\tfrac{23}{7}x_3 = -\tfrac{46}{7} \quad (\mathbf{r}_3' = \mathbf{r}_3 - \tfrac{2}{7}\mathbf{r}_2)$$

Step 3. Using rule (ii) above, reduce the coefficients of x_2 and x_3 in the second and third equations above to 1:

$$x_1 + 2x_2 + x_3 = 1,$$
$$x_2 + \tfrac{1}{7}x_3 = -\tfrac{5}{7} \quad (\mathbf{r}_2' = \tfrac{1}{7}\mathbf{r}_2),$$
$$x_3 = 2 \quad (\mathbf{r}_3' = -\tfrac{7}{23}\mathbf{r}_3).$$

Step 4. Starting from the third equation, we can now solve the equations by **back subsitution**. Since $x_3 = 2$, from the second equation,

$$x_2 = -\tfrac{5}{7} - \tfrac{1}{7}x_3 = -\tfrac{5}{7} - \tfrac{1}{7} \times 2 = -1,$$

from the first equation,

$$x_1 = 1 - 2x_2 - x_3 = 1 + 2 - 2 = 1.$$

Thus the solution is

$$x_1 = 1, \qquad x_2 = -1, \qquad x_3 = 2.$$

The method is known as **Gaussian elimination**.

In fact, we need not write down the equations for x_1, x_2, x_3 at each stage, since all the information in (10.3) is given by the 3×4 matrix

$$\begin{bmatrix} 1 & 2 & 1 & 1 \\ -2 & 3 & -1 & -7 \\ 1 & 4 & -2 & -7 \end{bmatrix},$$

which is known as the **augmented matrix** for the system of equations. The elementary operations referred to previously become **elementary row operations on the matrix**. We can reproduce the steps above by the following more compact procedure:

$$\begin{bmatrix} \mathbf{1} & 2 & 1 & 1 \\ -2 & 3 & -1 & -7 \\ 1 & 4 & -2 & -7 \end{bmatrix} \rightarrow \begin{bmatrix} 1 & 2 & 1 & 1 \\ 0 & 7 & 1 & -5 \\ 0 & 2 & -3 & -8 \end{bmatrix} \quad \begin{pmatrix} \mathbf{r}_2' = \mathbf{r}_2 + 2\mathbf{r}_1 \\ \mathbf{r}_3' = \mathbf{r}_3 - \mathbf{r}_1 \end{pmatrix}$$

$$\rightarrow \begin{bmatrix} 1 & 2 & 1 & 1 \\ 0 & 7 & 1 & -5 \\ 0 & 0 & -\frac{23}{7} & -\frac{46}{7} \end{bmatrix} \quad (\mathbf{r}_3' = \mathbf{r}_3 - \tfrac{2}{7}\mathbf{r}_2)$$

$$\rightarrow \begin{bmatrix} 1 & 2 & 1 & 1 \\ 0 & 1 & \frac{1}{7} & -\frac{5}{7} \\ 0 & 0 & 1 & 2 \end{bmatrix} \quad \begin{pmatrix} \mathbf{r}_2' = \tfrac{1}{7}\mathbf{r}_2 \\ \mathbf{r}_3' = -\tfrac{7}{23}\mathbf{r}_3 \end{pmatrix},$$

where the arrow '→' means 'is transformed into'. The final matrix is said to be in **echelon** form, that is, it has zeros below the diagonal elements starting from the top left. We can now solve the equations by back substitution as before.

The elements in bold type are known as **pivots** and they must be nonzero. They are used to clear the elements in the column below them. If any pivot turns out to be zero as the method progresses, then that equation or row is replaced by the first row **below** which has a nonzero coefficient in the column. If there are no further nonzero coefficients, then the pivot moves across to the next column.

It is possible to complete the Gaussian elimination by using further row operations on the echelon matrix. Thus, continuing from the echelon form above

$$\begin{bmatrix} 1 & 2 & 1 & 1 \\ 0 & 2 & \frac{1}{7} & -\frac{5}{7} \\ 0 & 0 & \mathbf{1} & 2 \end{bmatrix} \rightarrow \begin{bmatrix} 1 & 2 & 0 & -1 \\ 0 & \mathbf{1} & 0 & -1 \\ 0 & 0 & 1 & 2 \end{bmatrix} \quad \begin{pmatrix} \mathbf{r}_1' = \mathbf{r}_1 - \mathbf{r}_3 \\ \mathbf{r}_2' = \mathbf{r}_2 - \tfrac{1}{7}\mathbf{r}_3 \end{pmatrix}$$

$$\rightarrow \begin{bmatrix} 1 & 0 & 0 & 1 \\ 0 & 1 & 0 & -1 \\ 0 & 0 & 1 & 2 \end{bmatrix} \quad (\mathbf{r}_1' = \mathbf{r}_1 - 2\mathbf{r}_2),$$

where the pivots are bold again. The final matrix now represents the solution set $x_1 = 1$, $x_2 = -1$, $x_3 = 2$.

Example 7.1. Using Gaussian elimination and back substitution, solve the set of equations

$$\begin{aligned} x_1 + x_2 + 2x_3 \phantom{{}+ x_4} &= 4, \\ 2x_1 + 2x_2 + x_3 - x_4 &= -1, \\ x_2 + x_3 + x_4 &= 6, \\ x_2 - x_3 + 2x_4 &= 5. \end{aligned}$$

We first perform the pivotal row operations on the augmented matrix as follows:

$$
\begin{bmatrix}
1 & 1 & 2 & 0 & 4 \\
2 & 2 & 1 & -1 & -1 \\
0 & 1 & 1 & 1 & 6 \\
0 & 1 & -1 & 2 & 5
\end{bmatrix}
\rightarrow
\begin{bmatrix}
1 & 1 & 2 & 0 & 4 \\
0 & 0 & -3 & -1 & -9 \\
0 & 1 & 1 & 1 & 6 \\
0 & 1 & -1 & 2 & 5
\end{bmatrix}
$$
$$(\mathbf{r}_2' = \mathbf{r}_2 - 2\mathbf{r}_1)$$

$$
\rightarrow
\begin{bmatrix}
1 & 1 & 2 & 0 & 4 \\
0 & 1 & 1 & 1 & 6 \\
0 & 0 & -3 & -1 & -9 \\
0 & 1 & -1 & 2 & 5
\end{bmatrix}
\quad (\mathbf{r}_2 \leftrightarrow \mathbf{r}_3)
$$

$$
\rightarrow
\begin{bmatrix}
1 & 1 & 2 & 0 & 4 \\
0 & 1 & 1 & 1 & 6 \\
0 & 0 & -3 & -1 & -9 \\
0 & 0 & -2 & 1 & -1
\end{bmatrix}
$$
$$(\mathbf{r}_4' = \mathbf{r}_4 - \mathbf{r}_2)$$

$$
\rightarrow
\begin{bmatrix}
1 & 1 & 2 & 0 & 4 \\
0 & 1 & 1 & 1 & 6 \\
0 & 0 & -3 & -1 & -9 \\
0 & 0 & 0 & \frac{5}{3} & 5
\end{bmatrix}
$$
$$(\mathbf{r}_4' = \mathbf{r}_4 - \tfrac{2}{3}\mathbf{r}_3)$$

$$
\rightarrow
\begin{bmatrix}
1 & 1 & 2 & 0 & 4 \\
0 & 1 & 1 & 1 & 6 \\
0 & 0 & 1 & \frac{1}{3} & 3 \\
0 & 0 & 0 & 1 & 3
\end{bmatrix}
\quad
\begin{pmatrix}
\mathbf{r}_3' = -\tfrac{1}{3}\mathbf{r}_3 \\
\mathbf{r}_4' = \tfrac{3}{5}\mathbf{r}_4
\end{pmatrix}.
$$

(Note the row change $\mathbf{r}_2 \leftrightarrow \mathbf{r}_3$ because of the zero pivot.) Back substitution now gives

$$x_4 = 3, \qquad x_3 = 3 - \tfrac{1}{3}x_4 = 2, \qquad x_2 = 6 - x_3 - x_4 = 1,$$
$$x_1 = 4 - x_2 - 2x_3 = -1.$$

The solution can now be written as

$$
\begin{bmatrix}
x_1 \\
x_2 \\
x_3 \\
x_4
\end{bmatrix}
=
\begin{bmatrix}
-1 \\
1 \\
2 \\
3
\end{bmatrix}.
$$

10.2 The inverse matrix by Gaussian elimination

We can also use Gaussian elimination to find the inverse matrix. In solving the set of equations $Ax = d$, we obtain the solution $x = A^{-1}d$. In other words:

> Use elementary row operations to transform A into the identity I, and use the same operations to transform I into A^{-1}. **(10.5)**

Suppose that we require the inverse of

$$A = \begin{bmatrix} 0 & 1 & 0 & 2 \\ 1 & 0 & 1 & 0 \\ 0 & 1 & 0 & 1 \\ 1 & 0 & 2 & 0 \end{bmatrix}.$$

We reduce A to I_4 and perform the same row operations on I_4. Thus, we can write down the steps **in parallel** as follows:

$$A = \begin{bmatrix} 0 & 1 & 0 & 2 \\ 1 & 0 & 1 & 0 \\ 0 & 1 & 0 & 1 \\ 1 & 0 & 2 & 0 \end{bmatrix} \qquad I_4 = \begin{bmatrix} 1 & 0 & 0 & 0 \\ 0 & 1 & 0 & 0 \\ 0 & 0 & 1 & 0 \\ 0 & 0 & 0 & 1 \end{bmatrix}$$

$$\rightarrow \begin{bmatrix} 1 & 0 & 1 & 0 \\ 0 & 1 & 0 & 2 \\ 0 & 1 & 0 & 1 \\ 1 & 0 & 2 & 0 \end{bmatrix} \quad (r_1 \leftrightarrow r_2) \quad \rightarrow \begin{bmatrix} 0 & 1 & 0 & 0 \\ 1 & 0 & 0 & 0 \\ 0 & 0 & 1 & 0 \\ 0 & 0 & 0 & 1 \end{bmatrix}$$

$$\rightarrow \begin{bmatrix} 1 & 0 & 1 & 0 \\ 0 & 1 & 0 & 2 \\ 0 & 1 & 0 & 1 \\ 0 & 0 & 1 & 0 \end{bmatrix} \quad (r_4' = r_4 - r_1) \quad \rightarrow \begin{bmatrix} 0 & 1 & 0 & 0 \\ 1 & 0 & 0 & 0 \\ 0 & 0 & 1 & 0 \\ 0 & -1 & 0 & 1 \end{bmatrix}$$

$$\rightarrow \begin{bmatrix} 1 & 0 & 1 & 0 \\ 0 & 1 & 0 & 2 \\ 0 & 0 & 0 & -1 \\ 0 & 0 & 1 & 0 \end{bmatrix} \quad (r_3' = r_3 - r_2) \quad \rightarrow \begin{bmatrix} 0 & 1 & 0 & 0 \\ 1 & 0 & 0 & 0 \\ -1 & 0 & 1 & 0 \\ 0 & -1 & 0 & 1 \end{bmatrix}$$

$$\rightarrow \begin{bmatrix} 1 & 0 & 1 & 0 \\ 0 & 1 & 0 & 2 \\ 0 & 0 & 1 & 0 \\ 0 & 0 & 0 & -1 \end{bmatrix} \quad (r_3 \leftrightarrow r_4) \quad \rightarrow \begin{bmatrix} 0 & 1 & 0 & 0 \\ 1 & 0 & 0 & 0 \\ 0 & -1 & 0 & 1 \\ -1 & 0 & 1 & 0 \end{bmatrix}$$

$$\rightarrow \begin{bmatrix} 1 & 0 & 1 & 0 \\ 0 & 1 & 0 & 2 \\ 0 & 0 & 1 & 0 \\ 0 & 0 & 0 & 1 \end{bmatrix} \quad (\mathbf{r}_4' = -\mathbf{r}_4) \quad \rightarrow \begin{bmatrix} 0 & 1 & 0 & 0 \\ 1 & 0 & 0 & 0 \\ 0 & -1 & 0 & 1 \\ 1 & 0 & -1 & 0 \end{bmatrix}$$

$$\rightarrow \begin{bmatrix} 1 & 0 & 1 & 0 \\ 0 & 1 & 0 & 0 \\ 0 & 0 & 1 & 0 \\ 0 & 0 & 0 & 1 \end{bmatrix} \quad (\mathbf{r}_2' = \mathbf{r}_2 - 2\mathbf{r}_4) \quad \rightarrow \begin{bmatrix} 0 & 1 & 0 & 0 \\ -1 & 0 & 2 & 0 \\ 0 & -1 & 2 & 0 \\ 1 & 0 & -1 & 0 \end{bmatrix}$$

$$\rightarrow \begin{bmatrix} 1 & 0 & 0 & 0 \\ 0 & 1 & 0 & 0 \\ 0 & 0 & 1 & 0 \\ 0 & 0 & 0 & 1 \end{bmatrix} \quad (\mathbf{r}_1' = \mathbf{r}_1 - \mathbf{r}_3) \quad \rightarrow \begin{bmatrix} 0 & 2 & 0 & -1 \\ -1 & 0 & 2 & 0 \\ 0 & -1 & 0 & 1 \\ 1 & 0 & -1 & 0 \end{bmatrix}$$

$$= \mathbf{I}_4 \qquad\qquad\qquad\qquad = A^{-1}.$$

We conclude that

$$A^{-1} = \begin{bmatrix} 0 & 2 & 0 & -1 \\ -1 & 0 & 2 & 0 \\ 0 & -1 & 0 & 1 \\ 1 & 0 & -1 & 0 \end{bmatrix}.$$

10.3 Compatible and incompatible sets of equations

Not all sets of equations have solutions. For example $x + y = 1$ and $x + y = 2$ has no solutions. Consider the set of equations

$$x + y - \cdot z = 3,$$

$$3x - y + 3z = 5,$$

$$x - y + 2z = 2.$$

We can sense that there might be a problem by first evaluating the determinant of the coefficients of x, y, and z. Thus

$$\begin{vmatrix} 1 & 1 & -1 \\ 3 & -1 & 3 \\ 1 & -1 & 2 \end{vmatrix} = \begin{vmatrix} 1 & 1 & -1 \\ 4 & 0 & 2 \\ 2 & 0 & 1 \end{vmatrix} \quad \begin{pmatrix} \mathbf{r}_2' = \mathbf{r}_2 + \mathbf{r}_1 \\ \mathbf{r}_3' = \mathbf{r}_3 + \mathbf{r}_1 \end{pmatrix}$$

$$= - \begin{vmatrix} 4 & 2 \\ 2 & 1 \end{vmatrix} = 0.$$

Thus Cramer's rule will fail, although there still may be solutions. We can detect the existence of solutions more readily by using Gaussian elimination. In this case the application of row operations on the augmented matrix leads to

$$
\begin{bmatrix} 1 & 1 & -1 & 3 \\ 3 & -1 & 3 & 5 \\ 1 & -1 & 2 & 2 \end{bmatrix} \rightarrow \begin{bmatrix} 1 & 1 & -1 & 3 \\ 0 & -4 & 6 & -4 \\ 0 & -2 & 3 & -1 \end{bmatrix} \begin{pmatrix} \mathbf{r}_2' = \mathbf{r}_2 - 3\mathbf{r}_1 \\ \mathbf{r}_3' = \mathbf{r}_3 - \mathbf{r}_1 \end{pmatrix}
$$

$$
\rightarrow \begin{bmatrix} 1 & 1 & -1 & 3 \\ 0 & -4 & -6 & -4 \\ 0 & 0 & 0 & 1 \end{bmatrix} (\mathbf{r}_3' = \mathbf{r}_3 - \tfrac{1}{2}\mathbf{r}_2),
$$

which is the echelon form for this set of equations. However, row 3 is inconsistent since $0 \neq 1$. Hence these equations can have no solutions.

On the other hand, consider the following set:

$$x + y - z = 1,$$

$$3x - y + 3z = 5,$$

$$x - y + 2z = 2,$$

(this is the previous set with one change to the first equation). Gaussian elimination now gives

$$
\begin{bmatrix} 1 & 1 & -1 & 1 \\ 3 & -1 & 3 & 5 \\ 1 & -1 & 2 & 2 \end{bmatrix} \rightarrow \begin{bmatrix} 1 & 1 & -1 & 1 \\ 0 & -4 & 6 & 2 \\ 0 & -2 & 3 & 1 \end{bmatrix} \begin{pmatrix} \mathbf{r}_2' = \mathbf{r}_2 - 3\mathbf{r}_1 \\ \mathbf{r}_3' = \mathbf{r}_3 - \mathbf{r}_1 \end{pmatrix}
$$

$$
\rightarrow \begin{bmatrix} 1 & 1 & -1 & 1 \\ 0 & -4 & 6 & 2 \\ 0 & 0 & 0 & 0 \end{bmatrix} (\mathbf{r}_3' = \mathbf{r}_3 - \tfrac{1}{2}\mathbf{r}_2).
$$

Row 3 is now consistent, and row 2 is $-4y + 6z = 2$. Hence

$$y = -\tfrac{1}{4}(2 - 6z)$$

and, from row 1,

$$x = 1 - y + z = \tfrac{3}{2} - \tfrac{1}{2}z.$$

Thus z can take any value, say λ, so the full solution set is

$$
\begin{bmatrix} x \\ y \\ z \end{bmatrix} = \begin{bmatrix} \tfrac{3}{2} - \tfrac{1}{2}\lambda \\ -\tfrac{1}{2} + \tfrac{3}{2}\lambda \\ \lambda \end{bmatrix}
$$

for any value of λ. It can be seen in this case that there exists an infinite number of solutions, a different one for each different value of λ.

Geometrically, in three dimensions, it can be seen why equations can have a unique solution, no solution, or an infinite set of solutions. Any equation such as

$$ax + by + cz = d$$

represents a plane in $\mathbb{R}^3$. Three equations represent three planes, and we need only visualize how they might intersect or not. The coordinates of any point of intersection of the planes in the solution of the equations. The three diagrams in Fig. 10.1 show how three planes can intersect in a single point, no point, or a line of points.

(a)

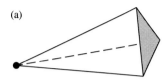

(b)

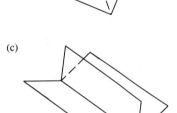

(c)

Fig. 10.1
(a) unique solution, (b) no solution, (c) line of solutions.

Example 10.2. Determine the complete sets of values for a and b which make the equations

$$x - 2y + 3z = 2,$$

$$2x - y + 2z = 3,$$

$$x + y + az = b,$$

have (i) a unique solution, (ii) no solutions, (iii) an infinite set of solutions.

Reduce the augmented matrix to echelon form using pivots to clear each column successively:

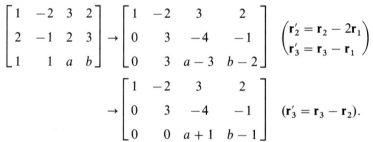

$$\begin{bmatrix} 1 & -2 & 3 & 2 \\ 2 & -1 & 2 & 3 \\ 1 & 1 & a & b \end{bmatrix} \rightarrow \begin{bmatrix} 1 & -2 & 3 & 2 \\ 0 & 3 & -4 & -1 \\ 0 & 3 & a-3 & b-2 \end{bmatrix} \quad \left(\begin{array}{l} \mathbf{r}_2' = \mathbf{r}_2 - 2\mathbf{r}_1 \\ \mathbf{r}_3' = \mathbf{r}_3 - \mathbf{r}_1 \end{array} \right)$$

$$\rightarrow \begin{bmatrix} 1 & -2 & 3 & 2 \\ 0 & 3 & -4 & -1 \\ 0 & 0 & a+1 & b-1 \end{bmatrix} \quad (\mathbf{r}_3' = \mathbf{r}_3 - \mathbf{r}_2).$$

We can now interpret the echelon matrix.
　(i) If $a \neq -1$, then z has the unique solution

$$z = \frac{b-1}{a+1}.$$

Also y and x can be found by back substitution.
　(ii) If $a = -1$, and $b \neq 1$ then row 3 will lead to an inconsistency. Hence there are no solutions of the equations.
　(iii) If $a = -1$, and $b = 1$ then row 3 implies $z = \lambda$ for any number λ. Also x and y can be found by back substitution.

　This example illustrates the advantage of Gaussian elimination over the formula-based method of Cramer's rule. The Gaussian method can still be used if the number of equations differs from the number of unknowns. Consider the following example.

Example 10.3. Investigate all solutions of

$$x + y - \ z = 1$$

$$3x - y + 3z = 5,$$

$$x - y + 2z = 2,$$

$$x \quad\quad + \ z = 3.$$

The augmented matrix is

$$\begin{bmatrix} 1 & 1 & -1 & 1 \\ 3 & -1 & 3 & 5 \\ 3 & -1 & 2 & 2 \\ 1 & 0 & 1 & 3 \end{bmatrix} \rightarrow \begin{bmatrix} 1 & 1 & -1 & 1 \\ 0 & -4 & 6 & 2 \\ 0 & -2 & 3 & 1 \\ 0 & -1 & 2 & 2 \end{bmatrix} \begin{pmatrix} \mathbf{r}_2' = \mathbf{r}_2 - 3\mathbf{r}_1 \\ \mathbf{r}_3' = \mathbf{r}_3 - \mathbf{r}_1 \\ \mathbf{r}_4' = \mathbf{r}_4 - \mathbf{r}_1 \end{pmatrix}$$

$$\rightarrow \begin{bmatrix} 1 & 1 & -1 & 1 \\ 0 & -4 & 6 & 2 \\ 0 & 0 & 0 & 0 \\ 0 & 0 & \frac{1}{2} & \frac{3}{2} \end{bmatrix} \begin{pmatrix} \mathbf{r}_3' = \mathbf{r}_3 - \frac{1}{2}\mathbf{r}_2 \\ \mathbf{r}_4' = \mathbf{r}_4 - \frac{1}{4}\mathbf{r}_2 \end{pmatrix}$$

$$\rightarrow \begin{bmatrix} 1 & 1 & -1 & 1 \\ 0 & -4 & 6 & 2 \\ 0 & 0 & \frac{1}{2} & \frac{3}{2} \\ 0 & 0 & 0 & 0 \end{bmatrix} (\mathbf{r}_3 \leftrightarrow \mathbf{r}_4).$$

Row 4 is consistent, while row 3 implies $z = 3$. Then y and x can be found by back substitution in rows 2 and 1.

On the other hand, there may be more variables than equations, as in the following example.

Example 10.4. Show that the following equations are inconsistent:

$$x_1 + \ x_2 + \ x_3 + 2x_4 = 1,$$

$$x_1 - 2x_2 + 3x_3 - \ x_4 = 4,$$

$$3x_1 - 3x_2 + 7x_3 \quad\quad = 7.$$

Proceed as before, and successively reduce the augmented matrix by pivots. Thus

$$\begin{bmatrix} 1 & 1 & 1 & 2 & 1 \\ 1 & -2 & 3 & -1 & 4 \\ 3 & -3 & 7 & 0 & 7 \end{bmatrix} \rightarrow \begin{bmatrix} 1 & 1 & 12 & 1 & 1 \\ 0 & -3 & 2 & -3 & 3 \\ 0 & -6 & 4 & -6 & 4 \end{bmatrix}$$

$$\begin{pmatrix} \mathbf{r}_2' = \mathbf{r}_2 - \mathbf{r}_1 \\ \mathbf{r}_3' = \mathbf{r}_3 - 3\mathbf{r}_1 \end{pmatrix}$$

$$\rightarrow \begin{bmatrix} 1 & 1 & 12 & 1 & 1 \\ 0 & -3 & 2 & -3 & 3 \\ 0 & 0 & 0 & 0 & -2 \end{bmatrix}$$

$$(\mathbf{r}_3' = \mathbf{r}_3 - 2\mathbf{r}_2).$$

Since $0 \neq 2$, row 3 indicates an inconsistency, so the equations are incompatible.

10.4 Homogeneous sets of equations

Any set of equations $A\mathbf{x} = \mathbf{0}$ is known as a **homogeneous** set; it is a set of linear equations with zero right-hand sides. Clearly, the equations always have the so-called **trivial** solution $\mathbf{x} = \mathbf{0}$, but there may exist **nontrivial** solutions. What are the conditions for their existence? Consider the following example.

Example 10.5. Find the value of a for which the following equations have nontrivial solutions:

$$x + y + z = 0,$$
$$x + 2y = 0,$$
$$x - 3y + az = 0.$$

Proceed in the usual way using Gaussian reduction. Thus

$$\begin{bmatrix} 1 & 1 & 1 & 0 \\ 1 & 2 & 0 & 0 \\ 1 & -3 & a & 0 \end{bmatrix} \rightarrow \begin{bmatrix} 1 & 1 & 1 & 0 \\ 0 & 1 & -1 & 0 \\ 0 & -4 & a-1 & 0 \end{bmatrix} \begin{pmatrix} \mathbf{r}_2' = \mathbf{r}_2 - \mathbf{r}_1 \\ \mathbf{r}_3' = \mathbf{r}_3 - \mathbf{r}_1 \end{pmatrix}$$

$$\rightarrow \begin{bmatrix} 1 & 1 & 1 & 0 \\ 0 & 1 & -1 & 0 \\ 0 & 0 & a-5 & 0 \end{bmatrix} \quad (\mathbf{r}_3' = \mathbf{r}_3 + 4\mathbf{r}_2).$$

Hence z can only be nonzero if $a = 5$. Hence, nontrivial solutions exist if, and only if, $a = 5$. Back substituting, we find that the solution is

$$z = \lambda, \qquad y = z = \lambda, \qquad x = -y - z = -2\lambda,$$

for any λ.

If A is a square matrix, then Cramer's rule (Section 10.1) implies that

$$x \det A = 0.$$

Hence x is a nonzero column vector if and only if $\det A = 0$, which is a test for nontrivial solutions for systems in which the number of variables is the same as the number of equations.

> **Homogeneous equations** $Ax = 0$, **where** A **is square.**
>
> If $\det A = 0$, there is an infinite number of nontrivial solutions. If $\det A \neq 0$, the only solution is $x = 0$. **(10.6)**

Example 10.6. Find all conditions on the constants a, b, and c in order that

$$x + y + z = 0,$$
$$ax + by + cz = 0,$$
$$a^2 x + b^2 y + c^2 z = 0,$$

should have nontrivial solutions. Find the solutions in the cases (a) $a = 1$, $b = 1$, $c = 2$; (b) $a = 1$, $b = 1$, $c = 1$.

This system of equations will have nontrivial solutions for x, y, and z if, and only if,

$$D = \begin{vmatrix} 1 & 1 & 1 \\ a & b & c \\ a^2 & b^2 & c^2 \end{vmatrix} = 0.$$

Thus

$$D = \begin{vmatrix} 1 & 0 & 0 \\ a & b-a & c-a \\ a^2 & b^2-a^2 & c^2-a^2 \end{vmatrix} \quad \begin{pmatrix} \mathbf{c}_2' = \mathbf{c}_2 - \mathbf{c}_1 \\ \mathbf{c}_3' = \mathbf{c}_3 - \mathbf{c}_1 \end{pmatrix}$$

$$= (b-a)(c-a) \begin{vmatrix} 1 & 0 & 0 \\ a & 1 & 1 \\ a^2 & b+a & c+a \end{vmatrix}$$

$$= (b-a)(c-a)(c+a-b-a)$$

$$= (b-c)(c-a)(a-b).$$

Hence, nontrivial solutions exist if $b = c$, $c = a$, or $a = b$.

(a) ($a = 1$, $b = 1$, $c = 2$) The equations become

$$x + y + z = 0,$$
$$x + y + 2z = 0,$$
$$x + y + 4z = 0.$$

The augmented matrix is

$$\begin{bmatrix} 1 & 1 & 1 & 0 \\ 1 & 1 & 2 & 0 \\ 1 & 1 & 4 & 0 \end{bmatrix} \rightarrow \begin{bmatrix} 1 & 1 & 1 & 0 \\ 0 & 0 & 1 & 0 \\ 0 & 0 & 3 & 0 \end{bmatrix} \begin{pmatrix} \mathbf{r}_2' = \mathbf{r}_2 - \mathbf{r}_1 \\ \mathbf{r}_3' = \mathbf{r}_3 - \mathbf{r}_1 \end{pmatrix}$$

$$\rightarrow \begin{bmatrix} 1 & 1 & 1 & 0 \\ 0 & 0 & 1 & 0 \\ 0 & 0 & 0 & 0 \end{bmatrix} \quad (\mathbf{r}_3' = \mathbf{r}_3 - 3\mathbf{r}_2).$$

Row 2 implies $z = 0$, while row 1 implies $x = -y$. Let $y = \lambda$, say. Then the solution set is

$$\begin{bmatrix} x \\ y \\ z \end{bmatrix} = \begin{bmatrix} -\lambda \\ \lambda \\ 0 \end{bmatrix} = \begin{bmatrix} -1 \\ 1 \\ 0 \end{bmatrix} \lambda,$$

for any λ.

(b) ($a = 1$, $b = 1$, $c = 1$) Applying Gaussian elimination, we find that

$$\begin{bmatrix} 1 & 1 & 1 & 0 \\ 1 & 1 & 1 & 0 \\ 1 & 1 & 1 & 0 \end{bmatrix} \rightarrow \begin{bmatrix} 1 & 1 & 1 & 0 \\ 0 & 0 & 0 & 0 \\ 0 & 0 & 0 & 0 \end{bmatrix} \begin{pmatrix} \mathbf{r}_2' = \mathbf{r}_2 - \mathbf{r}_1 \\ \mathbf{r}_3' = \mathbf{r}_3 - \mathbf{r}_1 \end{pmatrix}.$$

Hence, we are left with row 1, which implies

$$x + y + z = 0.$$

Let $z = \lambda$ and $y = \mu$. Then $x = -\lambda - \mu$. Hence the solution is

$$\begin{bmatrix} x \\ y \\ z \end{bmatrix} = \begin{bmatrix} -\lambda - \mu \\ \mu \\ \lambda \end{bmatrix} = \begin{bmatrix} -1 \\ 0 \\ 1 \end{bmatrix} \lambda + \begin{bmatrix} -1 \\ 1 \\ 0 \end{bmatrix} \mu$$

for any λ and any μ. Note that this is a two-parameter solution set.

10.5 Gauss–Seidel iterative method of solution

The method of Gaussian elimination described in Section 10.1 is not a practical approach, by hand, for a large system with perhaps 30 equations in 30 unknowns. Whatever method is employed, they will have to be solved by computer. But decisions about what scheme may be the best for a given set of equations are not always easy. There are many direct iterative methods in addition to the row-operation method described in Section 10.1. Two such methods will be briefly explained here.

Consider the equations

$$3x_1 + x_2 + x_3 = -1,$$
$$-x_1 + 4x_2 + x_3 = -8,$$
$$2x_1 + x_2 + 5x_3 = -14.$$

Write the equations as

$$x_1 = \tfrac{1}{3}(-x_2 - x_3 - 1), \tag{10.7}$$
$$x_2 = \tfrac{1}{4}(x_1 - x_3 - 8), \tag{10.8}$$
$$x_3 = \tfrac{1}{5}(-2x_1 - x_2 - 14), \tag{10.9}$$

where x_1, x_2, and x_3 are now the subjects of the three equations.

To start the iteration choose initial solutions for x_2 and x_3, say, $x_2^{(0)} = 0$ and $x_3^{(0)} = 0$, without thinking about the equations. Calculate $x_1^{(1)}$ from equation (10.7) as

$$x_1^{(1)} = \tfrac{1}{3}(-x_2^{(0)} - x_3^{(0)} - 1). \tag{10.10}$$

Use $x_1^{(1)}$ in (10.8) with $x_3^{(0)}$. Thus

$$x_2^{(1)} = \tfrac{1}{4}(x_1^{(1)} - x_3^{(0)}). \tag{10.11}$$

Finally, use the update $x_2^{(1)}$ in (10.9) to find $x_3^{(1)}$:

$$x_3^{(1)} = \tfrac{1}{5}(2x_1^{(1)} - x_2^{(1)} - 14). \tag{10.12}$$

Hence we have calculated a new approximate solution given by $x_1^{(1)}$, $x_2^{(1)}$, $x_3^{(2)}$. Now repeat the calculations starting with $x_2^{(1)}$ and $x_3^{(1)}$ as the new initial values to obtain $x_1^{(2)}$, $x_2^{(2)}$, $x_2^{(3)}$. The output from these **iterations** is shown in the following table.

i	0	1	2	3	4	5	6	7
$x_1^{(i)}$	–	−0.3333	1.1110	1.0570	1.0030	0.9984	0.9997	1.0000
$x_2^{(i)}$	0	−2.0830	−1.1600	−0.9825	−0.9926	−0.9997	−1.0000	−1.0000
$x_3^{(i)}$	0	−2.2500	−3.0130	−3.0260	−3.0030	−2.9990	−3.0000	−3.0000

All solutions are quoted to four decimal places. It can be seen that the exact solution, which is $x_1 = 1$, $x_2 = -1$, $x_3 = -3$, can be achieved to this accuracy after seven steps for this example. This is known as the **Gauss–Seidel scheme** for numerical solution of linear equations.

An alternative method without updating given by

$$x_1^{(1)} = \tfrac{1}{3}(-x_2^{(0)} - x_3^{(0)} - 1),$$
$$x_2^{(1)} = \tfrac{1}{4}(x_1^{(0)} - x_3^{(0)} - 8),$$
$$x_3^{(1)} = \tfrac{1}{5}(-2x_1^{(0)} - x_2^{(0)} - 14),$$

is known as **Jacobi's method**. However, convergence to the exact solution is slower with this scheme.

These methods do not always converge to the solution. For example, the Gauss–Seidel scheme applied to the system (10.2) fails since the iterates continue to increase in size. It can be shown that the Gauss–Seidel scheme converges if the magnitude of each leading diagonal element exceeds the sum of the magnitudes of the remaining elements in the same row of the matrix of coefficients. This is the case for the system given by (10.8a, b, c). Here the matrix of coefficients is

$$\begin{bmatrix} 3 & 1 & 1 \\ -1 & 4 & 1 \\ 2 & 1 & 5 \end{bmatrix}.$$

Each of the diagonal elements dominates the remaining elements in that row, since

$$3 \geqslant 1 + 1 = 2, \qquad 4 \geqslant |-1| + 1 = 2, \qquad 5 \geqslant 2 + 1 = 3.$$

This property of the system of equations is known as **diagonal dominance**. If the matrix is not diagonally dominant, then the scheme may or may not converge. Usually a few steps will indicate whether this is likely to be the case.

The schemes for both these methods can be expressed in matrix form as follows. Let the system of equations

$$Ax = d,$$

where $A = [a_{ij}]$ is an $n \times n$ matrix. Let

$$A = A_{\mathrm{L}} + D + A_{\mathrm{U}},$$

where A_{L}, D, and A_{U} are respectively the lower triangular, diagonal, and upper triangular matrices given by

$$A_{\mathrm{L}} = \begin{bmatrix} 0 & 0 & \dots & 0 \\ a_{21} & 0 & \dots & 0 \\ a_{31} & a_{32} & \dots & 0 \\ \dots & \dots & \dots & \dots \\ a_{n1} & a_{n2} & \dots & 0 \end{bmatrix}, \quad D = \begin{bmatrix} a_{11} & 0 & \dots & 0 \\ 0 & a_{22} & \dots & 0 \\ \dots & \dots & \dots & \dots \\ 0 & 0 & \dots & a_{nn} \end{bmatrix},$$

$$A_{\mathrm{U}} = \begin{bmatrix} 0 & a_{12} & a_{13} & \dots & a_{1n} \\ 0 & 0 & a_{23} & \dots & a_{2n} \\ \dots & \dots & \dots & \dots \\ 0 & 0 & 0 & \dots & 0 \end{bmatrix}.$$

The matrix equation becomes

$$A_{\mathrm{L}}x + Dx + A_{\mathrm{U}}x = d.$$

It is easy to find x from an equation of the form $Dx = \cdots$, since D is a diagonal matrix with a simple inverse. This is the matrix which is updated by Jacobi's method. Assuming that $x^{(0)} = [x_1^{(0)}, \dots, x_n^{(0)}]^{\mathrm{T}}$ is the given initial estimate, the approximate solution at step r will be computed from

$$Dx^{(r)} = -A_{\mathrm{U}}x^{(r-1)} - A_{\mathrm{L}}x^{(r-1)} + d.$$

On the other hand, in the Gauss–Seidel scheme, we take advantage of the observation that $x_1^{(r)}, x_2^{(r)}, \ldots$ are successively computed from rows $1, 2, \ldots$. Hence they can be used in the rows that follow. Thus the Gauss–Seidel iterations are given by

$$Dx^{(r)} = -A_L x^{(r)} - A_U x^{(r-1)} + d.$$

Problems, Chapter 10

10.1. (Section 10.1). Solve the following systems of linear equations using Cramer's rule:

(a)
$$x_1 \qquad + x_3 = \quad 1,$$
$$x_2 - x_3 = \quad 3,$$
$$2x_1 + x_2 \qquad = -1;$$

(b)
$$x_1 + 7x_2 + \quad x_3 = \quad 1,$$
$$x_2 - \quad x_3 = \quad 3,$$
$$2x_1 + \quad x_2 + 10x_3 = -1;$$

(c)
$$x_1 + 5x_2 - x_3 = \quad 1,$$
$$-3x_2 + \quad x_2 - x_3 = \quad 1,$$
$$3x_1 + \quad x_2 + x_3 = -3;$$

(d)
$$x_1 + \quad x_2 + \quad x_3 = 1,$$
$$ax_1 + \quad bx_2 + \quad cx_3 = d,$$
$$a^2 x_1 + b^2 x_2 + c^2 x_3 = d^2;$$

(e)
$$x_1 + 5x_2 \qquad + 2x_4 = 1,$$
$$-3x_2 \qquad - x_4 = 1,$$
$$3x_2 + x_3 + \quad x_4 = 1,$$
$$2x_2 + x_3 + \quad x_4 = 2.$$

10.2. (Section 10.1). The currents i_1, i_2, i_3 (in amps) flow in parts of a circuit which contains a variable resistor of resistance R (in ohms). The equations for the currents are given by

$$4i_1 - \quad i_2 - \quad i_3 = \quad 12,$$
$$- i_1 + Ri_2 \qquad = \quad 24,$$
$$i_1 \qquad + 5i_3 = -12,$$

in terms of the voltages on the right-hand side. For design reasons, the current i_3 should be 2 amps. How many ohms should the resistance R be?

10.3. (Section 10.3) Show that the following sets of equations are inconsistent.

(a)
$$x_1 + 2x_2 + \quad x_3 = 3,$$
$$x_1 - 3x_2 + 2x_3 = 4,$$
$$5x_1 + 5x_2 + 6x_3 = 1;$$

(b)
$$x_1 + x_2 + \quad x_3 \qquad = 2,$$
$$x_1 \qquad + \quad x_3 + 2x_4 = 3,$$
$$x_1 + x_2 \qquad + \quad x_4 = 4,$$
$$- x_2 + 2x_3 \qquad = 2;$$

(c)
$$x_1 + \quad x_2 \qquad\qquad\qquad = 1,$$
$$x_2 + \quad x_3 \qquad\qquad = 1,$$
$$x_3 + \quad x_4 \qquad = 1,$$
$$x_4 + \quad x_5 = 1,$$
$$x_1 + 3x_2 + 5x_3 + 7x_4 + 4x_5 = 1.$$

10.4. (Section 10.3). Determine the complete set of values for a and b that make the equations

$$x + \quad y - \quad z = 2,$$
$$2x + 3y + \quad z = 3,$$
$$5x + 7y + az = b,$$

have (i) a unique solution, (ii) no solutions, (iii) an infinite set of solutions.

10.5. (Section 10.3). Investigate all solutions of the system

$$x - \quad y + 2z = 1,$$
$$x + \quad y + 3z = 2,$$
$$x + 2y - \quad z = 3,$$
$$x - 2y + 6z = 0.$$

10.6. (Section 10.3). Show that the following equations are inconsistent.

$$x_1 + \quad x_2 + \quad x_3 - x_4 = 10,$$
$$x_1 - \quad x_2 - \quad x_3 \qquad = 1,$$
$$4x_1 - 2x_2 - 2x_3 - x_4 = 5.$$

10.7. (Section 10.1). Solve the following equations by

Gaussian elimination.

$$x_2 + 2x_3 - x_4 = 11,$$

$$x_1 + x_2 + x_3 + x_4 = 1,$$

$$2x_1 + x_2 - x_3 + 4x_4 = 0,$$

$$x_1 - x_2 + x_3 - 2x_4 = 2.$$

10.8. (Section 10.3). Find the value of a for which the linear equations

$$ax - y + 2z = 1,$$

$$x + 2y - az = 2,$$

$$4x + y - 2z = 2,$$

have no solutions.

10.9. (Section 10.2). Find the inverses of the following matrices.

(a) $\begin{bmatrix} 6 & -3 & 6 \\ 3 & 6 & 6 \\ -12 & -3 & 6 \end{bmatrix}$;

(b) $\begin{bmatrix} 1 & -1 & 2 \\ 1 & 2 & 1 \\ -4 & -1 & 2 \end{bmatrix}$;

(c) $\begin{bmatrix} 2 & -1 & 2 & 0 \\ 1 & 0 & -1 & 2 \\ 0 & 0 & -1 & 2 \\ -1 & 0 & 1 & 0 \end{bmatrix}$;

(d) $\begin{bmatrix} 1 & 1 & 0 & 0 & 0 & 0 \\ 0 & 1 & 0 & 0 & 0 & 0 \\ 0 & 0 & 1 & 0 & 0 & 0 \\ 0 & 0 & 0 & 1 & 0 & 0 \\ 0 & 0 & 0 & 0 & 1 & 1 \\ 0 & 0 & 0 & 0 & 0 & 1 \end{bmatrix}$;

(e) $\begin{bmatrix} 1 & 0 & 0 & 0 & 0 \\ 1 & 1 & 0 & 0 & 0 \\ 1 & 1 & 1 & 0 & 0 \\ 1 & 1 & 1 & 1 & 0 \\ 1 & 1 & 1 & 1 & 1 \end{bmatrix}$.

10.10. Show that

$$\begin{bmatrix} 1 & 0 & 1 & 0 & 1 \\ 0 & 1 & 0 & 1 & 0 \\ 1 & 0 & 1 & 0 & 1 \\ 0 & 1 & 0 & 1 & 0 \\ 1 & 0 & 1 & 0 & 1 \end{bmatrix}$$

is a singular matrix.

10.11. (Section 10.1). The four planes

$$6x - 3y - z = -3,$$

$$2x - y + 5z = 15,$$

$$y + z = 1,$$

$$2x + y - z = 1,$$

are the faces of a tetrahedron. Find the coordinates of all its vertices.

10.12. A light source is situated at the point $P:(2, 2, 2)$. A triangle has the points $A:(1, 1, 1)$, $B:(1, 0, 1)$, $C:(2, 1, 1)$ as vertices. Find the coordinates of the vertices of the shadow of the triangle on the coordinate planes $x = 0$, $y = 0$, and $z = 0$.

10.13. The parabola

$$y = \alpha + \beta x + \gamma x^2$$

is required which passes through the three points (x_1, y_1), (x_2, y_2), and (x_3, y_3). When solutions exist, find α, β, and γ, and discuss the cases where there are no solutions.

10.14. Find all values of the constants λ and μ in order that the equations

$$x + y + z = 4,$$

$$x - y + z = 2,$$

$$2x + y - \lambda z = \mu,$$

may have (a) just one solution, (b) no solutions, (c) an infinite set of solutions.

10.15. (Section 10.2). For each of the sets of equations below, set up the augmented matrix and, using elementary row operations, decide on the consistency of the equations. If they are consistent, obtain all solutions in each case.

(a) $x + y + z = 3,$

$$3x + 5y + z = -1,$$

$$x + 2y = 0;$$

(b)
$$y + z = 1,$$
$$x + y + 2z = 3,$$
$$x - y \quad\;\; = 1;$$

(c)
$$x + 2y + z = \quad 4,$$
$$x + y \quad\;\; = -1,$$
$$3x + 4y - z = 12.$$

10.16. (Section 10.4). Find all solutions of the determinant equation

$$\begin{vmatrix} 1-k & 2 & -1 \\ 2 & 1-k & -1 \\ -1 & -1 & 2-k \end{vmatrix} = 0.$$

What are the values of k for which the following set of equations has nontrivial solutions?

$$(1-k)x + \quad 2y \quad\quad - z = 0,$$
$$2x + (1-k)y \quad\quad - z = 0,$$
$$-x - \quad\quad y + (2-k)z = 0.$$

10.17. (Section 10.4). Show that

$$\begin{vmatrix} a^2+t & ab & ca \\ ab & b^2+t & bc \\ ca & bc & c^2+t \end{vmatrix} = t^2(t + a^2 + b^2 + c^2).$$

For what values of t do the equations

$$(1+t)x + \quad 2y + \quad\;\; 3z = 0,$$
$$2x + (4+t)y \quad\;\; + 6z = 0,$$
$$3x + \quad\;\; 6y + (9+t)z = 0,$$

have nontrivial solutions? Find all solutions in each case.

10.18. (Section 10.3). For what values of k do the equations

$$kx_1 + 4x_2 - x_3 + 3x_4 = 0,$$
$$4x_1 + x_2 - x_3 + 3x_4 = 0,$$
$$4x_1 - x_2 + kx_3 + 3x_4 = 0,$$
$$4x_1 - x_2 + 3x_3 + kx_4 = 0,$$

have nontrivial solutions?

10.19. (Section 10.3). Show that the equations

$$x_1 + 2x_2 + 3x_3 \quad\quad = 4,$$
$$2x_1 + 3x_2 + 8x_3 - x_4 = 20,$$
$$2x_1 + 5x_2 + 4x_3 + x_4 = 5,$$

are inconsistent.

10.20. (Section 10.2). Find the inverses of

$$\begin{bmatrix} 1 & \lambda & 0 \\ 0 & 1 & \lambda \\ 0 & 0 & 1 \end{bmatrix} \quad \text{and} \quad \begin{bmatrix} 1 & 0 & 0 \\ \mu & 1 & 0 \\ 0 & \mu & 1 \end{bmatrix}.$$

Hence find the inverse of

$$\begin{bmatrix} 1+\lambda\mu & \lambda & 0 \\ \mu & 1+\lambda\mu & \lambda \\ 0 & \mu & 1 \end{bmatrix}.$$

Find the inverse of

$$\begin{bmatrix} 13 & 3 & 0 \\ 4 & 13 & 3 \\ 0 & 4 & 1 \end{bmatrix}.$$

10.21. (Section 9.4). Express the determinant

$$\begin{vmatrix} 1 & 1 & 1 \\ a^2 & b^2 & c^2 \\ a(b+c) & b(c+a) & c(a+b) \end{vmatrix}$$

as the product of factors.

Obtain the values of a, b, and c for which nontrivial solutions of

$$x + a^2y + a(b+c)z = 0,$$
$$x + b^2y + b(c+a)z = 0,$$
$$x + c^2y + c(a+b)z = 0,$$

exist. Find the complete solution in the case $a + b = -c$.

10.22. (Section 10.5). Using the Gauss–Seidel iterative scheme solve the system of equations:

$$3x_1 + \;\; x_2 + \;\; x_3 = 5,$$
$$6x_2 - 2x_3 + 3x_4 = 6,$$
$$x_1 + 4x_3 - 2x_4 = 1,$$
$$x_2 + 2x_3 - 4x_4 = 2.$$

Show that the iterations converge to a solution accurate to four significant figures within eleven steps, starting from $(0, 1, 0, 0)$. Confirm also that the matrix of coefficients is diagonally dominant.

10.23. (Section 10.5). Show that the Gauss–Seidel scheme fails for the system

$$x_1 - 2x_2 + \;\; x_3 = 4,$$
$$x_1 - \;\; x_2 - \;\; x_3 = 1,$$
$$2x_1 + 3x_2 - 4x_3 = 4,$$

(the matrix in this case is not diagonally dominant.)

10.24. (Section 10.5). Show that one row of the matrix of coefficients fails to be dominant in the system

$$6x_1 - x_2 + x_3 = 2,$$
$$3x_1 + 2x_2 + x_3 = 1,$$
$$x_1 - x_2 + 4x_3 = 5.$$

However, confirm that the Gauss–Seidel scheme delivers a solution accurate to four significant figures after ten iterations starting at $(0, 0, 1)$.

10.25. For comparison purposes with the Gauss–Seidel method, solve the equations (10.7)–(10.9), namely

$$x_1 = \tfrac{1}{3}(-x_2 - x_3 - 1),$$
$$x_2 = \tfrac{1}{4}(x_1 - x_3 - 8),$$
$$x_3 = \tfrac{1}{5}(-2x_1 - x_2 - 14),$$

using the Jacobi method. How many steps are required to achieve the same accuracy as that in the table in Section 10.5, i.e. to five significant figures?

11

Eigenvalues and eigenvectors

11.1 Eigenvalues of a matrix

With any square matrix A, we can associate a set of homogeneous linear equations $Ax = 0$. As we saw in Section 10.4 of the previous chapter, such a set of equations will only have a nontrivial solution set if $\det A = 0$. Consider now the $n \times n$ set of equations

$$Ax = \lambda x, \quad \text{or} \quad (A - \lambda I_n)x = 0, \tag{11.1}$$

where λ is a parameter, and I_n is the unit matrix (see Section 7.3(5)). In order for these equations to have nontrivial solutions, we must have

$$\det(A - \lambda I_n) = 0.$$

This can only be satisfied if λ takes certain values. These are called the **eigenvalues** of the matrix A, and the equation they satisfy is called the **characteristic equation** of A. The characteristic equation is a polynomial equation, of degree n in λ. We usually list the eigenvalues as λ_1, λ_2, and so on.

Example 11.1. Find the eigenvalues of

$$A = \begin{bmatrix} 1 & 3 \\ 2 & 2 \end{bmatrix}.$$

The eigenvalues of A are given by the determinant equation

$$\det(A - \lambda I_2) = \begin{vmatrix} 1 - \lambda & 3 \\ 2 & 2 - \lambda \end{vmatrix} = 0,$$

which can be expanded into

$$(1 - \lambda)(2 - \lambda) - 6 = 0, \quad \text{or} \quad \lambda^2 - 3\lambda - 4 = 0.$$

This factorizes into $(\lambda - 4)(\lambda + 1) = 0$: hence the eigenvalues are $\lambda_1 = -1$, $\lambda_2 = 4$.

Example 11.2. Find the eigenvalues of

$$A = \begin{bmatrix} 2 & -2 \\ 1 & 4 \end{bmatrix}.$$

In this case

$$\det(A - \lambda I_2) = \begin{vmatrix} 2 - \lambda & -2 \\ 1 & 4 - \lambda \end{vmatrix} = (2 - \lambda)(4 - \lambda) + 2$$

$$= \lambda^2 - 6\lambda + 10 = 0,$$

and the quadratic equation has the roots

$$\lambda = \tfrac{1}{2}[6 \pm \sqrt{(36 - 40)}] = 3 \pm j.$$

Thus real matrices can have **complex eigenvalues**.

Example 11.3. Find the eigenvalues of

$$A = \begin{vmatrix} 1 & 2 & 1 \\ 2 & 1 & 1 \\ 1 & 1 & 2 \end{vmatrix}.$$

Here

$$\det(A - \lambda I_3) = \begin{vmatrix} 1 - \lambda & 2 & 1 \\ 2 & 1 - \lambda & 1 \\ 1 & 1 & 2 - \lambda \end{vmatrix}$$

$$= \begin{vmatrix} 4 - \lambda & 4 - \lambda & 4 - \lambda \\ 2 & 1 - \lambda & 1 \\ 1 & 1 & 2 - \lambda \end{vmatrix} \quad (\mathbf{r}_1' = \mathbf{r}_1 + \mathbf{r}_2 + \mathbf{r}_3)$$

$$= (4 - \lambda) \begin{vmatrix} 1 & 1 & 1 \\ 2 & 1 - \lambda & 1 \\ 1 & 1 & 2 - \lambda \end{vmatrix}$$

$$= (4 - \lambda) \begin{vmatrix} 1 & 0 & 0 \\ 2 & -1 - \lambda & -1 \\ 1 & 0 & 1 - \lambda \end{vmatrix} \quad \begin{pmatrix} \mathbf{c}_2' = \mathbf{c}_2 - \mathbf{c}_1 \\ \mathbf{c}_3' = \mathbf{c}_3 - \mathbf{c}_1 \end{pmatrix}$$

$$= (4 - \lambda)(-1 - \lambda)(1 - \lambda),$$

$$= 0,$$

if $\lambda = 4$ or ± 1. Hence the eigenvalues are

$$\lambda_1 = 4, \quad \lambda_2 = 1, \quad \lambda_3 = -1.$$

Eigenvalues

The eigenvalues of the $n \times n$ square matrix A are the
solutions λ of the determinant equation   **(11.2)**

$$\det(A - \lambda I_n) = 0.$$

11.2 Eigenvectors

Associated with each eigenvalue λ of A, there will be a nontrivial solution of the equation $(A - \lambda I_n)x = 0$. These are called the **eigenvectors of** A corresponding to the eigenvalue λ, and are generally denoted in this text by s. Thus, if λ is an eigenvalue of A, then there will exist a corresponding eigenvector $s \neq 0$ of

$$(A - \lambda I_n)s = 0.$$

The solution of this set of linear equations can be found by Gaussian elimination.

Example 11.4. Find the eigenvectors of

$$A = \begin{bmatrix} 1 & 3 \\ 2 & 2 \end{bmatrix}.$$

From Example 11.1, the eigenvalues are $\lambda_1 = 4$ and $\lambda_2 = -1$. Let the corresponding eigenvectors be

$$s_1 = \begin{bmatrix} a_1 \\ b_1 \end{bmatrix}, \qquad s_2 = \begin{bmatrix} a_2 \\ b_2 \end{bmatrix}.$$

Thus $(A - \lambda_1 I_2)s_1 = 0$ becomes

$$\begin{bmatrix} 1-4 & 3 \\ 2 & 2-4 \end{bmatrix}\begin{bmatrix} a_1 \\ b_1 \end{bmatrix} = \begin{bmatrix} 0 \\ 0 \end{bmatrix}, \quad \text{or} \quad \begin{cases} -3a_1 + 3b_1 = 0 \\ 2a_1 - 2b_1 = 0 \end{cases}.$$

Solution is easy in this case, and the solutions can be expressed as $a_1 = b_1 = \alpha$ for any α. If we put $\alpha = 1$, then an eigenvector is

$$s_1 = \begin{bmatrix} 1 \\ 1 \end{bmatrix}.$$

Any nonzero value of α will give an eigenvector; we usually choose a convenient value for the parameter to give one solution. The others are multiples of this.

Similarly $(A - \lambda_2 I_2)s_2 = 0$ implies

$$\begin{bmatrix} 1+1 & 3 \\ 2 & 2+1 \end{bmatrix}\begin{bmatrix} a_2 \\ b_2 \end{bmatrix} = 0, \quad \text{or} \quad \begin{cases} 2a_2 + 3b_2 = 0 \\ 2a_2 + 3b_2 = 0 \end{cases}.$$

An eigenvector in this case could be

$$s_2 = \begin{bmatrix} \beta \\ -\frac{2}{3}\beta \end{bmatrix},$$

for any nonzero β. As before, we choose a particular value of β which makes the eigenvector specific and simple. In this case we could put $\beta = 3$ to give the eigenvector

$$s_2 = \begin{bmatrix} 3 \\ -2 \end{bmatrix}.$$

Example 11.5. Find the eigenvectors of

$$A = \begin{bmatrix} 1 & 2 & 1 \\ 2 & 1 & 1 \\ 1 & 1 & 2 \end{bmatrix}.$$

The eigenvalues of A are $\lambda_1 = 4$, $\lambda_2 = 1$, $\lambda_3 = -1$ (see Example 11.3). Let the corresponding eigenvectors be

$$s_i = \begin{bmatrix} a_i \\ b_i \\ c_i \end{bmatrix} \quad (i = 1, 2, 3).$$

In each case, we need to solve $(A - \lambda_i I_3)s_i = 0$. If $\lambda_1 = 4$, then

$$-3a_1 + 2b_1 + c_1 = 0,$$
$$2a_1 - 3b_1 + c_1 = 0,$$
$$a_1 + b_1 - 2c_1 = 0.$$

Gaussian elimination leads to

$$\begin{bmatrix} -3 & 2 & 1 & 0 \\ 2 & -3 & 1 & 0 \\ 1 & 1 & -2 & 0 \end{bmatrix} \rightarrow \begin{bmatrix} -3 & 2 & 1 & 0 \\ 0 & -\frac{5}{3} & \frac{5}{3} & 0 \\ 0 & \frac{5}{3} & -\frac{5}{3} & 0 \end{bmatrix} \quad \begin{pmatrix} \mathbf{r}_2' = \mathbf{r}_2 + \frac{2}{3}\mathbf{r}_1 \\ \mathbf{r}_3' = \mathbf{r}_3 + \frac{1}{3}\mathbf{r}_1 \end{pmatrix}$$

$$\rightarrow \begin{bmatrix} -3 & 2 & 1 & 0 \\ 0 & -\frac{5}{3} & \frac{5}{3} & 0 \\ 0 & 0 & 0 & 0 \end{bmatrix} \quad (\mathbf{r}_3' = \mathbf{r}_3 + \mathbf{r}_2).$$

By back substitution, $c_1 = \alpha$, $b_1 = c_1 = \alpha$, $a_1 = \frac{1}{3}(2b_1 + c_1) = \alpha$. Thus, with $\alpha = 1$, an eigenvector is

$$s_1 = \begin{bmatrix} 1 \\ 1 \\ 1 \end{bmatrix}.$$

The other eigenvectors corresponding to λ_1 are simply multiples of s_1. Using the same procedure shows that the two eigenvalues corresponding respectively to λ_1 and λ_2 can be chosen to be

$$s_2 = \begin{bmatrix} -1 \\ -1 \\ 2 \end{bmatrix}, \qquad s_2 = \begin{bmatrix} 1 \\ -1 \\ 0 \end{bmatrix}.$$

Example 11.6. Find the eigenvalues and eigenvectors of

$$A = \begin{bmatrix} 1 & 2 & -1 \\ 1 & 2 & -1 \\ 2 & 2 & -1 \end{bmatrix}.$$

In this example,

$$\det(A - \lambda I_3) = \begin{vmatrix} 1 - \lambda & 2 & -1 \\ 1 & 2 - \lambda & -1 \\ 2 & 2 & -1 - \lambda \end{vmatrix}$$

$$= \begin{vmatrix} -\lambda & \lambda & 0 \\ 1 & 2 - \lambda & -1 \\ 2 & 2 & -1 - \lambda \end{vmatrix} \quad (\mathbf{r}_1' = \mathbf{r}_1 - \mathbf{r}_2)$$

$$= \begin{vmatrix} -\lambda & 0 & 0 \\ 1 & 3 - \lambda & -1 \\ 2 & 4 & -1 - \lambda \end{vmatrix} \quad (\mathbf{c}_2' = \mathbf{c}_2 + \mathbf{c}_1)$$

$$= -\lambda[(3 - \lambda)(-1 - \lambda) + 4]$$

$$= -\lambda(\lambda - 1)^2.$$

This particular matrix has an eigenvalue 0 and a **repeated** eigenvalue 1. How does this affect the eigenvectors? Let the eigenvectors be, for $\lambda_1 = 0$ and $\lambda_2 = 1$,

$$s_i = \begin{bmatrix} a_i \\ b_i \\ c_i \end{bmatrix} \quad (i = 1, 2).$$

For $\lambda_1 = 0$,

$$a_1 + 2b_1 - c_1 = 0,$$
$$a_1 + 2b_1 - c_1 = 0,$$
$$2a_1 + 2b_1 - c_1 = 0.$$

Hence $a_1 = 0$, $b_1 = \alpha$, $c_1 = 2\alpha$, for any α. An eigenvector is

$$s_1 = \begin{bmatrix} 0 \\ 1 \\ 2 \end{bmatrix}.$$

For $\lambda_2 = 1$,

$$2b_2 - c_2 = 0,$$
$$a_2 + b_2 - c_2 = 0,$$
$$2a_2 + 2b_2 - 2c_2 = 0.$$

If we let $b_2 = \beta$, then $c_2 = 2\beta$ and $a_2 = c_2 - b_2 = \beta$. Hence we can associate with $\lambda_2 = 1$ the eigenvector

$$s_2 = \begin{bmatrix} 1 \\ 1 \\ 2 \end{bmatrix},$$

by putting $\beta = 1$. There are only two independent eigenvectors in this case.

Note that **if A has a zero eigenvalue, then A must be a singular matrix** since det $A = 0$. And conversely, if A is singular, then A has at least one zero eigenvalue.

The matrix in Example 11.6 has two eigenvalues (one repeated) and two eigenvectors. The meaning of this reduced eigenvector set will be illustrated in the context of coordinate transformations in Section 11.4. As the next example illustrates, a matrix can have a repeated eigenvalue but still retain a full set of independent eigenvectors.

Example 11.7. Find the eigenvalues and eigenvectors of

$$A = \begin{bmatrix} 3 & 0 & -1 \\ 0 & 1 & 0 \\ 2 & 0 & 0 \end{bmatrix}.$$

Thus

$$\det(A - \lambda I_3) = \begin{vmatrix} 3 - \lambda & 0 & -1 \\ 0 & 1 - \lambda & 0 \\ 2 & 0 & -\lambda \end{vmatrix}$$

$$= (3 - \lambda)(1 - \lambda)(-\lambda) + (-1)(-2)(1 - \lambda)$$

$$= (1 - \lambda)[-3\lambda + \lambda^2 + 2]$$

$$= -(\lambda - 2)(\lambda - 1)^2.$$

Let $\lambda_1 = 2$ and $\lambda_2 = 1$ with corresponding eigenvectors

$$s_i = \begin{bmatrix} a_i \\ b_i \\ c_i \end{bmatrix} \quad (i = 1, 2).$$

For $\lambda_1 = 2$,

$$a_1 \qquad - c_1 = 0,$$
$$- b_1 \qquad = 0,$$
$$2a_1 \qquad - 2c_1 = 0.$$

We can let $b_1 = 0$, $c_1 = \alpha$, $a_1 = \alpha$. Hence we can choose

$$s_1 = \begin{bmatrix} 1 \\ 0 \\ 1 \end{bmatrix}.$$

For $\lambda_2 = 1$,

$$2a_1 - c_1 = 0,$$
$$0 = 0,$$
$$2a_1 - c_1 = 0.$$

If $a_1 = \beta$, then $c_1 = 2\beta$ but b_1 can then take any value γ, say. Hence, the eigenvector set is

$$
s_2 = \begin{bmatrix} \beta \\ \gamma \\ 2\beta \end{bmatrix} = \beta \begin{bmatrix} 1 \\ 0 \\ 2 \end{bmatrix} + \gamma \begin{bmatrix} 0 \\ 1 \\ 0 \end{bmatrix},
$$

that is, it contains *two* parameters β and γ. The choices of $\beta = 1$ with $\gamma = 0$, and $\beta = 0$ with $\gamma = 1$, say, give two independent eigenvectors

$$
\begin{bmatrix} 1 \\ 0 \\ 2 \end{bmatrix} \quad \text{and} \quad \begin{bmatrix} 0 \\ 1 \\ 0 \end{bmatrix}.
$$

Unlike in the previous example, we can associate three distinct eigenvectors with this matrix even though the matrix has only two eigenvalues. We shall take up this point again in connection with the diagonalization of matrices.

> **Eigenvectors**
> The eigenvectors of a square matrix A are the non-trivial solutions s of the homogeneous equations   **(11.3)**
> $$(A - \lambda_r I_n)s = 0, \quad \text{for each eigenvalue } \lambda_r.$$

11.3 Linear dependence

It is useful in mathematics to gather, in a collection or set, elements which have common features. For example, we might consider the set of all integers, the set of all fractions, or the set of all real numbers. In a similar way, we can gather all $m \times n$ matrices. They all obey certain rules, and are said to form a vector space. We shall not consider the general case here, but restrict ourselves to the set of *all* $m \times 1$ column vectors: this set is called an m-dimensional **vector space** V_m. These vectors obey the rules of matrix algebra. Thus if

$$
s_1 = \begin{bmatrix} a_1 \\ a_2 \\ \vdots \\ a_m \end{bmatrix}, \qquad s_2 = \begin{bmatrix} b_1 \\ b_2 \\ \vdots \\ b_m \end{bmatrix},
$$

then s_1 and s_2 belong to V_m, and so does $\alpha s_1 + \beta s_2$ for any constants α and β.

An important set of vectors in $\mathcal{V}_m$ is the set of **base vectors**

$$e_1 = \begin{bmatrix} 1 \\ 0 \\ 0 \\ \vdots \\ 0 \end{bmatrix}, \quad e_2 = \begin{bmatrix} 0 \\ 1 \\ 0 \\ \vdots \\ 0 \end{bmatrix}, \dots, e_m = \begin{bmatrix} 0 \\ 0 \\ 0 \\ \vdots \\ 1 \end{bmatrix}.$$

Any vector in $\mathcal{V}_m$ can be expressed as a **linear combination** of these vectors. Thus

$$x_1 = \begin{bmatrix} a_1 \\ a_2 \\ \vdots \\ a_m \end{bmatrix} = a_1 e_1 + a_2 e_2 + \cdots + a_m e_m.$$

The set of vectors $\{e_1, e_2, \dots, e_m\}$ is said, therefore, to form a **basis** of $\mathcal{V}_m$. None of the vectors $e_1, e_2, \dots, e_m$ can be expressed as a linear combination of the others, so that they are said to be **linearly independent**. A set of n column vectors $s_1, s_2, \dots, s_n$ is said to be **linearly dependent** if there exist constants $\alpha_1, \alpha_2, \dots, \alpha_n$, **not all zero**, such that

$$\alpha_1 s_1 + \alpha_2 s_2 + \cdots + \alpha_n s_n = \mathbf{0}.$$

If the above equation holds *only* when $\alpha_1 = \alpha_2 = \cdots = \alpha_n = 0$, then the vectors are linearly independent. It can be proved that any set of linearly m independent vectors can form a basis of the vector space $\mathcal{V}_m$.

11.4 Point mappings
Any matrix can be interpreted in terms of a transformation between coordinates. For example if (x, y, z) and (X, Y, Z) are coordinates in two three-dimensional cartesian spaces and A is a 3×3 matrix, then the mapping or transformation

$$X = Ax, \quad X = \begin{bmatrix} X \\ Y \\ Z \end{bmatrix}, \quad x = \begin{bmatrix} x \\ y \\ z \end{bmatrix},$$

maps points in the (x, y, z) space into points in the (X, Y, Z) space as shown in Fig. 11.1. The inverse mapping, if it exists, is given by

$$x = A^{-1}X,$$

in which case, all points in the (X, Y, Z) space can be mapped into points in the (x, y, z) space. While straight lines map into straight lines, other quantities such as the distances between points and the angles between lines will generally not be preserved in such mappings.

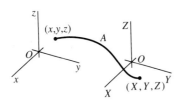

Fig. 11.1

What happens if A^{-1} does not exist? It means that there are points in the (X, Y, Z) space which cannot be reached from any point in (x, y, z). For example, if

$$A = \begin{bmatrix} 1 & 2 & -1 \\ 3 & 4 & -2 \\ 2 & 2 & -1 \end{bmatrix},$$

then

$$X = x + 2y - z, \qquad Y = 3x + 4y - 2z, \qquad Z = 2x + 2y - z.$$

A is a singular matrix and has no inverse, since det $A = 0$. Solve the equations for x, y, and z by Gaussian elimination. Thus

$$\begin{bmatrix} 1 & 2 & -1 & X \\ 3 & 4 & -2 & Y \\ 2 & 2 & -1 & Z \end{bmatrix} \rightarrow \begin{bmatrix} 1 & 2 & -1 & X \\ 0 & -2 & 1 & Y - 3X \\ 0 & -2 & 1 & Z - 2X \end{bmatrix} \quad \begin{pmatrix} \mathbf{r}_2' = \mathbf{r}_2 - 3\mathbf{r}_1 \\ \mathbf{r}_3' = \mathbf{r}_3 - 2\mathbf{r}_1 \end{pmatrix}$$

$$\rightarrow \begin{bmatrix} 1 & 2 & 1 & X \\ 0 & -2 & 1 & Y - 3X \\ 0 & 0 & 0 & Z + X - Y \end{bmatrix} \quad (\mathbf{r}_3' = \mathbf{r}_3 - \mathbf{r}_2).$$

The equations will only be consistent if $Z + X - Y = 0$. Hence *all* points in the (x, y, z) space map on to the *plane* $X - Y + Z = 0$.

A similar situation arises if the matrix in Example 11.6 is interpreted as a transformation matrix. The zero eigenvalue indicates a **singular** transformation.

It is possible for all points in the (x, y, z) space to map on to a *line* in (X, Y, Z). For example, if

$$A = \begin{bmatrix} 1 & 2 & 3 \\ 2 & 4 & 6 \\ 3 & 6 & 9 \end{bmatrix},$$

then in an obvious way, since the rows satisfy $\mathbf{r}_2 = 2\mathbf{r}_1$ and $\mathbf{r}_3 = 3\mathbf{r}_1$, all transformed points lie on the line given by $X = \alpha$, $Y = 2\alpha$, $Z = 3\alpha$, where α is a parameter.

11.5 Diagonalization of a matrix

We will take a constructive approach to this problem for a 3×3 matrix. Consider the matrix of Examples 11.3 and 11.5, namely

$$A = \begin{bmatrix} 1 & 2 & 1 \\ 2 & 1 & 1 \\ 1 & 1 & 2 \end{bmatrix}$$

which has the eigenvalues $\lambda_1 = 4$, $\lambda_2 = 1$, $\lambda_3 = -1$ and eigenvectors

$$s_1 = \begin{bmatrix} 1 \\ 1 \\ 1 \end{bmatrix} \qquad s_2 = \begin{bmatrix} -1 \\ -1 \\ 2 \end{bmatrix} \qquad s_3 = \begin{bmatrix} 1 \\ -1 \\ 0 \end{bmatrix}.$$

Construct a matrix C which has these eigenvectors as its columns:

$$C = [s_1 \quad s_2 \quad s_3] = \begin{bmatrix} 1 & -1 & 1 \\ 1 & -1 & -1 \\ 1 & 2 & 0 \end{bmatrix}.$$

Form the product

$$AC = A[s_1 \quad s_2 \quad s_3] = [As_1 \quad As_2 \quad As_3] = [\lambda_1 s_1 \quad \lambda_2 s_2 \quad \lambda_3 s_3],$$

the last equality holding since the eigenvector s_i is defined as a nonzero solution of $As_i = \lambda_i s_i$. Hence

$$AC = [s_1 \quad s_2 \quad s_3]D,$$

where

$$D = \begin{bmatrix} \lambda_1 & 0 & 0 \\ 0 & \lambda_2 & 0 \\ 0 & 0 & \lambda_3 \end{bmatrix},$$

that is, D is a diagonal matrix with eigenvalue elements. Thus

$$AC = [s_1 \quad s_2 \quad s_3]D = CD.$$

If we premultiply this equation by C^{-1}, then

$$C^{-1}AC = C^{-1}CD = I_3 D = D.$$

Effectively the matrix C has **diagonalized** the matrix A. In the example,

$$C^{-1} = \begin{bmatrix} 1 & -1 & 1 \\ 1 & -1 & -1 \\ 1 & 2 & 0 \end{bmatrix}^{-1} = \begin{bmatrix} \frac{1}{3} & \frac{1}{3} & \frac{1}{3} \\ -\frac{1}{6} & -\frac{1}{6} & \frac{1}{3} \\ \frac{1}{2} & -\frac{1}{2} & 0 \end{bmatrix}.$$

Finally, it can be checked that

$$\begin{bmatrix} \frac{1}{3} & \frac{1}{3} & \frac{1}{3} \\ -\frac{1}{6} & -\frac{1}{6} & \frac{1}{3} \\ \frac{1}{2} & -\frac{1}{2} & 0 \end{bmatrix} \begin{bmatrix} 1 & 2 & 1 \\ 2 & 1 & 1 \\ 1 & 2 & 0 \end{bmatrix} \begin{bmatrix} 1 & -1 & 1 \\ 1 & -1 & -1 \\ 1 & 2 & 0 \end{bmatrix} = \begin{bmatrix} 4 & 0 & 0 \\ 0 & 1 & 0 \\ 0 & 0 & -1 \end{bmatrix}.$$

It might appear at first sight that there is not a unique answer for D since C is not uniquely defined. However, if C is replaced by kC, with $k \neq 0$, then $(kC)^{-1} = k^{-1}C^{-1}$, from which it still follows that

$$(kC)^{-1}A(kC) = \frac{1}{k}C^{-1}ACk = C^{-1}AC = D,$$

since the k always cancels out. Individual columns in C can also be multiplied by different factors, depending on the choice of eigenvector, without changing the outcome.

Example 11.8. Use the eigenvalues and eigenvectors of

$$\begin{bmatrix} 3 & 0 & -1 \\ 0 & 1 & 0 \\ 2 & 0 & 0 \end{bmatrix}$$

obtained in Example 11.7 to construct a transformation which diagonalizes A, and verify that the diagonalized matrix is

$$D = \begin{bmatrix} 2 & 0 & 0 \\ 0 & 1 & 0 \\ 0 & 0 & 1 \end{bmatrix}.$$

From Example 11.7, we see that A has the eigenvalues $\lambda_1 = 2$ and $\lambda_2 = \lambda_3 = 1$. However, we can associate two linearly independent eigenvectors with the repeated eigenvalue. Thus, we can define C by

$$C = [s_1 \quad s_2 \quad s_3] = \begin{bmatrix} 1 & 1 & 0 \\ 0 & 0 & 1 \\ 1 & 2 & 0 \end{bmatrix}.$$

Its inverse is

$$C^{-1} = \begin{bmatrix} 2 & 0 & -1 \\ -1 & 0 & 1 \\ 0 & 1 & 0 \end{bmatrix}.$$

Finally it can be verified that

$$C^{-1}AC = \begin{bmatrix} 2 & 0 & -1 \\ -1 & 0 & 1 \\ 0 & 1 & 0 \end{bmatrix} \begin{bmatrix} 3 & 0 & -1 \\ 0 & 1 & 0 \\ 2 & 0 & 0 \end{bmatrix} \begin{bmatrix} 1 & 1 & 0 \\ 0 & 0 & 1 \\ 1 & 2 & 0 \end{bmatrix}$$

$$= \begin{bmatrix} 2 & 0 & 0 \\ 0 & 1 & 0 \\ 0 & 0 & 1 \end{bmatrix} = D.$$

Following the remarks just before this example

$$C = [2s_1 \quad 3s_2 \quad -s_3] = \begin{bmatrix} 2 & 3 & 0 \\ 0 & 0 & -1 \\ 2 & 6 & 0 \end{bmatrix}.$$

would equally well be an acceptable matrix in the diagonalization.

Example 11.9. Find a transformation which diagonalizes the matrix

$$A = \begin{bmatrix} 2 & -2 \\ 1 & 4 \end{bmatrix}.$$

From Example 11.2, the eigenvalues are $\lambda_1 = 3 + j$, $\lambda_2 = 3 - j$. Corresponding eigenvectors are

$$s_1 = \begin{bmatrix} -1 + j \\ 1 \end{bmatrix}, \qquad s_2 = \begin{bmatrix} -1 - j \\ 1 \end{bmatrix}.$$

The eigenvalues and eigenvectors are complex-valued but this does not affect the method. The matrix C becomes

$$C = [s_1 \quad s_2] = \begin{bmatrix} -1 + j & -1 - j \\ 1 & 1 \end{bmatrix}.$$

Its inverse is

$$C^{-1} = \frac{1}{\det C} \begin{bmatrix} 1 & 1 + j \\ -1 & -1 + j \end{bmatrix} = \frac{1}{2j} \begin{bmatrix} 1 & 1 + j \\ -1 & -1 + j \end{bmatrix}.$$

Finally check that

$$C^{-1}AC = \frac{1}{2j} \begin{bmatrix} 1 & 1 + j \\ -1 & -1 + j \end{bmatrix} \begin{bmatrix} 2 & -2 \\ 1 & 4 \end{bmatrix} \begin{bmatrix} -1 + j & -1 - j \\ 1 & 1 \end{bmatrix}$$

$$= \begin{bmatrix} 3 + j & 0 \\ 0 & 3 - j \end{bmatrix}.$$

Diagonalizing a matrix.

To diagonalize a matrix A:

   (i) find the eigenvalues of A;
   (ii) find n linearly independent eigenvectors s_n of A **(11.4)**
      (if they exist);
   (iii) construct the matrix C of eigenvectors;
   (iv) calculate the inverse C^{-1} of C;
   (v) compute $C^{-1}AC$.

Not all matrices can be diagonalized in this way. In Example 11.6 where

$$A = \begin{bmatrix} 1 & 2 & -1 \\ 1 & 2 & -1 \\ 2 & 2 & -1 \end{bmatrix},$$

we can associate only two linearly independent eigenvectors with the eigenvalue 0 and the repeated eigenvalue 1, and no diagonalizing matrix C can be constructed.

11.6 Powers of matrices

The transformation C of the previous section can be used to obtain a formula for calculating powers of square matrices. This follows since it is a simple matter to find powers of diagonal matrices. Thus if

$$D = \begin{bmatrix} \lambda_1 & 0 & 0 \\ 0 & \lambda_2 & 0 \\ 0 & 0 & \lambda_3 \end{bmatrix},$$

then

$$D^2 = \begin{bmatrix} \lambda_1 & 0 & 0 \\ 0 & \lambda_2 & 0 \\ 0 & 0 & \lambda_3 \end{bmatrix}\begin{bmatrix} \lambda_1 & 0 & 0 \\ 0 & \lambda_2 & 0 \\ 0 & 0 & \lambda_3 \end{bmatrix} = \begin{bmatrix} \lambda_1^2 & 0 & 0 \\ 0 & \lambda_2^2 & 0 \\ 0 & 0 & \lambda_3^2 \end{bmatrix},$$

and, in general,

$$D^n = \begin{bmatrix} \lambda_1^n & 0 & 0 \\ 0 & \lambda_2^n & 0 \\ 0 & 0 & \lambda_3^n \end{bmatrix}.$$

In the previous section we showed that, if a 3×3 matrix A has three linearly independent eigenvectors, then we can find a matrix C such that

$$AC = CD,$$

where D is a diagonal matrix, its elements consisting of the eigenvalues of A. Thus by multiplying on the right by C^{-1} we find that

$$A = CDC^{-1}.$$

Hence

$$A^2 = CDC^{-1}CDC^{-1} = CDI_3DC^{-1} = CD^2C^{-1},$$

since $C^{-1}C = I_3$. Continuing this process, we find that

$$A^3 = A^2A = CD^2C^{-1}CDC^{-1} = CD^3C^{-1},$$

and, in general,

$$A^n = CD^nC^{-1}.$$

Example 11.10. Find a formula for A^n, where

$$A = \begin{bmatrix} 1 & 2 & 1 \\ 2 & 1 & 1 \\ 1 & 1 & 2 \end{bmatrix}.$$

(See Examples 11.3 and 11.5 and Section 11.5.)

The eigenvalues of A are $\lambda_1 = 4$, $\lambda_2 = 1$, $\lambda_3 = -1$; and the diagonalizing transformation, with its inverse, is

$$C = \begin{bmatrix} 1 & -1 & 1 \\ 1 & -1 & -1 \\ 1 & 2 & 0 \end{bmatrix}, \qquad C^{-1} = \begin{bmatrix} \frac{1}{3} & \frac{1}{3} & \frac{1}{3} \\ -\frac{1}{6} & -\frac{1}{6} & \frac{1}{3} \\ \frac{1}{2} & -\frac{1}{2} & 0 \end{bmatrix}.$$

Hence

$$A^n = CD^nC^{-1} = \begin{bmatrix} 1 & -1 & 1 \\ 1 & -1 & -1 \\ 1 & 2 & 0 \end{bmatrix} \begin{bmatrix} 4 & 0 & 0 \\ 0 & 1 & 0 \\ 0 & 0 & -1 \end{bmatrix}^n \begin{bmatrix} \frac{1}{3} & \frac{1}{3} & \frac{1}{3} \\ -\frac{1}{6} & -\frac{1}{6} & \frac{1}{3} \\ \frac{1}{2} & -\frac{1}{2} & 0 \end{bmatrix}$$

$$= \begin{bmatrix} 1 & -1 & 1 \\ 1 & -1 & -1 \\ 1 & 2 & 0 \end{bmatrix} \begin{bmatrix} 4^n & 0 & 0 \\ 0 & 1^n & 0 \\ 0 & 0 & (-1)^n \end{bmatrix} \begin{bmatrix} \frac{1}{3} & \frac{1}{3} & \frac{1}{3} \\ -\frac{1}{6} & -\frac{1}{6} & \frac{1}{3} \\ \frac{1}{2} & -\frac{1}{2} & 0 \end{bmatrix}$$

$$= \begin{bmatrix} 4^n & -1 & (-1)^n \\ 4^n & -1 & -(-1)^n \\ 4^n & 2 & 0 \end{bmatrix} \begin{bmatrix} \frac{1}{3} & \frac{1}{3} & \frac{1}{3} \\ -\frac{1}{6} & -\frac{1}{6} & \frac{1}{3} \\ \frac{1}{2} & -\frac{1}{2} & 0 \end{bmatrix}$$

$$= \frac{4^n}{3} \begin{bmatrix} 1 & 1 & 1 \\ 1 & 1 & 1 \\ 1 & 1 & 1 \end{bmatrix} + \frac{1}{6} \begin{bmatrix} 1 & 1 & -2 \\ 1 & 1 & -2 \\ -2 & -2 & 4 \end{bmatrix} + \frac{(-1)^n}{2} \begin{bmatrix} 1 & -1 & 0 \\ -1 & 1 & 0 \\ 0 & 0 & 0 \end{bmatrix}.$$

Example 11.11. Let

$$P = \begin{bmatrix} 1 - \alpha & \alpha \\ \beta & 1 - \beta \end{bmatrix},$$

where $0 < \alpha, \beta < 1$. Find P^n and $\lim_{n \to \infty} P^n$.

The matrix P is an example of a **row-stochastic** matrix, that is, all elements are non-negative and the sum of the elements in each row is 1. The eigenvalues of P are given by

$$\begin{vmatrix} 1 - \alpha - \lambda & \alpha \\ \beta & 1 - \beta - \lambda \end{vmatrix} = 0.$$

Hence

$$(1 - \alpha - \lambda)(1 - \beta - \lambda) - \alpha\beta = 0,$$

or

$$\lambda^2 - \lambda(2 - \alpha - \beta) + 1 - \alpha - \beta = 0.$$

The roots $\lambda_1 = 1$, $\lambda_2 = 1 - \alpha - \beta = p$, say. Choose the corresponding eigenvectors

$$s_1 = \begin{bmatrix} 1 \\ 1 \end{bmatrix}, \qquad s_2 = \begin{bmatrix} -\alpha \\ \beta \end{bmatrix}.$$

Let

$$C = [s_1 \quad s_2] = \begin{bmatrix} 1 & -\alpha \\ 1 & \beta \end{bmatrix}.$$

Its inverse is given by

$$C^{-1} = \frac{1}{\alpha + \beta} \begin{bmatrix} \beta & \alpha \\ -1 & 1 \end{bmatrix}.$$

Thus

$$P^n = CD^nC^{-1} = \begin{bmatrix} 1 & -\alpha \\ 1 & \beta \end{bmatrix} \begin{bmatrix} 1 & 0 \\ 0 & p^n \end{bmatrix} \begin{bmatrix} \beta & \alpha \\ -1 & 1 \end{bmatrix} \frac{1}{\alpha + \beta}$$

$$= \frac{1}{\alpha + \beta} \begin{bmatrix} 1 & -\alpha p^n \\ 1 & \beta p^n \end{bmatrix} \begin{bmatrix} \beta & \alpha \\ -1 & 1 \end{bmatrix}$$

$$= \frac{1}{\alpha + \beta} \begin{bmatrix} \beta + \alpha p^n & \alpha - \alpha p^n \\ \beta - \beta p^n & \alpha + \beta p^n \end{bmatrix}$$

$$= \frac{1}{\alpha + \beta} \begin{bmatrix} \beta & \alpha \\ \beta & \alpha \end{bmatrix} + \frac{p^n}{\alpha + \beta} \begin{bmatrix} \alpha & -\alpha \\ -\beta & \beta \end{bmatrix}.$$

Since $0 < \alpha < 1$ and $0 < \beta < 1$, it follows that

$$p = 1 - \alpha - \beta < 1 \quad \text{and} \quad p = 1 - \alpha - \beta > 1 - 1 - 1 = -1,$$

that is, $|p| < 1$. As $n \to \infty$, then $p^n \to 0$ and

$$P^n \to \frac{1}{\alpha + \beta} \begin{bmatrix} \beta & \alpha \\ \beta & \alpha \end{bmatrix}.$$

Powers of a square matrix.

To find the power A^n of a diagonalizable matrix A:

  (i) find the eigenvalues and eigenvectors of A;

  (ii) construct a matrix C such that $C^{-1}AC = D$,   **(11.5)** where D is the diagonal matrix of eigenvalues;

  (iii) the required answer is

    $A^n = CD^nC^{-1}.$

11.7 Quadratic forms

Suppose that $x = [x_1, x_2, \ldots, x_n]^T$, an n-dimensional column vector with elements $x_1, x_2, \ldots, x_n$. Any polynomial function of these elements in which every term is of degree two in them is known as a **quadratic form**. Thus, if $n = 3$, then

$$x_1^2 + 8x_1x_2 + x_2^2 + 6x_2x_3 + x_3^2 \quad .$$

is an example of a quadratic form. Quadratic forms can always be expressed as a matrix product of the form

$$x^T A x.$$

The example above can be written as

$$[x_1 \quad x_2 \quad x_3] \begin{bmatrix} 1 & 4 & 0 \\ 4 & 1 & 3 \\ 0 & 3 & 1 \end{bmatrix} \begin{bmatrix} x_1 \\ x_2 \\ x_3 \end{bmatrix}. \tag{11.6}$$

In this representation, A is required to be a *symmetric* matrix. Nonsymmetric representations are possible with, for example,

$$A = \begin{bmatrix} 1 & 0 & 0 \\ 8 & 1 & 2 \\ 0 & 4 & 1 \end{bmatrix},$$

but the symmetric form is standard.

Let us find the eigenvalues of the symmetric matrix in (11.6) in the usual way by solving

$$\begin{vmatrix} 1-\lambda & 4 & 0 \\ 4 & 1-\lambda & 3 \\ 0 & 3 & 1-\lambda \end{vmatrix} = 0.$$

Hence

$$(1-\lambda)[(1-\lambda)^2 - 9] - 4 \cdot 4 \cdot (1-\lambda) = 0$$

or

$$(1-\lambda)[(1-\lambda)^2 - 25] = 0.$$

It follows that the eigenvalues are $\lambda_1 = 1$, $\lambda_2 = -4$, $\lambda_3 = 6$. It can be shown by the methods previously explained that corresponding eigenvectors are

$$s_1 = \begin{bmatrix} 3 \\ 0 \\ -4 \end{bmatrix}, \qquad s_2 = \begin{bmatrix} -4 \\ 5 \\ -3 \end{bmatrix}, \qquad s_3 = \begin{bmatrix} 4 \\ 5 \\ 3 \end{bmatrix}.$$

If a and b are two column vectors, and

$$a^T b = 0,$$

then a and b are said to be **orthogonal**. If we examine the eigenvectors s_1, s_2, and s_3 above, then it is easy to see that

$$s_1^T s_2 = [3 \quad 0 \quad -4] \begin{bmatrix} -4 \\ 5 \\ -3 \end{bmatrix} = -12 + 0 + 12 = 0,$$

and similarly that $s_2^T s_3 = 0$ and $s_3^T s_1 = 0$. Thus the three eigenvectors are **mutually orthogonal**: regarded as ordinary vectors in the sense of Chapter 9, they are **mutually perpendicular**.

It will be shown that this property of the eigenvalues follows from the symmetry of the matrix of the quadratic form. However, we first show that the eigenvectors of a symmetric matrix must be real numbers.

Theorem 11.1. If A is a symmetric real matrix, then its eigenvalues are real.

Proof. Suppose that $\lambda_1 = \alpha + j\beta$ is an eigenvalue. Since the equation $\det(A - \lambda I_n) = 0$ is a real polynomial in λ, it must also have an eigenvalue $\bar{\lambda}_1 = \alpha - j\beta$. Let s_1 and $\bar{s}_1$ be the eigenvectors corresponding to λ_1 and its conjugate $\bar{\lambda}_1$. Thus

$$As_1 = \lambda_1 s_1, \qquad A\bar{s}_1 = \bar{\lambda}_1\bar{s}_1. \tag{11.7}$$

Since A is symmetric, it follows that $(A\bar{s}_1)^{\mathrm{T}} = \bar{s}_1^{\mathrm{T}}A^{\mathrm{T}} = \bar{s}_1^{\mathrm{T}}A$, and we can replace (11.7) by

$$As_1 = \lambda_1 s_1, \qquad \bar{s}_1^{\mathrm{T}}A = \bar{\lambda}_1\bar{s}_1^{\mathrm{T}}. \tag{11.8}$$

Multiply the first equation in (11.8) on the left by $\bar{s}_1^{\mathrm{T}}$, and the second equation on the right by s_1. Thus

$$\bar{s}_1^{\mathrm{T}}As_1 = \lambda_1\bar{s}_1^{\mathrm{T}}s_1, \qquad \bar{s}_1^{\mathrm{T}}As_1 = \bar{\lambda}_1\bar{s}_1^{\mathrm{T}}s_1.$$

Elimination of $\bar{s}_1^{\mathrm{T}}As_1$ leads to

$$(\lambda - \bar{\lambda}_1)\bar{s}_1^{\mathrm{T}}s_1 = 0.$$

Since $\bar{a}_n a_n = |a_n|^2$,

$$\bar{s}_1^{\mathrm{T}}s_1 = [\bar{a}_1 \quad \bar{a}_2 \cdots \bar{a}_n]\begin{bmatrix} a_1 \\ a_2 \\ \vdots \\ a_n \end{bmatrix} = |a_1|^2 + |a_2|^2 + \cdots + |a_n|^2 > 0,$$

where

$$s_1 = \begin{bmatrix} a_1 \\ a_2 \\ \vdots \\ a_n \end{bmatrix},$$

it follows that $\lambda_1 = \bar{\lambda}_1$ or $\alpha + j\beta = \alpha - j\beta$, from which we conclude that $\beta = 0$.

Theorem 11.2. If A is a symmetric matrix, then the eigenvectors associated with two *distinct* eigenvalues are orthogonal.

Proof. Let λ_1 and λ_2 be the distinct eigenvalues, and s_1 and s_2 their corresponding eigenvectors. Then

$$As_1 = \lambda_1 s_1, \qquad As_2 = \lambda_2 s_2.$$

Transpose the second equation so that the equations become

$$As_1 = \lambda_1 s_1, \qquad s_2^T A = \lambda_2 s_2^T. \tag{11.9}$$

Multiply the first equation in (11.9) by s_2^T on the left, and the second equation by s_1 on the right. Hence

$$s_2^T A s_1 = \lambda_1 s_2^T s_1, \qquad s_2^T A s_1 = \lambda_2 s_2^T s_1.$$

Eliminate $s_2^T A s_1$ between these equations, leaving

$$\lambda_1 s_2^T s_1 = \lambda_2 s_2^T s_1, \quad \text{or} \quad (\lambda_1 - \lambda_2) s_2^T s_1 = 0.$$

Since $\lambda_1 \neq \lambda_2$, it follows that $s_2^T s_1 = 0$; namely s_1 and s_2 are orthogonal.

11.8 Positive-definite matrices

A quadratic form $x^T A x$ is said to be **positive-definite** if $x^T A x > 0$ for all $x \neq 0$. If this is true, we simply describe the matrix A as **positive-definite**. Suppose that we consider the particular case in which A is a 3×3 matrix. Let $\lambda_1, \lambda_2, \lambda_3$ be its eigenvalues with corresponding eigenvectors s_1, s_2, s_3 **which are chosen so that they are all unit vectors**, that is, $s_1^T s_1 = s_2^T s_2 = s_3^T s_3 = 1$.

Suppose that A is symmetric. As we saw in Section 11.5, we can diagonalize A by using the matrix

$$C = [s_1 \quad s_2 \quad s_3],$$

so that

$$C^{-1}AC = D = \begin{bmatrix} \lambda_1 & 0 & 0 \\ 0 & \lambda_2 & 0 \\ 0 & 0 & \lambda_3 \end{bmatrix}.$$

For a symmetric matrix, the eigenvectors are orthogonal (Theorem 11.2). Hence

$$s_1^T C = s_1^T [s_1 \quad s_2 \quad s_3],$$
$$= [s_1^T s_1 \quad s_1^T s_2 \quad s_1^T s_3],$$
$$= [1 \quad 0 \quad 0],$$

since s_1 is a unit vector. In a similar way,

$$s_2^T C = [0 \quad 1 \quad 0], \qquad s_3^T C = [0 \quad 0 \quad 1].$$

Hence, if we construct a matrix with s_1^T, s_2^T, s_3^T as its rows, then

$$C^T C = \begin{bmatrix} s_1^T \\ s_2^T \\ s_3^T \end{bmatrix} C = \begin{bmatrix} 1 & 0 & 0 \\ 0 & 1 & 0 \\ 0 & 0 & 1 \end{bmatrix} = I_3.$$

In other words, **the transpose of C is equal to the inverse of C:**

$$C^T = \begin{bmatrix} s_1^T \\ s_2^T \\ s_3^T \end{bmatrix}$$

is the inverse of C, that is, $C^T = C^{-1}$. Square matrices with this property are said to be **orthogonal** matrices.

Suppose that we now definite a transformation by $x = CX$, where C is an orthogonal matrix. Then, in terms of X, the quadratic form becomes

$$x^T A x = (CX)^T A C X$$
$$= X^T C^T A C X$$
$$= X^T D X = \lambda_1 X_1^2 + \lambda_2 X_2^2 + \lambda_3 X_3^2.$$

It follows from this result, for 3×3 matrices, and by implication for higher order, that a **quadratic form is positive-definite if and only if all its eigenvalues are positive.**

Example 11.12. Find an orthogonal matrix C which transforms the quadratic form $x^T A x$ where

$$A = \begin{bmatrix} 3 & -1 & 0 \\ -1 & 3 & 0 \\ 0 & 0 & 1 \end{bmatrix}$$

into a diagonal quadratic form $x^T D x$.

The eigenvalues of A are given by $\det(A - \lambda I_3) = 0$, where

$$\det(A - \lambda I_3) = \begin{vmatrix} 3 - \lambda & -1 & 0 \\ -1 & 3 - \lambda & 0 \\ 0 & 0 & 1 - \lambda \end{vmatrix},$$
$$= ((3 - \lambda)^2 - 1)(1 - \lambda),$$
$$= (\lambda - 2)(\lambda - 4)(1 - \lambda).$$

Hence the eigenvalues are $\lambda_1 = 1$, $\lambda_2 = 2$, $\lambda_3 = 4$. Since all the eigenvalues are positive, it follows that the quadratic form is positive-definite. The corresponding eigenvectors are

$$s_1 = \begin{bmatrix} 0 \\ 0 \\ 1 \end{bmatrix}, \qquad s_2 = \begin{bmatrix} 1/\sqrt{2} \\ 1/\sqrt{2} \\ 0 \end{bmatrix}, \qquad s_3 = \begin{bmatrix} -1/\sqrt{2} \\ 1/\sqrt{2} \\ 0 \end{bmatrix}.$$

Hence the required orthogonal matrix C is

$$C = [s_1 \quad s_2 \quad s_3] = \begin{bmatrix} 0 & 1/\sqrt{2} & -1/\sqrt{2} \\ 0 & 1/\sqrt{2} & 1/\sqrt{2} \\ 1 & 0 & 0 \end{bmatrix}.$$

The relation between the coordinates (x, y, z) and (X, Y, Z) in the transformation

$$x = CX = [s_1 \quad s_2 \quad s_3]X$$

where the eigenvectors s_1, s_2, s_3 are orthogonal unit vectors, can be seen as follows. Put $X = 1$, $Y = 0$, $Z = 0$, which is a point on the X axis. Since

$$X = \begin{bmatrix} 1 \\ 0 \\ 0 \end{bmatrix},$$

it follows that the corresponding point in the x frame is $x = s_1$. In other words the elements (a_1, b_1, c_1) of s_1 are the coordinates in the x space of the point $A_1 : (1, 0, 0)$ in the X space. Similarly, the elements of s_2 and s_3 are respectively the coordinates of $A_2 : (0, 1, 0)$ and $A_3 : (0, 0, 1)$ in the X space (see Fig. 11.2).

We know that the eigenvectors are mutually orthogonal, that is $s_i^T s_j = 0$ $(i \neq j)$. We want to show that this implies that the new axes $OXYZ$ are also mutually perpendicular. Consider the triangle OA_1A_2: we want to show that $A_1\widehat{O}A_2$ is a right angle, so that the triangle is subject to Pythagoras's theorem:

$$A_1A_2^2 - OA_1^2 - OA_2^2 = (a_1 - a_2)^2 + (b_1 - b_2)^2 + (c_1 - c_2)^2$$
$$- (a_1^2 + b_1^2 + c_1^2) - (a_2^2 + b_2^2 + c_2^2)$$
$$= -2(a_1a_2 + b_1b_2 + c_1c_2) = -2s_1^T s_2 = 0,$$

since the eigenvectors are unit vectors and orthogonal. Hence, by the Pythagorean equality, $A_1\widehat{O}A_2$ is a right angle. Similarly, the other angles $A_2\widehat{O}A_3$ and $A_3\widehat{O}A_1$ are right angles. Hence the new axes are mutually perpendicular. It can be shown that the det $C = \pm 1$. If det $C = 1$, then the X coordinates can be obtained from the x coordinates by a rotation about the origin O. If det $C = -1$, then a reflection and rotation are required.

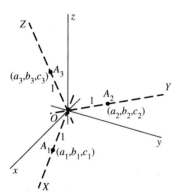

Fig. 11.2

Orthogonal mapping between axes. The coordinates (a_1, b_1, c_1), (a_2, b_2, c_2), (a_3, b_3, c_3) are measured in the x space.

Example 11.13. Show that

$$C = \tfrac{1}{3} \begin{bmatrix} 1 & 2 & 2 \\ 2 & 1 & -2 \\ 2 & -2 & 1 \end{bmatrix}$$

is an orthogonal matrix. If $x = CX$, what does the point $x = 1$, $y = 2$, $z = -1$ map into in the (X, Y, Z) coordinates?

In this example,

$$s_1 = \tfrac{1}{3}\begin{bmatrix} 1 \\ 2 \\ 2 \end{bmatrix}, \qquad s_2 = \tfrac{1}{3}\begin{bmatrix} 2 \\ 1 \\ -2 \end{bmatrix}, \qquad s_3 = \tfrac{1}{3}\begin{bmatrix} 2 \\ -2 \\ 1 \end{bmatrix}.$$

Clearly

$$s_1^{\mathrm{T}} s_1 = \tfrac{1}{9}[1 \quad 2 \quad 2]\begin{bmatrix} 1 \\ 2 \\ 2 \end{bmatrix} = 1.$$

Similarly $s_2^{\mathrm{T}} s_2 = 1$ and $s_3^{\mathrm{T}} s_3 = 1$. Also

$$s_1^{\mathrm{T}} s_2 = \tfrac{1}{9}[1 \quad 2 \quad 2]\begin{bmatrix} 2 \\ 1 \\ -2 \end{bmatrix}$$

$$= \tfrac{1}{9}[1 \times 2 + 2 \times 1 + 2 \times (-2)] = 0.$$

Similarly, $s_2^{\mathrm{T}} s_3 = 0$ and $s_3^{\mathrm{T}} s_1 = 0$. We need to invert the transformation so that

$$X = C^{-1} x = C^{\mathrm{T}} x$$

$$= \begin{bmatrix} s_1^{\mathrm{T}} \\ s_2^{\mathrm{T}} \\ s_3^{\mathrm{T}} \end{bmatrix}\begin{bmatrix} 1 \\ 2 \\ -1 \end{bmatrix} = \tfrac{1}{3}\begin{bmatrix} 1 \times 1 + 2 \times 2 + 2 \times (-1) \\ 2 \times 1 + 1 \times 2 + (-2) \times (-1) \\ 2 \times 1 + (-2) \times 2 + 1 \times (-1) \end{bmatrix}$$

$$= [1 \quad 2 \quad -1]^{\mathrm{T}}.$$

Hence $X = 1$, $Y = 2$, $Z = -1$.

11.9 An application to a vibrating system

Positive-definite matrices occur frequently in applications. For example, consider the system consisting of two particles of equal mass m and three equal springs stretched in a straight line between two supports as shown in Fig. 11.3. Suppose that, in equilibrium, the springs are unstretched, each of length a. The mechanical system vibrates longitudinally so that the displacements of the particles are x and y as shown.

If a spring is stretched or compressed from equilibrium by a length x, then its potential energy stored is $\tfrac{1}{2}kx^2$ where k is a constant known as the **stiffness** of the spring, which measures its reaction to being stretched or compressed. The total potential energy of the

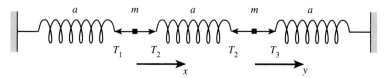

Fig. 11.3
Longitudinal oscillations.

system is
$$V = \tfrac{1}{2}kx^2 + \tfrac{1}{2}k(y - x)^2 + \tfrac{1}{2}y^2.$$
Note that the extension of the middle spring is $y - x$. Thus
$$V = \tfrac{1}{2}kx^2 + \tfrac{1}{2}ky^2 - kyx + \tfrac{1}{2}kx^2 + \tfrac{1}{2}ky^2,$$
$$= kx^2 - kxy + ky^2,$$
$$= \tfrac{1}{2}x^\mathrm{T}Kx,$$
where
$$x = \begin{bmatrix} x \\ y \end{bmatrix}, \qquad K = \begin{bmatrix} 2k & -k \\ -k & 2k \end{bmatrix}.$$
The eigenvalues of K are given by $\det(K - \lambda I_2) = 0$, that is,
$$\begin{vmatrix} 2k - \lambda & -k \\ -k & 2k - \lambda \end{vmatrix} = 0, \quad \text{or} \quad (2k - \lambda)^2 - k^2 = 0.$$
Hence, the eigenvalues are $\lambda_1 = k$ and $\lambda_2 = 3k$, which are both positive, implying that the potential energy is a positive-definite quadratic form. This is not surprising, since we might expect the potential energy to take a minimum value in equilibrium. The corresponding eigenvectors are
$$s_1 = \frac{1}{\sqrt{2}} \begin{bmatrix} 1 \\ 1 \end{bmatrix}, \qquad s_2 = \frac{1}{\sqrt{2}} \begin{bmatrix} 1 \\ -1 \end{bmatrix},$$
normalized as unit vectors. The matrix of eigenvectors, C, is given by
$$C = [s_1 \; s_2] = \frac{1}{\sqrt{2}} \begin{bmatrix} 1 & 1 \\ 1 & -1 \end{bmatrix}.$$
The transformation $x = CX$ introduces the coordinates $X^\mathrm{T} = (X, Y)$ in which
$$V = \tfrac{1}{2}X^\mathrm{T} \begin{bmatrix} k & 0 \\ 0 & 3k \end{bmatrix} X = \tfrac{1}{2}(kX^2 + 3kY^2).$$
These are known as the **normal coordinates** of the system, and are related to x and y by
$$\begin{bmatrix} x \\ y \end{bmatrix} = C \begin{bmatrix} X \\ Y \end{bmatrix} = \begin{bmatrix} (X + Y)/\sqrt{2} \\ (X - Y)/\sqrt{2} \end{bmatrix}.$$
Normal coordinates are often more convenient coordinates to use.

For the same problem, we can also derive the equations of motion of each particle. If T_1, T_2, and T_3 are the tensions in each of the springs, then applying Newton's law (force equals mass times acceleration) for each particle gives the **differential equations**
$$T_2 - T_1 = m\ddot{x}, \tag{11.10}$$
$$T_3 - T_2 = m\ddot{y}. \tag{11.11}$$
where $\ddot{x}$ and $\ddot{y}$ stand for $\mathrm{d}^2x/\mathrm{d}t^2$ and $\mathrm{d}^2y/\mathrm{d}t^2$ respectively. The tension in a spring is k times the extension, by Hooke's law, where k is the stiffness of the spring. Thus
$$T_1 = kx, \qquad T_2 = k(y - x), \qquad T_3 = -y.$$

Substitution into (11.10) and (11.11) yields

$$k(2y - x) = m\ddot{x}, \tag{11.12}$$

$$k(-2y + x) = m\ddot{y}, \tag{11.13}$$

In matrix form, these equations can be combined into the vector equation

$$\ddot{x} + Ax = 0,$$

where

$$\ddot{x} = \begin{bmatrix} \ddot{x} \\ \ddot{y} \end{bmatrix}, \qquad A = \begin{bmatrix} k & -2k \\ -k & 2k \end{bmatrix}.$$

If we use the normal coordinates, then $x = CX$ implies

$$C\ddot{X} + ACX = 0.$$

Multiply on the left by $C^{-1} = C^{T}$:

$$\ddot{X} + C^{T}ACX = 0,$$

or

$$\ddot{X} + DX = 0, \tag{11.14}$$

where

$$D = \begin{bmatrix} \lambda_1 & 0 \\ 0 & \lambda_2 \end{bmatrix} = \begin{bmatrix} k & 0 \\ 0 & 3k \end{bmatrix}.$$

Equation (11.14) now separates into the two differential equations

$$\ddot{X} + kX = 0,$$

$$\ddot{Y} + 3kY = 0,$$

which, unlike (11.12) and (11.13) are now no longer *simultaneous* equations, but uncouple into two equations which can be solved separately and independently for X and Y. We say more about the solution of differential equations in Chapter 16.

Problems, Chapter 11

11.1. (Sections 11.1, 2). Find the eigenvalues and eigenvectors of the following matrices:

(a) $\begin{bmatrix} 2 & 3 \\ 4 & 6 \end{bmatrix}$; (b) $\begin{bmatrix} 6 & 3 \\ 2 & 7 \end{bmatrix}$; (c) $\begin{bmatrix} 2 & 1 \\ 4 & 6 \end{bmatrix}$;

(d) $\begin{bmatrix} 1 & 1 \\ 4 & 5 \end{bmatrix}$; (e) $\begin{bmatrix} 1 & 2 \\ 14 & 5 \end{bmatrix}$; (f) $\begin{bmatrix} 2 & -2 \\ 4 & 6 \end{bmatrix}$.

11.2. (Section 11.1). Show that the eigenvalues of the *symmetric* matrix

$$A = \begin{bmatrix} a & b \\ b & c \end{bmatrix},$$

where a, b, and c are real numbers, are real.

11.3. (Section 11.1). Find the eigenvalues of

$$A = \begin{bmatrix} 6 & 3 \\ 2 & 7 \end{bmatrix},$$

(see Problem 11.1b). Find the inverse of A and find its eigenvalues. What relationship, would you guess, exists between the eigenvalues of A and those of A^{-1}? Find the eigenvalues of A^2. How do they relate to those of A?

11.4. (Sections 11.1, 2). Find the eigenvalues and eigenvectors of

(a) $\begin{bmatrix} 1 & 1 & 2 \\ 1 & 2 & 1 \\ 2 & 1 & 1 \end{bmatrix}$ (b) $\begin{bmatrix} 2 & 1 & 2 \\ 1 & 2 & 2 \\ 2 & 1 & 2 \end{bmatrix}$;

(c) $\begin{bmatrix} 2 & 0 & 0 \\ 0 & 2 & 2 \\ 0 & 2 & -1 \end{bmatrix}$; (d) $\begin{bmatrix} 6 & 5 & 5 \\ 5 & 6 & 5 \\ 5 & 5 & 6 \end{bmatrix}$.

11.5. (Sections 11.1, 2). Find the eigenvalues and eigenvectors of

$$\begin{bmatrix} 1 & 2 & 0 & 0 \\ 3 & 2 & 0 & 0 \\ 0 & 0 & 3 & 1 \\ 0 & 0 & 1 & 3 \end{bmatrix}$$

11.6. (Sections 11.1, 2). Show that

$$A = \begin{bmatrix} 1 & 0 & 0 \\ 0 & 2 & 2 \\ 0 & 2 & 5 \end{bmatrix}$$

has a repeated eigenvalue. Find the corresponding eigenvectors. How many linearly independent eigenvectors are there?

11.7. (Sections 11.1, 2). Show that the matrix

$$A = \begin{bmatrix} -a & -1 & a+1 \\ a+1 & -a & -1 \\ -a & a+1 & -a \end{bmatrix}$$

has a zero eigenvalue. For design reasons, a second eigenvalue must be 3. What is the value of the parameter a for this to occur? Find the third eigenvalue for this value of a, and also all the eigenvectors.

11.8. (Sections 11.1, 2). A matrix is said to be **idempotent** if $A^2 = A$. Explain why all eigenvalues of A must be either 0 or 1. Show that

$$A = \begin{bmatrix} 1 & 0 & 0 \\ 0 & 3 & 6 \\ 0 & -1 & -2 \end{bmatrix}$$

is idempotent. Find the eigenvalues and eigenvectors of A and A^2 and confirm the above result.

11.9. (Sections, 11.1, 2). Let

$$A = \tfrac{1}{2}\begin{bmatrix} 1 & 1 & 1 & 1 \\ 1 & 1 & -1 & -1 \\ 1 & -1 & 1 & -1 \\ 1 & -1 & -1 & 1 \end{bmatrix}.$$

Show that $A^2 = I_4$. Explain why the eigenvalues of A must be either 1 or -1. Can A be diagonalized?

11.10. Find the eigenvalues $\lambda_1, \lambda_2, \lambda_3$ of

$$A = \begin{bmatrix} 1 & 2 & 1 \\ 2 & 1 & 1 \\ 1 & 1 & 2 \end{bmatrix}.$$

The **trace** of a square matrix is the sum of the elements in the leading diagonal. Thus if $B = [b_{ij}]$ is an $n \times n$ matrix, then

$$\text{trace } B = b_{11} + b_{22} + \cdots + b_{nn}.$$

Confirm for A above that trace $A = \lambda_1 + \lambda_2 + \lambda_3$. Also verify that $\det A = \lambda_1\lambda_2\lambda_3$.

11.11. (Section 11.3). Show that the vectors

$$s_1 = \begin{bmatrix} 1 \\ 2 \\ 1 \end{bmatrix}, \quad s_2 = \begin{bmatrix} 2 \\ -1 \\ 3 \end{bmatrix}, \quad s_3 = \begin{bmatrix} 4 \\ 3 \\ 5 \end{bmatrix}$$

are linearly dependent.

11.12. (Section 11.4). Find a linear transformation which maps the three points $(1, 2, -1)$, $(1, 0, 1)$, $(-1, 1, 0)$ in the (x, y, z) space respectively into the points $(2, 1, -1)$, $(-1, 0, 2)$, $(1, 1, 1)$ in the (X, Y, Z) space. What does the point $(3, 1, 2)$ map into? Is there a point which maps into itself?

11.13. (Section 11.4). Given that $X = Ax$, where

$$A = \begin{bmatrix} 1 & -1 & 2 \\ 2 & 3 & -1 \\ 4 & 1 & 3 \end{bmatrix},$$

find the plane in the X space into which all points in the x space are mapped.

11.14. (Section 11.5). Let

$$A = \begin{bmatrix} -4 & 1 & -2 \\ 2 & -2 & 1 \\ 0 & 1 & 0 \end{bmatrix}.$$

Find the eigenvalues of A and a set of corresponding eigenvectors. Hence construct a matrix C which makes $C^{-1}AC$ a diagonal matrix.

11.15. Show that the matrix

$$\begin{bmatrix} 0 & -1 \\ 1 & 0 \end{bmatrix}$$

cannot be diagonalized.

11.16. (Section 11.5). Find a matrix C which diagonalizes

$$A = \begin{bmatrix} 2 & 0 & 0 \\ 0 & 2 & 2 \\ 0 & 2 & -1 \end{bmatrix}$$

Verify that $C^{-1}AC = D$, where D is the diagonal matrix of eigenvalues.

11.17. Using the diagonalization result

$$C^{-1}AC = D$$

for a matrix A which has a n linearly independent eigenvectors, show that

$$\det A = \lambda_1 \lambda_2 \cdots \lambda_n,$$

where $\lambda_1, \lambda_2, \ldots, \lambda_n$ are the eigenvectors of A. (Hint: use the result $\det AB = \det A \det B$ for square matrices.)

11.18. (Section 11.6). Find the eigenvalues and eigenvectors of the row-stochastic matrix

$$A = \begin{bmatrix} \frac{1}{4} & \frac{1}{2} & \frac{1}{4} \\ \frac{1}{2} & \frac{1}{4} & \frac{1}{4} \\ \frac{1}{4} & \frac{1}{4} & \frac{1}{2} \end{bmatrix}.$$

Find a formula for A^n. How does A behave as $n \to \infty$?

11.19. Show that

$$A = \begin{bmatrix} 1 & 0 & 0 \\ 0 & \cos \alpha & -\sin \alpha \\ 0 & \sin \alpha & \cos \alpha \end{bmatrix},$$

is an orthogonal matrix. Describe the mapping defined by

$$X = Ax.$$

Which set of points remains unaffected by the mapping?

11.20. (Section 11.8). Show that

$$A = \frac{1}{2} \begin{bmatrix} 1 & -1 & 1 & -1 \\ 1 & -1 & -1 & 1 \\ 1 & 1 & -1 & -1 \\ 1 & 1 & 1 & 1 \end{bmatrix}$$

is an orthogonal matrix.

11.21. Show that, in the transformation

$$\begin{bmatrix} X \\ Y \end{bmatrix} = \begin{bmatrix} \cos \alpha & -\sin \alpha \\ \sin \alpha & \cos \alpha \end{bmatrix} \begin{bmatrix} x \\ y \end{bmatrix},$$

the angle between the two sets of axes is α. What do the axes of x and y become in the (X, Y) plane?

11.22. Show that the nonzero eigenvalues of the skew-symmetric matrix

$$A = \begin{bmatrix} 0 & a & b \\ -a & 0 & c \\ -b & -c & 0 \end{bmatrix}$$

are imaginary for a, b, c real.

11.23. Let

$$A = \begin{bmatrix} 1 & 2 & 1 \\ 2 & 1 & 1 \\ 1 & 1 & 2 \end{bmatrix}.$$

Show that

$$\det(A - \lambda I_3) = -\lambda^3 + 4\lambda^2 + \lambda - 4.$$

Verify that

$$-A^3 + 4A^2 + A - 4 = 0.$$

In other words, the matrix A satisfies its own characteristic equation. This is known as the **Cayley–Hamilton** theorem, and holds generally for square matrices. Use the result to find the inverse matrix A^{-1}.

11.24. Find the eigenvalues and eigenvectors of

$$A = \begin{bmatrix} 5 & -1 & -3 & 3 \\ -1 & 5 & 3 & -3 \\ -3 & 3 & 5 & -1 \\ 3 & -3 & -1 & 5 \end{bmatrix}.$$

Construct a matrix C such that $C^{-1}AC$ is the diagonal matrix of eigenvalues. Write down $\det A$.

11.25. (Section 11.7). Express the following quadratic forms in the form $x^T Ax$, where A is a 3×3 symmetric matrix:
(a) $x_1^2 + x_2^2 + x_3^2 + 4x_1 x_2 - 4x_1 x_3 + 4x_2 x_3$;
(b) $x_1 x_2 - x_1 x_3 + x_2 x_3$.
  Find eigenvalues of A in each case, and find also a matrix C which transforms each into the form $\lambda_1 X_1^2 + \lambda_2 X_2^2 + \lambda_3 X_3^2$.

11.26. (Section 11.7). Which of each of the following quadratic forms is positive-definite?
(a) $4x_1^2 + x_2^2 - 4x_1x_2$;
(b) $x_1^2 + x_2^2 + 2x_3^2 + 2x_2x_3 + 2x_3x_1 + 4x_1x_2$;
(d) $6x_1^2 + 2x_2^2 - x_3x_1$.

11.27. (Section 11.9). Consider three particles, each of mass m, and four equal springs stretched in a straight line between fixed supports distance $4a$ apart by four springs each with unstretched lengths a (as in Fig. 11.3, but with three particles). Consider longitudinal oscillations of the systems and let x, y, z be the extensions of the springs, assuming Hooke's law with stiffness k for the tension in each spring, show that x, y, z satisfy the differential equations

$$k(-2x + y) = m\ddot{x},$$

$$k(x - 2y + z) = m\ddot{y},$$

$$k(y - 2z) = m\ddot{z}.$$

Express the equations in the matrix form

$$\ddot{x} + Ax = 0.$$

Find the eigenvalues and eigenvectors of A. Construct a matrix C such that $C^T A C$ is diagonal. Obtain differential equations for the normal coordinates X, Y, Z.

11.28. Let

$$A = \begin{bmatrix} 0 & 1 & 0 \\ 0 & 0 & 1 \\ 1 & 0 & 0 \end{bmatrix}.$$

Calculate A^2 and A^3. Find a general formula for A^n.
   Show that A has two complex eigenvalues, and find the corresponding eigenvectors. Construct a matrix C which diagonalizes A and find a formula for A^n. Compare this result with the *ad hoc* method above.

11.29. Let A be a square matrix, and let S represent the sum of the powers of A from A up to A^n:

$$S = A + A^2 + A^3 + \cdots + A^n.$$

By multiplying the equation by A and subtraction, show that

$$S = A(I + A^n)(I - A)^{-1},$$

and state any cases for which this method fails.
   If

$$A = \begin{bmatrix} 1 & 3 \\ 2 & 2 \end{bmatrix},$$

(see Examples 11.1 and 11.3), find a formula for A^m and the sum

$$S = \sum_{m=1}^{n} A^m.$$

11.30. Let

$$A = \begin{bmatrix} 1 & 2 & 1 \\ 2 & 1 & 1 \\ 1 & 1 & 2 \end{bmatrix}.$$

Find the eigenvectors of A and confirm that

$$s_1 = \begin{bmatrix} 2 \\ 2 \\ 2 \end{bmatrix}, \quad s_2 = \begin{bmatrix} -1 \\ -1 \\ 2 \end{bmatrix}, \quad s_3 = \begin{bmatrix} -3 \\ 3 \\ 0 \end{bmatrix}$$

are eigenvectors of A. Construct the matrix C and verify that

$$D = C^{-1} A C,$$

where D is the diagonal matrix of eigenvalues. (This is a reworking of the problem at the beginning of Section 8.5, but with different eigenvectors.)

12 Antidifferentiation and area

12.1 Reversing differentiation

Compare the following two problems:

Problem A: $\dfrac{d}{dx} \sin x = f(x)$; what is $f(x)$?

Problem B: $\dfrac{d}{dx} F(x) = \cos x$; what is $F(x)$?

We know already that, in Problem A,

$$f(x) = \cos x.$$

This provides *one* answer to Problem B, which is solved by

$$F(x) = \sin x.$$

Since $\cos x$ is the **derivative** of $\sin x$, we say that $\sin x$ is an **antiderivative** of $\cos x$ (we say *an* antiderivative because it is not the only one; for example, $\sin x + 1$ is also an antiderivative).

The **antidifferentiation** problem can be expressed in various ways; for example

(a) What must be differentiated to get $\cos x$?
(b) What curves have slope equal to $\cos x$ at every point?
(c) Find y as a function of x if $dy/dx = \cos x$.

Finding antiderivatives is the opposite or **inverse** process to that of finding derivatives.

The following examples show that **a function $f(x)$ has an infinite number of antiderivatives:** there is an infinite number of functions whose derivatives are $f(x)$. However, they are all very simple variants on a single function.

Example 12.1. Find y as a function of x if $dy/dx = 2x$.

One solution is $y = x^2$, because its derivative is $2x$. But the derivatives of $x^2 + 3$, $x^2 - \frac{1}{2}$, and so on are also equal to $2x$.

In fact

$$y = x^2 + C$$

is an antiderivative of $2x$ for any constant C.

Some of these solutions are shown in Fig. 12.1. Different choices for C just shift the graph bodily up or down parallel to itself. Therefore, for any particular value of x, such as is represented by the vertical line PQR, the slopes are all the same, independently of the value of C.

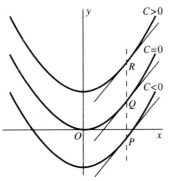

Fig. 12.1

Evidently the same thing will happen whatever function we start with: if we find one solution, we can add constants to obtain more.

Example 12.2. Find a collection of antiderivatives of $\sin 2x$.

We want y such that $dy/dx = \sin 2x$. If we differentiate a cosine we get something involving a sine, so first of all test whether $y = \cos 2x$ is close to being an antiderivative of $\sin 2x$. We find that $dy/dx = -2 \sin 2x$. This contains an unwanted factor (-2). It can be eliminated by choosing instead

$$y = \frac{1}{-2} \cos 2x = -\tfrac{1}{2} \cos 2x,$$

for then we have $dy/dx = -\tfrac{1}{2}(-2 \sin 2x) = \sin 2x$, which is right. Therefore, one antiderivative is $-\tfrac{1}{2} \cos 2x$, and the rest are of the form

$$y = -\tfrac{1}{2} \cos 2x + C \quad (C \text{ is any constant}).$$

Example 12.3. Solve the equation $dy/dx = e^{-3x}$ (that is to say, find a collection of antiderivatives of e^{-3x}.)

Try $y = e^{-3x}$; then $dy/dx = -3\,e^{-3x}$. To avoid the unwanted factor (-3) we should have taken

$$y = \frac{1}{(-3)} e^{-3x} = -\tfrac{1}{3} e^{-3x}$$

From this we construct an infinite collection of antiderivatives:

$$-\tfrac{1}{3} e^{-3x} + C \quad (C \text{ any constant}).$$

It can be proved that the above process, of finding a particular antiderivative of a function and adding constants, generates **all possible antiderivatives** for that function.

> **Antiderivatives of** $f(x)$
>
> A function $F(x)$ is called an **antiderivative** of $f(x)$ if
> $$\frac{d}{dx} F(x) = f(x).$$
> If $F(x)$ is any particular antiderivative of $f(x)$, then **all the antiderivatives** are given by
> $$F(x) + C,$$
> where C can be any constant. (Therefore, **any two antiderivatives differ by a constant**.)

(12.1)

An antiderivative of a function is also more usually called an **indefinite integral** of the function, and the process of getting it is called **integration**. If you know the term already, it is perfectly safe to use it. We shall change over to it in Chapter 13.

Example 12.4. Find all the antiderivatives of x^3.

We have to find the y which fits the equation $dy/dx = x^3$. Differentiation reduces a power of x by unity, so try $y = x^4$:

$$dy/dx = 4x^3.$$

The factor 4 is unwanted; we needed $\frac{1}{4} x^4$ to give x^3. Therefore all antiderivatives are given by

$$y = \tfrac{1}{4} x^4 + C,$$

where C is any constant.

Sums of terms and constant multipliers are treated in the same way as in differentiation: the multipliers stay as multipliers and each term is treated separately, as in the next example.

Example 12.5. Obtain all the antiderivatives of $2 e^{-3x} - \frac{1}{2}x^3 + 2$.

From the previous examples, one antiderivative of e^{-3x} is $-\frac{1}{3}e^{-3x}$, and one for x^3 is $\frac{1}{4}x^4$. Also, one antiderivative of 2 is obviously $2x$. Therefore one antiderivative of the given expression is

$$2(-\tfrac{1}{3}e^{-3x}) - \tfrac{1}{2}(\tfrac{1}{4}x^4) + 2x,$$

and all its antiderivatives are of the form

$$-\tfrac{2}{3}e^{-3x} - \tfrac{1}{8}x^4 + 2x + C,$$

where C is any constant.

The following examples show the importance in practice of including the constant C.

Example 12.6. A point is at $x = 2$ on the x axis at time $t = 0$, then moves with velocity $v = t - t^2$. Find where it is at time $t = 3$.

Velocity is the rate at which displacement changes with time: $v = \mathrm{d}x/\mathrm{d}t$. In this case

$$v = \mathrm{d}x/\mathrm{d}t = t - t^2.$$

Therefore x is some antiderivative of $t - t^2$. All of its antiderivatives are included in

$$x = \tfrac{1}{2}t^2 - \tfrac{1}{3}t^3 + C,$$

where C is any constant.

   To find what value C must take in this case, we obviously have to take the starting point into consideration: $x = 2$ when $t = 0$. To obtain the value of C, substitute these values into our expression:

$$2 = 0 - 0 + C.$$

Therefore $C = 2$, so the position at any time is given by

$$x = \tfrac{1}{2}t^2 - \tfrac{1}{3}t^3 + 2.$$

   Finally, when $t = 3$, we have $x = -\tfrac{5}{2}$.

Example 12.7. Find the equation of the curve which passes through the point $(\pi, -1)$ and whose slope is given by $\mathrm{d}y/\mathrm{d}x = \sin 2x$.

Since the required y is an antiderivative of $\sin 2x$, the equation of the curve must take the form

$$y = -\tfrac{1}{2}\cos 2x + C,$$

where C is *some* (not 'any') constant. Since also we know that the curve passes through the point $x = \pi$, $y = -1$, we must require

$$-1 = -\tfrac{1}{2}\cos 2\pi + C = -\tfrac{1}{2} + C,$$

so $C = -\tfrac{1}{2}$. Finally the required curve is

$$y = -\tfrac{1}{2}\cos 2x - \tfrac{1}{2}.$$

Example 12.8. Obtain the antiderivatives of $(3x - 2)^3$.

As in the earlier examples, we try to guess the structure of y, given that $\mathrm{d}y/\mathrm{d}x = (3x - 2)^3$. There is not much to go on, so try an analogy with x^3; it would lead us to try something like $y = (3x - 2)^4$. To check this, differentiate using the chain rule with $u = 3x - 2$ and $y = u^4$:

$$\frac{\mathrm{d}y}{\mathrm{d}x} = 4(3x - 2)^3 \cdot 3 = 12(3x - 2)^3.$$

The factor 12 is unwanted; we really needed $y = \tfrac{1}{12}(3x - 2)^4$. Therefore all the antiderivatives are given by $y = \tfrac{1}{12}(3x - 2)^4 + C$.

The technique used in the previous examples can be used for functions like $(ax + b)^n$, e^{ax+b}, $\cos(ax + b)$, and $\sin(ax + b)$. However, it would not work in this simple way for a function such as $(2x^2 - 3)^2$ or $\sin(2x^2 - 3)$: the antiderivative of $(2x^2 - 3)^2$ is *not* equal to $\frac{1}{3}(2x^2 - 3)^3$, because x^2 is present rather than x (try it, using the Chain Rule).

12.2 Constructing a table of antiderivatives

Since antidifferentiation is the inverse of differentiation, any table of derivatives can be read backwards in order to provide antiderivatives. Suppose that two typical entries in a table of derivatives are as follows:

Given function	Derivative
$F(x)$	$f(x) = \dfrac{\mathrm{d}}{\mathrm{d}x} F(x)$
$\sin ax$	$a \cos ax$
e^{ax}	$a\, e^{ax}$

By interchanging the columns and modifying the headings, we get two entries in a possible table of antiderivatives:

Given function $f(x)$	One antiderivative $F(x)$
$a \cos ax$	$\sin ax$
$a\, e^{ax}$	e^{ax}

However, these entries are not yet in the form we should like them. For example, for the first entry we would prefer to have $\cos ax$ in the left column, instead of $a \cos ax$. Therefore divide both entries by the constant a, remember to introduce the arbitrary constant C to register *all* the antiderivatives, and we have a more convenient table:

Given function $f(x)$	Antiderivatives $F(x)$
$\cos ax$	$\dfrac{1}{a}\sin ax + C$
e^{ax}	$\dfrac{1}{a}e^{ax} + C$

By such means the short table (12.2) is produced. To verify any entry, differentiate the function in the right-hand column; the result should be the entry on the left. The letter C stands for 'any constant' or 'an arbitrary constant'.

A short table of antiderivatives

Given function $f(x)$	Antiderivatives $F(x)$	
a (constant)	$ax + C$	
* x^m (unless $m = -1$)	$\dfrac{1}{m+1}x^{m+1} + C$	
** x^{-1} $\left(\text{i.e. } \dfrac{1}{x}\right)$	$\begin{cases} \ln x + C & \text{if } x > 0 \\ \ln(-x) + C & \text{if } x < 0 \end{cases}$	**(12.2)**
	or $\ln|x| + C$ (any x)	
e^{ax}	$\dfrac{1}{a}e^{ax} + C$	
$\cos ax$	$\dfrac{1}{a}\sin ax + C$	
$\sin ax$	$-\dfrac{1}{a}\cos ax + C$	

Notice particularly the two starred entries. The formula * covers most cases, but it does not produce antiderivatives of the function x^{-1} (i.e. of $1/x$). Here $m = -1$, so the entry on the right becomes infinite, and therefore meaningless. Therefore the antiderivatives of x^{-1} must be given by some different formula, and this is shown under **. All we have to do is to verify the formula ** as in the following example. (The **modulus** or **absolute value** notation $|x|$ is explained in Section 1.1.)

Example 12.9. Confirm that the antiderivatives of x^{-1} (i.e. $1/x$) are given by $\ln x + C$ if x is positive and by $\ln(-x) + C$ if x is negative, and that $\ln|x| + C$ covers both cases.

(Remember that $\ln x$ does not have a meaning if x is negative or zero). All we have to do to verify the correctness of the formulae is to differentiate the proposed antiderivatives. Since $(d/dx)\ln x = x^{-1}$, the result is right when x is positive.

Suppose now that x is negative. Then $-x$ is positive, so $\ln(-x)$ has a meaning. Using the chain rule (3.3) with $u = -x$,

$$\frac{d}{dx}\ln(-x) = \frac{1}{-x}(-1) = \frac{1}{x},$$

so the second result is confirmed.

But (see Section 1.1) $|x| = x$ if $x > 0$ and $|x| = -x$ when $x < 0$, so $\ln|x|$ is an antiderivative whether x is positive or negative.

Example 12.10. Find the antiderivatives of $(2x - 3)^{-1}$.

The power $m = -1$ is the starred case in the table 12.2, so try $y = \ln(2x - 3)$, supposing initially that $2x - 3 > 0$. Then

$$\frac{dy}{dx} = \frac{2}{2x - 3}, \quad \text{or} \quad 2(2x - 3)^{-1}.$$

The unwanted factor 2 will not appear if we try again with $y = \frac{1}{2}\ln(2x - 3)$. Also $2x - 3$ might be negative, so we introduce a modulus sign. Finally we have

$$y = \tfrac{1}{2}\ln|2x - 3| + C.$$

12.3 Signed area generated by a graph

Figure 12.2 shows the graph of a function $y = f(x)$ between $x = a$ and $x = b$, in which we assume that **the x and y scales are the same**. Divide the range as shown into N sections so that in any section y is either positive only, or negative only.

Let $A_1, A_2, \ldots$ denote the **geometrical areas** of these segments,

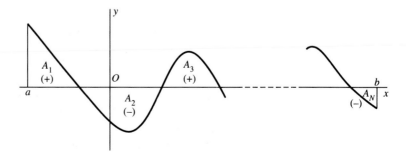

Fig. 12.2

and A the sum of these. *Geometrical* area is always positive, so A_1, $A_2, \ldots$ are all **positive numbers**. Then

$$A = A_1 + A_2 + A_3 + \cdots + A_N. \tag{12.3}$$

is naturally called 'the geometrical area between the curve and the x axis'.

We require a different quantity, $\mathcal{A}$, called the **signed area** between the curve and the x axis. This is defined by

$$\mathcal{A} = A_1 - A_2 + A_3 - \cdots - A_N. \tag{12.4}$$

In forming $\mathcal{A}$, we use the rule: **If y is positive, the contribution takes a positive sign; if y is negative, the contribution takes a negative sign**. This quantity has a far more useful range of applications than has geometrical area. For example, suppose that a point is moving on a straight line; then the signed displacement from its starting-point is equal to the *signed* area of its velocity-time graph.

We show how to calculate the signed area $\mathcal{A}$ of the graph of $y = f(x)$ between two given points, $x = a$ and $x = b$ (Fig. 12.3a). Let $\mathcal{A}(x)$ represent the signed area between a and a variable point with coordinate x (Fig. 12.3a). Increase x by a small step δx; the signed area from a to $x + \delta x$ is $\mathcal{A}(x + \delta x)$. The change in signed area, $\delta\mathcal{A} = \mathcal{A}(x + \delta x) - \mathcal{A}(x)$ (positive or negative), is equal to the signed area of $PQRS$ in Fig. 12.3a and b. This is very nearly equal to the signed area of the rectangle $PQRN$ in Fig. 12.3b (in this case the required sign is negative) so

$$\delta\mathcal{A} \approx f(x)\,\delta x$$

which automatically takes the right sign. Therefore

$$\frac{\delta\mathcal{A}}{\delta x} \approx f(x).$$

Now let $\delta x \to 0$; '$\approx$' becomes '$=$', and $\delta\mathcal{A}/\delta x$ becomes $\mathrm{d}\mathcal{A}/\mathrm{d}x$, so that

$$\frac{\mathrm{d}\mathcal{A}}{\mathrm{d}x} = f(x). \tag{12.5}$$

From (12.5) $\mathcal{A}(x)$ must be *one of the antiderivatives* of $f(x)$. To find which one, choose *any particular antiderivative* and call it $F(x)$. Then $\mathcal{A}(x)$ can differ from $F(x)$ only by a constant, k say, so

$$\mathcal{A}(x) = F(x) + k. \tag{12.6}$$

To determine the value of k, use the fact that $\mathcal{A}(x) = 0$ at $x = a$, because the starting point is then the same as the end point; that is to say,

$$\mathcal{A}(a) = 0.$$

Therefore, from (12.6)

$$\mathcal{A}(a) = 0 = F(a) + k,$$
or
$$k = -F(a), \tag{12.7}$$

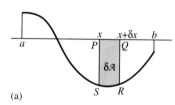

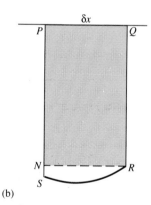

(a)

(b)

Fig. 12.3

a known quantity, since we selected the antiderivative $F(x)$ of $f(x)$ ourselves. The required area $\mathcal{A}$ between a and b is given by

$$\mathcal{A} = \mathcal{A}(b) = F(b) - F(a),$$

by putting $x = b$ into (12.6), with (12.7) as the value of k.

> **The signed area $\mathcal{A}$ of $f(x)$ between $x = a$ and b** is given by
> $$\mathcal{A} = F(b) - F(a),$$
> where $F(x)$ is *any antiderivative* of $f(x)$.
>
> (12.8)

In practice we naturally use the simplest antiderivative, in which the C in the table is zero. Any nonzero choice of C will cancel out and disappear, since it will be present in both $F(a)$ and $F(b)$.

Example 12.10. The signed area of $y = x^2$ from $x = -1$ to $x = 2$.

(This happens to be the same as the geometrical area, because y is never negative.) Here $a = -1$ and $b = 2$. Also, the simplest antiderivative of x is

$$F(x) = \tfrac{1}{3} x^3.$$

Therefore, from (12.8),

$$\mathcal{A} = F(b) - F(a) = \tfrac{1}{3} (2)^3 - \tfrac{1}{3} (-1)^3 = 3.$$

There is a special notation: the **square-bracket notation**, which we shall use generally from now onward.

> **Square-bracket notation**
> $\big[F(x)\big]_a^b$ stands for $F(b) - F(a)$.
>
> (12.9)

Example 12.11. Find (a) the signed area, and (b) the geometrical area, between $y = \sin x$ and the x axis from $x = 0$ to $x = 2\pi$.

(a) $f(x) = \sin x$, so $F(x) = -\cos x$ is an antiderivative. From (12.8) and (12.9), with $a = 0$ and $b = 2\pi$, the signed area $\mathcal{A}$ is given by

$$\mathcal{A} = \big[-\cos x\big]_0^{2\pi} = -\big[\cos x\big]_0^{2\pi} = -(\cos 2\pi - \cos 0) = 0,$$

as is expected from Fig. 12.4: the positive and negative sections cancel.

(b) The geometrical area A can be obtained by splitting the range into a positive section 0 to π, and a negative section from π to 2π (see Fig. 12.4). The negatively-signed section π to 2π must have its

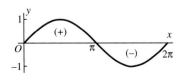

Fig. 12.4

sign reversed in order to give the *geometrical* area:

$$A = [\text{geometrical area of 1st loop}]$$
$$+ [\text{geometrical area of 2nd loop}]$$
$$= [\text{signed area of 1st loop}] - [\text{signed area of second loop}].$$

This is equal to

$$[F(x)]_0^\pi - [F(x)]_\pi^{2\pi} = [-\cos x]_0^\pi - [-\cos x]_\pi^{2\pi}$$
$$= (-\cos \pi + \cos 0) - (-\cos 2\pi + \cos \pi)$$
$$= (1 + 1) - (-1 + (-1)) = 2 + 2 = 4.$$

Problems, Chapter 12
Note: In case you have already met the term 'indefinite integral', the term 'antiderivative' has the same meaning.

12.1. Obtain all the antiderivatives of the following functions, and check their correctness by differentiating your results.
(a) x^5; $3x^4$; $2x^3$; $\frac{1}{3}x^2$; $6x$; $f(x) = 3$; $f(x) = 0$.
(b) $-\frac{1}{2}x^{-3}$; $2x^{-2}$; $3x^{-1}$ when $x > 0$ (if in doubt, see (12.2)).
(c) $x^{\frac{3}{2}}$; $x^{\frac{1}{2}}$; $x^{-\frac{1}{2}}$; $x^{\frac{4}{3}}$; $x^{-\frac{1}{3}}$.
(d) $1/x^2$ (write as x^{-2}); $1/x^4$; $1/x$ when $x < 0$ (see (12.2)).
(e) $\sqrt{x}(= x^{\frac{1}{2}})$; $1/\sqrt{x}$; $1/x^{\frac{3}{2}}$.
(f) $3x$; $\frac{1}{2}x^2$; $1/3x^2$; $3/4x^{\frac{1}{4}}$.
(g) e^x; e^{-x}; $5e^{2x}$; $e^{-\frac{1}{2}x}$; $3e^{-2x}$.
(h) $\cos x$; $\cos 3x$; $\sin x$; $\sin 3x$;
(i) $1 - 3x$; $1 + 2x - 3x^2$; $3x^4 - 4x^2 + 5$.
(j) $x(x + 1)$(expand by removing the brackets); $(1 + 2x)(1 - 2x)$; $(x + 1)^2$; $(1 + x)(1 - 1/x)$; $x^2(x + x^2)$.
(k) $(x + 1)/x$ (turn it into the sum of two terms); $(2\sqrt{x} - 1)/\sqrt{x}$ (put $\sqrt{x} = x^{\frac{1}{2}}$ and $1/\sqrt{x} = x^{-\frac{1}{2}}$, then simplify as the sum of two terms); $(x + 1)^2/x^3$.
(l) $e^x + e^{-x}$; $2e^{2x} - 3e^{3x}$; $e^{\frac{1}{2}x}(1 + e^{-\frac{1}{2}x})$; $1/e^{2x}(= e^{-2x})$; $(e^{2x} - e^{-2x})/e^{2x}$.
(m) $2 \cos 2x$; $3 \sin \frac{1}{2}x - 4 \cos \frac{1}{3}x$; $2 + \sin 2x$.

12.2. Find all the antiderivatives of the following by trial and error, as explained in the text. Confirm your answers by differentiation.
(a) $(x + 1)^3$ (start by trying $(x + 1)^4$); $(3x + 1)^3$; $(3x - 8)^3$.
(b) $(1 - x)^4$; $(8 - 3x)^{\frac{1}{2}}$; $(1 - x)^{\frac{1}{3}}$.
(c) $(2x + 1)^{-2}$; $(1 - x)^{-\frac{1}{2}}$; $2/(3x + 1)^3$; $1/4(1 - x)^{\frac{1}{4}}$.
(d) $2 \cos(3x - 2)$ (try first $\sin(3x - 2)$); $3 \sin(1 - x)$; $2 \sin(2 - 3x)$.

12.3. (See Example 12.10). Find the antiderivatives of the following.
(a) $1/(x + 1)$; $1/(x - 1)$; $3/(3x - 2)$; $2/(5x - 4)$.
(b) $1/(1 - x)$; $1/(4 - 5x)$.
(c) $x/(x + 1)$ (it can be written as $1 - 1/(x + 1)$);
(d) $(x + 1)/(x - 1)$ (compare (c)).

12.4. Use the identities $\cos^2 A = \frac{1}{2}(1 + \cos 2A)$, $\sin^2 A = \frac{1}{2}(1 - \cos 2A)$, and $\sin A \cos A = \frac{1}{2} \sin 2A$ to get rid of the squares and products in the following expressions, and in that way obtain the antiderivatives.
(a) $\cos^2 x$; $\sin^2 x$; $\sin x \cos x$.
(b) $3 \cos^2 2x$; $\sin^2 3x$; $\sin 2x \cos 2x$.
(c) $\cos^4 x$ (you will have to use the identities twice).

12.5. (a) Show that $(d/dx)(x e^x) = e^x + x e^x$. By rearranging the terms, show that the antiderivatives of $x e^x$ are $e^x(x - 1) + C$ (use the fact that e^x can be written as $(d/dx) e^x$). Confirm the result by differentiation.
(b) Differentiate $x^2 e^x$. By rearranging the terms and using the result in (a), find the antiderivatives of $x^2 e^x$.

12.6. Use the result (12.8) to obtain the signed areas between the given graphs and the x axis. By roughly sketching the graphs of the functions for which you obtain zero, explain this fact.
(a) $y = x$, $0 \leqslant x \leqslant 2$;
(b) $y = x$, $-1 \leqslant x \leqslant 1$;
(c) $y = -x^2$, $0 \leqslant x \leqslant 1$;
(d) $y = \cos x$, $-\pi \leqslant x \leqslant \pi$;
(e) $y = \cos x - 1$, $0 \leqslant x \leqslant 2\pi$;
(f) $y = x^{-1}$, $-2 \leqslant x \leqslant -1$ (note that x is negative in this range);
(g) $y = \sin 3x$, $0 \leqslant x \leqslant \frac{2}{3}\pi$;
(h) $y = 1/(1 - x)$, $2 \leqslant x \leqslant 3$ (note: $1 - x$ is negative over this range, so make sure you understand Example 12.10; alternatively, write $1/(1 - x) = -1/(x - 1)$).

12.7. Obtain the geometric area between the graph and the x axis in each of the following cases. It is necessary to treat each positive or negative section separately.
(a) $y = -3$, $0 \leqslant x \leqslant 1$ (this is negative all the way);
(b) $y = x^3$, $-1 \leqslant x \leqslant 1$;
(c) $y = 4 - x^2$, $-1 \leqslant x \leqslant 3$;
(d) $y = \cos x$, $0 \leqslant x \leqslant 2\pi$.

12.8. Find the most general function which satisfies the following equations. (Note: $\dfrac{d^2 x}{dt^2} = \dfrac{d}{dt}\dfrac{dx}{dt}$,

$\dfrac{d^3 x}{dt^3} = \dfrac{d}{dt}\dfrac{d^2 x}{dt^2}$ etc. Work in several steps, finding the next lowest derivative in each step.)

(a) $\dfrac{d^2 x}{dt^2} = 0$;   (b) $\dfrac{d^2 x}{dt^2} = t$;   (c) $\dfrac{d^2 x}{dt^2} = \sin t$;

(d) $\dfrac{d^3 x}{dt^3} = 0$;   (e) $\dfrac{d^3 x}{dt^3} = \cos t$;

(f) $\dfrac{d^2 x}{dt^2} = g$ (g is a constant);

(g) $\dfrac{d^4 y}{dx^4} = w_0$ (w_0 is constant; this relates to the displacements $y(x)$ of a bending beam.).

13 The definite and indefinite integral

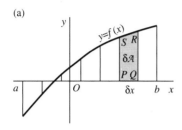

(a)

(b)

Fig. 13.1

13.1 Signed area as the sum of strips

Consider signed area from another point of view. Figure 13.1a represents the graph of a function $y = f(x)$ between $x = a$ and $x = b$. Since we are going to talk about area, assume that the x **and** y **scales are the same**. Divide the interval a to b into N small equal steps each of length

$$\delta x = \frac{b - a}{N}.$$

To any step PQ there is a signed area element $PQRS$ which we call $\delta \mathcal{A}$. The total signed area $\mathcal{A}$ is equal to the sum of all these:

$$\mathcal{A} = \sum_{x=a}^{x=b} \delta \mathcal{A}$$

which signifies 'the sum of all the elements $\delta \mathcal{A}$ between a and b'. The typical area element $PQRS$ is shown magnified in Fig. 13.1b. When δx is small, the signed area $\delta \mathcal{A}$ is nearly that of the shaded rectangle $PQNS$. Therefore, for δx small, we have

$$\mathcal{A} = \sum_{x=a}^{x=b} \delta \mathcal{A} \approx \sum_{x=a}^{x=b} f(x)\, \delta x.$$

When $\delta x \to 0$ (with N increasing correspondingly), the approximation approaches perfection and we have

---

Signed area as a sum

Signed area $\mathcal{A}$ of $y = f(x)$, $x = a$ to b: (13.1)

$$\mathcal{A} = \lim_{\delta x \to 0} \sum_{x=a}^{x=b} f(x)\, \delta x.$$

13.2 Numerical illustration of the sum formula

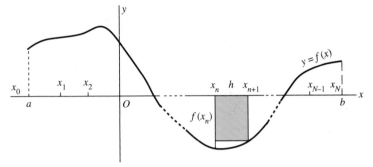

Fig. 13.2

We shall specify the sum in (13.1) in more detail, with the idea of obtaining a specific **algorithm** for actually calculating such a sum on a computer. In Fig. 13.2 we show the graph $y = f(x)$. There are N equal subdivisions: we shall call the **length of a subdivision** h rather than δx as in Fig. 13.1, since this is conventional when making numerical calculations:

$$h = \frac{b - a}{N}.$$

The relevant points of subdivision are labelled x_0 to x_{N-1}:

$$x_0 = a, \quad x_1 = a + h, \quad x_2 = a + 2h, \, \ldots \, ,$$
$$x_{N-1} = a + (N - 1)h,$$

which can be expressed as

$$x_n = a + nh \quad \text{for } n = 0, 1, 2, \ldots, N - 1.$$

(The point $x_N = b$ is not wanted for a sum of type (13.1).) Then the area of the nth approximating rectangle in (13.1) is $f(x_n)h$, and the approximating sum in (13.1) becomes

$$\mathcal{A} \approx f(a)h + f(a + h)h + \cdots + f(a + (N - 1)h)h$$
$$= h \sum_{n=0}^{N-1} f(x_n) \text{ where } x_n = a + nh, \text{ with } n = 0, 1, 2, \ldots, N - 1.$$

Computation of approximating sums (rectangle rule)

If $y = f(x)$ with range $x = a$ to b, the signed-area approximation with N subdivisions is

$$\mathcal{A} \approx h \sum_{n=0}^{N-1} f(x_n),$$

(13.2)

where $h = (b - a)/N$ and $x_n = a + nh$.

When we take larger and larger N, and smaller and smaller h correspondingly, we expect that the approximation will approach the exact value. The following example illustrates this for the very simple case of the signed area associated with a straight line. The **algorithm** (13.2) is very easy to program on a computer for any function $f(x)$. It is called the rectangle rule.

Example 13.1. Calculate the sum in (13.2) when $f(x) = x$, $a = -1$, $b = 2$, for $N = 30, 300, 3000, \ldots$, showing how the results approach the exact value 1.5.

The graph $y = x$ is a straight line, from which it is very easy to see that the signed area is exactly 1.5. The computed results are as follows ($b - a = 3$, and so $h = 3/N$):

N	30	300	3000	30,000	$\cdots$
h	0.1	0.01	0.001	0.0001	$\cdots$
$\mathcal{A} \approx$	1.05	1.455	1.4955	1.49955	$\cdots$

The approximations are approaching 1.5, though very slowly. We shall see in Section 14.3 how to improve such calculations.

13.3 The definite integral and area

The expression $\lim\limits_{\delta x \to 0} \sum\limits_{x=a}^{x=b} f(x)\,\delta x$ of (12.1), which is equal to the signed area, has a very important brief notation:

Definite-integral notation

$$\lim_{\delta x \to 0} \sum_{x=a}^{x=b} f(x)\,\delta x \text{ is denoted by } \int_a^b f(x)\,\mathrm{d}x. \qquad (13.3)$$

(Historically, a large letter S for 'sum' used to be printed instead of $\sum$: the sign $\int$ is really just an extended letter S.) The expression $\int_a^b f(x)\,\mathrm{d}x$ is called a **definite integral**, to be read: '**the integral of** $f(x)\,\mathrm{d}x$ **from** a **to** b'. Here $f(x)$ is the **integrand**, or the function to be **integrated**. The letter x is the name of the **variable of integration**, while a is the **lower limit** and b the **upper limit** for the **integration** process.

We already found a way to obtain the signed area by using an antiderivative: see (12.8). In the new notation, (12.8) is expressed as follows.

Signed area expressed as a definite integral

The signed area $\mathcal{A}$ of $f(x)$ between $x = a$ and b is given by

$$\int_a^b f(x)\,dx = [F(x)]_a^b = F(b) - F(a) \qquad \textbf{(13.4)}$$

where $F(x)$ is any antiderivative of $f(x)$.

Notice that in a **definite integral any letter can be used for the variable of integration**, because the letter itself disappears in the course of evaluation; for example

$$\int_0^1 x\,dx = \left[\tfrac{1}{2}x^2\right]_0^1 = \left[\tfrac{1}{2}x^2\right]_{x=0}^{x=1} = \tfrac{1}{2};$$

$$\int_0^1 t\,dt = \left[\tfrac{1}{2}t^2\right]_0^1 = \left[\tfrac{1}{2}t^2\right]_{t=0}^{t=1} = \tfrac{1}{2}; \quad \text{and so on.}$$

13.4 The indefinite-integral notation

The symbol

$$\int f(x)\,dx,$$

with no limits of integration specified, is called an **indefinite integral of** $f(x)$, and has exactly the **same meaning as the word 'antiderivative'** that we have used up until now, and which we have denoted by $F(x)$.

Indefinite-integral notation

$\int f(x)\,dx$, with no limits specified, stands for any $\qquad$ **(13.5)**

antiderivative of $f(x)$.

The expression $\int_a^b f(x)\,dx$ is called a definite integral because it takes a definite value: it represents a specified signed area and there is no arbitrary constant on the right. However, an indefinite integral $\int f(x)\,dx$ does not stand for a number; it represents an antiderivative, which is a function. This function is to be written in terms of the current variable of integration, so the name of the variable is usually significant. Also there will be a disposable, or arbitrary, constant, in the usual way. For example

$$\int x^2 \, dx = \tfrac{1}{3}x^3 + C, \qquad \int e^{2t} \, dt = \tfrac{1}{2} e^{2t} + C,$$

$$\int \cos u \, du = \sin u + C$$

and so on, where C is a constant. In some problems we shall assign or discover a definite value for C; in others we might want to keep C as an arbitrary constant in order to express every possible antiderivative (hence, 'indefinite' integral).

Example 13.2. Find the signed area $\mathcal{A}$ associated with the graph $y = 3e^{2x}$ from $x = 1$ to $x = 3$ using the new notation.

We shall need an antiderivative $F(x)$ (i.e. an indefinite integral) of $3e^{2x}$. Using the notation (13.5), we may write

$$F(x) = \int 3e^{2x} \, dx = \tfrac{3}{2}e^{2x},$$

(for this purpose any antiderivative will do, so we have put $C = 0$). Then, from (13.4),

$$\mathcal{A} = \int_1^3 3e^{2x} \, dx = \left[\tfrac{3}{2}e^{2x}\right]_1^3$$

$$= \tfrac{3}{2}\left[e^{2x}\right]_1^3 = \tfrac{3}{2}(e^6 - e^2).$$

In the last example, we might as well have written

$$\int_1^3 3e^{2x} \, dx = \left[\int 3e^{2x} \, dx\right]_1^3$$

in the first place, without ever introducing $F(x)$. If we do this, we get another version of (13.4) which it is often convenient to use:

Signed area, using the notation for definite and indefinite integrals

The signed area of $f(x)$ from a to b is

$$\int_a^b f(x) \, dx = \left[\int f(x) \, dx\right]_a^b, \qquad (13.6)$$

where $\int f(x) \, dx$ is any indefinite integral (antiderivative) of $f(x)$.

13.5 Integrals unrelated to area

Integrals arise constantly in applications, but only seldom is there any direct connection with area. The following example starts by giving information that seems to have nothing to do with area. However, we show that the problem can be thought of in terms of an area, and therefore can be solved in terms of a definite integral.

Example 13.3. A small object P is pushed steadily along the x axis from $x = 0$ to $x = 1$, against a resistive force $f(x) = x^2$. Find the work done against the resistance.

Divide the range $x = 0$ to 1 into a large number of short steps of length δx. In general, if the resistive force is constant the work done over a distance is (force) × (distance moved). Although the force on P is not constant, over a short distance δx the work δW done by the applied force is given approximately by

$$\delta W \approx f(x)\,\delta x = x^2\,\delta x.$$

The total work W is given by

$$W = \sum_{x=0}^{x=1} \delta W \approx \sum_{x=0}^{x=1} x^2\,\delta x.$$

Letting $\delta x \to 0$, we obtain exactly

$$W = \lim_{\delta x \to 0} \sum_{x=0}^{x=1} x^2\,\delta x. \tag{13.7}$$

But this expression matches equation (13.1): *it represents the signed area of the curve* $y = x^2$ *between* $x = 0$ *and* 1. Consequently we can say immediately that

$$W = \int_0^1 x^2\,dx = \left[\tfrac{1}{3}x^3\right]_0^1 = \tfrac{1}{3}. \tag{13.8}$$

The reader should think very carefully about the step from (13.7) to (13.8), because it can be generalized to apply to any similar problem. Suppose that there arises, in any context whatever, a sum of the type

$$\lim_{\delta x \to 0} \sum_{x=a}^{x=b} f(x)\,\delta x.$$

Then such a *sum can always be interpreted as representing a certain signed area* (namely the signed area of $y = f(x)$ between $x = a$ and $x = b$) so *it can always be represented by the definite integral* $\int_a^b f(x)\,dx$. We do not have to repeat this argument every time we encounter such a sum; from now on, we call on the general statement:

The limit of a sum represented by a definite integral

In all cases

$$\lim_{\delta x \to 0} \sum_{x=a}^{x=b} f(x)\,\delta x = \int_a^b f(x)\,dx. \tag{13.9}$$

The integral is then evaluated using (13.4) or (13.6).

The variable occurring need not be denoted by x, as the following examples demonstrate.

Example 13.4. An object is driven along a straight line with velocity $v(t) = e^{\frac{1}{2}t}$ between times $t = 0$ and $t = 2$. There is a resistive force $g(v) = 3v^2$, where v is velocity. Find the total work done against the resistance.

In a short interval between times t and $t + \delta t$, the distance travelled, δx, is given approximately by

$$\delta x \approx v(t)\,\delta t = e^{\frac{1}{2}t}\,\delta t.$$

The work δW done in this time interval is approximated by

$$\delta W \approx g(v)\,\delta x = 3v^2\,\delta x$$

$$\approx 3v^2(e^{\frac{1}{2}t}\,\delta t) = 3e^t(e^{\frac{1}{2}t}\,\delta t) = 3e^{\frac{3}{2}t}\,\delta t.$$

Therefore the total work W required is given by

$$W = \lim_{\delta t \to 0} \sum_{t=0}^{t=2} 3e^{\frac{3}{2}t}\,\delta t$$

$$= \int_0^2 3e^{\frac{3}{2}t}\,dt = 2[e^{\frac{3}{2}t}]_0^2 = 2(e^3 - 1).$$

Example 13.5. During a rainy period extending from $t = 0$ to $t = 10$ days, the rainfall rate r from moment to moment in units of centimetres per day, is found to be $r(t) = \frac{3}{5}t - \frac{3}{50}t^2$. Find the total depth of rainfall, R, for the period.

Take a short time interval from t to $t + \delta t$ (expressed as a fraction of a day). During this period, the rainfall δR is given approximately by

$$\delta R \approx r(t)\,\delta t = (\tfrac{3}{5}t - \tfrac{3}{50}t^2)\,\delta t$$

('approximately' because the rate of fall r varies a little even through a short time). The total rainfall from $t = 0$ to 10 days is equal to the sum of all the contributions as the steps δt tend to zero (while becoming proportionately more numerous):

$$R = \lim_{\delta t \to 0} \sum_{t=0}^{t=10} (\tfrac{3}{5}t - \tfrac{3}{50}t^2)\, \delta t$$

$$= \int_{0}^{10} (\tfrac{3}{5}t - \tfrac{3}{50}t^2)\, dt \quad \text{(by (13.9))}$$

$$= [\tfrac{3}{5}(\tfrac{1}{2}t^2) - \tfrac{3}{50}(\tfrac{1}{3}t^3)] = \tfrac{3}{10}(10)^2 - \tfrac{1}{50}(10)^3 = 10 \quad \text{(cm).}$$

Example 13.6. Suppose that, in Example 13.5, the rainfall rate is given by $r(t) = t^{\frac{1}{2}} e^{-t}$ (cm per day). Obtain the total rainfall R between $t = 0$ and 10 days.

Proceeding as before, the total rainfall is given by

$$R = \int_{0}^{10} t^{\frac{1}{2}} e^{-t}\, dt.$$

We cannot find an indefinite integral to enable R to be evaluated. However, we know from (13.9) that R *must be equal to the area under the (r, t) graph*, which can be *computed numerically* by using the numerical method of equation (13.2).

Divide the range $t = 0$ to 10 into N strips (so that $\delta t = 10/N$), then the approximation corresponding to (13.2) becomes

$$R \approx \frac{10}{N} \sum_{n=0}^{N-1} t_n^{\frac{1}{2}} e^{-t_n}, \quad \text{where} \quad t_n = n\left(\frac{10}{N}\right).$$

The following computed values show how the exact result is approached when we take N larger and larger:

N	5	10	100	1000
δx	2.00	1.00	0.10	0.100
R	0.4701	0.7070	0.8796	0.8859

The exact answer is $0.88587 \cdots$.

We shall show in Chapter 14 that we are not tied to equation (13.2) for calculating signed area, but can find far better computing formulae.

13.6 Improper integrals

If a definite integral has an **infinite range**, or the **integrand becomes infinite** at some point in its range, the integral is said to be **improper**. Usually these present no particular problem.

Example 13.7. Evaluate $\displaystyle\int_0^\infty e^{-2x}\,dx$

Putting $\displaystyle\int e^{-2x}\,dx = -\tfrac{1}{2}e^{-2x}$, we have

$$\int_0^\infty e^{-2x}\,dx = -\tfrac{1}{2}\big[e^{-2x}\big]_0^\infty = -\tfrac{1}{2}(0-1) = \tfrac{1}{2}.$$

Example 13.8. Evaluate $\displaystyle\int_1^\infty \frac{dx}{x^2}$.

$$\int_1^\infty x^{-2}\,dx = \big[-x^{-1}\big]_1^\infty = [0-(-1)] = 1.$$

In Examples 13.7 and 13.8 we have (see Fig. 13.3) two cases of

Fig. 13.3

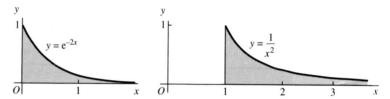

an infinitely long figure which encloses a finite area. This cannot always happen, even if the integrand goes to zero when $x \to \infty$.

Example 13.9. Consider $\displaystyle\int_1^\infty \frac{dx}{x}$.

We have

$$\int_1^\infty x^{-1}\,dx = [\ln x]_1^\infty.$$

The logarithm becomes infinite as x becomes infinite, so the integral is meaningless. The function x^{-1} does not tend to zero fast enough to keep the area finite as we extend the range to infinity.

The case when the **integrand becomes infinite** at some point in its range has similar features:

Example 13.10. Consider (a) $\displaystyle\int_0^1 x^{-\frac{1}{2}}\,dx$; (b) $\displaystyle\int_0^1 x^{-1}\,dx$.

Notice that $x^{-\frac{1}{2}}$ and x^{-1} are infinite at $x = 0$.

(a) $\displaystyle\int_0^1 x^{-\frac{1}{2}}\,dx = 2\big[x^{\frac{1}{2}}\big]_0^1 = 2[1-0] = 2.$

Therefore the integral gives no problem; it is again a case of an infinitely extended figure (extended in the y direction this time) containing a finite area.

(b) On the other hand,

$$\int_0^1 x^{-1}\, dx = \left[\ln x\right]_0^1,$$

and the integral this time is infinite, because $\ln 0$ is $(-\infty)$.

There are improper integrals which do not work out for a different reason:

Example 13.11. Consider the integrals

(a) $\displaystyle\int_0^X \cos x\, dx,$ (b) $\displaystyle\int_0^\infty \cos x\, dx.$

(a) We have

$$\int_0^X \cos x\, dx = \left[\sin x\right]_0^X = \sin X.$$

So long as X is finite, there is therefore no problem.

(b) However, for $\displaystyle\int_0^\infty \cos x\, dx$ we would, straightforwardly, have a term $\sin \infty$ to interpret. The only sensible meaning that we could attach to $\sin \infty$ is that it stands for $\lim_{X \to \infty} \sin X$. But $\sin X$ has no definite limit as $X \to \infty$; it goes up and down between ± 1 for ever.

Improper integrals which give a definite finite result are said to **converge**. If not, they are said to **diverge**.

13.7 Integration of complex functions: a new type of integral

To differentiate or integrate a function containing the 'imaginary' element j, simply treat j like an ordinary real constant. Thus, for example,

$$\frac{d}{dx} e^{jx} = j e^{jx}$$

and

$$\int e^{jx}\, dx = \frac{1}{j} e^{jx} + C = -j e^{jx} + C,$$

where C is an arbitrary constant (which in this context we would allow to be itself a complex number). Suppose that a and b are *real* numbers, and that

$$c = a + jb.$$

Then

$$\frac{d}{dx} e^{cx} = c\,e^{cx},$$

and

$$\int e^{cx}\,dx = \frac{1}{c} e^{cx} + C. \tag{13.10}$$

We use (13.10) with $c = a + jb$ to work out two integrals U and V which frequently occur in practice:

$$U = \int e^{ax} \cos bx\,dx \quad \text{and} \quad V = \int e^{ax} \sin bx\,dx.$$

Omit the arbitrary constant C for the moment. Observe that

$$U + jV = \int e^{ax} \cos bx\,dx + j \int e^{ax} \sin bx\,dx$$

$$= \int e^{ax}(\cos bx + j \sin bx)\,dx = \int e^{ax} e^{jbx}\,dx$$

$$= \int e^{(a + jb)x}\,dx$$

$$= \frac{1}{a + jb} e^{(a + jb)x} \quad \text{(from 13.10)}$$

$$= \frac{a - jb}{a^2 + b^2} e^{ax}(\cos bx + j \sin bx)$$

$$= \frac{1}{a^2 + b^2} e^{ax}[(a \cos bx + b \sin bx) + j(-b \cos bx + a \sin bx)].$$

Equate this last expression to $U + jV$: the real and imaginary parts must separately be equal; so, after introducing the arbitrary constant, we have

$$\text{(a)} \quad \int e^{ax} \cos bx\,dx$$

$$= \frac{1}{a^2 + b^2} e^{ax}(a \cos bx + b \sin bx) + C, \tag{13.11}$$

$$\text{(b)} \quad \int e^{ax} \sin bx\,dx$$

$$= \frac{1}{a^2 + b^2} e^{ax}(-b \cos bx + a \sin bx) + C.$$

The integrals can be expressed more simply in terms of a phase angle (see Section 18.1). Put

$$\frac{a}{(a^2 + b^2)^{\frac{1}{2}}} = \cos\phi \quad \text{and} \quad \frac{b}{(a^2 + b^2)^{\frac{1}{2}}} = -\sin\phi$$

into (13.11a), and

$$\frac{-b}{(a^2 + b^2)^{\frac{1}{2}}} = \cos\theta \quad \text{and} \quad \frac{a}{(a^2 + b^2)^{\frac{1}{2}}} = -\sin\theta$$

into (13.11b). Notice also that ϕ and θ must therefore be related by

$$\theta = \phi - \tfrac{1}{2}\pi.$$

Then (13.11) becomes

$$\text{(a)} \int e^{ax} \cos bx \, dx = \frac{1}{(a^2 + b^2)^{\frac{1}{2}}} e^{ax} \cos(bx + \phi),$$

$$\text{(b)} \int e^{ax} \sin bx \, dx = \frac{1}{(a^2 + b^2)^{\frac{1}{2}}} e^{ax} \cos(bx + \phi - \tfrac{1}{2}\pi) \qquad \textbf{(13.12)}$$

where $\cos\phi = a/(a^2 + b^2)^{\frac{1}{2}}$, $\sin\phi = -b/(a^2 + b^2)^{\frac{1}{2}}$.

Example 13.12. Evaluate $I = \displaystyle\int_0^\infty e^{-x} \cos 2x \, dx$.

Equation (13.11a) or (13.12a) can be used directly with $a = -1$, $b = 2$. However, we will go through the working from first principles, but express the argument differently. Remember that

$$e^{\alpha + j\beta} = e^\alpha e^{j\beta} = e^\alpha(\cos\beta + j\sin\beta);$$

then we have $e^{-x} \cos 2x = \text{Re } e^{(-1 + 2j)x}$. Therefore

$$I = \int_0^\infty e^{-x} \cos 2x \, dx = \text{Re} \int_0^\infty e^{(-1 + 2j)x} \, dx$$

$$= \text{Re}\left[\frac{1}{-1 + 2j} e^{(-1 + 2j)x} \right]_0^\infty$$

$$= \text{Re}\left(0 - \frac{1}{-1 + 2j} \right) = \text{Re} \frac{1 + 2j}{5} = \tfrac{1}{5}.$$

13.8 The area analogy for a definite integral

A signed area can be represented as a definite integral as in (13.4). Conversely, any definite integral $\int_a^b f(x) \, dx$, whatever it represents, can be interpreted as representing the signed area of the graph $y = f(x)$ between a and b. The connection with area means that we have a picture of an integral which can often give useful information without the need to evaluate the integral, which might in any case be impossible. One example of this is the simple numerical method described in Section 13.2. We restate the connection, calling it the **area analogy**.

The area analogy

The definite integral $\displaystyle\int_a^b f(x)\,\mathrm{d}x$ always represents the   **(13.13)**

signed area of the graph $y = f(x)$ from $x = a$ to b.

The following section illustrates the use of the principle (13.13): it will also be referred to in later chapters.

13.9 Using the area analogy

In this section we shall use t instead of x, and x instead of y, so that we consider functions of the form

$$x = f(t).$$

The reason is that time t is commonly the physical variable in contexts where these techniques are found useful.

Sometimes, by using the area analogy (13.13), the graph of the integrand of $\int_a^b f(t)\,\mathrm{d}t$ makes it obvious that the value of a definite integral is zero. Figure 13.4 shows some simple cases.

The range of the integrals on the two sides of the origin are equal, and because of the special symmetry the positive and negative contributions cancel out. Such functions are called **odd functions**, or **functions odd about the origin**. They have the following property (see Section 1.4):

Odd functions $f(t)$

satisfy the condition $f(-t) = -f(t).$ **(13.14)**

Some basic odd functions are

$$t, t^3, t^5, \ldots, \text{ and their reciprocals;}$$

$\sin at, \sin^3 at, \ldots,$ and $\tan at, \tan^3 at, \ldots,$ where a is constant.

Symmetrical integrals over functions which are odd about the origin

If $f(-t) = -f(t)$ then $\displaystyle\int_{-c}^{c} f(t)\,\mathrm{d}t = 0.$ **(13.15)**

(a)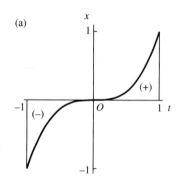

(b)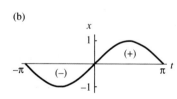

(c)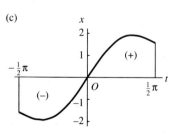

Fig. 13.4
(a) $x = t^3$; $\int_{-1}^{1} t^3\,\mathrm{d}t = 0$.
(b) $x = \sin t$; $\int_{-\pi}^{\pi} \sin t\,\mathrm{d}t = 0$.
(c) $x = t + \sin 2t$;
$\int_{-\frac{1}{2}\pi}^{\frac{1}{2}\pi} (t + \sin 2t)\,\mathrm{d}t = 0$.

Another useful class are **even functions**, which are symmetrical about the x axis:

> **Even functions $f(t)$**
>
> $f(t)$ is even if $f(-t) = f(t)$.   (13.16)

Some basic even functions are

$$t^2, t^4, t^6, \ldots, \text{ and their reciprocals;}$$

$$\cos at \text{ and } \cos^n at \text{ } (a \text{ is a constant});$$

also even powers of any odd function, such as $\tan^6 at$. It is also useful to realize that

$$(\text{odd function}) \times (\text{even function}) = (\text{odd function}),$$

$$(\text{odd function}) \times (\text{odd function}) = (\text{even function}).$$

Some even functions are shown in Fig. 13.5.

(a)
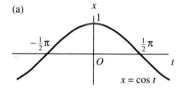

Example 13.13. Show that the following integrals are zero:

(a) $\displaystyle\int_{-\frac{1}{2}\pi}^{\frac{1}{2}\pi} t^4 \sin 3t \, dt;$ (b) $\displaystyle\int_{-\pi}^{\pi} t^5 \cos 3t \cos \tfrac{1}{2}t \, dt;$

(c) $\displaystyle\int_{-1}^{1} (e^{2t} - e^{-2t}) \, dt$

(b)
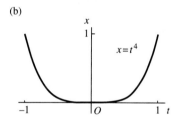

(a) The function t^4 is even and $\sin 3t$ is odd, so the integrand is odd. Since the range is symmetrical about the origin, the integral is zero.

(b) t^5 is odd, $\cos 3t$ is even, and $\cos \tfrac{1}{2}t$ is even, so the integrand is odd and the integral is zero as in (a).

(c) $e^{2t} - e^{-2t}$ is odd (put $-t$ in place of t in the function—it just changes its sign). Therefore the integral is zero.

(c)
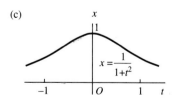

Fig. 13.5

If we integrate an even function between $\pm c$, the graph shows that we get the same contribution from both sies of the origin, which gives the following result.

> **Symmetrical integrals of even functions**
>
> If $f(t)$ is even, then $\displaystyle\int_{-c}^{c} f(t) \, dt = 2 \int_{0}^{c} f(t) \, dt.$ (13.17)

These ideas may also be useful if there is special symmetry about some point other than the origin.

(a)

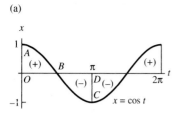

(b)

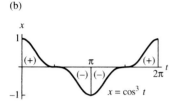

Fig. 13.6

Example 13.14. Show that $\displaystyle\int_0^{2\pi} \cos^3 t \, dt = 0$.

In the graph of $x = \cos t$ (Fig. 13.6) the parts OBA and DBC are congruent, and similarly for the other pair of divisions; in fact all four divisions are congruent. For the graph of $x = \cos^3 t$ the shape is changed, but the four pieces remain congruent and retain their original sign. The resulting cancellation gives zero for the integral.

13.10 Definite integrals having variable limits

Integrals of the following type occur rather frequently in applications:

$$I(x) = \int_c^x f(t) \, dt, \tag{13.18}$$

where c is a constant. Although $I(x)$ depends on x, this is still a **definite integral** because it has limits of integration; no arbitrary constant occurs. Notice that we **avoid using x as the variable of integration when a limit of integration involves** x: the same letter would be serving two totally different purposes. Therefore we have changed the variable of integration to t.

Suppose that $F(t)$ represents any particular indefinite integral, or antiderivative, of $f(t)$. Then, as always,

$$I(x) = \big[F(t)\big]_c^x = F(x) - F(c). \tag{13.19}$$

Since $F(c)$ is constant,

$$\frac{dI(x)}{dx} = \frac{d}{dx}\int_c^x f(t) \, dt = \frac{dF(x)}{dx} = f(x).$$

Similarly we can obtain

$$\frac{d}{dx}\int_x^c f(t) \, dt = -f(x).$$

Now consider the more complicated case

$$K(x) = \int_{u(x)}^{v(x)} f(t) \, dt = F(v(x)) - F(u(x)).$$

By using the chain rule, we have

$$\frac{d}{dx} F(v(x)) = \frac{dF(v)}{dv}\frac{dv(x)}{dx} = f(v(x)) \frac{dv(x)}{dx},$$

with a similar result for $F(u(x))$, and finally we have the results

Differentiation of integrals

(a) $\dfrac{d}{dx} \displaystyle\int_{u(x)}^{v(x)} f(t)\,dt = f(v(x))\dfrac{dv(x)}{dx} - f(u(x))\dfrac{du(x)}{dx}.$

(b) (Special cases):

$$\dfrac{d}{dx} \int_{c}^{x} f(t)\,dt = f(x) \qquad\qquad \textbf{(13.20)}$$

and

$$\dfrac{d}{dx} \int_{x}^{c} f(t)\,dt = -f(x).$$

(The results (13.20b) are simply (13.20a) in the respective cases $v(x) = x$, $u(x) = c$, or $v(x) = c$, $u(x) = x$.) It is worth noticing that (13.20) does not require you to integrate anything!

Example 13.15. Obtain dI/dx when $I(x) = \displaystyle\int_{x^2}^{3x^2} e^t\,dt$.

Here $f(t) = e^t$, $u(x) = x^2$, $v(x) = 3x^2$.

Therefore $\dfrac{dI(x)}{dx} = e^{3x^2}\, 6x - e^{x^2}\, 2x = 2x(3e^{3x^2} - e^{x^2}).$

Equation (13.19) shows that $I(x) = \displaystyle\int_{c}^{x} f(t)\,dt$ is an antiderivative of $f(x)$. It might be thought that, by choosing various values for c, we could reproduce *all* the antiderivatives $F(x)$ corresponding to $f(x)$. However, this expectation is only sometimes correct.

Example 13.16. Let $f(x) = \cos x$, with antiderivatives $\sin x + C$, where C is an arbitrary constant. Demonstrate that for the integral $\int_{c}^{x} f(t)\,dt$ it is not possible to find a value of c which will reproduce the antiderivative $\sin x + 1000$.

We have

$$\int_{c}^{x} \cos t\,dt = \big[\sin t\big]_{c}^{x} = \sin x - \sin c.$$

But $-1 \leqslant \sin c \leqslant 1$ no matter what value of c we take, so we could never make the integral equal to $\sin x + 1000$.

(a)

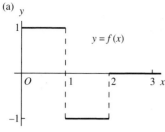

(b)

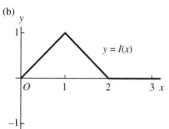

Fig. 13.7

Example 13.17. The function shown in Fig. 13.7a is described by $f(x) = 1$ when $0 \leqslant x < 1$, $f(x) = -1$, when $1 \leqslant x < 2$, and $f(x) = 0$ when $x \geqslant 2$. Sketch a graph of the function

$$I(x) = \int_0^x f(t)\, dt.$$

Range $0 \leqslant x < 1$. $I(x) = \displaystyle\int_0^x 1\, dt = x$ **(i)**

Range $1 \leqslant x < 2$. In this range, $f(t)$ is described by a different expression, -1 instead of 1, so we must split up the integral:

$$I(x) = \int_0^x f(t)\, dt = \int_0^1 f(t)\, dt + \int_1^x f(t)\, dt$$

$$= 1 + \int_1^x (-1)\, dt \quad \text{(after using (i) at } x = 1)$$

$$= 1 - (x - 1) = 2 - x. \qquad \textbf{(ii)}$$

Range $x > 2$. Split the integral again:

$$I(x) = \int_0^2 f(t)\, dt + \int_2^x f(t)\, dt$$

$$= 0 + \int_2^x 0\, dt = 0, \qquad \textbf{(iii)}$$

where we used the value of (ii) at $x = 2$. The resulting graph of $I(x)$ is shown in Fig. 13.7b.

Problems, Chapter 13

13.1. Sketch each of the following curves; then express the signed areas under them firstly as the sums of strips, as in (13.1), and secondly as definite integrals, as in (13.3); and finally evaluate them by (13.4).
(a) $y = x^3$, $-1 \leqslant x \leqslant 2$;   (b) $y = x^5$, $-1 \leqslant x \leqslant 1$;
(c) $y = \sin x$, $-\pi \leqslant x \leqslant 0$;   (d) $y = e^{-2x}$, $0 \leqslant x \leqslant 1$.

13.2. Evaluate the following indefinite integrals (remember the arbitrary constant).

(a) $\displaystyle\int x^{\frac{1}{2}}\, dx$;   (b) $\displaystyle\int (x+1)^{\frac{1}{2}}\, dx$;   (c) $\displaystyle\int e^{\frac{1}{2}x}\, dx$;

(d) $\displaystyle\int \sin x\, dx$;   (e) $\displaystyle\int (\cos x - 2 \sin 2x)\, dx$;

(f) $\displaystyle\int t^{-\frac{1}{2}}\, dt$;   (g) $\displaystyle\int \cos 2u\, du$;   (h) $\displaystyle\int 3e^{-\frac{1}{2}y}\, dy$;

(i) $\displaystyle\int (1 + 3t^2 - 2t)\, dt$;   (j) $\displaystyle\int (1 + 4 \cos 4w)\, dw$;

(k) $\displaystyle\int (-x)^{\frac{1}{2}}\, dx$ when x is negative (you will have to experiment to find a valid antiderivative).

13.3. Evaluate the following definite integrals.

(a) $\displaystyle\int_{-1}^{1} x^3\, dx$;   (b) $\displaystyle\int_{-1}^{1} x^2\, dx$;   (c) $\displaystyle\int_0^2 dx$;

(d) $\displaystyle\int_0^4 x^{\frac{1}{2}}\, dx$;   (e) $\displaystyle\int_{-1}^{1} (1 - 3x + 2x^2)\, dx$;

(f) $\displaystyle\int_1^2 (x^{-3} + x^{-2})\, dx$;   (g) $\displaystyle\int_1^2 x^{-2}\, dx$;

(h) $\displaystyle\int_{-2}^{-1} x^{-1}\, dx$ (take care: the x values are negative);

(i) $\int_{-2}^{-1} (-x)^{\frac{1}{2}}\,dx$ (see the remark in Problem 13.2k);

(j) $\int_{0}^{1} e^{-3x}\,dx$; (k) $\int_{0}^{\frac{1}{4}\pi} \sin 4x\,dx$;

(l) $\int_{0}^{2\pi} \sin\tfrac{1}{2}x\,dx$; (m) $\int_{0}^{2\pi} \cos\tfrac{1}{2}x\,dx$.

13.4. Evaluate the following integrals, using the notation of (13.4).

(a) $\int_{0}^{1} x(x^2 + x + 1)\,dx$; (b) $\int_{-1}^{1} (x-1)(x+1)\,dx$;

(c) $\int_{0}^{2} x(x^2 - 1)\,dx$; (d) $\int_{1}^{2} \dfrac{x + x^2}{x^3}\,dx$;

(e) $\int_{1}^{2} \dfrac{t(t+1)}{t^{\frac{1}{2}}}\,dt$; (f) $\int_{1}^{4} \dfrac{\sqrt{u}-1}{u}\,du$;

(g) $\int_{-1}^{0} \dfrac{dw}{2w + 3}$; (h) $\int_{-2}^{-1} \dfrac{x}{x-1}\,dx$;

(i) $\int_{0}^{\pi} \cos^2 3t\,dt$ $(\cos^2 A = \tfrac{1}{2}(1 + \cos 2A))$.

13.5. Evaluate the following infinite integrals.

(a) $\int_{1}^{\infty} e^{-3t}\,dt$; (b) $\int_{0}^{\infty} e^{-\frac{1}{2}v}\,dv$; (e) $\int_{1}^{\infty} \dfrac{dx}{x^3}$;

(d) $\int_{0}^{\infty} \dfrac{dx}{(2x+3)^2}$; (e) $\int_{0}^{1} \dfrac{ds}{s^{\frac{1}{4}}}$; (f) $\int_{1}^{2} \dfrac{dt}{(t-1)^{\frac{1}{2}}}$;

(g) $\int_{0}^{\infty} e^{-2t} \sin 3t\,dt$ (see Section 13.7);

(h) $\int_{0}^{\infty} e^{-\frac{1}{4}t} \cos 2t\,dt$ (see Section 13.7).

13.6. The **mean** or **average value** of $f(t)$ over an interval $0 \leqslant t \leqslant T$ is the quantity $\dfrac{1}{T}\int_{0}^{T} f(t)\,dt$. Find the mean values of the following over the intervals given.

(a) $f(t) = t,\ 0 \leqslant t \leqslant 1$; (b) $f(t) = t,\ -1 \leqslant t \leqslant 1$;
(c) $f(t) = \sin t,\ 0 \leqslant t \leqslant \pi$;
(d) $f(t) = \sin t,\ 0 \leqslant t \leqslant 2\pi$; (e) $f(t) = t^{-2},\ 1 \leqslant t \leqslant T$;
(f) $f(t) = e^{-t} \cos t\,dt,\ 0 \leqslant t \leqslant 2\pi$;
(g) $f(t) = e^{-2t} \sin t\,dt,\ 0 \leqslant t < \infty$;
(h) $f(t) = 1 - e^{-t}$ (work out the mean value over $0 \leqslant t \leqslant T$ for several increasing values of T, and deduce the value of the mean over $0 \leqslant t < \infty$; if you put $T = \infty$ into the integral directly, it turns out to be infinite, or 'diverges' (see Section 13.6), so no conclusion can be drawn from this approach).

(i) $f(t) = t^{-1},\ 1 \leqslant t < \infty$ (it is necessary to follow the procedure in the previous question, for the same reason).

13.7. Use the even/odd properties of the integrands (see Section 13.9) to prove the following results.

(a) $\int_{-\pi}^{\pi} \sin^4 t\,dt = 2\int_{0}^{\pi} \sin^4 t\,dt$;

(b) $\int_{-1}^{1} \dfrac{t^3}{(1 + t^4)}\,dt = 0$;

(c) $\int_{-\pi}^{\pi} \dfrac{t \cos t}{1 + t^2}\,dt = 0$; (d) $\int_{-\frac{1}{4}\pi}^{\frac{1}{4}\pi} t^2 \sin(t^3)\,dt = 0$.

13.8. (Computational: See Section 13.2 and Examples 13.1 and 13.6). Write a simple program based on the algorithm (13.2) to evaluate a definite integral $\int_{a}^{b} f(x)\,dx$. Assume that you have a subroutine for evaluating $f(x)$, and that you input a, b, and N (the number of subdivisions); also either a permissible error E, or a parameter M which determines the number of iterations. If you use E, the process might be written to print out when two successive iterations are within E of each other. Check the correctness of the program by using a function such as x^2 as integrand.

Estimate the values of the following integrals.

(a) $\int_{1}^{2} \dfrac{e^{-x}}{x}\,dx$; (b) $\int_{0}^{\pi} \sin x^2\,dx$; (c) $\int_{0}^{1} \cos e^{-x}\,dx$.

13.9. (Computational). (a) Convince yourself that $e^{-x^2} < e^{-x}$ when $x > 1$. Use the area analogy (13.13) to show that, if $b > 1$, then $\int_{b}^{\infty} e^{-x^2}\,dx < \int_{b}^{\infty} e^{-x}\,dx$.

Deduce that, if E is a positive number and $E < 1$, then $\int_{b}^{\infty} e^{-x^2}\,dx < E$ for $b > -\ln E$.

(b) Use the program written for Problem 13.8 to evaluate the *improper* integral $\int_{0}^{\infty} e^{-x^2}\,dx$ to within two decimal places, in the following way. You have to stop the integral somewhere: the program cannot deal with $b = \infty$. Take a permissible error $E = 0.001$, say, to leave some leeway. Referring to (a), choose $b > -\ln E$, and compute the integral $\int_{0}^{b} e^{-x^2}\,dx$. The part of the original integrand between b and infinity will then be negligible.

13.10. (Section 13.10). Find dI/dx where $I(x)$ is given by the following integrals.

(a) $\displaystyle\int_0^x t^2 \, dt;$ (b) $\displaystyle\int_0^x \sin^5 t \, dt;$ (c) $\displaystyle\int_0^x \frac{e^t}{1+t} \, dt;$

(d) $\displaystyle\int_0^{e^x} t \ln t \, dt;$ (e) $\displaystyle\int_{\sqrt{x}}^{\sqrt{(x+1)}} \sin(t^2) \, dt.$

13.11. (See Example 13.16: it is necessary to split up the integrals.) Obtain $\displaystyle\int_0^x f(t) \, dt$ where $f(x)$ is defined by the following.

(a) $f(x) = \begin{cases} 0 & \text{if } x < -1 \\ x & \text{if } -1 \leqslant x \leqslant 1 \\ 0 & \text{if } x > 1 \end{cases}$; consider positive

and negative values of x.

(b) $f(x) = \begin{cases} x & \text{if } 0 \leqslant x < 1 \\ 2 - x & \text{if } 1 \leqslant x \leqslant \frac{3}{2} \\ 0 & \text{if } x > \frac{3}{2} \end{cases}$; consider positive x only.

13.12. An 'RL' circuit has a constant current I_0 flowing, produced by a constant applied voltage. A switch cuts off the voltage and closes the circuit again at time $t = t_0 > 0$. For $t > t_0$, the current is given by $I(t) = I_0 e^{-R(t-t_0)/L}$. Obtain expressions for $Q(t)$ for $t \geqslant 0$, where

$$Q(t) = \int_0^t I(u) \, du.$$

14

Applications involving the integral as a sum

Reminder: The short table (12.2) provides all the indefinite integrals (antiderivatives) required for this chapter.

14.1 Examples of integrals arising from a sum
The examples which follow show typical cases where integrals arise from sums of the type (13.9).

Example 14.1. The tension T in an elastic string is given by $T = 0.01x$ (kg m s^{-2}), where x is the extension beyond the natural length. Find the work done on the string to stretch it 2 metres beyond its natural length.

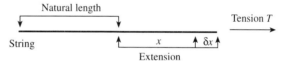

Fig. 14.1

To stretch it from extension x to $x + \delta x$, the work δW required is approximated by

$$\delta W \approx \text{force} \times \text{distance} = T\,\delta x = 0.01x\,\delta x.$$

The total work W is approximated by

$$W = \sum_{x=0}^{x=2} \delta W \approx \sum_{x=0}^{x=2} 0.01x\,\delta x.$$

Now let $\delta x \to 0$. Then we obtain

$$W = \lim_{\delta x \to 0} \sum_{x=0}^{x=2} 0.01x\,\delta x = \int_0^2 0.01x\,dx \quad \text{(from (13.9))}$$

$$= \left[\int 0.01x\,dx\right]_0^2 = 0.01\left[\tfrac{1}{2}x^2\right]_0^2 = 0.02 \quad (\text{kg m}^2\,\text{s}^{-2}).$$

Example 14.2. A car runs from rest to rest in 1 hour, its velocity v being given by $v = 200t(1 - t)$ (in kilometres per hour). The rate of fuel consumption, f (in litres per kilometre), is related to the velocity by $f = 10^{-4} v^2$. Find (a) the distance travelled and (b) the fuel used.

(a) In time δt it travels a distance δx, where

$$\delta x \approx v \, \delta t.$$

The total displacement x (which is equal to the distance travelled since v is always positive) is therefore

$$x = \lim_{\delta t \to 0} \sum_{t=0}^{t=1} v \, \delta t = \int_0^1 v \, dt$$

$$= \int_0^1 200t(1 - t) \, dt = 200 \int_0^1 (t - t^2) \, dt$$

$$= 200 \left[\tfrac{1}{2} t^2 - \tfrac{1}{3} t^3 \right]_0^1 = 33\tfrac{1}{3} \quad \text{(km)}$$

(b) In distance δx, it uses an amount of fuel δF approximated by

$$\delta F \approx f \, \delta x \approx fv \, \delta t = 10^{-4} v^2 (v \, \delta t) = 800t^3 (1 - t)^3 \, \delta t.$$

The total fuel used, F, is given by

$$F = \lim_{\delta t \to 0} \sum_{t=0}^{t=1} 800t^3 (1 - t)^3 \, \delta t = \int_0^1 800t^3 (1 - t)^3 \, dt$$

$$= 800 \int_0^1 (t^3 - 3t^4 + 3t^5 - t^6) \, dt$$

$$= 800 \left[\tfrac{1}{4} t^4 - \tfrac{3}{5} t^5 + \tfrac{1}{2} t^6 - \tfrac{1}{7} t^7 \right]_0^1 = 5.71 \quad \text{(litres).}$$

Example 14.3. The straight line $y = (r/h)x$, between $x = 0$ and $x = h$, is rotated around the x axis to sweep out a solid cone of height h and circular base radius r. Obtain an expression for its volume.

Divide the interval OH into a large number of equal small steps δx (Fig. 14.2) Consider the step PQ between x and $x + \delta x$. This identifies a thin slice of the cone, like a slice of bread. Its volume δV is nearly that of a cylinder of radius y and thickness δx, so

$$\delta V \approx \pi y^2 \, \delta x.$$

The total volume is obtained by adding all the δV and then letting the slices tend to zero thickness (at the same time becoming proportionately more numerous):

$$V = \lim_{\delta x \to 0} \sum_{x=0}^{x=h} \pi y^2 \, \delta x = \int_0^h \pi y^2 \, dx$$

$$= \int_0^h \frac{\pi r^2}{h^2} x^2 \, dx = \frac{\pi r^2}{h^2} \left[\tfrac{1}{3} x^3 \right]_0^h$$

$$= \frac{\pi r^2}{h^2} \frac{h^3}{3} = \tfrac{1}{3} \pi r^2 h.$$

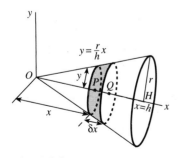

Fig. 14.2

The **volume** of any **solid of revolution** between $x = a$ and $x = b$, formed by rotating a **profile** $y = f(x)$ around the x axis, can be found in exactly the same way:

> **Volume of a solid of revolution around the x axis**
>
> Profile $y = f(x)$, $a \leqslant x \leqslant b$.
>
> Volume $V = \displaystyle\int_a^b \pi y^2 \, dx$.
>
> (14.1)

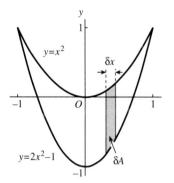

Fig. 14.3

Example 14.4. Find the geometrical area enclosed between the curves $y = 2x^2 - 1$ and $y = x^2$.

This problem is complicated if we have to think all the time about the difference between *signed* and *geometrical* area as in Chapter 13.

Here it will be done in a different way. Divide the interval $-1 \leqslant x \leqslant 1$ into short steps of length δx and consider the area elements indicated in Fig. 14.3. They are nearly rectangular, and the **geometrical** (positive) area δA of each is given by

$$\delta A \approx |x^2 - (2x^2 - 1)| \, \delta x = (-x^2 + 1) \, \delta x$$

(we may drop the modulus signs since $-x^2 + 1 \geqslant 0$ in the given range).

The total geometrical area A is therefore given by

$$A = \lim_{\delta x \to 0} \sum_{x=-1}^{x=1} (-x^2 + 1) \, \delta x = \int_{-1}^{1} (-x^2 + 1) \, dx$$

$$= \left[-\tfrac{1}{3}x^3 + x \right]_{-1}^{1} = (-\tfrac{1}{3} + 1) - (\tfrac{1}{3} - 1) = \tfrac{4}{3}.$$

14.2 Geometrical area in polar coordinates

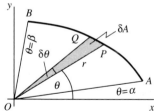

Fig. 14.4

In Fig. 14.4, $\overset{\frown}{AB}$ represents part of a curve which is described in polar coordinates by

$$r = f(\theta), \qquad \alpha \leqslant \theta \leqslant \beta.$$

Form a new, nonrectangular type of area element δA by dividing the θ range, $\theta = \alpha$ to $\theta = \beta$, into small angular steps $\delta\theta$, expressed in **radians**. (We use A rather than $\mathcal{A}$ because, in polar coordinates, we always regard r as being positive, and we shall count the area elements as positive.) A typical area element has the shape OPQ.

When $\delta\theta$ is small, OPQ has very nearly the same area as a narrow *circular* sector of radius r and angle $\delta\theta$ radians. Its area is therefore a fraction $\delta\theta/2\pi$ of a complete circle of radius r and area πr^2:

$$\delta A \approx \frac{\delta\theta}{2\pi} \pi r^2 = \tfrac{1}{2} r^2 \, \delta\theta.$$

The total area is obtained by adding all the elements and letting

$\delta\theta$ tend to zero:

$$A = \lim_{\delta\theta \to 0} \sum_{\theta=\alpha}^{\theta=\beta} \tfrac{1}{2}r^2\,\delta\theta = \int_\alpha^\beta \tfrac{1}{2}r^2\,\mathrm{d}\theta.$$

where $r = f(\theta)$.

Area of a sector in polar coordinates

For a sector $r = f(\theta)$, with $\alpha \leqslant \theta \leqslant \beta$,

$$A = \tfrac{1}{2}\int_\alpha^\beta r^2\,\mathrm{d}\theta.$$

(14.2)

Example 14.5. Find the area of the loop of the curve $r = 3\sin 2\theta$ in the first quadrant.

For the loop shown in Fig. 14.5, the range of θ is $0 \leqslant \theta \leqslant \tfrac{1}{2}\pi$. Thus in (14.2)

$$f(\theta) = 3\sin 2\theta, \qquad \alpha = 0, \qquad \beta = \tfrac{1}{2}\pi.$$

The area is therefore given by

$$A = \tfrac{1}{2}\int_0^{\frac{1}{2}\pi} (3\sin 2\theta)^2\,\mathrm{d}\theta = \tfrac{9}{2}\int_0^{\frac{1}{2}\pi} \sin^2 2\theta\,\mathrm{d}\theta.$$

But, for any angle B, $\sin^2 B = \tfrac{1}{2}(1 - \cos 2B)$; so

$$A = \tfrac{9}{4}\int_0^{\frac{1}{2}\pi} (1 - \cos 4\theta)\,\mathrm{d}\theta$$

$$= \tfrac{9}{4}\left[\theta - \sin 4\theta\right]_0^{\frac{1}{2}\pi} = \tfrac{9}{8}\pi.$$

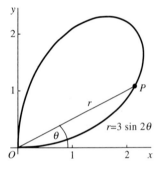

Fig. 14.5

14.3 The numerical estimation of integrals

Practical problems often give rise to integrals which the investigator cannot evaluate or find in a dictionary of integrals. Indeed, sometimes integrals which are very simple-looking cannot, *in principle*, be expressed in terms of ordinary 'formulae' at all. However, numerical approximations to definite integrals can usually be obtained to any required degree of accuracy by using numerical methods in conjunction with a computer. We will mention some very simple methods which call directly on the **area analogy** (13.13), which we repeat here:

The area analogy

The definite integral $\displaystyle\int_a^b f(x)\,\mathrm{d}x$ is equal to the signed area between $y = f(x)$ and the x axis from $x = a$ to $x = b$.

(14.3)

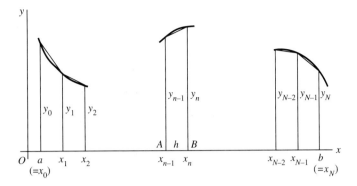

Fig. 14.6

In Examples (13.1) and (13.4), we illustrated the use of the area analogy (13.13) using as the area approximation the sum in (13.1), which had been introduced only for the purpose of establishing the principle. It only gives close approximations if we use very small step lengths; but, now that the area analogy is established, we can look for approximation methods that will be more efficient.

An improved area approximation is shown in Fig. 14.6, where the curve $y = f(x)$ is 'fitted' by a polygonal curve. The approximation to the area of each strip individually is obviously better in general than we would get from a rectangle. Divide the interval $x = a$ to $x = b$ into N steps. We shall denote the length of each step by h (instead of δx, because h is conventional in numerical analysis). Then

$$h = \frac{b - a}{N}.$$

Number the $N + 1$ points of division $0, 1, 2, \ldots, N$: the x values are

$$x_0 \; (=a), \; x_1, \; x_2, \; \ldots, \; x_{N-1}, \; x_N \; (=b)$$

and the y values $y_0, y_1, y_2, \ldots, y_{N-1}, y_N$.

Each of the approximating area elements is a trapezium. The signed area $\delta \mathcal{A}_n$ of the nth area element is given by

$$\delta \mathcal{A}_n \approx \tfrac{1}{2}(y_{n-1} + y_n)h = \frac{b - a}{2N}(y_{n-1} + y_n).$$

The total area $\mathcal{A}$ is approximated by the sum of these:

$$\mathcal{A} \approx \sum_{n=1}^{N} \frac{b - a}{2N}(y_{n-1} + y_n)$$

$$= \frac{b - a}{2N}\left[(y_0 + y_1) + (y_1 + y_2) + \cdots + (y_{N-1} + y_N)\right]$$

$$= \frac{b - a}{N}\left[\tfrac{1}{2}y_0 + (y_1 + y_2 + \cdots + y_{N-1}) + \tfrac{1}{2}y_N\right].$$

This is called the **trapezium rule**.

Trapezium rule

$$\int_a^b f(x)\,dx \approx \frac{b-a}{N}\left[\tfrac{1}{2}y_0 + (y_1 + y_2 + \cdots y_{N-1}) + \tfrac{1}{2}y_N\right]$$

(14.4)

The interval is divided into N equal steps:

$x_0\ (=a),\ x_1,\ldots,\ x_N\ (=b)$ are the division points; and $y_n = f(x_n)$ $(n = 0, 1, 2, \ldots, N)$.

In the following example, we compare the trapezium rule (14.4) with the **rectangle rule** (13.2), which we can recast for comparison as

$$\int_a^b f(x)\,dx \approx \frac{b-a}{N}(y_0 + y_1 + \cdots + y_{N-1}).$$

Example 14.6. Compare the efficiency of the trapezium rule (14.4) with the rectangle rule (13.2) for approximating to $\int_0^1 e^{-x}\,dx$.

We set out the results in the following table.

N	10	100	1000
h	0.1	0.01	0.001
Rectangle rule	0.66	0.635	0.6324
Trapezium rule	0.632657	0.632125	0.632120

The exact value is 0.6321205.... For three-decimal accuracy, the rectangle rule requires about 1000 divisions and the trapezium rule only about 12. There are many formulae which are far more efficient than even the trapezium rule – one of the best of these, for combining simplicity with accuracy, being Simpson's rule (see Problem 14.21). The reader should look at books on numerical analysis for others.

14.4 Centre of mass, moment of inertia

Suppose that there are N particles attached to a weightless plane sheet (Fig. 14.7), the nth particle being at $P : (x_n, y_n)$ and having mass m_n, where $n = 1, 2, \ldots, N$.

Let $G : (\bar{x}, \bar{y})$ be the **centre of mass**. It is the balancing point of the assembly; the point such that the total moment of the particles about any axis through G is zero. Consider in particular the axes CGD and AGB, parallel to the y and x axes and passing through G. Then

$$\sum_{n=1}^N m_n(x_n - \bar{x}) = 0, \qquad \sum_{n=1}^N m_n(y_n - \bar{y}) = 0.$$

(14.5)

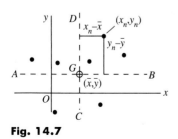

Fig. 14.7

These can be written

$$\sum_{n=1}^{N} m_n x_n - \bar{x} \sum_{n=1}^{N} m_n = 0, \qquad \sum_{n=1}^{N} m_n y_n - \bar{y} \sum_{n=1}^{N} m_n = 0.$$

Let $\sum_{n=1}^{N} m_n = M$, the total mass; then these equations give

$$\bar{x} = \frac{1}{M} \sum_{n=1}^{N} m_n x_n, \qquad \bar{y} = \frac{1}{M} \sum_{n=1}^{N} m_n y_n.$$

If instead of a number of particles there is a solid plate, then this too has a balancing point. Assume that the **plate is uniform** so that its mass per unit area, μ (greek mu), is the same everywhere on it.

We also assume that the shape of the plate is such that no vertical or horizontal line cuts across the boundary more than twice: once going in and again going out. If the shape does not have this property, then the process as explained here has to be modified.

Suppose that the centre of mass G is at $(\bar{x}, \bar{y})$. Divide the area into narrow vertical strips of width δx (Figure 14.8a). Let the

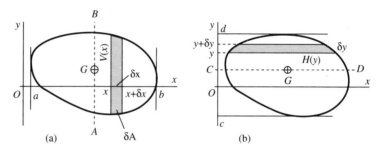

Fig. 14.8   (a)   (b)

total length, or height, of a representative strip as shown by $V(x)$. Then its geometrical area δA is nearly equal to $V \delta x$, and its mass δm is nearly $\mu V \delta x$. Therefore the moment about a vertical axis AB through $G : (x, y)$ is approximately equal to $(x - \bar{x}) \delta m \approx (x - \bar{x}) \mu V \delta x$. The sum of all the elementary moments must be zero, since G is the mass centre, So, in the limit as δx tends to 0, we have

$$\lim_{\delta x \to 0} \sum_{x=a}^{x=b} (x - \bar{x}) V(x) \mu \, \delta x = 0,$$

where $x = a$ and $x = b$ represent the extreme left and right limits of the plate. Since μ and $\bar{x}$ are constants, this is the same as

$$\mu \lim_{\delta x \to 0} \sum_{x=a}^{x=b} x V(x) \, \delta x = \mu \bar{x} \lim_{\delta x \to 0} \sum_{x=a}^{x=b} V(x) \, \delta x = \mu A \bar{x},$$

where A is the area of the plate, equal to $\lim_{\delta x \to 0} \sum_{x=a}^{x=b} V(x) \, \delta x$.

Cancelling μ, we obtain

$$\bar{x} = \frac{1}{A} \lim_{\delta x \to 0} \sum_{x=a}^{x=b} x V(x) \, \delta x = \frac{1}{A} \int_a^b x V(x) \, \mathrm{d}x.$$

Similarly, by dividing the y axis into steps δy, and considering the moments of horizontal strips of length $H(y)$ (see Fig. 14.8b) about a horizontal axis CD through G, we obtain

$$\bar{y} = \frac{1}{A} \int_c^d yH(y)\,\mathrm{d}y,$$

where $y = c$ and $y = d$ are the extreme lower and upper limits of the plate.

In these expressions, all reference to mass has gone (μ is no longer present). Therefore the centre of mass of a **uniform plate** is also called the **centroid** of the figure representing the plate, and it depends only on its shape and size.

In fact the moments about *every* line through G are zero, not simply the moments about AB and CD parallel to the x and y axes that we used to find G.

Centre of mass of a uniform convex plate, or centroid of a convex area, $G : (\bar{x}, \bar{y})$

$$\bar{x} = \frac{1}{A} \int_a^b xV(x)\,\mathrm{d}x; \qquad \bar{y} = \frac{1}{A} \int_c^d yH(y)\,\mathrm{d}y, \qquad \textbf{(14.6)}$$

where A is the area; here, respectively, $V(x)$ and $H(y)$ are the lengths of the vertical and horizontal strips, and $x = a, b$ (resp. $y = c, d$) are the extreme horizontal (resp. vertical) boundaries of the figure.

Example 14.7. Find the position of the centroid or centre of mass of an isosceles triangle of height h and base b.

It is a great advantage to choose axes which make the job as simple as possible. In this case, use the axes shown in Fig. 14.9.

From the symmetry of the isosceles triangle about the x axis, the centroid must lie on this axis, so $\bar{y} = 0$ without any calculations.

The sides have equations

$$y = \pm \frac{b}{2h} x;$$

therefore the length of the strip at x is given by

$$V(x) = \frac{b}{h} x.$$

Also the area A is given by

$$A = \tfrac{1}{2} bh.$$

Therefore by (14.6),

$$\bar{x} = \frac{2}{bh} \int_0^h x\left(\frac{b}{h} x\right) \mathrm{d}x = \frac{2}{h^2} \int_0^h x^2 \,\mathrm{d}x = \tfrac{2}{3} h.$$

(In these coordinates, $\bar{x}$ is independent of the base b.)

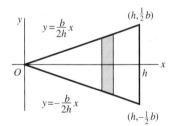

Fig. 14.9

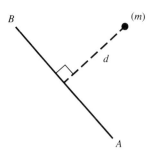

Fig. 14.10

The **moment of inertia** is important for problems in mechanics involving rotation: it plays a part similar to that of mass in nonrotational problems. The moment of inertia of a single particle of mass m about any axis AB is defined to be md^2, where d is its perpendicular distance from AB (see Fig. 14.10). For the moment of inertia of an assemblage of particles, the individual contributions are added. For a solid plate the contributions of small area elements are likewise added, as if they were particles, and in the limit we obtain a definite integral. It is important to select axes and suitably-shaped area elements to make a particular problem manageable.

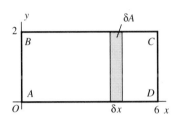

Fig. 14.11

Example 14.8. Find the moment of inertia I of a uniform rectangular plate $ABCD$ about the edge AB when $AB = 2$, $BC = 6$, and the mass per unit area is 2.

Set up axes parallel to the sides, and area elements which are vertical strips of height 2 and width δx, as shown in Fig. 14.11. The axis of rotation is the y axis. The mass δm of each strip is given by

$$\delta m = \text{(surface density)} \times \text{(area)} = 2\,\delta A = 2 \times 2 \times \delta x = 4\,\delta x.$$

The moment of inertia of the strip distant x from the y axis is therefore

$$x^2\,\delta m = 4x^2\,\delta x.$$

The total moment of inertia I is given by

$$I = \lim_{\delta x \to 0} \sum_{x=0}^{x=6} 4x^2\,\delta x = 4\int_0^6 x^2\,dx$$

$$= 4\left[\tfrac{1}{3}x^3\right]_0^6 = 288.$$

Example 14.9. Find an expression for the moment of inertia I of an isosceles triangle ABC about its base AB, when $AB = b$, its height is h, and its mass is M.

The axes and the representative strip at x are shown in Fig. 14.12. The equations of BC and AC are

$$y = \pm\left(-\frac{b}{2h}x + \tfrac{1}{2}b\right)$$

respectively, so the length $V(x)$ to be assigned to the strip is

$$V(x) = -\frac{b}{h}x + b,$$

and the area δA is approximated by

$$\delta A \approx \left(-\frac{b}{h}x + b\right)\delta x.$$

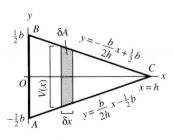

Fig. 14.12

Since the plate is uniform, the mass per unit area is (total mass)/(area), or $M/\frac{1}{2}bh$, so the mass element δm is approximated by

$$\delta m \approx \frac{M}{\frac{1}{2}bh}\left(-\frac{b}{h}x + b\right)\delta x = \frac{2M}{h}\left(1 - \frac{x}{h}\right)\delta x.$$

Therefore the moment of inertia I is given by

$$I = \lim_{\delta x \to 0}\sum_{x=0}^{x=h} x^2\,\delta m = \lim_{\delta x \to 0}\sum_{x=0}^{x=h} x^2\,\frac{2M}{h}\left(1 - \frac{x}{h}\right)\delta x$$

$$= \frac{2M}{h}\int_0^h x^2\left(1 - \frac{x}{h}\right)dx = \frac{2M}{h}\int_0^h\left(x^2 - \frac{1}{h}x^3\right)dx$$

$$= \frac{2M}{h}\left[\tfrac{1}{3}x^3 - \frac{1}{4h}x^4\right]_0^h = \frac{2M}{h}\frac{h^2}{12} = \tfrac{1}{6}Mh^2.$$

Example 14.10. Find the moment of inertia of a circular disc of radius R and mass M about an exis through its centre and perpendicular to the plane of the disc.

The usual (x, y) coordinates are not natural to this problem. In Fig. 14.13, the polar coordinate r ranges from 0 to R. Break this range into ring-shaped steps as shown, the representative ring or **annulus** having inner radius r and thickness δr. These constitute the area elements δA.

We have $\delta A \approx 2\pi r\,\delta r$, and the mass per unit area is $M/\pi R^2$, so that the mass of the ring δm is approximately

$$\delta m \approx \frac{M}{\pi R^2}2\pi r\,\delta r.$$

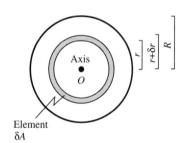

Element
δA

Fig. 14.13

The moment of inertia of the ring must be equal to that of a suitable distribution of closely-spaced particles along its circumference. The contribution of each of these imaginary particles to the moment of inertia of the ring is equal to its mass times r^2. Since r is constant on the ring, its moment of inertia δI is equal to the total mass of the ring times r^2:

$$\delta I \approx r^2\,\delta m \approx \frac{2M}{R^2}r^3\,\delta r.$$

Finally

$$I = \lim_{\delta r \to 0}\sum_{r=0}^{r=R}\frac{2M}{R^2}r^3\,\delta r = \frac{2M}{R^2}\int_0^R r^3\,dr$$

$$= \frac{2M}{R^2}\cdot\tfrac{1}{4}R^4 = \tfrac{1}{2}MR^2.$$

Problems, Chapter 14
(Units are kilogram, metre, second where they are unstated).

14.1. The resistance R of a compression spring is given by $R = 100x + 1000x^2$, where x is the displacement from its natural length. Find the work done in compressing it through a distance of 0.01.

14.2. The velocity v of a point moving along the x axis is $v = 20 - 10t$, where t is the time. The displacement x taking place in a short time δt is approximated by $\delta x \simeq v\,\delta t$. Express the displacement which takes place between $t = 2$ and $t = 4$ as a definite integral, and evaluate it. What is its x coordinate at $t = 4$ if it was at $x = 3$ when $t = 2$?

14.3. Each of the following curves is the profile of a solid of revolution which has the x axis as its central axis. Find the volume in each case (see (14.1)–you should briefly go through the whole argument until you understand it, not simply quote the formula):
(a) $y = e^{-x}, 0 \leqslant x \leqslant 1$; (b) $y = 1/x, 1 \leqslant x \leqslant 2$;
(c) $y = x(1 - x), 0 \leqslant x \leqslant 1$;
(d) $y = \sin x, 0 \leqslant x \leqslant \pi$; (e) $y = x^3, -1 \leqslant x \leqslant 1$(the fact that x^3 is negative over part of its range does not have to be taken into account: the volume elements are always positive, unlike area elements);
(f) $y = x(1 - x), 0 \leqslant x \leqslant 2$ (see the note in (e));
(g) $y = x^{-1}, 1 \leqslant x < \infty$ (contrast Example 13.9b, for area); (h) $y = x^{\frac{1}{4}}, 0 \leqslant x \leqslant 1$.

14.4. Show that the volume of a sphere of radius R is $\frac{4}{3}\pi R^3$. (a sphere is a solid of revolution.)

14.5. (a) Find the volume of the ellipsoid obtained by rotating the elliptical profile $x^2/a^2 + y^2/b^2 = 1$ about the x axis.
(b) If the x and y scales of the profile ellipse in (a) are contracted or expanded by suitable factors, it becomes a unit circle. Deduce from this fact the formula for the volume of the ellipsoid of revolution.

14.6. The curve $y = \frac{1}{2}x$ between $y = 1$ and $y = 2$ is rotated about the y axis to profile a vertical spindle, or truncated cone. Find its volume.

14.7. A uniform beam AB of length L has mass m per unit length. It is cemented horizontally at A into a wall at the end A. Sum the moments about A of elements of length δx, form a definite integral, and so find the moment supporting the beam at A.

14.8. A 'beam' in the shape of a circular spindle made of material of density 500 is fixed to a vertical wall at the end A with its axis of symmetry horizontal. Its cross-sectional area (perpendicular to its axis) is $4 \times 10^{-4}(1 + 0.4x^2)$, where x is measured from A. Its length is 1. Find the moment at A required to support it under gravity.

Suppose that the data are the same, except that the cross-section is square, or possibly irregular in shape. Does this affect the answer? Suppose that the axis is bent, but that x still measures the perpendicular distance from the wall: is the calculation affected?

14.9. A narrow tube of length 10 cm and cross-section 0.1 cm² contains a chemical solution, with concentration $c(x) = 0.04\,e^{-\frac{1}{4}x}$ gm cm^{-3}, where x is the distance from one end. Find the total mass of solute in the tube.

14.10. The water clock in Fig. 14.14 has depth 0.5 m,

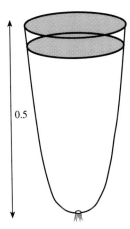

0.5

Fig. 14.14

and its profile is given by $r(h) = 0.39h^{\frac{1}{4}}$, where $r(h)$ is the radius at height h from the outlet in the bottom. The size of the outlet hole is such as to drain the water at a rate given by

$$\frac{dV}{dt} = -0.003h^{\frac{1}{4}} \text{ m}^3 \text{ hr}^{-1}.$$

where V is the volume of water remaining. Show that the water level falls at a uniform rate, and find how long it runs. (Consider the change δh in level which occurs in a short time δt.)

14.11. An alternating current $i = i_0 \cos \omega t$ flows through a resistor R. The instantaneous rate of heat generation is Ri^2 heat units per unit time. Find the heat generated in a complete cycle of the current, that is, in a period $2\pi/\omega$. Does is make any difference at what instant you regard the period as starting? (To carry out the integration, you will need the identity $\cos^2 A = \frac{1}{2}(1 + \cos 2A)$.)

14.12. Find the geometric area enclosed between the curves $y = -x$ and $y = x(x - 1)$ on the interval $0 \leqslant x \leqslant 2$, by considering vertical strips between the curves of width δx.

14.13. Find the geometric area enclosed between the curves $y = -x$ and $y = x^3$ between $x = -1$ and $x = 1$ by considering vertical strips of width δx connecting the curves. (Be careful about signs: these curves cross.)

14.14. For the angular ranges specified, sketch the curves given in polar coordinates below and find the sectorial areas.
(a) $r = \theta, 0 \leqslant \theta \leqslant 2\pi$ (a spiral arc);
(b) $r = 2 \cos \theta, -\frac{1}{2}\pi \leqslant \theta \leqslant \frac{1}{2}\pi$ (a circle);
(c) $r = e^{\theta/2\pi}, 0 \leqslant \theta \leqslant \pi$ (spiral arc);
(d) $r = \sin 2\theta, 0 \leqslant \theta \leqslant \frac{1}{2}\pi$.
(Remember the identities $\cos^2 A = \frac{1}{2}(1 + \cos 2A)$, $\sin^2 A = \frac{1}{2}(1 - \cos 2A)$.)

14.15. The end of a water trough is a rectangle of height H and width L. Find the total force and moment on the end when the trough is full. (The pressure, meaning the force per unit area acting perpendicularly on any surface, at depth y is $\rho g y$, where ρ is density and g the gravitation constant.)

14.16. Determine the position of the centre of mass of a symmetrical cone of circular cross-section which has height H and base radius R.

14.17. Find the moment of inertia of a rectangle, having sides a and b, about an axis through its centre, parallel to the sides of length b.

14.18. Obtain the moment of inertia of an isosceles triangle of height H and base B about an axis through its vertex which is (a) parallel to the base, and (b) perpendicular to the base.

14.19. Use the trapezium rule (14.4) to evaluate the following integrals to 1% accuracy. (The exact value can be obtained by evaluating the integrals in the usual way.)

(a) $\displaystyle\int_0^1 e^{\frac{1}{2}x}\, dx$;   (b) $\displaystyle\int_0^\pi \sin x\, dx$;   (c) $\displaystyle\int_{-\frac{1}{2}\pi}^{\frac{1}{2}\pi} \cos x\, dx$.

14.20. The following integrals are either difficult or impossible to evaluate directly. Estimate them by using the trapezium rule (14.4). (Since you cannot know the exact answer in advance, you can proceed by running the program using increasingly fine divisions until you get no change in some predetermined number of decimal places.)

(a) $\displaystyle\int_0^{\frac{1}{2}\pi} \sin^{\frac{1}{2}} x\, dx$;   (b) $\displaystyle\int_0^1 e^{-x^2}\, dx$;

(c) $\displaystyle\int_1^2 \frac{e^x\, dx}{1 + x^3}$;   (d) $\displaystyle\int_1^2 \frac{\sin x}{x}\, dx$.

14.21. The following is called **Simpson's rule** for numerical integration. It results from splitting the points of division into successive groups of three, then exactly fitting the corresponding groups of points on the graph by second-degree polynomials. For this purpose, N **must be an even number**:

$$\int_a^b y\, dx \approx \frac{b - a}{3N}(y_0 + 4y_1 + 2y_2 + 4y_3 \\ + 2y_4 + \cdots + 4y_{N-1} + y_N)$$

Show that $\displaystyle\int_0^1 e^{-x^2}\, dx$ is given correctly to four decimal places by using only four subdivisions. Compare the trapezium rule and the rectangle rule.

14.22. Consider the curve $y = f(x)$ for $a \leqslant x \leqslant b$. Show that the arc length δs associated with a short step δx is given by $\delta s \approx (\delta x + \delta y)^{\frac{1}{2}}$. Deduce that the total length s of the curve is given by

$$s = \int_a^b \left[1 + \left(\frac{dy}{dx}\right)^2\right]^{\frac{1}{2}} dx.$$

This type of integral is usually impossible to evaluate explicitly, but can be done numerically. Compute the lengths of the following curves. (Try the trapezium rule, Simpson's rule of Problem 14.21, and an integrating routine from a software package if you know how to use it: the interest lies in comparing them.)
(a) $y = \sin x, 0 \leqslant x \leqslant 1$;   (b) $y = x^2, 0 < x < 2$;
(c) $y = e^x, -1 < x < 1$.
(d) $y = (1 - x^2)^{\frac{1}{2}}, 0 \leqslant x \leqslant 1$ (a circle, so it can be done directly.)

15 Systematic techniques for integration

15.1 Substitution method for $\int f(ax + b)\,dx$

Consider the indefinite integral

$$\int (3x - 2)^3 \, dx.$$

We carried out this integration in Example 12.8 by starting with a guess that the result will resemble $(3x - 2)^4$. We now describe a method less dependent on trial and error.

We shall take up a clue suggested by the chain rule procedure (Section 3.3). Put

$$3x - 2 = u. \tag{15.1}$$

Then the integral becomes

$$\int u^3 \, dx.$$

Unfortunately this is *not* equal to $\frac{1}{4}u^4 + C$, because dx, not du, is present: the variable of integration is still x. Thinking in terms of an integral as a sum, δx is not the same size as δu; in fact from (15.1) $\delta u = 3 \, \delta x$, which suggests what to do with the new integral.

From (15.1), $du/dx = 3$, which we write as

$$dx = \tfrac{1}{3} \, du.$$

Put this into the integral, and it works through straightforwardly:

$$\int (3x - 2)^3 \, dx = \int u^3 (\tfrac{1}{3} \, du) = \tfrac{1}{12} u^4 + C.$$

Now use (15.1) to change back to x:

$$\int (3x - 2)^3 \, dx = \tfrac{1}{12}(3x - 2)^4 + C,$$

and this is correct. In checking its correctness by differentiation, we use the chain rule with $u = 3x - 2$, and find we are simply reversing the order of the operations that we just went through.

Example 15.1. Use a substitution to obtain $\int \dfrac{\mathrm{d}x}{2x-1}$.

Try
$$u = 2x - 1.$$
We shall need to express $\mathrm{d}x$ in terms of u. Since $\mathrm{d}u/\mathrm{d}x = 2$, we have
$$\mathrm{d}x = \tfrac{1}{2}\,\mathrm{d}u.$$
The integral therefore becomes, in terms of u,
$$\int \frac{\mathrm{d}x}{2x-1} = \int \frac{(\tfrac{1}{2}\,\mathrm{d}u)}{u} = \tfrac{1}{2}\ln|u| + C$$
$$= \tfrac{1}{2}\ln|2x-1| + C.$$

Example 15.2. Evaluate $\int \sin(3x+2)\,\mathrm{d}x$.

Put
$$u = 3x + 2,$$
then $\mathrm{d}u/\mathrm{d}x = 3$, so $\mathrm{d}u = 3\,\mathrm{d}x$, or $\mathrm{d}x = \tfrac{1}{3}\,\mathrm{d}u$. The integral becomes
$$\int \sin(3x+2)\,\mathrm{d}x = \int \sin u \cdot (\tfrac{1}{3}\,\mathrm{d}u)$$
$$= (-\tfrac{1}{3}\cos u) + C = -\tfrac{1}{3}\cos(3x+2) + C.$$

The essence of the matter is that the **change of variable** or **substitution** led to a simpler integral than the one we started with. In general, for integrals of this type, we have the following result.

Type $\int f(ax+b)\,\mathrm{d}x$

Put $u = ax + b$; then $\dfrac{\mathrm{d}u}{\mathrm{d}x} = a$, or $\mathrm{d}x = \dfrac{1}{a}\,\mathrm{d}u$. The $\qquad$ **(15.2)**

integral transforms to $\dfrac{1}{a}\displaystyle\int f(u)\,\mathrm{d}u$.

It is worthwhile to try this substitution in more general cases, even if it is not obvious that a simplification will take place.

Example 15.3. Evaluate $\int x(2x-1)^3\,\mathrm{d}x$.

This is not quite of the form (15.2) because of the presence of the loose x. Nevertheless, put
$$u = 2x - 1,$$
with the object of simplifying at least the most complicated part.

Then
$$du = 2\,dx, \quad \text{or} \quad dx = \tfrac{1}{2}\,du.$$
We also need to express x in terms of u, using $u = 2x - 1$:
$$x = \tfrac{1}{2}(u + 1).$$
Now we have
$$\int x(2x - 1)^3\,dx = \int \tfrac{1}{2}(u + 1)u^3(\tfrac{1}{2}\,du)$$
$$= \tfrac{1}{4}\int (u^4 + u^3)\,du = \tfrac{1}{20}u^5 + \tfrac{1}{16}u^4 + C$$
$$= \tfrac{1}{20}(2x - 1)^5 + \tfrac{1}{16}(2x - 1)^4 + C.$$

Do not miss the possibility of making a substitution in **simple cases**; for example:
$$\int e^{-3x}\,dx\colon \text{ put } u = -3x,\ dx = -\tfrac{1}{3}\,du;$$
$$\int \sin 3x\,dx\colon \text{ put } u = 3x,\ dx = \tfrac{1}{3}\,du;$$
$$\int \frac{1 + x}{1 - x}\,dx\colon \text{ put } u = 1 - x,\ dx = -du.$$

15.2 Substitution method for $\displaystyle\int f(ax^2 + b)x\,dx$

Example 15.3. Evaluate $\displaystyle\int x\,e^{x^2}\,dx$.

Try putting
$$u = x^2,$$
with the objective of simplifying the unfamiliar-looking term e^{x^2}. It is then necessary to deal with x and dx in the integral. We have
$$\frac{du}{dx} = 2x,$$
which we can write as $du = 2x\,dx$, or
$$x\,dx = \tfrac{1}{2}\,du.$$
In this way we have translated the whole group $(x\,dx)$ into terms of u, instead of having to deal separately with x and dx. Therefore
$$\int x\,e^{x^2}\,dx = \int e^{x^2}(x\,dx) = \int e^u(\tfrac{1}{2}\,du)$$
$$= \tfrac{1}{2}\int e^u\,du = \tfrac{1}{2}e^u + C = \tfrac{1}{2}e^{x^2} + C,$$

where C is an arbitrary constant. The correctness of the result can be checked by differentiating it.

Example 15.4. Evaluate $\displaystyle\int \frac{x\,\mathrm{d}x}{3x^2+2}$.

Notice that the integral can be written in the form

$$\int \frac{1}{3x^2+2}\,(x\,\mathrm{d}x).$$

The integrand contains a function of x^2 and the combination $x\,\mathrm{d}x$ which appeared in Example 15.3. This suggests putting $u=x^2$ to give a simpler integral. However, we can do even better than this.

Put

$$u = 3x^2 + 2.$$

Then $\mathrm{d}u/\mathrm{d}x = 6x$, so that

$$x\,\mathrm{d}x = \tfrac{1}{6}\,\mathrm{d}u.$$

Therefore

$$\int \frac{1}{3x^2+2}\,(x\,\mathrm{d}x) = \int \frac{1}{u}\,(\tfrac{1}{6}\,\mathrm{d}u) = \tfrac{1}{6}\int \frac{\mathrm{d}u}{u}$$

$$= \tfrac{1}{6}\ln|u| + C = \tfrac{1}{6}\ln(3x^2+2) + C,$$

where C is an arbitrary constant. The modulus sign in the logarithm can be discarded because $3x^2+2$ is always positive.

The general result is as follows:

Integrals of type $I = \displaystyle\int xf(ax^2+b)\,\mathrm{d}x$

Put $u = ax^2 + b$; then $x\,\mathrm{d}x = \dfrac{1}{2a}\,\mathrm{d}u$; so   (15.3)

$$I = \frac{1}{2a}\int f(u)\,\mathrm{d}u.$$

15.3 Substitution method for $\displaystyle\int \cos^m ax \sin^n ax \,\mathrm{d}x$ (m or n odd)

Example 15.5. Evaluate $\displaystyle\int \sin^3 x \cos x \,\mathrm{d}x$.

Aim to simplify the worst term by putting

$$u = \sin x.$$

Then $\sin^3 x$ becomes u^3, and we must deal with $\cos x\,\mathrm{d}x$. As always, begin with $\mathrm{d}u/\mathrm{d}x = \cos x$. Therefore

$$\mathrm{d}u = \cos x\,\mathrm{d}x,$$

so, by good fortune, $\cos x\, dx$ appears in one piece. Then we have

$$\int \sin^3 x \cdot (\cos x\, dx) = \int u^3\, du = \tfrac{1}{4}u^4 + C = \tfrac{1}{4}\sin^4 x + C,$$

with C arbitrary. (Check by differentiating.)

Example 15.6. Evaluate $\displaystyle\int \tan x\, dx$.

$$\int \tan x\, dx = \int \frac{\sin x}{\cos x}\, dx = \int \frac{1}{\cos x}(\sin x\, dx).$$

This time, put

$$u = \cos x,$$

so that $du/dx = -\sin x$. From this we obtain

$$du = -\sin x\, dx,$$

so, apart from the sign, we have exactly the combination required for the rest of the integrand. Then

$$\int \frac{1}{\cos x}(\sin x\, dx) = \int \frac{1}{u}(-du) = -\ln|u| + C = -\ln|\cos x| + C,$$

where C is arbitrary. This often appears as $\ln|\sec x| + C$ in tables of integrals.

This technique can be used for products $\cos^m ax \sin^n ax$, when either m or n (or both) are odd numbers, either positive or negative, and for certain other cases as well.

Example 15.7. Evaluate $\displaystyle\int \cos^3 x\, dx$. (This is the case $m = 3$, $n = 0$.)

Write $\displaystyle\int \cos^3 x\, dx = \int \cos^2 x \cdot (\cos x\, dx)$, and put

$$u = \sin x$$

(not $\cos x$ as possibly expected). Then $du/dx = \cos x$, so that

$$du = \cos x\, dx.$$

The remaining part of the integrand is $\cos^2 x$, and we can transform this by writing

$$\cos^2 x = 1 - \sin^2 x = 1 - u^2.$$

Then we have

$$\int \cos^2 x \cdot (\cos x\, dx) = \int (1 - u^2)\, du = u - \tfrac{1}{3}u^3 + C$$

$$= \sin x - \tfrac{1}{3}\sin^2 x + C,$$

where C is arbitrary.

The reader should try also the substitution $u = \cos x$. It leads to an integral in terms of u that is correct but worse than the original.

Example 15.8. Evaluate $I = \int \cos^3 2x \sin^3 2x \, dx$.

(Here $m = 3$, $n = 3$.) The technique requires us to decompose the term whose power is odd. Here both powers are odd, so either will do. We shall split the integrand like this:

$$I = \int \cos^3 2x \sin^2 2x(\sin 2x \, dx).$$

Put $u = \cos 2x$ so that

$$\sin 2x \, dx = -\tfrac{1}{2} \, du.$$

Since $\sin^2 2x = 1 - \cos^2 2x$, the integral becomes

$$I = \int u^3(1 - u^2)(-\tfrac{1}{2} \, du)$$

$$= -\tfrac{1}{2} \int (u^3 - u^5) \, du = -\tfrac{1}{8}u^4 + \tfrac{1}{12}u^6 + C$$

$$= -\tfrac{1}{8} \cos^4 2x + \tfrac{1}{12} \cos^6 2x + C,$$

with C arbitrary.

The general rule is as follows.

Integrals of type $I = \int \cos^m ax \sin^n ax \, dx$

(a) If n is odd, put $I = \int \cos^m ax \sin^{n-1} ax \, (\sin ax \, dx)$;

then $u = \cos ax$, $\sin ax \, dx = -\dfrac{1}{a} \, du$, and $\sin^2 ax = 1 - \cos^2 ax$.

(b) If m is odd, write   **(15.4)**

$$I = \int \cos^{m-1} ax \sin^n ax(\cos ax \, dx);$$

then $u = \sin ax$, $\cos ax \, dx = \dfrac{1}{a} \, du$, and $\cos^2 ax = 1 - \sin^2 ax$.

(c) If n and m are odd, use either of (a) and (b).

15.4 Definite integrals and change of variable

For the previous examples involving indefinite integrals, we changed the variable to u, carried out the integration, and then expressed the result back in terms of x. For a *definite* integral, it is often more convenient to **express the limits of integration in terms of** u, as well as the integrand, and in that way work with u right up to the end. In the following example, both procedures are illustrated.

Example 15.9. Evaluate $I = \displaystyle\int_0^{\frac{1}{2}\pi} \cos^3 x \, dx$ in two ways.

(a) (First finding an indefinite integral in terms of x.) As in Example 15.7, put $u = \sin x$, $du = \cos x \, dx$;

$$\int \cos^3 x \, dx = \int (1 - u^2) \, du = u - \tfrac{1}{3}u^3$$

$$= \sin x - \tfrac{1}{3}\sin^3 x,$$

taking the simplest case with $C = 0$. Then

$$I = \left[\sin x - \tfrac{1}{3}\sin x \right]_0^{\frac{1}{2}\pi} = (1 - \tfrac{1}{3}) - 0 = \tfrac{2}{3}.$$

(b) (Working with u throughout.) Put $u = \sin x$ into I. In order to express the limits of integration in terms of u, note that $u = 0$ when $x = 0$, and $u = 1$ when $x = \tfrac{1}{2}\pi$. Then (writing the limits so as to make them more explicit)

$$\int_{x=0}^{x=\frac{1}{2}\pi} \cos^3 x \, dx = \int_{u=0}^{u=1} (1 - u^2) \, du$$

$$= \left[u - \tfrac{1}{3}u^3 \right]_{u=0}^{u=1} = (1 - \tfrac{1}{3}) - 0 = \tfrac{2}{3}.$$

It would have been *wrong* to write the integral in the form

$$\int_0^{\frac{1}{2}\pi} (1 - u^2) \, du.$$

This would imply that we were going to put u equal to 0 and $\tfrac{1}{2}\pi$ after integrating.

Example 15.10. Find the centroid (centre of mass) of the uniform semicircular plate shown in Fig. 15.1.

The symmetry shows that the centroid G lies on the x axis. From (15.6), the x coordinate of G is given by

$$\bar{x} = \frac{1}{\frac{1}{2}\pi R^2} \int_0^R V(x) x \, dx.$$

Since $x^2 + y^2 = R^2$, we have $V(x) = 2(R^2 - x^2)^{\frac{1}{2}}$, so that

$$\bar{x} = \frac{4}{\pi R^2} \int_0^R (R^2 - x^2)^{\frac{1}{2}} x \, dx.$$

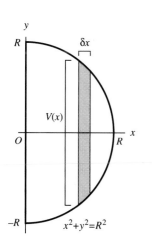

Fig. 15.1

This is an integral of the type of (15.3). To simplify it, put $u = R^2 - x^2$, so that $du/dx = -2x$ and $x\,dx = -\frac{1}{2}\,du$. Also, $u = R^2$ ^{when} $x = 0$, and $u = 0$ when $x = R$. Therefore

$$\bar{x} = \frac{4}{\pi R^2} \int_{R^2}^{0} u^{\frac{1}{2}}(-\tfrac{1}{2}\,du)$$

$$= -\frac{2}{\pi R^2} \tfrac{2}{3}[u^{\frac{3}{2}}]_{R^2}^{0} = -\frac{4}{3\pi R^2}[0 - R^3] = \frac{4}{3\pi}R.$$

15.5 Occasional substitutions

Finding an advantageous new variable u is often a process of trial and error. Frequently the possible usefulness of a substitution is more easy to see in the form

$$x = f(u)$$

rather than u as a function of x as in the previous work.

Example 15.11. Find a substitution to evaluate $\displaystyle\int \frac{dx}{(1 - x^2)^{\frac{1}{2}}}$.

Try to simplify $(1 - x^2)^{\frac{1}{2}}$ first, hoping that dx will work out conveniently. To do this try

$$x = \sin u; \tag{15.5}$$

then $(1 - x^2)^{\frac{1}{2}} = (1 - \sin^2 u)^{\frac{1}{2}} = \cos u$. Also $dx/du = \cos u$, so

$$dx = \cos u\,du.$$

Therefore

$$\int \frac{dx}{(1 - x^2)^{\frac{1}{2}}} = \int \frac{\cos u\,du}{\cos u} = u + C = \arcsin x + C,$$

a result which can be confirmed from the table of derivatives in Appendix D. You might try putting $u = 1 - x^2$ instead: the resulting integral is different from, but no better than, the original.

The entries in Appendix E can be obtained from the basic forms such as those in (12.2) by means of substitutions, as in the following examples.

Example 15.12. From Example 15.11, we know that $\displaystyle\int \frac{dx}{(1 - x^2)^{\frac{1}{2}}} =$ $\arcsin x + C$. Use this result to obtain $\displaystyle\int \frac{dx}{(4 - x^2)^{\frac{1}{2}}}$.

Aim to convert $(4 - x^2)^{\frac{1}{2}}$ into something like $(1 - u^2)^{\frac{1}{2}}$, so as to be able to use the result in the table.

$$(4 - x^2)^{\frac{1}{2}} = 2(1 - \tfrac{1}{4}x^2)^{\frac{1}{2}} = 2[1 - (\tfrac{1}{2}x)^2]^{\frac{1}{2}},$$

and make the substitution

$$u = \tfrac{1}{2}x.$$

Then $du/dx = \tfrac{1}{2}$, so that $dx = 2\,du$. Therefore

$$\int \frac{dx}{(4 - x^2)^{\frac{1}{2}}} = \int \frac{2\,du}{2(1 - u^2)^{\frac{1}{2}}} = \int \frac{du}{(1 - u^2)^{\frac{1}{2}}}$$
$$= \arcsin u + C = \arcsin \tfrac{1}{2}x + C.$$

Example 15.13. From Appendix E, $\displaystyle\int \frac{dx}{1 + x^2} = \arctan x + C$. Use this result to evaluate $\displaystyle\int \frac{dx}{1 + 9x^2}$.

We want to transform $1 + 9x^2$ to a form close to $1 + u^2$, so put

$$u = 3x,$$

so that $dx = \tfrac{1}{3}\,du$. Then

$$\int \frac{dx}{1 + 9x^2} = \int \frac{\tfrac{1}{3}\,du}{1 + u^2} = \tfrac{1}{3}\arctan u + C = \tfrac{1}{3}\arctan 3x + C.$$

If the required integral does not seem to be similar to one that is already known, then one has in effect to guess a suitable substitution:

Example 15.14. Evaluate $\displaystyle\int \frac{\ln^3 x}{x}\,dx$.

We can simplify the logarithm (at the risk of extra complexity elsewhere) by putting

$$x = e^u$$

so that $\ln x = \ln e^u = u$. Since $dx/du = e^u$, we have $dx = e^u\,du$. Therefore

$$\int \frac{\ln^3 x}{x}\,dx = \int \frac{u^3}{e^u} e^u\,du = \int u^3\,du$$
$$= \tfrac{1}{4}u^4 + C = \tfrac{1}{4}\ln^4 x + C.$$

The general shape of the integrand sometimes suggests a substitution that is sure to simplify it. Suppose we notice that $f(x)$ takes the special form

$$f(x) = cg(u)\frac{du}{dx},$$

where c is a constant and u is a function of x. For example,

$$(x^4 + 1)^7 x^3 = c(x^4 + 1)^7 \frac{d}{dx}(x^4 + 1);$$

and, in this case, $c = \frac{1}{4}$, $u(x) = x^4 + 1$, and $g(u) = u^7$. In the general case,

$$\int f(x)\,dx = c\int g(u)\frac{du}{dx}\,dx = c\int g(u)\,du.$$

(Any $f(x)$ can in principle be written in this form: the question is only whether it is easy to see how it breaks up.) Having observed the form of $u(x)$, the substitution should be made in the usual way.

Example 15.15. Evaluate $\int (x^{\frac{1}{2}} + 1)^{\frac{1}{3}} x^{-\frac{1}{2}}\,dx.$

The important thing is to spot that $(d/dx)(x^{\frac{1}{2}} + 1)$ is like the remaining factor, $x^{-\frac{1}{2}}$. This suggests that $u = x^{\frac{1}{2}} + 1$ is the right substitution. Specifically, put $u = x^{\frac{1}{2}} + 1$; then

$$\frac{du}{dx} = \tfrac{1}{2}x^{-\frac{1}{2}} \text{ and so } x^{-\frac{1}{2}}\,dx = 2\,du.$$

The integral becomes

$$2\int u^{\frac{1}{3}}\,du = \tfrac{3}{2}u^{\frac{4}{3}} + C = \tfrac{3}{2}(x^{\frac{1}{2}} + 1)^{\frac{4}{3}} + C.$$

15.6 Partial fractions for integration

In Section 1.10, it was shown how a **rational function** $P(x)/Q(x)$, where $P(x)$ and $Q(x)$ are polynomials, $P(x)$ is of lower degree than $Q(x)$, and $Q(x)$ factorizes into real factors, can be expressed as the sum of simpler **partial fractions**. This provides a method for integrating rational functions.

Example 15.15. Evaluate $\displaystyle\int \frac{dx}{x^2 - 1}.$

By the methods of Section 1.10, we find that

$$\frac{1}{x^2 - 1} = \frac{1}{(x - 1)(x + 1)} = \tfrac{1}{2}\frac{1}{x - 1} - \tfrac{1}{2}\frac{1}{x + 1}.$$

Therefore

$$\int \frac{dx}{x^2 - 1} = \tfrac{1}{2}\int \frac{dx}{x - 1} - \tfrac{1}{2}\int \frac{dx}{x + 1}$$

$$= \tfrac{1}{2}\ln|x - 1| - \tfrac{1}{2}\ln|x + 1| + C.$$

Other equivalent forms are $\tfrac{1}{2}\ln\left|\dfrac{x - 1}{x + 1}\right| + C$, $\ln\left|\dfrac{x - 1}{x + 1}\right|^{\frac{1}{2}} + C$ and $\tfrac{1}{2}\ln\left|B\,\dfrac{x - 1}{x + 1}\right|$, where C and B are arbitrary.

As a result of expanding in partial fractions we may encounter integrands of the type

$$\frac{cx + d}{px^2 + qx + r}$$

in which the equation $px^2 + qx + r = 0$ has no real roots (that is, the denominator has no real factors). The following Example shows how to evaluate them by 'completing the square' in the denominator.

Example 15.16. Evaluate $I = \int \dfrac{(x + 1)\, dx}{x^2 + 4x + 8}$.

The quadratic form $x^2 + 4x + 8$ has no real factors. 'Completing the square' in the denominator consists of writing $x^2 + 4x + 8$ in the form $(x + a)^2 + b$. The first two terms, $x^2 + 4x$, can be written

$$x + 4x = (x + 2)^2 - 4,$$

so

$$x^2 + 4x + 8 = (x + 2)^2 - 4 + 8 = (x + 2)^2 + 4.$$

The integral becomes

$$I = \int \frac{(x + 1)\, dx}{(x + 2)^2 + 4} = \tfrac{1}{4} \int \frac{(x + 1)\, dx}{[\tfrac{1}{2}(x + 2)]^2 + 1}.$$

Now put $u = \tfrac{1}{2}(x + 2)$, or $x = 2u - 2$, from which

$$dx = 2\, du.$$

Then

$$I = \tfrac{1}{2} \int \frac{2u - 1}{u^2 + 1}\, du = \int \frac{u}{u^2 + 1}\, du - \tfrac{1}{2} \int \frac{1}{u^2 + 1}\, du.$$

To evaluate the first integral, use the substitution $v = u^2 + 1$, as in Section 15.2; the second is a standard integral. We obtain

$$I = \tfrac{1}{2} \ln(x^2 + 4x + 8) - \tfrac{1}{2} \arctan \tfrac{1}{2}(x + 2) + C.$$

15.7 Integration by parts

This method is totally unrelated to the techniques we have so far described, and can be used to integrate special types of **product**. It is needed very frequently for obtaining fundamental general results.

Suppose that we are given any $u(x)$ and $v(x)$. Then, by the product rule,

$$\frac{d}{dx}(uv) = u\frac{dv}{dx} + v\frac{du}{dx}.$$

Since both sides are equal, their indefinite integrals can only differ by a constant, so

$$\int \frac{d}{dx}(uv)\,dx = \int u\frac{dv}{dx}\,dx + \int v\frac{du}{dx}\,dx + C. \qquad (15.6)$$

Look at the integral on the left. It means 'an antiderivative of $(d/dx)[u(x)v(x)]$'. But, from the definition (12.1), $u(x)v(x)$, is an antiderivative. Therefore (15.6) becomes

$$uv = \int u\frac{dv}{dx}\,dx + \int v\frac{du}{dx}\,dx + C.$$

Now rearrange the terms to obtain

$$\int u\frac{dv}{dx}\,dx = uv - \int v\frac{du}{dx}\,dx + C.$$

This is the formula for **integration by parts**.

Integration by parts

$$\int u\frac{dv}{dx}\,dx = uv - \int v\frac{du}{dx}\,dx + C, \qquad (15.7)$$

where C is an arbitrary constant.

It is not at first obvious how this complicated result could be of any use, but the point of it is that the right-hand integral might be simpler than the one on the left. The process was once called 'partial integration', because the uv part is already integrated out. (For the effect of missing out C, see Problem 15.19.)

Example 15.17. Evaluate $\int x\,e^x\,dx$ by integrating by parts.

First observe that the integrand consists of the **product** of two factors, x and e^x, both of which we can integrate and differentiate any number of times. We relate this fact to (15.7) by identifying them with u and dv/dx respectively: put

$$u = x \quad \text{and} \quad \frac{dv}{dx} = e^x. \qquad (i)$$

Then

$$\frac{du}{dx} = 1 \quad \text{and} \quad v = \int e^x\,dx = e^x, \qquad (ii)$$

where we have chosen v to be the simplest antiderivative of e^x. Nothing would ultimately be changed by introducing an arbitrary constant C into v: *any antiderivative will do* (see Problem 15.18).

Fill in the right-hand side of (15.7) by picking out $u, v, du/dx$ from (i) and (ii), and introduce the constant C:

$$\int x\,e^x\,dx = x\,e^x - \int (e^x)(1)\,dx + C$$

$$= x\,e^x - \int e^x\,dx + C = x\,e^x - e^x + C,$$

where C is arbitrary.

We obtained a simplification because we chose x, rather than e^x, to be assigned to u. Since du/dx is simpler than u, it seemed possible that the right side of (15.7) might be simpler than the left. (To see what happens when we put $u = e^x$, $dv/dx = x$, see Example 15.19.)

As in Example 15.17, you should always write out stages (i) and (ii) in full. and do the subsequent working in full, or you will make mistakes.

Example 15.18. Evaluate $\int x \cos 2x\,dx$.

Put $u = x$, $dv/dx = \cos 2x$. Then

$$\frac{du}{dx} = 1, \qquad v = \int \cos 2x\,dx = \tfrac{1}{2}\sin 2x.$$

Substituting these functions into the right-hand side of (15.7):

$$\int x \cos 2x\,dx = x(\tfrac{1}{2}\sin 2x) - \int (\tfrac{1}{2}\sin 2x)(1)\,dx + C$$

$$= \tfrac{1}{2}x \sin 2x - \tfrac{1}{2}(-\tfrac{1}{2}\cos 2x) + C$$

$$= \tfrac{1}{2}x \sin 2x + \tfrac{1}{4}\cos 2x + C.$$

Example 15.19. For $\int x\,e^x\,dx$ (see Example 15.17), try the effect of assigning x and e^x to u and dv/dx the 'wrong way round'.

In Example 15.17, we successfully put $u = x$ and $dv/dx = e^x$. Now try instead

$$u = e^x, \qquad \frac{dv}{dx} = x,$$

then

$$\frac{du}{dx} = e^x, \qquad v = \int x\,dx = \tfrac{1}{2}x^2.$$

The integration-by-parts formula becomes

$$\int x\,e^x\,dx = e^x(\tfrac{1}{2}x^2) - \int (\tfrac{1}{2}x^2)\,e^x\,dx + C$$

$$= \tfrac{1}{2}x^2\,e^x - \tfrac{1}{2}\int x^2\,e^x\,dx + C,$$

which is a true result, but the transformed integral is worse than the original.

Sometimes it is not immediately obvious that the method can be made to work, as in the following.

Example 15.20. Evaluate $\int \ln x \, dx$.

Write $\ln x = (\ln x)(1)$, so that the integral becomes

$$\int (\ln x)(1) \, dx.$$

We can now put $u = \ln x$ and $dv/dx = 1$, so that

$$\frac{du}{dx} = \frac{1}{x}, \qquad v = x.$$

Then

$$\int (\ln x)(1) \, dx = (\ln x)(x) - \int (x)\left(\frac{1}{x}\right) dx + C$$

$$= x \ln x - \int dx + C = x \ln x - x + C,$$

where C is an arbitrary constant.

The **integrals of other inverse functions**, such as arcsin x and arctan x, respond to the same technique.

Example 15.21. Evaluate $\int x^2 \sin x \, dx$.

It is necessary in this problem to integrate by parts twice. Put

$$u = x^2, \qquad \frac{dv}{dx} = \sin x;$$

then

$$\frac{du}{dx} = 2x, \qquad v = -\cos x.$$

From (15.7),

$$\int x^2 \sin x \, dx = -x^2 \cos x + 2 \int x \cos x \, dx + C.$$

Integrate the integral on the right by parts; put

$$u = x, \qquad \frac{dv}{dx} = \cos x,$$

so that $du/dx = 1$ and $v = \sin x$. From (15.7), we obtain finally

$$\int x^2 \sin x \, dx = -x^2 \cos x + 2x \sin x + 2 \cos x + C.$$

15.8 Integration by parts: definite integrals

The integration-by-parts formula (15.7) expresses a relation between indefinite integrals, or antiderivatives. Suppose that we have a **definite integral** of the form

$$\int_a^b u \frac{dv}{dx} \, dx,$$

which we expect to integrate by parts. Then, from (15.7),

$$\int_a^b u \frac{dv}{dx} \, dx = \left[uv - \int v \frac{du}{dx} \, dx \right]_a^b.$$

The operation $[\,\cdots\,]$ applies to the two terms separately, so we have:

Integration by parts (definite integrals)

$$\int_a^b f(x) \, dx = \int_a^b u \frac{dv}{dx} \, dx = \left[uv \right]_a^b - \int_a^b v \frac{du}{dx} \, dx. \qquad \text{(15.8)}$$

This can sometimes considerably simplify the working, especially if more than one integration by parts is needed.

Example 15.22. Evaluate $\displaystyle\int_0^{\frac{1}{2}\pi} x^2 \sin x \, dx$.

As in Example 15.21, put $u = x^2$ and $dv/dx = \sin x$. Then

$$\frac{du}{dx} = 2x, \qquad v = -\cos x.$$

From (15.8),

$$\int_0^{\frac{1}{2}\pi} x^2 \sin x \, dx = \left[x(-\cos x) \right]_0^{\frac{1}{2}\pi} - \int_0^{\frac{1}{2}\pi} (-\cos x)(2x) \, dx$$

$$= 2 \int_0^{\frac{1}{2}\pi} x \cos x \, dx,$$

because the bracketed term is zero; we did not have to wait to the end of the calculation to see it go. To evaluate the remaining integral, integrate by parts again, putting $u = x$ and $dv/dx = \cos x$; we have

$$\frac{du}{dx} = 1, \qquad v = \sin x.$$

Use (15.8) again:

$$2 \int_0^{\frac{1}{2}\pi} x \cos x \, dx = 2\left(\left[x \sin x \right]_0^{\frac{1}{2}\pi} - \int_0^{\frac{1}{2}\pi} \sin x \, dx \right)$$

$$= 2\left(\tfrac{1}{2}\pi + \left[\cos x \right]_0^{\frac{1}{2}\pi}\right) = 2\left[\tfrac{1}{2}\pi + \left(0 - 1\right)\right]$$

$$= \pi - 2.$$

The following result is important for Chapter 22, and involves the use of (15.8):

$$\int_0^\infty e^{-t} t^N \, dt = N!$$

when $N = 0, 1, 2, 3, \ldots,$.
(0! is defined to be 1)

(15.9)

Here $N!$ stands for the **factorial**:

$$N! = N(N - 1)(N - 2) \cdots 3 \cdot 2 \cdot 1.$$

The symbol 0!, which is apparently arbitrarily given the value 1, does not fit this pattern; it should be regarded at this stage as being just a useful convention.

To prove (15.9), let k represent any of the numbers $0, 1, 2, \ldots,$ and write

$$\int_0^\infty e^{-t} t^k \, dt = F(k),$$

to indicate the integral's dependence on the parameter k; for example $\int_0^\infty e^{-t} t^3 \, dt$ is denoted by $F(3)$. Notice in particular that

$$F(0) = \int_0^\infty e^{-t} \, dt = \left[-e^{-t} \right]_0^\infty = 1. \tag{15.10}$$

For $k = 1, 2, \ldots,$ integrate by parts. Put $u = t^k$ and $dv/dt = e^{-t}$; then

$$F(k) = \int_0^\infty e^{-t} t^k \, dt = [t^k(-e^{-t})]_0^\infty - \int_0^\infty (kt^{k-1})(-e^{-t}) \, dt$$

$$= k \int_0^\infty e^{-t} t^{k-1} \, dt$$

(the bracket is zero because $k \geqslant 1$). The integral is $F(k - 1)$, so

$$F(k) = kF(k - 1) \quad \text{for } k = 1, 2, \ldots. \tag{15.11}$$

By integrating by parts, we have reduced the degree of t by unity. We could evaluate $F(N)$, where N is given, by integrating by parts again and again until we reach $F(0)$, given as 1 by (15.10). But we do not have to integrate by parts any more: equation (15.11) does it for us. Put $k = 0, 1, 2, 3, \ldots,$ successively: we obtain

$F(0) = 1$ (by (15.10)),

$F(1) = 1F(0)$ (by (15.11)) $= 1 \cdot 1 = 1!,$

$F(2) = 2F(1)$ (by (15.11)) $= 2(1!) = 2!,$

$F(3) = 3F(2)$ (by (15.11)) $= 3(2!) = 3!,$

and so on (each line uses the result of the previous line). So, if we are given N, we shall reach $F(N)$ after N lines, and find that

$$F(N) = N!,$$

The argument above can be expressed in a different way. Using (15.11) repeatedly we have

$$F(N) = NF(N-1) = N(N-1)F(N-2) = \cdots,$$

until we arrive at $F(0)$, which is 1, and we are left with $N!$ on the right.

Equation (15.11) is an example of a **reduction formula**, by which an integral can be systematically reduced, one step at a time, to progressively simpler integrals. (See the problems at the end of this chapter.)

Problems, Chapter 15

15.1. (Section 15.1). Obtain $\int f(x)\,dx$ when the $f(x)$ are as follows.
(a) $\sin 3x$; (b) $\cos 4x$; (c) e^{-3x};
(d) $(1+x)^{10}$; (e) $(1-x)^9$; (f) $(3-2x)^5$;
(g) $(1+2x)^n$; (h) $x(x-1)^4$; (i) $(1-x)^{\frac{1}{2}}$;
(j) $(2x-3)^{-\frac{1}{2}}$ for $x > \frac{3}{2}$; (k) $1/(3x+2)^2$;
(l) $1/(1-x)^4$; (m) $1/(1+x)$; (n) $1/(2x+3)$;
(o) $x/(1-x)^2$; (p) $(1+x)/(1-x)$;
(q) $x/(x-1)^{\frac{1}{2}}$ for $x > 1$; (r) $\cos(1-2x)$;
(s) $\sin(2x-3)$.

15.2. (Section 15.1). Evaluate the following indefinite integrals.

(a) $\int (2t-5)^5\,dt$; (b) $\int \sin\tfrac{1}{2}(3t-1)\,dt$;

(c) $\int \dfrac{1}{(2w+1)^2}\,dw$; (d) $\int e^{-3r}\,dr$;

(e) $\int (-t)^{\frac{1}{2}}\,dt$ if $t < 0$; (f) $\int \dfrac{s\,ds}{(1-s)^3}$;

(f) $\int \cos(\omega t - \phi)\,dt$.

15.3. (Section 15.2). Obtain $\int f(x)\,dx$ when the $f(x)$ are as follows.
(a) $x\,e^{-x^2}$; (b) $x\sin x^2$; (c) $x\cos x^2$;
(d) $x\cos(x^2+3)$; (e) $x\cos(1-3x^2)$;
(f) $x(x^2-1)^4$; (g) $x(3x^2+4)^3$;
(h) $x/(1+2x^2)$; (i) $x^3(1-x^2)^3$ (note: $x^3 = xx^2$);
(j) $x/(1+x^2)$; (k) $x/(3x^2-2)$.

15.4. (Section 15.3). Find $\int f(x)\,dx$ where the $f(x)$ are as follows.
(a) $\sin x \cos x$; (b) $\sin^2 x \cos x$; (c) $\sin^2 2x \cos 2x$;
(d) $\cos^2 x \sin x$; (e) $\cos^2 3x \sin 3x$; (f) $\sin^3 x \cos x$;
(g) $\cot 2x$; (h) $\tan\tfrac{1}{2}x$; (i) $(\sin^3 x)/\cos x$;
(j) $\sin^3 x$ ($=\sin^2 x \sin x$);
(k) $\tan^3 x$ (compare (j)); (l) $\cos^3 x$ (compare (j)).

15.5. Evaluate the following definite integrals by using any necessary substitutions.

(a) $\displaystyle\int_{-1}^{1} (1+x)^7\,dx$; (b) $\displaystyle\int_{-1}^{1} (1-\tfrac{1}{2}x)^7\,dx$;

(c) $\displaystyle\int_{0}^{1} x(1-x^2)^3\,dx$; (d) $\displaystyle\int_{0}^{1} \dfrac{x\,dx}{2x+3}$;

(e) $\displaystyle\int_{-3}^{-2} \dfrac{dx}{1+x}$ (note: $x < -1$); (f) $\displaystyle\int_{3}^{4} \dfrac{dx}{2-3x}$;

(g) $\displaystyle\int_{0}^{1} x^3(1-x^2)^3\,dx$; (h) $\displaystyle\int_{0}^{\frac{1}{4}\pi} \tan t\,dt$;

(i) $\displaystyle\int_{\frac{1}{12}\pi}^{\frac{1}{6}\pi} \cot 3w\,dw$; (j) $\displaystyle\int_{0}^{\frac{1}{2}\pi} \sin u \cos u\,du$;

(k) $\displaystyle\int_{0}^{\pi} (\sin v)^{\frac{1}{2}}\cos v\,dv$; (l) $\displaystyle\int_{-\frac{1}{4}\pi}^{\frac{1}{4}\pi} \cos^3\theta\,d\theta$;

(m) $\displaystyle\int_{0}^{\frac{1}{4}\pi} \sin 2t\,dt$; (n) $\displaystyle\int_{-\frac{1}{4}\pi/\omega}^{\frac{1}{4}\pi/\omega} \cos(\omega t + \phi)\,dt$.

15.6. Use the identities $\cos^2 A = \frac{1}{2}(1 + \cos 2A)$, $\sin^2 A = \frac{1}{2}(1 - \cos 2A)$, or $\sin A \cos A = \frac{1}{2}\sin 2A$ to evaluate the following.

(a) $\displaystyle\int_0^\pi \sin^2 t\, dt$;   (b) $\displaystyle\int_0^\pi \cos^2 t\, dt$;

(c) $\displaystyle\int_0^{\frac{1}{2}\pi} \sin^2 2t\, dt$;   (d) $\displaystyle\int_0^{\frac{1}{2}\pi} \cos^2 \tfrac{1}{2}t\, dt$;

(e) $\displaystyle\int_{-\pi}^\pi \sin^2 3t \cos 3t\, dt$;   (f) $\displaystyle\int_0^\pi \cos^4 u\, du$.

15.7. Use the substitutions suggested to evaluate $\displaystyle\int f(x)\, dx$ for the following $f(x)$. (In several of the questions the identity $1 + \tan^2 A = 1/\cos^2 A$ is needed. You may also have to refer to the table, Appendix E.)
(a) $\ln x/x$ (put $x = e^u$);
(b) $x(1 - x^2)^{\frac{1}{2}}$ (try (i) $u = 1 - x^2$, (ii) $x = \sin u$);
(c) $1/(e^x + e^{-x})$ (put $u = e^x$);
(d) $1/(1 - x^2)^{\frac{1}{2}}$ (try (i) $x = \sin u$, (ii) $x = \cos u$; why do the results seem to be different?);
(e) $\tan^2 x$ (put $u = \tan x$);
(f) $1/x^2(1 + x^2)$ (put $x = 1/u$, followed by another process);
(g) $1/(1 + x^2)$ (put $x = \tan u$);
(h) $1/\cos^2 x$ (put $u = \tan x$);
(i) $\dfrac{1}{t^{\frac{1}{2}}(1 + t)}$ (put $t = u^2$);   (j) $\dfrac{1}{t^2}\sin\dfrac{1}{t}$ (put $t = 1/u$);
(k) $(1 - x^2)^{\frac{1}{2}}$ (put $x = \sin u$);
(l) $1/(1 + x^2)^{\frac{1}{2}}$ (put $x = \tan u$).

15.8. Use partial fractions to evaluate $\displaystyle\int f(x)\, dx$ for the following $f(x)$.
(a) $1/(x^2 - 4)$;   (b) $1/x(x + 2)$;
(c) $1/x^2(x - 1)$;   (d) $x/(2x + 1)(x + 1)$;
(e) $(x + 1)/(4x^2 - 9)$;   (f) $1/x(x^2 + 1)$;
(g) $x/(2x^2 + 3x + 1)$;   (h) $1/x^2(2x + 1)$;
(i) $1/\cos x$ (first put $u = \sin x$);
(j) $1/\sin x$ (first put $u = \cos x$).

15.9. Obtain $\displaystyle\int f(x)\, dx$ for each of the following $f(x)$, noting that they take the form $cg(u)\, du/dx$ (see the remark at the end of Section 15.5), so that (a), for example, will respond to the substitution $u = x^3 - 1$.
(a) $x^2(x^3 - 1)^5$;   (b) $(x - 1)(x^2 - 2x + 3)^{-1}$;
(c) $1/(x \ln^2 x)$;   (d) $x^{\frac{1}{2}}(3x^{\frac{3}{2}} + 2)^{\frac{1}{2}}$;
(e) $(e^x - e^{-x})/(e^x + e^{-x})$;   (f) $1/x^{\frac{1}{2}}(x^{\frac{1}{2}} + 1)$;
(g) $x^2/(x^3 + 1)$.

15.10. Use integration by parts (Section 15.7) to obtain $\displaystyle\int f(x)\, dx$ for each of the following $f(x)$.

(a) $x e^{-x}$;   (b) $x e^{3x}$;   (c) $x e^{-3x}$;
(d) $x \cos x$;   (e) $x \sin x$;   (f) $x \cos \frac{1}{2}x$;
(g) $x \sin 2x$;   (h) $x(1 - x)^{10}$;   (i) $x \ln x$;
(j) $x^n \ln x$, $n \neq -1$;   (k) $(\ln x)/x$ (the method might seem to have failed; but look again).

15.11. Use integration by parts (see Example 15.20), writing the integrand as $f(x)(1)$, to obtain $\displaystyle\int f(x)\, dx$ for each of the following $f(x)$.
(a) $\ln^2 x$;   (b) $\arcsin x$;   (c) $\arccos x$;
(d) $\arctan x$.

15.12. To evaluate $\displaystyle\int f(x)\, dx$ for the following $f(x)$, integrate by parts twice; then look closely at your result. (If it does not work out you have probably made a mistake with a sign.) Compare your results with (13.11).
(a) $e^x \sin x$;   (b) $e^{-x} \sin x$ (c) $e^{-x} \cos x$.

15.13. (Integration by parts: definite integrals, Section 15.8). Evaluate the following.

(a) $\displaystyle\int_0^{\frac{1}{2}\pi} x \cos x\, dx$;   (b) $\displaystyle\int_0^\pi x \cos 2x\, dx$;

(c) $\displaystyle\int_0^\pi x^2 \cos x\, dx$;

(d) $\displaystyle\int_0^\infty e^{-x} \sin x\, dx$ (integrate by parts twice);

(e) $\displaystyle\int_0^\infty e^{-x} \cos x\, dx$ (integrate by parts twice);

(f) $\displaystyle\int_1^2 \frac{\ln x\, dx}{x}$;   (g) $\displaystyle\int_0^1 \arcsin x\, dx$;

(h) $\displaystyle\int_{-1}^1 \arccos x\, dx$;   (i) $\displaystyle\int_0^\infty \arctan x\, dx$;

(j) $\displaystyle\int_1^2 \ln x\, dx$.

15.14. (Compare (15.9)). Denote $\displaystyle\int_0^1 x^k e^x\, dx$ by $F(k)$ for $k = 0, 1, 2, \ldots$. Integrate by parts to obtain the reduction formula

$$F(k) = e - kF(k - 1),$$

(provided that k is positive). By applying it four times, show that

$$F(4) = \int_0^1 x^4 e^x\, dx = -15e + 24F(0) = 9e - 24.$$

15.15. Denote $\int_0^{\frac{1}{2}\pi} \cos^k x \, dx$ by $F(k)$ when $k = 0, 1, 2,$ .... Integrate by parts twice to show that

$$F(k) = \frac{k-1}{k} F(k-2) \quad \text{for } k = 2, 3, \dots.$$

Evaluate $F(0)$ and $F(1)$. Use the reduction formula repeatedly, together with $F(0)$ and $F(1)$, to evaluate

$$\int_0^{\frac{1}{2}\pi} \cos^4 x \, dx \quad \text{and} \quad \int_0^{\frac{1}{2}\pi} \cos^5 x \, dx.$$

15.16. Follow the lines of Problems 15.14 and 15.15 to obtain the following reduction formulae, and to integrate the special cases given. The letter k is an integer as specified for each case.

(a) Let

$$F(k) = \int_1^2 (\ln x)^k \, dx \quad (k \geq 0).$$

Show that $F(k) = 2 \ln^k 2 - kF(k-1)$ for $k \geq 1$, and evaluate $\int_1^2 (\ln x)^3 \, dx$.

(b) Let $F(k) = \int_0^\pi x^k \sin x \, dx \quad (k \geq 0)$. Integrate by parts twice to show that

$$F(k) = -\pi^k - k(k-1)F(k-2)$$

for $k \geq 2$. Evaluate $\int_0^\pi x^4 \sin x \, dx$ and $\int_0^\pi x^5 \sin x \, dx$.

(c) Obtain a reduction formula for

$$F(k) = \int_0^{\frac{1}{2}\pi} \sin^k x \, dx,$$

and use it to evaluate $\int_0^{\frac{1}{2}\pi} \sin^4 x \, dx$ and $\int_0^{\frac{1}{2}\pi} \sin^5 x \, dx$.

15.17. (Change of variable etc.). Denote the integral $\int_1^c \frac{dx}{x}$ for $c > 0$ by $F(c)$. Deduce the properties (a) to (d) below. $F(c)$ is obviously equal to $\ln c$, but *do not use any of the known properties of the logarithm*; pretend that this is the first time you have ever seen the integral.

(a) $F(a^{-1}) = -F(a)$ if $a > 0$. (Hint: put $c = a^{-1}$ in the definition; then change the variable to u where $u = x^{-1}$.)

(b) $F(ab) = F(a) + F(b)$ if a and $b > 0$. (Hint: change the variable appropriately; break the integral into two parts; then use (a).)

(c) $F(a/b) = F(a) - F(b)$, where a and $b > 0$.

(d) $F(a^n) = nF(a)$ if $a > 0$ and n has any value.

15.18. (Integration by parts). It is stated in Example 15.17 that, in obtaining v from dv/dx, we may take any antiderivative (so naturally we always take the simplest one, with $C = 0$ in the tables). Confirm that this is true for Example 15.17, in which $u = x$ and $dv/dx = e^x$, by choosing $v(x) = e^x + A$ instead.

Prove that the truth of (15.7) is always unaffected by the choice of antiderivative for $v(x)$.

15.19. (Integration by parts: an apparent paradox). Consider the following calculation.

$$\int x^{-1} \, dx = \int x^{-1}(1) \, dx$$

$$= x^{-1}x - \int (-x^{-2})x \, dx = 1 + \int x^{-1} \, dx.$$

Therefore $0 = 1$. How is this to be resolved?

16

Unforced linear differential equations with constant coefficients

16.1 Differential equations and their solutions

Suppose that we have a problem in which a quantity x that we are studying depends on the time t; that is to say, x is a function of t, which we will write as $x(t)$. From the physics and geometry of the problem we can often obtain an *indirect* relation between x and t, called an *equation* for x. The equation might be an ordinary algebraic equation such as $x^2 + 2xt = 1$, but it might contain dx/dt or d^2x/dt^2, as in the equation $d^2x/dt^2 = g$ for a falling body, where g is the gravitational acceleration. This is a simple example of a **differential equation**, and we can solve it by the methods of earlier chapters (compare Problem 12.8f).

The equation

$$\frac{dx}{dt} = 3x$$

is also a differential equation, but we do not yet know how to find an explicit solution for x in terms of t. Obviously not just anything will do; if for instance we try $x = t^2$ it does not work, because then $dx/dt = 2t$, but $3x = 3t^2$, and these are quite different.

A clue is given by interpreting the equation: it says that a quantity x always grows at a rate proportional to the amount of x already present. This is a property of the exponential function (see Section 1.10), so we might try exponential functions of t. In fact,

$$x = e^{3t}$$

solves the equation, because then $dx/dt = 3\,e^{3t}$, and this is equal to $3x$, as required. However, it is not the only solution, because

$$x = A\,e^{3t},$$

where A is any constant, also solves the equation.

In general, a **differential equation for x as a function of t** is an equation involving at least the first derivative dx/dt as well as, possibly, x and t separately. Some examples are

$$\frac{dx}{dt} + 2xt = 1, \qquad \frac{d^2x}{dt^2} + \frac{dx}{dt} + x = 0, \qquad \frac{d^3x}{dt^3} = \frac{x^2}{t^2}.$$

In such equations, t is called the **independent variable** and x the **dependent variable**. An equation is called **first-order, second-order**, and so on, according to the **order of the highest derivative** in it: dx/dt, d^2x/dt^2, and so on.

Problems in science and engineering are often most easily formulated in terms of differential equations. Suppose for example that in the RL circuit of Fig. 16.1 the switch is closed at time $t = 0$, and that subsequently the voltage applied is $E(t)$. Then the current $x(t)$ is found by solving the differential equation

$$L \frac{dx}{dt} + Rx = E(t).$$

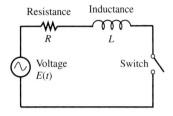

Fig. 16.1

Here we have collected all the terms that involve x (including dx/dt) on the left side and have put the term that does not involve x, namely $E(t)$, on the right. This is the conventional arrangement. The term independent of x which comes on the right is then called the **forcing term**, the reason being obvious in this case.

The differential equation with the same left-hand side, but with a **zero forcing term** on the right, plays a key role in obtaining solutions of the original equation. Such equations are called **unforced differential equations**, or sometimes **homogeneous equations**, and are the subject of this chapter. Also, for the present, we shall further restrict ourselves to **linear equations with constant coefficients**, which have the form:

Linear unforced differential equations with constant coefficients

(a) First-order:

$$\frac{dx}{dt} + cx = 0 \ (c \text{ constant}). \tag{16.1}$$

(b) Second-order:

$$\frac{d^2x}{dt^2} + b\frac{dx}{dt} + cx = 0 \ (b, c \text{ constants}).$$

These are called **linear** because there are no squares, products, etc. involving x and its derivatives. Such equations have comparatively simple characteristics. The simplest instance of all is

$$\frac{dx}{dt} = 0.$$

It has solutions $x = A$, where A is any constant. There is therefore **an infinity of solutions**, and we must expect this to be true in more general cases too.

A **solution of a differential equation** is any function $x(t)$ which fits, or satisfies, the equation. This is illustrated in the next two examples.

Example 16.1. For the differential equation $dx/dt + 2x = 0$, verify that (a) $x = e^{2t}$ is not a solution, (b) $x = 2 e^{-2t}$ is a solution.

(a) Test $x = e^{2t}$. Then $dx/dt = 2 e^{2t}$ and so

$$\frac{dx}{dt} + 2x = 2 e^{2t} + 2 e^{2t} = 4 e^{2t}.$$

This is not zero, so e^{2t} is not a solution.

(b) Test $x = 2 e^{-2t}$. Then $\dfrac{dx}{dt} = -4 e^{-2t}$ and so

$$\frac{dx}{dt} + 2x = -4 e^{-2}t + 4 e^{-2t} = 0.$$

The zero value is what the equation requires, so $2 e^{-2t}$ is a solution.

Incidentally, we can confirm in the same way that $x = A e^{-2t}$, where A is any constant, is always a solution. We have

$$\frac{dx}{dt} + 2x = -2A e^{-2t} + 2A e^{-2t} = 0,$$

as it should be. This is the infinity of solutions we were expecting.

Example 16.2. Verify that the following functions are solutions of the second-order equation $d^2x/dt^2 + 4x = 0$: (a) $x = \cos 2t$, (b) $x = \sin 2t$, (c) $x = A \cos 2t + B \sin 2t$, where A and B are any constants.

Note that 'verify' means 'try out': you are not expected to show how the solutions were obtained.

(a) If $x = \cos 2t$, then $dx/dt = -2 \sin 2t$, and $d^2x/dt^2 = -4 \cos 2t$. Therefore

$$\frac{d^2x}{dt^2} + 4x = -4 \cos 2t + 4 \cos 2t = 0$$

as required.

(b) Similarly, if $x = \sin 2t$, then

$$\frac{d^2x}{dt^2} + 4x = -4 \sin 2t + 4 \sin 2t = 0.$$

(c) Confirmation is straightforward, but the underlying reason why the previous solutions can be combined into a new solution in this way is made clearer by organizing the calculation as follows.

$$\frac{d^2x}{dt^2} + 4x = \frac{d^2}{dt^2}(A\cos 2t + B\sin 2t) + 4(A\cos 2t + B\sin 2t)$$

$$= A\left(\frac{d^2}{dt^2}\cos 2t + 4\cos 2t\right)$$

$$+ B\left(\frac{d^2}{dt^2}\sin 2t + 4\sin 2t\right),$$

by rearranging the terms. We already know that the two bracketed expressions are zero, so the whole expression is zero as required.

The separation of $d^2x/dt^2 + 4x$ into an 'A' part and a 'B' part in this way is possible only because the equation is *linear*.

16.2 Solving first-order linear unforced equations

Consider the equation

$$\frac{dx}{dt} + cx = 0 \quad (c \text{ a fixed constant}). \tag{16.2}$$

If we write it in the form

$$\frac{dx}{dt} = (-c)x,$$

it can be seen to describe the variation of a quantity $x(t)$ which decays (if c is positive) or grows (if c is negative) at a rate proportional to the amount of x already present. From Section 1.10, we know that exponential functions have this property. We shall therefore test for solutions of the form

$$x(t) = A e^{mt} \tag{16.3}$$

where A and m are unknown constants which we shall try to adjust to fit the equation. From (16.3),

$$\frac{dx}{dt} + cx = Am e^{mt} + cA e^{mt} = A(m + c) e^{mt}.$$

This quantity must be *zero for all values of t in* order to fit the differential equation (16.2). Ignoring the possibility $A = 0$, which gives us the so-called **trivial solution** $x(t) = 0$, we must have

$$m = -c,$$

and in that case it does not matter what value is given to A. We have therefore found a collection of solutions $x(t) = A e^{-ct}$, where A is an arbitrary constant. It can be proved that there are no other solutions, and so we call the solutions we have found the **general solution** of the equation.

> **The general solution of**
>
> $$\frac{dx}{dt} + cx = 0$$
>
> where c is a given constant, is
>
> $$x = A\,e^{-ct},$$
>
> where A is any constant.

(16.4)

Example 16.3. Find the general solution of $dx/dt - 4x = 0$.

We will rework the theory. Look for solutions of the form $x = A\,e^{mt}$:

$$\frac{dx}{dt} - 4x = Am\,e^{mt} - 4A\,e^{mt} = A\,e^{mt}(m - 4).$$

This is zero for all time if $m = 4$, whatever the value of A. Therefore the general solution (which includes the trivial one mentioned above) is

$$x = A\,e^{4t}, \quad \text{with } A \text{ an arbitrary constant.}$$

Figure 16.2 depicts several of these solutions, corresponding to various values of the arbitrary constant A.

Each value of A gives a different curve, and these **solution curves** fill the whole plane. Also the curves do not cross, so there is one and only one curve through every point. This corresponds to the fact that the slope dx/dt has one and only value at every point, namely the value prescribed by the differential equation $dx/dt = 4x$ taken at the point. This is all strong evidence that we have found *all the solutions*. More is said about the graphical way of understanding differential equations in Chapters 20 and 21.

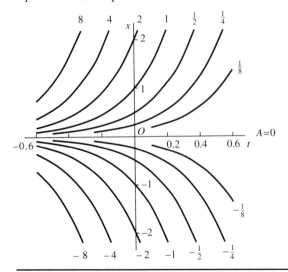

Fig. 16.2
The values of A are indicated on the curves

Example 16.4. Find all the solutions of $3\dfrac{dx}{dt} + 2x = 0$.

We could carry out the full calculation as in the previous example. However, if instead we want to quote the *formula*, (16.4), we must first write the equation in the form

$$\frac{dx}{dt} + \tfrac{2}{3}x = 0.$$

Therefore $c = \tfrac{2}{3}$ (not 2), and the general solution is

$$x = A\,e^{-\frac{2}{3}t}, \quad \text{with } A \text{ any constant.}$$

It is worthwhile to memorize the formula (16.4).

In practical cases we do not usually need all the solutions, but only the one which satisfies some further condition of the problem. Frequently the condition supplied describes the condition prevailing at the start of the action, or at some other time, as in the following.

Example 16.5. Find the solution of $\dfrac{dx}{dt} - 4x = 0$ for which $x = 2$ when $t = 1$.

Other ways of saying this are 'find the solution curve which passes through the point $(1, 2)$', or 'find a solution $x(t)$ so that $x(1) = 2$'.

From Example 16.3, all the possible solutions are given by

$$x = A\,e^{4t}.$$

Since $x = 2$ when $t = 1$, we must have $2 = A\,e^{4}$. Therefore

$$A = 2\,e^{-4}$$

and the single solution picked out is

$$x = (2\,e^{-4})\,e^{4t} = 2\,e^{4(t-1)}.$$

An extra condition of this type is called an **initial condition**. It describes the **state of the system** at a given time. The differential equation together with its initial condition is called an **initial-value problem**.

Initial-value problem, first-order equation

(a) Differential equation: $\dfrac{dx}{dt} + cx = 0$.

(b) Initial condition: $x = x_0$ at $t = t_0$,

(or $x(t_0) = x_0$), with x_0 and t_0 specified.

(16.5)

16.3 Solving second-order linear unforced equations

For second-order differential equations of the type (16.1), we use a similar technique.

Example 16.6. Find some solutions of the equation

$$\frac{d^2x}{dt^2} + \frac{dx}{dt} - 2x = 0.$$

We will look first for absolutely basic solutions. Test whether there are any solutions of the form $x(t) = e^{mt}$, where m is constant. Because $dx/dt = m\,e^{mt}$ and $d^2x/dt^2 = m^2\,e^{mt}$, we have

$$\frac{d^2x}{dt^2} + \frac{dx}{dt} - 2x = m^2\,e^{mt} + m\,e^{mt} - 2\,e^{mt}$$

$$= e^{mt}(m^2 + m - 2).$$

This is zero for all time if $m^2 + m - 2 = 0$, or if

$$m = 1 \text{ or } -2.$$

This gives us two solutions, namely

$$x(t) = e^t, \quad \text{and} \quad x(t) = e^{-2t}.$$

From this **basis**, we can obtain more solutions. Guided by Example 16.2c, we show that also

$$x(t) = A\,e^t + B\,e^{-2t},$$

where A and B are arbitrary constants, is a solution. By substituting into the equation and sorting the terms into those with coefficient A and those with coefficient B, we obtain

$$A\left(\frac{d^2}{dt^2}e^t + \frac{d}{dt}e^t - 2\,e^t\right) + B\left(\frac{d^2}{dt^2}e^{-2t} + \frac{d}{dt}e^{-2t} - 2\,e^{-2t}\right) = 0,$$

because e^t and e^{-2t} are known already to be solutions, so both of the bracketed expressions are zero.

This is the principle, but consider now the general case

$$\frac{d^2x}{dt^2} + b\frac{dx}{dt} + cx = 0.$$

Look for solutions of the form $x = e^{mt}$. Then

$$\frac{d^2x}{dt^2} + b\frac{dx}{dt} + cx = e^{mt}(m^2 + bm + c).$$

This will be zero for all t, as required by the differential equation, if

$$m^2 + bm + c = 0. \tag{16.6}$$

(16.6) is called the **characteristic equation**. Being quadratic, it may have two real solutions, exactly one real solution, or two complex solutions, depending on the coefficients.

Roots m_1 and m_2 of the characteristic equation real and different.
    In this case,

$$x(t) = e^{m_1 t} \quad \text{and} \quad x(t) = e^{m_2 t}$$

are solutions of the differential equation, and from these we can construct a whole **family of solutions**

$$x(t) = A\,e^{m_1 t} + B\,e^{m_2 t}$$

where A and B are arbitrary. It can be proved that there are no more solutions: this gives the **general solution**. The pair of functions $(e^{m_1 t}, e^{m_2 t})$ is called a **basis** for the general solution.

$$\frac{d^2 x}{dt^2} + b\,\frac{dx}{dt} + cx = 0; \ \textbf{roots } m_1 \textbf{ and } m_2 \textbf{ of}$$

$m^2 + bm + c = 0$ **real and different**   (16.7)

Basis of solutions: $e^{m_1 t}$, $e^{m_2 t}$.

General solution: $A\,e^{m_1 t} + B\,e^{m_2 t}$ (A, B arbitrary).

Example 16.7. Find the general solution of $2\dfrac{d^2 x}{dt^2} - \dfrac{dx}{dt} - x = 0$.

To correspond with the standard form, (16.7), we should have to write the equation in the form $d^2 x/dt^2 - \frac{1}{2}\,dx/dt - \frac{1}{2}\,x = 0$, but there is no need to do this if we directly test for solutions of the form $x = e^{mt}$. The characteristic equation then takes the form $2m^2 - m - 1 = 0$, or $(2m + 1)(m - 1) = 0$, so that $m_1 = -\frac{1}{2}$, $m_2 = 1$. Therefore the basis for the general solution is the solution pair $(e^{-\frac{1}{2}t}, e^t)$, and the general solution is

$$x(t) = A\,e^{-\frac{1}{2}t} + B\,e^t, \quad A \text{ and } B \text{ arbitrary.}$$

The roots m_1 and m_2 of the characteristic equation are equal.
    Suppose that $m_1 = m_2 = m_0$, say. We have then only one function in our basis instead of two, and we might expect the general solution to be $A\,e^{m_0 t}$. However, all we know is that *there is essentially only one solution of the form e^{mt}* (ignoring simple multiples of e^{mt}), but we shall see in the next example that there does exist another solution not of this form, namely

$$x(t) = t\,e^{m_0 t}. \tag{16.8}$$

We might therefore think there will be no end to it: if $t\,e^{m_0 t}$ is a solution, then why not $t^2\,e^{m_0 t}$, or some function of great complication? However, it can be proved that **every second-order linear differential equation has two linearly independent solutions** (i.e. they are not just multiples of each other); also that **these form a basis of solutions:** we do not need any others to construct the most general solution. Formally:

Basis and general solution of

$$\frac{d^2x}{dt^2} + b\frac{dx}{dt} + cx = 0$$

(a) There exist two linearly independent solutions.
(b) If $u(t)$ and $v(t)$ are any two linearly independent   **(16.9)**
solutions, these form a basis for the general solution;
that is to say, the general solution is given by

$$x(t) = Au(t) + Bv(t),$$

where A and B are arbitrary constants.

Example 16.8. Find the general solution of $\dfrac{d^2x}{dt^2} + 4\dfrac{dx}{dt} + 4x = 0$.

The characteristic equation, formed by substituting $x(t) = e^{mt}$, is

$$m^2 + 4m + 4 = (m + 2)^2 = 0,$$

and the only value of m that we find is $m = -2$. It corresponds to
the basic solution e^{-2t}.

The theorem (16.9) guarantees there is another independent solution, and it does not matter how we find it. Test the truth of (16.8),
which proposes an independent solution having the form

$$x(t) = t\,e^{-2t}.$$

Then

$$\frac{dx}{dt} = (1 - 2t)\,e^{-2t},$$

and

$$\frac{d^2x}{dt^2} = (-4 + 4t)\,e^{-2t}.$$

Therefore

$$\frac{d^2x}{dt^2} + 4\frac{dx}{dt} + 4x = [(-4 + 4t) + 4(1 - 2t) + 4t]\,e^{-2t},$$

which is zero, so $x(t) = t\,e^{-2t}$ is a second solution, and it is
independent of the first. By (16.9), the solution basis is therefore

$$(e^{-2t}, t\,e^{-2t}),$$

and the general solution is

$$x(t) = A\,e^{-2t} + Bt\,e^{-2t}, \quad A \text{ and } B \text{ arbitrary.}$$

The second solution always takes the same form (see Problem
16.8):

Characteristic equation: coincident roots

If $\dfrac{\mathrm{d}^2 x}{\mathrm{d}t^2} + b\,\dfrac{\mathrm{d}x}{\mathrm{d}t} + cx = 0$, in which $b^2 - 4c = 0$ (for coincident roots), and m_0 is the single solution of the
characteristic equation $m^2 + bm + c = 0$, then the sol-
ution basis is $(\mathrm{e}^{m_0 t},\ t\,\mathrm{e}^{m_0 t})$ and the general solution is
$x(t) = A\,\mathrm{e}^{m_0 t} + Bt\,\mathrm{e}^{m_0 t}$ (A and B arbitrary constants). **(16.10)**

16.4 Complex roots of the characteristic equation

If $b^2 < 4c$, the roots m_1 and m_2 of the characteristic equation
$m^2 + bm + c = 0$ for the differential equation $\mathrm{d}^2 x/\mathrm{d}t^2 + b\,\mathrm{d}x/\mathrm{d}t + cx = 0$ are complex. Since they are roots of a quadratic equation,
they must be **complex conjugate**, so put

$$m_1 = \alpha + \mathrm{j}\beta, \qquad m_2 = \alpha - \mathrm{j}\beta,$$

where α and β are real numbers. The corresponding functions

$$\mathrm{e}^{(\alpha + \mathrm{j}\beta)t} \quad \text{and} \quad \mathrm{e}^{(\alpha - \mathrm{j}\beta)t} \tag{16.11}$$

are genuine solutions of the differential equation. They are complex
functions, so we call (16.11) a **complex basis** for solutions of the
differential equation. If we are interested in complex as well as real
solutions, then we can allow the arbitrary constants A and B to be
complex as well, in an all-inclusive **general complex solution**

$$x(t) = A\,\mathrm{e}^{(\alpha + \mathrm{j}\beta)t} + B\,\mathrm{e}^{(\alpha - \mathrm{j}\beta)t}.$$

Suppose, however, that we want the general solution to consist
only of real functions. Then a **basis of real solutions** can be got
from (16.11) in the following way. Choose *either one* of the complex
solutions (16.11), say $\mathrm{e}^{(\alpha + \mathrm{j}\beta)t}$. Then (see (6.8))

$$\mathrm{e}^{(\alpha + \mathrm{j}\beta)t} = \mathrm{e}^{\alpha t}\,\mathrm{e}^{\mathrm{j}\beta t} = \mathrm{e}^{\alpha t}\cos \beta t + \mathrm{j}\,\mathrm{e}^{\alpha t}\sin \beta t.$$

This function solves the differential equation, so its real and
imaginary parts separately must also solve it. Therefore

$$(\mathrm{e}^{\alpha t}\cos \beta t,\ \mathrm{e}^{\alpha t}\sin \beta t)$$

is a **real basis** for the general (real) solution

$$x(t) = A\,\mathrm{e}^{\alpha t}\cos \beta t + B\,\mathrm{e}^{\alpha t}\sin \beta t, \tag{16.12}$$

where A and B are arbitrary (but real, of course).

Equation (16.12) can be written in a different form. Using the
identity (1.19), we have

$$A\cos \beta t + B\sin \beta t = C\cos (\beta t + \varphi),$$

where C and φ are constants related to A and B. Therefore (16.12)
can be written

$$x(t) = C\,\mathrm{e}^{\alpha t}\cos (\beta t + \varphi).$$

Since A and B are arbitrary, so are C and φ.

Example 16.9. Find the general solution of $\dfrac{d^2x}{dt^2} + 4x = 0$.

The characteristic equation is $m^2 + 4 = 0$. Its solutions are $m = \pm 2j$. Therefore the complex solution basis is (e^{2jt}, e^{-2jt}). But

$$e^{2jt} = \cos 2t + j \sin 2t,$$

and the real and imaginary parts give a basis for the real solutions:

$$(\cos 2t, \sin 2t).$$

Therefore the general solution is

$$x(t) = A \cos 2t + B \sin 2t \quad (A, B \text{ arbitrary}).$$

Example 16.10. Find the general solution of $\dfrac{d^2x}{dt^2} + 2\dfrac{dx}{dt} + 2x = 0$.

Setting $x = e^{mt}$ gives the characteristic equation $m^2 + 2m + 2 = 0$, so that $m = -1 \pm j$. Therefore

$$(e^{(-1+j)t}, e^{(-1-j)t})$$

is a basis for complex solutions. But

$$e^{(-1+j)t} = e^{-t}(\cos t + j \sin t),$$

whose real and imaginary parts are

$$e^{-t}\cos t, \quad e^{-t}\sin t.$$

These form the basis for the real solutions. The general solution is

$$x(t) = A e^{-t}\cos t + B e^{-t}\sin t.$$

If we chose instead to take the real and imaginary parts of $e^{(-1-j)t}$, we would obtain $(e^{-t}\cos t, -e^{-t}\sin t)$ as a basis. The minus sign will be absorbed into the *arbitrary* constant B: no new solutions appear.

The general solution method can be summed up as follows:

$\dfrac{d^2x}{dt^2} + b\dfrac{dx}{dt} + cx = 0$, **when** $m^2 + bm + c = 0$ **has**

complex roots $m_1, m_2 = \alpha \pm j\beta$ **(i.e.** $b^2 < 4c$)

Complex basis: $e^{(\alpha + j\beta)t}, e^{(\alpha - j\beta)t}$.

Real basis: $e^{\alpha t}\cos\beta t, e^{\alpha t}\sin\beta t$.   **(16.13)**

General solution:

(a) $x(t) = A e^{\alpha t}\cos\beta t + B e^{\alpha t}\sin\beta t$
         (A and B arbitrary);

or

(b) $x(t) = C e^{\alpha t}\cos(\beta t + \phi)$ (C and ϕ arbitrary).

A very important case is when $b = 0$ and $c > 0$, illustrated by Example 16.9. In that case, $\alpha = 0$. In conventional notation, putting $c = \omega^2$, we obtain the following result:

$$\frac{d^2x}{dt^2} + \omega^2 x = 0$$

Characteristic equation: $m^2 + \omega^2 = 0$; $m_1, m_2 = \pm j\omega$.

Complex basis: $e^{j\omega t}, e^{-j\omega t}$.

Real basis: $\cos \omega t, \sin \omega t$.

General solution: (a) $x(t) = A \cos \omega t + B \sin \omega t$,

or

(b) $\qquad\qquad x(t) = C \cos (\omega t + \phi)$.

(16.14)

In the special case (16.14), the alternative solution form

$$x(t) = C \cos(\omega t + \phi)$$

shows that the solutions oscillate regularly, swinging above and below the t axis to an extent governed by the amplitude C. In the general case (16.13),

$$x(t) = C e^{\alpha t} \cos(\beta t + \phi),$$

the solutions oscillate, but the amplitude is governed by the factor $C e^{\alpha t}$. If α is positive, the oscillation constantly grows; if α is negative, it dies away to zero. This is fully discussed in Chapter 18, but Fig. 16.3 shows a particular case.

The **damped unforced linear oscillator** is the simplest linear model of an oscillating mechanical or electrical system which has a small amount of friction or some other form of energy-loss mechanism (see Chapter 18 for a full discussion). In a customary notation the equation is

$$\frac{d^2x}{dt^2} + 2k\frac{dx}{dt} + \omega^2 x = 0.$$

The term $2k\, dx/dt$ expresses the energy-absorbing property. Assume

$$k^2 < \omega^2.$$

The characteristic equation is $m^2 + 2km + \omega^2 = 0$, so that

$$m = -k \pm (k^2 - \omega^2)^{\frac{1}{2}} = -k \pm j(\omega^2 - k^2)^{\frac{1}{2}},$$

since $k^2 < \omega^2$. From (16.13), $\alpha = -k$ and $\beta = (\omega^2 - k^2)^{\frac{1}{2}}$, so finally:

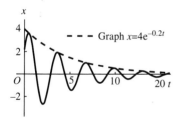

Fig. 16.3
Graph of
$x(t) = 4 e^{-0.2t} \cos(2t - 1)$

$$\frac{d^2x}{dt^2} + 2k\frac{dx}{dt} + \omega^2 x = 0 \text{ where } k^2 < \omega^2$$

General solution:

(a) $\quad x(t) = A\,e^{-kt}\cos(\omega^2 - k^2)^{\frac{1}{2}}t$

$$+ B\,e^{-kt}\sin(\omega^2 - k^2)^{\frac{1}{2}}t \qquad \text{(16.15)}$$

(A and B arbitrary constants); or

(b) $\quad x(t) = C\,e^{-kt}\cos[(\omega^2 - k^2)^{\frac{1}{2}}t + \phi]$

(C and ϕ arbitrary).

16.5 Initial conditions for second-order equations

The general solution of a second-order differential equations involves two arbitrary constants, and the solutions are therefore an order of magnitude more numerous than in the first-order case. Unlike the first-order case, the solution curves may cross – in fact, there is an infinite number of solution curves through any point on the (x, t) plane, as indicated in Fig. 16.4a.

To pick out a particular solution, we need to determine the two arbitrary constants. Two pieces of information are necessary. These may consist of **two initial conditions**: conditions which define the **state of the system** at some starting time t_0: the values of $x(t)$ and the slope dx/dt at $t = t_0$ are given (see Fig. 16.4b). For example, the equation $d^2x/dt^2 + \omega_0^2 x = 0$ describes the oscillations of a particle on a spring; the initial conditions tell us its position and velocity (i.e. its **state**) when it starts off. We then have an **initial-value problem**:

(a)

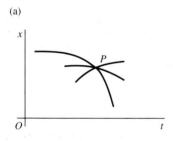

(b)

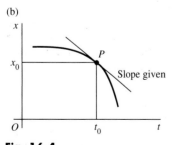

Fig. 16.4
(a) An infinite number of curves pass through each point. (b) Selection of a solution given P and the slope at P.

Initial-value problem

(i) Equation: $\dfrac{d^2x}{dt^2} + b\dfrac{dx}{dt} + cx = 0.$

(ii) Initial conditions:

$$x = x_0 \text{ and } \frac{dx}{dt} = x_1 \text{ at } t = t_0, \qquad \text{(16.16)}$$

which may be expressed alternatively as

$$x(t_0) = x_0, \qquad x'(t_0) = x_1,$$

where x_0 and x_1 are given.

Example 16.11. Find the solution of $d^2x/dt^2 + 4x = 0$ for which $x = 1$ and $dx/dt = 2$ at $t = 0$ (that is, $x(0) = 1$, $x'(0) = 2$).

First we need all the solutions. From Example 16.9, these are $x(t) = A \cos 2t + B \sin 2t$, where A and B may take any values. Since $x = 1$ at $t = 0$,

$$1 = A + 0, \quad \text{so} \quad A = 1.$$

For the other condition, we first need $x'(t)$ in general:

$$x'(t) = -2A \sin 2t + 2B \cos 2t.$$

At $t = 0$, we are given that $x'(t) = 2$, so the last equation becomes

$$2 = 0 + 2B, \quad \text{or} \quad B = 1.$$

The required solution is therefore $x(t) = \cos 2t + \sin 2t$.

Problems, Chapter 16
(For the 'dash' notation $x'(t) = dx/dt$ **etc., see** (4.1).)

16.1. Say which of the following equations are linear, unforced, with constant coefficients, i.e. can be rearranged to conform with (16.1a).
(a) $x' = 3t$; (b) $x' = \frac{1}{2}x$; (c) $x' + tx = 0$;
(d) $3x' - 2x^2 = 0$; (e) $x' - x = 0$; (f) $x' = 0$;
(g) $\dfrac{x'}{x^2} = 3$; (h) $\dfrac{dy}{dx} + \frac{1}{2}y = 1$; (i) $\dfrac{1}{y}\dfrac{dy}{dx} = 2$;
(j) $L\dfrac{dI}{dt} + RI = 0$; (k) $\dfrac{v' + v + v^2}{v' - v + v^2} = 1$.

16.2. Write down all the solutions of the following equations. Check one or two of them by substitution into the differential equation.
(a) $x' + 5x = 0$; (b) $x' - \frac{1}{2}x = 0$;
(c) $x' - x = 0$; (d) $x' + 3x = 0$;
(e) $3x' + 4x = 0$; (f) $x' = 2x$; (g) $x' = 3x$;
(h) $x'/x = -3$; (i) $(x' + 1)/(x + 1) = 1$.

16.3. Solve the following initial-value problems.
(a) $x' + 2x = 0$, $x = 3$ when $t = 0$;
(b) $3x' - x = 0$, $x = 1$ when $t = 1$;
(c) $y' - 2y = 0$, $y = 2$ when $x = -3$;
(d) $x' + x = 0$, $x(-1) = 10$;
(e) $2y' - 3y = 0$, $y(0) = 1$;
(f) Find the curve whose slope at any point (x, y) is equal to $5x$, and which passes through the point $(1, -2)$.

16.4. Suppose that the generator in Fig. 16.1 is short-circuited and cut out at a moment when the current in the circuit is I_0. Find an expression for the current subsequently. Show that the ratio L/R provides a measure of the time it takes for the current to die away.

16.5. A radioactive element disintegrates at a rate proportional to the amount of the original element still remaining. Show that if $A(t)$ represents the activity of the element at time t, then

$$\frac{dA}{dt} + kA = 0,$$

where k is a positive constant.
(a) Solve the initial-value problem for A if $A = A_0$ (given) at time $t = 0$.
(b) The time taken for the activity to drop to half of the starting value is called the half-life period. For Uranium 232, it is found that 17.5% has decayed after 20 years. Show that its half-life period is about 72 years.

16.6. (Götterdämmerung) Once upon a time, rabbits in Elysium reached maturity instantly and bred with a birthrate of 20 rabbits per year per couple. No rabbit ever died. At the start of the experiment Zeus released 50 male and 50 female rabbits.

By treating the number of rabbits as a continuously varying quantity and considering the number born in a short time δt, construct a differential equation and then an initial-value problem for $R(t)$, the rabbit population. Find how many rabbits there were at the end of Year 4.

Appalled by this result and assisted by Pluto, Zeus launched another similar experiment, in which any rabbit was allowed to live for one year only. Construct the differential equation for the population. Did this alleviate the situation appreciably?

16.7. Obtain all solutions of the following equations. (The characteristic equations all have real roots, not necessarily distinct.)
(a) $x'' - 3x' + 2x = 0$; (b) $x'' + x' - 2x = 0$;
(c) $x'' - x = 0$; (d) $x'' - 4x = 0$;
(e) $3x'' - \frac{1}{4}x = 0$; (f) $x'' - 9x = 0$;
(g) $x'' + 2x' - x = 0$; (h) $x'' - 2x' - 2x = 0$;
(i) $2x'' + 2x' - x = 0$; (j) $3x'' - x' - 2x = 0$;
(k) $x'' + 4x' + 4x = 0$; (l) $x'' + 6x' + 9x = 0$;
(m) $4x'' + 4x' + x = 0$; (n) $x'' = 0$.

16.8. Verify that, when the characteristic equation corresponding to $x'' + bx' + cx = 0$ has coincident roots, $m_1 = m_2 = m_0$, say, then the function $x(t) = t\,e^{m_0 t}$ provides a second solution for the basis of the general solution. (For coincident roots, $b^2 = 4c$.)

16.9. Solve the following initial-value problems.
(a) $x'' - 4x = 0$, $x(0) = 1$, $x'(0) = 0$;
(b) $x'' + x' - 2x = 0$, $x(0) = 0$, $x'(0) = 2$;
(c) $y'' - 4y' + 4y = 0$, $y(0) = 0$, $y'(0) = -1$;
(d) $y'' + 2y' + y = 0$, $y(1) = 0$, $y'(1) = 1$;
(e) $x'' - 9x = 0$, $x(1) = 1$, $x'(1) = 1$;
(f) $x'' - 4x' = 0$, $x(1) = 1$, $x'(1) = 0$.

16.10. Obtain all solutions of the following equations. (The roots of the characteristic equations are complex.)
(a) $x'' + x = 0$; (b) $x'' + 9x = 0$;
(c) $x'' + \frac{1}{4}x = 0$; (d) $x + \omega_0^2 x = 0$;
(e) $x'' + 2x' + 2x = 0$; (f) $y'' - 2y' + 2y = 0$;
(g) $y'' + y' + y = 0$; (h) $2x'' + 2x' + x = 0$;
(i) $3x'' + 4x' + 2x = 0$; (j) $3x'' - 4x' + 2x = 0$.

16.11. Solve the following initial-value problems.
(a) $x'' + x = 0$, $x(0) = 0$, $x'(0) = 1$;
(b) $x'' + 4x = 0$, $x(0) = 1$, $x'(0) = 0$;
(c) $x'' + \omega_0^2 x = 0$, $x(0) = a$, $x'(0) = b$;
(d) $x'' + 2kx' + x = 0$, $x(0) = 0$, $x'(0) = b$ for the cases $k^2 > 1$, $k^2 < 1$ and $k^2 = 1$.

(Use the A, B form): finding the constants C and ϕ in (16.14b) for an initial-value problem can be comparatively difficult.)

16.12. The approximate equation for small swings of a pendulum is

$$\frac{d^2\theta}{dt^2} + \frac{g}{l}\theta = 0$$

where θ is the inclination from the vertical (in radians), l is the length and g the gravitational acceleration. The pendulum is held still at an angle α, and is then passively released. Find the subsequent motion.

16.13. The pendulum in Problem 16.12 is hanging at rest; then the bob is given a small velocity v in the direction of θ increasing. Find the subsequent motion.

16.14. If there is a little friction in the pendulum of Problem 16.12, the equation of motion takes the form

$$\frac{d^2\theta}{dt^2} + K\frac{d\theta}{dt} + \frac{g}{l}\theta = 0,$$

where K is an additional positive constant which takes account of the friction (assumed to be proportional to the angular velocity). In a particular case (mks units), $g = 9.7$, $l = 20$, $K = 0.066$. The pendulum is at rest at first, hanging freely. It is then pushed so as to give the bob a velocity of 1 metre per second. Find the subsequent motion.

16.15. Consider the third-order differential equation

$$\frac{d^3y}{dx^3} - y = 0.$$

Proceed by analogy with the method of Section 16.3: by substituting $y = e^{mx}$, and obtaining a characteristic equation for m (a cubic equation), find three distinct basic solutions of this type. By introducing arbitrary constants A, B, C, find as wide a variety of solutions as you can (in fact, this is the general solution).

16.16. By proceeding as in Problem 15.15, find a wide variety of solutions of the equation

$$\frac{d^3y}{dx^3} + y = 0.$$

16.17. By proceeding with the equation

$$\frac{d^4y}{dx^4} - y = 0$$

as in Problem 16.15, obtain the collection of solutions

$$y(x) = A\,e^x + B\,e^{-x} + C\cos x + D\sin x,$$

where A, B, C, D are arbitrary constants.

16.18. A tapered concrete column of height H metres is to support a statue of mass M (i.e. weight Mg force units, where g is the gravitational acceleration) at the top. Pressure (force per unit area) may not exceed P. Show that the most economical construction for the column is for its cross-sectional area $A(y)$, where y is distance above the ground, to satisfy the equation

$$A(y) = \frac{Mg}{P} + \frac{\rho g}{P} \int_y^H A(u)\, du,$$

where ρ is the density of concrete. By differentiating this expression (see Section 13.10), obtain a differential equation for $A(y)$, and an initial condition for the equation, and solve it.

17

Forced linear differential equations

17.1 Particular solutions for standard forcing terms

Consider the equation

$$\frac{d^2x}{dt^2} + b\frac{dx}{dt} + cx = f(t).$$

The function $f(t)$ is called the **forcing term**; it represents physically the external **input** to the physical system that the equation describes, and the system will respond with an **output** $x(t)$ which depends on the input $f(t)$.

If $f(t)$ is an **exponential function** $K\,e^{\alpha t}$, a **sine** or **cosine function** $K\sin\beta t$ or $K\cos\beta t$, or a **polynomial**, then we can find an individual **particular solution** by trial.

Example 17.1. Find a particular solution of

$$\frac{d^2x}{dt^2} + \frac{dx}{dt} - 2x = 3\,e^{2t}.$$

Try for a solution containing the same exponential as that on the right-hand side of the equation:

$$x(t) = p\,e^{2t},$$

where p is some constant – not an arbitrary constant, but one whose value we shall settle by substitution: only one value will do. Then

$$\frac{dx}{dt} = 2p\,e^{2t}, \quad \text{and} \quad \frac{d^2x}{dt^2} = 4p\,e^{2t},$$

so that

$$\frac{d^2x}{dt^2} + \frac{dx}{dt} - 2x = 4p\,e^{2t} + 2p\,e^{2t} - 2p\,e^{2t}$$

$$= e^{2t}(4 + 2 - 2)p = 4p\,e^{2t}.$$

This must equal the given right-hand side, $3\,e^{2t}$ *for all values of t*, which is only possible if $4p = 3$, or

$$p = \tfrac{3}{4}.$$

Therefore one particular solution is

$$x(t) = \tfrac{3}{4} e^{2t}.$$

There are many other solutions, as we shall see in Section 17.5, but they are all based on this particular solution.

Example 17.2. Find a particular solution of $\dfrac{d^2 x}{dt^2} + 4x = 2 \cos 3t$.

Guess that there might be a solution of the form

$$x = p \cos 3t.$$

Then

$$\frac{dx}{dt} = -3p \sin 3t, \quad \text{and} \quad \frac{d^2 x}{dt^2} = -9p \cos 3t,$$

so that

$$\frac{d^2 x}{dt^2} + 4x = -9p \cos 3t + 4p \cos 3t = -5p \cos 3t.$$

This must be the same as the right-hand side of the equation in order for the guessed function to be a solution, so

$$-5p \cos 3t = 2 \cos 3t.$$

Therefore $p = -\tfrac{2}{5}$, and the required solution is

$$x(t) = -\tfrac{2}{5} \cos 3t.$$

In most cases when the right-hand side is a sine or cosine it will not work so simply, as is illustrated in the following example.

Example 17.3. Find a particular solution of

$$\frac{d^2 x}{dt^2} + \frac{dx}{dt} - 2x = 2 \cos 3t.$$

The form $p \cos 3t$ cannot be made to fit this equation because the dx/dt term on the left produces a $\sin 3t$ term, making the left and right sides impossible to match for all t. Try instead

$$x = p \cos 3t + q \sin 3t.$$

Then

$$\frac{dx}{dt} = -3p \sin 3t + 3q \cos 3t,$$

and

$$\frac{d^2 x}{dt^2} = -9p \cos 3t - 9q \cos 3t.$$

Therefore

$$\frac{d^2x}{dt^2} + \frac{dx}{dt} - 2x = (-9p + 3q - 2p)\cos 3t$$
$$+ (-9q - 3p - 2q)\sin 3t$$
$$= (-11p + 3q)\cos 3t + (-3p - 11q)\sin 3t$$
$$= 2\cos 3t$$

(the last expression being the right-hand side of the equation) for all t.

The only way to satisfy this condition is to require both

$$-11p + 3q = 2, \qquad -3p - 11q = 0.$$

The solution of these two simultaneous equations for p and q is

$$p = -\tfrac{11}{65}, \qquad q = \tfrac{3}{65}.$$

Therefore a particular solution is

$$x = -\tfrac{11}{65}\cos 3t + \tfrac{3}{65}\sin 3t.$$

It will be necessary in nearly all cases to take both sine and cosine terms into account.

The case when $f(t)$ is a **constant** often occurs.

Example 17.4. Find a particular solution of $\dfrac{d^2x}{dt^2} - 2\dfrac{dx}{dt} + 4x = 3$.

Test whether there is a constant solution

$$x(t) = p \quad \text{(a constant)}.$$

By substituting this in the differential equation, we get

$$0 + 0 + 4p = 3,$$

so that $p = \tfrac{3}{4}$, and the particular solution is just $x(t) = \tfrac{3}{4}$, which is obvious after it has been worked out.

If the right-hand side is a **polynomial** (the sum of one or more powers of x), then the solution will be a polynomial.

Example 17.5. Find a solution of $\dfrac{d^2x}{dt^2} - \dfrac{dx}{dt} + x = 3 + 2t^2$.

Try a solution of the form $x(t) = p + qt + rt^2$, where p, q, and r are constants. It is normally necessary to try a polynomial of the same degree as the forcing term, which in this case has degree 2, and to include all the lower-degree terms in the trial solution.

Since

$$\frac{dx}{dt} = q + 2rt, \quad \text{and} \quad \frac{d^2x}{dt^2} = 2r,$$

we must have

$$2r - (q + 2rt) + (p + qt + rt^2) = 3 + 2t^2.$$

Match up the coefficients of the three powers of t; we find that

$$r = 2, \qquad -2r + q = 0, \qquad 2r - q + p = 3.$$

The equations are easy to solve and lead to the solution

$$x(t) = 3 + 4t + 2t^2.$$

Particular solutions of

$$\frac{d^2x}{dt^2} + b\frac{dx}{dt} + cx = f(t) \textbf{ and } \frac{dx}{dt} + cx = f(t)$$

(a) $f(t) = K\,e^{\alpha t}$: try a solution $x(t) = p\,e^{\alpha t}$. (17.1)

(b) $f(t) = K\cos\beta t$ or $K\sin\beta t$: try a solution

$$x(t) = p\cos\beta t + q\sin\beta t.$$

(c) $f(t)$ is a polynomial of degree N: try a polynomial of the same degree, with all its terms present.

There are **exceptional cases** where these substitutions have to be modified. For example, $d^2x/dt^2 = t$ has a polynomial solution of degree 3, not degree 1. These cases are treated in Section 17.3.

If the forcing term on the right-hand side consists of the sum of several constituent terms, then obtain a particular solution for each one, and add them, as in the following example.

Example 17.6. Obtain a particular solution of

$$\frac{d^2x}{dt^2} + 4x = 1 + e^{-t}.$$

Solve $\dfrac{d^2x_1}{dt^2} + 4x_1 = 1$ for $x_1(t)$ and $\dfrac{d^2x_2}{dt^2} + 4x_2 = e^{-t}$ for $x_2(t)$; then

$$x(t) = x_1(t) + x_2(t)$$

will be a particular solution of the original equation.

For $x_1(t)$, try for a constant solution $x_1(t) = p$: it is found that $p = \frac{1}{4}$, so $x_1(t) = \frac{1}{4}$.

For $x_2(t)$, try $x_2(t) = q\,e^{-t}$ (following the method of Example 17.1). The substitution gives $q\,e^{-t} + 4q\,e^{-t} = e^{-t}$, so that $q = \frac{1}{5}$, and the solution is $x_2(t) = \frac{1}{5}\,e^{-t}$.

Therefore, a particular solution $x(t)$ of the original equation is

$$x(t) = x_1(t) + x_2(t) = \tfrac{1}{4} + \tfrac{1}{5}\,e^{-t}.$$

The method just described is another consequence of the linearity of the class of equations considered. It is also called the **super-position principle**.

These methods apply equally to **first-order equations**.

Example 17.7. Obtain a particular solution of $\dfrac{dx}{dt} + x = 3\cos 2t$.

Remembering Example 17.3, we expect the solution will have to contain both cosine and sine terms, so try

$$x(t) = p\cos 2t + q\sin 2t.$$

The substitution gives $(p + 2q)\cos 2t + (-2p + q)\sin 2t = 3\cos 2t$, so that $p = \frac{3}{5}$, $q = \frac{6}{5}$, and the solution is

$$x(t) = \tfrac{3}{5}\cos 2t + \tfrac{6}{5}\sin 2t.$$

17.2 Harmonic forcing term by using complex solutions

In Example 17.3, we solved a second-order equation with the term $\cos 2t$ on the right by choosing constants p and q so that the expression $p\cos 2t + q\sin 2t$ would fit the equation. We shall explain another important method for obtaining solutions, which derives the required real solutions from complex solutions of a related equation.

First of all, consider the general differential equation

$$\frac{d^2 X}{dt^2} + b\,\frac{dX}{dt} + cX = a\,e^{j\beta t}, \tag{17.2}$$

where b, c, a, and β are real constants, and j is the complex element. Since the forcing term is an exponential, we shall test for a particular solution of (17.2) having the form

$$X(t) = P\,e^{j\beta t}$$

as in Example 17.1 (but this time we must expect that P will be a complex constant).

Example 17.8. Find a particular (complex) solution of the complex differential equation $\dfrac{d^2 X}{dt^2} + \dfrac{dX}{dt} + X = 3\,e^{2jt}$.

Look for a solution of the form $X(t) = P\,e^{2jt}$. To find P, substitute this expression into the left-hand side of the differential equation;

$$(2j)^2 P\,e^{2jt} + (2j)P\,e^{2jt} + P\,e^{2jt} = P(-4 + 2j + 1)\,e^{2jt}$$

$$= P(-3 + 2j)\,e^{2jt}.$$

This must be the same as the right-hand side of the equation, $3\,e^{2jt}$, for all values of t. Therefore, $P(-3 + 2j) = 3$, so

$$P = \frac{3}{-3 + 2j} = \frac{3(-3 - 2j)}{(-3)^2 + 2^2} = -\tfrac{3}{13}(3 + 2j).$$

Therefore $X(t) = -\tfrac{3}{15}(3 + 2j)\,e^{2jt}$ is a particular solution. When expanded, it becomes

$$X(t) = -\tfrac{3}{15}(3 + 2j)(\cos 2t + j \sin 2t)$$
$$= -\tfrac{3}{15}(3 \cos 2t - 2 \sin 2t) + j\{-\tfrac{3}{15}(2 \cos 2t + 3 \sin 2t)\}.$$

Consider next the real equation for $x(t)$:

$$\frac{d^2 x}{dt^2} + b\frac{dx}{dt} + cx = a \cos \beta t, \qquad (17.3)$$

where b, c, a, β are all real. We know that

$$\cos \beta t = \text{Re } e^{j\beta t} \qquad (17.4)$$

(see (6.8)). Therefore, if we can find a particular solution $X(t)$ of the complex equation (17.2), its real part will solve the corresponding real equation (17.3).

Example 17.9. Find a particular solution of the equation

$$\frac{d^2 x}{dt^2} + \frac{dx}{dt} - 2x = 2 \cos 3t.$$

This is the same problem as Example 17.3, reworked so that the methods can be compared. Since $\cos 3t = \text{Re } e^{3jt}$, the corresponding complex equation for $X(t)$ is

$$\frac{d^2 X}{dt^2} + \frac{dX}{dt} - 2X = 2\,e^{3jt}.$$

To find a particular solution of this new equation, try $X(t) = P\,e^{3jt}$:

$$\frac{d^2 X}{dt^2} + \frac{dX}{dt} - 2X = 9j^2 P\,e^{3jt} + 3jP\,e^{3jt} - 2P\,e^{3jt}$$
$$= (9j^2 + 3j - 2)P\,e^{3jt} = (-11 + 3j)P\,e^{3jt}.$$

This must equal $2\,e^{3jt}$ for all values of t, so

$$P = \frac{2}{-11 + 3j} = \frac{2(-11 - 3j)}{(-11)^2 + 3^2} = -(\tfrac{11}{65} + \tfrac{3}{65}j).$$

Therefore we have a complex solution of the complex equation:

$$X(t) = -(\tfrac{11}{65} + \tfrac{3}{65}j)\,e^{3jt} = -(\tfrac{11}{65} + \tfrac{3}{65}j)(\cos 3t + j \sin 3t).$$

For $x(t)$, we require only the real part of this expression:

$$x(t) = \text{Re } X(t) = -\tfrac{11}{65}\cos 3t + \tfrac{3}{65}\sin 3t,$$

which is what we obtained in Example 17.3 for the same problem.

In the case when the right-hand side of the equation has the form $a \sin \omega t$, the calculation is the same, but the imaginary part of the complex solution must be extracted instead of the real part. The following example demonstrates also how right-hand sides of the form

$$a\, e^{\alpha t} \cos \beta t, \qquad a\, e^{\alpha t} \sin \beta t$$

can be handled in the same way.

Example 17.10. Find a solution of $\dfrac{d^2 x}{dt^2} + x = e^{-2t} \sin 3t$.

Use the fact that

$$e^{-2t} \sin 3t = \operatorname{Im} (e^{-2t}\, e^{3jt}) = \operatorname{Im} e^{(-2+3j)t}.$$

Therefore, consider the corresponding complex equation

$$\frac{d^2 X}{dt^2} + X = e^{(-2+3j)t}.$$

To find a solution, try the form

$$X(t) = P\, e^{(-2+3j)t}.$$

We find in the usual way that

$$(-2 + 3j)^2 P\, e^{(-2+3j)t} + P\, e^{(-2+3j)t} = e^{(-2+3j)t}$$

for all values of t. Therefore

$$P = \frac{1}{(-2+3j)^2 + 1} = \frac{-1}{4(1+3j)} = -\tfrac{1}{40}(1 - 3j)$$

and

$$X(t) = -\tfrac{1}{40}(1 - 3j)\, e^{(-2+3j)t}.$$

If we take the imaginary part of $X(t)$, we obtain a solution of the original equation:

$$x(t) = \operatorname{Im}[-\tfrac{1}{40}(1 - 3j)(e^{-2t}\, e^{3jt})]$$
$$= -\tfrac{1}{40}\, e^{-2t}(-3 \cos 3t + \sin 3t).$$

The same result could be obtained by substituting

$$x(t) = p\, e^{-2t} \cos 3t + q\, e^{-2t} \sin 3t,$$

but this would be a very laborious and error-prone process.

The method is particularly advantageous when the coefficients are general constants. The following equation will be important in Chapter 18.

Example 17.11. Find a particular solution of

$$\frac{d^2 x}{dt^2} + 2k\frac{dx}{dt} + \omega_0^2 x = a \cos \omega t, \quad \text{where } a > 0.$$

Since $\cos \omega t = \mathrm{Re}\, \mathrm{e}^{\mathrm{j}\omega t}$, first find a solution of

$$\frac{\mathrm{d}^2 X}{\mathrm{d}t^2} + 2k \frac{\mathrm{d}X}{\mathrm{d}t} + \omega_0^2 X = a\, \mathrm{e}^{\mathrm{j}\omega t}.$$

By substituting $X(t) = P\, \mathrm{e}^{\mathrm{j}\omega t}$, we find that

$$P = a/[(\omega_0^2 - \omega^2) + \mathrm{j}(2k\omega)].$$

It is easier if we put P into polar coordinates: $P = |P|\, \mathrm{e}^{\mathrm{j}\phi}$, where

$$|P| = |a/[(\omega_0^2 - \omega^2) + \mathrm{j}(2k\omega)]|$$
$$= a/[(\omega_0^2 - \omega^2)^2 + (2k\omega)^2]^{\frac{1}{2}},$$

and

$$\phi = \arg[(\omega_0^2 - \omega^2) - \mathrm{j}(2k\omega)],$$

since $a > 0$. Then we obtain

$$x(t) = \mathrm{Re}(P\, \mathrm{e}^{\mathrm{j}\omega t}) = \frac{a \cos(\omega t + \phi)}{[(\omega_0^2 - \omega^2)^2 + (2k\omega)^2]^{\frac{1}{2}}},$$

where ϕ is the polar angle of the point $((\omega_0^2 - \omega^2), -2k\omega)$ on an Argand diagram.

Particular solution of $\dfrac{\mathrm{d}^2 x}{\mathrm{d}t^2} + b \dfrac{\mathrm{d}x}{\mathrm{d}t} + cx = f(t)$

(a) $f(t) = a \cos \beta t$ or $a \sin \beta t$
(i.e. $\alpha = 0$). Put $X(t) = P\, \mathrm{e}^{\mathrm{j}\beta t}$ to solve

$$X'' + bX' + cX = a\, \mathrm{e}^{\mathrm{j}\beta t}. \tag{17.5}$$

Then $x(t) = \mathrm{Re}\, X(t)$ or $\mathrm{Im}\, X(t)$, corresponding to $\cos \beta t$ or $\sin \beta t$ respectively.

(b) $f(t) = a\, \mathrm{e}^{\alpha t} \cos \beta t$ or $a\, \mathrm{e}^{\alpha t} \sin \beta t$.
Solve $X'' + bX' + cX = a\, \mathrm{e}^{(\alpha + \mathrm{j}\beta)t}$, and continue as in (a).

17.3 Particular solutions: exceptional cases

There are exceptional cases for each of the three rules (17.1), when the suggested substitution does not give any result because the trial function delivers zero when it is substituted into the left-hand side. This means (as with the similar exceptional case of a single solution of the characteristic equation, (16.10)) that the trial solution must have a different form.

The most important exception is the case of the equation

$$\frac{\mathrm{d}^2 x}{\mathrm{d}t^2} + \beta^2 x = a \cos \beta t \quad (\text{or } a \sin \beta t).$$

Note that β occurs on both sides of the equation. This is a special case of Example 17.11 in which $k = 0$ and $\omega^2 = \omega_0^2 = \beta^2$. The rule in (17.1) suggests substituting

$$x(t) = p \cos \beta t + q \sin \beta t,$$

and choosing p and q so that the two sides match. But we already know that this is a solution of the unforced equation $d^2x/dt^2 + \beta^2 x = 0$, and the inevitable zero that we get on making the substitution cannot be matched to $a \cos \beta t$ on the right.

In this case, the solution is quite different. The following results can be confirmed by direct substitution.

Particular solutions: two exceptional cases

(a) $\dfrac{d^2x}{dt^2} + \beta^2 x = a \cos \beta t$: solution

$$x(t) = \frac{a}{2\beta} t \sin \beta t. \tag{17.6}$$

(b) $\dfrac{d^2x}{dt^2} + \beta^2 x = a \sin \beta t$: solution

$$x(t) = -\frac{a}{2\beta} t \cos \beta t.$$

Example 17.12. Find a particular solution of $\dfrac{d^2x}{dt^2} + 9x = 5 \sin 3t$.

Here, $x = p \cos 3t$ and $x = q \sin 3t$ both give $d^2x/dt^2 + 9x = 0$, so the standard solution form does not work. From (17.6), with $\beta = 3$ and $K = 5$, the required solution is

$$x(t) = -\frac{5}{2 \times 3} t \cos 3t = -\tfrac{5}{6} t \cos 3t.$$

This solution is sketched in Fig. 17.1. Unlike the ordinary sine and cosine type solutions, it grows indefinitely. Such solutions have an important physical significance described in Chapter 18.

There are other exceptional cases that are not so frequently encountered; some examples are given among the problems at the end of the chapter.

17.4 The general solution of forced equations
Consider the equation

$$\frac{d^2x}{dt^2} - x = -2 \cos t. \tag{17.7}$$

A particular solution, $x_p(t)$, say, is

$$x_p(t) = \cos t. \tag{17.8}$$

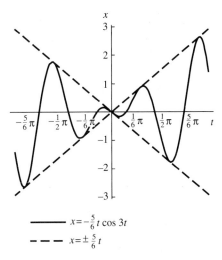

$$\text{———} \quad x = -\tfrac{5}{6} t \cos 3t$$
$$\text{- - -} \quad x = \pm \tfrac{5}{6} t$$

Fig. 17.1

From earlier experience, we should expect other solutions. In order to find some, consider what happens when we substitute various functions $x(t)$ in the expression

$$\frac{\mathrm{d}^2}{\mathrm{d}t^2} x(t) - x(t). \tag{17.9}$$

For example, when we put $x(t) = \cos t$, we obtain

$$\frac{\mathrm{d}^2}{\mathrm{d}t^2} \cos t - \cos t = -2 \cos t,$$

as demanded by (17.7).

Suppose now that we can find another function, $x(t) = x_c(t)$ say, which produces *zero* out of (17.9). For example, $x(t) = x_c(t) = \mathrm{e}^t$ makes $\mathrm{d}^2 x / \mathrm{d}t^2 - x$ equal to zero. It is then obvious that if we put

$$x(t) = x_p(t) + x_c(t) = \cos t + \mathrm{e}^t$$

into (17.9), we again obtain $(-2 \cos t)$ on the right: that is to say, we have found another solution of (17.7).

But we already know, from Chapter 16, *all the functions* $x_c(t)$ that give zero when they are put into (17.9): they are the solutions of the equation

$$\frac{\mathrm{d}^2 x}{\mathrm{d}t^2} - x = 0, \tag{17.10}$$

and are given by

$$x(t) = x_c(t) = A \,\mathrm{e}^t + B \,\mathrm{e}^{-t},$$

where A and B are any constants. Therefore

$$x(t) = \cos t + A \,\mathrm{e}^t + B \,\mathrm{e}^{-t} \tag{17.11}$$

is always a solution of (17.7). The differential equation (17.10) is called the **unforced equation corresponding to the original equation** (17.7), and its solutions $x(t)$ are called the **complementary functions** of the problem (they complement or extend the particular solution of (17.7) that we obtained).

To show that we have obtained all possible solutions of (17.7), take the particular solution $\cos t$ that we obtained, and suppose that $x_p(t)$ is any other solution of (17.7). Evidently the function $x(t) = x_p(t) - \cos t$ satisfies (17.10), so $x(t)$ must be a complementary function. Therefore

$$x_p(t) = \cos t + \text{(a complementary function)},$$

so $x_p(t)$ must be one of the solutions already expressed by (17.11). Therefore (17.11) is the **general solution** of (17.7). Exactly the same argument would have applied in the general case:

General solution of $\dfrac{d^2x}{dt^2} + b\dfrac{dx}{dt} + cx = f(t)$

(i) Obtain any particular solution, $x_p(t)$.
(ii) Obtain all the solutions $Ax_{c1}(t) + Bx_{c2}(t)$ of the corresponding unforced equation

$$\frac{d^2x}{dt^2} + b\frac{dx}{dt} + cx = 0 \qquad\qquad \textbf{(17.12)}$$

(the complementary functions).
The sum of these gives the general solution:

$$x(t) = x_p(t) + Ax_{c1}(t) + Bx_{c2}(t).$$

The theory and method is exactly the same for **linear equations** of the first order, and of any order, whether the coefficients are constant or not.

Example 17.13. Find the general solution of $\dfrac{d^2x}{dt^2} + 4x = 3\cos 5t$.

Particular solution $x_p(t)$. Looking forward into the calculation, it can be seen that the solution needs no $\sin 5t$ term. Therefore try

$$x_p(t) = p\cos 5t.$$

The substitution into the equation gives

$$p(-25\cos 5t) + 4p\cos 5t = 3\cos 5t$$

for all t, so $p = -\frac{1}{7}$. Therefore

$$x(t) = -\tfrac{1}{7}\cos 5t.$$

Complementary functions $x_c(t)$. We require the solutions $x_c(t)$ of the corresponding unforced equation $d^2x/dt^2 + 4x_c = 0$. Try for solutions of the form $x_c(t) = p\,e^{mt}$. The substitution produces the characteristic equation $m^2 + 4 = 0$. Therefore $m = \pm 2j$, so a pair of solutions

$$(e^{2jt}, e^{-2jt})$$

constitutes a complex basis. To get a real basis, choose either one, say e^{2jt}, and find its real and imaginary parts. These are

$$\cos 2t, \quad \sin 2t,$$

and this is the required real basis. Therefore, all the complementary functions are given by

$$x_c(t) = A \cos 2t + B \sin 2t \quad (A \text{ and } B \text{ arbitrary constants}).$$

General solution. This is the sum of the two:

$$x(t) = -\tfrac{1}{7} \cos 5t + A \cos 2t + B \sin 2t.$$

As explained in Section 17.3, the straightforward trial method for the complementary functions fails if the forcing term on the right is already a complementary function, so it can be a useful tactic to look at the complementary functions first. The following example contains this feature, and is also an initial-value problem.

Example 17.14. (a) Obtain the general solution of

$$\frac{d^2 x}{dt^2} + 4x = 3 + 2 \cos 2t.$$

(b) Find the particular solution for which $x = 0$ and $dx/dt = 0$ when $t = 0$.

(a) *Complementary functions* $x_c(t)$. These are the solutions of $d^2 x_c/dt^2 + 4x_c = 0$. We found them in Example 17.13: they are

$$x_c(t) = A \cos 2t + B \sin 2t \quad \text{with } A \text{ and } B \text{ arbitrary}.$$

Particular solution $x_p(t)$. There are two terms on the right, so find a particular solution for each term separately, and add them.

Therefore for a solution, $x_{p1}(t)$ say, of $d^2 x_{p1}/dt^2 + 4x_{p1} = 3$, we can obviously take

$$x_{p1}(t) = \tfrac{3}{4}.$$

Corresponding to the other term, we need a solution, $x_{p2}(t)$ say, of $\dfrac{d^2 x_{p2}}{dt^2} + 4x_{p2} = 2 \cos 2t$. We should normally expect a solution of the form $p \cos 2t + q \sin 2t$. However, looking at the complementary functions we found, this function is already a complementary function; so we have the exceptional case (17.6), which gives

$$x_{p2}(t) = \tfrac{1}{2}t \sin 2t.$$

Therefore a particular solution of the original equation is

$$x_p(t) = x_{p1}(t) + x_{p2}(t) = \tfrac{3}{4} + \tfrac{1}{2}t \sin 2t.$$

General solution.

$$x(t) = A \cos 2t + B \sin 2t + \tfrac{3}{4} + \tfrac{1}{2}t \sin 2t.$$

(b) *Initial-value problem.* We require also dx/dt:

$$\frac{dx}{dt} = -2A \sin 2t + 2B \cos 2t + \tfrac{1}{2} \sin 2t + t \cos 2t.$$

The initial conditions prescribe $x(0) = 0$, or

$$A + \tfrac{3}{4} = 0,$$

and $x'(0) = 0$, or

$$2B = 0.$$

Therefore $A = -\tfrac{3}{4}$ and $B = 0$, so the required particular solution is

$$x(t) = -\tfrac{3}{4} \cos 2t + \tfrac{3}{4} + t \sin 2t.$$

17.5 First-order linear equations with a variable coefficient

So far, the coefficient c in the equation $dx/dt + cx = f(t)$ has been a constant. We shall now suppose c to be variable; call it $g(t)$:

$$\frac{dx}{dt} + g(t)x = f(t). \tag{17.13}$$

The equation is of linear type (no squares, products, etc. between terms involving x are present), and the idea of obtaining a general solution by adding complementary functions to any particular solution still holds good. However, it is nearly impossible to guess suitable trial functions so we need a new approach to finding solutions.

If we could express the left-hand side $dx/dt + g(t)x$ of the equation as

$$\frac{d}{dt} (\text{something}),$$

then the equation would be easy to solve. This cannot be done, but we can instead do the next best thing. This is to obtain a certain function $I(t)$, called an **integrating factor**, such that

$$I(t)\left(\frac{dx}{dt} + g(t)x\right) = \frac{d}{dt}[I(t)x] \tag{17.14}$$

identically (that is, for every function $x(t)$ and for all values of t).

The following example shows the meaning of this idea and the way it is used.

Example 17.15. (a) Show that $I(t) = e^t$ is an integrating factor for the expression $dx/dt + x$. (b) Use it to find the general solution of the equation $dx/dt + x = e^{2t}$.

(a) We have to confirm (17.14), that

$$e^t\left(\frac{dx}{dt} + x\right) = \frac{d}{dt}(e^t x).$$

Work from the right-hand side. Differentiate the product $e^t x$:

$$\frac{d}{dt}(e^t x) = e^t \frac{dx}{dt} + e^t x = e^t \left(\frac{dx}{dt} + x \right),$$

which is the same as the left-hand side, so e^t is an integrating factor.

(b) Multiply both sides of the differential equation by e^t:

$$e^t \left(\frac{dx}{dt} + x \right) = e^t e^{2t} = e^{3t}.$$

Because of the result in (a), we can write this as

$$\frac{d}{dt}(e^t x) = e^{3t}.$$

Therefore

$$e^t x = \int e^{3t} \, dt = \tfrac{1}{3} e^{3t} + A \quad (A \text{ arbitrary}),$$

or

$$x = \tfrac{1}{3} e^{2t} + A e^{-t}.$$

To find a general expression for an integrating factor, refer back to the definition (17.14); the property of the integrating factor $I(t)$ is

$$I(t) \left(\frac{dx}{dt} + g(t)x \right) = \frac{d}{dt}[I(t)x].$$

This is the same as

$$I(t) \frac{dx}{dt} + I(t)g(t)x = I(t) \frac{dx}{dt} + x \frac{dI(t)}{dt},$$

or

$$I(t)g(t) = \frac{dI(t)}{dt}$$

(after cancelling $I(t) \dfrac{dx}{dt}$, and dividing through by x). This can be written

$$\frac{1}{I(t)} \frac{dI(t)}{dt} = g(t)$$

or

$$\frac{d \ln I(t)}{dt} = g(t).$$

Therefore

$$\ln I(t) = \int^t g(t) \, dt,$$

or

$$I(t) = e^{\int g(t) \, dt}.$$

(In the case of Example 17.15, we had $g(t) = 1$, and the present formula gives $I(t) = e^{\int dt} = e^{t+C}$; the choice $C = 0$ gives the integrating factor suggested – any other choice would do.)

Integrating factor for the equation

$$\frac{dx}{dt} + g(t)x = f(t).$$

Put $I(t) = e^{\int g(t)\,dt}$;

then $I(t)\left(\dfrac{dx}{dt} + g(t)x\right) \equiv \dfrac{d}{dt}[I(t)x].$

(17.15)

Solution of $\dfrac{dx}{dt} + g(t)x = f(t).$

Multiply both sides by $I(t)$ (see (17.15)): the equation becomes

$$\frac{d}{dt}[I(t)x(t)] = I(t)f(t);$$

then $I(t)x(t) = \displaystyle\int I(t)f(t)\,dt + C$, giving $x(t)$.

(17.16)

Example 17.16. Find the general solution of $\dfrac{dx}{dt} - \dfrac{1}{t}x = t^3$, for $t > 0$.

Here $g(t) = -1/t$. Then $\displaystyle\int g(t)\,dt = C - \ln t$ (we need only consider t positive), so that

$$I(t) = e^{-\ln t} = t^{-1},$$

where we have chosen $C = 0$ for convenience. Multiply both sides by $I(t) = t^{-1}$:

$$t^{-1}\left(\frac{dx}{dt} - \frac{1}{t}x\right) = t^{-1}t^3 = t^2.$$

By (17.15), this can be written

$$\frac{d}{dt}(t^{-1}x) = t^2.$$

Therefore

$$t^{-1}x = \int t^2\,dt = \tfrac{1}{3}t^3 + C,$$

so that

$$x(t) = \tfrac{1}{3}t^4 + Ct.$$

The solution obviously falls into the shape

(particular solution) + (complementary function).

We should have gained nothing by considering negative t, or by adding an arbitrary constant, when working out $\int^1 \frac{1}{t} \, dt$: we only need *any* integrating factor, not all possible ones.

Notice particularly that, in the examples, we did not need to calculate or check the truth of a statement like

$$t^{-1}\left(\frac{dx}{dt} - \frac{1}{t}x\right) = \frac{d}{dt}(t^{-1}x).$$

We already know that t^{-1} is an integrating factor, and this is the very property that an integrating factor is designed to possess.

Be prepared to recognize this type of equation in disguised form, or when different letters are involved; for example,

$$\frac{dy}{dx} = \frac{x+y}{x+1}$$

is the same as

$$\frac{dy}{dx} - \frac{1}{x+1}y = \frac{x}{x+1}.$$

Problems, Chapter 17

17.1. Find a particular solution of each of the following equations by trial as in Section 17.1.
(a) $x' + x = 3\,e^{2t}$; (b) $x' - 3x = t^3 + 1$;
(c) $2x' + 3x = t + 3\,e^t$; (d) $x'' + x = 3\,e^{2t}$;
(e) $x'' - \frac{1}{4}x = 2\,e^t + 3\,e^{-t}$; (f) $x'' - 2x' + x = 3$;
(g) $x'' + 4x' - x = 3t^2 - t$;
(h) $x'' - x = 2\cos t$; (i) $2x'' + 3x = 2\sin 3t$;
(j) $2x'' + x' = \sin t - \cos t$;
(k) $x'' + 2x' + x = \cos 2t$;
(l) $\dfrac{d^2y}{dx^2} - y = 1 - 3\,e^{2x}$;
(m) $\dfrac{d^2y}{dx^2} - \dfrac{dy}{dx} + 2y = 3\sin 2x$.

17.2. Use the method of Section 17.2 to find a particular solution of the following.
(a) $x'' - x = 3\cos 2t$; (b) $x'' + x = 2\sin 3t$;
(c) $x'' + 2x' + x = 3\sin t$; (d) $x'' - x' - x = 3\cos t$;
(e) $2x'' + x' + 2x = 2\cos 2t$;
(f) $3x'' + 2x' + x = 2\sin 2t$;
(g) $x'' - 4x = e^{-t}\cos t$ (note: $e^{-t}\cos t = \operatorname{Re} e^{(-1+j)t}$);
(h) $x'' - 4x = 3\,e^t \sin 2t$ (note: $e^t \sin 2t = \operatorname{Im} e^{(1+2j)t}$);

(i) Show that a solution of $x'' + x' + 4x = 5\cos 3t$ is $(6/\sqrt{34})\cos(3t + \phi)$, where $\phi = -\arctan \frac{3}{5}$.

17.3. The following differential equations are examples of the exceptional cases treated in Section 17.3. Find a particular solution in each case.
(a) $x'' + x = 3\cos t$; (b) $x'' + 4x = 3\sin 2t$;
(c) $x'' + 4x = 1 + 3\cos 2t$; (d) $\dfrac{d^2y}{dx^2} + 9y = 2\sin 3x$.
(e) $\dfrac{d^2y}{dx^3} - 2\dfrac{dy}{dx} + 2y = e^x \cos x$.

17.4. The following are exceptional cases of types not described in Section 17.3. Find a particular solution for each.
(a) $x'' - x = e^t$; try a solution of the form $pt\,e^t$.
(b) $x'' - 2x' + x = e^t$; try a solution of the form $pt^2\,e^t$. (In this case, both e^t and $t\,e^t$ are complementary functions, so the form in (a) will not work.)
(c) Consider the simple differential equation $d^2x/dt^2 = t$. A first try with the form $pt + q$ suggested by (17.1) does not lead to a result. Try polynomials of higher degree than 1.

(d) $\dfrac{d^2y}{dx^2} + \dfrac{dy}{dx} = x$. The absence of a term in y causes
the second-degree trial function $px^2 + qx + r$ to fail.
Try a third-degree of polynomial instead.

(e) $x'' - 2x' + 2x = e^t \cos t$. Try $t\,e^t(p \cos t + q \sin t)$,
or modify the complex-number approach of Section
17.2 to obtain a particular solution.

(f) First-order equations also have exceptional cases.
Consider the equation $\dfrac{dx}{dx} - y = e^x$. (If you have read
as far as Section 17.5, you can also handle it by using
an integrating factor.)

17.5. Find the general solution of the following equations.

(a) $x'' + 9x = 3\,e^{2t}$;   (b) $x'' - 4x = 2\,e^{-t}$;
(c) $4x'' - x = 1 + 3 \cos 2t$;

(d) $\dfrac{d^2y}{dx^2} + 2\dfrac{dy}{dx} + 2y = 3$;

(e) $x'' - 2x' + 2x = 3 \sin 2t$;
(f) $4x'' - 2x' - 2x = 3t^2$.
(g) $x'' + x' = 2 - 3\,e^{-t} \cos t$;
(h) $2x'' + x' - x = \frac{1}{2}t + 3\,e^{-t}$;

(i) $\dfrac{d^2y}{dx^2} + y = 1 + 2\,e^{3x} + x^2$;

(j) $\dfrac{d^2y}{dx^2} + 2\dfrac{dy}{dx} + y = 3 \cos 2x + \sin 2x$;

(k) $\dfrac{d^2y}{dx^2} + 4\dfrac{dy}{dx} + 5y = e^{-x} \sin x$.

17.6. Use an integrating factor (Section 17.6) to find
the general solution of the following equations.

(a) $x' - 3x = 0$;   (b) $x' + 2x = 3$;
(c) $x' - 2tx = t$;   (d) $x' - t^{-1}x = t + t\,e^{-t}$;
(e) $x' - t^{-1}x = t - 1$;   (f) $tx' - 2x + 3 = 0$;

(g) $\dfrac{dy}{dx} + \dfrac{1}{x+1}\,y = \sin x$ (you will need to use integration by parts to perform the integration);

(h) $3\dfrac{dy}{dx} + \dfrac{1}{x}\,y = x$;   (i) $(x - 1)\dfrac{dy}{dx} - y = (x - 1)^2$;

(j) $x' - \dfrac{1}{t}\,x = \ln t$;   (k) $tx' - x = 1 + t$;

(l) $\dfrac{dy}{dx} = \dfrac{x + y}{x + 1}$;   (m) $x' + x \cos t = \cos t$;

(n) $x\dfrac{dy}{dx} = \dfrac{1 - y}{1 - x}$;   (o) $(1 - t^2)x' + tx = t$.

17.7. Show that the general solution of

$$\dfrac{dy}{dx} + \dfrac{1}{x}\,y = f(x)$$

is given by

$$y(x) = \dfrac{1}{x}\int xf(x)\,dx + \dfrac{C}{x},$$

where C is any constant. Find the solution of the
equation

$$\dfrac{dy}{dx} + \dfrac{1}{x}\,y = \ln x$$

for $x > 0$, for which $y = 0$ when $x = 1$.

17.8. (a) Use an integrating factor to show that the
general solution of

$$\dfrac{dy}{dx} + y = f(x)$$

is

$$y(x) = e^{-x}\int e^x f(x)\,dx + C\,e^{-x},$$

where C is an arbitrary constant.

(b) Show that the particular solution for which
$y = y_0$ when $x = 0$ is given by

$$y(x) = y_0 + e^{-x}\int_0^x e^u f(u)\,du.$$

17.9. (Newton cooling). An object is heated or cooled
above or below the ambient air temperature T_0.
Under certain physical assumptions, the body temperature T satisfies the equation

$$dT/dt = -k(T - T_0),$$

where k is a positive constant. Find the general
solution of the equation.

The body is at $100°\,C$ in an atmosphere at $40°\,C$.
After 3 minutes, its temperature is $85°\,C$. Find the
value of k, and determine when the body will reach
$60°\,C$.

18 Harmonic functions and the harmonic oscillator

18.1 Harmonic oscillations

Consider the equation

$$\frac{d^2x}{dt^2} + \omega^2 x = 0,$$

where we assume $\omega > 0$. Its solutions (see (16.14b)) are

$$x(t) = C \cos(\omega t + \phi) \tag{18.1}$$

where C and ϕ are any constants. We can write

$$x(t) = C \cos(\omega t + \phi) = C \cos \omega(t + \phi/\omega).$$

Therefore the graph of (18.1) is merely the graph $x = C \cos \omega t$ shifted, or **translated**, a distance ϕ/ω along the t axis; to the left if ϕ is positive, to the right if ϕ is negative. Sine functions are included in the collection (18.1), because $\sin \omega t = \cos(\omega t - \frac{1}{2}\pi)$. These functions are spoken of generally as **harmonic functions**.

In applications it is usual to adjust ϕ so that

$$C > 0 \qquad \text{and} \qquad -\pi < \phi \leqslant \pi \tag{18.2}$$

which can always be done without changing the function values described by the expression (18.1). We say then that the function (18.1) is in **standard form**.

Example 18.1. Express $-2 \cos(3t - \frac{7}{3}\pi)$ in standard form.

Note that $\cos(A + \pi) = -\cos A$. Therefore

$$-2 \cos(3t - \tfrac{7}{3}\pi) = 2 \cos(3t - \tfrac{7}{3}\pi + \pi) = 2 \cos(3t - \tfrac{4}{3}\pi).$$

We now have positive C, but ϕ is still out of range according to (18.2). To bring it within range increase it by 2π, which alters nothing:

$$2 \cos(3t - \tfrac{4}{3}\pi) = 2 \cos(3t - \tfrac{4}{3}\pi + 2\pi) = 2 \cos(3t + \tfrac{2}{3}\pi)$$

which is now in standard form.

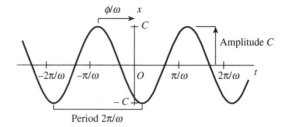

Fig. 18.1
$x = C \cos(\omega t + \phi)$; $c > 0$,
$-\pi < \phi \leqslant \pi$.

The features of the function $x(t) = C \cos(\omega t + \phi)$ are shown in Fig. 18.1. Assume that the expression is in standard form (18.2). The graph swings between $\pm C$, and C is its **amplitude**. It is periodic (see Section 1.6), repeating itself at intervals of length $2\pi/\omega$, which is its minimum **period**. The number of complete oscillations per unit time is the **frequency** (e.g. in cycles per second, or Hertz units), and

$$\text{Frequency} = (\text{period})^{-1} = \omega/2\pi. \qquad (18.3)$$

The parameter ω is **angular frequency**, often shortened merely to 'frequency'. The parameter ϕ is the **phase** or **phase angle**. As explained above, ϕ/ω represents the distance that the graph $x = C \cos \omega t$ has to be shifted to coincide with (18.1).

Frequently the independent variable represents **length** x instead of time t, as in a form such as $y = C \cos(\omega x + \phi)$. Then $2\pi/\omega$ is called **wavelength** rather than 'period', and $\omega/2\pi$ the **wave number** rather than 'frequency'.

Graphs of harmonic functions are often displayed by plotting x against the dimensionless variable

$$\tau = \omega t$$

(τ is the Greek letter 'tau') rather than against t. Thus τ will be the name of the new time-like axis, so that

$$x = C \cos(\tau + \phi)$$

The x, τ graph is drawn in Fig. 18.2.

The new graph has τ period 2π: it repeats itself when τ increases by 2π. It is the same as the graph of $x = C \cos \tau$ displaced through an interval in τ of length ϕ. Expressed in terms of the period or wavelength in Fig. 18.2, it is clear that $\phi = \pi$ represents a displace-

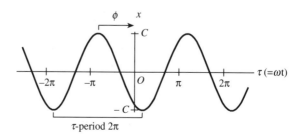

Fig. 18.2

ment of half a wavelength, $\phi = \frac{1}{2}\pi$ represents a displacement of a quarter of a wavelength, and so on.

18.2 Phase difference: lead and lag
Suppose that two oscillations have the same angular frequency ω, but are out of step because they have a different phase:
$$x_1(t) = C_1 \cos(\omega t + \phi_1), \qquad x_2(t) = C_2 \cos(\omega t + \phi_2).$$
Then they are said to be **out of phase** by an angle $\phi_2 - \phi_1$, or $\phi_1 - \phi_2$. More specifically, the following terminology is widely used in science and engineering applications:

> **Phase difference—lead and lag**
> $x(t) = C_1 \cos(\omega t + \phi_x)$, and $y(t) = C_2 \cos(\omega t + \phi_y)$
> are harmonic functions in standard form with the $\qquad$ **(18.4)**
> same circular frequency ω. If $\phi_1 > \phi_2$, then x is said
> to **lead** y, or y is said to **lag** x, by an angle $\phi_1 - \phi_2$.

The reason for these terms is illustrated in Example 18.2.

Example 18.2. If a voltage $v = v_0 \cos \omega t$ is applied to a coil having self-inductance L, the resulting current $i = \dfrac{v_0}{\omega L} \cos(\omega t - \frac{1}{2}\pi)$, so that v leads i, or i lags v, by $\frac{1}{2}\pi$ (or 90°). Illustrate the sense of the terms graphically for this case.

The curves in Fig. 18.3 are plotted against the variable $\tau = \omega t$ and represent
$$v = v_0 \cos \tau \qquad \text{and} \qquad i = \frac{v_0}{\omega L} \cos(\tau - \frac{1}{2}\pi).$$
Choose one of the curves, say the v curve, and select a prominent feature, say the maximum at B. Now search an interval within $\pm \pi$ of B (that is to say, within half a period on either side of B) for the corresponding feature of i. This is the maximum of i at A.

Now, as we move from left to right (time increasing) through the interval, B appears before A—that is to say, at a quarter period ($\frac{1}{2}\pi$) earlier than A. This will be true for any feature of v within its own *symmetrical* corresponding interval of $\pm \pi$. It is equivalent to saying that, when the two variables to be compared are in standard form, the one with the greater phase leads, and the other lags, by the phase difference (taken positively).

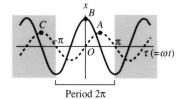

Fig. 18.3
——— $v = v_0 \cos \tau$;
- - - - - $i = (v_0/\omega L)\cos(\tau - \frac{1}{2}\pi)$.
The period inspected is symmetrical about the chosen feature at B.

In Example 18.2 it is **essential to limit the search to the prescribed single period**. Otherwise we could argue (see Fig. 18.3) that because, say, C appears before B, therefore i leads v, which is contrary to the definition. Also, notice that if one entity leads another it does not in the least imply that the first is to be taken as the cause of the second.

Suppose that two oscillations, having the same amplitude and frequency, differ in phase by π so they are displaced by half a period. If the oscillations are added together, there is total cancellation, as shown in Fig. 18.4. The following example shows what happens when the phase difference is less extreme.

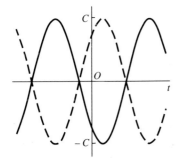

Fig. 18.4

——— $x = C\cos(\omega t + \phi)$;

- - - - - $x = C\cos(\omega t + \phi \pm \pi)$.

Example 18.3. Two waves described by $C\cos\omega t$ and $C\cos(\omega t + \phi)$ are superimposed (added). Show that the result is a harmonic wave of the same frequency, and show how the amplitude varies as ϕ varies between $\pm\pi$.

From Appendix B the sum can be written

$$C[\cos\omega t + \cos(\omega t + \phi)] = 2C\cos\tfrac{1}{2}\phi\cos(\omega t + \tfrac{1}{2}\phi).$$

This is a harmonic oscillation with angular frequency ω, phase $\tfrac{1}{2}\phi$, and amplitude $2C\cos\tfrac{1}{2}\phi$. As ϕ goes from $-\pi$ through zero to π, the amplitude goes from zero (cancellation) through the value $2C$ and back to zero.

This type of superposition is of importance in describing **interference** and **diffraction** phenomena. If the amplitudes of the components are not the same, a similar calculation applies (see Problem 18.4 at the end of this chapter).

18.3 Physical models of a differential equation

Figure 18.5a shows a piston of mass m running in a cylinder, controlled by a spring which obeys Hooke's law (a linear spring) and has stiffness s, acted on by an external force $F(t)$. The displacement of the piston from its equilibrium position is $x(t)$. Assume also that there is a frictional resistance proportional to the velocity:

$$\text{(Frictional resistance)} = K\frac{dx}{dt}, \qquad (K > 0).$$

The equation of motion, force equals mass times acceleration, becomes

$$F(t) - K\frac{dx}{dt} - sx = m\frac{d^2x}{dt^2},$$

or

$$\frac{d^2x}{dt^2} + \frac{K}{m}\frac{dx}{dt} + \frac{s}{m}x = \frac{F(t)}{m}. \tag{18.5}$$

This equation is of the type discussed in Chapter 17:

$$\frac{d^2x}{dt^2} + b\frac{dx}{dt} + cx = f(t), \tag{18.6}$$

in which

$$b = \frac{K}{m}, \qquad c = \frac{s}{m}, \qquad \text{and} \qquad f(t) = \frac{F(t)}{m}.$$

(a)

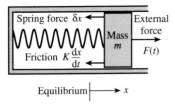

(b)

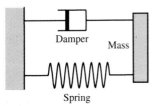

Fig. 18.5

(a) Mass–spring system. The arrows indicate the actual direction of the forces when $F(t)$, sx, and $K\,dx/dt$ take positive values. (b) Schematic representation: the spring and the frictional element must be in parallel.

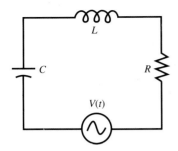

Fig. 18.6

Figure 18.6 represents an LCR circuit driven by a voltage source of zero impedance, $V(t)$. If Q is the charge on the capacitor, then

$$L\frac{d^2Q}{dt^2} + R\frac{dQ}{dt} + \frac{1}{C}Q = V(t),$$

or $\quad \dfrac{d^2Q}{dt^2} + \dfrac{R}{L}\dfrac{dQ}{dt} + \dfrac{1}{LC}Q = \dfrac{1}{L}V(t).$ (18.7)

Again, this is an equation of the type (18.6), with

$$x = Q, \qquad b = \frac{R}{L}, \qquad c = \frac{1}{LC}, \qquad f(t) = \frac{1}{L}V(t).$$

These two physical systems serve as **models** of the differential equation (18.6). They are also models of each other, for by choosing the same coefficients b and c and the same forcing term $f(t)$ the circuit would serve as a precise **analogue** of the piston and mimic its behaviour exactly. A vast number of systems share the governing equation (18.6), at least approximately. Such a system is called a **linear oscillator**.

18.4 Free oscillations of a linear oscillator

Suppose that in the piston system there is no external force acting, so that $F(t) = 0$ for all t. We shall choose a conventional notation that simplifies the algebra a little. Equation (18.6) will be written

$$\frac{d^2x}{dt^2} + 2k\frac{dx}{dt} + \omega_0^2 x = 0$$ (18.8)

(in which we have put $K/m = 2k$, $s/m = \omega_0^2$, $F(t) = 0$). This equation describes the **free oscillations** of the mass–spring system.

The parameter k is a measure of the amount of friction in the system. We shall consider the case when k is 'small'. This is not very meaningful because k is not dimensionless, so we could change our units so as to make it as large as we wished. The only thing that makes sense is to compare it with another parameter having the same dimensions. We specify that

$$k^2 < \omega_0^2.$$ (18.9)

We have already worked out this problem (see (16.15)). The solutions of (18.8) subject to (18.9) are given by

$$x(t) = C\,e^{-kt}\cos[(\omega_0^2 - k^2)^{\frac{1}{2}}t + \phi],$$ (18.10)

where C and ϕ are arbitrary. These are called the **free oscillations** or **natural oscillations** of the system represented by the equation.

If the friction, or so-called **damping**, is zero then $k = 0$ and the equation for the free oscillations becomes

$$\frac{d^2x}{dt^2} + \omega_0^2 x = 0$$ (18.11)

with solutions

$$x(t) = C \cos(\omega_0 t + \phi) \tag{18.12}$$

which are harmonic functions with circular frequency ω.

The friction, or damping, changes (18.12) into (18.10). The frequency is changed from ω_0 to $(\omega_0^2 - k^2)^{\frac{1}{2}}$, which is a small change if k is small, and the regular oscillations of (18.12) are caused to die away through the factor e^{-kt} in (18.10). The general effect is shown in Fig. 18.7. We say that the oscillation **decays exponentially** down to zero, when all the initial energy is used up on friction.

If $k^2 > \omega_0^2$, then there is a comparatively large amount of friction, and the form of the solution is different from (18.10) (see Problem 18.7). There are no oscillations; the $x(t)$ curve dies away without crossing the t axis more than once, as in a **dead-beat** electrical instrument or shock absorber.

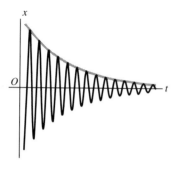

Fig. 18.7

18.5 Forced oscillations and transients

Return to the equation (18.5) for the mass–spring system with a nonzero external force $F(t)$ acting. As before, put $K/m = 2k$, $s/m = \omega_0^2$, and $F(t)/m = f(t)$, so that we get

$$\frac{d^2 x}{dt^2} + 2k \frac{dx}{dt} + \omega_0^2 x = f(t),$$

which is a forced equation of the type considered in Chapter 17.

We shall consider only the case when $f(t) = K \cos \omega t$:

$$\frac{d^2 x}{dt^2} + 2k \frac{dx}{dt} + \omega_0^2 x = K \cos \omega t, \tag{18.13}$$

and suppose as before that the friction (or resistance) is 'small':

$$k^2 < \omega_0^2.$$

The mass in the piston system is now subject to competing stimuli. Left to itself it would oscillate as in (18.10) with circular frequency $(\omega_0^2 - k^2)^{\frac{1}{2}}$, and finally come to rest. However, the forcing term is trying to make it oscillate with a different circular frequency ω. The result is described by the general solution of (18.13). This is equal to the sum of a particular solution (already worked out in Example 17.11 and the complementary functions which are the free oscillations described by (18.10):

General solution of the forced linear oscillator equation

$$\frac{d^2 x}{dt^2} + 2k \frac{dx}{dt} + \omega_0^2 x = K \cos \omega t.$$

$$x(t) = \frac{K}{[(\omega_0^2 - \omega^2)^2 + 4k^2\omega^2]^{\frac{1}{2}}} \cos(\omega t + \Phi) \tag{18.14}$$
$$+ C e^{-kt} \cos[(\omega_0^2 - k^2)^{\frac{1}{2}} t + \phi],$$

in which Φ is the polar angle of the point $(\omega_0^2 - \omega^2, -2k\omega)$, and C and ϕ are arbitrary.

The structure of (18.14) is very important: the general features are summarized in (18.15) below.

Forced oscillations of a linear oscillator

(A) The forced oscillation (first term of (18.14)) co-exists with a free oscillation (second term). The free oscillation proceeds as if no forcing term were present.

(B) The term representing the forced oscillation is harmonic, with the same frequency as the forcing term, but a different phase and amplitude. The term is invariable; initial conditions can have no effect on it since it contains no adjustable constants.   (18.15)

(C) The free oscillation term adjusts to any initial conditions by means of the constants C and ϕ.

(D) If k is positive, that is to say if there is any friction (or resistance in the case of a circuit), the free oscillations die away to zero due to the factor e^{-kt}. Therefore all solutions ultimately settle into the same steady oscillation, independently of the initial conditions.

On account of (D) the free oscillation is called a **transient** oscillation, and may show itself, for example, by a brief irregularity in the voltage or current upon switching an electrical apparatus.

Example 18.5. The circuit shown is initially quiescent and uncharged. Find the charge $Q(t)$ on the capacitor after switching the circuit on.

We shall rework the problem from first principles. The equation is

$$10^{-3}\frac{d^2Q}{dt^2} + 8\cdot10^{-3}\frac{dQ}{dt} + 10Q = 2\cos 90t,$$

or $\quad \dfrac{d^2Q}{dt^2} + 8\dfrac{dQ}{dt} + 10^4Q = 2\cdot10^3\cos 90t.$

Complementary functions Q_c (*natural oscillation*). The characteristic equation is $m^2 + 8m + 10^4 = 0$, so that $m = -4 + 91.92j$ and the complementary functions are $Q_c = B\,e^{-4t}\cos(91.92t + \phi)$, where B and ϕ are arbitrary.

Particular solution Q_p (*forced oscillation*). Look for a solution to the corresponding complex equation

$$\frac{d^2X}{dt^2} + 8\frac{dX}{dt} + 10^4X = 2\times10^3\,e^{90jt},$$

and take its real part. By trying a solution of the form $X(t) = P\,e^{90jt}$ we obtain $P = 0.9205 - 0.3488j$. In polar coordinates this becomes $P = 0.9843\,e^{-0.3622j}$. The corresponding complex solution, in polar

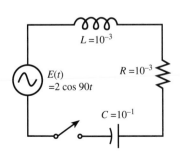

Fig. 18.8

$L = 10^{-3}$

$E(t)$
$= 2\cos 90t$

$R = 10^{-3}$

$C = 10^{-1}$

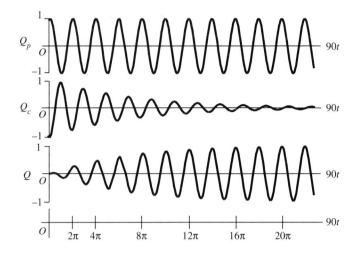

Fig. 18.9
(a) Forced oscillation,
$Q_p = 0.984 \cos(90t - 0.362)$.
(b) Transient, $Q_c =$
$0.996 \, e^{-4t} \cos(91.92t + 2.749)$.
(c) Total oscillation,
$Q = Q_p + Q_c$.

coordinates, is

$$X(t) = 0.9843 \, e^{(90t - 0.3622)j}.$$

Therefore the particular solution is

$$Q_p(t) = \text{Re } X(t) = 0.9843 \cos(90t - 0.362).$$

The general solution. This is

$$Q(t) = 0.984 \cos(90t - 0.362) + B \, e^{-4t} \cos(91.92t + \phi).$$

Initial conditions. At $t = 0$, Q and dQ/dt are zero. After obtaining dQ/dt and substituting $t = 0$ into Q and dQ/dt, we obtain the equations

$$B \cos \phi = -0.9204,$$

$$4 \cos \phi + 91.92B \sin \phi = 31.389.$$

The solution is $B = 0.996$, $\phi = 2.749$, so $Q(t)$ is given by

$$Q(t) = 0.984 \cos(90t - 0.362) + 0.996 \, e^{-4t} \cos(91.92t + 2.749).$$

Figure 18.9 shows the individual contributions of the two terms.

18.6 Resonance
Return to equation (18.14) for the linear oscillator and its solutions and examine the forced oscillation, which is all that is left after the transient has died away. Its amplitude, A say, is given by

$$A = \frac{K}{[(\omega_0^2 - \omega^2)^2 + 4k^2\omega^2]^{\frac{1}{2}}}.$$

Different values for the forcing frequency ω will produce different amplitudes; some values of ω will be more effective than others in generating a large amplitude.

Regard ω_0 and k as representing the fixed characteristics of some kind of system, and consider an experiment in which we try to excite it with a controllable input $K \cos \omega t$, keeping K constant but trying various values of ω. The amplitude A will be greatest when $(\omega_0^2 - \omega^2)^2 + 4k^2\omega^2 = g(\omega)$ say, is a minimum with respect to the variable ω. It is found by solving $\mathrm{d}g/\mathrm{d}\omega = 0$ (see Problem 18.15), that the minimum occurs when

$$\omega^2 = \omega_0^2 - 2k^2.$$

When ω takes that value the amplitude A will take its greatest possible value for the given K and ω, given by

$$A = \frac{K}{2k(\omega_0^2 - k^2)^{\frac{1}{2}}}.$$

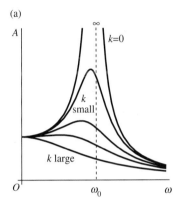

(a)

Figure 18.10 shows schematically how the amplitude A, and also the phase Φ in (18.14), vary with forcing frequency ω. Different curves are obtained according to the amount of friction or damping (or resistance in the case of a circuit) in the system, measured by the size of k; as the damping decreases, the maximum increases. When the condition for a maximum is satisfied, the system is said to be in a state of **resonance**.

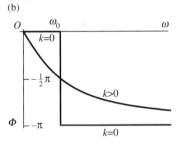

(b)

Fig. 18.10

Resonating system

$$\frac{\mathrm{d}^2 x}{\mathrm{d}t^2} + 2k\frac{\mathrm{d}x}{\mathrm{d}t} + \omega_0^2 x = K \cos \omega t.$$

Forced amplitude $A = \dfrac{K}{[(\omega_0^2 - \omega^2)^2 + 4k^2\omega^2]^{\frac{1}{2}}}.$ (18.16)

Resonant frequency $\omega^2 = \omega_0^2 - 2k^2.$

Resonance amplitude $\dfrac{K}{2k(\omega_0^2 - k^2)^{\frac{1}{2}}}$

A physical feeling for the buildup of a large amplitude can be obtained by thinking of a child being pushed on a swing by two people, one on either side of the swing. The method is to push the swing the way it wants to go, and not to work against it. This is best done by pushing it, forward and backward alternately, when it is at the bottom of its path. The driving frequency is then the same as the natural frequency of the swing, and the driving cycle is a quarter of a period out of phase with the swing's cycle, because the force is a maximum when the displacement is a minimum. In terms of (18.16), k is assumed small, so that $\omega^2 = \omega_0^2$ very nearly, and the phase difference Φ is nearly $\frac{1}{2}\pi$ or a quarter of a period, the forcing term leading the response by this amount.

Suppose next that there is **zero friction**;

$$k = 0.$$

so that

$$\frac{d^2x}{dt^2} + \omega_0^2 x = K \cos \omega t, \tag{18.17}$$

and from (18.16) the forced amplitude A is

$$A = \frac{K}{\omega_0^2 - \omega^2}. \tag{18.18}$$

The natural frequency of this system is exactly ω_0. When ω (the forcing frequency) gets close to ω_0, the amplitude A can become very large, approaching infinity as ω approaches ω_0: see Fig. 18.10(a).

When $\omega = \omega_0$ the equation becomes

$$\frac{d^2x}{dt^2} + \omega_0^2 x = K \cos \omega_0 t, \tag{18.19}$$

and apparently $A = \infty$. This result cannot be said to describe a steady solution of (18.19), but must be reconcilable with (18.19) in some way. In fact it is the 'exceptional case' of equation (17.6(a)), and has a solution

$$x(t) = \frac{K}{2\omega_0} t \sin \omega_0 t, \tag{18.20}$$

shown in Fig. 17.1. This particular solution conveniently satisfies the initial conditions

$$x(0) = 0, \qquad x'(0) = 0, \tag{18.21}$$

that is to say, the conditions for initial quiescence. It therefore represents a system without friction and in a state of resonance, which starts up from rest. Its oscillations grow steadily to infinity due to the factor t in (18.20). The equation does not have any solutions corresponding to steady forced oscillations, such as we found earlier in systems having even a small amount of friction.

18.7 Nearly linear systems
Consider the pendulum of Fig. 18.11. It consists of a weightless rod of length l, pivoted at the top and carrying a point mass m at the lower end. It makes an angle $\theta(t)$ with the vertical. The equation of motion is

$$\frac{d^2\theta}{dt^2} + \frac{g}{l} \sin \theta = 0, \tag{18.22}$$

where g is the gravitational constant.

This equation is **nonlinear** since $\sin \theta$ is not of the form $a\theta + b$, so the methods of Chapter 17 do not apply to it. However, the

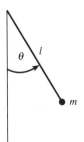

Fig. 18.11

Taylor series for $\sin \theta$ begins:

$$\sin \theta = \theta - \tfrac{1}{6}\theta^3 + \cdots,$$

(θ in radians), so provided that θ remains small enough we can approximate $\sin \theta$ by

$$\sin \theta \approx \theta.$$

The error is about 10% when $\theta = 45°$, and 0.1% at $5°$. Put this into (18.22); we obtain the **approximate linearized equation**

$$\frac{\mathrm{d}^2\theta}{\mathrm{d}t^2} + \frac{g}{l}\theta = 0. \tag{18.23}$$

The general solution is $\theta(t) = C \cos[(g/l)^{\frac{1}{2}}t + \phi]$. The values of C and ϕ will depend on how it was set going; the initial conditions amount to prescribing the position and angular velocity at $t = 0$. However, C must be small for the approximation to be justified.

Exactly linear equations are uncommon. Most frequently they occur as the result of a simplifying approximation such as we carried out for the pendulum. Usually some function in the equation is linearized at the expense of a restriction on the dependent variable.

Example 18.6. A mass m is fixed at the midpoint of a piece of elastic having natural length l and stiffness s. The elastic is stretched between two points a distance $L > l$ apart. Find the period of small lateral vibrations.

(See Fig. 18.12.) The extension e of the branch AC is $AC - \tfrac{1}{2}l$, so the tension T in either branch is

$$T = se = s(AC - \tfrac{1}{2}l).$$

The total restoring force F is $2T \sin \theta$, and

$$\sin \theta = NC/AC = x/AC,$$

so $\quad F = 2s(AC - \tfrac{1}{2}l)x/AC = 2s(1 - l/2AC)x.$

The equation of motion is

$$m\frac{\mathrm{d}^2x}{\mathrm{d}t^2} = -F,$$

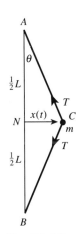

Fig. 18.12
A and B are fixed a distance L apart where $L > l$; the mass m at C is displaced from equilibrium at N by a distance $x(t)$.

so we must put AC in terms of x. Now

$$AC = (\tfrac{1}{4}L^2 + x^2)^{\frac{1}{2}},$$

so the equation will be nonlinear. However, if the oscillations which we expect are of small amplitude compared with L, we can put

$$AC \simeq \tfrac{1}{2}L$$

with an error of something like $2x^2/L^2$, and the approximation to the restoring force becomes

$$F \simeq 2s(1 - l/L)x.$$

The equation of motion becomes approximately

$$m \frac{d^2x}{dt^2} = -2s(1 - l/L)x,$$

or $$\frac{d^2x}{dt^2} + \frac{2s}{m}(1 - l/L)x = 0.$$

This is the linearized equation, good for small amplitudes. It has solutions

$$x(t) = C \cos\left\{ \left[\frac{2s}{m} \left(1 - \frac{l}{L} \right) \right]^{\frac{1}{2}} t + \phi \right\},$$

where C and ϕ are arbitrary. The approximate period is

$$2\pi \bigg/ \left[\frac{2s}{m} \left(1 - \frac{l}{L} \right) \right]^{\frac{1}{2}}.$$

It is interesting to consider the case when the string is unstretched in the equilibrium position, so that $L = l$. The resulting equation cannot be straightforwardly linearized.

Problems

18.1. Express the following in standard amplitude–phase form $C \cos(\omega t + \phi)$, with $C > 0$ and $-\pi < \phi \leqslant \pi$.

(a) $3 \cos(3t + \frac{3}{2}\pi)$;   (b) $3 \cos(\omega t - 3\pi)$;
(c) $2 \sin 3t$;   (d) $3 \sin(2t + \frac{1}{2}\pi)$;
(e) $-3 \cos(2t - \frac{1}{2}\pi)$;   (f) $-4 \cos(2t + \frac{1}{4}\pi)$;
(g) $-\sin t$;   (h) $3 \cos 2t + 4 \sin 2t$;
(i) $\cos 2t + \cos(2t - \pi)$;
(j) $\cos(2t - \frac{3}{2}\pi) - \cos(2t + \frac{3}{2}\pi)$.

18.2. State whether x leads or lags y in the following cases, and by how much.
(a) $x = 4 \cos 3t$, $y = 3 \cos(3t - \frac{1}{2}\pi)$.
(b) $x = 2 \cos(2t + \frac{1}{4}\pi)$, $y = 3 \cos(2t + \frac{9}{2}\pi)$.
(c) $x = -3 \cos 2t$, $y = 4 \cos 2t$.
(d) $x = \cos 3t$, $y = \sin 3t$.
(e) $x = 2 \cos 3t$, $y = \cos(3t - \frac{9}{4}\pi)$.

18.3. Obtain the free oscillations of the following in the form $C \cos(\omega t + \phi)$. State (i) the natural frequency if the damping coefficient is put to zero; (ii) the frequency that actually occurs in the cosine term of the solution; (iii) the number of complete cycles needed for the amplitude to drop to 0.1 of its value at $t = 0$.

(a) $x'' + 20x' + (2.5 \times 10^5)x = 0.$
(b) $x'' + 0.5x' + 4x = 0.$
(c) $x'' + 0.15x' + 3x = 0.$
(d) $x'' + x' + 20x = 0.$

18.4. Express $A \cos \omega t + B \sin(\omega t + \frac{1}{4}\pi)$ in the standard form $C \cos(\omega t + \phi)$ when (a) $A = 3^{\frac{1}{2}}$, $B = 1$; (b) $A = 3^{\frac{1}{2}}$, $B = -1$; (c) $A = -3^{\frac{1}{2}}$, $B = 1$; (d) $A = -3^{\frac{1}{2}}$, $B = -1$.

18.5. (a) Show that the maxima and minima of $x(t) = C e^{-kt} \cos(\omega t + \phi)$ occur at times T_N given by

$$\omega T_N + \phi = -\arctan \frac{k}{\omega} + N\pi,$$

where N is any integer.
   (b) Show that the values of $x(t)$ at these points are given by

$$x(T_N) = \frac{(-1)^N \omega C e^{-kT_N}}{(\omega^2 + k^2)^{\frac{1}{2}}}.$$

18.6. Consider an expression of the form

$$x(t) = e^{-t/T} g(t),$$

where T is a constant, and $g(t)$ itself does not have any term in it like $e^{\pm kt}$ (for example, $g(t)$ might be a constant, or $\cos t$, or even t^3, but it must not be, for example, $e^{-2t} \cos t$). Then T is called the time constant of $f(t)$.

(a) State the time constant for Q_c in Example 18.5.

(b) Describe how T provides a measure of the rate of exponential decay of $f(t)$, rather like the half-life period of a radioactive substance.

18.7. (Heavy damping.) Find the general solution of the equation

$$x'' + 2kx' + \omega_0^2 x = 0$$

when $k^2 > \omega^2$. Describe the general character of the solutions, contrasting them with the case when $k^2 < \omega^2$.

18.8. Solve the equation

$$x'' + 10x' + 24x = 0$$

subject to the initial conditions $x(0) = -3$, $x'(0) = 16$. Show that the solution curve crosses the t axis only once, at the point $t = \frac{1}{4} \ln 2$.

18.9. ('Critical damping'.) Find the general solution of the equation

$$x'' + 2kx' + \omega_0^2 x = 0$$

for the case when $k^2 = \omega^2$.

18.10. The following equation could represent the damped vertical motion of a mass supported by a spring and subjected to an external periodic force:

$$x'' + x' + 36x = 10 \cos \omega t, \qquad \text{for } t > 0,$$

the system being in equilibrium under no force for $t \leqslant 0$.

(a) Find the period of the free (damped) oscillations. Show that any free oscillations stimulated at startup are reduced by a factor of about 10 after five periods of oscillation.

(b) Obtain expressions in terms of ω for the amplitude and phase of the forced oscillation.

(c) Find the condition for resonance.

(d) Plot curves of amplitude and phase against ω for a range $4 \leqslant \omega \leqslant 8$.

18.11. A particle rolls to and fro under gravity at the bottom of a parabolic cylinder having vertical cross-section $y = ax^2$. There is negligible friction. The equation of motion in terms of horizontal displacement x is then

$$x'' + 2ax(g + 2ax'^2)/(1 + 4a^2 x^2) = 0.$$

Show that for small oscillations the period is $2^{\frac{1}{2}}\pi/(ag)^{\frac{1}{2}}$.

18.12. A particle is balanced at the topmost point, $x = y = 0$, of an inverted parabolic cylinder whose shape is described by $y = -ax^2$, y being measured vertically upward. Its equation of motion is

$$x'' + 2ax(2ax'^2 - g)/(1 + 4a^2 x^2) = 0.$$

By linearizing the equation show that, if the particle is slightly disturbed, it starts to move away from its initial position $(0, 0)$ at an increasing rate. (This condition is called **unstable equilibrium**.)

18.13. The equation for the displacement $x(t)$ of an electrical circuit fixed on springs and influenced by a current-carrying conductor is

$$x'' + 4[x - 2/(3 - x)] = 0.$$

(a) Show that there are two positions x at which the circuit could theoretically be in equilibrium. ('**Equilibrium**' means that

$$x(t) = \text{constant}$$

is a solution of the equation.)

(b) Call the equilibrium positions $x = a$ and $x = b$. To investigate the state of affairs near $x = a$, put

$$x = a + u$$

into the equation, so as to obtain an equation for $u(t)$, which is the distance from a. Then do the same thing near $x = b$ by putting

$$x = b + v,$$

where v is distance from b. Tidy the equations as far as possible.

(c) Suppose that u in one case, and v in the other, are small, and linearize the equations in each case.

(d) Show that in one case small oscillations take place, but that in the other the displacement tends to increase. (One is called a **stable** equilibrium state, the other **unstable**.)

18.14. A particle moves in a plane under a central attractive force γ/r^α per unit mass, where r and θ are its plane polar coordinates relative to an origin in the attracting body. Its equation of motion can be expressed in the form

$$\frac{\mathrm{d}^2 u}{\mathrm{d}\theta^2} + u - \frac{\gamma}{H^2} u^{\alpha - 2} = 0,$$

where $u = r$ and H is its (constant) angular momentum.

Show that the equation has a *constant* solution $u = u_0$, which is equivalent to a circular orbit. Does it stay close to this orbit if its position u is slightly changed from u_0, while H keeps its original value? (Hint: put $u = u_0 + x$ and linearize the equation for small x. You may assume that for small values of x/u (see (5.4d)).

$$(u_0 + x)^{\alpha - 2} \approx u_0^{\alpha - 2}\left(1 + (\alpha - 2)\frac{x}{u_0}\right)$$

18.15. Given the expression for the forced amplitude A in equation (18.16), deduce the expressions for the resonant frequency and resonant amplitude.

19 Steady forced oscillations: phasors, impedance, transfer functions

19.1 Phasors

We shall consider circuits driven by an applied harmonically alternating voltage, with resistances placed so that that any free oscillations set up by switching on the circuit die away, leaving only a periodic forced oscillation, as described in Section 18.5. It is only this remaining, **steadily-oscillating state** that is discussed here.

Let $x(t)$ represent any variable in the circuit, such as the current in a particular branch. If the frequency of the applied voltage is $\omega/2\pi$, then all these possible variables $x(t)$ share the **same frequency** $\omega/2\pi$ once the transients have died away, though in general the **phases and amplitudes of different variables are different**. Here we adopt the standardized amplitude/phase form of (18.2), assuming that

$$x(t) = c \cos(\omega t + \phi), \quad \text{with } c > 0 \text{ and } -\pi < \phi \leqslant \pi. \quad \textbf{(19.1)}$$

We can write $x(t)$ in a complex form instead:

$$x(t) = \mathrm{Re}(c\, \mathrm{e}^{\mathrm{j}(\omega t + \phi)}) = \mathrm{Re}(c\, \mathrm{e}^{\mathrm{j}\phi} \cdot \mathrm{e}^{\mathrm{j}\omega t}).$$

The complex coefficient $c\, \mathrm{e}^{\mathrm{j}\phi}$ that multiplies $\mathrm{e}^{\mathrm{j}\omega t}$ is called the **phasor corresponding to** $x(t)$, and it is independent of time t. **Every variable $x(t)$ will have its own phasor**, but the factor $\mathrm{e}^{\mathrm{j}\omega t}$ is the same for each one. In a circuit, ϕ and c usually depend on ω, so the values of the corresponding phasors will depend on ω.

Corresponding to each variable denoted by a lowercase letter, we use a bold capital letter to denote the phasor. This style is traditional, and emphasizes that phasors, being complex numbers, can be treated as vectors in the Argand diagram.

Phasor of a harmonic oscillation

The phasor of $x(t) = c \cos(\omega t + \phi)$ is the complex number $X = c\, \mathrm{e}^{\mathrm{j}\phi}$. $\quad$ **(19.2)**

In engineering applications, phasors $c\, \mathrm{e}^{\mathrm{j}\phi}$ are often written in the form $c\underline{/\phi}$, and ϕ may be expressed in degrees. Thus, if $X = 3\,e^{-\frac{1}{2}\pi}$, we can write.

$$X = 3\underline{/-\tfrac{1}{2}\pi} = 3\underline{/-45°}.$$

The two numbers displayed are the polar coordinates of the point which represents the phasor on an Argand diagram, in this case the point $(3/\sqrt{2}, -3/\sqrt{2})$ corresponding to

$$X = 3\cos(-45°) + j3\sin(-45°) = \frac{3}{\sqrt{2}} - j\,\frac{3}{\sqrt{2}}.$$

It is often convenient to express a phasor in the form $a + jb$ rather than in the polar form $c\,e^{j\phi}$.

Example 19.1. Find the phasor of $x(t) = -3\cos(2t + \frac{1}{2}\pi)$.

In standard form (19.1), $x(t) = 3\cos(2t - \frac{1}{2}\pi)$. The phasor X is therefore given by $X = 3\,e^{-\frac{1}{2}\pi}$ or $3\underline{/-90°}$.

Example 19.2. Given that the prevailing angular frequency is $\omega = 10^4$, find the functions $x(t)$ having the following phasors. (a) $X = 1/(-1 + j)$, (b) $X = (1 - \sqrt{3}j)/(-1 + j)$.

(a) Put X into polar form:

$$X = \frac{1}{-1 + j} = \frac{-1 - j}{(-1)^2 + 1^2} = -\tfrac{1}{2} - \tfrac{1}{2}j = \frac{1}{\sqrt{2}}\,e^{-\frac{3}{4}\pi j}.$$

Therefore

$$x(t) = \frac{1}{\sqrt{2}}\cos(10^4 t - \tfrac{3}{4}\pi).$$

(b) $1 - \sqrt{3}j = 2\,e^{-\frac{1}{3}\pi j}$ (as can be seen by putting the point $1 - \sqrt{3}j$ on an Argand diagram). Therefore, using (a),

$$X = (2\,e^{-\frac{1}{3}\pi j})\left(\frac{1}{\sqrt{2}}\,e^{-\frac{3}{4}\pi j}\right) = \sqrt{2}\,e^{-\frac{13}{12}\pi j}.$$

The phase $(-\frac{13}{12}\pi)$ is out of the standard range (19.1), so add 2π to it, leaving X unchanged. We obtain $X = \sqrt{2}\,e^{\frac{11}{12}\pi j}$, so

$$x(t) = \sqrt{2}\cos(10^4 t + \tfrac{11}{12}\pi)$$

Example 19.3. Let $x(t) = \sqrt{3}\cos\omega t - \sin\omega t$. Find the corresponding phasor.

Take the terms separately:

$$\sqrt{3}\cos\omega t = \mathrm{Re}(\sqrt{3}\,e^{j\omega t});$$
$$\sin\omega t = \cos(\omega t - \tfrac{1}{2}\pi) = \mathrm{Re}(e^{-\frac{1}{2}\pi j}\,e^{j\omega t}).$$

Combining them, we obtain

$$x(t) = \mathrm{Re}[(\sqrt{3} - e^{-\frac{1}{2}\pi j})\,e^{j\omega t}].$$

Therefore

$$X = \sqrt{3} - e^{-\frac{1}{2}\pi j} = \sqrt{3} - (-j)$$
$$= 2\,e^{\frac{1}{6}\pi j} \text{ or } 2\underline{/30°}.$$

19.2 Algebra of phasors

As seen in Example 19.3, when oscillations associated with the same value of ω combine by addition, so do their phasors. Suppose, for instance, that $u(t)$ and $v(t)$ have the same angular frequency ω, and that their phasors are U and V. Then $u(t) = \text{Re}(U\,e^{j\omega t})$ and $v(t) = \text{Re}(V\,e^{j\omega t})$, so

$$u(t) + v(t) = \text{Re}[(U + V)\,e^{j\omega t}],$$

whose phasor is $U + V$. The addition holds similarly if there are more terms present.

Addition principle for phasors

If $u(t)$, $v(t)$, ... have a common frequency, and $z(t) = u(t) + v(t) + \cdots$, then $Z = U + V + \cdots$, where Z, U, V, ... are the corresponding phasors.　　　(19.3)

Differentiation and integration give important results. If $x(t) = \text{Re}(X\,e^{j\omega t})$, where X is the phasor, then $dx/dt = \text{Re}(j\omega X\,e^{j\omega t})$, so that the phasor of dx/dt is $j\omega X$. Differentiate again, and a further factor $j\omega$ is introduced, so that the phasor of d^2x/dt^2 is $(j\omega)^2 X$, and so on.

For $\int x(t)\,dt$, we find in the same way that the phasor is $X/j\omega$. Here the arbitrary constant has been put to zero because, in normal use, all the variables that occur oscillate.

Phasors of derivatives and integrals

Variable:	x	$\dfrac{dx}{dt}$	$\dfrac{d^2x}{dt^2}$	$\displaystyle\int x\,dt$
Phasor:	$X = c\,e^{j\phi}$	$j\omega X$	$-\omega^2 X$	$\dfrac{1}{j\omega}X$

(19.4)

Example 19.4. Obtain the phasor of the expressions

(a) $L\dfrac{d^2q}{dt^2} + R\dfrac{dq}{dt} + \dfrac{q}{C}$; (b) $L\dfrac{di}{dt} + \dfrac{1}{C}\displaystyle\int i\,dt$; in terms of the phasors

Q of $q(t)$ and I of $i(t)$. (L, R, and C are constants, and the prevailing frequency is ω.)

(a) From (19.4) and the addition principle (19.3), the phasor is

$$L(j\omega)^2 Q + R(j\omega)Q + (1/C)Q = [(1/C - L\omega^2) + jR\omega]Q.$$

(b) The phasor is

$$L(j\omega)I + (1/C\, j\omega)I = j(L\omega - 1/C\omega)I.$$

Example 19.5. Find the steady-state solution of

$$\frac{d^2 x}{dt^2} + 8\frac{dx}{dt} + 10^4 x = 2.10^3 \cos 90t.$$

This is equivalent to the circuit equation in Example 18.5, with $x(t)$ in place of $q(t)$. The prevailing value of ω is 90. Let X be the phasor of $x(t)$. The phasor of the right-hand side is $2 \cdot 10^3$, so by using (19.4) we obtain

$$[(90j)^2 + 8(90j) + 10^4]X = 2 \cdot 10^3,$$

or

$$(1900 + 720j)X = 2 \cdot 10^3,$$

from which X can be found.

$$X = \frac{2 \cdot 10^3}{1900 + 720j} = \frac{1}{0.95 + 0.36j}$$

$$= \frac{1}{1.0159\, e^{0.3622j}} = 0.984\, e^{-0.362j}.$$

Therefore

$$x(t) = \text{Re}[0.984\, e^{-0.362j}\, e^{90jt}]$$

$$= 0.984 \cos(90t - 0.362),$$

as we found in Example 18.5 for the forced oscillation.

19.3 Phasor diagrams

Complex numbers can be represented by vectors in an Argand diagram (see Section 6.2), and they are added in the same way as the corresponding vectors. Phasors are just complex numbers, so they **add like vectors** too. This fact can be used to show pictorially how a number of superposed oscillations which are not in phase with each other contribute to the sum. The diagrams concerned are called **phasor diagrams**.

Example 19.6. Let $u(t) = 2\cos 10t$, $v(t) = \cos(10t - \frac{1}{2}\pi)$, and $w(t) = 3\cos(10t + \frac{1}{4}\pi)$. Find $p(t) = u(t) + v(t) + w(t)$ by means of a phasor diagram.

The phasors corresponding to u, v, and w are $U = 2$, $V = e^{-\frac{1}{2}\pi j}$, and $W = 3\, e^{\frac{1}{4}\pi j}$. In the polar-coordinate notation they are $U = 2/\underline{0}$, $V = 1/\underline{-90°}$, $W = 3/\underline{45°}$. They are shown as position vectors in Fig. 19.1a, and in Fig. 19.1b they are strung together as usual for

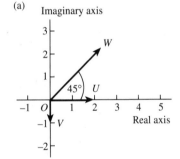

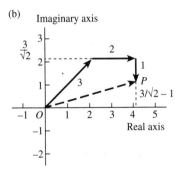

Fig. 19.1
(a) Argand diagram showing U, V, W. (b) The sum $U + V + W = \overline{OP}$.

addition. The vector $\overline{OP}$ can be measured off from the diagram, or calculated using the dimensions shown. We have

$$|\overline{OP}| = [(3/\sqrt{2} + 1)^2 + (3/\sqrt{2} - 1)^2]^{\frac{1}{2}} = 4.27,$$

$$\phi = \arctan \frac{3/\sqrt{2} - 1}{3/\sqrt{2} + 1} = 0.266 \text{ (radians)}.$$

Therefore $p(t) = 4.27 \cos(10t + 0.266)$.

19.4 Phasors and complex impedance
In the following table, an electric current

$$i(t) = c \cos(\omega t + \phi),$$

with phasor

$$I = c \, e^{j\phi}$$

is caused to pass through a resistor, an inductor, and a capacitor, separately. The resulting voltage drop $v(t)$ associated with each is shown, together with its phasor V. It is the unique steadily oscillating state that is being described by the phasors.

	Resistor	**Inductor**	**Capacitor**
Voltage drop:	$v = Ri$	$v = L \dfrac{di}{dt}$	$v = \dfrac{1}{C} \displaystyle\int i \, dt$
Voltage phasor V:	$Rc \, e^{j\phi} = RI$	$j\omega L I$	$\dfrac{1}{j\omega C} I$
Voltage phase:	ϕ (in phase)	$\phi + \frac{1}{2}\pi$ (v leads i)	$\phi - \frac{1}{2}\pi$ (v lags i)

(19.5)

A similar table can be constructed if the voltage rather than current is prescribed. The entries can be read from the table above; for example, if the phasor of the voltage applied to an inductor is V, the phasor of the resulting current is $V/j\omega L$.

Discussion of circuits in terms of phasors is said to take place in the **frequency domain**, rather than the **time domain** associated with the differential equations of the circuits.

Each of the three cases in the table can be written in the form

$$V = ZI,$$

where Z is either R, $j\omega L$, or $(j\omega C)^{-1}$. The quantity Z is called the **complex impedance** of these elements. There is a plain analogy with Ohm's law for direct current through a resistance. We have

> **Complex impedance** Z
>
> Resistor $Z = R$
> Inductor $Z = j\omega L$ **(19.6)**
>
> Capacitor $Z = \dfrac{1}{j\omega C}$.

By stringing elements of this type together in series, we can form composite units. The combined unit has a complex impedance which is the sum of the complex impedances of the individual elements:

Example 19.7. Show that the complex impedance Z of two elements in series, whose complex impedances are Z_1 and Z_2, is given by

$$Z = Z_1 + Z_2.$$

Suppose that the impedance of the unit is Z; we mean by this that, if V is the phasor of the voltage drop across the unit and I is the phasor of the current through it (see Fig. 19.2), then

$$V = ZI.$$

From Fig. 19.2, $v = v_1 + v_2$; therefore, by (19.3), the corresponding phasors satisfy

$$V = V_1 + V_2.$$

But i, and therefore I, is the same for Z_1 and Z_2, so

$$V_1 = Z_1I, \qquad V_2 = Z_2I.$$

Therefore $V = Z_1I + Z_2I = ZI$, or

$$Z = Z_1 + Z_2.$$

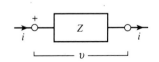

Fig. 19.2

If the two impedances are in parallel, the analogy with Ohm's law again exists:

Example 19.8. Show that the complex impedance Z of any two elements Z_1 and Z_2 in parallel is given by $\dfrac{1}{Z} = \dfrac{1}{Z_1} + \dfrac{1}{Z_2}$.

From Fig. 19.3, $i = i_1 + i_2$; so, by (19.3), $I = I_1 + I_2$. The voltage drop is the same for both branches, so

$$I_1 = V/Z_1, \quad I_2 = V/Z_2, \quad I = V/Z.$$

Therefore $I = (1/Z_1 + 1/Z_2)V = (1/Z)V$, from which the result follows.

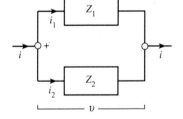

Fig. 19.3
Two impedances in parallel and their combined impedance.

It is easy to extend these two results to encompass more elements, and therefore we have the following general result.

> **Complex impedance Z of series and parallel circuits**
>
> (a) Impedances $Z_1, Z_2, \ldots$, in series:
> $$Z = Z_1 + Z_2 + \cdots. \tag{19.7}$$
> (b) Impedances $Z_1, Z_2, \ldots$, in parallel:
> $$\frac{1}{Z} = \frac{1}{Z_1} + \frac{1}{Z_2} + \cdots.$$

The analogy with resistive circuits, evident from these formulae, goes much further. The general rules which govern voltages and currents in a passive linear circuit are Kirchhoff's laws: (i) that the algebraic sum of the voltages around any closed circuit is zero; (ii) that the resultant current entering any junction is zero. There is also a linear voltage/current relation for each branch. In terms of phasors and complex impedances for a circuit in a state of steady harmonic oscillation, these conditions become the following.

> | | |
> |---|---|
> | Around any closed circuit, | $\sum V = 0.$ |
> | At any junction, | $\sum I = 0.$ |
> | On any branch, | $V = ZI.$ |
>
> (19.8)

These rules have the same form as the rules for resistive direct-current circuits, with V, I, and Z appearing in them in place of v, i, and R. It follows that **general rules applicable to DC circuits may be borrowed** for the purpose of the circuits we have been considering. Such rules are the Wheatstone bridge rules, Thévenin's theorem, and the structure of equivalent circuits. However, the restriction to steady harmonic oscillation must be remembered: many circuits can be made to 'balance' like a Wheatstone bridge for steady oscillations, but not for more general disturbances.

Example 19.9. Find the steady AC current in the circuit shown.

The unit comprising R and C consists of two complex impedances in parallel, R and $(j\omega C)^{-1}$. If Z is the combined impedance, then

$$\frac{1}{Z} = \frac{1}{R} + \frac{1}{(j\omega C)^{-1}},$$

which gives

$$Z = \frac{R}{1 + j\omega RC}.$$

Z is in series with the other impedance, $j\omega L$, so the impedance of

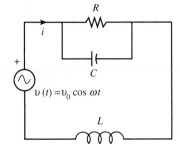

Fig. 19.4

the circuit is given by

$$Z = \frac{R}{1 + j\omega RC} + j\omega L = \frac{R(1 - \omega^2 LC) + j\omega L}{1 + j\omega RC}.$$

Since $I = V/Z$, and $V = v_0$, we obtain

$$I = \frac{v_0(1 + j\omega RC)}{R(1 - \omega^2 LC) + j\omega L} = \frac{v_0(1 + \omega^2 R^2 C^2)^{\frac{1}{2}}}{[R^2(1 - \omega^2 LC)^2 + \omega^2 L^2]^{\frac{1}{2}}} e^{j(\phi_1 - \phi_2)},$$

where

$$\phi_1 = \arctan \omega RC, \qquad \phi_2 = \arctan \frac{\omega L}{R(1 - \omega^2 LC)}.$$

Finally,

$$i(t) = \mathrm{Re}(I\, e^{j\omega t})$$

$$= \frac{v_0(1 + \omega^2 R^2 C^2)^{\frac{1}{2}}}{[R^2(1 - \omega^2 LC)^2 + \omega^2 L^2]^{\frac{1}{2}}} \cos(\omega t + \phi_1 - \phi_2).$$

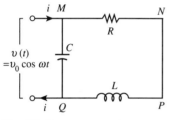

Fig. 19.5

Example 19.10. Find the steady AC current entering the circuit shown in Fig. 19.5.

The phasor of the voltage source is $V = v_0$. By (19.6) the impedance of $MNPQ$ is $R + j\omega L$, and that of MCP is $1/j\omega C$. These are in parallel, so by (19.7) the impedance Z of the circuit viewed between M and P is given by

$$\frac{1}{Z} = \frac{1}{R + j\omega L} + \frac{1}{1/j\omega C}.$$

Therefore

$$I = \frac{V}{Z} = v_0 \left(\frac{1}{R + j\omega L} + j\omega C \right).$$

The simplest way to get an expression for $i(t)$ is to treat the two terms in the parentheses on the right separately (though this does not give the answer in standard form). We obtain

$$i(t) = \frac{v_0}{(R^2 + \omega^2 L^2)^{\frac{1}{2}}} \cos\left(\omega t - \arctan \frac{\omega L}{R} \right)$$

$$+ v_0 \omega C \cos(\omega t + \tfrac{1}{2}\pi).$$

Example 19.11. (Balanced bridge circuit.) (a) For Fig. 19.6a, show that (i) if $i(t) = 0$, then $Z_1/Z_2 = Z_3/Z_4$, (ii) if $Z_1/Z_2 = Z_3/Z_4$, then $i(t) = 0$. (b) Check that $i(t) = 0$ in the circuit of Fig. 19.6b.

(a) The analogy (19.8) between resistive and general circuits for steady harmonic oscillations enables us to borrow ordinary Wheatstone-bridge theory, substituting current and voltage phasors and complex impedances for the usual constant currents, voltages, and resistances. We can therefore say immediately that the circuit

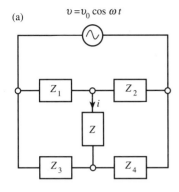

(a) $v = v_0 \cos \omega t$

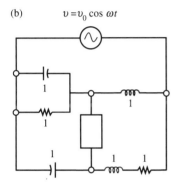

(b) $v = v_0 \cos \omega t$

Fig. 19.6

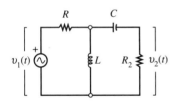

Fig. 19.7

is balanced ($i(t) = 0$) if, and only if,

$$Z_1/Z_2 = Z_3/Z_4.$$

(b) Z_1 consists of a capacitor and resistor in parallel; so, by (19.6) and (19.7),

$$\frac{1}{Z_1} = \frac{1}{(j/\omega)} + \frac{1}{1} \quad \text{or} \quad Z_1 = \frac{1}{1 + j\omega}.$$

Also

$$Z_2 = j\omega, \qquad Z_3 = \frac{1}{j\omega}, \qquad Z_4 = 1 + j\omega.$$

Therefore

$$\frac{Z_1}{Z_2} = \frac{1}{j\omega(1 + j\omega)} \quad \text{and} \quad \frac{Z_3}{Z_4} = \frac{1}{j\omega(1 + j\omega)};$$

so, from (a), $i(t) = 0$ and the bridge is balanced.

19.5 Transfer functions in the frequency domain
Consider the circuit of Fig. 19.7, in which the applied voltage is $v_1(t)$, with phasor $V_1 = c_1 \, e^{j\phi_1}$. Suppose that the voltage drop $v_2(t)$ across R has a phasor $V_2 = c_2 \, e^{j\phi_2}$.

Consider the ratio of these two phasors, denoting it by G_{12} (G standing for **voltage gain**):

$$G_{12} = \frac{V_2}{V_1} = \frac{c_2 \, e^{j\phi_2}}{c_1 \, e^{j\phi_1}} = \frac{c_2}{c_1} \, e^{j(\phi_2 - \phi_1)}.$$

Then

$$|G| = \frac{c_2}{c_1},$$

which is the ratio of the peak voltages, or amplitudes, of $v_2(t)$ and $v_1(t)$. The argument (polar angle) of G_{12} is the phase difference between them. If instead we are interested in the current $i_2(t)$ through R_2 produced by $v_1(t)$, then we need the ratio

$$Z_{12} = V_1/I_2,$$

where I_2 is the phasor of $i_2(t)$. This quantity, a voltage divided by a current, is called a **transfer impedance**. Alternatively, we could consider the ratio

$$Y_{21} = I_2/V_1,$$

in which Y_{21} is called a **transfer admittance** (whose parallel is conductance in DC theory).

In general, the ratio of an **output** (such as a current in a selected branch) to an **input** (such as a voltage driving a network) is called a **transfer function in the frequency domain**. A different class of transfer functions is discussed in Chapter 23, on Laplace transforms.

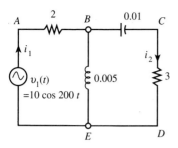

Fig. 19.8

Example 19.12. Find the transfer impedance $Z_{12} = V_1/I_2$ for the circuit of Fig. 19.8 when the prevailing angular frequency ω is 200.

The currents indicated take account of Kirchhoff's second rule (19.8), that the sum of the currents entering a junction is zero. The first law expressed in terms of the phasors (see (19.8)), that the sum of the voltage drops round closed circuits is zero, gives for the circuits $ABCDEA$ and $BCDEB$ respectively:

$$2I_1 + \left(\frac{1}{200 \times 0.01j} + 3\right)I_2 = V_1,$$

and

$$\left(\frac{1}{200 \times 0.01j} + 3\right)I_2 - (200 \times 0.005j)(I_2 - I_1) = 0.$$

After simplification, these become

$$2I_1 + (3 - \tfrac{1}{2}j)I_2 = V_1,$$

$$jI_1 + (3 - \tfrac{3}{2}j)I_2 = 0.$$

The solution for I_2 is

$$I_2 = \frac{j}{-\tfrac{11}{2} + 6j}\, V.$$

The transfer function required is

$$Z_{12} = V_1/I_2 = 6 + \tfrac{11}{2}j.$$

The amplitude of $i_2(t)$ is given by

$$|I_2| = |V_1|/|I_2| = 10/8.14 = 1.23.$$

Its phase is:

$$(\text{phase of } V_1) - (\text{phase of } Z_{12}) = 0 - 0.74 = 0.74.$$

The current leads the voltage by this amount.

The methods described in this chapter were invented in the late nineteenth century to assist engineers working with alternating current to interpret and make calculations on their circuits. So long as only steady harmonic oscillations had to be considered, there was no need to solve differential equations: only algebraic equations are involved and these are much simpler to manipulate. Since that time, the methods have been extensively developed so as to permit computer calculation for circuits of any degree of complexity, using matrix algebra, graph theory, and other sophisticated techniques. In Section 22.16, another method for algebrizing circuit equations is described, using Laplace transforms.

Problems, Chapter 19

19.1. Write down the phasors X corresponding to the oscillations $x(t)$ given below, in polar and $a + jb$ form.
(a) $2 \cos(10t + \frac{1}{2}\pi)$; (b) $-2 \cos(10t + \frac{1}{2}\pi)$;
(c) $3 \sin \omega t$; (d) $-4 \sin(3t - \frac{1}{4}\pi)$.

19.2. Write the following phasors X in polar form, and give the corresponding oscillations $x(t)$ when the angular frequency is ω.
(a) $1 - j$; (b) $2j$; (c) $-3j$; (d) $-\sqrt{2} - \sqrt{2}j$;
(e) $-2\sqrt{3} - 2j$; (f) $-1 - \sqrt{3}j$; (g) $1/(1 - 2j)$;
(h) $j/(1 - 2j)$; (i) $(2 + 3j)/(2 - 3j)$; (j) $1/3j + 2j$.

19.3. Write down the phasors corresponding to the following oscillators. State the amplitude and phase.
(a) $\cos 2t + \cos(2t - \frac{1}{4}\pi)$; (b) $\cos 3t - \sin 3t$;
(c) $\sin 3t + 2 \cos 3t$.

19.4. Either algebraically, or by calculation or measurement based on a phasor diagram, give the phasors of the following functions.
(a) $-\cos 2t + \cos(2t + \frac{1}{4}\pi) + \cos(2t - \frac{1}{2}\pi)$;
(b) $\cos 1760t - 3 \cos(1760t - \frac{1}{2}\pi) + \cos(1760t + \frac{1}{2}\pi)$.

19.5. Show that the point on an Argand diagram corresponding to $c\,e^{j(\omega t + \phi)}$ moves on a circle, centre the origin and radius c, with constant angular velocity ω, and that its projection on the x axis is given by $x = c \cos(\omega t + \phi)$. (A phase diagram such as Fig. 19.1 is a snapshot of conditions at $t = 0$ in such a representation: the whole diagram rotates unaltered.)

19.6. Obtain the complex impedance of the following circuit branches.

(a)

(b)

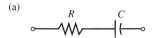

(c)

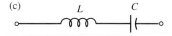

(d)

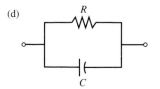

(e)

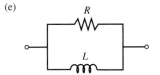

(f)

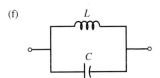

(g)

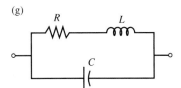

(h)

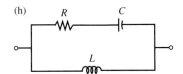

(i)

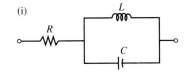

(j)

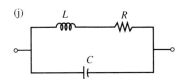

(k)

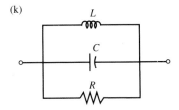

(l)

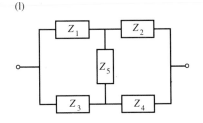

19.7. A voltage $v = 2 \cos \omega t$ is applied to each of the terminals in Problem 19.6. Find the amplitude and phase of the current passing through the circuit.

19.8. In Fig. 19.10, numerical values are given to the complex impedances (the units are usually ohms, although the quantities may be complex). A voltage with phasor V_0 is applied. Obtain the phasor V_1 as indicated, the corresponding voltage gain V_1/V_0, and the transfer impedance V_0/I_1.

(b)

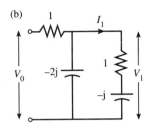

(a)

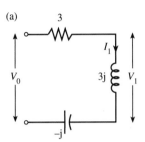

(c)

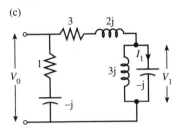

Fig. 19.10

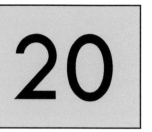

20 Graphical, numerical and other aspects of first-order equations

20.1 Graphical features of first-order equations

In this section we shall use x for the independent variable (instead of t), and y as the dependent variable (instead of x), and consider differential equations of the form

$$\frac{\mathrm{d}y}{\mathrm{d}x} = f(x, y), \tag{20.1}$$

where $f(x, y)$ is unrestricted. If $f(x, y)$ happens to take the form $g(x) + h(x)y$, the equation is **linear** and can be handled by the method of Section 17.6; otherwise none of the methods so far discussed will work.

However, we can always obtain a rough picture of the **solution curves** by using a simple fact. Choose any point $x = a$, $y = b$, on the (x, y) plane. Then equation (20.1) says that the slope of the solution curve which passes through (a, b) must be equal to $f(a, b)$, and this has a definite numerical value that we can work out. So take a large number of points (a, b) on the (x, y) plane. For each of them, work out $f(a, b)$, and draw through the point a short line whose slope is equal to $f(a, b)$, as is done in Fig. 20.1a for the special case $f(x, y) = xy$. These are called **direction indicators**. Given enough of these, it is possible to draw a **family of curves** which follow their directions smoothly, as in Fig. 20.1b. Each of the curves represents a solution of (20.1), because its slope, or derivative, is correctly reproduced at each point on it. The picture is called a **lineal-element diagram** or a **direction field**. The technique can in principle be used for first-order equations however complicated they may be.

Rather than to draw the direction indicators at grid points as in Fig. 20.1, it is often better to look for curves, called **isoclines**, along which the slope is constant, as in the following example.

(a)

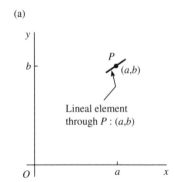

(b)

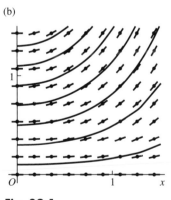

Fig. 20.1
Lineal-element diagram indicating solution curves for $\mathrm{d}y/\mathrm{d}x = xy$.

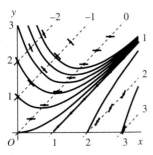

Fig. 20.2

Solution curves of
$dy/dx = x - y$ in the first
quadrant. – – – – – isoclines;
values of K indicated.
——— solution curves.

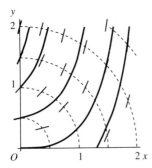

Fig. 20.3

Pattern of solution curves of
$dy/dx = x^2 + y^2$
in the first quadrant.
– – – – – isoclines,
——— solution curves.

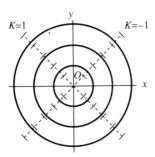

Fig. 20.4

Sketch the solution curves
$dy/dx = -x/y$.
– – – – – isoclines,
——— solution curves.

Example 20.1. Sketch the solution curves of $\dfrac{dy}{dt} = x - y$. Here dy/dx takes *constant* values K on the isoclines $x - y = K$, or $y = x - K$. For example, $dy/dx = 0$ on $y = x$, $dy/dx = 1$ on $y = x - 1$, and so on. If we draw the line $y = x - K$, then the indicators along it are all parallel, with slope K, so it is easier to draw a large number of them. Figure 20.2 is constructed in this way. (This equation is in fact linear with constant coefficients, its solutions being $y = x - 1 + C e^{-x}$.)

Example 20.2. Sketch the solution curves of $\dfrac{dy}{dx} = x^2 + y^2$.

The isoclines are the circles $x^2 + y^2 = K$ (see Fig. 20.3), on each of which the slope is equal to K (which must be a positive number here).

Closed curves can occur, as in the following example.

Example 20.3. Sketch the solution curves of $\dfrac{dy}{dx} = -\dfrac{x}{y}$.

The isoclines are the radial straight lines $-x/y = K$ (see Fig. 20.4), or

$$y = -\frac{1}{K} x,$$

on which the slopes are equal to K. On $y = 0$, the slope K must be infinite so the direction indicators are vertical.

The method illustrates why there is always an infinite number of solutions: there will be **a single solution curve through every point where** $f(x, y)$ **has a definite value**. The type of exception that might arise is illustrated by the case $f(x, y) = (xy)^{\frac{1}{2}}$, which only has a meaning when x and y have the same sign; there are solutions in only the first and third quadrants of the (x, y) plane. Again, there can be points from which several solution curves emanate: this occurs at the origin in Example 20.3, where $f(x, y)$ takes the indeterminate form $0/0$; nothing can be taken for granted at such a point, but elsewhere the curves do not intersect.

To prescribe a point (a, b) through which a curve must pass is equivalent to imposing an initial condition on the solution: the corresponding initial condition would read 'Find the solution for which $y = b$ when $x = a$'. Therefore, it can be seen that, even when an equation is not linear, an initial condition of this sort will give exactly one solution; points where $f(a, b)$ is indeterminate being excepted.

20.2 A method for numerical solution

For the equation

$$\frac{dy}{dx} = f(x, y),$$

consider an adaptation of the graphical method described in the previous section. Start at any point $P_0 : (x_0, y_0)$, and draw an indicator with slope $f(x_0, y_0)$ from P_0 to $P_1 : (x_1, y_1)$ a short distance away (see Fig. 20.5). Then P_1 will lie close to the solution curve through P_0. Do the same thing starting with P_1, and so on, continuing as far as is necessary. It is also possible to proceed backwards from P_0. Provided that the steps are small enough, it seems likely that $P_0, P_1, P_2, \ldots$, will be close to the solution curve through P_0, so we have an approximate solution to the initial-value problem: Find the solution of $dy/dx = f(x, y)$ for which $y = y_0$ when $x = x_0$.

Obviously, to draw the solution curve in this way is not really practicable; but we need not actually draw it, because the same process can be carried out numerically as follows.

As shown in Fig. 20.6, choose a small, constant step length h in x for going from point to point: $P_0 \to P_1 \to P_2 \to \cdots$, where the vectors $\overrightarrow{P_0 P_1}$, $\overrightarrow{P_1 P_2}$, $\overrightarrow{P_2 P_3}, \ldots$ point in the direction of the indicators at their respective starting points $P_0, P_1, P_2, \ldots$. Corresponding to the x steps, call the y steps $k_1, k_2, k_3, \ldots$:

$$y_1 = y_0 + k_1, \quad y_2 = y_1 + k_2, \quad y_3 = y_2 + k_3, \ldots,$$

where

$$k_1 = hy'(x_0) = hf(x_0, y_0), \quad k_2 = hy'(x_1) = hf(x_1, y_1), \quad \ldots$$

Therefore:

for P_1: $\quad x_1 = x_0 + h, \quad y_1 = y_0 + hf(x_0, y_0)$;

for P_2: $\quad x_2 = x_1 + h, \quad y_2 = y_1 + hf(x_1, y_1)$;

and, in general, with $n = 1, 2, 3, \ldots$, in turn,

for P_n: $\quad x_n = x_{n-1} + h, \quad y_n = y_{n-1} + hf(x_{n-1}, y_{n-1})$.

We expect that the points will be close to the solution curve. This is the **Euler method** for approximating to the solution of the differential equation.

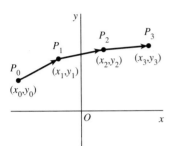

Fig. 20.5
Step-by-step use of direction indicators along a particular solution curve.

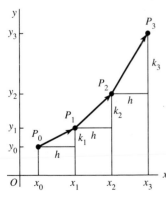

Fig. 20.6
Three steps in the numerical solution of $dy/dx = f(x, y)$, starting at the point $P : (x_0, y_0)$.

Euler method for initial-value problems

Differential equation: $\dfrac{dy}{dx} = f(x, y)$.

Initial condition: $y = b$ when $x = a$.   **(20.2)**

Approximate solution: Put $x_0 = a$, $y_0 = b$; then
$x_n = x_{n-1} + h, \quad y_n = y_{n-1} + f(x_{n-1}, y_{n-1})h$
for $n = 1, 2, \ldots$ successively.

This recipe, or **algorithm**, describes a step-by-step repetitive process, or **iteration**; essentially the same thing has to be done over and over again. The procedure which produces a new (x, y) from the preceding (x, y) is called a **recurrence relation** (compare Newton's method for solving equations in Chapter 3). Such a process is easy to program on a computer, and Fig. 20.7 is the skeleton of a flow diagram. The program should contain a method for stopping itself when x has gone far enough; also, since small intervals h are usually necessary, it is useful to include a means of recording only the results for preset values of x to avoid voluminous output.

Fig. 20.7
Flow diagram for the initial-value problem.

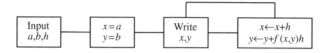

Example 20.4. Use the Euler method to obtain a solution of the initial-value problem

$$\frac{dy}{dx} = xy^2, \text{ with } y = 1 \text{ at } x = 0,$$

between $x = 0$ and $x = 1$. Compare the result with the exact solution $y = (1 - \frac{1}{2}x^2)^{-1}$ when steps of $h = 0.2, 0.1, 0.01,$ and 0.001 are adopted. The following results are obtained.

x	0	0.2	0.4	0.6	0.8	1.0
Exact y	1.0000	1.0204	1.0870	1.2195	1.4706	2.0000
$h = 0.2$	1.0000	1.0401	1.1265	1.2788	1.5405	2.0151
$h = 0.1$	1.0000	1.0304	1.1074	1.2507	1.5081	2.0062
$h = 0.01$	1.0000	1.0214	1.0891	1.2228	1.4746	2.0003
$h = 0.001$	1.0000	1.0205	1.0872	1.2198	1.4710	2.0000

Euler's method is very simple; but it is usually good enough to provide reasonable accuracy over a finite range, provided that small enough intervals are used. The simplest way of checking accuracy is to experiment with successively smaller intervals h, noting when further reduction in h does not change the values of y obtained at the number of decimal places required. Several problems on these lines are given at the end of the chapter.

There exist, however, far more sophisticated algorithms which will give great accuracy over long ranges without having to use minute values of h (which can introduce problems of its own). The computer programs for such methods can be found in libraries of computer routines. The theoretical side of the subject is called numerical analysis; mathematical theory makes it possible, for example, to estimate the size of interval required without carrying out trials.

20.3 Nonlinear equations of separable type
The equation

$$\frac{dy}{dx} = \frac{y^2}{x^2}$$

is nonlinear (note the y^2 term), and none of the theory of Chapters 16 and 17 can be adapted to solve it. Write it in the form

$$\frac{dy}{y^2} = \frac{dx}{x^2}.$$

On the left only y appears, and on the right only x appears. The form looks like an invitation to integrate both sides:

$$\int \frac{dy}{y^2} = \int \frac{dx}{x^2} + C,$$

so $-1/y = -1/x + C$. Therefore

$$y = \frac{x}{1 - Cx},$$

where C is arbitrary. The reader should consider checking that these really are solutions by substituting into the equation. Notice that there is no sign of the complementary functions and particular solutions found for linear equations: an arbitrary constant C does occur, but it is imbedded deep in the expression. The solution curves are shown in Fig. 20.8: each curve has its individual asymptotes, namely the lines $y = -C^{-1}$, $x = C^{-1}$.

The method is called **separation of variables**. It can be applied to equations which are **separable**, that is to say, ones that can be arranged in the form

$$\frac{dy}{dx} = g(x)h(y);$$

where the right-hand side is the product of two terms, one a function of x only, and the other a function of y only. Alternatively, you might see it more easily as an equation which can be put into the form

$$Y(y)\,dy = X(x)\,dx,$$

$X(x)$ and $Y(y)$ being functions respectively of x only and y only. For example the equations

$$\frac{dy}{dx} = x^{\frac{1}{2}}y^{\frac{1}{2}}, \qquad \frac{dy}{dx} = e^x \sin y, \qquad \frac{y\,dy}{x\,dx} = \cos(y^2)$$

are of the right type.

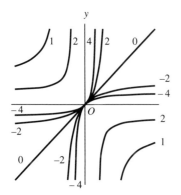

Fig. 20.8
Solution curves
$y = x/(1 - Cx)$ for
$dy/dx = y^2/x^2$. Values of C
indicated on the curves.

Separation of variables

Equation type: $\dfrac{dy}{dx} = g(x)h(y)$.

Separate the terms: $\dfrac{dy}{h(y)} = g(x)\,dx$.   (20.3)

Integrate: $\displaystyle\int \dfrac{dy}{h(y)} = \int g(x)\,dx + C$, so that y is
expressed as function of x (usually an implicit function).
C may take a range of values.

Example 20.5. Find solutions of the equation $y\dfrac{dy}{dx} = \cos x$.

This can be written $y\,dy = \cos x\,dx$. By integrating both sides, we
obtain $\frac{1}{2}y^2 = \sin x + C$, giving $y = \pm 2^{\frac{1}{2}}(\sin x + C)^{\frac{1}{2}}$, where C is
only to a certain extent arbitrary. It cannot be completely arbitrary;
for example, if $C = -100$, then $\sin x + C$ will *always* be negative
(because $-1 \leqslant \sin x \leqslant 1$), so the square root never has a real value.
We must have $C > -1$ to get any real solution. If C lies between
-1 and $+1$, there will be intervals in x which make $\sin x + C$
negative; this fact leads to the oval curves in Fig. 20.9: you should
experiment with, say, $y = \pm 2^{\frac{1}{2}}(\sin x + \frac{1}{2})^{\frac{1}{2}}$ from this point of view.
Since $\sin x$ has period 2π, so has the solution picture.

Fig. 20.9
Solution curves for the
equation

$$y\frac{dy}{dx} = \cos x,$$

which are the functions
$y = \pm 2^{\frac{1}{2}}(\sin x + C)^{\frac{1}{2}}$.

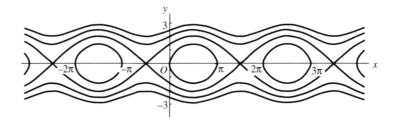

Example 20.6. Find solutions of $\dfrac{dy}{dx} = \dfrac{y(x+1)}{x(y+1)}$.

After separating, we have

$$\int \frac{1+y}{y}\,dy = \int \frac{1+x}{x}\,dx + C,$$

or

$$\int \left(1 + \frac{1}{y}\right) dy = \int \left(1 + \frac{1}{x}\right) dx + C.$$

Therefore $y + \ln|y| = x + \ln|x| + C$, or $y\,e^y = Ax\,e^x$, where A is
arbitrary.

We cannot further reduce this 'solution' to express y explicitly in terms of x. The 'answer' is nearly as obscure as the original equation, for we could sketch solutions curves or compute solutions directly by the methods of Sections 17.1 and 17.2 (see Problem 20.3a).

The separation-of-variables techniques requires initiative, even in the simplest cases, as can be seen from the following example.

Example 20.7. Find solutions of $\dfrac{dy}{dx} = 2y^{\frac{1}{2}}$.

After separating, we have $\frac{1}{2} \displaystyle\int y^{-\frac{1}{2}}\, dy = \int dx$, or

$$y^{\frac{1}{2}} = x + C. \tag{20.4}$$

To express y in terms of x, square both sides. We get

$$y = (x + C)^2. \tag{20.5}$$

This represents a family of parabolas, as shown in Fig. 20.10a. But it cannot be right: the curves cross at every point, although dy/dx has only one value at any point. In fact, since $y^{\frac{1}{2}} > 0$, only the positive value of $y'(x)$ is legitimate, and this gives the right-hand branches of Fig. 20.10b.

We also lost a solution, namely $y(x) = 0$. This is connected with the fact that, for this solution, we in effect divided by zero when we first separated the equation.

The production of **pseudosolutions** and the nonappearance of certain **singular solutions** in the final formula, as in Example 20.7, is a problem endemic to techniques for nonlinear differential equations.

20.4 Differentials and the solution of first-order equations

In this section, it is important to distinguish between an **identity** such as $d(y^2)/dx = 2y\, dy/dx$, which is true for any $y(x)$ (it is just a special case of the chain rule), and an **equation** such as $dy/dx = xy$, which will only be true for special functions $y(x)$.

Take an identity such as

$$\frac{d}{dx} x^2 = 2x, \tag{20.7a}$$

and consider another way of writing it:

$$d(x^2) = 2x\, dx. \tag{20.7b}$$

It is as if we multiply (20.7a) by dx and cancel the dx on the left to obtain (20.7b). Conversely, if we divide (20.7b) by dx, we

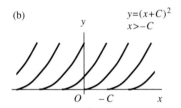

(a)

(b)

Fig. 20.10
(a) $y = (x + C)^2$ for various C. (b) The solutions of the differential equation, consisting only of the right-hand branches of the parabolas. (Note: $y(x) = 0$ is also a valid solution.)

recover (20.7a). We already used this process to help to change the variable in an integral in Section 15.1.

Now consider a more complicated identity, obtained from the product rule for differentiation ($y(x)$ represents any function of x):

$$\frac{d}{dx}(xy) = y + x\frac{dy}{dx}. \tag{20.8a}$$

A parallel expression of the same identity, obtained as before, is

$$d(xy) = y\,dx + x\,dy. \tag{20.8b}$$

Given either one of them, we can immediately construct the other, so we shall regard such pairs of expressions as being simply different ways of writing the same thing. In effect we did a similar thing when carrying out the separation-of-variables process for differential equations in Section 20.3, and we are leading up to a generalization of this method.

In general, a **differential expression** or **differential form** has the shape

$$P(x, y)\,dx + Q(x, y)\,dy, \tag{20.9}$$

where $P(x, y)$ and $Q(x, y)$ are two functions of x and y. In (20.7b), we had $P(x, y) = 2x$ and $Q(x, y) = 0$; in (20.8b), we had $P(x, y) = y$ and $Q(x, y) = x$. The symbols on the left of (20.7b) and (20.8b), $d(x^2)$ and $d(xy)$, are called the **differentials** of x^2 and xy respectively.

The table (20.10) (below) gives a list of useful identities written in the usual form and the differential form for comparison.

Standard form	Differential form	
$\dfrac{d}{dx}(C) = 0 \quad (C \text{ constant})$	$dC = 0$	
$\dfrac{d}{dx}(x^2) = 2x$	$d(x^2) = 2x\,dx$	
$\dfrac{d}{dx}(y^2) = 2y\dfrac{dy}{dx}$	$d(y^2) = 2y\,dy$	
$\dfrac{d}{dx}(xy) = x\dfrac{dy}{dx} + y$	$d(xy) = y\,dx + x\,dy$	(20.10)
$\dfrac{d}{dx}\left(\dfrac{y}{x}\right) = \dfrac{1}{x^2}\left(x\dfrac{dy}{dx} - y\right)$	$d\left(\dfrac{y}{x}\right) = -\dfrac{1}{x^2}(y\,dx - x\,dy)$	
$\dfrac{d}{dx}\left(\dfrac{x}{y}\right) = \dfrac{1}{y^2}\left(y - x\dfrac{dy}{dx}\right)$	$d\left(\dfrac{x}{y}\right) = \dfrac{1}{y^2}(y\,dx - x\,dy)$	
$\dfrac{d}{dx}\left(\ln\dfrac{y}{x}\right) = \dfrac{1}{xy}\left(x\dfrac{dy}{dx} - y\right)$	$d\left(\ln\dfrac{y}{x}\right) = -\dfrac{1}{xy}(y\,dx - x\,dy)$	

Differential forms can be manipulated. For example:

(i) $2x \, dx - y \, dy = d(x^2) - d(\tfrac{1}{2}y^2) = d(x^2 - \tfrac{1}{2}y^2)$.

(ii) $(x + y) \, dx + x \, dy = x \, dx + (y \, dx + x \, dy)$
$$= d(\tfrac{1}{2}x^2) + d(xy) = d(\tfrac{1}{2}x^2 + xy).$$

(iii) If $u(x)$ and $v(x)$ are two functions, then
$$d(uv) = u \, dv + v \, du;$$

which is the product rule for derivatives in differential form. These results are all identities; i.e. true for all functions $y(x)$, $u(x)$, $v(x)$.

Example 20.8. Put $d(x^3 + x \sin y)$ into the form $P \, dx + Q \, dy$. We have
$$d(x^3 + x \sin y) = d(x^3) + d(x \sin y),$$

for the first term, $(d/dx)x^3 = 3x^2$, and in differential form this becomes $d(x^3) = 3x^2 \, dx$. For the second term, we can use the product rule in the form (iii) above (or write it in standard form first):
$$d(x \sin y) = \sin y \, dx + x \, d(\sin y)$$

followed by the chain rule
$$d(x \sin y) = \sin y \, dx + x \cos y \, dy.$$

Finally,
$$d(x^3 + x \sin y) = (3x^2 + \sin y) \, dx + (x \cos y) \, dy,$$

so that, in (20.9), $P(x, y) = 3x^2 + \sin y$ and $Q(x, y) = x \cos y$.

First order **differential equations** can also be written as differential forms. The simplest is the equation
$$\frac{dy}{dx} = 0,$$

which has solutions $y(x) = C$, where C is any constant. In differential form, the equation becomes
$$dy = 0,$$

and the solutions are compatible with the first entry in the table above: $dC = 0$ when C is a constant.

Example 20.9. Find solutions of the equation $\dfrac{dy}{dx} = -\dfrac{x^3}{y^2}$.

In differential form, this becomes
$$x^3 \, dx + y^2 \, dy = 0.$$

But
$$x^3 \, dx + y^2 \, dy = d(\tfrac{1}{4}x^4) + d(\tfrac{1}{3}y^3) = d(\tfrac{1}{4}x^4 + \tfrac{1}{3}y^3).$$

This will be zero if

$$\tfrac{1}{4}x^4 + \tfrac{1}{3}y^3 = C,$$

where C is, in this case, any constant. The equation is in fact separable; you should compare this with the process in Section 20.3.

Example 20.10. Find solutions of $\dfrac{dy}{dx} = \dfrac{x - y}{x + y}$.

In differential form, this becomes

$$0 = (x - y)\,dx - (x + y)\,dy$$
$$= x\,dx - y\,dx - x\,dy - y\,dy.$$

Try to rearrange it so that recognizable forms appear:

$$0 = x\,dx - y\,dy - (y\,dx + x\,dy)$$
$$= d(\tfrac{1}{2}x^2) - d(\tfrac{1}{2}y^2) - d(xy)$$
$$= d(\tfrac{1}{2}x^2 - \tfrac{1}{2}y^2 - xy).$$

This differential will be zero as required if

$$\tfrac{1}{2}x^2 - \tfrac{1}{2}y^2 - xy = C,$$

where C is the 'variable constant', or **parameter**, which will generate a whole family of solutions.

In the previous example, the terms were rearranged in a search for a group like $y\,dx + x\,dy$ that would simplify, in that case to $d(xy)$. If a differential form can be expressed *identically* (that is to say, for all $y(x)$) in the form of a single differential:

$$P(x, y)\,dx + Q(x, y)\,dy \equiv dF(x, y)$$

(a fixed function of x and y), then it is called a **perfect differential form**, or a **perfect differential**. Usually this is impossible. For example, consider the differential form $y\,dx$. It can be proved that there *does not exist* any fixed function $F(x, y)$ such that

$$y(x)\,dx = dF(x, y(x)) \quad \text{for every } y(x)$$

(try looking for one).

To solve differential equations by this method, we search for a perfect differential in order to be able to conclude with the steps

$$`dF(x, y) = 0; \quad \text{therefore } F(x, y) = C',$$

as in the examples. If a perfect differential is not already present, we might be able to produce one by multiplying through by a suitable function, called an **integrating factor** for the expression. For example, $y\,dx - x\,dy$ is not a perfect differential; but, from (20.10),

$$\frac{1}{x^2}(y\,dx - x\,dy) = d\left(-\frac{y}{x}\right),$$

so the new expression is a perfect differential.

> **Perfect differential forms**
>
> Let y be an arbitrary function of x. Then $P(x, y)\,dx +$
> $Q(x, y)\,dy$ is a perfect differential if it can be written
> as
> $$P(x, y)\,dx + Q(x, y)\,dy \equiv dF(x, y),$$
> where $F(x, y)$ is a fixed function of x and y.
>
> **(20.11)**

> **Integrating factor for differential forms**
>
> A function $I(x, y)$ is an integrating factor for the
> differential form $P\,dx + Q\,dy$ if $I(P\,dx + Q\,dy)$ is a
> perfect differential.
>
> **(20.12)**

It can be proved that every differential form has an integrating factor, but only occasionally is it easy to see one.

Example 20.11. Find a family of solutions of $x\dfrac{dy}{dx} = y$.

(This is a linear equation, which is also separable, so we have other methods for solving it.) In differential form:

$$y\,dx - x\,dy = 0,$$

and we cannot do anything with the left-hand side as it stands. However, the remark above suggests we multiply by $1/x^2$, obtaining

$$0 = (1/x^2)(y\,dx - x\,dy) = d(-y/x).$$

Therefore

$$-y/x = C, \quad \text{or} \quad y = -Cx,$$

are the solutions, as is easily confirmed.

There are other possibilities; for example (see (20.10)) we might divide by y^2 or xy. In the end, these lead to the same set of solutions.

Note that, in Example 20.11, the equation is linear:

$$\frac{dy}{dx} + \left(-\frac{1}{x}\right)y = 0,$$

and so we can alternatively use the method of Section 17.6. An 'integrating factor' $I(x)$ is also used there: it is

$$I(x) = e^{-\int x^{-1}\,dx} = e^{-\ln x + C} = 1/x$$

(where we choose $C = 0$ for simplicity), which is different from, though related to, the ones which work for the differential form $y\,dx - x\,dy$ above.

Example 20.12. Find a set of solutions of $x\dfrac{dy}{dx} = y + y^2 x$.

Equivalently, $y \, dx - x \, dy + y^2 x \, dx = 0$. The first two terms cannot be written as $dF(x, y)$. The table (20.10) offers three integrating factors, x^{-2}, y^{-2}, and $(xy)^{-1}$ to choose from. It is, however, also necessary to be able to manage the remaining term, $y^2 x \, dx$, after multiplying by the integrating factor, so we choose y^{-2}, which gives

$$0 = (1/y^2)(y \, dx - x \, dy) + x \, dx$$
$$= d(x/y) + d(\tfrac{1}{2}x^2) = d(x/y + \tfrac{1}{2}x^2).$$

Therefore $x/y + \tfrac{1}{2}x^2 = C$, or $y = x/(C - \tfrac{1}{2}x^2)$, are solutions.

20.5 Change of variable in a differential equation

Occasionally we can find a **change of variable**, or **substitution**, which will simplify a differential equation. Some general types are given here for illustration.

> **Equations not involving** y
>
> If a differential equation contains
> $$x, \quad dy/dx, \quad d^2y/dx^2, \quad \dots,$$
> but not y, substitute the independent variable $w = dy/dx$ in place of y, producing an equation for w of lower order.   (20.13)

To see how this works, consider the following example.

Example 20.13. Find solutions of $\dfrac{d^2y}{dx^2} + \left(\dfrac{dy}{dx}\right)^2 = 0$.

The variable y is not independently present, so put

$$w = \frac{dy}{dx}.$$

Then $d^2y/dx^2 = dw/dx$, so that the equation becomes

$$\frac{dw}{dx} + w^2 = 0.$$

This is a separable equation, and the method of Section 17.3 gives

$$-\int \frac{dw}{w^2} = \int dx, \quad \text{or} \quad \frac{1}{w} = x + A,$$

where A is constant, so that

$$w = \frac{1}{x + A}.$$

For the second stage, remember that $w = dy/dx$, so we have

$$\frac{dy}{dx} = \frac{1}{x + A}.$$

Therefore

$$y = \ln|x + A| + B,$$

where A and B are constants which we see, in retrospect, may be chosen entirely arbitrarily.

Sometimes it is possible to change the independent variable y into something else to obtain a more manageable equation:

Equation of the form $\dfrac{dy}{dx} = f\left(\dfrac{y}{x}\right)$.

Change to a new dependent variable v by

$$v = y/x \hspace{4cm} \textbf{(20.14)}$$

and solve the resulting separable equation.
(To make the change write $y = xv$, so that

$$dy/dx = x\, dv/dx + v.)$$

Example 20.14. Find solutions of $\dfrac{dy}{dx} = \dfrac{3y - x}{3x - y}$.

This equation can be written in the form

$$\frac{dy}{dx} = \frac{3y/x - 1}{3 - y/x},$$

which has the form $f(y/x)$, so change the dependent variable from y to $v = y/x$. To obtain dy/dx in terms of v, write $y(x) = xv(x)$. Then $dy/dx = x\, dv/dx + v$, and in terms of v the equation becomes

$$x\frac{dv}{dx} + v = \frac{3v - 1}{3 - v}, \quad \text{or} \quad x\frac{dv}{dx} = \frac{v^2 - 1}{3 - v}.$$

This new equation is separable (Section 20.3) (a separable equation will always be obtained at this stage). Following (20.3):

$$\int \frac{1}{x}\, dx = \int \frac{3 - v}{v^2 - 1}\, dv = \int \left(\frac{1}{v - 1} - \frac{2}{v + 1}\right) dv.$$

Therefore $\ln|[(v - 1)/(v + 1)^2]| = \ln|x| + C$, where C is an arbitrary constant. After returning to y and simplifying, we have

$$(y - x)/(y + x)^2 = c, \hspace{3cm} \textbf{(20.15)}$$

where $c = \pm e^C$. The solution curves are shown in Fig. 20.11, plotted by working directly from the differential equation and using a numerical method (see Section 20.2).

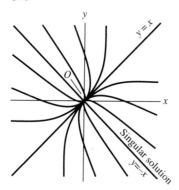

Fig. 20.11
Solution curves for

$$\frac{dy}{dx} = \frac{3y - x}{3x - y}.$$

Notice the special solutions
$y = \pm x$.

The methods of separation of variables, differentials, substitutions, etc. used to solve nonlinear equations are rather hazardous. In Example 20.14 there are two solutions which do not appear in (20.15), represented by the straight lines $y = \pm x$ in Fig. 20.11 and corresponding to the limiting cases $C \to \pm \infty$. Therefore (20.15) is not a truly general solution. These extra **singular solutions** can be found by trying the form $y = mx$ in the equation and solving the resulting quadratic for m.

Singular solutions are important in the theory of vibrations, population problems, and other nonlinear fields. In general, they represent limiting cases of the ordinary solutions; in this case, $y = x$ is an **envelope** of the ordinary solutions: that is, a curve that is tangential to each ordinary solution (see Example 27.10). This was the case with the singular solution $y(x) = 0$ of Example 20.7.

The following example illustrates another way of transforming a differential equation. In a mechanical context, the transformation of the derivative d^2x/dt^2 involved is called the **energy transformation**.

Example 20.15. The acceleration of a vehicle is constrained by its velocity: $d^2x/dt^2 = Kv^{-2}$, where $v = dx/dy$ and K is a constant. Find the velocity as a function of distance if $x(0) = 0$ and $v(0) = 0$.

Transform the acceleration by the chain rule, using x as the intermediate variable:

$$\frac{d^2y}{dt^2} = \frac{dv}{dt} = \frac{dv}{dx}\frac{dx}{dt} = v\frac{dv}{dx} = \tfrac{1}{2}\frac{d(v^2)}{dx}.$$

The differential equation now relates v to x:

$$\tfrac{1}{2}\frac{d(v^2)}{dx} = Kv^{-2}.$$

We could put $v^2 = u$ and separate, but for the sake of adventurousness we shall leave v^2 as it is and turn the equation upside down:

$$2\frac{dx}{d(v^2)} = K^{-1}v^2.$$

Now integrate both sides, courageously using v^2 as the variable:

$$2x = K^{-1}\int v^2\, d(v^2) = \tfrac{1}{2}K^{-1}(v^2)^2 + C,$$

or

$$v^4 = 4Kx - 2CK.$$

The initial condition $x(0) = v(0) = 0$ then gives

$$v(x) = (4Kx)^{\frac{1}{4}}.$$

Problems, Chapter 20

In these problems, y' means dy/dx.

20.1. Sketch a lineal-element diagram for the solution curves of each of the following.

(a) $y' = -y$; (b) $y' = x - y$; (c) $y' = x/y$;

(d) $y' = xy$; (e) $y' = -y/x$; (f) $y' = y/x$;

(g) $y' = (x - 1)y$; (h) $y' = \dfrac{1}{x^2 + y^2}$;

(i) $y' = \dfrac{1}{x^2 + y^2 - 1}$; (j) $y' = (1 - y^2)^{\frac{1}{2}}$;

(k) $y' = (y/x)^{\frac{1}{2}}$. Make sure that not all your curves lie in the first quadrant.

20.2. (Computational). Use the Euler method to compute approximate solutions to the following initial-value problems. Try various values of the step h. Compare the results with the exact solutions provided.
(a) $y' = -\frac{1}{2}y$ with $y = 1$ at $x = 0$, over the range $0 \leqslant x \leqslant 2$. (The exact solution is $y = e^{-\frac{1}{2}x}$.)
(b) $y' = -x/y$ with $y = -1$ at $x = -1$, over the range $-1 \leqslant x \leqslant 1$. (The exact solution is $y = -(2 - x^2)^{\frac{1}{2}}$. Try to extend your results forward and backward (using negative h) to the range $-2 \leqslant x \leqslant 2$.)
(c) $y' = (1 - y^2)^{\frac{1}{2}}$, with $y = 0$ at $x = 0$, over the range $0 \leqslant x \leqslant 3.1416$. (The exact solution is $y = \sin x$.)

20.3. (Computational). Use the Euler method to calculate a few representative solution curves in the following cases. Each curve will have a different initial condition. You should follow each curve fowards, and probably backwards as well (negative h), sufficiently far to get a clear idea of how it is behaving. It might be advantageous to use smaller intervals h over some sections than over others.
(a) $y' = y(x + 1)/x(y + 1)$. This refers to Example 20.6, with the 'solutions' $y e^y = Ax e^x$, where A is an arbitrary constant.
(b) $y' = 2y^{\frac{1}{2}}$. This refers to Example 20.7.
(c) $y' = (y/x)^{\frac{1}{2}}$ has solution curves in the first and *third* quadrants. Sketch a lineal element diagram to obtain the broad pattern, then compute a few representative curves. (The general solution is

$$|y| = (|x|^{\frac{1}{2}} - C)^{\frac{1}{2}} \quad \text{for } |x|^{\frac{1}{2}} > C. \;)$$

20.4. (Separation of variables). Obtain solutions of the following equations.
(a) $y' = x/y$; (b) $y' = 2x/y$;
(c) $y' = x/(y + 2)$; (d) $y' = (x + 3)/(y + 2)$;
(e) $y' = x^2/y^2$; (f) $y' = -x^2/y^2$;
(g) $y' = y^2/x^2$; (h) $y' = -y^2/x^2$;
(i) $2xy' = y^2$; (j) $yy' + x = 1$;
(k) $\dfrac{dx}{dt} = 3t^2x^2$; (l) $(\sin x)\dfrac{dx}{dt} = t$;

(m) $e^{x+y}\dfrac{dy}{dx} = 1$; (n) $(1 + x^2)\dfrac{dy}{dx} + (1 + y^2) = 0$, with $y(0) = -1$.

20.5. Show that the solution of the initial-value problem

$$\frac{dy}{dx} = -\frac{x}{y},$$

where $y = 1$ when $x = 2$, is obtained from the equation

$$\int_2^x u \, du = -\int_1^y v \, dv.$$

Generalize this technique to apply to the initial-value problem

$$\frac{dy}{dx} = g(x)h(y)$$

where $y = b$ when $x = a$.

20.6. Solve the following equations and sketch the solution curves. Take care to avoid spurious solutions as in Example 20.7. Look out for solutions you might have lost in the process: these are usually suggested by the sketch.
(a) $x\dfrac{dy}{dx} = 2y^{\frac{1}{2}}$; (b) $\dfrac{dy}{dx} = xy^{\frac{1}{2}}$;

(c) $\dfrac{dy}{dx} = (1 - y^2)^{\frac{1}{2}}$; (d) $x\dfrac{dy}{dx} = (1 - y^2)^{\frac{1}{2}}$.

20.7. (Differential method). Obtain a family of solutions of the following equations. (Usually they must be left in implicit form.)
(a) $\dfrac{dy}{dx} = \dfrac{2x - y}{x + 2y}$ (check whether there is also a solution of the form $y = mx$);

(b) $\dfrac{dy}{dx} = \dfrac{y}{y^2 - x}$;

(c) $\dfrac{dy}{dx} = \dfrac{x^2 - y}{x + y}$;   (d) $\dfrac{dy}{dx} = \dfrac{2x - y}{x - 2y}$;

(e) $\dfrac{dy}{dx} = \dfrac{x - 2xy}{x^2 - y}$;   (f) $\dfrac{dy}{dx} = \dfrac{3x^2}{3y^2 + 1}$;

(g) $\dfrac{dy}{dx} + \dfrac{2xy}{x^2 - 1} = 0$ (this is also a linear equation);

(h) $(1 - \sin y)\dfrac{dy}{dx} + \cos x = 0$;

(i) $(1 + 3\,e^{3y})\dfrac{dy}{dx} = 2\,e^{2x} - 1$;

(j) $(e^{x+y} + 1)\dfrac{dy}{dx} + (e^{x+y} - 1) = 0$;

(k) $\dfrac{dy}{dx} = \dfrac{1 + \cos x \sin y}{1 + \sin x \cos y}$.

20.8. (Differential method). Some of these need an integrating factor (see equation (20.12)) such as the ones suggested.

(a) $\dfrac{dy}{dx} = \dfrac{y}{x}\dfrac{y - 2x}{x - 2y}$ (Check also for solutions of the form $y = mx$)

(b) $\dfrac{dy}{dx} = \dfrac{y(1 - x^2)}{x(1 + x^2)}$ (divide by x^2);

(c) $\dfrac{dy}{dx} = \dfrac{y^2}{y^2 - 1}$ (divide by y^2);

(d) $\dfrac{dy}{dx} = \dfrac{y(y + 1)}{y^2 - x}$ (divide by y^2);

(e) $\dfrac{dy}{dx} = \dfrac{y(x^2 + y^2 - y)}{x(x^2 + y^2)}$ (divide by $x^2 y^2$);

(f) $\dfrac{dy}{dx} = \dfrac{y\,x^3 - y}{x\,x^3 + y}$ (show that this reduces to

$x^3 y^2\, d(x/y) = y\, d(xy)$;

now put $u = xy$ and $v = x/y$).

20.9. The 'logistic equation' $dP/dt = aP - bP^2$, where $a > 0$ and $b > 0$, represents the growth of a population $P(t) > 0$ in which unrestricted growth is prevented by the term $-bP^2$ representing pressure on the means of subsistence. Solve the equation, sketch the solution curves, and show that in all circumstances $P(t) \to a/b$ as t tends to infinity.

20.10. (Computational). A population $P(t)$ of protozoa is assumed to increase according to the equation $dP/dt = aP - bP^2$, where a and b are constants. Starting with 10 protozoa, they are observed to increase by 150% per day while the numbers are still low, and to reach a fairly steady level ($dP/dt = 0$) of 25,000 after a few weeks. Find an approximation to a and b.

Use a numerical method to compute the population curve for the first 10 days.

Compare the curve obtained from the law $dP/dt = aP - bP^4$.

20.11. (See Example 20.15). (a) A falling body of mass m is subject to air resistance equal to Kv^α, where v is its speed of fall: its equation of motion is then

$$\dfrac{d^2 x}{dt^2} = g - \dfrac{K}{m}\left(\dfrac{dx}{dt}\right)^\alpha,$$

where g is the gravitational acceleration and x represents its position measured vertically downwards. Without solving the equation, show that the limiting speed of fall is equal to $(mg/K)^{1/\alpha}$.

(b) Substitute $v = dx/dt$ to obtain an equation for v^2 of the form

$$\dfrac{d(v^2)}{dt} = 2\left(g - \dfrac{K}{m}(v^2)^{\frac{1}{2}\alpha}\right).$$

(c) Assume that (in mks units) $K = 4$, $m = 80$, $\alpha = 1.2$, $g = 10$, and that the mass is dropped from rest. Use Euler's method (20.2) to obtain v^2, and hence v, over a sufficient distance to compare with the limiting speed of fall.

20.12. (See Section 17.5). Solve the following (implicitly) by putting $y = xw$.
(a) $dy/dx = (x^2 - xy + y^2)/xy$;
(b) $dy/dx + (x^2 + y^2)/xy = 0$;
(c) $dy/dx + (x - y)/(3x + y) = 0$;
(d) $dy/dx = 2xy/(3x^2 - 4y^2)$;
(e) $dy/dx + 2(2x^2 + y^2)/xy = 0$.

20.13. Show that, if the substitution $w = y^{1-n}$ is made in the equation $y' + g(x)y = h(x)y^n$ (called the Bernoulli equation), we obtain the linear equation $w' + (1 - n)g(x)w = (1 - n)h(x)$.

Use this result to solve (a) $y' + y = y^4$,
(b) $y' + y = y^{-\frac{1}{2}}$.

20.14. The equation

$$d^2y/dx^2 + (b/x)\,dy/dx + (c/x^2)y = 0$$

is called **equidimensional** (the dx^2, $x\,dx$, and x^2 in the denominators are considered to have the same dimensions). It is a linear equation with zero on the right, so we expect a general solution of the form $Ay_1(x) + By_2(x)$, where A and B are arbitrary. Show how a basis of solutions $(y_1(x), y_2(x))$ can be obtained in two ways:

(a) Look for solutions having the form $y = x^M$, where M is an unknown constant. Note that M might be complex; in that case, to obtain *real* solutions, use

$$x^{\alpha + j\beta} = e^{(\alpha + j\beta)\ln x} = e^{\alpha \ln x}\,e^{j\beta \ln x}$$
$$= x^{\alpha}[\cos(\beta \ln x) + j \sin(\beta \ln x)],$$

as in Section 16.5.

(b) Change the independent variable to t, where $t = \ln x$, or $x = e^t$. The new equation has constant coefficients.

(c) Use either method to find the solutions of the equations

(i) $d^2y/dx^2 - (2/x)\,dy/dx + (2/x^2)y = 0$;
(ii) $d^2y/dx^2 - (1/x)\,dy/dx + 1/x^2 = 0$;
(iii) $d^2y/dx^2 + (3/x)\,dy/dx + (2/x^2)y = 0$.

20.15. (Computational). A boat enters a river at O, and tries to reach the point A on the other bank, directly opposite O and distant H from O, by keeping its nose pointed towards A, at an angle θ from OA (see Fig. 20.12). The speed of the boat in still water is V, and the uniform stream speed is $v < V$.

Show that, when the boat is at (x, y),

$$dx/dt = V\cos\theta, \qquad dy/dt = v - V\sin\theta$$

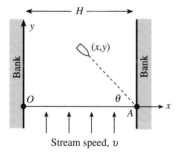

Fig. 20.12

where t is time. By dividing one equation by the other find a differential equation for y in terms of x. Given the values $v = 1$ m s^{-1}, $V = 4$ m s^{-1}, $H = 30$ m, compute the path of the boat. (As you approach close to A, you will encounter a problem with dy/dx.)

20.16. (Computational). As in Problem 20.15, but construct a differential equation for a stream having a parabolic distribution of velocity, greatest in the middle and zero att the banks, of the form

$$v(x) = ax(H - x).$$

Put in plausible values for V, v, H, and a, and compute the path.

20.17. A mouse M enters a room at O and rushes to its hole at H with speed v, pursued by the cat C, who

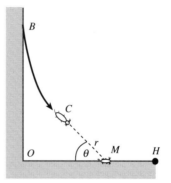

Fig. 20.13

starts from B at the same moment as the mouse appears. The cat runs with speed $V > v$, always directly towards the mouse. Show that

$$dr/dt = v\cos\theta - V \quad \text{and} \quad d\theta/dt = -(v\sin\theta)/r,$$

where r and θ are polar coordinates for the cat relative to the (moving) mouse. Construct a differential equation for r in terms of θ, and solve it (it is really only a question of integration).

21 Introduction to the phase plane

However many methods may be invented for solving differential equations, there will always remain equations beyond their scope. But this does not mean that nothing can be done with them. The important van der Pol equation, which models a type of electrical oscillator:

$$\frac{d^2x}{dt^2} + c(x^2 - 1)\frac{dx}{dt} + x = 0,$$

where $c > 0$, cannot be solved explicitly. However, there are still comparatively simple methods which enable us to demonstrate its really important feature, which is that every solution, no matter how the device is started off, settles down into the same regular period oscillation.

Techniques enabling such conclusions to be drawn without actually solving the equation are called **qualitative methods**. This chapter outlines a way of looking at differential equations which is at the basis of many of these techniques. Qualitative methods do not consist of a collection of fixed results, and tend to be exploratory. Therefore computation is important. In the final section, a simple computing method is described which is easy to program, but is effective enough to analyse realistic physical and biological models.

We shall take t (time) as the independent variable. For derivatives with respect to time, we use the conventional **dot notation** (just like the dash notation (4.1)):

$$\dot{x} = \frac{dx}{dt}, \qquad \ddot{x} = \frac{d^2x}{dt^2}.$$

21.1 Autonomous second-order equations
Let the independent variable be t and the dependent variable x. We shall only discuss equations which can be written in the form

$$\ddot{x}(t) = Q\big(x(t), \dot{x}(t)\big),$$

in which t **does not appear independently under Q on the right-hand side.** Such equations are called **autonomous**. For example, the equation $\ddot{x} - x\dot{x} + 1 = 0$ is autonomous, but $t\ddot{x} - x\dot{x} + 1 = 0$ is not autonomous. If startup conditions are supplied so as to specify an **initial-value problem**:

$$\ddot{x} = Q(x, \dot{x}), \qquad x(t_0) = x_0, \quad \dot{x}(t_0) = y_0 \qquad \text{(21.1)}$$

(there is a special reason for using the symbol y_0 here), we expect that the initial conditions will select exactly one solution.

Suppose that the equation represents an electrical system, and that a graph of x against t for $t \geqslant t_0$ can be plotted automatically. If we find the clock in the plotter has been wrongly set, then it will not make the graph unusable; only its starting time will be wrong. Similarly, if we do one experiment starting at $t = 8.00$ hr and repeat it at $t = 13.00$ hr, the graphs plotted will be the same shape although the starting times are different (see Fig. 21.1). Intuition suggests that for autonomous equations, namely those in which t does not occur independently, it will not be an arbitrary clock time t_0 assigned to startup that counts, but the time elapsed from startup, $t - t_0$.

This intuition is correct. The mathematical reason is that a change of time scale from t to $t - t_0$ does not change the form of the differential equation, so the same phenomena follow. Put

$$T = t - t_0, \quad \text{so that} \quad x(t) = X(T).$$

Then $\mathrm{d}X/\mathrm{d}T = \mathrm{d}x/\mathrm{d}t$ and $\mathrm{d}^2X/\mathrm{d}T^2 = \mathrm{d}^2x/\mathrm{d}t^2$. Also $t = t_0$ becomes $T = 0$, so that the new initial-value problem is

$$\ddot{X} = Q(X, \dot{X}), \qquad X(0) = x_0, \quad \dot{X}(0) = y_0. \qquad \text{(21.2)}$$

The equation is unchanged, but the starting time is assigned the value zero. Suppose (21.2) is solved in terms of T. Restore t by putting $T = t - t_0$. The solution $x(t)$ of (21.1) is then a function only of $t - t_0$, so it depends only on the time elapsed from startup.

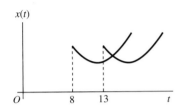

$x(t)$

O 8   13 t

Fig. 21.1
Two experiments, with a device described by an autonomous equation, which start at different times.

Example 21.1. Solve the initial-value problems $\ddot{x} + \omega^2 x = 0$, with $x(t_0) = x_0$ and $\dot{x}(t_0) = y_0$.

The general solution is $x(t) = A \cos \omega t + B \sin \omega t$. The process of finding A and B from the equations obtained by substituting the expressions for $x(t)$ and $\dot{x}(t)$ into the initial conditions is quite complicated (try it). Instead, put

$$T = t - t_0, \qquad x(t) = X(T).$$

Then

$$\ddot{X} + \omega^2 X = 0, \qquad X(0) = x_0, \quad \dot{X}(0) = y_0.$$

The solution of this system is simple:

$$X(T) = x_0 \cos \omega T + \omega^{-1} y_0 \sin \omega T.$$

Put $T = t - t_0$; then the required solution is

$$x(t) = x_0 \cos \omega(t - t_0) + \omega^{-1} y_0 \sin \omega(t - t_0),$$

that is to say, x is a function only of the elapsed time $t - t_0$.

> **Autonomous second-order equations**
> The solution of the initial-value problem
>
> $$\ddot{x} = Q(x, \dot{x}) \tag{21.3}$$
>
> with $x(t_0) = x_0$ and $\dot{x}(t_0) = y_0$ is a function only of elapsed time $t - t_0$.

21.2 Constructing a phase diagram for $(x, \dot{x})$

Consider again the initial-value problem of Example 21.1:

$$\ddot{x} + \omega^2 x = 0, \qquad x(t_0) = x_0, \quad \dot{x}(t_0) = y_0. \tag{21.4}$$

This problem could arise in connection with a mass oscillating on a spring. The initial conditions imply that the position and velocity are prescribed at the start, $t = t_0$. If asked what the system was doing at $t = t_0$, specification of the position and velocity seems to constitute an adequate description of its state. It is in fact a perfect description, since it is exactly what is required to determine the whole future of the system. It is therefore reasonable to call the pair of numbers

$$(x_0, y_0)$$

the **state of the system** at t_0.

Subsequently the system moves smoothly through a succession of states: x and $\dot{x}$ will vary in time. Catch the system at any moment t_1; then the state $\big(x(t_1), \dot{x}(t_1)\big)$ serves as fresh initial conditions for all the subsequent motion, but there can never be any conflict with what was predicted from the original initial conditions. It is the **succession of states** which is the subject of this chapter; the precise time that the states occur takes a secondary place.

To track the succession of states $\big(x(t), \dot{x}(t)\big)$ for the initial-value problem (21.4), we could in principle begin by finding its solution. The solution (Example 21.1) is

$$x(t) = x_0 \cos \omega t + \omega^{-1} y_0 \sin \omega t,$$

so

$$\dot{x}(t) = -\omega x_0 \sin \omega t + y_0 \cos \omega t.$$

In effect, these equations specify the states **parametrically** (with parameter t). However, the expressions do not clearly reveal the association of x with $\dot{x}$, which is what we actually *observe* from moment to moment. Moreover, we need a method for equations we cannot solve.

We shall take a different route to the states $(x, \dot{x})$ which does not require that we solve the differential equation. Write

$$\dot{x} = y. \tag{21.5a}$$

Since $\ddot{x} = -\omega^2 x$, and $\ddot{x} = (\mathrm{d}/\mathrm{d}t)\dot{x}$, we have

$$\dot{y} = -\omega^2 x. \tag{21.5b}$$

These two **simultaneous *first-order* differential equations** are equivalent to the second-order differential equation (21.4). Divide (21.5b) by (21.5a), and use (3.8):

$$\frac{dy}{dt}\bigg/\frac{dx}{dt} = \frac{dy}{dx} = -\omega^2\frac{x}{y}. \tag{21.6a}$$

Time has disappeared from the problem, and we now have a single first-order equation connecting x and y (i.e. connecting x and $\dot{x}$). Separate the variables in (21.6a), and we obtain

$$\int y\, dy = -\omega^2 \int x\, dx,$$

or

$$\omega^2 x^2 + y^2 = C, \tag{21.6b}$$

where C is a positive, but otherwise arbitrary, constant.

The motive for introducing y as the symbol for $\dot{x}$ now becomes clear. Set up a pair of axes x and y as in Fig. 21.2. This framework is called the $(x, \dot{x})$ **phase plane**. The solution (21.6b) represents the family of ellipses displayed in the figure, and is called an $(x, \dot{x})$ **phase diagram for the differential equation** $\ddot{x} + \omega^2 x = 0$.

Any point on the diagram, say $A : (x_0, y_0)$, represents an initial state. If we follow the curve passing through A, we obtain the sequence of states for the corresponding solution. Equation (21.6a) alone does not tell us which way to travel along the curve; for this purpose, we momentarily resurrect t. The arrows indicate the directions that correspond to time going forwards rather than backwards. We defined y by

$$\frac{dx}{dt} = \dot{x} = y.$$

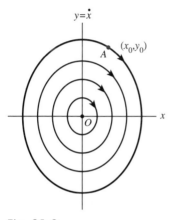

Fig. 21.2

$(x, \dot{x})$ phase diagram for $\ddot{x} + \omega^2 x = 0$, displaying $y = \dot{x}$ against x.

If we are in the **upper half plane**, then $y > 0$, so dx/dt is positive. Therefore $x(t)$ **is increasing**, and the directive arrow points **from left to right**. By a similar argument with $y < 0$ we find that in the **lower half plane** the arrow points **from right to left**. We must follow the arrow. Supplied with arrows, the state curves are called **phase paths,** or **trajectories**, or **orbits** for the differential equation $\ddot{x} + \omega^2 x = 0$.

Starting from $A : (x_0, y_0)$, follow the phase path. In going round, we can pick out a new feature in passing: that $\dot{x}$ is zero when x is at a minimum or maximum, and *vice versa*. Eventually we get back to A, renewing the initial state. Continue to follow the path around, duplicating the first circuit; the succession of states is repeated time after time.

This repetition does not itself establish that this is a truly *periodic* process. When we meet A again at the end of the first circuit, it is at a later time, t_1 say, so the initial conditions for the original equation are to this extent changed. Even though the system must follow the same path, perhaps it takes twice as long to go around the second time. However, from the discussion in Section 21.1, the time to complete any circuit, or to go repeatedly between any two fixed points on the circuit, is invariable because the equation is autonomous. This argument does not depend on what equation we started with, so we can say in general: **any closed phase path represents a periodic oscillation**.

Finally notice in Fig. 21.2 the bullet at the origin. This point represents a true solution, namely

$$x(t) = 0, \qquad y(t) = 0.$$

It is a special case of an **equilibrium point**, meaning a **constant solution**

$$x(t) = k, \qquad \dot{x}(t) = 0,$$

where k is a constant. Equilibrium points are of great importance in phase diagrams. A zero solution does not mean the same as no solution. Because the constant k is zero, this solution is called 'trivial'; but this does not lessen its significance. An equilibrium point surrounded by closed curves is called a **centre**. It represents periodic oscillations about equilibrium. (Note, however, that oscillations of different amplitudes do not usually have the same period.).

Since we chose a simple case, we have not discovered anything we did not know already, so consider the following example.

Example 21.2. Sketch an $(x, \dot{x})$ phase plane for the equation $\ddot{x} + cx^3 = 0$, where $c > 0$.

This represents small *lateral* oscillations $x(t)$ of a mass attached to the middle of an elastic string that is fixed at the ends and is *unextended* when $x = 0$. It is the same as Example 18.6, with $l = L$, and $c = 4s/ml$. We can regard explicit solutions as being unobtainable.

Put $y = \dot{x}$. Then $\ddot{x} = \dot{y}$, and we have two first-order equations, together equivalent to the original equation:

$$\dot{x} = y, \qquad \dot{y} = -cx^3,$$

Therefore

$$\frac{dy}{dx} = \frac{dy}{dt} \Big/ \frac{dx}{dt} = -c\frac{x^3}{y}.$$

By separating the variables, we obtain

$$\int y \, dy = -c \int x^3 \, dx,$$

so that $\frac{1}{2}y^2 = -\frac{1}{4}cx^4 + C$, where C is an arbitrary constant. Therefore

$$y = \pm(c/2)^{\frac{1}{2}}(A - x^4)^{\frac{1}{2}},$$

where A is arbitrary. This is the equation for the phase paths shown in Fig. 21.3.

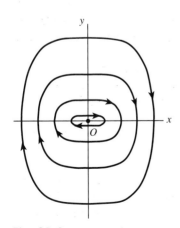

Fig. 21.3

The phase diagram of Fig. 21.3 consists entirely of closed curves; so, by exactly the same reasoning as in Example 21.1, we can deduce that every solution of the differential equation is a periodic oscillation. (However, we cannot say that they all have the same period: in fact they do not.) This phase diagram has therefore revealed an important fact about an equation that we could not solve.

21.3 $(x, \dot{x})$ **phase diagrams for other linear equations; stability**

For the moment, we shall stay with equations that are familiar from Chapter 16.

Example 21.3. Construct the phase diagram for $\ddot{x} - \omega^2 x = 0$.

Put $\dot{x} = y$; then $\dot{y} = \omega^2 x$. To eliminate t, form

$$\frac{\dot{y}}{\dot{x}} = \frac{dy}{dx} = \omega^2 \frac{x}{y}.$$

Separate the variables:

$$\int y \, dy = \omega^2 \int x \, dx,$$

or

$$y^2 - \omega^2 x^2 = A,$$

where A is arbitrary. This represents the family of hyperbolas in Fig. 21.4, having asymptotes $y = \pm \omega x$. The directions of the arrows follow the rule in Section 21.2: left to right in the upper half plane.

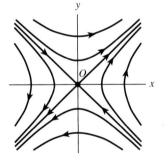

Fig. 21.4

Differential equations usually describe the behaviour of some circuit, machine, ecosystem, or something else in the real world, and our concern is with interpreting the phase diagrams in such a way as to bring to light features of practical importance. One important question is that of **stability of equilibrium** in a system. In our phase diagrams, the question turns into that of the **stability of an equilibrium point**, such as the origin in Examples 21.1 to 21.3.

In the real world, systems are never in perfect equilibrium, but are always subject to small external disturbances and internal fluctuations. For a system to work, it is important that small causes give small effects, and that the effects do not commence to grow catastrophically. In this case, a system is said to be **stable** with respect to small disturbances; otherwise it is **unstable**. The precise criterion for tolerable behaviour will depend on what we require from the particular system.

An equilibrium point surrounded by a structure of curves resembling those shown in Fig. 21.4 is called a **saddle**. It could hardly be classed as anything but an **unstable equilibrium point**. Apart from two special directions, if equilibrium is disturbed – even by a hairsbreadth – the system will find itself on one of the hyperbolas, and so it will be swept further and further away from equilibrium.

On the other hand a **centre**, exemplified in Figs 21.2 and 21.3, would often be called a **stable equilibrium point**. If equilibrium is disturbed by a small amount, the system does not go wild; it simply oscillates around its equilibrium position. However, a car which behaved like that would be regarded as very unstable. Subject to a continual battering, it would vibrate objectionably, and the vibrations would never die away; therefore the stronger sort of stability illustrated in the following two examples is preferable.

Example 21.4. Construct a phase diagram for $\ddot{x} + \frac{1}{4}\dot{x} + x = 0$.

The equivalent pair of first-order equations is

$$\dot{x} = y, \qquad \dot{y} = -\tfrac{1}{4}y - x. \tag{i}$$

Therefore

$$\frac{\mathrm{d}y}{\mathrm{d}x} = \frac{-\frac{1}{4}y - x}{y}. \tag{ii}$$

This is hard to solve. To produce a phase diagram, we could solve the original equation for $x(t)$ by the methods of Chapter 16, and then obtain $y = \dot{x}(t)$. These are parametric equations for (x, y) curves. However, this would not be in the spirit of this chapter, because almost never are we able to solve the original equation. Instead, the Euler method of Section 20.2 can be used to obtain solution curves for (ii). (As we shall see later, one can better work from (i) using Section 21.10.) We obtain the pattern of spiral curves surrounding the origin shown in Fig. 21.5.

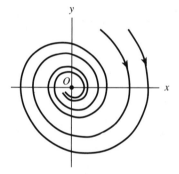

Fig. 21.5

Example 21.4 is a case of a linear oscillator with small damping, discussed in Section 16.4. The phase paths show that, from any starting point, the origin is approached via a sequence of diminishing spirals. Therefore any initial disturbance from equilibrium dies away. The equilibrium point is called a **stable spiral**. For the equation

$$\ddot{x} - \tfrac{1}{4}\dot{x} + x = 0,$$

the pattern of curves is the same, but the arrows are reversed, so we obtain an **unstable spiral**.

Example 21.5. Construct a phase diagram for $2\ddot{x} + 7\dot{x} + 3x = 0$.

This is a case of heavy damping: We obtain

$$\dot{x} = y, \qquad \dot{y} = -\tfrac{7}{2}y - \tfrac{3}{2}x,$$

or

$$\frac{\mathrm{d}y}{\mathrm{d}x} = \frac{-\frac{7}{2}y - \frac{3}{2}x}{y}.$$

The phase paths, calculated numerically, are as shown in Fig. 21.6.

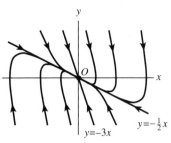

Fig. 21.6

In Fig. 21.6, the origin is called a **stable node**. All solutions fall straight into the origin without any oscillations: the system is **deadbeat**. Notice the structure of the node. There are two straight-line solutions to the equation

$$\frac{dy}{dx} = \frac{-\frac{7}{2}y - \frac{3}{2}x}{y},$$

which can be found by trying for solutions of the form $y = mx$. Then $dy/dx = m = (-\frac{7}{2}m - \frac{3}{2})/m$, or $2m^2 + 7m + 3 = 0$. Therefore $m = -\frac{1}{2}$ or $m = -3$, and the two linear solutions are $y = -\frac{1}{2}x$, $y = -3x$. They divide the plane into four sectors which contain curved phase paths. Each of the curves has the property that it is tangential to $y = -\frac{1}{2}x$ at the origin, and parallel to $y = -3x$ at infinity. This behaviour is characteristic of nodes arising from linear equations, and the mutual tangency at the origin is common to all nodes, even those arising from nonlinear equations.

The technique for second-order differential equations can be summed up as follows.

$(x, \dot{x})$ **phase plane for** $\ddot{x} = Q(x, \dot{x})$

(a) *Phase path equations*:

$$\dot{x} = y, \qquad \dot{y} = Q(x, y):$$

(b) *The direction* of a phase path (x, y) is left to right if $y > 0$, and right to left if $y < 0$.

(c) *Equilibrium points* (constant solutions) are at $(x, 0)$, where x is any solution of $Q(x, 0) = 0$.

(d) *Alternative equation* $dy/dx = Q(x, y)/y$. (Shows that phase paths meet only at equilibrium points or other points where $Q(x, y)/y$ is undefined, and that paths cross the x axis at right angles.)

(21.8)

21.4 The pendulum equation

The equation for a pendulum (Fig. 21.7) consisting of a light rod AB of length l freely pivoted at A and carrying a mass at B is

$$\ddot{x} + \omega^2 \sin x = 0,$$

where x is the angle of inclination and $\omega^2 = g/l$; here g is the gravitational acceleration. This equation can be solved, but only with difficulty, and by using recondite functions. From (21.7), the equations for the phase paths are

$$\dot{x} = y, \qquad \dot{y} = -\omega^2 \sin x. \tag{21.9}$$

The equilibrium points solve the equation $\sin x = 0$, so

$$x = 0, \pm\pi, \pm 2\pi, \ldots, \quad \text{with} \quad y = 0. \tag{21.10a}$$

If

$$x = 0, \pm 2\pi, \pm 4\pi, \ldots, \tag{21.10b}$$

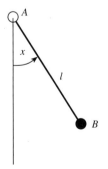

Fig. 21.7

the pendulum is hanging vertically from its pivot in equilibrium. These values of x all represent the same *observed* state, though on the phase plane they correspond to different points. Similarly the values

$$x = \pm\pi, \pm 3\pi, \ldots, \tag{21.10c}$$

represent a state which is not usually thought of in connection with a pendulum: the pendulum rod is perched vertically upwards (and insecurely) on its pivot A.

Consider $x = 0$ as being representative of the freely-hanging state, the equilibrium points (21.10b). See what happens when the displacement from $x = 0$ is small, but the pendulum is no longer in equilibrium. We can then put

$$\sin x \approx x.$$

The original equation becomes $\ddot{x} + \omega^2 x = 0$ (approximately), with solutions $x = C\cos(\omega t + \phi)$, where C (small) and ϕ are arbitrary. This is the familiar condition of small, isochronous oscillations. We have already solved the same problem for the phase plane in Section 21.2 and we found a centre: the family of ellipses shown in Fig. 21.2:

$$\omega^2 x^2 + y^2 = C,$$

where C is an arbitrary non-negative constant. This family will be repeated (for small C) around $x = \pm 2\pi, \pm 4\pi, \ldots$ in a progressively developing phase diagram; see Fig. 21.8 below.

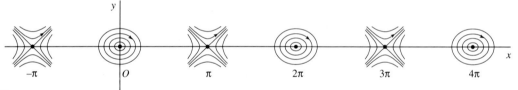

Fig. 21.8

Next, consider the case when the pendulum stands vertically: we choose as representative of (21.10c) the case $x = \pi$. To find what happens when the state is slightly displaced from the point $(\pi, 0)$, put

$$x = \pi + X,$$

where the new variable X is going to be small. Then

$$\sin x = \sin(\pi + X) = \sin\pi \cos X + \cos\pi \sin X$$
$$= -\sin X \approx -X.$$

From (21.8), with X instead of $\sin x$, the (approximate) equation for the displacement X from the equilibrium point is

$$\ddot{X} - \omega^2 X = 0.$$

From Example 21.3, this is equivalent to a saddle point on the phase plane at $X = 0$ (i.e. at $x = \pi$), and all the other equilibrium points in (21.10c) will have the identical structure, which implies instability.

Fig. 21.9
Phase diagram $(x, \dot{x})$ for the pendulum equation $\ddot{x} + \omega^2 \sin x = 0$, given by $y = \pm\sqrt{2\omega}(\cos x - A)^{\frac{1}{2}}$ with $A \leqslant 1$. The figure extends with period 2π. The undulating curves (for $A < -1$) represent a whirling motion. The separatrices correspond to $A = 1$. There are centres at $x = 0, \pm 2\pi, \ldots$, and saddles at $x = \pm\pi, \pm 3\pi, \ldots$.

The state of affairs around the equilibrium points is shown in Fig. 21.8. The rest of the phase diagram could be computed from (21.9), but in this case it is not difficult to sketch it in its entirety as

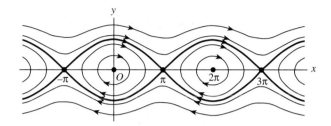

in Fig. 21.9. From (21.9),

$$\frac{dy}{dx} = \frac{\omega^2 \sin x}{y}.$$

By separating the variables, we obtain the equations of the paths, which can be written in the form

$$y = \pm\sqrt{2\omega}(\cos x - A)^{\frac{1}{2}}, \tag{21.11}$$

where A is (to an extent) arbitrary. Since $\cos x$ has period 2π, the repetitious nature of Fig. 21.8 is explained. Notice that $(\cos x - A)^{\frac{1}{2}}$ is real only when $\cos x \geqslant A$. Therefore $A \leqslant 1$. With that limitation, there are two main ranges of A which give significantly different patterns of (x, y) curves: $-1 \leqslant A \leqslant 1$ and $A < -1$. The centres correspond to $A = 1$, and the special curves joining the saddles, called the **separatrices**, correspond to $A = -1$. Notice the regular whirling motions which occur if $y = \dot{x}$ is large enough.

21.5 The general phase plane
There exists a great field of problems which, right from the start, take the form of simultaneous first-order differential equations:

$$\dot{x} = P(x, y), \qquad \dot{y} = Q(x, y). \tag{21.12}$$

Example 21.6. A community of foxes and rabbits lives in uneasy harmony on an island. The rabbit population is $x(t)$, and they eat grass. The fox population is $y(t)$; they eat rabbits. Construct a differential equation model for the population variation.

In a short time δt there is a rabbit population increase $ax\,\delta t$ $(a > 0)$ due to births and natural deaths, and a decrease $-bxy\,\delta t$ due to meetings with foxes, the frequency of which we suppose to be jointly proportional to the population densities of these animals. The net change in time δt is therefore

$$\delta x = ax\,\delta t - bxy\,\delta t.$$

Divide by δt and let $\delta t \to 0$; we obtain

$$\frac{\mathrm{d}x}{\mathrm{d}t} = ax - bxy. \tag{i}$$

For the foxes, assume that a shortage of prey causes a death rate c from starvation, offset by a fecundity factor dxy among those who get something to eat. Then, in time δt,

$$\delta y = -cy\,\delta t + dxy\,\delta t.$$

Divide by δt and let $\delta t \to 0$:

$$\frac{\mathrm{d}y}{\mathrm{d}t} = -cy + dxy. \tag{ii}$$

(i) and (ii) form a simultaneous (nonlinear) first-order system:

$$\dot{x} = ax - bxy, \qquad \dot{y} = -cy + dxy, \tag{21.13}$$

with $a, b, c, d, > 0$.

We shall now show how important characteristics of the solutions $x(t)$ and $y(t)$ of the equations (21.13) resulting from this example can be revealed on a **general phase plane** by plotting x and y. (Notice that $\dot{x}$ is no longer equal to y, so this is not the same as the $(x, \dot{x})$ phase plane that we had before.) The pair of values (x, y) will be called a **state**. Although we do not prove it, the values of x and y at a particular t constitute **initial conditions** determining the solution for all subsequent time $t > t_0$, and because the equations are **autonomous**, the solutions are functions only of $t - t_0$.

Firstly, $(x, y) = (k, l)$ is a **constant solution** provided that it satisfies the differential equations: that is to say, provided that

$$0 = x(a - by) \quad \text{and} \quad 0 = -y(c - dx).$$

Therefore $(x, y) = (0, 0)$ and $(x, y) = (c/d, a/b)$ are the constant solutions: on the (x, y) phase plane, they are the **equilibrium points**.

As before, divide the two equations, obtaining

$$\frac{\mathrm{d}y}{\mathrm{d}t} \bigg/ \frac{\mathrm{d}x}{\mathrm{d}t} = \frac{\mathrm{d}y}{\mathrm{d}x} = -\frac{(c - dx)y}{(a - by)x}. \tag{21.14}$$

This is a separable equation; after separation it becomes

$$\int \left(\frac{a}{y} - b\right) \mathrm{d}y = -\int \left(\frac{c}{x} - d\right) \mathrm{d}x,$$

or

$$a \ln y - by + c \ln x - dx = C. \tag{21.15}$$

Equation (21.15) represents the closed curves shown in Fig. 21.10. It is possible to see in advance that they are closed: the reason will be given shortly.

The direction arrows on the figure do not obey the rule (21.8b) for the $(x, \dot{x})$ phase plane. Each case has to be treated separately.

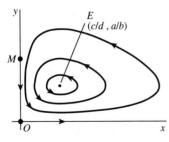

Fig. 21.10

The principle is easy: we have to find the **direction at a single point**, and the directions elsewhere are settled by **continuity of direction** – we expect adjacent curves to have the same direction. We might take the point $M : (0, m)$ in Fig. 21.10. At this point, the second equation of (21.13) gives $y = -cm < 0$, so y is decreasing at M. Once this direction is settled, the directions on the other curves follow by continuity.

There is a **centre** at the equilibrium point $E : (c/d, a/b)$. If the rabbit/fox populations take the values at E, the equations predict that the state will be permanent. A bad season for grass, or a disease amongst the foxes, will put the population state somewhere else, and thereafter the populations will undergo periodic oscillations. If foxes feast and thrive, rabbits languish, eventually starving the foxes; therefore rabbits prosper again; and so on.

The equilibrium point at O is unstable. If rabbits are introduced into a desert island paradise the population increases indefinitely, following the x axis arrow. If foxes are introduced to control the rabbits, a great periodic cycle is set up which goes on for as long as nothing else changes. Clearly the model is imperfect; but, provided that we have a program which will plot general phase paths, the complexity of the model is really a matter of no importance.

Earlier we said that the implicit equation (21.15) for the phase paths gives closed curves. It is useful to be able to recognize this feature.

Condition that $f(x) + g(y) = C$ (C arbitrary) represents a centre

If $f(x)$ has a minimum at $x = \alpha$, and $g(y)$ has a minimum at $y = \beta$, then there is an equilibrium point at (α, β) which is a centre (or substitute 'maximum' for 'minimum' in both places).   **(21.16)**

To prove this you need to look forward at Section 25.1. In three dimensions, x, y, z, the surface $z = f(x) + g(y)$ is bowl-shaped, with a minimum or maximum (α, β). The paths are the curves cut out by intersection with the horizontal planes $z = C$. The functions $c \ln x - dx$ and $a \ln y - by$ have maxima at $x = c/d$, $y = a/b$.

Finally, for a general system

$$x = P(x, y), \qquad y = Q(x, y),$$

the equilibrium points are where

$$P(x, y) = Q(x, y) = 0,$$

and might therefore appear anywhere in the phase plane, not just on the x axis as with the $(x, \dot{x})$ plane. The following statements recall the main features encountered in this section.

> **General phase plane**
>
> (a) *Phase-path equations*: $\dot{x} = P(x, y)$, $\dot{y} = Q(x, y)$.
> (b) *Equilibrium points*: the solutions of
> $P(x, y) = Q(x, y) = 0$.   **(21.17)**
> (c) *Phase-path direction*: find the direction at one
> point and use continuity for other paths.
> (d) *Alternative equation*: $\mathrm{d}y/\mathrm{d}x = Q(x, y)/P(x, y)$.

21.6 Approximate linearization

In connection with the pendulum problem of Section 21.4, we were
able to analyse equilibrium points by using a linear approximation
valid near the points. In the general case, suppose that $\dot{x} = P(x, y)$,
$\dot{y} = Q(x, y)$, and that (k, l) is an equilibrium point:

$$P(k, l) = Q(k, l) = 0. \tag{21.18}$$

We shall obtain a linear approximation to $P(x, y)$ and $Q(x, y)$ valid
near (k, l). Put

$$x = k + X, \qquad y = l + Y, \tag{21.19}$$

where we suppose that X and Y are small. Then, because of (21.18),
the approximations will take the form

$$P(x, y) = aX + bY, \qquad Q(x, y) = cX + dY, \tag{21.20}$$

(with no constant term present).

Equations (21.20) should provide information about the phase
paths near the equilibrium point $(x, y) = (k, l)$, which has become
$(X, Y) = (0, 0)$, and tell us at least whether they are stable or
unstable. This is often true, but not always: if, to take an extreme
case, $a = b = c = d = 0$, then we would hardly want to rely on it.

The single equation corresponding to (21.17d) which connects X
and Y in the approximation is

$$\frac{\mathrm{d}Y}{\mathrm{d}X} = \frac{cX + dY}{aX + bY} \tag{21.21}$$

The algebra involved in classifying this equation with respect to the
coefficients is very complicated, and we merely summarize the
results, omitting some special cases. For general reference, we will
express the results in the notation (x, y) instead of (X, Y).

> **Equilibrium point** $(0, 0)$ **of the linear system**
> $\dot{x} = ax + by$, $\dot{y} = cx + dy$
> Put $p = a + d$, $q = ad - bc$, $\Delta = p^2 - 4q$.
> (a) If $q > 0$ and $\Delta > 0$: a node ($p < 0$, stable; $p > 0$,
> unstable.)

> (b) If $q > 0$ and $\Delta < 0$: a spiral ($p < 0$, stable; $p > 0$, unstable.) **(21.22)**
> (c) If $q < 0$: a saddle.
> (d) If $p = 0$ and $q > 0$: a centre.
> (e) Path directions: investigate one point.

Since these equations are **linear**, the phase diagram centred on $(0, 0)$ is **self-similar**: the pattern of paths is the same if viewed through a microscope or seen over an immense field, so we are not restricted to small x and y.

In applying (21.22), do not be too ready to decide that the *original* equations have a centre just because the linear ones do: the small difference on changing back from the linear approximation may be all that is necessary to change a centre into a spiral.

Example 21.7. Classify the equilibrium points of the system
$$\dot{x} = x - y, \qquad \dot{y} = 1 - xy.$$
The equilibrium points are where $x - y = 0$, $1 - xy = 0$; that is, at $(1, 1)$ and $(-1, -1)$.

Near $(1, 1)$. Put $x = 1 + X$, $y = 1 + Y$. Then $x - y = X - Y$, and
$$1 - xy = 1 - (1 + X)(1 + Y) \approx -X - Y$$
for X and Y small. Therefore, in (21.21),
$$a = 1, \quad b = -1, \quad c = -1, \quad d = -1;$$
so $p = 0$, $q = -2$, $\Delta = 8$. According to (21.22), this is a saddle point (which is an unstable equilibrium point).

Near $(-1, -1)$. Put $x = -1 + X$, $y = -1 + Y$; then we obtain
$$x - y = X - Y, \qquad 1 - xy \approx -X - Y,$$
so that $a = 1$, $b = -1$, $c = 1$, $d = 1$. Therefore $q = 2 > 0$, $p = 2 > 0$, $\Delta = -4 < 0$; so, by (21.22), the point is an unstable spiral. The phase diagram is shown in Fig. 21.11.

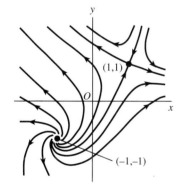

Fig. 21.11

21.7 Limit cycles

Spirals, centres, etc. occur for both linear and nonlinear systems, but a **limit cycle** is a feature only of nonlinear systems. When it occurs it usually represents the most important phenomenon in the phase plane. The following example includes a limit cycle.

Example 21.8. Sketch a phase diagram for
$$\ddot{x} + (x^2 + \dot{x}^2 - 1)\dot{x} + x = 0.$$
Put
$$\dot{x} = y, \qquad \dot{y} = (1 - x^2 - y^2)y - x, \tag{i}$$

It is possible to express the phase paths in polar coordinates r, θ:
$$r^2 = x^2 + y^2 \quad \text{and} \quad \tan \theta = y/x.$$
Differentiate these equations with respect to t;
$$r\dot{r} = x\dot{x} + y\dot{y},$$
$$\dot{\theta}/\cos^2 \theta = (x\dot{y} - y\dot{x})/x^2. \tag{ii}$$
Substitute (i) into (ii): remember that $\dot{x} = y$, and put $x = r \cos \theta$ and $y = r \sin \theta$ as necessary. Then $r(t)$ and $\theta(t)$ are found to satisfy
$$\dot{r} = -r(r^2 - 1) \sin^2 \theta, \tag{iii}$$
$$\dot{\theta} = -1 - (r^2 - 1) \sin \theta \cos \theta. \tag{iv}$$
A particular solution of (iii) and (iv) is $r = 1$, with $\dot{\theta} = -1$. This indicates a path consisting of the circle $r = 1$, followed around in the clockwise direction with unit angular velocity.

Also, from (iii),
$$\dot{r} \begin{cases} > 0 & \text{if } r < 1, \\ < 0 & \text{if } r > 1, \end{cases} \tag{v}$$

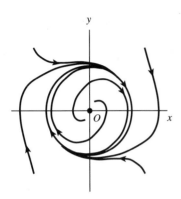

Fig. 21.12

so the circle is approached from points inside by means of expanding spirals, and from points outside by contracting spirals. The phase diagram is shown in Fig. 21.12.

If we start from any initial conditions except for the equilibrium point $(0, 0)$, the system settles down gradually to the regular oscillation represented by the circle. This behaviour has a physical explanation. The 'coefficient' $x^2 + \dot{x}^2 - 1$, although variable, serves the purpose of a damping coefficient. Outside the circle, when $x^2 + y^2 - 1 > 1$ (remember $y = \dot{x}$), energy is lost and the paths tend to drift inwards. When $x^2 + y^2 - 1 < 1$ there is negative damping; energy is being supplied, so the amplitude of paths within the circle increases. For points on the circle $x^2 + y^2 - 1 = 0$ the damping is zero, so the motion is harmonic (the solutions are $x = \cos(t + \phi)$, with ϕ any constant), consistent with the circular path.

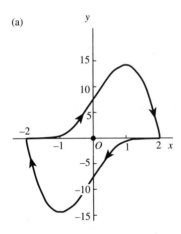

(a)

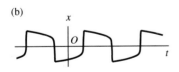

(b)

Fig. 21.13
(a) Limit cycle for
$\ddot{x} + 10(x^2 - 1)\dot{x} + x = 0$.
(b) The solution $x(t)$
corresponding to the limit
cycle.

The circular path $r = 1$ in Example 21.8 is an example of a **limit cycle**, which is defined generally as **an isolated closed phase path**. If the paths approach it spirally (in a broad sense) from both sides, it is called a **stable limit cycle**. It then represents a **stable oscillation**: if we disturb, or **perturb**, the oscillation by a small amount, it simply creeps back into the original oscillation. If the paths on one or both sides point away from the limit cycle, it is called **unstable** and is unlikely ever to be observed in practice.

To show how strange a limit cycle can be, we return to the van der Pol equation for the special case $\ddot{x} + 10(x^2 - 1)\dot{x} + x = 0$. Figure 21.13 shows its phase diagram together with the solution represented by the limit cycle.

21.8 A numerical method for $\dot{x} = P$, $\dot{y} = Q$
We shall show a numerical method, related to Euler's method of

Section 20.2, for plotting phase paths of the system

$$\dot{x}(t) = P\big(x(t), y(t)\big), \qquad \dot{y}(t) = Q\big(x(t), y(t)\big). \qquad \textbf{(21.23)}$$

Essentially we use t as a parameter in a step-by-step solution. Start from an initial point $P : (x_0, y_0)$ at time t_0. The choice of t_0 does not affect the path constructed because the equations are autonomous. Take short time steps of length h. Then we proceed from point to point in the diagram;

$$P_0 : (x_0, y_0) \to P_1 : (x_1, y_1) \to P_2 : (x_2, y_2) \to \cdots.$$

Since, approximately,

$$x_{n+1} - x_n = h\dot{x}(t_n) = hP(x_n, y_n),$$

and similarly for $y_{n+1} - y_n$, the rule for getting from P_n to P_{n+1} is as follows.

Euler's method for $\dot{x} = P(x, y)$, $\dot{y} = Q(x, y)$

$$x_{n+1} = x_n + hP(x_n, y_n), \qquad y_{n+1} = y_n + hQ(x_n, y_n); \qquad \textbf{(21.24)}$$

Compute for $n = 0, 1, 2, \ldots$ successively.

This process gives rise to rather unevenly spaced points on a phase path, widely spaced when P and Q are large, and very closely spaced near an equilibrium point, where P and Q are inevitably small. However, they have the advantage that, if necessary, regular **time** indications can be marked on the path while it is being computed.

If **evenly spaced** points are wanted, the parameter can be changed from time t to arc length s. We have $\delta s^2 = \delta x^2 + \delta y^2$; so

$$\frac{\mathrm{d}s}{\mathrm{d}t} = \left[\left(\frac{\mathrm{d}x}{\mathrm{d}t}\right)^2 + \left(\frac{\mathrm{d}y}{\mathrm{d}t}\right)^2\right]^{\frac{1}{2}} = (P^2 + Q^2)^{\frac{1}{2}}.$$

Therefore

$$\frac{\mathrm{d}x}{\mathrm{d}s} = \frac{\mathrm{d}x}{\mathrm{d}t} \Big/ \frac{\mathrm{d}s}{\mathrm{d}t} = \frac{P}{(P^2 + Q^2)^{\frac{1}{2}}} \quad \text{and} \quad \frac{\mathrm{d}y}{\mathrm{d}s} = \frac{Q}{(P^2 + Q^2)^{\frac{1}{2}}}$$

are equivalent equations for the path, in terms of arc length s. This gives the following method.

To compute the paths of the system $\dot{x} = P(x, y)$, $\dot{y} = Q(x, y)$, at evenly spaced points

Apply (21.24) to the equivalent system

$$\frac{\mathrm{d}x}{\mathrm{d}s} = \tilde{P}(x, y), \qquad \frac{\mathrm{d}y}{\mathrm{d}s} = \tilde{Q}(x, y), \qquad \textbf{(21.25)}$$

where

$$\tilde{P} = P/(P^2 + Q^2)^{\frac{1}{2}}, \quad \tilde{Q} = Q/(P^2 + Q^2)^{\frac{1}{2}}.$$

The step length h is the distance along the path.

Problems, Chapter 21

Many of the problems involve computation. The method of Section 21.8 is sufficient, but a high-accuracy computer library routine would allow the use of much larger values of h, and therefore be more efficient.

21.1. (Computation.) Practice computing a phase diagram in the following cases. Information is given for checking, but imagine that you do not have it. The equilibrium points are at $(0, 0)$. Take different starting points; you may have to work backwards as well as forwards, by changing the sign of h.

(a) $\dot{x} = y$, $\dot{y} = -4x$. (A centre, $4x^2 + y^2 = C$. If your paths do not nearly close, try a smaller interval h.)

(b) $\dot{x} = y$, $\dot{y} = x$. (A saddle, $x^2 - y^2 = C$. Find the asymptotes by trying $y = mx$ in the equations: they are $y = \pm x$. It is difficult to make sense of the diagram without this information.)

(c) $\dot{x} = y$, $\dot{y} = -2x - 3y$. (Stable node. Find the two solutions $y = -x$, $y = 2x$, which are radial straight lines as in (b). These represent *four* paths since they are interrupted by the origin.)

(d) $\dot{x} = y$, $\dot{y} = -3x - y$. (Stable spiral.)

(e) $\dot{x} = y$, $\dot{y} = -2x + y$. (Unstable spiral.)

(f) Recompute (a), marking off a time scale on each of the paths, showing intervals in t of around 0.3.

(g) Recompute (b) with a time scale as in (f).

(h) $\dot{x} = y$, $\dot{y} = -2y$. A different type: what is the second-order equation that it comes from?

21.2. Sketch the phase paths for the following equations by first solving for them: form dy/dx and separate the variables.

(a) $\dot{x} = y$, $\dot{y} = x$;   (b) $\dot{x} = x$, $\dot{y} = y$;
(c) $\dot{x} = -y$, $\dot{y} = x$;   (d) $\dot{x} = -x$, $\dot{y} = y$;
(e) $\dot{x} = 2y$, $\dot{y} = x$;   (f) $\dot{x} = -2y$, $\dot{y} = x$.

21.3. Solve the following by using the **energy transformation** $d^2x/dt^2 = \frac{1}{2} d(\dot{x}^2)/dx$ (Example 20.15), and sketch the $(x, \dot{x})$ phase diagrams.

(a) $\ddot{x} = e^x$;
(b) $\ddot{x} + \dot{x}^2 + x = 0$ the transformed equation is linear in y^2);
(c) $\ddot{x} - 8x\dot{x} = 0$;
(d) $\ddot{x} = e^x - e^{-x}$ (the Poisson–Boltzmann equation).

21.4. Classify the equilibrium point $(0, 0)$ for each of the following linear equations by using (21.22). Sketch the phase diagram: in cases where it is appropriate you should first obtain the radial straight paths $y = mx$ by substitution. State which are unstable.

(a) $\dot{x} = x - 5y$, $\dot{y} = x - y$;
(b) $\dot{x} = x + y$, $\dot{y} = x - 2y$;
(c) $\dot{x} = -4x + 2y$, $\dot{y} = 3x - 2y$;
(d) $\dot{x} = x + 2y$, $\dot{y} = 2x + 2y$;
(e) $\dot{x} = 4x - 2y$, $\dot{y} = 3x - y$;
(f) $\dot{x} = 2x + 3y$, $\dot{y} = -3x - 3y$.

21.5. For the equations given: find any equilibrium points; obtain a linear approximation at each equilibrium point by the method of Section 21.6; classify it from (21.22) (finding the straight line paths in the case of nodes and saddles); and put the sketches on a phase diagram. Guess how the diagram away from the equilibrium points is filled in (isoclines, Section 17.1, might help here). Then turn to Problem 21.6.

(a) $\dot{x} = x - y$, $\dot{y} = x + y - 2xy$;
(b) $\dot{x} = 1 - xy$, $\dot{y} = (x - 1)y$;
(c) $\dot{x} = x - y$, $\dot{y} = x^2 - 1$;
(d) $\ddot{x} + x - x^3 = 0$ (with $\dot{x} = y$);
(e) $\dot{x} = 4x - 2xy$, $\dot{y} = -2y + xy$, for $x > 0$ and $y > 0$ (foxes and rabbits, Example 21.6: classify $(0, 0)$ as if x and y could be negative).

21.6. (Computational). Check some of the phase diagrams you sketched in Problem 21.5 by computing representative phase paths. Look out for separatrices, which end at equilibrium points.

21.7. Sketch possible phase diagrams from the information given. If a phase path ends in mid air, or if you have a closed curve without an equilibrium point inside, then there is something wrong. There are often several possibilities; for example, a path might either join two equilibrium points or split, forming two branches going to infinity. Suppose that the only equilibrium points at a finite distance are those given in the following cases.

(a) centre at $(0, 0)$, saddle at $(1, 0)$;
(b) centre at $(0, 0)$, saddles at $(\pm 1, 0)$;
(c) unstable node at $(0, 0)$, stable node at $(1, 0)$;
(d) centres at $(\pm 1, 0)$.

21.8. (Computational). Obtain a phase diagram for the following (in some of these the linear approximation point is zero, so it gives no information):

(a) $\ddot{x} + |\dot{x}|\dot{x} + x = 0$;   (b) $\ddot{x} + |\dot{x}|\dot{x} + x^3 = 0$;
(c) $\dot{x} = x^4 - x^2$;   (d) $\dot{x} = 2xy$, $\dot{y} = y^2 - x^2$;
(e) $\dot{x} = 2xy$, $\dot{y} = x^2 - y^2$;

(f) $\ddot{x} + \dot{x}(x^2 + \dot{x}^2) + x = 0$ (notice that the origin is a spiral, although the linear approximation has a centre – see the remark following (21.22)).

21.9. From the Taylor series (5.4b), $\sin x \approx x - \frac{1}{6}x^3$ for small x, so the pendulum equation (21.8) is approximated by $\ddot{x} + \omega^2(x - \frac{1}{6}x^3) = 0$ (the Duffing equation). Sketch or compute the phase diagram, and comment on the differences from Fig. 21.9, for the exact equation.

21.10. (Computational). For a modified form of the predator–prey problem (compare Example 21.6), in a special case, the equations are

$$x = 4x - 2xy - x^2, \qquad y = -2y + xy - 2y^2.$$

The additional terms in x^2 and y^2 are meant to account for competition for resources among rabbits and among foxes. Use a linear approximation at the equilibrium points in order to classify them, then compute the phase diagram.

21.11. A model for $H(t)$ hosts supporting $P(t)$ dangerous parasites is $\dot{H} = (a - bP)H$, $\dot{P} = (c - dP/H)P$, where a, b, c, d, are positive. Analyse the system in the (H, P) plane.

21.12. Figure 21.14 represents a spring of stiffness s and natural length l, pivoted at A at a height h above a smooth wire CD. At B is a mass m, attached to the spring and sliding on the wire. The equation of motion is

$$\ddot{x} + \frac{s}{m}\left(1 - \frac{l}{(h^2 + x^2)^{\frac{1}{2}}}\right)x = 0.$$

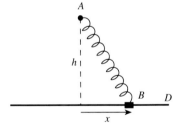

Fig. 21.14

Classify the equilibrium points when $l < h$, $l = h$, and $l > h$.

21.13. (Computational). Solve Problem 21.12 modified so that there is friction between the bead and the wire equal to $k\dot{x}$. Classify the equilibrium points and construct the phase diagrams.

21.14. (Computational). Construct phase diagrams for the equation $\ddot{x} + k\dot{x} - x + x^2 = 0$. Consider various values of k.

21.15. (Computational). Construct a phase diagram for the following equations. (They each contain a limit cycle.)
(a) $\ddot{x} + \frac{1}{2}(x^2 + \dot{x}^2 - 1)\dot{x} + x = 0$;
(b) $\ddot{x} + \frac{1}{5}(x^2 - 1)\dot{x} + x = 0$;
(c) $\ddot{x} + \frac{1}{5}(\frac{1}{3}\dot{x}^2 - 1)\dot{x} + x = 0$;
(d) $\ddot{x} + 5(x^2 - 1)\dot{x} + x = 0$.

21.16. As in Problem 21.7, sketch phase diagrams for the general (x, y) phase plane compatible with the following information. The equilibrium points and limit cycles specified are the only ones allowed.
(a) $(0, 0)$ is a spiral and $x^2 + y^2 = 1$ is a stable limit cycle.
(b) $(0, 0)$ is a spiral, $x^2 + y^2 = 1$ a stable limit cycle, and $x^2 + y^2 = 4$ another limit cycle.
(c) $(\pm 1, 0)$ are saddles, $(0, 0)$ is a centre, and $x^2 + y^2 = 4$ is a stable limit cycle.
(d) $(\pm 1, 0)$ are centres, $(0, 0)$ is a saddle, and $x^2 + y^2 = 4$ a stable limit cycle.
(e) $(0, 0)$ is a centre; the only closed path with $x^2 + y^2 > 1$ is the stable limit cycle $x^2 + y^2 = 4$.

21.17. Show that, in polar coordinates, the system

$$\dot{x} = -y + x(1 - x^2 - y^2),$$
$$\dot{y} = x + y(1 - x^2 - y^2)$$

becomes

$$\dot{r} = r(1 - r^2), \qquad \dot{\theta} = 1.$$

By investigating the sign of $\dot{r}$, explain why the system has just one limit cycle, which is stable. Sketch the phase diagram.

21.18. Find the locations of all the equilibrium points of

$$\dot{x} = (x^2 + y^2 - 1)y, \qquad \dot{y} = -(x^2 + y^2 - 1)x.$$

Explain why the circle $x^2 + y^2 = 1$ does not represent periodic motion.

21.19. Verify that the differential equation

$$\ddot{x} + \left(1 - x^2 - \frac{\dot{x}^2}{\omega^2}\right)\dot{x} + \omega^2 x = 0,$$

has the particular solution $x = \cos \omega(t - t_0)$ for any t_0. What is the corresponding phase path in the $(x, y = \dot{x})$ plane? Put further details on a sketch of the phase diagram.

21.20. Locate all equilibrium points of the system

$$\dot{x} = (x^2 - 1)y, \qquad \dot{y} = (y^2 - 1)x,$$

and sketch its phase diagram.

21.21. The linear system given by (21.22), namely

$$\dot{x} = ax + by, \qquad \dot{y} = cx + dy,$$

can be expressed in matrix form as

$$\dot{\mathbf{x}} = A\mathbf{x},$$

where

$$\mathbf{x} = \begin{bmatrix} x \\ y \end{bmatrix}, \qquad A = \begin{bmatrix} a & b \\ c & d \end{bmatrix}.$$

Try an exponential solution in the form

$$\mathbf{x} = \mathbf{C}\, e^{\lambda t},$$

where $\mathbf{C}$ is a constant column vector, and show that λ must satisfy

$$\det(A - \lambda I_2) = 0.$$

In other words, the solutions for λ are the *eigenvalues* of A (see Chapter 11). If λ_1 and λ_2 are *distinct* eigenvalues, show that the general solution is

$$\mathbf{x} = \mathbf{C}_1\, e^{\lambda_1 t} + \mathbf{C}_2\, e^{\lambda_2 t}.$$

What is the solution if $\lambda_1 = \lambda_2$?

Write down the roots of the quadratic equation, and discuss how the possible cases (e.g. real roots, imaginary roots, complex roots, etc.) fit in with the centre, saddle, node, and spiral, as classified in (21.22).

22

The Laplace transform

22.1 The Laplace transform

Suppose that $f(t)$ is a specified function, and that s is a real positive parameter (that is to say, a supplementary variable). Then the integral

$$\int_0^\infty e^{-st} f(t)\,dt = F(s)$$

is called the **Laplace transform** of $f(t)$: the integral **transforms** $f(t)$ into another function $F(s)$.

For example, suppose that $f(t) = e^{2t}$. Then

$$F(s) = \int_0^\infty e^{-st} e^{2t}\,dt = \int_0^\infty e^{-(s-2)t}\,dt$$

$$= \left[-\frac{1}{s-2} e^{-(s-2)t} \right]_0^\infty$$

$$= -\frac{1}{s-1}(e^{-\infty} - e^0) = -\frac{1}{s-2}(0-1) = \frac{1}{s-2}.$$

This result is true only if $s > 2$; otherwise the integral is infinite. We shall always assume that s is large enough to ensure that the integrals we encounter remain finite, or **converge** (see Section 13.6).

We also use the symbol L to stand for the 'Laplace transform of'. We have just proved that

$$F(s) = L\{e^{2t}\} = \frac{1}{s-2}.$$

Laplace transform of $f(t)$

$$L\{f(t)\} = F(s) = \int_0^\infty e^{-st} f(t)\,dt. \tag{22.1}$$

Another, very useful, notation is to indicate a transformed function by a **tilde sign**: $L\{f(t)\} = \tilde{f}(s)$, $L\{x(t)\} = \tilde{x}(s)$, and so on.

The letter p is often used for the parameter instead of s, especially in mainly theoretical texts.

22.2 Laplace transforms of t^n, $e^{\pm t}$, sin t, cos t

(a) *Positive, whole-number powers t^n, $n = 0, 1, 2, \ldots$.*

$$L\{t^n\} = \int_0^\infty e^{-st} t^n \, dt.$$

Simplify the integral by substituting $u = st$, so that

$$t = \frac{1}{s} u \quad \text{and} \quad dt = \frac{1}{s} du.$$

Since s is **positive**, the limits of integration $t = 0$ and ∞ correspond to $u = 0$ and ∞ respectively. Therefore

$$L\{t^n\} = \int_0^\infty e^{-u} \left(\frac{u}{s}\right)^n \frac{du}{s} = \frac{1}{s^{n+1}} \int_0^\infty e^{-u} u^n \, ds$$

$$= \frac{n!}{s^{n+1}}$$

for $n = 0, 1, 2, \ldots$ (from the standard integral, (15.9)). Note that $0!$ is to be interpreted as being equal to 1.

Laplace transform of powers

$$L\{t^n\} = \frac{n!}{s^{n+1}} \quad \text{for } n = 0, 1, 2, \ldots.$$

Special cases:

$$L\{1\} = \frac{1}{s}, \quad L\{t\} = \frac{1}{s^2}, \quad L\{t^2\} = \frac{2!}{s^3}, \quad L\{t^3\} = \frac{3!}{s^4}.$$

(22.2)

Example 22.1. Find the Laplace transform $F(s)$ of $f(t)$ when

$$f(t) = 1 - t + \frac{1}{2!} t^2 - \frac{1}{3!} t^3.$$

Composite expressions are dealt with in the following way.

$$F(s) \text{ or } L\{t\} = L\left\{1 - t + \frac{1}{2!} t^2 - \frac{1}{3!} t^3\right\},$$

$$= L\{1\} - L\{t\} + \frac{1}{2} L\{t^2\} - \frac{1}{3!} L\{t^3\},$$

(which follows from the fact that each $L\{\cdots\}$ stands for the integral (22.1))

$$= \frac{1}{s} - \frac{1}{s^2} + \frac{1}{2!} \frac{2!}{s^3} - \frac{1}{3!} \frac{3!}{s^4},$$

$$= \frac{1}{s} - \frac{1}{s^2} + \frac{1}{s^3} - \frac{1}{s^4}.$$

(b) *Exponential* $e^{\pm t}$.

$$L\{e^{\pm t}\} = \int_0^\infty e^{-st} e^{\pm t} \, dt = \int_0^\infty e^{-(s \mp 1)t} \, dt$$

$$= -\frac{1}{s \mp 1} \left[e^{-(s \mp 1)t} \right]_0^\infty.$$

$s \mp 1$ are both positive if we take $s > 1$, in which case

$$L\{e^{\pm t}\} = -\frac{1}{s \mp 1} (0 - 1) = \frac{1}{s \mp 1}.$$

$$L\{e^t\} = \frac{1}{s - 1}, \qquad L\{e^{-t}\} = \frac{1}{s + 1}. \tag{22.3}$$

(c) *Sine and cosine.*

$$L\{\cos t\} = \frac{s}{s^2 + 1}, \qquad L\{\sin t\} = \frac{1}{s^2 + 1}. \tag{22.4}$$

Since $\cos t + j \sin t = e^{jt}$, both of these can be verified at the same time by working out $L\{e^{jt}\}$ and then separating the real and imaginary parts.

$$L\{e^{jt}\} = \int_0^\infty e^{-st} e^{jt} \, dt = \int_0^\infty e^{-(s - j)t} \, dt$$

$$= -\frac{1}{s - j} \left[e^{-(s - j)t} \right]_0^\infty = -\frac{1}{s - j} (0 - 1)$$

(since s is positive)

$$= \frac{1}{s - j} = \frac{s + j}{s^2 + 1}.$$

Therefore, as in (22.4),

$$\int_0^\infty e^{-st} \cos t \, dt = \text{Re} \int_0^\infty e^{-st} e^{jt} \, dt = \frac{s}{s^2 + 1};$$

$$\int_0^\infty e^{-st} \sin t \, dt = \text{Im} \int_0^\infty e^{-st} e^{jt} \, dt = \frac{1}{s^2 + 1}.$$

Example 22.2. Find the Laplace transform of $3t^2 + 2e^{-t} - 5 \cos t$.

$$L\{3t^2 + 2e^{-t} - 5 \cos t\} = 3L\{t^2\} + 2L\{e^{-t}\} - 5L\{\cos t\},$$

$$= 3 \frac{2!}{s^3} + 2 \frac{1}{s + 1} - 5 \frac{s}{s^2 + 1},$$

$$= \frac{6}{s^3} + \frac{2}{s + 1} - \frac{5s}{s^2 + 1}.$$

22.3 Scale rule; shift rule; factors t^n and e^{kt}

The following rules make it easy to derive more complicated transforms from the basic ones of Section 22.3.

Scale rule

If $\mathcal{L}\{f(t)\} = F(s)$, and $k > 0$, then

$$\mathcal{L}\{f(kt)\} = \frac{1}{k} F\left(\frac{s}{k}\right).$$

(22.5)

The proof is as follows.

$$\mathcal{L}\{f(kt)\} = \int_0^\infty e^{-st} f(kt) \, dt.$$

Change the variable by putting $u = kt$, so that $t = u/k$ and $dt = du/k$. The limits of integration $t = 0$ and ∞ go into $u = 0$ and ∞ respectively because $k > 0$. Therefore

$$\mathcal{L}\{f(kt)\} = \int_0^\infty e^{-s(u/k)} f(u) \left(\frac{du}{k}\right)$$

$$= \frac{1}{k} \int_0^\infty e^{-(s/k)t} f(u) \, du,$$

$$= \frac{1}{k} F\left(\frac{s}{k}\right),$$

since $F(s) = \int_0^\infty e^{-su} f(u) \, du$.

The following are special cases.

If k is any constant, positive or negative, then

(a) $\mathcal{L}\{e^{kt}\} = \dfrac{1}{s-k}$,

(b) $\mathcal{L}\{\cos kt\} = \dfrac{s}{s^2 + k^2}$,

(c) $\mathcal{L}\{\sin kt\} = \dfrac{k}{s^2 + k^2}$.

(22.6)

These are proved as follows.

(a) Suppose that m is a positive number; then combining (22.3) with the scale rule (22.5) gives

$$\mathcal{L}\{e^{\pm mt}\} = \frac{1}{m} \cdot \frac{1}{s/m \mp 1} = \frac{1}{s \mp m}.$$

The result (22.6a) therefore holds good for both positive and negative k ($k = \pm m$ for $m > 0$).

(b) From (22.4),

$$L\{\cos t\} = \frac{s}{s^2 + 1}.$$

Therefore, by the scale rule (14.5), if $k > 0$,

$$L\{\cos kt\} = \frac{1}{k} \frac{s/k}{(s/k)^2 + 1} = \frac{k}{s^2 + k^2}.$$

This is true also if k is negative, since it is equal to $\int_0^\infty e^{-st} \cos kt\, dt$ (see (22.1)).

(a) is similar to (b).

Example 22.3. Find the Laplace transform of $2\cos(3t + \frac{1}{4}\pi)$.

$$\cos(3t + \tfrac{1}{4}\pi) = \cos\tfrac{1}{4}\pi \cos 3t - \sin\tfrac{1}{4}\pi \sin 3t = (\cos 3t - \sin 3t)/\sqrt{2}.$$

Therefore by (22.6)

$$L\{\cos(3t + \tfrac{1}{4}\pi)\} = \frac{1}{\sqrt{2}}\left(\frac{s}{s^2 + 9} - \frac{3}{s^2 + 9}\right) = \frac{1}{\sqrt{2}}\frac{s - 3}{s^2 + 9}.$$

Suppose that we know the Laplace transform $F(s)$ of a function $f(t)$ already. Then the Laplace transform of $e^{kt} f(t)$ can immediately be written down

$$L\{e^{kt}f(t)\} = \int_0^\infty e^{-st} e^{kt} f(t)\, dt = \int_0^\infty e^{(s-k)t} f(t)\, dt.$$

But $\int_0^\infty e^{-st} f(t)\, dt = F(s)$, which is supposed to be known, and here we have $s - k$ in place of s. Therefore

$$L\{e^{kt} f(t)\} = F(s - k).$$

Shift rule (multiplication by e^{kt})

If $L\{f(t)\} = F(s)$ and k is any constant, then $\hspace{2cm}$ **(22.7)**
$$L\{e^{kt} f(t)\} = F(s - k).$$

The shift rule is so called because the transform function $F(s)$ is 'shifted' a distance k along the s axis by the presence of the factor e^{kt}.

Example 22.4. Find $L\{e^{-3t} \sin 2t\}$.

From (22.6),

$$L\{\sin 2t\} = \frac{2}{s^2 + 4}.$$

By the shift rule (22.7) with $k = -3$, we deduce that

$$L\{e^{-3t} \sin 2t\} = \frac{2}{(s + 3)^2 + 4} = \frac{2}{s^2 + 6s + 13}.$$

Example 22.5. Find $L\{t^3 \, e^{4t}\}$.

From (22.2), $L\{t^3\} = 3!/s^4$. The shift rule with $k = 4$ gives

$$L\{e^{4t} \, t^3\} = \frac{3!}{(s - 4)^4}.$$

There is a rule similar to (22.7) by which we can find the Laplace transform of $t^n f(t)$ when the transform of $f(t)$ is known:

Multiplication by t^n

If $L\{f(t)\} = F(s)$, then

$$L\{t^n f(t)\} = (-1)^n \frac{d^n F(s)}{ds^n}.$$

(22.8)

The simplest way to prove this is to start with the right-hand side. Since

$$\int_0^\infty e^{-st} \, f(t) \, dt = F(s),$$

then

$$\frac{dF(s)}{ds} = \frac{d}{ds} \int_0^\infty e^{-st} \, f(t) \, dt$$

$$= \int_0^\infty \frac{d(e^{-st})}{ds} \, f(t) \, dt$$

$$= \int_0^\infty (-t \, e^{-st}) f(t) \, dt = - \int_0^\infty e^{-st}(t f(t)) \, dt$$

$$= -L\{t f(t)\}.$$

Every time we differentiate, another factor t and another multiplication by -1 appears, which takes us to (22.8).

Example 22.6. Find $L\{t \cos 3t\}$.

Since, by (22.6b),

$$F(s) = L\{\cos 3t\} = \frac{s}{s^2 + 9},$$

then, by (22.8),

$$L\{t \cos 3t\} = -\frac{d}{ds} \frac{s}{s^2 + 9} = -\frac{9 - s^2}{(s^2 + 9)^2} = \frac{s^2 - 9}{(s^2 + 9)^2}.$$

Note the two following special cases, which occur frequently.

$$L\{t\cos kt\} = \frac{s^2 - k^2}{(s^2 + k^2)^2},$$

$$L\{t\sin kt\} = \frac{2ks}{(s^2 + k^2)^2}. \tag{22.9}$$

Example 22.7. Find $L\{t^3 e^{-3t}\}$ (a) by using the shift rule, (b) by using (22.8), (c) by working directly from the definition of the Laplace transform.

(a) From (22.2),

$$L\{t^3\} = \frac{6}{s^4}.$$

Therefore, using the shift rule (22.7a) with $k = -3$,

$$L\{e^{-3t} t^3\} = \frac{6}{(s+3)^4}.$$

(b) From (22.6) with $k = -3$

$$L\{e^{-3t}\} = \frac{1}{s+3}.$$

From (22.8) with $n = 3$,

$$L\{t^3 e^{-3t}\} = (-1)^3 \frac{d^3}{ds^3} \frac{1}{s+3} = (-1)^3 \frac{(-1)(-2)(-3)}{(s+3)^4} = \frac{6}{(s+3)^4}.$$

(c) From the definition, (22.1),

$$L\{t^3 e^{-3t}\} = \int_0^\infty e^{-st} t^3 e^{-3t} \, dt = \int_0^\infty e^{-(s+3)t} t^3 \, dt.$$

From (22.2), this is equal to

$$\frac{3!}{(s+3)^4}.$$

22.4 Inverting a Laplace transform

Given a function $f(t)$, we obtain its transform $F(s)$ by using the definition (22.1). Alternatively, if a function $F(s)$ is presented, then we can try to recover the original $f(t)$, from which $F(s)$ is obtained. This is called the **original** of $F(s)$. This second question is the **inverse problem** for the Laplace transform – to find '?' in the equation

$$L\{?\} = F(s).$$

We shall assume here that there is only one answer to this problem. The process of finding $f(t)$ from $F(s)$ is called **inversion** of $F(s)$.

The notation

$$f(t) \leftrightarrow F(s)$$

is a useful notation which underlines the two-way correspondence between $f(t)$ and $F(s)$.

We can open up a 'dictionary' for this purpose, as we did for derivatives and integrals. The most important results we have so far are given in the table (22.10) below.

$f(t)$ for $t > 0$		$F(s)$	
$\begin{cases} t^n & (n = 0, 1, \ldots), \\[2mm] \dfrac{1}{(m-1)!}\, t^{m-1} & (m = 1, 2, \ldots) \end{cases}$		$\left.\begin{matrix} \dfrac{n!}{s^{n+1}} \\[3mm] \dfrac{1}{s^m} \end{matrix}\right\}$	
e^{kt} (any k)		$\dfrac{1}{s-k}$	**(22.10)**
$\cos kt$ (any k)		$\dfrac{s}{s^2 + k^2}$	
$\begin{cases} \sin kt & \text{(any } k) \\[2mm] \dfrac{1}{k}\sin kt & \text{(any } k \neq 0) \end{cases}$		$\left.\begin{matrix} \dfrac{k}{s^2 + k^2} \\[3mm] \dfrac{1}{s^2 + k^2} \end{matrix}\right\}$	

A much fuller table which also includes the various rules can be found in Appendix F. Remember that everything we do with Laplace transforms refers to $t \geqslant 0$ only: the defining integral (22.1) calls only for values of $t \geqslant 0$.

Partial fractions are often useful for inverting transforms.

Example 22.8. Given the transform $1/s(s+1)$, find the original.

In partial fractions,

$$\frac{1}{s(s+1)} = \frac{1}{s} - \frac{1}{s+1}.$$

From the table above,

$$\frac{1}{s} \leftrightarrow 1 \quad \text{and} \quad \frac{1}{s+1} \leftrightarrow \mathrm{e}^{-t},$$

so that

$$\frac{1}{s(s+1)} \leftrightarrow 1 - \mathrm{e}^{-t}.$$

Example 22.9. Invert the Laplace transform

$$\frac{s+1}{s(s^2+4)}.$$

The partial-fraction rules require the form

$$\frac{s+1}{s(s^2+4)} = \frac{A}{s} + \frac{Bs+C}{s^2+4}.$$

When the constants are determined by the method of Section 1.12, we find that $A = \frac{1}{4}$, $B = -\frac{1}{4}$, $C = 1$, so that

$$\frac{s+1}{s(s^2+4)} = \frac{\frac{1}{4}}{s} + \frac{-\frac{1}{4}s+1}{s^2+4},$$

$$= \frac{\frac{1}{4}}{s} - \frac{1}{4}\frac{s}{s^2+4} + \frac{1}{s^2+4}.$$

From (22.2),

$$\frac{1}{s} \leftrightarrow 1.$$

From the Table (22.10),

$$\frac{s}{s^2+4} \leftrightarrow \cos 2t, \qquad \frac{1}{s^2+4} \leftrightarrow \tfrac{1}{2}\sin 2t.$$

Therefore

$$\frac{s+1}{s(s^2+4)} \leftrightarrow \tfrac{1}{4} - \tfrac{1}{4}\cos\ 2t + \tfrac{1}{2}\sin 2t.$$

Example 22.10. Invert the Laplace transform

$$\frac{3s+2}{s^2+2s+2}.$$

The quadratic denominator does not have real factors, so partial fractions are not available. Instead we complete the square:

$$s^2 + 2s + 2 = (s+1)^2 - 1 + 2 = (s+1)^2 + 1.$$

We aim to write the whole expression in terms of $s+1$ so that we can apply the shift rule (22.7). So put also

$$3s + 2 = 3(s+1) - 3 + 2 = 3(s+1) - 1,$$

and the transform becomes

$$\frac{3(s+1)-1}{(s+1)^2+1}.$$

If we had s instead of $s+1$, we could invert the transform:

$$\frac{3s-1}{s^2+1} = \frac{3s}{s^2+1} - \frac{1}{s^2+1} \leftrightarrow 3\cos t - \sin t.$$

Therefore, by the shift rule with $k = -1$,

$$\frac{3(s + 1) - 1}{(s + 1)^2 + 1} \leftrightarrow e^{-t}(3 \cos t - \sin t).$$

22.5 Laplace transforms of derivatives

Suppose that $L\{f(t)\} = F(s)$. Then the Laplace transforms of $df(t)/dt$, $d^2 f(t)/dt^2, \dots$ can be expressed in terms of $F(s)$.

In the definition,

$$\hat{L}\left\{\frac{df(t)}{dt}\right\} = \int_0^{\infty} e^{-st} \frac{df(t)}{dt} \, dt.$$

Integrate the right-hand side by parts. Using the notation of Section 15.7, put

$$u = e^{-st}, \qquad \frac{dv}{dt} = \frac{df(t)}{dt},$$

so that

$$\frac{du}{dt} = -s \, e^{-st}, \qquad v = f(t).$$

Then

$$L\left\{\frac{df(t)}{dt}\right\} = \int_0^{\infty} e^{-st} \frac{df(t)}{dt} \, dt$$

$$= [e^{-st} f(t)]_0^{\infty} - \int_0^{\infty} (-s \, e^{-st}) f(t) \, dt$$

$$= 0 - e^0 f(0) + s \int_0^{\infty} e^{-st} f(t) \, dt$$

$$= -f(0) + sL\{f(t)\}.$$

In other words, if $L\{f(t)\} = F(s)$, then

$$L\left\{\frac{df(t)}{dt}\right\} = sF(s) - f(0). \tag{22.11}$$

(Note that it is $f(0)$, not $F(0)$, that arises here.)

We can use (22.11) again and again to obtain successively

$$L\left\{\frac{d^2 f(t)}{dt^2}\right\} = L\left\{\frac{d}{dt}\frac{df(t)}{dt}\right\}$$

and higher derivatives, from which we obtain the following rule.

Laplace transform of derivatives

If $L\{f(t)\} = F(s)$, then

$$L\left\{\frac{\mathrm{d}f(t)}{\mathrm{d}t}\right\} = sF(s) - f(0),$$

$$L\left\{\frac{\mathrm{d}^2 f(t)}{\mathrm{d}t^2}\right\} = s^2 F(s) - sf(0) - f'(0), \qquad \textbf{(22.12)}$$

$$L\left\{\frac{\mathrm{d}^3 f(t)}{\mathrm{d}t^3}\right\} = s^3 F(s) - s^2 f(0) - sf'(0) - f''(0),$$

and so on.

Example 22.11. Obtain the transform of the expression

$$\frac{\mathrm{d}^2 x}{\mathrm{d}t^2} + 2\frac{\mathrm{d}x}{\mathrm{d}t} + 3x,$$

when $x = 4$ and $\mathrm{d}x/\mathrm{d}t = 5$ at $t = 0$.

Put $L\{x(t)\} = X(s)$. Then

$$L\left\{\frac{\mathrm{d}^2 x}{\mathrm{d}t^2} + 2\frac{\mathrm{d}x}{\mathrm{d}t} + 3x\right\} = L\left\{\frac{\mathrm{d}^2 x}{\mathrm{d}t^2}\right\} + 2L\left\{\frac{\mathrm{d}x}{\mathrm{d}t}\right\} + 3L\{x\}$$

$$= s^2 X - sx(0) - x'(0) + 2[sX - x(0)] + 3X$$

$$= s^2 X - 4s - 5 + 2(sX - 4) + 3X$$

$$= (s^2 + 2s + 3)X - 4s - 13.$$

22.6 Application to differential equations

The results (22.12) enable initial-value problems for differential equations to be solved.

Example 22.12. Find the solution of

$$\frac{\mathrm{d}x}{\mathrm{d}t} + 2x = \mathrm{e}^{-t}$$

for which $x = 3$ when $t = 0$.

Since

$$\frac{\mathrm{d}x}{\mathrm{d}t} + 2x = \mathrm{e}^{-t},$$

it is also true that

$$L\left\{\frac{\mathrm{d}x}{\mathrm{d}t}\right\} + 2L\{x\} = L\{\mathrm{e}^{-t}\}.$$

Write

$$\mathcal{L}\{x(t)\} = X(s).$$

By (22.12) the transformed equation becomes

$$sX - 3 + 2X = \frac{1}{s+1}$$

(where we put $x(0) = 3$ as specified by the initial condition). The transform $X(s)$ of $x(t)$ is therefore given by

$$X(s) = \frac{3s+4}{(s+1)(s+2)} = \frac{1}{s+1} + \frac{2}{s+2}$$

$$\leftrightarrow x(t) = e^{-t} + 2e^{-2t},$$

which is the required solution.

It can be seen that the terms involving $f(0), f'(0), \ldots$ in (22.12), far from being merely a nuisance, are exactly what is required to translate a differential equation together with initial conditions into a simpler problem in ordinary algebra. We do not have to match up arbitrary constants with the initial conditions; these conditions are built into the transformed equations.

In many physical situations, we want to know what happens when an inactive or **quiescent** system is 'switched on'. In such cases, we have **zero initial conditions** at $t = 0$. For a system described by a second-order differential equation, we assume by this that the variable and its first derivative are initially set to zero.

Example 22.13. A system is described by the equation

$$\frac{d^2x}{dt^2} + 2\frac{dx}{dt} + 4x = 1.$$

It is initially quiescent and is then switched on. Find the subsequent time variation of x.

We have $x(0) = x'(0) = 0$. Let

$$x(t) \leftrightarrow X(s).$$

Then the equation transforms to

$$s^2 X + 2sX + 4X = \frac{1}{s},$$

(notice the $1/s$) so that

$$X = \frac{1}{s(s^2 + 2s + 4)} = \frac{1}{4}\frac{1}{s} - \frac{1}{4}\frac{s+2}{s^2 + 2s + 4}.$$

The quadratic has no real factors; therefore the second term is rewritten in the manner of Example 22.10:

$$X = \tfrac{1}{4}\frac{1}{s} - \tfrac{1}{4}\frac{(s+1)+1}{(s+1)^2+3}$$

$$= \tfrac{1}{4}\frac{1}{s} - \tfrac{1}{4}\left(\frac{s+1}{(s+1)^2+3} + \frac{1}{(s+1)^2+3}\right).$$

To invert the last two terms: from (22.10)

$$\frac{s}{s^2+3} \leftrightarrow \cos\sqrt{3}t, \qquad \frac{1}{s^2+3} = \frac{1}{\sqrt{3}}\sin\sqrt{3}t.$$

By using the shift rule (22.7) with $k = -1$, we obtain

$$\frac{s+1}{(s+1)^2+3} \leftrightarrow e^{-t}\cos\sqrt{3}t, \qquad \frac{1}{(s+1)^2+3} \leftrightarrow \frac{1}{\sqrt{3}}e^{-t}\sin\sqrt{3}t.$$

Therefore

$$x(t) = \tfrac{1}{4} - \tfrac{1}{4}(e^{-t}\cos\sqrt{3}t + \tfrac{1}{3}\sqrt{3}\,e^{-t}\sin\sqrt{3}t).$$

Example 22.14. Solve the equation

$$\frac{d^2x}{dt^2} + \omega_0^2 x = a\cos\omega_0 t,$$

with $x(0) = x'(0) = 0$.

If we put $\mathcal{L}\{x(t)\} = X(s)$, then the equation transforms into

$$s^2 X + \omega_0^2 X = \frac{as}{s^2+\omega_0^2},$$

so that

$$X = \frac{as}{(s^2+\omega_0^2)^2}.$$

We can read off the inverse from (22.9) with $k = \omega_0$:

$$x(t) = \frac{a}{2\omega_0}\,t\sin\omega_0 t.$$

This equation is one of the exceptional resonant types discussed in Section 17.3. The advantage of using the Laplace transform is easy to see.

Example 22.15. Solve the simultaneous equations

$$\frac{dx}{dt} = x - y, \qquad \frac{dy}{dt} = x + y,$$

with the initial conditions $x(0) = 1$, $y(0) = 0$.

Let $\mathcal{L}\{x(t)\} = X(s)$ and $\mathcal{L}\{y(t)\} = Y(s)$. Then the transformed equations, including initial conditions, are

$$sX - 1 = X - Y, \qquad sY = X + Y.$$

Therefore

$$(1 - s)X - Y = -1,$$

$$X + (1 - s)Y = 0.$$

By solving these equations, we obtain

$$X = \frac{-1 + s}{s^2 - 2s + 2}, \qquad Y = \frac{1}{s^2 - 2s + 2}.$$

The denominators, $s^2 - 2s + 2$, have no real factors, so use the method of Example 22.10 to rewrite these expressions as

$$X = \frac{s - 1}{(s - 1)^2 + 1}, \qquad Y = \frac{1}{(s - 1)^2 + 1},$$

so that the shift rule (22.7) can be used to invert them. By (22.10),

$$\frac{s}{s^2 + 1} \leftrightarrow \cos t, \qquad \frac{1}{s^2 + 1} \leftrightarrow \sin t.$$

Therefore, by the shift rule with $k = 1$,

$$x(t) = e^t \cos t, \qquad y(t) = e^t \sin t.$$

22.7 The unit function and the delay rule

The Heaviside **unit function** $H(t)$ (or $U(t)$) was introduced in Section 1.4. Here is a reminder of its definition:

Unit function $H(t)$

$$H(t) = \begin{cases} 0 & \text{when } t \leqslant 0, \\ 1 & \text{when } t > 0. \end{cases} \qquad (22.13)$$

It is shown again in Fig. 22.1a. Figures 22.1b–e show how it can be used to describe various **step functions** and **switching functions**.

For example, the composition of the three segments of Fig. 22.1e is specified by:

$$e^t[H(t - 1) - H(t - 2)] = \begin{cases} e^t(0 - 0) = 0 & \text{if } t \leqslant 1, \\ e^t(1 - 0) = e^t & \text{if } 1 < t \leqslant 2, \\ e^t(1 - 1) = 0 & \text{if } t > 2. \end{cases}$$

Related Laplace transforms are given as follows.

(a)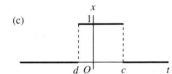

(b)

(c)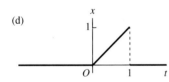

(d)

(e)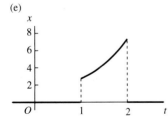

Fig. 22.1
(a) $x = H(t)$, (b) $x = H(t - c)$,
(c) $x = H(t - d) - H(t - c)$,
(d) $x = t[H(t - 1) - H(t)]$,
(e) $x = e^t[H(t - 1) - H(t - 2)]$.

Laplace transform for the unit function

$$L\{H(t)\} = \frac{1}{s}, \qquad L\{H(t-c)\} = \frac{e^{-cs}}{s} \quad (c \text{ positive}). \qquad \textbf{(22.14)}$$

The various combination rules such as the shift rule (22.7) work for $H(t)$ in the same way as for smooth functions $f(t)$.

Example 22.16. Find $L\{f(t)\}$ when $f(t) = e^t(H(t-2) - H(t-1))$.
This is the function shown in Fig. 22.1e. Then, from the definition,

$$L\{f(t)\} = \int_0^\infty e^{-st} e^t[H(t-2) - H(t-1)]\, dt,$$

$$= \int_1^2 e^{-(s-1)t}\, dt = -\frac{1}{s-1}\left[e^{-(s-1)t}\right]_1^2$$

$$= -\frac{1}{s-1}(e^{-2(s-1)} - e^{-(s-1)}).$$

Alternatively, we could use the shift rule (22.7), though it has no particular advantage.

Example 22.17. Find the Laplace transform of the function shown in Fig. 22.2.

By considering the segments one at a time and using Fig. 22.1c, we have

$$x(t) = [H(t) - H(t-1)] - [H(t-1) - H(t-2)]$$
$$+ [H(t-2) - H(t-3)] - \cdots,$$
$$= H(t) - 2H(t-1) + 2H(t-2) - 2H(t-3) + \cdots + .$$

From (22.14)

$$H(t-n) \leftrightarrow \frac{e^{-ns}}{s}.$$

Therefore

$$L\{x(t)\} = \frac{1}{s} - \frac{2}{s}(e^{-s} - e^{-2s} + e^{-3s} - \cdots).$$

The brackets contain an infinite geometric series with first term e^{-s} and common ratio $-e^{-s}$. Therefore

$$L(x(t)) = \frac{1}{s} - \frac{2}{s}\frac{e^{-s}}{1 + e^{-s}} = \frac{1 - 3\,e^{-s}}{s(1 + e^{-s})}.$$

Suppose that we have a function $g(t)$ which has a meaning for all positive t, such as $g(t) = e^{-t}$. Its Laplace transform is

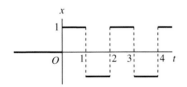

Fig. 22.2

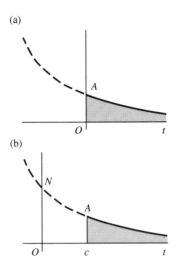

Fig. 22.3
(a) Graph of $g(t)$.
(b) Graph of $g(t-c)H(t-c)$.

$G(s) = \int_0^\infty e^{-st} g(t)\, dt$. All values of $g(t)$ for t positive are called on to contribute to this integral, but none of its values for negative t are called upon (Fig. 22.3a).

Now translate the function a distance c (positive) to the right as in Fig. 22.3b. The new graph represents $g(t-c)$. It brings with it a section NA which originally corresponded to negative values of t. We cannot expect that the Laplace transform of this new function $g(t-c)$ can be expressed in terms of $G(s)$, because none of these t values played any part in the calculation of $G(s)$.

Therefore we cut out the section NA by considering not $g(t-c)$, but $g(t-c)H(t-c)$, which is shaded in Fig. 22.3b, and is congruent to the shaded part of Fig. 22.3a.

Then

$$L\{g(t-c)H(t-c)\} = \int_0^\infty e^{-st}\, g(t-c)H(t-c)\, dt$$

$$= \int_c^\infty e^{-st}\, g(t-c)\, dt.$$

Put $t - c = u$, so that $t = u + c$ and $dt = du$. The integral becomes

$$\int_0^\infty e^{-s(u+c)}\, g(u)\, du = e^{-sc} \int_0^\infty e^{-su}\, g(u)\, du = e^{-sc}\, G(s).$$

This is the **second shift rule**, or the **delay rule**; so called because $g(t-c)H(t-c)$ does not start until $t = c$.

Delay rule

If $G(s) \leftrightarrow g(t)$ and $c > 0$, then   **(22.15)**

$$e^{-cs}\, G(s) \leftrightarrow g(t-c)H(t-c).$$

It is most often useful in inverting a Laplace transform.

Example 22.18. Find the Laplace inverse transform of e^{-2s}/s^2.
Put $G(s) = 1/s^2$. Then

$$G(s) = \frac{1}{s^2} \leftrightarrow g(t) = t.$$

By the delay rule,

$$\frac{e^{-2s}}{s^2} = e^{-2s}\, G(s) \leftrightarrow (t-2)H(t-2),$$

a function which suddenly takes off from zero at $t = 2$.

Example 22.19. Find the Laplace inverse transform of

$$\frac{e^{-2(s+1)}}{(s+1)(s+2)}.$$

Put

$$G(s) = \frac{1}{(s+1)(s+2)} = \frac{1}{s+1} - \frac{1}{s+2}$$

$$\leftrightarrow g(t) = e^{-t} - e^{-2t}.$$

We require the inverse transform of $e^{-2(s+1)}G(s)$. By the delay rule with $c = 2$, this is given by

$$e^{-2(s+1)}G(s) = e^{-2} e^{-2s} G(s)$$

$$\leftrightarrow e^{-2}(e^{-(t-2)} - e^{-2(t-2)})H(t-2) = (e^{-t} - e^{-2t+2})H(t-2).$$

Example 22.20. Solve the differential equation

$$\frac{dx}{dt} + 2x = f(t)$$

with $x(0) = 0$, where (Fig. 22.4)

$$f(t) = \begin{cases} 0 & \text{when } t < 1, \\ e^{-t} & \text{when } 1 \leqslant t \leqslant 2, \\ 0 & \text{when } t > 2. \end{cases}$$

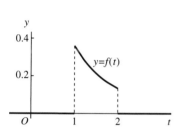

Fig. 22.4

Let $L\{x(t)\} = X(s)$. We need

$$L\{f(t)\} = \int_1^2 e^{-st} e^{-t} \, dt = \int_1^2 e^{-(s+1)t} \, dt$$

$$= \frac{1}{s+1} (e^{-(s+1)} - e^{-2(s+1)}) = F(s).$$

The transformed equation is then

$$sX + 2X = F(s), \quad \text{or} \quad X = F(s)/(s+2).$$

Therefore

$$X = (e^{-(s+1)} - e^{-2(s+1)}) \frac{1}{(s+1)(s+2)}$$

$$= (e^{-(s+1)} - e^{-2(s+1)}) \left(\frac{1}{s+1} - \frac{1}{s+2} \right)$$

$$= e^{-1} e^{-s} \left(\frac{1}{s+1} - \frac{1}{s+2} \right) - e^{-2} e^{-2s} \left(\frac{1}{s+1} - \frac{1}{s+2} \right).$$

Apply the delay rule with $c = 1$ and $c = 2$, noting that

$$\frac{1}{s+1} - \frac{1}{s+2} \leftrightarrow e^{-t} - e^{-2t}.$$

We obtain

$$x(t) = e^{-1}(e^{-(t-1)} - e^{-2(t-1)})H(t-1)$$

$$- e^{-2}(e^{-(t-2)} - e^{-2(t-2)})H(t-2)$$

$$= (e^{-t} - e^{1-2t})H(t-1) - (e^{-t} - e^{2-2t})H(t-2).$$

Both terms are zero before 'switch-on' at $t = 1$. Between $t = 1$ and 2, only the first term contributes. For $t > 2$ both terms are present, the second being stimulated by 'switching off'.

Problems, Chapter 22

The dot notation, $\dot{x} = dx/dt$, $\ddot{x} = d^2x/dt^2$, etc. is used in some of the questions.

22.1. Write down $\mathcal{L}\{x(t)\}$, where $x(t)$ is as follows.
(a) e^t; (b) $4\,e^{-t}$; (c) $3\,e^t - e^{-t}$;
(d) $3t^2 - 1$; (e) $\frac{1}{2}t^3 + 2t^2 - 3$; (f) $3 + 2t^4$;
(g) $3 \sin t - \cos t$; (h) $2(\cos t - \sin t)$;
(i) $1 + \dfrac{1}{1!}t + \dfrac{1}{2!}t + \cdots + \dfrac{1}{n!}t^n$ (you get a geometric series; see Section 1.13).

22.2. (Scale rule). Find $\mathcal{L}\{x(t)\}$ for the following cases of $x(t)$.
(a) e^{3t}; (b) $1 - 2\,e^{-2t}$; (c) $\sin \omega t$;
(d) $\cos \omega t$; (e) $3 \cos 2t - 2 \sin 2t$;
(f) $\cos^2 t$ (express it in terms of $\cos 2t$);
(g) $\sin^2 t$ (see (f)).

22.3. (See Section 22.3). Find $\mathcal{L}\{x(t)\}$ in the following cases of $x(t)$.
(a) $t^2\,e^t$ (easiest to start with t^2); (b) $t\,e^{-2t}$;
(c) $t^2\,e^{-t}$; (d) $e^{2t} \cos t$; (e) $e^{-t} \sin t$;
(f) $e^t \sin 3t$; (g) $e^{-2t} \sin 3t$; (h) $e^{-3t} \cos 2t$;
(i) $t \cos 3t$; (j) $t \sin 3t$; (k) $t^2 \sin t$;
(l) $t^4\,e^{-t}$ (compare the three methods: (i) start with t^4 and use the shift rule, (ii) start with e^{-t} and use (22.8), (iii) work directly from the definition (22.1)).

22.4. Obtain the Laplace transform for $\sin kt$ by differentiating that of $\cos kt$ with respect to k.

22.5. Invert the following Laplace transforms.
(a) $1/s^2$; (b) $1/s$; (c) $3/2s$; (d) $3/s^5$; (e) $1/(s-3)$;
(f) $1/(s+4)$; (g) $3/(2s-1)$; (h) $2/(2-3s)$;
(i) $1/s(s-1)$; (j) $1/(s^2+s-1)$; (k) $s/(s^2-1)$;
(l) $(2s-1)/(s^2-1)$; (m) $s/(s^2+1)$; (n) $1/(s^2+4)$;
(o) $(2s-1)/(s^2+4)$; (p) $(2s-1)/s(s-1)$;
(q) $(s^2-1)/s(s-1)(s+2)(s+3)$;
(r) $s/(s-1)(s^2+1)$; (s) $1/(s-1)^3$;
(t) $(2s+1)/(s^2-2s+2)$; (u) $s/(s^2+1)(s^2+4)$.

22.6. Find the Laplace transform of the following expressions involving $x(t)$, where $\mathcal{L}\{x(t)\} = X(s)$.
(a) $\dot{x}(t)$, where $x(0) = 6$; (b) $\dot{x}(t)$, where $x(0) = 0$;
(c) $\ddot{x}(t)$, where $x(0) = 3$, $\dot{x}(0) = 5$;
(d) $\ddot{x}(t)$, where $x(0) = 0$, $\dot{x}(0) = 0$;

(e) $2\ddot{x} + 3\dot{x} - 2x$, where $x(0) = 5$, $\dot{x}(0) = -2$;
(f) $3\ddot{x} - 5\dot{x} + x - 1$, where $x(0) = 0$, $\dot{x}(0) = 0$.

22.7. Use the Laplace transform to solve the following initial-value problems.
(a) $\ddot{x} + 3\dot{x} + 2x = 0$, $x(0) = 0$, $\dot{x}(0) = 1$;
(b) $\ddot{x} + \dot{x} - 2x = 0$, $x(0) = 3$, $\dot{x}(0) = 0$;
(c) $\ddot{x} + 4\dot{x} = 0$, $x(0) = x_0$, $\dot{x}(0) = y_0$;
(d) $\ddot{x} + \omega^2 x = 0$, $x(0) = c$, $\dot{x}(0) = 0$.
(e) $\ddot{x} + 2\dot{x} + 2x = 0$, $x(0) = 3$, $\dot{x}(0) = 0$;
(f) $d^4y/dx^4 - y = 0$, $y(0) = 1$, $y'(0) = 0$, $y''(0) = 0$, $y'''(0) = 0$ (use x instead of t as the variable in the Laplace transform).

22.8. Use the Laplace transform to solve the following initial-value problems.
(a) $\ddot{x} = 1 + t + e^t$, $x(0) = 0$, $\dot{x}(0) = 0$;
(b) $\ddot{x} + x = 3$, $x(0) = 0$, $\dot{x}(0) = 1$;
(c) $\ddot{x} + 2\dot{x} + 2x = 3$, $x(0) = 1$, $\dot{x}(0) = 0$;
(d) $\ddot{x} - x = e^{2t}$, $x(0) = 0$, $\dot{x}(0) = 1$;
(e) $\ddot{x} - x = t\,e^t$, $x(0) = 1$, $\dot{x}(0) = 1$;
(f) $\ddot{x} - 4x = 1 - e^{2t}$, $x(0) = 1$, $\dot{x}(0) = -1$;
(g) $\ddot{x} - 4x = e^{2t} + e^{-2t}$, $x(0) = 0$, $\dot{x}(0) = 0$;
(h) $\ddot{x} + \omega^2 x = C \cos \omega t$, $x(0) = x_0$, $\dot{x}(0) = y_0$;
(i) $\dddot{x} - 2\ddot{x} - \dot{x} + 2x = e^{-2t}$, $x(0) = 0$, $\dot{x}(0) = 0$, $\ddot{x}(0) = 2$ (look out for factors in the denominator of $X(s)$).

22.9. Solve the following simultaneous first-order differential equations, for the given initial values.
(a) $\dot{x} = x - y$, $\dot{y} = x + y$, $x(0) = 1$, $y(0) = 0$;
(b) $\dot{x} = 2x + 4y + e^{4t}$, $\dot{y} = x + 2y$, $x(0) = 1$, $y(0) = 0$;
(c) $\dot{x} = x - 4y$, $\dot{y} = x + 2y$, $x(0) = 2$, $y(0) = 1$.

22.10. Find the general solution of the following by putting $x(0) = A$, $\dot{x}(0) = B$, where A and B are arbitrary.
(a) $\ddot{x} + x = e^t$; (b) $\ddot{x} - x = 3$; (c) $\ddot{x} - 2\dot{x} + x = e$.

22.11. Find the general solution of $d^4y/dx^4 - y = e^t$, by putting $y(0) = A$, $y'(0) = B$, $y''(0) = C$, $y'''(0) = D$, where A, B, C, D are arbitrary. (Let the variable in the Laplace transform (22.1) be x instead of t.)

22.12. This is a system of first-order equations for

$x_0(t), x_1(t), \ldots, x_n(t)$:

$$\dot{x}_0 = -\beta x_1, \qquad \dot{x}_r = \beta(x_{r-1} - x_r)$$

for $r = 1, 2, \ldots, n$. Solve them by using the Laplace transform, showing that $x_r = \dfrac{1}{r!}(\beta t)^r \, e^{-\beta t}$.

22.13. Use the delay rule (22.15) to obtain the Laplace transform of $e^{-t}(t - 2) \cos(t - 2)H(t - 2)$.

22.14. Find the functions which give rise to the following Laplace transforms:
(a) $e^{-2s}/(s + 3)$;   (b) $(1 - s e^{-s})/(s^2 + 1)$;
(c) $e^{-2s}/(s - 4)$;   (d) $s e^{-s}/(s + 1)(s + 2)$;
(e) $e^{-s}/(s - 1)(s^2 - 2s + 2)$.

22.15. Solve the following differential equations assuming that the initial state is of quiescence $x(0) = \dot{x}(0) = 0$:

(a) $\ddot{x} + x = f(t)$, where
$$f(t) = \begin{cases} 1 & \text{for } 0 < t \leqslant 1, \\ 0 & \text{for } t > 1. \end{cases}$$

(b) $\ddot{x} - 4x = f(t)$, where
$$f(t) = \begin{cases} 1 & \text{for } 0 < t \leqslant 1, \\ 0 & \text{for } t > 1. \end{cases}$$

(c) $\ddot{x} - 4x = f(t)$, where
$$f(t) = \begin{cases} t & \text{for } 0 < t \leqslant 1, \\ 2 - t & \text{for } 1 < t \leqslant 2, \\ 0 & \text{for } t > 2. \end{cases}$$

(d) $\ddot{x} + x = f(t)$, where
$$f(t) = \begin{cases} \cos t & \text{for } 0 < t \leqslant \pi, \\ 0 & \text{for } t > \pi. \end{cases}$$

23 Applications of the Laplace transform

23.1 Division by s and integration

Multiplication by s is associated with differentiation (see (22.12)). Division by s is associated with integration, as follows.

Division rule

If $G(s) \leftrightarrow g(t)$, then $\dfrac{1}{s} G(s) \leftrightarrow \displaystyle\int_0^t g(\tau)\, d\tau$. $\qquad$ **(23.1)**

To prove this, put $(1/s)G(s) = F(s)$; then we must express $f(t)$ in terms of $g(t)$. Rewrite the relation between $F(s)$ and $G(s)$ in the form

$$sF(s) = G(s).$$

But from (22.12) we know that, in general,

$$\frac{df}{dt} \leftrightarrow sF(s), \quad \text{provided that } f(0) = 0;$$

so then we have $df/dt \leftrightarrow G(s)$. This is equivalent to the initial-value problem $df/dt = g(t)$, with $f(0) = 0$. The solution is

$$f(t) = \int_0^t g(\tau)\, d\tau.$$

Example 23.1. Find $f(t)$ when $F(s) = 1/s(s^2 + 1)$, (a) by using partial fractions, (b) by using (23.1).

(a) $\dfrac{1}{s(s^2 + 1)} = \dfrac{1}{s} - \dfrac{s}{s^2 + 1} \leftrightarrow 1 - \cos t$.

(b) In the notation of (23.1), put

$$G(s) = \frac{1}{s^2 + 1} \leftrightarrow \sin t.$$

Therefore

$$F(s) = \frac{1}{s(s^2 + 1)} = \frac{1}{s}\frac{1}{s^2 + 1}$$

$$\leftrightarrow \int_0^t \sin \tau \, dt = 1 - \cos t.$$

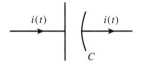

Fig. 23.1

Figure 23.1 shows a capacitor, of capacitance C, being charged by a current $i(t)$, the voltage drop across the plates being $v(t)$. Assume that the capacitor is uncharged at $t = 0$; then, at a later time t,

$$v(t) = \frac{1}{C} \int_0^t i(\tau) \, d\tau.$$

Therefore, according to (23.1), the relation between the Laplace transforms of $v(t)$ and $i(t)$ is

$$V(s) = \frac{1}{Cs} I(s). \tag{23.2}$$

We say that (23.2) describes the situation **in the s domain**, as we spoke of description in the frequency, or ω, domain in Section 19.3.

If the capacitor has an initial charge q_0, then

$$v(t) = \frac{1}{C}\left(\int_0^t i(\tau) \, d\tau + q_0 \right).$$

Since $q_0 \leftrightarrow s^{-1} q_0$, this transforms into

$$V(s) = \frac{1}{C}\frac{1}{s}[I(s) + q_0]. \tag{23.3}$$

We shall not be concerned with this case.

Example 23.2. The circuit shown is switched on at time $t = 0$. It is initially quiescent, and there is zero charge on the capacitor. Find the current for $t > 0$.

The circuit equation is

$$v_0 \cos \omega t = Ri(t) + \frac{1}{C} \int_0^t i(\tau) \, d\tau.$$

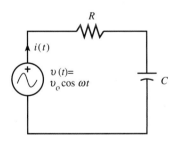

Fig. 23.2

Such an equation is called an **integral equation** for $i(t)$. The Laplace transform of the equation is

$$\frac{v_0 s}{s^2 + \omega^2} = RI(s) + \frac{1}{Cs} I(s),$$

so

$$I(s) = \frac{v_0}{R} \frac{s^2}{(s + 1/RC)(s^2 + \omega^2)}$$

$$= \frac{v_0}{R} \frac{1}{1 + (RC\omega)^2} \left(\frac{(RC\omega)^2 s}{s^2 + \omega^2} - \frac{RC\omega^2}{s^2 + \omega^2} + \frac{1}{s + 1/RC} \right)$$

after splitting into partial fractions. Therefore, for $t > 0$,

$$i(t) = \frac{v_0}{R} \frac{1}{1 + (RC\omega)^2} [(RC\omega)^2 \cos \omega t - RC\omega \sin \omega t + e^{-t/RC}].$$

The first two terms represent a steady forced oscillation and the final term is a transient.

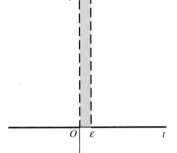

$x = \delta(t)$

Fig. 23.3

23.2 The impulse function

Figure 23.3a shows the graph of a function which is zero everywhere except for a tall, narrow rectangle with width ε and height $1/\varepsilon$, so that the area under the graph is equal to 1. Imagine that ε is a very small number, as small as we wish. This very tall and very narrow picture is a simplified version of the **impulse function or delta function**, usually denoted by $\delta(t)$. It is used in problems involving sudden and brief events, to represent (say) impulsive force between two bodies in collision; voltage from a lightning strike; or, if the variable is position rather than time, a point force.

The impulse or delta function $\delta(t)$

Informal definition: $\delta(t) = 1/\varepsilon$ for $0 < t < \varepsilon$, and $\delta(t) = 0$ elsewhere. The dimensions of $\delta(t)$ are those of t^{-1} (if t is time, that is $[T^{-1}]$).   **(23.4)**

In Figure 23.4, $\delta(t)$ is moved to the right so as to be at $t = c$; the vertical strip therefore represents $\delta(t - c)$. An ordinary function $f(t)$ crosses it at C. Consider the integral

$$\int_a^b f(t)\,\delta(t - c)\,\mathrm{d}t,$$

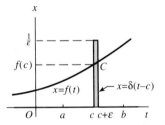

$x = f(t)$ $x = \delta(t-c)$

Fig. 23.4

where c lies between a and b. The integrand is zero except between c and $c + \varepsilon$; over this very narrow interval, $f(t)$ hardly changes from the value $f(c)$. Therefore (as closely as we wish)

$$\int_a^b f(t)\,\delta(t - c)\,\mathrm{d}t = \int_c^{c+\varepsilon} f(c)\varepsilon^{-1}\,\mathrm{d}t = f(c).$$

If c does not lie between a and b, then the integral is zero. The delta function is sometimes called a **sifting function** because of this property.

Sifting property of $\delta(t)$

$$\int_a^b f(t)\,\delta(t - c)\,\mathrm{d}t = \begin{cases} f(c) & \text{if } a \leqslant c < b, \\ 0 & \text{otherwise.} \end{cases}$$ **(23.5)**

We can obtain the Laplace transform of $\delta(t)$ from (23.5):

> **Laplace transform of** $\delta(t - c)$
>
> $$\mathcal{L}\{\delta(t - c)\} = \int_0^\infty e^{-st} \, \delta(t - c) \, dt = e^{-cs}, \tag{23.6}$$
>
> for $c \geqslant 0$. In particular, $\mathcal{L}\{\delta(t)\} = 1$.

Example 23.3. The equation $d^2x/dt^2 + \omega^2 x = f(t)$ represents the displacement x of a particle of unit mass on a spring of stiffness ω with external force $f(t)$. Find the motion for $t > 0$ if the particle is subjected to an impulse $I \, \delta(t - 1)$ at time $t = 1$, assuming equilibrium at $t = 0$. (I has dimensions [force $\times$ time].)

The equation is $d^2x/dt^2 + \omega^2 x = I \, \delta(t - 1)$. Its transform is

$$s^2 X + \omega^2 X = I \, e^{-s},$$

where $x(t) \leftrightarrow X(s)$. Therefore

$$X(s) = \frac{I}{s^2 + \omega^2} \, e^{-s}.$$

We know that

$$\frac{1}{s^2 + \omega^2} \leftrightarrow \frac{1}{\omega} \sin \omega t;$$

so, by the delay rule (22.15), we have

$$X(s) = \frac{I}{s^2 + \omega^2} \, e^{-s} \leftrightarrow x(t) = \frac{I}{\omega} \sin \omega(t - 1) \mathrm{H}(t - 1),$$

where H stands for the unit function (22.13). There is no motion until $t = 1$, when the impulse sets up free oscillations $(I/\omega) \sin \omega(t - 1)$.

Example 23.4. Find the current resulting from an impulsive voltage $I_v \, \delta(t)$ applied to the circuit of Fig. 23.5, the current being zero before application of the voltage. (The dimensions of I_v are [emf $\times$ time].)

The equation for the current is $L \, di/dt + Ri = I_v \, \delta(t)$. After transformation, with $i(0) = 0$, it becomes

$$LsI(s) + RI(s) = I_v.$$

Therefore

$$I(s) = \frac{I_v}{L(s + R/L)} \leftrightarrow i(t) = \frac{I_v}{\omega} e^{-Rt/L}.$$

The great, though brief, applied voltage gives only a finite current because of the counter-emf generated by the coil.

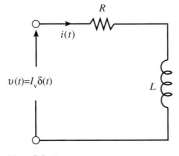

Fig. 23.5

The delta function can be regarded formally as the derivative of the unit function $\mathrm{H}(t)$. As in Fig. 23.6a, smooth out the transition

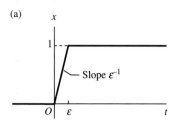

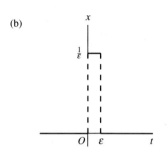

Fig. 23.6

of H(t), from zero to one, as t passes through the origin, by means of a sloping straight-line segment. The derivative of this function is equal to zero outside the transition interval $(0, \varepsilon)$ and equal to ε inside it; this specifies $\delta(t)$ as in (23.4).

Connection between H(t) and $\delta(t)$

$$\frac{d\mathrm{H}(t)}{dt} = \delta(t). \tag{23.7}$$

This only conforms with the Laplace-transform derivative rule (22.12),

$$s\left(\frac{1}{s}\right) - \mathrm{H}(0) = 1,$$

if we rather arbitrarily interpret H(0) as being zero. It should be understood that certain weaknesses result from treating the impulse function very informally; the real justification for its use is an elaborate mathematical subject called **distribution theory**.

23.3 Impedance in the s domain
In table (23.8) below three basic circuit elements are shown, together with their voltage-drop–current relations and the Laplace transforms of these relations, on the assumption that, **at $t = 0$, the current through the inductor and the charge on the capacitor are zero**. The expression 's domain' refers to transformed quantities.

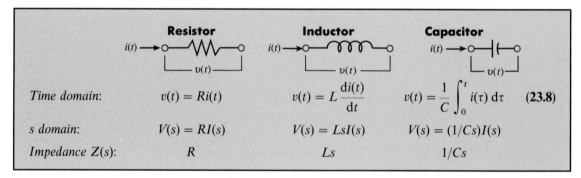

	Resistor	**Inductor**	**Capacitor**
Time domain:	$v(t) = Ri(t)$	$v(t) = L\dfrac{di(t)}{dt}$	$v(t) = \dfrac{1}{C}\displaystyle\int_0^t i(\tau)\, d\tau$ (23.8)
s domain:	$V(s) = RI(s)$	$V(s) = LsI(s)$	$V(s) = (1/Cs)I(s)$
Impedance Z(s):	R	Ls	$1/Cs$

Table (23.8) should be compared with the table (19.5) for the case of steady forced oscillations of frequency $\omega/2\pi$. The **impedances** $Z(s)$ in the s plane are analogous to the complex impedances R, $j\omega L$, and $1/j\omega C$ of (19.6) for the steady case. One can pass from one to the other by substituting $j\omega$ for s, or $-js$ for ω. However, the s forms allow arbitrary inputs to the circuit to be considered.

Impedances combine in series and parallel in the same way as do complex impedances (see (19.7)) in the frequency domain, but it is to be remembered that they refer to **zero initial conditions only**.

Combination of impedances $Z(s)$ in the s domain for zero initial state

Impedances in series

$$Z = Z_1 + Z_2 + \cdots.$$

Impedances in parallel

$$\frac{1}{Z} = \frac{1}{Z_1} + \frac{1}{Z_2} + \cdots$$

(23.9)

Example 23.5. The circuit shown in Fig. 23.7a is initially quiescent, with zero charge on the capacitor. The constant voltage v_0 is switched on at $t = 1$ and off at $t = 2$. Find the current $i(t)$.

The corresponding s domain impedances are shown in Fig. 23.7b, in which the elements R and C are grouped. They are in parallel, so (23.8) and (23.9) give

$$\frac{1}{Z_1} = \frac{1}{R} + \frac{1}{(Cs)^{-1}} = \frac{1}{3} + \frac{s}{12} = \frac{s+4}{12}$$

Hence

$$Z_1 = \frac{12}{s+4},$$

and also $Z_2 = Ls = 4s$. Then Z for the whole circuit is given by

$$Z = 4s + \frac{12}{s+4} = \frac{4(s+1)(s+3)}{s+4}.$$

Therefore

$$I(s) = \frac{s+4}{4(s+1)(s+3)} V(s).$$

Taking into account switch-on at $t = 1$ and switch-off at $t = 2$,

$$v(t) = v_0[\mathrm{H}(t-1) - \mathrm{H}(t-2)],$$

so

$$V(s) = v_0\left(\frac{1}{s}e^{-s} - \frac{1}{s}e^{-2s}\right).$$

Therefore

$$I(s) = \frac{v_0(s+4)}{4s(s+1)(s+3)}(e^{-s} - e^{-2s})$$

$$= v_0\left(\frac{1}{3s} - \frac{3}{8}\frac{1}{s+1} + \frac{1}{24}\frac{1}{s+3}\right)(e^{-s} - e^{-2s}).$$

(a)

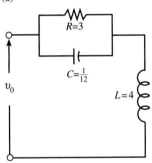

(b)

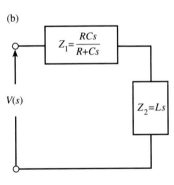

Fig. 23.7

The bracketed factor transforms back to

$$v_0(\tfrac{1}{3} - \tfrac{3}{8}e^{-t} + \tfrac{1}{24}e^{-3t}),$$

and, by using the delay rule (22.15) to deal with the exponentials,

$$i(t) = v_0(\tfrac{1}{3} - \tfrac{3}{8}e^{-(t-1)} + \tfrac{1}{24}e^{-3(t-1)})H(t-1)$$
$$- v_0(\tfrac{1}{3} - \tfrac{3}{8}e^{-(t-2)} + \tfrac{1}{24}e^{-3(t-2)})H(t-2).$$

Nothing happens until the system is switched on at $t = 1$, when the first term (only) is activated. At $t = 2$, when it is switched off, the second term comes in also; some current persists but it dies away to zero.

It must be emphasized that such a problem is considerably complicated when the initial conditions are not zero. For example, the expression (23.3) for an initially charged capacitor is not in the form of a voltage–impedance–current relationship. In such cases, it is necessary to start with the differential equations for individual branches.

23.4 Transfer functions in the s domain

The impedance $Z(s)$ which directly connects the current in a unit with the voltage drop across the same unit is a special case of a more general idea: to relate any two currents or voltages which occur in the network.

We suppose as before that we have a passive circuit consisting of linear resistors, capacitors, and inductors, and a single source of voltage which drives the circuit. Figure 23.8 represents such a network. We denote the driving voltage by $f(t)$, because much of what we say can be taken over into mechanical and other systems. The unknown voltages and currents we call the **variables**.

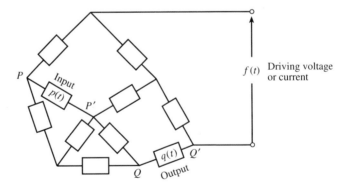

Fig. 23.8

Suppose that there are N currents and voltages to be determined, which we call $x_1(t), x_2(t), \ldots, x_N(t)$, with transforms $X_1(s), X_2(s), \ldots, X_N(s)$. A voltage $f(t)$, with transform $F(s)$, is applied somewhere in the network. Provided that the circuit is initially quiescent (zero initial conditions), each of the s domain equations for the currents and voltages takes one of only two possible forms:

$$\text{either} \quad a_1 X_1 + a_2 X_2 + \cdots + a_N X_N = 0$$
$$\text{or} \quad b_1 X_1 + b_2 X_2 + \cdots + b_N X_N = F,$$

where the coefficients are functions of s. Therefore the transforms $X_1, X_2, \ldots, X_N$ are all proportional to F:

$$X_n(s) = G_n(s)F(s),$$

for $n = 1, 2, \ldots, N$. The G_n are functions which **depend only on the circuit constants and not on the applied voltage**. They are called **transfer functions**. Nominate an arbitrary variable $p(t)$ in any branch as the **input**, and another variable $q(t)$ elsewhere as the **output**. The corresponding G_n are denoted by G_P and G_Q. The driving voltage $f(t)$ may serve as an input if we wish. Assuming that we start with **zero currents and charges**, which implies **zero initial conditions**, the transforms $P(s)$ and $Q(s)$ of $p(t)$ and $q(t)$ are related by

$$Q(s)/P(s) = G_Q F/G_P F = G_{PQ}(s),$$

say, where $G_{PQ}(s)$ is called the **transfer function from p to q**.

Transfer function $G_{PQ}(s)$ (zero initial conditions)

Let $p(t)$ (input) and $q(t)$ (output) be the voltage or current in any two branches. Then

$$Q(s)/P(s) = G_{PQ}(s),$$

where $G_{PQ}(s)$ is the transfer function from p to q. G_{PQ} depends only on the circuit parameters.

(23.10)

Transfer functions which connect different types of variable are given various names and conventional symbols in literature on systems. For example, in the s domain, voltage ÷ current is **impedance**; current ÷ voltage is **admittance**; voltage ÷ voltage is **voltage gain**, and so on.

Example 23.6. Find the transfer function $G(s)$ from the voltage transform $P(s)$ over R, regarded as the input, and the voltage transform $Q(s)$ over C, regarded as the output, in Fig. 23.9a.

Let the current $i(t)$ be as indicated. The impedances of the various groups are shown in Fig. 23.9b; these are in fact transfer functions between current and voltage for each unit. In terms of the transforms,

$$P(s) = RI(s), \qquad Q(s) = \frac{1}{Cs} I(s).$$

Therefore

$$G(s) = \frac{Q(s)}{P(s)} = \frac{1}{RCs}.$$

(a)

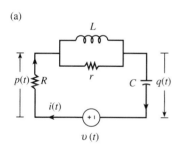

(b)

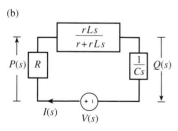

Fig. 23.9

Thus

$$Q(s) = \frac{1}{RC} \frac{1}{s} P(s),$$

(a)   circuit A

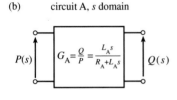

and so $q(t) = \dfrac{1}{RC} \displaystyle\int_0^t p(\tau)\, d\tau$, as expected.

(b)   circuit A, s domain

(c)

Suppose now that we have a circuit such as the one in Fig. 23.10a, called Circuit A, where $p(t)$ is the input voltage and $q(t)$ the output voltage. Figure 23.10b schematizes the arrangement and specifies the transfer function $G(s) = Q(s)/P(s)$ between p and q.

We could also symbolize the dependence of q on p by the scheme in Fig. 23.10c. However, this figure suggests the beginnings of some kind of series arrangement: it looks as if we could attach another circuit to the original one without altering the transfer function, and so get an easy calculation for the combined circuit. This is not true in general, but sometimes it is a useful approximation.

To illustrate this question, we will append to Circuit A another Circuit B. It is shown in Fig. 23.11a, together with its s domain representation and its transfer function. In Fig. 23.11b, A and B are connected across MN; here p, q, r, and their transforms P, Q, R represent the actual voltages across the terminals indicated. The question is, do the transfer functions written in the boxes still correctly give $Q(s)$ in terms of $P(s)$, and then $R(s)$ in terms of $Q(s)$?

Fig. 23.10

(a)

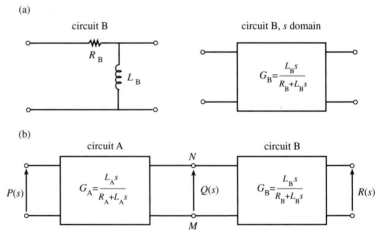

Fig. 23.11

If an appreciable amount of current passes between Circuits A and B after attachment, then $Q(s)$ must change, so the true transfer functions of both of the circuits will be changed, and the changes will not compensate each other. In special circumstances, however, the circuits may behave almost independently, or can be made to do so by means of technical arrangements such as feedback.

Example 23.7. The two circuits A and B shown in Fig. 23.12 are connected to form a composite circuit C. Show that

$$G(s) \approx G_A(s)G_B(s)$$

(where the $G(s)$ are the transfer functions for the voltages shown) if $1/R$ is much smaller than $1/r + 1/r_1$.

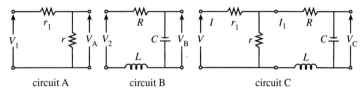

Fig. 23.12

circuit A   circuit B   circuit C

For Circuit A alone:

$$G_A(s) = \frac{V_A(s)}{V_1(s)} = \frac{r}{r + r_1}.$$

For Circuit B alone:

$$G_B(s) = \frac{V_B(s)}{V_2(s)} = \frac{\text{impedance of C}}{\text{total impedance}}$$

$$= \frac{1}{Cs} \cdot \frac{1}{R + Ls + 1/Cs}.$$

Therefore

$$G_A(s)G_B(s) = \frac{1}{Cs} \cdot \frac{r}{r + r_1} \cdot \frac{1}{R + Ls + 1/Cs}.$$

For Circuit C, by following the voltage drops around closed subcircuits as usual, we get

$$V = r_1 I + r(I - I_1),$$
$$0 = (R + Ls + 1/Cs)I_1 - r(I - I_1),$$

from which

$$I_1 = \frac{Vr}{(r + r_1)(r + R + Ls + 1/Cs) - r^2},$$

which represents the current 'leaking' between A and B. Therefore

$$G_C(s) = \frac{V_C(s)}{V(s)} = \frac{1}{Cs} \cdot \frac{r}{(r + r_1)(r + R + Ls + 1/Cs) - r^2},$$

which we have to compare with $G_A(s)G_B(s)$ above. Rewrite $G_C(s)$ in the form

$$G_C(s) = \frac{1}{Cs} \cdot \frac{r}{r + r_1} \cdot \frac{1}{R + Ls + 1/Cs - rr_1/(r + r_1)}.$$

It can be seen that $G_C(s) \approx G_A(s)G_B(s)$ if $rr_1/(r + r_1)$ is much smaller than R. But

$$\frac{rr_1}{r + r_1} = 1 \bigg/ \left(\frac{1}{r} + \frac{1}{r_1}\right),$$

and this is much smaller than R if $1/r + 1/r_1$ is much *greater* than $1/R$. The relation between the circuits could be represented in this case approximately by Fig. 23.13, as if they processed the voltage signals independently.

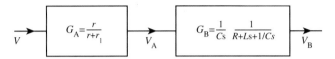

Fig. 23.13

Example 23.8. Figure 23.14 shows a chain of three systems, which act independently upon their inputs according to the transfer functions $G_A(s)$, $G_B(s)$, and $G_C(s)$ indicated in the boxes. Find the transfer function $G(s)$ between $F(s)$ and $F_C(s)$. Find $f_C(t)$ when $f(t) = H(t)$, for zero initial conditions.

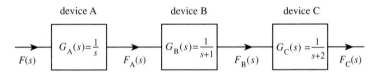

Fig. 23.14

We have

$$\frac{F_C}{F} = \frac{F_C}{F_B}\frac{F_B}{F_A}\frac{F_A}{F} = \frac{1}{s + 2}\frac{1}{s + 1}\frac{1}{s},$$

so

$$G(s) = \frac{1}{s(s + 1)(s + 2)}.$$

Now let $f(t) = H(t)$; then $F(s) = 1/s$. Therefore

$$F_C(s) = G(s)F(s) = \frac{1}{s(s + 1)(s + 2)}\frac{1}{s} = \frac{1}{s^2(s + 1)(s + 2)}.$$

In partial fractions,

$$F_C(s) = -\frac{3}{4}\frac{1}{s} + \frac{1}{2}\frac{1}{s^2} + \frac{1}{s + 1} - \frac{1}{4}\frac{1}{s + 2}.$$

Therefore $f(t) = -\frac{3}{4} + \frac{1}{2}t + e^{-t} - \frac{1}{4}e^{-2t}$ for $t > 0$.

It is possible to get an idea of important features of an output without going through the whole calculation:

Example 23.9. In a particular system, the output $X(s)$ in the s domain is related to the input $F(s)$ by $X(s) = G(s)F(s)$. Find the general character of $x(t)$ if $f(t) = \cos 2t$, $G(s) = s/(s + 1)(s^2 + 4s + 5)$.

Since $f(t) \leftrightarrow s/(s^2 + 4)$, we have

$$X(s) = \frac{s}{(s + 1)(s^2 + 4s + 5)} \frac{s}{s^2 + 4}.$$

If we expanded this in partial fractions, we should have terms of the types

$$\frac{1}{s + 1}, \frac{s}{(s + 2)^2 + 1}, \text{ and } \frac{1}{(s + 2)^2 + 1} \quad \text{(from } G\text{)},$$

and

$$\frac{s}{s^2 + 4} \text{ and } \frac{1}{s^2 + 4} \quad \text{(from } F\text{)}.$$

Therefore, in terms of time, we should obtain terms like

$$e^{-t}, e^{-2t} \sin t, \text{ and } e^{-2t} \cos t \text{ from } G \text{ (which are transients)},$$

$$\cos 2t \text{ and } \sin 2t \text{ from } F \text{ (a forced oscillation)}.$$

Finally we illustrate the relation between transfer functions in the s domain and complex transfer functions in the ω domain (Section 19.5).

Example 23.10. The transfer function between an input $F(s)$ and an output $X(s)$ is $1/(s^2 + 1)$. Find the amplitude and phase of the steady forced oscillation produced by an input $f(t) = 3 \sin 2t$.

As pointed out in Section 23.3, the complex impedance is simply the s domain impedance with $j\omega$ substituted for s. The same is true for any transfer function. In the ω domain representation, the input and output will be represented by phasors $F(\omega) = 3 e^{-\frac{1}{2}\pi j}$ and $X(\omega)$, corresponding to circular frequency $\omega = 2$ in this case. Then

$$X(\omega) = \frac{1}{(2j)^2 + 1} 3 e^{-\frac{1}{2}\pi j} = -e^{-\frac{1}{2}\pi j} = e^{\frac{1}{2}\pi j}.$$

The amplitude is the modulus of X, which is 1, and the phase is $\frac{1}{2}\pi$.

23.5 The convolution theorem
The following result enables us to interpret Laplace transforms which take the form of a **product** of two s functions.

Convolution theorem

Suppose $F(s) = G(s)H(s)$, and

$$G(s) \leftrightarrow g(t), \; H(s) \leftrightarrow h(t).$$

Then   (23.11)

$$f(t) = \int_0^t g(t - \tau)h(\tau)\, d\tau$$

(which is the same as $\int_0^t h(t - \tau)g(\tau)\, d\tau$).

This result will be proved in Chapter 29, Example 29.12. For the present we shall verify that it is true in some special cases.

Example 23.11. Find the inverse Laplace transform of

$$F(s) = \frac{1}{(s + 1)(s + 2)}.$$

Put $F(s) = G(s)H(s)$, where

$$G(s) = \frac{1}{s + 1}, \qquad H(s) = \frac{1}{s + 2};$$

then $g(t) = e^{-t}$ and $h(t) = e^{-2t}$. Equation (23.11) gives

$$F(s) \leftrightarrow f(t) = \int_0^t e^{-(t-\tau)} e^{-2\tau}\, d\tau$$

$$= \int_0^t e^{-t-\tau}\, d\tau = e^{-t} \int_0^t e^{-\tau}\, d\tau \text{ (note this step carefully)}$$

$$= e^{-t}(-e^{-t} + 1) = e^{-t} - e^{-2t}.$$

This result can be confirmed by using partial fractions instead:

$$\frac{1}{(s + 1)(s + 2)} = \frac{1}{s + 1} - \frac{1}{s + 2} \leftrightarrow e^{-t} - e^{-2t}.$$

Notice very carefully the distinction between t and τ in the integrals (23.11): τ is the variable of integration. **The variable t is a constant** so far as the integration process is concerned; so, for example, at one point we took e^{-t} outside the integral sign.

Example 23.12. Find the inverse transform of $1/s(s^2 + 1)$.
In Example 23.1, we showed in two different ways that

$$\frac{1}{s(s^2 + 1)} \leftrightarrow 1 - \cos t.$$

To confirm that (23.11) gives the same result, put

$$G(s) = \frac{1}{s^2 + 1} \quad \text{and} \quad H(s) = \frac{1}{s},$$

say. Then (for $t > 0$) $g(t) = \sin t$ and $h(t) = 1$, so

$$g(t - \tau) = \sin(t - \tau) \quad \text{and} \quad h(\tau) = 1.$$

Therefore, by the convolution theorem,

$$F(s) \leftrightarrow \int_0^t \sin(t - \tau)1 \, d\tau = \left[\cos(t - \tau)\right]_{\tau=0}^t = 1 - \cos t,$$

as expected.

Example 23.13. (See (23.11)). Confirm directly that

$$\int_0^t g(t - \tau)h(\tau) \, d\tau = \int_0^t h(t - \tau)g(\tau) \, d\tau.$$

In the first integral, change the variable, putting

$$u = t - \tau.$$

Then (remember t is to be treated like a constant) $du = -d\tau$.
Therefore

$$\int_0^t g(t - \tau)h(\tau) \, d\tau = \int_t^0 g(u)h(t - u)(-du)$$

$$= \int_0^t h(t - u)g(u) \, du,$$

which is the integral required, merely using u instead of τ for the variable of integration.

Example 23.14. Find an expression for the inverse transform of

$$F(s) = \frac{1}{s + 1} H(s)$$

in terms of $h(t)$, the inverse transform of $H(s)$.
Use the convolution theorem, (23.11), putting $G(s) = 1/(s + 1)$. Then

$$g(t) = e^{-t}.$$

We therefore obtain from (23.11)

$$f(t) = \int_0^t e^{-(t - \tau)} h(\tau) \, d\tau,$$

or its alternative form

$$f(t) = \int_0^t e^{-\tau} h(t - \tau) \, d\tau.$$

23.6 General response of a system from its impulsive response

We shall take an electrical network as our example, though what we say applies to linear mechanical systems as well. Suppose that it is activated by an applied voltage $f(t)$ (regarded as the input). Focus on any particular one of the currents or voltages in the circuit, and call it $x(t)$ (the output). The transfer function between input and output will be called $G(s)$. We have then

$$X(s) = G(s)F(s). \tag{23.12}$$

Suppose that we conduct an experiment in which we excite the circuit by means of a voltage impulse $I_v\,\delta(t)$, and record the result (the dimensions of I_v are [emf × time]: see (23.1)). Then

$$f(t) = I_v\,\delta(t), \quad \text{so that} \quad F(s) = I_v.$$

The current resulting from this special voltage (an impulsive input) will be called $x^*(t)$, with transform $X^*(s)$. Now put $F(s) = I_v$ and X^* for X into (23.12), and it becomes

$$X^*(s) = I_v G(s). \tag{23.13}$$

Such an experiment would therefore give us the corresponding transfer function $G(s)$ directly (we could even arrange for I_v to equal unity). Thus, even if the circuit is a 'black box' with its details unknown, we still know from (23.13) what to put into (23.12) for the case when $f(t)$ is any function at all:

$$X(s) = I_v^{-1}X^*(s)F(s).$$

Therefore, by the convolution theorem (23.11),

$$x(t) = I_v^{-1}\int_0^t x^*(t-\tau)f(\tau)\,\mathrm{d}\tau, \quad \text{or} \quad I_v^{-1}\int_0^t x^*(\tau)f(t-\tau)\,\mathrm{d}\tau.$$

This type of result applies to the other circuit variables such as voltages and charges, and to mechanical systems governed by linear differential equations. In terms of general outputs and inputs:

Output $x(t)$ from an input $f(t)$ to a quiescent linear system, in terms of the output $x^*(t)$ from an impulsive input $I\,\delta(t)$.

$$x(t) = I^{-1}\int_0^t x^*(t-\tau)f(\tau)\,\mathrm{d}\tau, \tag{23.14}$$

or

$$x(t) = I^{-1}\int_0^t x^*(\tau)f(t-\tau)\,\mathrm{d}\tau.$$

Example 23.15. The displacement $x^*(t)$ caused by an impulse $I\,\delta(t)$ applied to a certain mechanical linear system at rest is found to be $x^*(t) = \mathrm{e}^{-t} - \sin 2t$. Find the displacement $x(t)$ corresponding to an applied force $f(t) = \sin t$ starting at $t = 0$.

We have

$$x(t) = I^{-1} \int_0^t [e^{-(t-\tau)} - \sin 2(t-\tau)] \sin \tau \, d\tau \quad \text{(from (23.14))}$$

$$= I^{-1} e^{-t} \int_0^t e^\tau \sin \tau \, d\tau - I^{-1} \int_0^t \sin(2t - 2\tau) \sin \tau \, d\tau$$

$$= I^{-1} e^{-t} \int_0^t e^\tau \sin \tau \, d\tau$$

$$- \tfrac{1}{2} I^{-1} \int_0^t [\cos(2t - 3\tau) - \cos(2t - \tau)] \, d\tau,$$

by using the identity (1.18b). In the end, we find

$$x(t) = \tfrac{1}{2} I^{-1} e^{-t} + \tfrac{1}{3} I^{-1} \sin 2t - \tfrac{1}{6} I^{-1} (3 \cos t + \sin t)$$

for $t > 0$. The first term is a transient, and the second an induced free oscillation, while the third term represents the forced oscillation.

23.7 Convolution integral in terms of memory

An integral of the type

$$x(t) = \int_0^t g(t - \tau) f(\tau) \, d\tau,$$

such as arose in the convolution theorem (23.11), is called a **convolution integral**. Typically, f acts as some kind of 'cause', such as a driving force or voltage, and $x(t)$ stands for a certain 'effect' produced.

Choose a time t for observation; then divide the interval $\tau = 0$ to $\tau = t$ into a large number of equal time steps $\delta\tau$. We have

$$x(t) = \int_0^t g(t - \tau) f(\tau) \, d\tau \approx \sum_{\tau=0}^{\tau=t} g(t - \tau) f(\tau) \, \delta\tau.$$

Now choose any moment τ_1 between 0 and t: there was a force $f(\tau_1)$ applied at this moment, and its contribution to x at time $t > \tau_1$ is

$$g(t - \tau_1) f(\tau_1) \, \delta\tau.$$

The factor $g(t - \tau_1)$ takes into account the **time elapsed** between the cause and its effect – in some problems it would be appropriate to call $t - \tau_1$ the 'age' of $f(\tau_1)$ at the moment t of observation, and g an ageing factor. Depending on the type of problem, this factor might weaken or amplify the contribution of $f(\tau_1)$ to the integral as time t passes. The elapsed time is increased if either we take an earlier τ_1, or delay the time of observation by increasing t. Figure 23.15a shows a representative function $g(\alpha)$, where α stands for 'age', and Fig. 23.15b illustrates its effect on the influence of f at time τ_1 on x at a later time t.

(a)

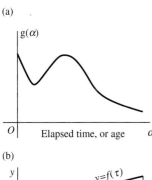

(b)

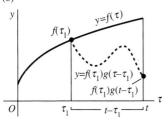

Fig. 23.15

Problems, Chapter 23

23.1. Invert the transforms (a) $1/s(s^2 + 1)$, (b) $1/s^2(s^2 + 1)$, (c) $1/s^3(s^2 + 1)$, by using (23.1).

23.2. The equation for the current $i(t)$ in an RLC circuit for zero initial charge is

$$L\frac{di}{dt} + Ri + C\int_0^t i(\tau)\,d\tau = v(t).$$

(a) Solve this equation when $L = 2$, $R = 3$, $C = \frac{1}{3}$, $v(t) = 3\cos t$ in conveniently scaled units, for zero initial current and charge.
(b) Adapt the equation to the case when $v(t) = 0$ and there is an initial charge q_0 on the capacitor, and solve it, given that $i(0) = 0$.
(c) The circuit in (a) is quiescent with zero charge; then, at $t = t_0$, a voltage of 300 units acts in it for 0.01 time units. Approximate the applied voltage by a suitable impulse function, and solve the equation for $i(t)$.

23.3. The displacement $x(t)$ of a mass on a spring with velocity damping and external force $f(t)$ per unit mass reduces to the conventional form $\ddot{x} + 2k\dot{x} + \omega^2 x = f(t)$. The initial conditions are $x(0) = 1$, $\dot{x}(0) = 1$. An impulse I is applied at $t = t_0$. Find the solution for $t > 0$ for $k^2 > \omega^2$.

23.4. A light plank of length l rests across a crevasse, and sags under the weight of a mountaineer of mass M standing at the centre. The displacement $u(x)$, where x is measured from one end, is determined in general by $K\,d^4u/dx^4 = f(x)$, where K is constant and $f(t)$ is force per unit length along the plank. The **boundary conditions**, which say that the plank merely rests on its ends, are

$$u(0) = u''(0) = u(l) = u''(l) = 0.$$

Treat the mountaineer as a point force and solve the problem using Laplace transforms. (Hint: Two conditions are prescribed at $x = 0$, but four are needed: call the missing ones A, B. Find A and B by requiring $u(l) = u''(l) = 0$.)

23.5. Find the impedances of the circuits shown.

(a)

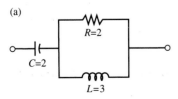

(continued)

(b)

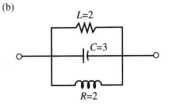

(c)

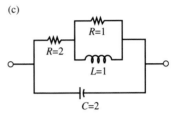

(d)

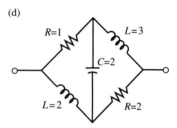

23.6. Find the transfer functions $V_2(s)/V_1(s)$ and $V_2(s)/I(s)$ in the following circuits.

(a)

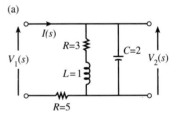

(b)

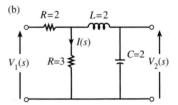

23.7. Evaluate the convolution integral

$$\int_0^t g(\tau)h(t - \tau)\,d\tau,$$

or

$$\int_0^t h(\tau)g(t-\tau)\,d\tau,$$

in the following cases. Sometimes it might be easier to invert the corresponding Laplace transform (23.11).
(a) $g(t) = e^t$, $h(t) = 1$;   (b) $g(t) = 1$, $h(t) = 1$;
(c) $g(t) = e^t$, $h(t) = e^t$;   (d) $g(t) = e^{-t}$, $h(t) = t$;
(e) $g(t) = t$, $h(t) = \sin t$;   (f) $g(t) = \cos t$, $h(t) = t$;
(g) $g(t) = \sin 3t$, $h(t) = e^{-2t}$;
(h) $g(t) = \sin t$, $h(t) = \sin t$;
(i) $g(t) = t^4$, $h(t) = \sin t$;   (j) $g(t) = t^n$, $h(t) = t^m$.

23.8. Use the convolution theorem (23.11) to obtain an expression, in the form of an integral, for a particular solution of the following equations.

(a) $\dfrac{d^2x}{dt^2} + \omega^2 x = f(t)$;   (b) $\dfrac{d^2x}{dt^2} - \omega^2 x = f(t)$.

23.9. Use the convolution theorem (23.11) to find a solution $x(t)$ of the following **Volterra-type integral equations**.

(a) $\displaystyle\int_0^t x(\tau)(t-\tau)\,d\tau = t^4$;

(b) $x(t) = 1 + \displaystyle\int_0^t x(\tau)(t-\tau)\,d\tau$;

(c) $x(t) = \sin t + \displaystyle\int_0^t x(\tau)\cos(t-\tau)\,d\tau$.

23.10. By following a similar argument to that leading up to (23.14), show that

$$x(t) = \frac{d}{dt}\int_0^t x^{**}(\tau)f(t-\tau)\,d\tau,$$

where $x^{**}(t)$ represents the response of a quiescent 'black box' to a unit-function input $H(t)$, and $x(t)$ is its response from quiescence to an input $f(t)$.

Suppose that the transform $X^{**}(s)$ of the unit-function response is given by $1/(s-1)(s+2)$ in a particular case. Obtain the response from zero initial conditions to an input $H(t)\sin\omega t$.

23.11. (a) a student learning a language aims to memorize 50 new words a day, starting at $t = 0$. She is successful in this but, after a time lapse α, remembers only a fraction $e^{-0.01\alpha}$ of those learned at any time. Express the number $N(t)$ of words still in her vocabulary in terms of a convolution integral, and evaluate it.

(b) The student decides to increase the number of words available by attempting $50 + 0.1t$ words per day. Find the new $N(t)$, assuming the same initial success and the same rate of forgetting.

23.12. A population $p(t)$ for $t > 0$ develops as follows. The p_0 individuals in existence at $t = 0$ die out on average via a factor $e^{-\gamma t}$, so that at time t only about $p_0\,e^{-\gamma t}$ are still in existence. For the rest, take any time $\tau < t$. The number born between τ and $\tau = \delta\tau$ is $bp(\tau)\,\delta\tau$, where b is the birthrate; these individuals die out through a factor $e^{-\beta(t-\tau)}$, where $\beta < \gamma$ and $t - \tau$ is the time elapsed from birth. Show that

$$p(t) = p_0\,e^{-\gamma t} + b\int_0^t p(\tau)\,e^{-\beta(t-\tau)}\,d\tau,$$

and solve the equation.

23.13. A simple harmonic oscillator with displacement x is subject to a constant force F_0 for $0 < t < t_0$, and allowed to oscillate freely for $t > t_0$. Its equation of motion is

$$m\ddot{x} + kx = F_0[H(t) - H(t - t_0)].$$

If the system starts from rest in equilibrium, show that the Laplace transform is

$$\mathcal{L}\{x(t)\} = \frac{F_0}{m}\frac{(1 - e^{-st})}{s(s^2 + \omega^2)},$$

where $\omega = \sqrt{(k/m)}$. Show that, for $0 < t < t_0$, the solution is

$$x(t) = \frac{F_0}{k}(1 - \cos\omega t),$$

and find the solution for $t > t_0$.

23.14. An equation of the form

$$\frac{dx(t)}{dt} = x(t-1) + t,$$

which relates the derivative at time t to the value of the function at an earlier time is an example of a **differential delay equation**. If $x(t) = 0$ for $t \leqslant 0$, show that the Laplace transform of the solution is

$$\mathcal{L}\{x(t)\} = \frac{1}{s^2(s + e^{-s})} = \frac{1}{s^3(1 + e^{-s}/s)}.$$

Expand $1/(1 + e^{-s}/s)$ in powers of e^{-s}/s using a binomial expansion, and show that

$$x(t) = 2\sum_{n=0}^{\lfloor t\rfloor} \frac{(t-n)^{n+2}}{(n+2)!},$$

where $\lfloor t\rfloor$ is the **integer floor function** (the largest integer less than or equal to t: for example, $\lfloor 2.3\rfloor = 2$, $\lfloor 3\rfloor = 3$, and $\lfloor -2.3\rfloor = -3$).

23.15. The equation

$$2 \int_0^t \cos(t - u) x(u) \, du = x(t) - t,$$

is an example of an **integral equation**. Note that the integral is of convolution type, which means that the Laplace transform of the equation is

$$2L\{\cos t\} X(s) = X(s) - \frac{1}{s^2}.$$

Show that the solution is

$$x(t) = 2(t - 1) e^t + t + 2.$$

23.16. The differential equation

$$\frac{d^2 x}{dt^2} + t \frac{dx}{dt} - x = 0,$$

does *not* have constant coefficients: the coefficient of dx/dt is t. Using the results (22.8) and (22.12), show that the transform of the differential equation subject to the conditions $x(0) = 0$ and $x'(0) = 1$ satisfies the first-order equation

$$-s \frac{dX(s)}{ds} + (s^2 - 2) X(s) = 1.$$

Verify that $X(s) = 1/s^2$ satisfies this equation, and hence obtain the required solution of the original equation.

23.17. Using the method outlined in Problem 23.16, solve the following variable-coefficient equations using Laplace transforms:
(a) $tx''(t) + (1 - t) x'(t) - x(t) = 0$, $x(0) = x'(0) = 1$;
(b) $x''(t) + tx'(t) - 2x(t) = 2$, $x(0) = x'(0) = 0$;
(c) $tx''(t) - x'(t) + tx(t) = \sin t$, $x(0) = 1$, $x'(0) = 0$.

Fourier series

24

Fundamental tone (250 Hz)

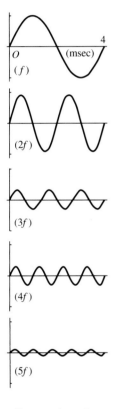

Compound sound

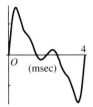

Fig. 24.1

24.1 The composition of vibrations

If a note on a piano is played, firstly by pressing the key and then by plucking the string, the sounds produced are very different, although the **pitch** or **fundamental frequency** heard is the same in both cases. The note produced by an instrument is not a pure tone or sinusoidal wave; it is a richer sound which contains other frequencies. These occur in different proportions when the same note is stimulated in different ways, or is sounded on different instruments.

A trained ear can detect some detail in these differences; the extra components can be distinguished and their pitch recognized, and they can be isolated by using resonators. The extra component frequencies of a note are all higher than the fundamental frequency, and related to it in a simple manner. If the fundamental frequency is f, then the **harmonics** present have frequencies

$$f, \quad 2f, \quad 3f, \quad 4f, \quad 5f, \dots,$$

the strength of the harmonics dropping off to zero as the frequency increases.

When these components are added, a profile for the composite wave is obtained. A particular note was found to have components as shown:

Order of harmonic:	1	2	3	4	5	...
Frequency:	f	$2f$	$3f$	$4f$	$5f$	...
Relative amplitude:	1.0	0.9	0.3	0.3	0.1	...

The shape and amplitude of the component harmonic waves, and of the composite wave, are shown in Fig. 24.1.

By means of an electronic synthesizer the proportions in which harmonics occur can be controlled and a great variety of sound quality generated, from flute to drum. Given any particular fundamental frequency f, it is plausible that we could generate a sound wave of any preassigned quality (that is to say, any shape) by adjusting the balance of the harmonics. This possibility is essentially what the theory of Fourier series is about, though in a wider context than that of sound waves.

24.2 Fourier series for a periodic function

The following symbols are used in connection with **periodic functions** (see Section 18.1):

f = frequency (cycles/time; if time is in seconds, the unit is the Hertz);

ω = angular frequency (radians/time): $\omega = 2\pi f = 2\pi/T$;

T = period or wavelength: $T = 1/f = 2\pi/\omega$.

A typical **periodic** function $P(t)$ with period T is shown in Fig. 24.2. Any full-period interval may be chosen for discussion; suppose it is the interval between $t = -\pi/\omega$ and $t = \pi/\omega$.

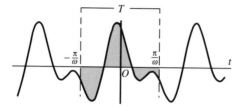

Fig. 24.2
A periodic function $P(t)$ with period $T = 2\pi/\omega$.

We shall express $P(t)$ over this interval, denoted by $[-\pi/\omega, \pi/\omega]$, as the sum of harmonic (sinusoidal) curves having frequencies $f, 2f, 3f, \ldots$, where $f = 1/T$; or, equivalently, angular frequencies $\omega, 2\omega, 3\omega, \ldots$, where $\omega = 2\pi/T$. A **constant term** is also needed, since the average value of $P(t)$ will not generally be zero. **Both sine and cosine terms** are needed, because if we involve only sines or only cosines, the sum will have a symmetry, odd or even (Section 13.9), which $P(t)$ might not have. Then we expect that

---

Standard form for a Fourier series of period T

$$P(t) = \tfrac{1}{2}a_0 + (a_1 \cos \omega t + b_1 \sin \omega t)$$
$$\quad + (a_2 \cos 2\omega t + b_2 \sin 2\omega t) + \cdots$$
$$= \tfrac{1}{2}a_0 + \sum_{n=1}^{\infty} (a_n \cos n\omega t + b_n \sin n\omega t),$$

where $\omega = 2\pi/T$.

(24.1)

Equation (24.1) is a **Fourier series** for $P(t)$, and the constants $a_0; a_1, b_1; a_2, b_2; \ldots$ are its **Fourier coefficients**. It will be shown how to determine the coefficients in Section 24.4: the factor $\tfrac{1}{2}$ in the constant term $\tfrac{1}{2}a_0$ is introduced to simplify the working.

We have spoken in terms of the one-period range $t = -\pi/\omega$ to $t = \pi/\omega$, but every term on the right of (24.1) is periodic with the same period $T = 2\pi/\omega$ as $P(t)$. Therefore **the series will describe** $P(t)$ **for every value of** t, not merely for t in the interval between $\pm\pi/\omega$.

24.3 Integrals of periodic functions

We prove two results needed for the next Section, in which the values of the coefficients in (24.1) are determined. Figure 24.3 represents a periodic function $P(t)$ with period T. Choose any value of t, say $t = t_0$, and compare the two integrals

$$\int_0^T P(t)\,dt \quad \text{and} \quad \int_{t_0}^{t_0+T} P(t)\,dt,$$

each of which is taken over a one-period interval of $P(t)$. The figure shows that the integrals are equal by virtue of the area analogy (13.13). The two shaded areas in Fig. 24.3a, b are assembled from identical elements which are simply added up in a different order.

(a)

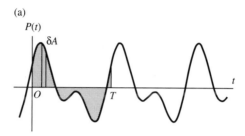

(b)

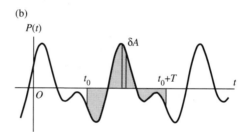

Fig. 24.3
Illustrating the area analogy for
(a) $\int_0^T P(t)\,dt$ and
(b) $\int_{t_0}^{t_0+T} P(t)\,dt$,
where $P(t)$ has period T.

> The integral over any one-period interval of a function $P(t)$ having period T,
>
> $$\int_{t_0}^{t_0+T} P(t)\,dt, \qquad\qquad (24.2)$$
>
> does not depend on t_0.

Example 24.1. Show that $\int_0^\pi \sin 2t \cos^2 t\,dt = 0$.

The period of $\cos^2 t$ is π, because

$$\cos^2 t = \tfrac{1}{2}(1 + \cos 2t)$$

and the period of $\cos 2t$ is π. By (24.2),

$$\int_0^\pi \sin 2t \cos^2 t\,dt = \int_{-\frac{1}{2}\pi}^{\frac{1}{2}\pi} \sin 2t \cos^2 t\,dt,$$

since the range $-\frac{1}{2}\pi$ to $\frac{1}{2}\pi$ also covers a period π. But the integrand is an odd function about the origin, so that the value of the last version is zero (Section 13.9).

The following special results can be proved by using the trigonometric identities in Appendix B which convert products to sums.

Trigonometric integrals over a one-period interval

(a) For n and $m = 0, 1, 2, \ldots$ with $n \neq m$,

$$\int_{-\pi/\omega}^{\pi/\omega} \cos n\omega t \cos m\omega t \, dt = 0,$$

$$\int_{-\pi/\omega}^{\pi/\omega} \sin n\omega t \sin m\omega t \, dt = 0,$$

$$\int_{-\pi/\omega}^{\pi/\omega} \cos n\omega t \sin m\omega t \, dt = 0.$$

(24.3)

(b) For $n = 1, 2, \ldots$

$$\int_{-\pi/\omega}^{\pi/\omega} \cos^2 n\omega t \, dt = \int_{-\pi/\omega}^{\pi/\omega} \sin^2 n\omega t \, dt = \pi/\omega.$$

For $n = 0$, we obtain

$$\int_{-\pi/\omega}^{\pi/\omega} dt = 2\pi/\omega \quad \text{and} \quad \int_{-\pi/\omega}^{\pi/\omega} 0 \, dt = 0.$$

(c) The range $-\pi/\omega$ to π/ω may be replaced by any interval of length $2\pi/\omega$.

24.4 Calculating the Fourier coefficients

From (24.1), we expect that any periodic function $P(t)$ having period $2\pi/\omega$ can be expressed in the form of a Fourier series

$$P(t) = \tfrac{1}{2}a_0 + \sum_{n=1}^{\infty} (a_n \cos n\omega t + b_n \sin \omega t). \tag{24.4}$$

To find a particular coefficient a_N, multiply both sides of (24.4) by $\cos N\omega t$:

$$P(t) \cos N\omega t = \tfrac{1}{2}a_0 \cos N\omega t$$

$$+ \sum_{n=1}^{\infty} (a_n \cos n\omega t \cos N\omega t + b_n \sin n\omega t \cos N\omega t).$$

Integrate both sides of this equation between $-\pi/\omega$ and π/ω:

$$\int_{-\pi/\omega}^{\pi/\omega} P(t)\cos N\omega t \, dt = \tfrac{1}{2}a_0 \int_{-\pi/\omega}^{\pi/\omega} \cos N\omega t \, dt$$

$$+ \sum_{n=1}^{\infty} \left(a_n \int_{-\pi/\omega}^{\pi/\omega} \cos n\omega t \cos N\omega t \, dt + b_n \int_{-\pi/\omega}^{\pi/\omega} \sin n\omega t \cos N\omega t \, dt \right).$$

$$(24.5)$$

Firstly consider the case $N = 0$. According to (24.3a), all terms under the summation sign in (24.5) are zero; so, after putting $\cos N\omega t = \cos 0 = 1$, we are left with

$$\int_{-\pi/\omega}^{\pi/\omega} P(t) \, dt = \tfrac{1}{2}a_0 \int_{-\pi/\omega}^{\pi/\omega} dt = \frac{\pi}{\omega} a_0.$$

Therefore

$$a_0 = \frac{\omega}{\pi} \int_{-\pi/\omega}^{\pi/\omega} P(t) \, dt. \qquad (24.6)$$

Now suppose that $N \neq 0$. By (24.3), all the integrals on the right of (24.6) are zero except the single one that involves a_N, so (24.6) reduces to

$$\int_{-\pi/\omega}^{\pi/\omega} P(t)\cos N\omega t \, dt = a_N \int_{-\pi/\omega}^{\pi/\omega} \cos^2 N\omega t \, dt = \frac{\pi}{\omega} a_N.$$

Therefore, for $N = 1, 2, 3, \ldots,$

$$a_N = \frac{\omega}{\pi} \int_{-\pi/\omega}^{\pi/\omega} P(t)\cos N\omega t \, dt. \qquad (24.7)$$

By comparing (24.7) with (24.6), it can be seen that a_0 and $a_1, a_2, \ldots$ are all given by the same formula. That is why the constant term in (24.4) is written as $\tfrac{1}{2}a_0$ instead of a_0.

To find b_N for $N = 1, 2, \ldots,$ multiply (24.4) by $\sin N\omega t$ and integrate. In a similar way to that described above, we find that, for $N = 1, 2, 3, \ldots,$

$$b_N = \frac{\omega}{\pi} \int_{-\pi/\omega}^{\pi/\omega} P(t)\sin N\omega t \, dt. \qquad (24.8)$$

Since $P(t)$ is a known function, the integrals in (24.6), (24.7), and (24.8) can be evaluated to give all the coefficients in the Fourier series (24.5).

In the following summary, the letter n is used in place of N to simplify the form of the results.

Fourier series for periodic functions

Function: $P(t)$, period $T = 2\pi/\omega$,

Fourier series: $P(t) = \frac{1}{2}a_0 + \sum_{n=1}^{\infty} (a_n \cos n\omega t + b_n \sin \omega t)$,

Fourier coefficients:

$$a_n = \frac{\omega}{\pi} \int_{-\pi/\omega}^{\pi/\omega} P(t) \cos n\omega t \, dt \quad (n = 0, 1, 2, \ldots), \qquad (24.9)$$

$$b_n = \frac{\omega}{\pi} \int_{-\pi/\omega}^{\pi/\omega} P(t) \sin n\omega t \, dt \quad (n = 1, 2, \ldots),$$

(in place of the range of integration, $-\pi/\omega$ to π/ω, *any other* one-period interval may be used).

It can be seen also that since

$$\frac{1}{2}a_0 = \frac{\omega}{2\pi} \int_{-\pi/\omega}^{\pi/\omega} P(t) \, dt = \frac{1}{T} \int_{-\frac{1}{2}T}^{\frac{1}{2}T} P(t) \, dt,$$

the following is true:

Average value of $P(t)$

The average value of $P(t)$ over a one-period interval   (21.10)
is equal to the constant term $\frac{1}{2}a_0$.

Notice the case of period 2π, which often occurs. In such cases $\omega = 2\pi/T = 1$:

Fourier series for functions $P(t)$ with period 2π

$$P(t) = \frac{1}{2}a_0 + \sum_{n=1}^{\infty} (a_n \cos nt + b_n \sin nt),$$

where

$$a_n = \frac{1}{\pi} \int_{-\pi}^{\pi} P(t) \cos nt \, dt, \qquad (24.11)$$

$$b_n = \frac{1}{\pi} \int_{-\pi}^{\pi} P(t) \sin nt \, dt.$$

(The integrals may be taken over any one-period interval instead of $[-\pi, \pi]$.)

24.5 Examples of Fourier series

The actual calculation of Fourier coefficients requires attention to detail, especially in respect of a_0.

Example 24.2. Find the Fourier series of the function $P(t)$ shown in Fig. 24.4.

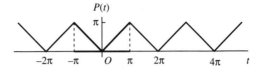

$P(t)$

Fig. 24.4

The period is 2π, so that $\omega = 2\pi/2\pi = 1$. Choosing the interval $-\pi$ to π as the basis of the calculation yields

$$P(t) = \begin{cases} -t & \text{if } -\pi \leqslant t \leqslant 0, \\ t & \text{if } 0 \leqslant t \leqslant \pi. \end{cases}$$

The coefficients can be obtained from (24.11):

Coefficients b_n. $P(t)$ is an even function about the origin (see Section 13.9), and $\sin nt$ is odd; therefore $P(t) \sin nt$ is odd. Hence the integrals defining b_n are all zero:

$$b_n = 0 \quad (n = 1, 2, \ldots). \tag{i}$$

Coefficients a_n. Since $P(t)$ is even and $\cos nt$ is even, $P(t) \cos nt$ is even; so (24.11) gives

$$a_n = \frac{2}{\pi} \int_0^\pi P(t) \cos nt \, dt = \frac{2}{\pi} \int_0^\pi t \cos nt \, dt,$$

$$= \frac{2}{\pi} \left(\left[\frac{t \sin nt}{n} \right]_0^\pi - \int_0^\pi \frac{\sin nt}{n} \, dt \right), \tag{ii}$$

after integrating by parts.

At this point it is seen that $n = 0$ is a case requiring separate treatment (basically because $\int \cos nt \, dt \neq n^{-1} \sin nt + C$ when $n = 0$). Postponing the question of $n = 0$, suppose firstly that $n = 1, 2, \ldots$. The formula becomes

$$a_n = \frac{2}{\pi n^2} \left[\cos nt \right]_0^\pi = \frac{2}{\pi} \frac{(-1)^n - 1}{n^2}.$$

Therefore

$$a_n = \begin{cases} 0 & \text{if } n \text{ is even}, \\ -4/\pi n^2 & \text{if } n \text{ is odd}. \end{cases} \tag{iii}$$

We still have to find a_0, which is given by (24.11) as

$$a_0 = \frac{2}{\pi} \int_0^\pi t \, dt = \pi. \tag{iv}$$

Collect the coefficients from (i), (iii), and (iv) and put them back into the Fourier series:

$$P(t) = \tfrac{1}{2}\pi - \frac{4}{\pi}\left(\frac{\cos t}{1^2} + \frac{\cos 3t}{3^2} + \frac{\cos 5t}{5^2} + \cdots\right).$$

(a)

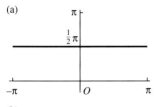

(b)

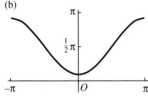

(c)

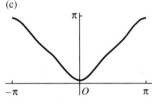

(d)

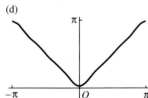

Fig. 24.5
(a) 1.571; (b) 1.571 − 1.273 cos t;
(c) 1.571 − 1.273 cos t − 0.141 cos 3t;
(d) 1.571 − 1.273 cos t −
0.141 cos 3t − 0.051 cos 5t.

Fig. 24.6

In Fig. 24.5, we show how $P(t)$ is gradually shaped as we take more and more terms of the Fourier series in Example 24.2. Here

$$P(t) = \tfrac{1}{2}\pi - \frac{4}{\pi}\left(\frac{\cos t}{1^2} + \frac{\cos 3t}{3^2} + \frac{\cos 5t}{5^2} + \cdots\right),$$

$$= 1.571 - 1.273\cos t - 0.141\cos 3t - 0.051\cos 5t - \cdots.$$

Example 24.3. Find the Fourier series for the function shown in Fig. 24.6.
The period is $T = 2\pi$, so that $\omega = 1$ and the Fourier series is

$$P(t) = \tfrac{1}{2}a_0 + \sum_{n=1}^{\infty}(a_n\cos nt + b_n\sin nt).$$

It makes no difference to the ease of calculation whether $-\pi$ to π or 0 to 2π is chosen as the basic interval. We will take 0 to 2π to remind the reader of the possibility. Then

$$P(t) = \begin{cases} t & (0 \leqslant t \leqslant \pi), \\ 0 & (\pi < t \leqslant 2\pi). \end{cases}$$

Coefficient a_n. From (24.11),

$$a_n = \frac{1}{\pi}\int_0^{2\pi} P(t)\cos nt\,\mathrm{d}t = \frac{1}{\pi}\int_0^{\pi} t\cos nt\,\mathrm{d}t.$$

Warned by Example 24.2, we deal first with the case $n = 1, 2, 3, \ldots$:

$$a_n = \frac{1}{\pi}\left[\frac{1}{n}t\sin nt + \frac{1}{n^2}\cos nt\right]_0^{\pi}$$

$$= \frac{1}{\pi}\left[\left(0 + \frac{1}{n^2}\cos n\pi\right) - \left(0 + \frac{1}{n^2}\right)\right] = \frac{1}{\pi n^2}[(-1)^n - 1].$$

The sequence has every even-order term zero:

$$a_1 = -\frac{2}{\pi}, \quad a_2 = 0, \quad a_3 = -\frac{2}{\pi 3^2}, \quad a_4 = 0, \quad a_5 = -\frac{2}{\pi 5^2}, \quad \cdots.$$

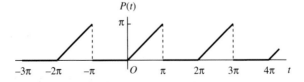

The case $n = 0$ is again special:

$$a_0 = \frac{1}{\pi}\int_0^{2\pi} P(t)\,\mathrm{d}t = \frac{1}{\pi}\left[\tfrac{1}{2}t^2\right]_0^{\pi} = \tfrac{1}{2}\pi.$$

Coefficient b_n.

$$b_n = \frac{1}{\pi} \int_0^{2\pi} P(t) \sin nt \, dt = \frac{1}{\pi} \int_0^{\pi} t \sin nt \, dt$$

$$= \frac{1}{\pi} \left[-\frac{t \cos nt}{n} + \frac{1}{n^2} \sin nt \right]_0^{\pi}$$

$$= \frac{1}{\pi} \left[-\frac{1}{n}(\pi \cos n\pi - 0) + \frac{1}{n^2}(0 - 0) \right] = -\frac{(-1)^n}{n}.$$

The series is difficult to write if the cosine and sine terms are kept together. By separating them, we obtain

$$P(t) = \tfrac{1}{4}\pi - \frac{2}{\pi}\left(\cos t + \frac{1}{3^2} \cos 3t + \frac{1}{5^2} \cos 5t + \cdots \right)$$

$$+ \left(\sin t - \frac{1}{2} \sin 2t + \frac{1}{3} \sin 3t - \cdots \right).$$

In Example 24.3, the function $P(t)$ jumps from π to zero at the points

$$t = \ldots, -\pi, \pi, 3\pi, \ldots.$$

To see what values are generated by the **series** at such points put, say $t = \pi$ into the series we obtained. All the cosine terms become (-1) and all the sine terms are zero, so that at $x = \pi$ the series delivers

$$\frac{1}{4}\pi + \frac{2}{\pi}\left(1 + \frac{1}{3^2} + \frac{1}{5^2} + \cdots \right).$$

A few minutes with a calculator makes it clear that this series for $P(\pi)$ cannot add up to π, and plainly it does not give zero either. In fact its sum is $\frac{1}{2}\pi$, half way between these values. The general rule is as follows.

> **Fourier series at a jump in value of a function**
>
> The sum of a Fourier series at a jump is equal to the   **(24.12)** average of the two function values on either side.

Fig. 24.7 shows how the function is fitted by the series when the seven terms up to $\cos 3t$ and $\sin 3t$ are taken.

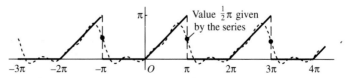

Fig. 24.7

24.6 Use of symmetry: sine and cosine series

In general, the Fourier series for a periodic function will contain both sine and cosine terms. However, the following results hold.

Even and odd functions $P(t)$

(a) If $P(t)$ is even about the origin, then

$$b_1 = b_2 = \cdots = 0.$$

(b) If $P(t)$ is odd about the origin, then

$$a_0 = a_1 = a_2 = \cdots = 0.$$

(24.13)

These results follow from (24.9) and (24.11), because $P(t) \sin n\omega t$ is odd if $P(t)$ is even, and $P(t) \cos n\omega t$ is odd if $P(t)$ is odd.

Example 24.4. Obtain the Fourier series for the switching function $P(t)$ shown in Fig. 24.8.

The period T is 2, so that $\omega = \pi$. Choose the basic interval to be $t = -1$ to 1. On this interval,

$$P(t) = \begin{cases} -1 & \text{for } -1 \leqslant t < 0, \\ 1 & \text{for } 0 \leqslant t \leqslant 1. \end{cases}$$

Since $P(t)$ is odd about the origin,

$$a_0 = a_1 = a_2 = \cdots = 0.$$

For the b_n, from (24.9), since the integrands are even functions,

$$b_n = \frac{\pi}{\pi} \int_{-1}^{1} P(t) \sin n\pi t \, dt = 2 \int_{0}^{1} \sin n\pi t \, dt = -\frac{2}{n\pi} \left[\cos n\pi t \right]_0^1$$

$$= -\frac{2}{n\pi} [(-1)^n - 1]$$

for $n = 1, 2, \ldots$. The sequence b_n is therefore

$$b_1 = \frac{4}{n}, \quad b_2 = 0, \quad b_3 = \frac{4}{\pi}\frac{1}{3}, \quad b_4 = 0, \ldots,$$

and the Fourier series is

$$P(t) = \frac{4}{\pi} \left(\sin \pi t + \frac{1}{3} \sin 3\pi t + \frac{1}{5} \sin 5\pi t + \cdots \right)$$

$$= \frac{4}{\pi} \sum_{r=1}^{\infty} \frac{\sin(2r-1)\pi t}{2r-1}.$$

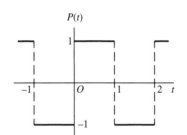

$P(t)$

Fig. 24.8

Example 24.5. Obtain the Fourier series for the switching function $P(t)$ shown in Fig. 24.9.

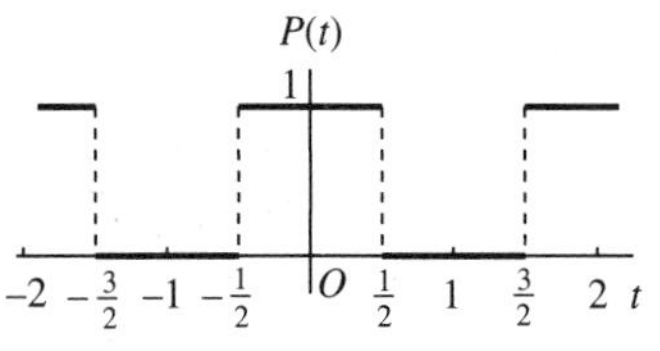

The period is 2, so that $\omega = \pi$. Choose $[-1, 1]$ as the representative interval; then

$$P(t) = \begin{cases} 1 & \text{if } -\frac{1}{2} \leqslant t \leqslant \frac{1}{2}, \\ 0 & \text{elsewhere on the interval.} \end{cases}$$

Since $P(t)$ is an even function, $b_1 = b_2 = b_3 = \cdots = 0$. The coefficients

$P(t)$

Fig. 24.9

a_n are given by

$$a_n = \int_{-1}^{1} P(t) \cos n\pi t \, dt = \int_{-\frac{1}{2}}^{\frac{1}{2}} \cos n\pi t \, dt$$

$$= \frac{1}{n\pi} \left[\sin n\pi t \right]_{-\frac{1}{2}}^{\frac{1}{2}} = \frac{2}{n\pi} \sin \tfrac{1}{2} n\pi.$$

As we have seen before, a_0 gives trouble since this formula is meaningless when $n = 0$. We have, in fact,

$$a_0 = \int_{-\frac{1}{2}}^{\frac{1}{2}} 1 \, dt = 1.$$

Then

$$a_0 = 1, \quad a_1 = 2/\pi, \quad a_2 = 0, \quad a_3 = -2/3\pi, \quad a_4 = 0, \dots,$$

so that the odd-order coefficients alternate in sign.

Finally the series is

$$P(t) = \tfrac{1}{2} + \frac{2}{\pi} \left(\cos \pi t - \frac{1}{3} \cos 3\pi t + \frac{1}{5} \cos 5\pi t - \cdots \right)$$

$$= \tfrac{1}{2} + \frac{2}{\pi} \sum_{r=1}^{\infty} \frac{(-1)^r \cos(2r - 1)\pi t}{2r - 1}.$$

24.7 Functions defined on a finite range: half-range series

It is often necessary to obtain a Fourier-type series for a function which is of interest only over some finite interval, and whose natural extension, if any, is not necessarily periodic. The Fourier *series* is invariably periodic, so that it cannot fit a nonperiodic function everywhere. For example, consider the problem of finding a Fourier series which will fit $f(t)$, where

$$f(t) = t \text{ between } t = 0 \text{ and } \pi,$$

when our *only* concern is whether the series fits $f(t)$ between 0 and π, the behaviour of the series elsewhere being a matter of indifference.

Figure 24.10 illustrates a technique for producing such series. We hold on to the given function inside the interval of interest, but **extend it by means of an artificial function which is periodic**. This extended function will have a Fourier series of its own, and it will agree with $f(t)$ on the interval 0 to π.

In Fig. 24.10b we have extended the nonperiodic function $f(t) = t$ on $0 \leqslant t \leqslant \pi$ to an artificial function $f_s(t)$ which has period 2π and is an *odd function*. Being odd, it has a Fourier series consisting of sine terms only, and this will correctly reproduce $f(t)$ on $0 \leqslant t \leqslant \pi$.

Alternatively, Fig. 24.10c shows how to get a series of cosine terms by an *even extension* $f_c(t)$, keeping $f_c(t) = f(t)$ on $0 \leqslant t \leqslant \pi$.

Again, Fig. 24.10d shows a fairly arbitrary extension of period 3π, which will have a Fourier series containing both sine and cosine

(a)

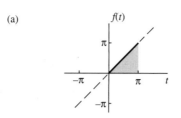

(b)

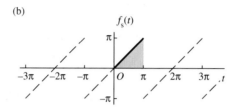

Fig. 24.10
(a) $f(t) = t$ on $0 \leqslant t \leqslant \pi$, with natural nonperiodic extension.
(b) $f_s(t) = t$ on $0 \leqslant t \leqslant \pi$, has period 2π and is an odd function.
(c) $f_c(t) = t$ on $0 \leqslant t \leqslant \pi$ has period 2π and is an even function.
(d) $f_a(t) = t$ on $0 \leqslant t \leqslant \pi$, and has an arbitrary extension of period 3π.

(c)

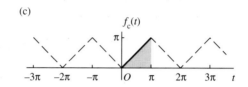

(d)

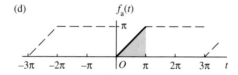

terms. Obviously there is an infinite number of possibilities, the most important being the so-called **half-range sine** and **cosine series,** corresponding to **odd and even extensions** respectively.

Example 24.6. Obtain a Fourier sine series for $f(t) = t$ on the interval $0 \leqslant t \leqslant \pi$.

Extend $f(t)$ on $0 \leqslant t \leqslant \pi$ as an odd function $f_s(t)$ with period 2π (not π) as shown in Fig. 24.10b. Then $\omega = 1$ in (24.9). Choose the interval $-\pi$ to π as basic. Then since $f_s(t)$ is odd, we know in advance from (24.13) that

$$f_s(t) = \sum_{n=1}^{\infty} b_n \sin nt$$

(sine terms only), where

$$b_n = \frac{1}{\pi} \int_{-\pi}^{\pi} f_s(t) \sin nt \, \mathrm{d}t$$

$$= \frac{2}{\pi} \int_{0}^{\pi} f_s(t) \sin nt \, \mathrm{d}t \quad \text{(since } f_s(t) \text{ is even; see (13.17))}$$

$$= \frac{2}{\pi} \int_0^\pi f(t) \sin nt \; dt$$

(since $f_s(t)$ agrees with $f(t)$ on $0 \leqslant t \leqslant \pi$)

$$= \frac{2}{\pi} \int_0^\pi t \sin nt \; dt = \frac{2}{\pi} \left[-\frac{1}{n} t \cos nt + \frac{1}{n^2} \sin nt \right]_0^\pi$$

$$= \frac{2}{\pi} \left(-\frac{\pi}{n} \cos n\pi \right) = \frac{2}{n} (-1)^{n+1}.$$

Therefore the required series is

$$\frac{2}{\pi} \left(\sin t - \frac{1}{2} \sin 2t + \frac{1}{3} \sin 3t - \cdots \right) = \frac{2}{\pi} \sum_{n=1}^\infty \frac{(-1)^{n+1} \sin nt}{n},$$

and this is equal to t on $0 \leqslant t \leqslant \pi$ (but nowhere else).

(a) **Half-range cosine series for** $0 \leqslant t \leqslant \pi$

$$f(t) = \tfrac{1}{2} a_0 + \sum_{n=1}^\infty a_n \cos nt, \qquad a_n = \frac{2}{\pi} \int_0^\pi f(t) \cos nt \; dt$$

(b) **Half-range sine series for** $0 \leqslant t \leqslant \pi$

(24.14)

$$f(t) = \sum_{n=1}^\infty b_n \sin nt, \qquad b_n = \frac{2}{\pi} \int_0^\pi f(t) \sin nt \; dt$$

Suppose that, more generally, a sine series representing $f(t)$ for $0 \leqslant t \leqslant t_0$ is required (Fig. 24.11). Extend $f(t)$ to an odd function

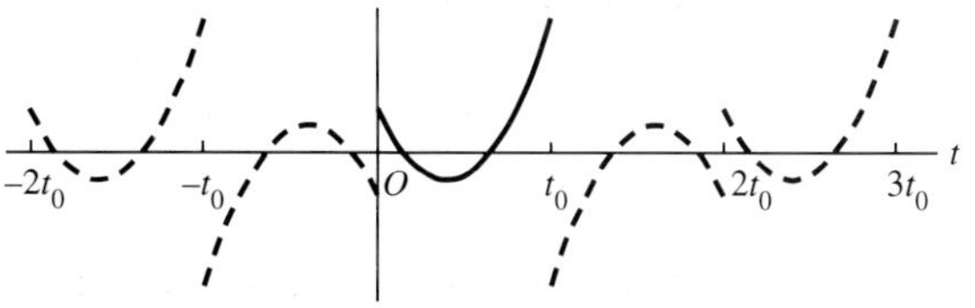

Fig. 24.11
$f(t)$, $0 \leqslant t \leqslant t_0$; odd extension, $f_s(t)$, period $2t_0$.

$f_s(t)$ having period $2t_0$. Then, in equation (24.9),

$$\omega = 2\pi/2t_0 = \pi/t_0,$$

and, since $f_s(t)$ is odd,

$$f_s(t) = \sum_{n=1}^\infty b_n \sin(n\pi t/t_0),$$

where

$$b_n = \frac{1}{t_0} \int_{-t_0}^{t_0} f_s(t) \sin \frac{n\pi t}{t_0} \; dt = \frac{2}{t_0} \int_0^{t_0} f(t) \sin \frac{n\pi t}{t_0} \; dt,$$

since $f_s(t) = f(t)$ on the interval 0 to t_0.

If the extension is carried out so as to produce an even periodic function, a similar calculation leads to a cosine expansion.

(a) Half-range cosine series for $0 \leqslant t \leqslant t_0$

$$f(t) = \tfrac{1}{2}a_0 + \sum_{n=1}^{\infty} a_n \cos \frac{n\pi t}{t_0},$$

$$a_n = \frac{2}{t_0} \int_0^{t_0} f(t) \cos \frac{n\pi t}{t_0} \, dt.$$

(24.15)

(b) Half-range sine series for $0 \leqslant t \leqslant t_0$

$$f(t) = \sum_{n=1}^{\infty} b_n \cos \frac{\pi t}{t_0}, \qquad b_n = \frac{2}{t_0} \int_0^{t_0} f(t) \sin \frac{n\pi t}{t_0} \, dt$$

24.8 Spectrum of a periodic function

Suppose that $P(t)$ is a periodic function with period T. The Fourier series has the form

$$P(t) = \tfrac{1}{2}a_0 + \sum_{n=1}^{\infty} (a_n \cos n\omega t + b_n \sin n\omega t)$$

where $\omega = 2\pi/T$. By the identity (1.19),

$$a_n \cos n\omega t + b_n \sin n\omega t = c_n \cos(n\omega t + \phi_n),$$

where ϕ_n is a phase angle, and

$$c_n = \sqrt{(a_n^2 + b_n^2)} \quad (n = 1, 2, \ldots),$$

which is the (positive) amplitude or strength of the nth term. For completeness, we include

$$c_0 = |a_0|.$$

The sequence $c_0, c_1, c_2, \ldots$ is called the **spectrum** of $P(t)$. If the series consists only of cosine (or sine) terms, then correspondingly $c_n = \sqrt{(a_n^2)} = |a_n|$ (or $|b_n|$) – the **spectral components** are always positive or zero.

The spectrum can be displayed as if it were a physical spectrum. Figure 24.12 shows the spectrum of the function worked out in Example 24.5. The property which makes the spectrum a useful concept is that **the spectrum is independent of the time origin of** t, although the Fourier series itself is not. If

$$P(t) = \tfrac{1}{2}a_0 + \sum_{n=1}^{\infty} c_n \cos(n\omega t + \phi_n),$$

then the series for $P(t - t_0)$, whose graph is the same shape as $P(t)$ but moved to the right a distance t_0, is

$$P(t - t_0) = \tfrac{1}{2}a_0 + \sum_{n=1}^{\infty} c_n \cos[n\omega(t - t_0) + \phi_n].$$

The c_n remain the same, and only the phase angle changes. Therefore it is only the shape of $P(t)$ which determines its spectrum, not its clock-timing. For this reason the spectral or harmonic composition of a piano note is always the same, independently of what time of day the note is played.

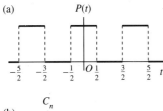

(a)

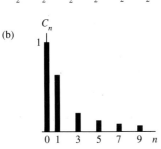

(b)

Fig. 24.12
See Example 24.5.
(a) $P(t) = \tfrac{1}{2} + (2/\pi)(\cos \pi t - \tfrac{1}{3} \cos 3\pi t + \tfrac{1}{5} \cos 5\pi t - \cdots)$.
(b) The spectral components are $\tfrac{1}{2}$, 0, $2/\pi$, 0, $2/3\pi$, 0, $2/5\pi$, ...

It is important to realize that the spectrum refers only to a *complete periodic function*, and not to an isolated segment such as those discussed in Section 24.7. For the functions shown in Fig. 24.10, there correspond different Fourier series which have different spectra.

24.9 Obtaining one Fourier series from another

There exist 'dictionaries' of Fourier series, but the entries cannot match exactly all the functions required in practice. If the broad shape of the dictionary entry is the same as that of the function whose series is needed, then scaling or translation along the *t* axis, or along the axis of $P(t)$, might be all that is required. The transition can require more than one stage.

The examples which follow are based on the standard form shown in Fig. 24.13, which was expressed as a Fourier series in Example 24.2:

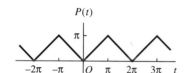

Fig. 24.13

$$P(t) = \tfrac{1}{2}\pi - \frac{4}{\pi}\left(\frac{\cos t}{1^2} + \frac{\cos 3t}{3^2} + \frac{\cos 5t}{5^2} + \cdots\right). \tag{24.16}$$

Example 24.7. Find the Fourier expansion of the function $Q(t)$ shown in Fig. 24.14.

This is the same as Fig. 24.13 except that the vertical dimension is reduced by a factor $1/\pi$. Therefore, from (24.16),

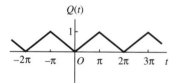

Fig. 24.14

$$Q(t) = \tfrac{1}{2} - \frac{4}{\pi^2}\left(\cos t + \frac{1}{3^2}\cos 3t + \cdots\right).$$

Example 24.8. Find the Fourier expansion of the function $Q(t)$ shown in Fig. 24.15.

Here the *t* scale is changed by a factor π. We obtain, from (24.16),

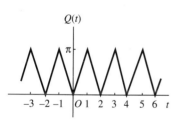

Fig. 24.15

$$Q(t) = \tfrac{1}{2}\pi - \frac{4}{\pi}\left(\cos \pi t + \frac{1}{3^2}\cos 3\pi t + \frac{1}{5^2}\cos 5\pi t + \cdots\right).$$

It is necessary to be careful here: it is not t/π but πt in the new series. Check the period: it is equal to 2, which is correct.

Example 24.9. Find the Fourier expansion of $Q(t)$ in Fig. 24.16.

The graph of $P(t)$ in Fig. 24.12 has been shifted a distance $\tfrac{1}{2}\pi$ to the left (see Fig. 24.16). Therefore

$$Q(t) = P(t + \tfrac{1}{2}\pi),$$

$$= \tfrac{1}{2}\pi - \frac{4}{\pi}\left(\cos(t + \tfrac{1}{2}\pi) + \frac{1}{3^2}\cos 3(t + \tfrac{1}{2}\pi) + \frac{1}{5^2}\cos 5(t + \tfrac{1}{2}\pi) + \cdots\right).$$

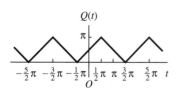

Fig. 24.16

As n goes through the sequence $1, 3, 5, 7, \ldots$, $\cos \tfrac{1}{2}n\pi = 0$ and $\sin \tfrac{1}{2}n\pi$

becomes the alternating sequence $1, -1, 1, -1, \ldots,$. Therefore

$$Q(t) = \tfrac{1}{2}\pi - \frac{4}{\pi}\left(\sin t - \frac{1}{3^2}\sin t + \frac{1}{5^2}\sin 5t - \cdots\right).$$

Problems, Chapter 24

24.1. Draw a sketch of the following odd 2π-periodic functions defined for $-\pi < t \leqslant \pi$, and find a general formula for their Fourier coefficients:

(a) $f(t) = \begin{cases} -1 & (-\pi < t < 0), \\ 1 & (0 \leqslant t \leqslant \pi); \end{cases}$

(b) $f(t) = t \quad (-\pi < t \leqslant \pi);$

(c) $f(t) = \begin{cases} -t^2 & (-\pi < t < 0), \\ t^2 & (0 \leqslant t \leqslant \pi); \end{cases}$

(d) $f(t) = \begin{cases} e^t - 1 & (-\pi < t < 0), \\ -(e^t - 1) & (0 \leqslant t \leqslant \pi); \end{cases}$

(e) $f(t) = \begin{cases} 1 & (-\pi < t \leqslant \tfrac{1}{2}\pi), \\ -1 & (-\tfrac{1}{2}\pi < t \leqslant 0), \\ 1 & (0 < t \leqslant \tfrac{1}{2}\pi), \\ -1 & (\tfrac{1}{2}\pi < t \leqslant \pi); \end{cases}$
(f) $f(t) = \sin \tfrac{1}{2}t.$

24.2. Draw a sketch of the following even 2π-periodic functions defined for $-\pi < t \leqslant \pi$, and find a general formula for their Fourier coefficients:

(a) $f(t) = \begin{cases} -1 & (-\pi < t \leqslant \tfrac{1}{2}\pi), \\ 1 & (-\tfrac{1}{2}\pi < t \leqslant \tfrac{1}{2}\pi), \\ -1 & (\tfrac{1}{2}\pi < t \leqslant \pi); \end{cases}$

(b) $f(t) = t^2;$ (c) $f(t) = \cos \tfrac{1}{2}t.$

24.3. Draw a sketch of the following 2π-periodic functions defined for $-\pi < t \leqslant \pi$, and find a general formula for their Fourier coefficients:

(a) $f(t) = \begin{cases} 0 & (-\pi < t < 0), \\ t & (0 \leqslant t \leqslant \pi); \end{cases}$

(b) $f(t) = \begin{cases} t + \pi & (-\pi < t < 0), \\ t & (0 < t \leqslant \pi). \end{cases}$

24.4. A half-rectified sine wave is given by the 2π-periodic function

$$f(t) = \begin{cases} 0 & (-\pi < t \leqslant 0), \\ \sin t & (0 < t \leqslant \pi). \end{cases}$$

Find the Fourier series of $f(t)$.

24.5. Show that the Fourier series of the 2π-periodic function

$$f(t) = \begin{cases} 0 & (-\pi < t \leqslant 0), \\ 1 & (0 < t \leqslant \pi), \end{cases}$$

is

$$\tfrac{1}{2} + \frac{2}{\pi}\left(\sin t + \frac{1}{3}\sin 3t + \frac{1}{5}\sin 5t + \cdots\right).$$

What value does the Fourier series take at $t = 0$? By choosing a particular value of t, find the sum of the series

$$1 - \frac{1}{3} + \frac{1}{5} - \frac{1}{7} + \cdots.$$

24.6. A signal $F\sin t$ with amplitude $F > 0$ is fully rectified into $F|\sin t|$. Find the Fourier series of the rectified signal. What is the amplitude of its first harmonic?

24.7. A Fourier series is given by

$$\sum_{n=1}^{\infty} \frac{n+a}{n^3 + an + 3}\sin nt,$$

where a is a design parameter in the system. Find a in order that the leading harmonics $n = 1$ and $n = 2$ have amplitudes in the ratio $2{:}1$. What is the amplitude of the next harmonic?

24.8. The two 2π-periodic signals shown in Fig. 24.17

(a)

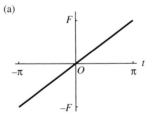

(b)

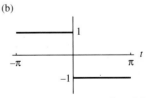

Fig. 24.17

are added. Find the Fourier series of the combined signal. What value should F take in order that the leading harmonic should disappear?

24.9. A T-periodic function is defined by

$$Q(t) = \tfrac{1}{4}T^2 - t^2 \quad \text{for } -\tfrac{1}{2}T \leqslant t \leqslant \tfrac{1}{2}T.$$

Find the Fourier series of $Q(t)$. What is the error between the sum of the first four terms of the series and $Q(t)$ at (a) $t = 0$, (b) $t = \tfrac{1}{4}T$?

24.10. A 2π-periodic function is defined by

$$f(t) = \begin{cases} \beta t(\pi - t) & (0 < t \leqslant \pi), \\ \beta t(\pi + t) & (-\pi < t \leqslant 0). \end{cases}$$

Find the Fourier series for $f(t)$. What is the ratio of the amplitudes of the third and first harmonics? Compare the values of $f(t)$ and the Fourier series up to and including the coefficient b_3 at $t = \tfrac{1}{2}\pi$.

24.11. Find the Fourier series of the 2π-periodic function defined by $f(t) = t(\pi^2 - t^2)$ for $-\pi < t \leqslant \pi$. Find the derivative of $f(t)$ for $-\pi < t \leqslant \pi$ and find its Fourier series. Confirm that the derivative of the Fourier series of $f(t)$ obtained by term-by-term differentiation is the same series as the Fourier series of $f'(t)$.

Consider now the function $g(t) = t^3$ defined for $-\pi < t \leqslant \pi$. Find the Fourier series of $g(t)$ and $g'(t)$. Confirm that the derivative of the Fourier series of $g(t)$ is *not* the same series as the Fourier series of $g'(t)$.

Comparing the functions of $f(t)$ and $g(t)$, what feature of $g(t)$ do you think causes the problem with its differentiated Fourier series?

24.12. Sketch the rectified sine wave defined by

$$P(t) = \begin{cases} 0 & (-\pi \leqslant t \leqslant 0), \\ |\sin 2t| & (0 < t \leqslant \pi), \end{cases}$$

extended so as to have period 2π. Find its Fourier series. (The identities

$$\sin A \cos B = \tfrac{1}{2}[\sin(A + B) + \sin(A - B)],$$
$$\sin A \sin B = \tfrac{1}{2}[-\cos(A + B) + \cos(A - B)],$$

will be needed.)

24.13. Show that

$$t = 2 \sum_{n=1}^{\infty} \frac{(-1)^{n-1}}{n} \sin nt$$

for $-\pi < t \leqslant \pi$. Integrate the terms from $t = 0$ to $t = x$, and rearrange them to show that

$$x^2 = 4 \sum_{n=1}^{\infty} \frac{(-1)^{n-1}}{n^2} - 4 \sum_{n=1}^{\infty} \frac{(-1)^{n-1}}{n^2} \cos nx.$$

Now use (24.10) to establish the value of the constant term in this Fourier series.

24.14. From Problem 24.13, or by direct means, obtain the Fourier series valid for $-\pi \leqslant t \leqslant \pi$:

$$t^2 = \tfrac{1}{3}\pi^2 + 4 \sum_{n=1}^{\infty} \frac{(-1)^n}{n^2} \cos nt.$$

By integrating all the terms in this expression from $t = 0$ to $t = x$, obtain a Fourier series for $x^3 - \pi^2 x$. (It is always valid to integrate a Fourier series in this way in order to obtain a new one, but differentiation of the terms does not always lead to a valid series.)

24.15. Obtain the Fourier series of the function having period T which is defined for $-\tfrac{1}{2}T \leqslant t < \tfrac{1}{2}T$ by

$$P(t) = \begin{cases} -2t & (-\tfrac{1}{2}T \leqslant t < 0), \\ 2t & (0 \leqslant t < \tfrac{1}{2}T). \end{cases}$$

Display its spectrum as in Fig. 24.12.

24.16. The function $f(t)$ is defined on the interval $0 \leqslant t \leqslant 1$ by $f(t) = 1$. Express $f(t)$ as a half-range Fourier series on $0 \leqslant t \leqslant 1$, (a) as a sine series, (b) as a cosine series. Sketch the sum of the *series* on $-\infty < t < \infty$ in both cases.

24.17. The function $f(t)$ is defined on $0 \leqslant t \leqslant 1$ by $f(t) = t$. Express $f(t)$ as a half-range Fourier series on $0 \leqslant t \leqslant 1$, (a) as a cosine series, (b) as a sine series.

24.18. Express $f(t) = \sin \omega t$ on $0 \leqslant t \leqslant \pi/\omega$ as a half-range cosine series. Sketch the sum of the *series* on $-\infty < t < \infty$.

24.19. Express $f(t) = \cos \omega t$ on $0 \leqslant t \leqslant \pi/\omega$ as a half-range sine series. Sketch the sum of the *series* on $-\infty < t < \infty$.

24.20. Express $f(t) = \cos t$ on $0 \leqslant t \leqslant 2\pi$ as a half-range sine series.

24.21. Express $f(t) = \cos t$ on $0 \leqslant t \leqslant 2\pi$ as a half-range cosine series.

24.22. Express the function $f(t)$, for $0 \leqslant t \leqslant \pi$, (a) as a half-range sine series, (b) as a half-range cosine series:

$$f(t) = \begin{cases} 1 & (0 \leqslant t < \tfrac{1}{2}\pi), \\ 0 & (\tfrac{1}{2}\pi \leqslant t \leqslant \pi). \end{cases}$$

24.23. The Fourier series for the function $P(t)$, period 2π, given by

$$P(t) = \begin{cases} -t & (-\pi \leqslant t \leqslant 0), \\ t & (0 \leqslant t \leqslant \pi), \end{cases}$$

is

$$P(t) = \tfrac{1}{2}\pi - \frac{4}{\pi}\left(\frac{\cos t}{1^2} + \frac{\cos 3t}{3^2} + \frac{\cos 5t}{5^2} + \cdots\right),$$

(see Example 24.2). Deduce from this the Fourier expansions of the following periodic functions.

(a) $Q(t)$, period 4, where

$$Q(t) = \begin{cases} -3t & (-2 \leqslant t \leqslant 0), \\ 3t & (0 \leqslant t \leqslant 2); \end{cases}$$

(b) $R(t)$, period 2, where

$$R(t) = \begin{cases} 1 + t & (-1 \leqslant t \leqslant 0), \\ 1 - t & (0 \leqslant t \leqslant 1). \end{cases}$$

(Sketch $R(t)$ to understand the connection with $P(t)$.)

(c) Check that $P(t)$, $Q(t)$, $R(t)$ have similar spectra.

24.24. The Fourier series for the function $P(t)$, period 2, given by

$$P(t) = \begin{cases} -1 & (-1 \leqslant t < 0), \\ 1 & (0 \leqslant t < 1), \end{cases}$$

for one period, is

$$P(t) = \frac{2}{\pi}\left(\sin \pi t + \frac{1}{3}\sin 3\pi t + \frac{1}{5}\sin 5\pi t + \cdots\right),$$

(see Example 24.4).

Deduce the expansion of the function $Q(t)$, period T, defined on one period by

$$Q(t) = \begin{cases} -a & (0 \leqslant t < \tfrac{1}{2}T), \\ a & (\tfrac{1}{2}T \leqslant t < T). \end{cases}$$

24.25. Find the Fourier series of the 2π-periodic sawtooth wave defined by

$$f(t) = t \quad (-\pi \leqslant t \leqslant \pi).$$

Determine the *forced* part of the solution of the second-order differential equation

$$\frac{d^2x}{dt^2} + \Omega^2 x = K \sin \omega t,$$

where $\omega \neq \pm\Omega$. Hence put together the periodic output of the forced system

$$\frac{d^2x}{dt^2} + \Omega^2 x = f(t),$$

where $f(t)$ is the sawtooth wave above. For what values of Ω does the system exhibit resonance?

24.26. The Fourier series of a function with period T is given by

$$f(t) = \tfrac{1}{2}a_0 + \sum_{n=1}^{\infty} (a_n \cos \omega t + b_n \sin \omega t),$$

where $T = 2\pi/\omega$. Multiply both sides of the equation by $f(t)$ and integrate between $-\tfrac{1}{2}T$ and $\tfrac{1}{2}T$, to obtain **Parseval's identity**

$$\frac{2}{T}\int_{-\frac{1}{2}T}^{\frac{1}{2}T} f(t)^2 \, dt = \tfrac{1}{2}a_0^2 + \sum_{n=1}^{\infty} (a_n^2 + b_n^2).$$

(a) Let $T = \pi$, and

$$f(t) = \begin{cases} -1 & (-\tfrac{1}{2}\pi < t \leqslant 0), \\ 1 & (0 < t \leqslant \tfrac{1}{2}\pi). \end{cases}$$

Show that

$$\sum_{n=1}^{\infty} \frac{1}{(2n+1)^2} = \frac{\pi^2}{8}.$$

(b) Let $f(t) = t \; (-\pi < t \leqslant \pi)$ be a 2π-periodic function. Find its Fourier series (see Problem 24.1b), and deduce the corresponding Parseval identity.

24.27. The function $f(t)$ with period T has the Fourier series

$$f(t) = \tfrac{1}{2}a_0 + \sum_{n=1}^{\infty} (a_n \cos n\omega t + b_n \sin n\omega t).$$

Find the Laplace transform of the function as the sum of a series of Laplace transforms of the trigonometric terms. Hence find the Laplace transform of the 2π-periodic function defined by

$$f(t) = \begin{cases} -t^2 & (-\pi < t < 0), \\ t^2 & (0 \leqslant t \leqslant \pi). \end{cases}$$

(See Problem 24.1c.)

24.28. A radio wave described by

$$x(t) = a \cos \omega t \cos \omega_0 t,$$

where ω_0 is very much greater than ω, represents a **carrier wave** subject to amplitude modulation by the comparatively slowly-varying term $\cos \omega t$, which represents a musical note. Roughly sketch the general character of $x(t)$.

(a) If $\omega = 500$ and $\omega_0 = 100,001$ (notice the 1 at the end), what is the period of $x(t)$?

(b) Let $\omega = p/q$ and $\omega_0 = r/s$, where p, q, r, s are whole numbers such that no two of them have a common divisor other than 1. What is the period? Express $x(t)$ as the sum of two waves with angular frequencies $\omega \pm \omega_0$ (these are called the *sidebands*). What is the Fourier cosine expansion based on this period?

(c) If you know about irrational numbers, show that $x_1(t) = \cos t \cos \sqrt{2}t$ never repeats itself *exactly*: it is not periodic. ∴

25 Differentiation of functions of two variables

25.1 Functions of more than one variable

Quantities in nature usually depend on, or are functions of, more than one variable. The elevation H of land above sea level depends on two map coordinates x and y; so H is a function of the two variables x and y, and we write $H(x, y)$. If we want to take account of geological changes, then time t becomes a consideration, and in that case H is a function of three variables x, y, t, and we write $H(x, y, t)$. It is easy to produce examples involving many variables; for example the distance between two points $P : (x_1, y_1, z_1)$ and $Q : (x_2, y_2, z_2)$ is a function of six variables. The state of the economy is a function of a multitude of variables. We alternatively speak of **a function in 1, 2, 3, ... dimensions**.

Suppose that a quantity z, called the **dependent variable**, depends on two **independent variables** x and y. The dependence can often be expressed by an explicit **formula** such as

$$z = x^3 + y^3, \qquad z = e^{x-2y}, \qquad z = |xy|,$$

and so on. To make statements which apply to all sorts of functions we use the notation

$$z = f(x, y),$$

or $z = g(x, y)$, etc. The letter f on its own signifies a **function** or **process**: a computer subroutine, a particular formula, or a set of rules which will generate a single number z when two numbers x and y are fed to it in the right order.

Thus, if

$$f(x, y) = 2x + y^2,$$

then

$$f(3, -2) = (2 \times 3) + (-2)^2 = 10,$$
$$f(-2, 3) = [2 \times (-2)] + 3^2 = 5,$$
$$f(a, b) = 2a + b^2, \qquad f(u^2, v) = 2u^2 + v^2,$$
$$f(-x, x) = -2x + x^2, \qquad f(y, x) = 2y + x^2.$$

Notice the last one particularly: it is different from $f(x, y)$.

25.2 Depiction of functions of two variables

Consider the particular function

$$f(x, y) = x^2 + y^2.$$

Set up x, y, z axes; put

$$z = x^2 + y^2,$$

and proceed as if plotting a graph. Take a large number of pairs (x, y), work out z for each, then put the point (x, y, z) in the axes. For example, if $x = 1$ and $y = 2$, then $z = 5$ and we 'plot' the point $(1, 2, 5)$ as shown in Fig. 25.1a. For Fig. 25.1b, a great number of points is supposed to have been plotted. They cover a surface shaped like an inverted bowl.

Fig. 25.1
Depicting the function
$f(x, y) = x^2 + y^2$.

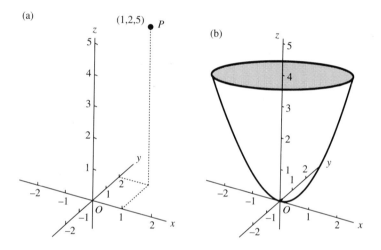

(a)

(b)

Every function has a characteristic **surface** shape, which is the analogue in three dimensions of the graphs used for functions of a single variable. Some other functions are depicted in Fig. 25.2.

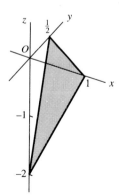

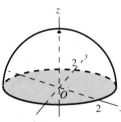

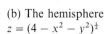

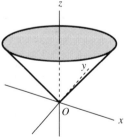

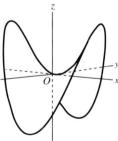

Fig. 25.2
(a) The plane
$z = 2x + 4y - 2$

(b) The hemisphere
$z = (4 - x^2 - y^2)^{\frac{1}{2}}$

(c) The cone
$z = (x^2 + y^2)^{\frac{1}{2}}$

(d) The saddle
$z = x^2 - y^2$

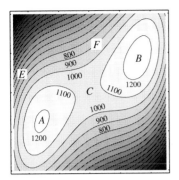

Fig. 25.3

(a)

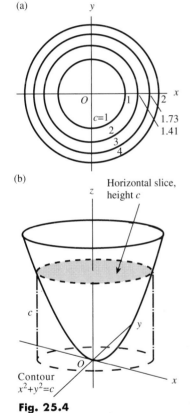

(b)

Fig. 25.4

Fig. 25.5

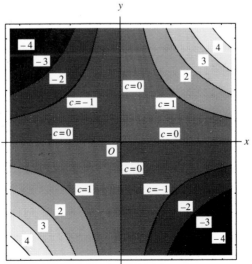

Another way of depicting a function is to sketch its **contour map** or its **level curves**. Fig. 25.3 shows a contour map of a patch of countryside. Along each contour the height is constant, and is indicated on the curve. The important features of the terrain are very easy to pick out; there are peaks at A and B, a pass at C (which is a 'saddle' as in Fig. 25.2d), valleys north-west and south-east of C, and ascents north-east and south-west of C. At E the contours are close together, so the slope is steep, and at F the contours are widely spaced so the slope is comparatively gentle.

Consider again the function $f(x, y) = x^2 + y^2$ depicted in Fig. 25.1. The contour of height c is the circle

$$x^2 + y^2 = c,$$

where $c > 0$, which is a circle of radius $c^{\frac{1}{2}}$, as shown in Fig. 25.4a. This can be visualized as in Fig. 25.4b, as a horizontal slice of the surface $z = x^2 + y^2$ at height c, projected on to the x, y plane.

Example 25.1. Sketch the contours of the function xy.

The contour of height c is given by the equation

$$z = xy = c,$$

or

$$y = c/x.$$

By varying c, taking positive and negative values, the contour map or level curves of Fig. 25.5 are obtained.

25.3 Partial derivatives

Suppose that $z = f(x, y)$ represents the height above sea level of a piece of countryside.

In Fig. 25.6a, an observer stands at the point $P : (x, y)$, facing east, in the direction of the x-axis. A short step forward takes him to $Q : (x + \delta x, y)$, up or down a slope. The altitude changes by an amount

$$\delta z = f(x + \delta x, y) - f(x, y).$$

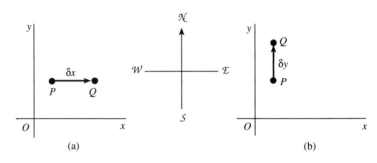

Fig. 25.6

(a)   (b)

The average slope in this direction over the step length δx is $\delta z/\delta x$, so the slope at P facing the observer is given by

$$\lim_{\delta x \to 0} \frac{\delta z}{\delta x} = \lim_{\delta x \to 0} \frac{f(x + \delta x, y) - f(x, y)}{\delta x}.$$

Since the variable y is constant during the step, this is **in effect an ordinary derivative, taken with respect to x only**. However, it is customary to signal that another variable is present, which is done by using the special sign: ∂ (still called 'dee') instead of the usual d for the derivative; writing

$$\frac{\partial f}{\partial x} \quad \text{or} \quad \frac{\partial z}{\partial x}$$

instead of df/dx or dz/dx. This is called the **partial derivative of** $f(x, y)$**, or of z, with respect to** x.

If the observer faces north and takes a step δy, as in Fig. 25.6b, then we obtain in the same way the slope $\partial f/\partial y$ or $\partial z/\partial y$ in the y direction.

Partial derivatives

If $z = f(x, y)$, then

$$\frac{\partial f}{\partial x} \text{ or } \frac{\partial z}{\partial x} = \lim_{\delta x \to 0} \frac{f(x + \delta x, y) - f(x, y)}{\delta x}, \tag{25.1}$$

$$\frac{\partial f}{\partial y} \text{ or } \frac{\partial z}{\partial y} = \lim_{\delta y \to 0} \frac{f(x, y + \delta y) - f(x, y)}{\delta y}.$$

Example 25.2. Find $\partial z/\partial x$ and $\partial z/\partial y$ at the point $x = 1$, $y = 3$ when

$$z = x^2 y + 2x^2 - 3y + 4.$$

For $\partial z/\partial x$, y has the status of a constant for the purpose of the differentiation, so

$$\frac{\partial z}{\partial x} = 2xy + 4x - 0 + 0 = 2xy + 4x.$$

At the point $(1, 3)$, $\partial z/\partial x = 10$.

For $\partial z/\partial y$, x is treated as constant, so

$$\frac{\partial z}{\partial y} = x^2 + 0 - 3 + 0 = x^2 - 3.$$

At $(1, 3)$, $\partial z/\partial y = -2$.

We often need to indicate the particular point at which a derivative is to be evaluated, like the point $(1, 3)$ in the previous Example. There are many notations in use for this purpose. We use

$$\left(\frac{\partial z}{\partial x}\right)_{(a,b)} \quad \text{and} \quad \left(\frac{\partial f}{\partial x}\right)_{(a,b)}$$

or

$$\left(\frac{\partial z}{\partial y}\right)_P \quad \text{and} \quad \left(\frac{\partial f}{\partial y}\right)_P$$

to mean the derivatives are to be evaluated at $P : (a, b)$. In this connection, the following definitions are equivalent to (25.1):

Partial derivatives at (a, b)

$$\left(\frac{\partial z}{\partial x}\right)_{(a,b)} \quad \text{or} \quad \left(\frac{\partial f}{\partial x}\right)_{(a,b)} = \lim_{x \to a} \frac{f(x, b) - f(a, b)}{x - a},$$

$$\left(\frac{\partial z}{\partial y}\right)_{(a,b)} \quad \text{or} \quad \left(\frac{\partial f}{\partial y}\right)_{(a,b)} = \lim_{y \to b} \frac{f(a, y) - f(a, b)}{y - b}. \tag{25.2}$$

Example 25.3. Obtain (a) $\dfrac{\partial}{\partial x}\left(\dfrac{x}{x + y}\right)$; (b) $\dfrac{\partial}{\partial y}\left(\dfrac{1}{(x^2 + y^2)^{\frac{1}{2}}}\right)$.

(a) We hold y constant and use the quotient rule (3.2):

$$\frac{\partial}{\partial x}\left(\frac{x}{x + y}\right) = \left((x + y)\frac{\partial x}{\partial x} - x\frac{\partial}{\partial x}(x + y)\right)\bigg/(x + y)^2 = \frac{y}{(x + y)^2},$$

since $\partial x/\partial x = 1$, and y is constant.

(b) x is held constant. Use the chain rule (3, 3), putting

$$u = x^2 + y^2, \qquad z = u^{-\frac{1}{2}};$$

then

$$\frac{\partial z}{\partial y} = \frac{dz}{du}\frac{\partial u}{\partial y}.$$

(We write $\partial u/\partial y$ instead of du/dy in the chain rule because both x and y are present in u, and x is being held constant.) Continuing, we have

$$\frac{\partial z}{\partial y} = -\tfrac{1}{2}u^{-\frac{3}{2}}(2x) = -x(x^2 + y^2)^{-\frac{3}{2}}$$

Example 25.4. The potential function $V(x, t) = A\,e^{-qt}\sin k(x - ct)$ represents an attenuating wave travelling to the right along a cable with speed c. Here A, q, k, c are constants. Find (a) the rate of change of V with time t at any fixed point x; (b) the 'potential gradient' $\partial V/\partial x$ along the wire at any moment.

(a) For $\partial V/\partial t$, use the product rule (3.1) with $u = A\,e^{-qt}$ and $v = \sin k(x - ct)$. We treat x as constant, so $\partial v/\partial t$ instead of dv/dt will be written into the product rule:

$$\frac{\partial V}{\partial t} = \frac{\partial(uv)}{\partial t} = u\frac{\partial v}{\partial t} + v\frac{du}{dt}$$

$$= A\,e^{-qt}[-kc\cos k(x - ct)] + (-qA\,e^{-qt})\sin k(x - ct)$$

$$= -A\,e^{-qt}[kc\cos k(x - ct) + q\sin k(x - ct)].$$

(b) $\dfrac{\partial V}{\partial x} = A\,e^{-qt}\dfrac{\partial}{\partial x}\sin k(x - ct) = kA\,e^{-qt}\cos k(x - ct),$

t being treated as constant.

It will be seen that **no new rules have to be learned** in order to obtain the partial derivatives of given functions. In fact the reader has always unconsciously carried out partial differentiation when differentiating expressions like $A\sin(\omega t + \phi)$, without worrying whether A, ω, ϕ were really constants or just to be treated as such while differentiating.

25.4 Higher derivatives
Having differentiated a function, we might want to differentiate it again. Thus, if $z = f(x, y)$, we can form $\dfrac{\partial z}{\partial x}$ and then $\dfrac{\partial}{\partial x}\left(\dfrac{\partial z}{\partial x}\right)$ or $\dfrac{\partial}{\partial y}\left(\dfrac{\partial z}{\partial x}\right)$, thus forming **second derivatives**, or derivatives higher than the second. There are four second derivatives, written as follows:

$$\frac{\partial}{\partial x}\left(\frac{\partial z}{\partial x}\right) = \frac{\partial^2 z}{\partial x^2}, \qquad \frac{\partial}{\partial y}\left(\frac{\partial z}{\partial x}\right) = \frac{\partial^2 z}{\partial y\,\partial x},$$

$$\frac{\partial}{\partial x}\left(\frac{\partial z}{\partial y}\right) = \frac{\partial^2 z}{\partial x\,\partial y}, \qquad \frac{\partial}{\partial y}\left(\frac{\partial z}{\partial y}\right) = \frac{\partial^2 z}{\partial y^2}.$$

Example 25.5. Obtain the four second derivatives when

$$z = x^3 y + xy^2 + x + y^2 + 1.$$

The first derivatives are

$$\frac{\partial z}{\partial x} = 3x^2 y + y^2 + 1, \qquad \frac{\partial z}{\partial y} = x^3 + 2xy + 2y.$$

Therefore

$$\frac{\partial^2 z}{\partial x^2} = \frac{\partial}{\partial x}\left(\frac{\partial z}{\partial x}\right) = 6xy,$$

$$\frac{\partial^2 z}{\partial y\,\partial x} = \frac{\partial}{\partial y}\left(\frac{\partial z}{\partial x}\right) = 3x^2 + 2y,$$

$$\frac{\partial^2 z}{\partial x\,\partial y} = \frac{\partial}{\partial x}\left(\frac{\partial z}{\partial y}\right) = 3x^2 + 2y,$$

$$\frac{\partial^2 z}{\partial y^2} = \frac{\partial}{\partial y}\left(\frac{\partial z}{\partial y}\right) = 2x + 2.$$

In the last example, we see that $\partial^2 z/\partial y\,\partial x = \partial^2 z/\partial x\,\partial y$. This is always true, although the proof in the general case is difficult:

For any function $f(x, y)$,

$$\frac{\partial^2 f}{\partial y\,\partial x} = \frac{\partial^2 f}{\partial x\,\partial y}. \qquad (25.3)$$

In higher derivatives, the ∂x and ∂y in the denominator may be arranged in any order.

For example, $\dfrac{\partial^3 f}{\partial x\,\partial y^2} = \dfrac{\partial^3 f}{\partial y^2\,\partial x} = \dfrac{\partial^3 f}{\partial y\,\partial x\,\partial y}$, and so on.

The next example shows how to manage a problem in notation. Often a function $f(x, y)$ is used in which the variables x and y only occur in a fixed combination $u = h(x, y)$, so that

$$f(x, y) = g(u), \quad \text{with} \quad u = h(x, y),$$

where g represents a general, unspecified, **function of a single variable**. To obtain a general formula for $\partial f/\partial x$ use the chain rule (3.3) (see also Example 4.2):

$$\frac{\partial f}{\partial x} = \frac{dg}{du}\frac{\partial u}{\partial x} = g'(u)\frac{\partial u}{\partial x} = \frac{\partial h}{\partial x}g'(h(x, y)).$$

(You must *not write* $\partial g/\partial x$ at this point.)

Thus suppose that $f(x, y) = g(5x - 3y)$; then

$$\frac{\partial f}{\partial x} = 5g'(5x - 3y).$$

If $z = g(u)$, where $u = h(x, y)$, then

$$\frac{\partial z}{\partial x} = \frac{dg}{du} \frac{\partial u}{\partial x} = g'(h(x, y)) \frac{\partial h}{\partial x},$$

$$\frac{\partial z}{\partial y} = \frac{dg}{du} \frac{\partial u}{\partial y} = g'(h(x, y)) \frac{\partial h}{\partial y}.$$

(25.4)

It is necessary to **work out** $g'(u)$ **first**, before substituting $u = h(x, y)$.

Example 25.6. Prove that if $z = \phi(x - ct)$, where ϕ is any function, then

$$\frac{\partial^2 z}{\partial x^2} = \frac{1}{c^2} \frac{\partial^2 z}{\partial t^2}.$$

Put $z = \phi(u)$ where $u = x - ct$. Then

$$\frac{\partial z}{\partial x} = \frac{d\phi}{du} \frac{\partial u}{\partial x} = \phi'(u).$$

By the chain rule again,

$$\frac{\partial^2 z}{\partial x^2} = \frac{\partial}{\partial x} \phi'(u) = \frac{d\phi'(u)}{du} \frac{\partial u}{\partial x} = \phi''(u).$$

Similarly

$$\frac{\partial z}{\partial t} = \frac{d\phi}{du} \frac{\partial u}{\partial t} = \phi'(u)(-c),$$

so

$$\frac{\partial^2 z}{\partial t^2} = \frac{\partial}{\partial t} [-c\phi'(u)] = \frac{d}{du} [-c\phi'(u)] \frac{\partial u}{\partial t} = (-c)^2 \phi''(u).$$

Therefore

$$\frac{\partial^2 z}{\partial x^2} = \frac{1}{c^2} \frac{\partial^2 z}{\partial t^2}.$$

The equation

$$\frac{\partial^2 z}{\partial x^2} = \frac{1}{c^2} \frac{\partial^2 z}{\partial t^2}$$

in Example 25.6 is called the **wave equation** in one space dimension. It is a **partial differential equation** as contrasted with the **ordinary differential equations** treated earlier in the book. We have verified that $\phi(x - ct)$ is always a solution, for any function ϕ. The general solution is

$$\phi(x - ct) + \psi(x + ct),$$

where ϕ and ψ are arbitrary functions. The general solution of partial differential equations involves arbitrary functions rather than the arbitrary constants which occur in ordinary differential equations: even the simple equation $\partial z/\partial x = 0$ has the general solution $z = f(y)$, where $f(y)$ is an arbitrary function.

25.5 Tangent plane and normal to a surface

The tangent plane to a surface $z = f(x, y)$ at a point Q on the surface plays the same role as the tangent line to a curve for functions of a single variable. The tangent plane is the plane that fits the surface near Q better than any other possible plane, as when a penny is pressed against a teapot at a particular point.

Suppose that the tangent plane at Q (Fig. 25.7) has the equation

$$z = Ax + Bx + C.$$

There are three constants to be determined, so we need three conditions to settle the values. Three conditions it is reasonable to expect the tangent plane to satisfy are

(i) It must pass through Q; so $c = Aa + Bb + C$.
(ii) In the x direction at Q, the slope A of the *plane* must be equal to the slope of the *surface*; so $A = \left(\dfrac{\partial f}{\partial x}\right)_Q$.
(iii) In the y direction at Q the slope B of the *plane* must be equal to the slope of the *surface*; so $B = \left(\dfrac{\partial f}{\partial y}\right)_Q$.

Therefore the required equation is

$$z = \left(\frac{\partial f}{\partial x}\right)_Q x + \left(\frac{\partial f}{\partial y}\right)_Q y + \left[c - \left(\frac{\partial f}{\partial x}\right)_Q a - \left(\frac{\partial f}{\partial y}\right)_Q b\right],$$

or, more tidily,

$$z - c = \left(\frac{\partial z}{\partial x}\right)_{(a,b)} (x - a) + \left(\frac{\partial z}{\partial y}\right)_{(a,b)} (y - b).$$

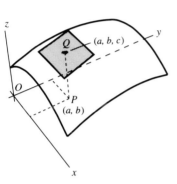

Fig. 25.7
The tangent plane to
$z = f(x, y)$ at $Q : (a, b, c)$,
where $c = f(a, b)$.

Tangent plane at $Q : (a, b, c)$ on the surface
$z = f(x, y)$

$$z - c = \left(\frac{\partial z}{\partial x}\right)_{(a,b)} (x - a) + \left(\frac{\partial z}{\partial y}\right)_{(a,b)} (y - b).$$

(25.5)

Example 25.7. Find the equation of the tangent plane at the point $Q: (2, 1, -2)$ on the sphere $x^2 + y^2 + z^2 = 9$.

Recast the equation into the form $z = f(x, y)$, noticing that Q is on the lower half of the sphere:

$$z = -(9 - x^2 - y^2)^{\frac{1}{2}}.$$

Then the chain rule gives:

$$\frac{\partial z}{\partial x} = -(-2x) \cdot \tfrac{1}{2}(9 - x^2 - y^2)^{-\frac{1}{2}}, \quad \text{and} \quad \left(\frac{\partial z}{\partial x}\right)_{(2,1)} = 1;$$

$$\frac{\partial z}{\partial y} = -(-2y) \cdot \tfrac{1}{2}(9 - x^2 - y^2)^{-\frac{1}{2}}, \quad \text{and} \quad \left(\frac{\partial z}{\partial y}\right)_{(2,1)} = \tfrac{1}{2}.$$

Therefore the equation of the tangent plane at Q is

$$z - (-2) = 1(x - 2) + \tfrac{1}{2}(y - 1),$$

or

$$z = x + \tfrac{1}{2}y - \tfrac{9}{2}.$$

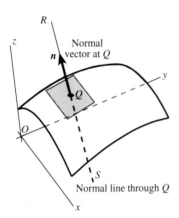

Fig. 25.8

A straight line SQR (Fig. 25.8) is said to be **normal** or perpendicular to the surface $z = f(x, y)$ at Q if it is perpendicular to its tangent plane at Q. The equation (25.5) for the tangent plane can be written in the form

$$\left(\frac{\partial z}{\partial x}\right)_Q x + \left(\frac{\partial z}{\partial y}\right)_Q y + (-1)z = C,$$

where C is a constant, so (see equation (9.17)) a triplet of direction ratios for the line normal to the surface at Q is

$$\left(\left(\frac{\partial z}{\partial x}\right)_Q, \left(\frac{\partial z}{\partial y}\right)_Q, -1\right). \tag{25.6}$$

Example 25.8. Find the cartesian (x, y, z) equation of the straight line normal to the surface $x^2 + y^2 + z^2 = 9$ at $(2, 1, -2)$.

From Example 25.5 (which has the same data), the direction ratios in (21.6) are

$$1, \quad \tfrac{1}{2}, \quad -1.$$

Therefore the equation of the normal line at Q is

$$\frac{x - 2}{1} = \frac{y - 1}{\tfrac{1}{2}} = \frac{z + 2}{-1}.$$

The triplet of direction ratios in (25.6) can be regarded as the three components of a vector parallel to the normal line. Such a vector is called a **normal vector** at Q, and is denoted usually by $\boldsymbol{n}$:

$$\boldsymbol{n} = \left(\left(\frac{\partial z}{\partial x}\right)_Q, \left(\frac{\partial z}{\partial y}\right)_Q, -1\right).$$

Any multiple of this vector is another normal vector, since it will be parallel to the same line. A normal vector placed at Q is shown in Fig. 25.8.

Normal vector n at $Q : (a, b, c)$ where $c = f(a, b)$, on the surface $z = f(x, y)$

$$n = \left(\frac{\partial z}{\partial x}\right)_Q \hat{\boldsymbol{i}} + \left(\frac{\partial z}{\partial y}\right)_Q \hat{\boldsymbol{j}} + (-1)\hat{\boldsymbol{k}}$$

$$= \left(\left(\frac{\partial z}{\partial x}\right)_Q, \left(\frac{\partial z}{\partial y}\right)_Q, -1\right),$$

(25.7)

or any multiple of this vector. Its components are direction ratios of the normal line at Q.

Example 25.9. Find several vectors normal to the sphere $x^2 + y^2 + z^2 = 9$ at the point $(2, 1, -2)$ on the sphere.

The data are again the same as in Example 25.5. The normal taken from (25.7) is $(1, \frac{1}{2}, -1)$. Another is $(-1, -\frac{1}{2}, 1)$, pointing in the opposite direction, while $(\frac{2}{3}, \frac{1}{3}, -\frac{2}{3})$ is a unit vector which is a normal.

25.6 Maxima, minima, and other stationary points

For a function of a single variable, a local maximum or minimum or a point of inflection occurs where the tangent line to the graph of the function is horizontal. For a function $f(x, y)$ of two variables, there are similar possibilities at **points where the tangent plane is horizontal.** Such points, or rather their (x, y) coordinates, are called **stationary points of** $f(x, y)$, because as we pass through them the function is momentarily neither increasing nor decreasing. Sometimes a stationary point is a local **minimum or maximum** as illustrated in Fig. 25.9a, b.

The condition for the tangent plane at Q on $z = f(x, y)$ to be horizontal is that the normal n at Q should be vertical, or parallel to the z axis. Therefore the x and y components of n in (25.7) must be zero:

$$\frac{\partial f}{\partial x} = 0, \qquad \frac{\partial f}{\partial y} = 0.$$

These constitute two simultaneous equations whose solutions (x, y) are the stationary points of $f(x, y)$.

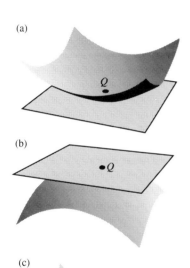

(a)

(b)

(c)

(d)

Fig. 25.9
(a) A local minimum.
(b) A local maximum.
(c) A saddle.
(d) A shoulder.

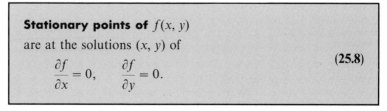

Stationary points of $f(x, y)$

are at the solutions (x, y) of

$$\frac{\partial f}{\partial x} = 0, \qquad \frac{\partial f}{\partial y} = 0.$$

(25.8)

We shall usually describe a stationary point of $f(x, y)$ as being 'at $P : (x, y)$' rather than 'at $Q : (x, y, z)$ on $z = f(x, y)$'. If necessary, the corresponding value of z can be worked out after finding (x, y).

(a)

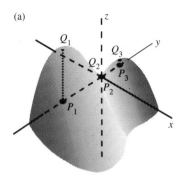

(b)

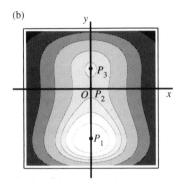

Fig. 25.10
(a) Local maxima at Q_1 and Q_3 and a saddle at Q_2.
(b) The contour map shows closed curves around the maxima.

Example 25.10. Find the stationary points of

$$f(x, y) = \tfrac{1}{3}x^3 - xy^2 - 2y,$$

and the value of $f(x, y)$ there.

Since $\partial f/\partial x = x^2 - y^2$ and $\partial f/\partial y = -2xy - 2$, stationary points occur where

$$x^2 - y^2 = 0, \qquad xy + 1 = 0.$$

The first equation is equivalent to $y = \pm x$. Consider these alternatives *separately*:

If $y = x$, the second equation becomes $x^2 + 1 = 0$, which has no solution. Therefore reject $y = x$.

If $y = -x$, the second equation becomes $-x^2 + 1 = 0$, which has solutions $x = \pm 1$. Corresponding to these we have

$$y = \mp x = \mp 1.$$

Therefore there are two stationary points, $(1, -1)$ and $(-1, 1)$. The values of $f(x, y)$ at these points are

$$f(1, -1) = \tfrac{4}{3}, \qquad f(-1, 1) = -\tfrac{4}{3}.$$

A stationary point at (a, b) is a **local maximum** if $f(a, b)$ is **greater than** $f(x, y)$ at all points in its immediate locality; it is a **local minimum** if the words 'less than' are substituted for 'greater than'. On a contour map, a maximum or minimum shows its presence by being surrounded by closed contours as in Fig. 25.10b.

As with functions of a single variable, the test for a maximum or minimum involves second derivatives. The following test enables maxima, minima and other stationary points to be distinguished in most cases, but we omit the proof, which is difficult.

Test for the character of a stationary point $P : (a, b)$ of $f(x, y)$

Suppose that $\partial f/\partial x = \partial f/\partial y = 0$ at P. Then P is

(a) a saddle if $\dfrac{\partial^2 f}{\partial x^2}\dfrac{\partial^2 f}{\partial y^2} - \left(\dfrac{\partial^2 f}{\partial x\, \partial y}\right)^2 < 0$ at P,

(b) a maximum if $\dfrac{\partial^2 f}{\partial x^2}\dfrac{\partial^2 f}{\partial y^2} - \left(\dfrac{\partial^2 f}{\partial x\, \partial y}\right)^2 > 0$ with $\dfrac{\partial^2 f}{\partial x^2} < 0 \left(\text{or } \dfrac{\partial^2 f}{\partial y^2} < 0\right)$ at P,

(c) a minimum if $\dfrac{\partial^2 f}{\partial x^2}\dfrac{\partial^2 f}{\partial y^2} - \left(\dfrac{\partial^2 f}{\partial x\, \partial y}\right)^2 > 0$ with $\dfrac{\partial^2 f}{\partial x^2} > 0 \left(\text{or } \dfrac{\partial^2 f}{\partial y^2} > 0\right)$ at P.

(d) If none of these apply, the point might be any type.

(25.9)

Example 25.11. Find and classify the stationary points of
$$f(x, y) = \tfrac{1}{3}x^3 + \tfrac{1}{3}y^3 - x^2 - y^2.$$
The stationary points are the solutions of $\partial f/\partial x = 0$, $\partial f/\partial y = 0$, or
$$x^2 - 2x = 0, \qquad y^2 - 2y = 0.$$
From the first, we obtain $x = 0$ or $x = 2$. From the second, $y = 0$ or $y = 2$. Therefore there are stationary points at $(0, 0)$, $(0, 2)$, $(2, 0)$, $(2, 2)$. To test them, we need the second derivatives at a general point:
$$\frac{\partial^2 f}{\partial x^2} = 2x - 2, \qquad \frac{\partial^2 f}{\partial y^2} = 2y - 2, \qquad \frac{\partial^2 f}{\partial x\, \partial y} = 0.$$
At $(0, 0)$, these become respectively $-2, -2, 0$. Then
$$\frac{\partial^2 f}{\partial x^2}\frac{\partial^2 f}{\partial y^2} - \left(\frac{\partial^2 f}{\partial x\, \partial y}\right)^2 = 4 > 0, \qquad \frac{\partial^2 f}{\partial x^2} = \frac{\partial^2 f}{\partial y^2} = -4 < 0.$$
Since the conditions of (25.9b) apply, the point is a maximum.
At $(0, 2)$ and $(2, 0)$,
$$\frac{\partial^2 f}{\partial x^2}\frac{\partial^2 f}{\partial y^2} - \left(\frac{\partial^2 f}{\partial x\, \partial y}\right)^2 = -4 < 0;$$
so, by (25.9a), both points are saddles.
At $(2, 2)$,
$$\frac{\partial^2 f}{\partial x^2}\frac{\partial^2 f}{\partial y^2} - \left(\frac{\partial^2 f}{\partial x\, \partial y}\right)^2 = 4 > 0, \qquad \frac{\partial^2 f}{\partial x^2} = \frac{\partial^2 f}{\partial y^2} = 4 > 0.$$
Therefore, by (25.9c), the point is a minimum.

25.7 The method of least squares

Suppose that a succession of experiments is performed in which we vary one quantity x, such as voltage applied to a circuit, and measure the corresponding value of another variable y, say the resulting current. The values recorded for one or both of the variables might be subject to random errors of measurement; on a graph of the results, this will show up as scatter among the points, as in Fig. 25.11.

We might have reason to believe that the underlying relation between x and y is a straight line. There is no way of deducing this line with certainty, but the following method is often used to obtain a convincing straight-line fit to the points.

Suppose that there are N points altogether; call them
$$(x_1, y_1), \quad (x_2, y_2), \quad \ldots, \quad (x_N, y_N).$$
The general point is called (x_n, y_n). Figure 25.11 shows a candidate for the best-fitting straight line,
$$y = ax + b,$$
and we have to adjust the constants a and b to obtain a good fit. The vertical deviation e_n of a point (x_n, y_n) from the line is shown:
$$e_n = y_n - (ax_n + b).$$

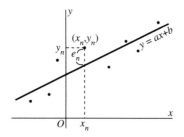

Fig. 25.11

The criterion we shall use to determine the best straight line is to choose a and b so that $\sum_{n=1}^{N} e_n^2$ is as small as possible; that is to say, we want to minimize

$$\sum_{n=1}^{N} e_n^2 = \sum_{n=1}^{N} (y_n - ax_n - b)^2 = f(a, b) \quad \text{(say)}.$$

Therefore a and b are the variables in this problem, and everything else has fixed values.

For a minimum, we require at least that

$$\frac{\partial f}{\partial a} = \frac{\partial f}{\partial b} = 0.$$

The derivatives are given by

$$\frac{\partial f}{\partial a} = \sum_{n=1}^{N} 2(-x_n)(y_n - ax_n - b) = 2 \sum_{n=1}^{N} (ax_n^2 + bx_n - x_n y_n),$$

$$\frac{\partial f}{\partial b} = \sum_{n=1}^{N} (-2)(y_n - ax_n - b) = 2 \sum_{n=1}^{N} (ax_n + b - y_n).$$

Noting that $\sum_{n=1}^{N} b = b + b + \cdots + b = Nb$, we find the conditions for a minimum as the following pair of simultaneous equations for a and b:

Method of least squares

To fit a straight line $y = ax + b$ to the N points (x_n, y_n) $(n = 1, 2, \ldots, N)$:

find a and b by solving the simultaneous equations   **(25.10)**

$$a \sum_{n=1}^{N} x_n^2 + b \sum_{n=1}^{N} x_n = \sum_{n=1}^{N} x_n y_n,$$

$$a \sum_{n=1}^{N} x_n + bN = \sum_{n=1}^{N} y_n.$$

We shall not prove that the stationary point of $f(a, b)$ found by this method is actually a minimum (see Problem 25.21).

Example 25.12. Find the straight line which best fits the data:

$$x_n \quad 0.0 \quad 1.1 \quad 3.2 \quad 3.9 \quad 7.1 \quad 8.9$$
$$y_n \quad 1.1 \quad 1.6 \quad 1.6 \quad 2.8 \quad 2.9 \quad 3.8.$$

Here $N = 6$, and the coefficients in (25.10) are

$$\sum_{n=1}^{6} x_n = 24.2, \qquad \sum_{n=1}^{6} y_n = 13.8,$$

$$\sum_{n=1}^{6} x_n^2 = 156.28, \qquad \sum_{n=1}^{6} x_n y_n = 72.21.$$

The equations for a and b therefore become

$$156.28a + 24.2b = 72.21,$$
$$24.2a + 6b = 13.8.$$

By solving these we find that $a = 0.28$, $b = 1.16$, so the required line is $y = 0.28x + 1.16$.

The equations for a and b are sometimes **ill-conditioned**, meaning that the solutions are very sensitive to small changes in the coefficients. It is therefore advisable to retain all the significant figures given by the data while solving them, despite the fact that we know they already embody the errors of measurement.

25.8 Differentiating an integral with respect to a parameter

Suppose that we have an integral whose integrand contains a parameter α as well as the variable of integration – for example,

$$\int_0^1 e^{\alpha t} \, dt, \qquad \int_{-\infty}^{\infty} g(x)h(x + \alpha) \, dx, \qquad \int \frac{dx}{x + \alpha}.$$

We shall consider a definite integral, though the process works in the same way for indefinite integrals. Indicate the dependence on α in the general case by

$$I(\alpha) = \int_a^b f(t, \alpha) \, dt.$$

Then $dI(\alpha)/d\alpha$ can be obtained by the following rule:

> **Differentiating an integral with respect to a parameter**
>
> If $\displaystyle\int_a^b f(t, \alpha) \, dt = I(\alpha)$, then   (25.11)
>
> $$\frac{dI(\alpha)}{d\alpha} = \int_a^b \frac{\partial f(t, \alpha)}{\partial \alpha} \, dt.$$

This process is also called **differentiation under the integral sign**. To prove (25.11), change α to $\alpha + \delta\alpha$; then $I(\alpha)$ changes to $I(\alpha + \delta\alpha)$. Put

$$I(\alpha + \delta\alpha) - I(\alpha) = \delta I(\alpha).$$

Then

$$\frac{\delta I(\alpha)}{\delta\alpha} = \frac{I(\alpha + \delta\alpha) - I(\alpha)}{\delta\alpha}$$

$$= \frac{1}{\delta\alpha}\left(\int_a^b f(t, \alpha + \delta\alpha) \, dt - \int_a^b f(t, \alpha) \, dt \right)$$

$$= \int_a^b \frac{f(t, \alpha + \delta\alpha) - f(t, \alpha)}{\delta\alpha} \, dt.$$

Now let $\delta\alpha \to 0$. Then $\delta I(\alpha)/\delta\alpha$ becomes $\mathrm{d}I(\alpha)/\mathrm{d}\alpha$, and the integrand becomes $\partial f(t, \alpha)/\partial\alpha$, which is the result (25.11).

Example 25.13. Evaluate $I(\alpha) = \displaystyle\int_0^\infty \frac{\mathrm{d}t}{t^2 + \alpha^2}$, where $\alpha > 0$, and

use (25.11) to evaluate $J(\alpha) = \displaystyle\int_0^\infty \frac{\mathrm{d}t}{(t^2 + \alpha^2)^2}$.

From Appendix E,

$$I(\alpha) = \int_0^\infty \frac{\mathrm{d}t}{t^2 + \alpha^2} = \left[\alpha^{-1}\arctan(t/\alpha)\right]_0^\infty = \frac{\pi}{2\alpha}.$$

By (25.11),

$$\frac{\mathrm{d}I}{\mathrm{d}\alpha} = \int_0^\infty \frac{\partial}{\partial\alpha}\frac{1}{t^2 + \alpha^2}\,\mathrm{d}t = \frac{\mathrm{d}}{\mathrm{d}\alpha}\frac{\pi}{2\alpha},$$

or

$$\int_0^\infty \frac{-2\alpha}{(t^2 + \alpha^2)^2}\,\mathrm{d}t = -\frac{\pi}{2\alpha^2}.$$

Therefore

$$J(\alpha) = \int_0^\infty \frac{\mathrm{d}t}{(t^2 + \alpha^2)^2} = \frac{\pi}{4\alpha^3}.$$

Problems, Chapter 25

25.1. Sketch contour maps of the following functions:
(a) $2x - 3y + 4$;   (b) $-x + 2y - 1$;
(c) $(x - 1)(y - 1)$;   (d) $x^2 + \frac{1}{4}y^2 - 1$;
(e) $x^2 + 2x + y^2$ (complete the square in x);
(f) y/x;   (g) $y^2 - x^2$;   (h) y/x^3;
(i) $x^3 + 4y^2$;   (j) $y/(x + y)$.

25.2. By sketching rough contour maps, indicate the paths of steepest ascent (the paths on which z increases most rapidly), starting at the point $(1, 1)$:
(a) $z = 2x - 3y + 4$;   (b) $z = x - y$;
(c) $z = x^2y^2$;   (d) $z = (x - 1)^2 + \frac{1}{4}(y - 1)^2$.

25.3. Obtain $\partial f/\partial x$ and $\partial f/\partial y$ at the point $(2, 1)$ for the following functions.
(a) $3x + 7y - 2$;   (b) $-2x + 3y + 4$;
(c) $2x^2 - 3y^2 - 2xy - x - y + 1$;
(d) $\frac{1}{8}x^3 + y^3 - 2y - 1$;   (e) $x^4y^2 - 1$;
(f) $(x - 1)(y - 2)$;   (g) $1/(xy)$;
(h) x/y;   (i) $\dfrac{x - y}{x + y}$;   (j) $\dfrac{3}{x^2 + y^2}$;
(k) $(x^2 + y^2)^{\frac{1}{2}}$;   (l) $(2x - 3y + 2)^3$;

(m) $e^{x^2 + y^2}$;   (n) $\cos(x^2 - y^2)$;
(o) $\sin(x/y)$;   (p) $\arctan(y/x)$.

25.4. (a) Let $z = g(ax + by)$, where a and b are constants. Express $\partial z/\partial x$ and $\partial z/\partial y$ in terms of $g'(ax + by)$ (which means $g'(u)$ when u is subsequently put equal to $ax + by$). Check your result for the cases when $g(u) = \cos u$ and $g(u) = e^u$.

(b) Let $z = g(\sin xy)$. Express $\partial z/\partial x$ and $\partial z/\partial y$ in terms of x, y, and $g'(\sin xy)$. Check the result by differentiating $e^{\sin xy}$ directly.

(c) A certain physical quantity V is a function only of the radial coordinate r in plane polar coordinates: $V = g(r)$, where $r = (x^2 + y^2)^{\frac{1}{2}}$. Express $\partial V/\partial x$ and $\partial V/\partial y$, firstly in terms of x and y, then in terms of r and θ.

25.5. In plane polar coordinates (r, θ) in the first quadrant, $r = (x^2 + y^2)^{\frac{1}{2}}$ and $x = r\cos\theta$. Form $\partial r/\partial x$ and $\partial x/\partial r$, and show that

$$\frac{\partial r}{\partial x}\frac{\partial x}{\partial r} \neq 1.$$

By considering the meaning of the derivatives

$\partial r/\partial x$ and $\partial x/\partial r$ near a particular point P in the manner of Fig. 25.6, show why it is not to be expected that the product should equal 1. (In the case of a single variable and ordinary derivatives, we often get true results by formally cancelling out symbols like dx, du, etc., as in the chain rule. This almost never works when more variables are present: see e.g. the next problem.)

25.6. (a) Let $z = \sin(x - y)$; show that $\dfrac{\partial z}{\partial x} \Big/ \dfrac{\partial z}{\partial y} = -1$.

(b) Let $z = g(x - y)$; show that $\dfrac{\partial z}{\partial x} \Big/ \dfrac{\partial z}{\partial y} = -1$.

25.7. Show that, if $z = g(x/y)$, then

$$x\frac{\partial z}{\partial x} + y\frac{\partial z}{\partial y} = 0,$$

and check the result in the case $z = \sin x/y$.

25.8. Find $\dfrac{\partial^2 f}{\partial x^2}, \dfrac{\partial^2 f}{\partial y^2}, \dfrac{\partial^2 f}{\partial y\,\partial x}, \dfrac{\partial^2 f}{\partial x\,\partial y}$ in each of the following cases (note (25.3)).
(a) $ax + by + c$; (b) $x^2 + 2y^2 + 3xy - x + 1$;
(c) $\sin(x - y)$; (d) y/x; (e) e^{2x+3y};
(f) $1/x + 1/y$; (g) $\sin 3x + \cos 2y$; (h) $(3x - 4y)^4$;
(i) $1/(x + y)$; (j) $\ln xy$; (k) $1/(x^2 + y^2)^{\frac{1}{2}}$.

25.9. Confirm that, if $r = (x^2 + y^2)^{\frac{1}{2}}$ and $z = \ln r$, then

$$\frac{\partial z}{\partial x} = \frac{x}{r^2} \quad \text{and} \quad \frac{\partial^2 z}{\partial x^2} = \frac{1}{r^2} - \frac{2x^2}{r^4}.$$

Show that $z = \ln r$ is a solution of the equation

$$\frac{\partial^2 z}{\partial x^2} + \frac{\partial^2 z}{\partial y^2} = 0.$$

(This is called Laplace's partial differential equation in two dimensions.)

25.10. Obtain the tangent plane and a normal vector for the following surfaces at the points given.
(a) $z = x^2 + y^2$ at $(1, 1, 2)$; (b) $z = xy$ at $(2, 2, 4)$;
(c) $z = x/y$ at $(2, 1, 2)$;
(d) $z = (29 - x^2 - y^2)^{\frac{1}{2}}$ at $(3, 4, 2)$;
(e) $z = x^2 + y^2 - 2x - 2y$ at $(1, 1, -2)$;
(f) $z = e^{xy}$ at $(0, 0, 1)$.

25.11. The two surfaces $z = x^2 + y^2$ and $z = x - y + 2$ intersect at the point $Q : (1, 1, 2)$. Find normal vectors at Q to each of the two surfaces, $\mathbf{n}_1$ to the first and $\mathbf{n}_2$ to the second. By considering the scalar product $\mathbf{n}_1 \cdot \mathbf{n}_2$, find the angle between the normals and hence the angle at which the surfaces cut at Q.

25.12. Find the stationary points of the following functions, and classify them using (25.9).
(a) $(x - 1)(y + 2)$; (b) $x^2 + y^2 - 2x + 2y$;
(c) $\frac{1}{3}x^3 - \frac{1}{3}y^3 - x + y + 3$;
(d) $\cos x + \cos y$; (e) $\ln(x^2 + x) + \ln(y^2 + y)$;
(f) $e^{x^2+y^2-2x+2y}$; (g) $xy + 1/x + 1/y$;
(h) $x^3 + y^3 - 3xy + 1$; (i) $\sin x + \sin y$;
(j) $xy^2 - x^2y + x - y + 1$;
(k) $(x^2 - y^2)(x + y)$; (l) $(2 - x^2 - y^2)^2$;
(m) $x^4 + y^4 + y - x$;
(n) $x^4 + y^4$ (this eludes the test (25.9) – the point is obviously a minimum).

25.13. Classify the stationary point of $ax^2 + 2hxy + by^2$ at $(0, 0)$ for various relations between a, b, and h.

25.14. Find positive numbers a, b, c so that
(a) $a + b + c = 21$ and abc is a maximum.
(b) $abc = 64$ and $a + b + c$ is a minimum.

25.15. Find the **absolute maximum** value of $(2 - x^2 - y^2)^2$ in the 'box' $-1 \leqslant x \leqslant 2$, $-1 \leqslant y \leqslant 2$. (It will be necessary to investigate the function on the four edges of the box separately, since the *absolute* maximum will not be revealed by the conditions (25.9) if it is on the edges.)

25.16. Find the shortest distance between the straight lines $x = y = z$ and $2x = y = z + 2$, by using a simple parametrization of each line. (Use different letters for the two parameters: these will be the new variables for the minimization.)

25.17. N points $(x_1, y_1), (x_2, y_2), \ldots, (x_N, y_N)$ are given in a plane, and $P : (x, y)$ is a general point. Find P so that the sum of the squares of its distances from the N given points is as small as possible.

25.18. (a) A rectangular box with a lid must hold a given volume V, and have the smallest possible surface area. Show that it must be a cube. (Call the lengths of two of its sides x and y.)

(b) An open-topped rectangular box must have a given volume V and its surface area must be as small as possible. Find its dimensions.

(c) A circular-cylindrical box must have a fixed volume V and minimum surface area. Find its dimensions (i) if it has a lid, (ii) if it has no lid.

(d) A rectangular container is required to have total surface area S, and a volume as large as possible. Find its dimensions (i) if it has a lid, (ii) if it does not have a lid.

25.19. Find the straight line which best fits the

experimental data in the sense of Section 21.7:

x	1	2	3	4	5
y	3.1	2.1	2.0	1.8	1.2.

25.20. The population P of a fast-breeding rodent was observed over a period of 12 months, and the following estimates obtained:

t (months)	0	2	3	5	8	10	12
P (pop'n)	12	23	26	60	170	300	690.

Assume that the underlying growth law takes the form (see Section 1.10)

$$P = A\,e^{bt},$$

where A and b are constants.

To estimate A and b, take the logarithm of this expression and treat $y = \ln P$ as a variable in the least-squares method of Section 25.7.

25.21. For the least-squares method of Section 25.7, use the test (25.9) to show that the values of a and b obtained do minimize the sum of squares. (This is, of course, rather obvious intuitively.)

25.22. Using Laplace transforms with respect to t, solve the partial differential equation

$$\frac{\partial z}{\partial t} + x\frac{\partial z}{\partial x} + z = 2x,$$

for $x > 0$ and $t > 0$, where $z(0, t) = 0$ and $z(x, 0) = 0$.

26 Functions of two variables: geometry and formulae

26.1 The incremental approximation

It was explained in Section 25.5 that the tangent plane at a point is the plane that best fits a surface at and around the point. The formula for the tangent plane to a surface $z = f(x, y)$ at $Q : (a, b, c)$, where $c = f(a, b)$, is

$$z - b = \left(\frac{\partial f}{\partial x}\right)_{(a,b)} (x - a) + \left(\frac{\partial f}{\partial y}\right)_{(a,b)} (y - b)$$

(see (25.5)). We will set up new axes with origin at Q, parallel to the old ones, and call them δx, δy, δz (see Fig. 26.1), anticipating that we shall be concerned with *small* distances from Q. Then

$$\delta x = x - a, \qquad \delta y = y - b, \qquad \delta z = z - c.$$

In the new coordinates, the equation of the tangent plane is

$$\delta z = \left(\frac{\partial f}{\partial x}\right)_{(a,b)} \delta y + \left(\frac{\partial f}{\partial y}\right)_{(a,b)} \delta y.$$

Now consider the quantity δf, where

$$\delta f = f(x, y) - f(a, b).$$

This is the change in z *on the surface* $z = f(x, y)$ from its value at Q. The tangent plane is the best-fitting plane to the surface at Q, so the formula

$$f(x, y) - f(a, b) = \delta f \approx \left(\frac{\partial f}{\partial x}\right)_{(a,b)} \delta x + \left(\frac{\partial f}{\partial y}\right)_{(a,b)} \delta y$$

must give the **best-fitting linear approximation** to δf near $x = a$, $y = b$:

Best linear approximation to $f(x, y)$ **near** (a, b).

$$\delta f \approx \left(\frac{\partial f}{\partial x}\right)_{(a,b)} \delta x + \left(\frac{\partial f}{\partial y}\right)_{(a,b)} \delta y, \qquad (26.1)$$

where $\delta f = f(x, y) - f(a, b), \delta x = x - a, \delta y = y - b$.

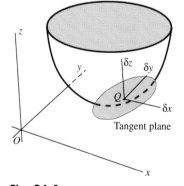

Fig. 26.1

The reader is more likely to **remember** the formula obtained by calling the general point (x, y) instead of (a, b), and putting z in place of f. Also the approximation will be good enough to be useful only when δx and δy are 'small' (how small will depend on circumstances):

Incremental approximation for $f(x, y)$
(mnemonic version)

For small enough increments δx and δy:

$$f(x + \delta x, y + \delta y) - f(x, y) \approx \frac{\partial f}{\partial x}\delta x + \frac{\partial f}{\partial y}\delta y \qquad (26.2)$$

If we put $z = f(x, y)$, this can be written

$$\delta z = \frac{\partial z}{\partial x}\delta x + \frac{\partial z}{\partial y}\delta y \quad \text{(approximately)}.$$

This will be the source of almost all our results from now on.

Example 26.1. Let $z = x^2 + 3y^2$. Find an approximation to δz in terms of δx and δy near the points (a) $x = 2$, $y = 1$; (b) $x = 3$, $y = 2$; (c) $x = 0$, $y = 0$. (d) Find the exact value of δz in case (a) and compare it with the approximate values in the three cases when δx and δy both take the values 0.1, 0.01, and 0.001.

In general, $\dfrac{\partial z}{\partial x} = 2x$ and $\dfrac{\partial z}{\partial y} = 6y$.

(a) At $(2, 1)$, $\partial z/\partial x = 4$ and $\partial z/\partial y = 6$. Therefore, from (26.2)

$$\delta z = 4\,\delta x + 6\,\delta y \quad \text{approximately}.$$

(b) At $(3, 2)$, $\partial z/\partial x = 6$ and $\partial z/\partial y = 12$; so

$$\delta z = 6\,\delta x + 12\,\delta y \quad \text{approximately}.$$

(c) At $(0, 0)$, $\partial z/\partial x = \partial z/\partial y = 0$; so the formula predicts

$$\delta z = 0 \quad \text{approximately}.$$

The reason is that $(0, 0)$ is a stationary point, so z hardly changes when we move a short distance from $(0, 0)$.

(d) From (a), the approximation near $(2, 1)$ when $\delta x = \delta y = 0.1$ is

$$\delta z = (4 \times 0.1) + (6 \times 0.1) = 1.0.$$

The exact value is given by

$$\delta z = f(2.1, 1.1) - f(2, 1) = 1.04,$$

so the error in estimating δz is -4%. If $\delta x = \delta y = 0.01$, the error is -0.4%; if $\delta x = \delta y = 0.001$, it is -0.04%.

We see from (d) in the example that the **approximation improves percentagewise as δx and δy get smaller**: it is not merely that the error decreases because δx, δy, δz all go to zero together. The following example shows the reason for this.

Example 26.2. Find the exact algebraic form of the error incurred by using (26.2) to estimate δz at $(2, 1)$ when $z = x^2 + 3y^2$ (see Example 26.1a).

Put $x = 2 + \delta x$ and $y = 1 + \delta y$. Then

$$\delta z = f(2 + \delta x, 1 + \delta y) - f(2, 1)$$
$$= (2 + \delta x)^2 + 3(1 + \delta y)^2 - 7$$
$$= (4\,\delta x + 6\,\delta y) + (\delta x^2 + 3\,\delta y^2).$$

The first two terms represent the linear approximation obtained in Example 26.1(a). The remainder is the error incurred; the part we ignore in the approximation. The error consists only of **higher powers of δx and δy**, and this will always be the case. Therefore **the error is an order of magnitude smaller than the linear terms retained in the incremental approximation (26.2)**.

26.2 Small changes and errors

The incremental approximation (26.1) or (26.2) can be used to estimate the effect of making **small changes** in the values of variables in a formula.

Example 26.3. Estimate the change in the value of

$$z = \frac{1}{(x^2 + y^2)^{\frac{1}{2}}}$$

when (x, y) change from $(3, 4)$ to $(3.1, 3.8)$.

Using (26.2), put $(x, y) = (3, 4)$, $\delta x = 0.1$, $\delta y = -0.2$. We require

$$\left(\frac{\partial z}{\partial x}\right)_{(3,4)} = [-x(x^2 + y^2)^{-\frac{3}{2}}]_{(3,4)} = -\tfrac{3}{125},$$

$$\left(\frac{\partial z}{\partial y}\right)_{(3,4)} = [-y(x + y)^{-\frac{3}{2}}]_{(3,4)} = -\tfrac{4}{125},$$

Therefore, approximately,

$$\delta z = (-\tfrac{3}{125})(0.1) + (-\tfrac{4}{125})(-0.2) = 0.004.$$

(The exact value of δz is $0.00391\ldots$.)

Example 26.4. The period T of the swings of a pendulum is equal to $2\pi(l/g)^{\frac{1}{2}}$, where l is the length and g the gravitation constant. Estimate the error in calculating T if, instead of using closely correct values $l = 1.015$ and $g = 9.812$ in the formula, we use the rounded values $l = 1$ and $g = 10$.

The formula corresponding to (26.12) is

$$\delta T \approx \frac{\partial T}{\partial l}\,\delta l + \frac{\partial T}{\partial g}\,\delta g.$$

Suppose for simplicity we decide to substitute the *rounded values* $l = 1$ and $g = 10$ into the coefficients: we obtain

$$\frac{\partial T}{\partial l} = (\pi l^{-\frac{1}{2}}g^{-\frac{1}{2}})_{(1,10)} = 0.993,$$

$$\frac{\partial T}{\partial g} = (-\pi l^{\frac{1}{2}}g^{-\frac{3}{2}})_{(1,10)} = -0.099.$$

Equation (26.2) then requires that we put

$$\delta l = (\text{true value}) - (\text{rounded value}) = 0.015,$$

$$\delta g = (\text{true value}) - (\text{rounded value}) = -0.188.$$

Then

$$\delta T \approx (\text{true value}) - (\text{rounded value})$$

$$\approx (0.993)(0.015) + (-0.099)(-0.188) = 0.333.$$

But this is not the *error*: for that we need

$$(\text{error}) = (\text{rounded value}) - (\text{true value}) = -\delta T,$$

so the *error* is about -0.033. (The exact error is $-0.339\ldots$.)

In the last example, we substituted the rounded (erroneous) values into $\partial T/\partial l$ and $\partial T/\partial g$, which led to a complication we might have avoided. However, usually there is no choice, the exact values being unknown. Let $z = f(x, y)$, and suppose that we want to estimate the error in z which could arise from using measured (i.e. approximate) values for x and y. The error Δx in x is

$$\Delta x = (\text{measured value of } x) - (\text{exact value of } x),$$

and similarly for Δy and Δz.

Usually we only know a range of possible error, not the errors themselves. For example we might say that a parcel weighed $1430(\pm 15)$ g, meaning that we think it is between 1415 and 1445 g. Therefore, the values of Δx and Δy are unknown, so **the exact values of x and y are unknown**, and are not available to go into (26.2) in place of (x, y). Suppose we put instead

$$x \text{ and } y = (\text{measured values}).$$

To correspond with this, the definition of δx, δy, δz in (26.2) requires

$$\delta x,\ \delta y,\ \delta z = \text{(true values)} - \text{(measured values)},$$

Therefore

$$\delta x = -\Delta x, \qquad \delta y = -\Delta y, \qquad \delta z = -\Delta z$$

go into (26.2). Every term has then a negative sign, so the formula in terms of Δx, Δy, Δz has the same shape as the incremental formula:

Small-error formula

If $z = f(x, y)$, then

$$\Delta z = \frac{\partial z}{\partial x} \Delta x + \frac{\partial z}{\partial y} \Delta y \quad \text{(approximately)}, \qquad (26.3)$$

where x and y are measured values, and Δ stands for

$$\text{error} = \text{(measured value)} - \text{(exact value)}.$$

This is used in the following way.

Example 26.5. In a triangle ABC, the side BC has length a given by

$$a = \frac{c \sin A}{\sin(A + B)}.$$

Suppose that $c = 10$ (exactly), and angles A and B are measured to $5°$ accuracy: $A = 45(\pm 5)°$, $B = 30(\pm 5)°$. Estimate the largest possible resulting error in a.

Put

$$a = f(A, B) = 10 \frac{\sin A}{\sin(A + B)}.$$

Then

$$\frac{\partial a}{\partial A} = 10 \frac{\sin(A + B) \cos A - \cos(A + B) \sin A}{\sin^2(A + B)} = 10 \frac{\sin B}{\sin^2(A + B)}.$$

Similarly

$$\frac{\partial a}{\partial B} = -10 \frac{\sin A}{\sin^2(A + B)}.$$

When taken at the measured values $A = 45°$ and $B = 30°$ (these are the only values available), we get $\partial a / \partial A = 5.36$ and $\partial a / \partial B = -7.58$. The error formula (26.3) becomes

$$\Delta a = 5.36\, \Delta A - 7.58\, \Delta B$$

approximately, where ΔA and ΔB *must be measured in radians*.

The greatest possible magnitude occurs if ΔA and ΔB happen to have the opposite signs and their greatest possible magnitudes; that is, if

$\Delta A = -\Delta B = \pm 0.087$ radians. In that case, $\Delta a = \pm 1.13$. Therefore

$$a = 7.32(\pm 1.13),$$

showing a possible error of 15%.

Example 26.6. One solution of the equation $x^2 + bx + c = 0$ is $x = \frac{1}{2}[-b + (b^2 - 4c)^{\frac{1}{2}}]$. (a) Find an approximate expression for the error Δx arising from small errors Δb and Δc in b and c. (b) Estimate the maximum possible error in the solution x if b and c are rounded to one decimal to give $b \simeq 3.1$, $c \simeq 2.1$.

(a) We have $\Delta x = (\partial x/\partial b)\, \Delta b + (\partial x/\partial c)\, \Delta c$, in which we must put

$$\frac{\partial x}{\partial b} = \tfrac{1}{2}[-1 + b(b - 4c)^{-\frac{1}{2}}], \qquad \frac{\partial x}{\partial c} = -(b^2 - 4c)^{-\frac{1}{2}}.$$

(b) Since b and c are rounded numbers, all that we know about them is that

$$b = 3.1(\pm 0.05), \qquad c = 2.1(\pm 0.05),$$

meaning that the error might be *anywhere* in the range indicated. Putting the face values $b = 3.1$ and $c = 2.1$ into (a), we obtain

$$\frac{\partial x}{\partial b} = 0.909, \qquad \frac{\partial x}{\partial c} = -0.909;$$

so, by (26.3),

$$\Delta x = 0.909\, \Delta b - 0.909\, \Delta c.$$

This takes its greatest possible magnitude when Δb and Δc take their maximum values and have *opposite* sign: that is, when

$$\Delta b = \pm 0.05, \qquad \Delta c = \mp 0.05.$$

In that case $\Delta x = \pm 0.909(0.05 + 0.05) = \pm 0.091$.

The value of x estimated from the rounded coefficients is $x = -1.095$. Although the rounding error is only at most 2.5%, the error in the solution could be as large as $\pm 8.3\%$.

26.3 The derivative in any direction

The plane in Fig. 26.2 is a map of a surface $z = f(x, y)$ with all detail omitted. At $P : (x, y)$ we see a slope $\partial z/\partial x$ if we look east, a slope $\partial z/\partial y$ looking north, and other slopes in other directions. We can find the slopes in other directions in terms of $\partial z/\partial x$ and $\partial z/\partial y$. It might seem that we could make the intermediate slopes equal to anything we liked, but if the surface at P is smooth enough to have a tangent plane, this is not so. In effect, the slopes we see are the slopes of the tangent plane in the various directions.

Consider the direction $\overline{PQ}$ which makes an angle θ with the positive x axis, the direction for positive angles being anticlockwise as with polar coordinates. Let the length $PQ = \delta s$, a short step, and let δx and δy be as shown. Then, by (26.2), the change in elevation

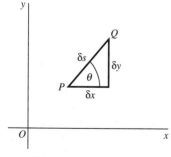

Fig. 26.2

in this direction is given approximately by

$$\delta z \approx \frac{\partial z}{\partial x}\,\delta x + \frac{\partial z}{\partial y}\,\delta y.$$

Divide by δs; we obtain

$$\frac{\delta z}{\delta s} \approx \frac{\partial z}{\partial x}\frac{\delta x}{\delta s} + \frac{\partial z}{\partial y}\frac{\delta y}{\delta s} = \frac{\partial z}{\partial x}\cos\theta + \frac{\partial z}{\partial y}\sin\theta$$

from Fig. 26.2. Now let $\delta s \to 0$; the approximation becomes exact, and we have an expression for the slope in any direction. Using the notation for the **directional derivative**,

$$\lim_{\delta s \to 0} \frac{\delta z}{\delta s} = \frac{dz}{ds},$$

we have the following formula.

Directional derivative

The slope of $z = f(x, y)$ at P in direction θ:

$$\frac{dz}{ds} = \left(\frac{\partial z}{\partial x}\right)_P \cos\theta + \left(\frac{\partial z}{\partial y}\right)_P \sin\theta.$$

(26.4)

Example 26.7. Find the slopes of the surface $z = xy + x^2$ at $P : (2, 3)$ in the direction $-120°$.

The direction is shown on Fig. 26.3.

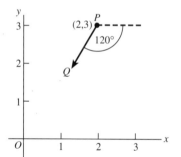

Fig. 26.3

$$\left(\frac{\partial z}{\partial x}\right)_{(2,3)} = (y + 2x)_{(2,3)} = 7,$$

$$\left(\frac{\partial z}{\partial y}\right)_{(2,3)} = (x)_{(2,3)} = 2.$$

Also

$$\cos(-120°) = -\sin 30° = -\tfrac{1}{2}$$

and

$$\sin(-120°) = -\cos 30° = -\tfrac{1}{2}\sqrt{3};$$

so

$$\frac{dz}{ds} = 7(-\tfrac{1}{2}) + 2(-\tfrac{1}{2}\sqrt{3}) = -\tfrac{1}{2}(7 + 2\sqrt{3}).$$

Example 26.8. The temperature distribution in a plate heated at the point $(0, 0)$ is given by $T = 1/(x^2 + y^2)^{\frac{1}{2}}$. (a) Find the temperature gradient at the point $(3, 3)$ in a direction of $45°$ to the positive x axis. (b) In polar coordinates, $T = 1/r$. Show that the result (a) is the same as $\partial T/\partial r$ taken at any point where $r = (3^2 + 3^2)^{\frac{1}{2}}$.

(a) $\left(\dfrac{\partial T}{\partial x}\right)_{(3,3)} = \left(-\dfrac{x}{(x^2 + y^2)^{\frac{3}{2}}}\right)_{(3,3)} = -\dfrac{1}{18\sqrt{2}}$,

$\left(\dfrac{\partial T}{\partial y}\right)_{(3,3)} = \left(-\dfrac{y}{(x^2 + y^2)^{\frac{3}{2}}}\right)_{(3,3)} = -\dfrac{1}{18\sqrt{2}}$.

Also $\cos\theta = 1/\sqrt{2}$ and $\sin\theta = 1/\sqrt{2}$. Therefore the temperature gradient at (3, 3) in the given direction is

$$\dfrac{\mathrm{d}T}{\mathrm{d}s} = -\tfrac{1}{18}.$$

(b) $T = 1/r$, so $\partial T/\partial r = -1/r^2$. At the given point, $r = 3\sqrt{2}$, so the result is the same.

Example 26.9. At any point on the plane $z = \sqrt{3}x - y + 4$, find (a) an expression for the slope $\mathrm{d}z/\mathrm{d}s$ in every direction, (b) the directions in which $\mathrm{d}z/\mathrm{d}s = 0$, (c) the directions in which $\mathrm{d}z/\mathrm{d}s$ is a maximum and a minimum.

(a) $\partial z/\partial x = \sqrt{3}$ and $\partial z/\partial y = -1$; and these are the same at every point. By (26.4),

$$\dfrac{\mathrm{d}z}{\mathrm{d}s} = \sqrt{3}\cos\theta - \sin\theta.$$

(b) $\mathrm{d}z/\mathrm{d}s = 0$ where $\sqrt{3}\cos\theta - \sin\theta = 0$, or $\tan\theta = \sqrt{3}$. Therefore $\theta = 60°$ or $\theta = -120°$. These directions are opposed: see Fig. 26.4. They give the direction of the contour through any point.

(c) $\mathrm{d}z/\mathrm{d}s$ is a maximum (direction of **steepest ascent**), or a minimum (**steepest descent**), in directions such that

$$\dfrac{\mathrm{d}}{\mathrm{d}\theta}\left(\dfrac{\mathrm{d}z}{\mathrm{d}s}\right) = 0,$$

or $-\sqrt{3}\sin\theta - \cos\theta = 0$, or $\tan\theta = -1/\sqrt{3}$. Therefore $\theta = -30'$ or $\theta = 150°$, these directions being directly opposed: see Fig. 26.4. By considering the sign of

$$\dfrac{\mathrm{d}^2}{\mathrm{d}\theta^2}\left(\dfrac{\mathrm{d}z}{\mathrm{d}s}\right),$$

or just by thinking about it, it can be seen that the directions of steepest ascent and descent are as shown.

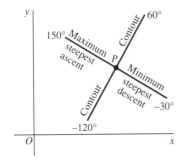

Fig. 26.4

In the last example, the directions of steepest ascent/descent at any point are perpendicular to the directions of the contours; we shall now show that this is true for all surfaces. On the contour map of $z = f(x, y)$, the slope in the direction θ at a point $P : (x, y)$ has the form (26.4):

$$\dfrac{\mathrm{d}z}{\mathrm{d}s} = A\cos\theta + B\sin\theta,$$

where A and B are the values of $\partial z/\partial x$ and $\partial x/\partial y$ at P. This is zero in the directions θ_1 where

$$\tan \theta_1 = -A/B.$$

The two directions θ_1 which satisfy this equation differ by π, so they indicate smooth passage of the contour through P. The gradient dz/ds is a maximum or minimum when

$$\frac{d}{d\theta}\left(\frac{dz}{ds}\right) = 0,$$

or in directions θ_2 where

$$\tan \theta_2 = B/A,$$

which give the directions of steepest ascent/descent. Since

$$\tan \theta_1 \tan \theta_2 = -1,$$

these directions are perpendicular, a fact known intuitively by any hill walker.

> **At each point on the map of $z = f(x, y)$, the direction of steepest ascent/descent is perpendicular to the contour.**   (26.5)

The two systems of curves, consisting of the contours and the curves which follow directions of steepest ascent or descent, are perpendicular wherever they cross, so they are called **orthogonal systems** of curves (see Section 27.4).

26.4 Implicit differentiation

An equation of the type

$$f(x, y) = c,$$

where c is a constant, describes a curve or curves in the (x, y) plane, since we can imagine solving it to obtain y as a function of x. For example, $x^2 + y^2 = 4$ represents the two semicircles $y = \pm(4 - x^2)^{\frac{1}{2}}$. Another interpretation is that the equation $f(x, y) = c$ describes the contour $z = c$ of a surface $z = f(x, y)$, projected into the (x, y) plane, as in Fig. 25.4b.

Although it is usually impossible in practice to solve for y in terms of x, it is always possible to obtain an expression for the slope dy/dx of the curve in terms of x and y. Choose any point $P : (x, y)$ on the curve (Fig. 26.5), and move along it a short distance to $Q : (x + \delta x, y + \delta y)$. Then dy/dx on the curve is given by

$$\frac{dy}{dx} = \lim_{\delta x \to 0} \frac{\delta y}{\delta x}.$$

Since P and Q both lie on the curve, $\delta f = 0$; so the incremental y

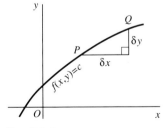

Fig. 26.5

approximation (26.2) gives

$$\frac{\partial f}{\partial x} \delta x + \frac{\partial f}{\partial y} \delta y \approx 0,$$

or

$$\frac{\delta y}{\delta x} \approx -\frac{\partial f}{\partial x} \bigg/ \frac{\partial f}{\partial y}.$$

Now let $\delta x \to 0$. The '$\approx$' becomes '$=$', and $\delta y / \delta x$ becomes dy/dx, from which we obtain:

The implicit-differentiation formula

The slope of $f(x, y) = c$ at any point (x, y) on the curve is given by   (26.6)

$$\frac{dy}{dx} = -\frac{\partial f}{\partial x} \bigg/ \frac{\partial f}{\partial y}.$$

The process is called **implicit differentiation** because $f(x, y) = c$ gives y in terms of x only 'implicitly', not explicitly.

Example 26.10. Find an expression for dy/dx at a general point (x, y) on the circle $x^2 + y^2 = 4$.

Here $f(x, y) = x^2 + y^2 = 4$, and so

$$\frac{\partial f}{\partial x} = 2x, \qquad \frac{\partial f}{\partial y} = 2y.$$

Therefore, by (26.7),

$$\frac{dy}{dx} = -\frac{2x}{2y} = -\frac{x}{y},$$

(provided that (x, y) is actually a point on the given circle.)

In the last example, we would have obtained exactly the same result for the circle $x^2 + y^2 = 1$, or $x^2 + y^2 = 100$. It is the numerical values of x and y to be put in the right-hand side which will distinguish the circle under discussion from all the other circles. In fact the equation we obtained,

$$\frac{dy}{dx} = -\frac{x}{y},$$

can be thought of as a differential equation. Its solutions (obtained by the method of Section 20.3) are $x^2 + y^2 = C$, which includes the given circle and all the others as well.

Example 26.11. Find dy/dx on the curve $x^3y - xy^3 = 6$ at the point $(2, 1)$.

(You can check that the point $(2, 1)$ is really on the curve.) Putting $f(x, y) = x^3y - xy^3$, we have

$$\frac{\partial f}{\partial x} = 3x^2y - y^3, \qquad \frac{\partial f}{\partial y} = x^3 - 3xy^2.$$

Therefore, at any point (x, y) on the curve,

$$\frac{dy}{dx} = -\frac{3x^2y - y^3}{x^3 - 3xy^2}.$$

At $(2, 1)$, the slope is

$$\left(\frac{dy}{dx}\right)_{(2, 1)} = -\tfrac{11}{2}.$$

(This is *not* a differential equation: it is a *numerical value* which holds at only a *single point*.)

The link with differential equations can be used in many ways, as in the following example.

Example 26.12. Find the family of curves which is orthogonal (perpendicular) to the curves $xy = C$.

The curves $xy = C$ are the contours of the function $f(x, y) = xy$, and the new family will be the curves of steepest ascent/descent on the contour map of xy.

The differential equation of the family $xy = c$ is, from the implicit-differentiation formula (26.6)

$$\frac{dy}{dx} = -\frac{y}{x}.$$

Wherever the new curves intersect with these they must cut at a right angle, so the product of their slopes at any intersection must be equal to -1. Therefore the new family must have the differential equation

$$\frac{dy}{dx} = \frac{x}{y}$$

because $(-y/x)_P(x/y)_P = -1$ at any point P. This equation can be solved by separating the variables (Section 20.3), which gives

$$\int y \, dy = \int x \, dx$$

or

$$y^2 - x^2 = B,$$

where B is an arbitrary constant. This is another family of hyperbolas. A small region of the x, y plane is shown in Fig. 26.6.

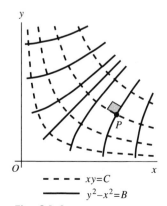

- - - - $xy = C$
—— $y^2 - x^2 = B$

Fig. 26.6

26.5 Normal to a curve

The slope at any point P on the curve $f(x, y) = c$ is equal to

$$-\left(\frac{\partial f}{\partial x}\right)_P \bigg/ \left(\frac{\partial f}{\partial y}\right)_P \quad \text{(by (26.6))}.$$

We shall obtain a vector $\mathbf{n}$ perpendicular or **normal** to the curve at P. A straight line through P perpendicular to the curve must have slope

$$\left(\frac{\partial f}{\partial y}\right)_P \bigg/ \left(\frac{\partial f}{\partial x}\right)_P,$$

because the product of the slopes must be equal to -1. A vector with components (a, b) has slope b/a, so one **normal vector** $\mathbf{n}$ is

$$\mathbf{n} = \left(\left(\frac{\partial f}{\partial x}\right)_P, \left(\frac{\partial f}{\partial y}\right)_P\right).$$

Any multiple of this $\mathbf{n}$ is also a normal at the point. Dropping the suffix P, we have the following result.

> **Normal vector $\mathbf{n}$ at the point (x, y) on the curve**
> $f(x, y) = c$
> $$\mathbf{n} = \left(\frac{\partial f}{\partial x}, \frac{\partial f}{\partial y}\right) = \frac{\partial f}{\partial x}\hat{\imath} + \frac{\partial f}{\partial y}\hat{\jmath}. \qquad (26.7)$$

Example 26.13. Find several normal vectors at the point $(2, 1)$ on the curve $x^2 + y^2 = 5$.

Putting $f(x, y) = x^2 + y^2$, we have $\partial f/\partial x = 2x$ and $\partial f/\partial y = 2y$; so

$$\left(\frac{\partial f}{\partial x}\right)_{(2,1)} = 4, \qquad \left(\frac{\partial f}{\partial y}\right)_{(2,1)} = 2.$$

Therefore one vector normal to the circle at $(2, 1)$ is

$$\mathbf{n} = (4, 2)$$

and from this any number of other normal vectors can be constructed by taking multiples. For example, $(2, 1)$, $(-2, -1)$, and $(\frac{2}{5}, \frac{1}{5})$ are also normals, the last one being a **unit normal** (one having unit length), which is often important.

Example 26.14. Find the angle of intersection between the curves $x^2 + y^2 = 5$ and $x^2 - y^2 = 3$ at the point $(2, 1)$.

In the last example, we showed that $\mathbf{n} = (4, 2)$ is normal to $x^2 + y^2 = 5$ at $(2, 1)$. Similarly the vector $\mathbf{n} = (4, -2)$ is normal to $x^2 - y^2 = 3$ at the point. From Fig. 26.7, it can be seen that the acute angle θ between the normals is equal to one of the angles

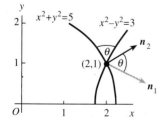

Fig. 26.7

between the curves (the other is $\pi - \theta$). From (9.5),

$$\cos\theta = \frac{\boldsymbol{n}_1\cdot\boldsymbol{n}_2}{|\boldsymbol{n}_1||\boldsymbol{n}_2|} = \frac{(4,2)\cdot(4,-2)}{\sqrt{20}\sqrt{20}} = \tfrac{3}{5},$$

so $\theta = 72.5°$.

26.6 Gradient vector in two dimensions

It is familiar that the value of a quantity such as pressure or temperature depends on, or is a function of, position (x, y). These are **scalar functions**: the values they take up are ordinary numbers. There are also vector quantities that depend on position. Figure (26.8) shows some streamlines for a fluid flowing over a long cylinder (assuming that the flow is always in the plane of the paper). The velocity $\boldsymbol{v}$ is a **vector** which varies from point to point, so we can write $\boldsymbol{v} = \boldsymbol{v}(x, y)$. Gravitational, magnetic, and electric fields are other instances or **vector functions of position** or **vector fields**. Associated with any scalar function, there is an important vector function which arises as follows.

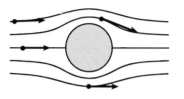

Fig. 26.8

We repeatedly produce formulae involving the pair of elements $\partial f/\partial x$ and $\partial f/\partial y$ in combinations of the form $U\,\partial f/\partial x + V\,\partial f/\partial y$, where U and V are constants or functions; for example as in (26.7), (26.2), and (26.4). We can manipulate this pair as a unit by regarding

$$\frac{\partial f}{\partial x}\,\hat{\boldsymbol{i}} + \frac{\partial f}{\partial y}\,\hat{\boldsymbol{j}} \quad\text{or}\quad \left(\frac{\partial f}{\partial x}, \frac{\partial f}{\partial y}\right)$$

as a vector function. We call this vector function the **gradient of** f and denote it by

grad f or ∇f

(∇ is pronounced 'del'). We shall see that it works rather like an ordinary derivative, but in two dimensions; hence its name.

Alternatively we can regard the symbol **grad** or ∇ standing alone as an **operator** (compare $\mathrm{d}/\mathrm{d}x$): it operates on scalar functions $f(x, y)$, instructing us to carry out the operation $\hat{\boldsymbol{i}}\,\partial/\partial x + \hat{\boldsymbol{j}}\,\partial/\partial y$ or $(\partial/\partial x, \partial/\partial y)$ on $f(x, y)$:

$$\textbf{grad } f(x, y) = \left(\hat{\boldsymbol{i}}\frac{\partial}{\partial x} + \hat{\boldsymbol{j}}\frac{\partial}{\partial y}\right)f(x, y).$$

Gradient in two dimensions

Given a scalar function $f(x, y)$, **grad** f or ∇f stands for

$$\hat{\boldsymbol{i}}\frac{\partial f}{\partial x} + \hat{\boldsymbol{j}}\frac{\partial f}{\partial y} \quad\text{or}\quad \left(\frac{\partial f}{\partial x}, \frac{\partial f}{\partial y}\right). \tag{26.8}$$

Alternatively, **grad** or ∇ stands for the operator

$$\hat{\boldsymbol{i}}\frac{\partial}{\partial x} + \hat{\boldsymbol{j}}\frac{\partial}{\partial y} \quad\text{or}\quad \left(\frac{\partial}{\partial x}, \frac{\partial}{\partial y}\right).$$

Example 26.15. Let $f(x, y) = x^2 + y^2$. Obtain (a) the vector function **grad** f: (b) the value of **grad** f at the point $(1, 2)$; (c) an expression for the magnitude, or length, of **grad** f at (x, y).

(a) **grad** $f = \dfrac{\partial f}{\partial x}\mathbf{i} + \dfrac{\partial f}{\partial y}\mathbf{j} = 2x\mathbf{i} + 2y\mathbf{j}$;

or we can use the alternative notations, and even the operator viewpoint:

$$\mathbf{V}f = \left(\frac{\partial}{\partial x}, \frac{\partial}{\partial y}\right)(x^2 + y^2) = (2x, 2y).$$

(b) At $x = 1$, $y = 2$, we have

grad $f = (2, 4)$.

(c) The magnitude or length of a vector $\mathbf{v} = (a, b)$ is $|\mathbf{v}| = (a^2 + b^2)^{\frac{1}{2}}$, so

$$|\textbf{grad } f| = [(2x)^2 + (2y)^2]^{\frac{1}{2}} = 2(x^2 + y^2)^{\frac{1}{2}}.$$

We can re-express some earlier results in terms of **grad**. For example, we may write (26.7) immediately as follows.

> **A normal vector n at the point (x, y) on the curve $f(x, y) = c$ is**
>
> $n = $ **grad** $f(x, y)$. $\hspace{2cm}$ (26.9)

As we remarked earlier, expressions occurring in physical theory frequently take the form

$$U\frac{\partial f}{\partial x} + V\frac{\partial f}{\partial y},$$

where U and V may be constants, or various functions. Then we can write such expressions as a scalar ('dot') product by inventing a new vector function $\mathbf{S} = U\mathbf{i} + V\mathbf{j}$:

> If $\mathbf{S} = U\mathbf{i} + V\mathbf{j}$, then
>
> $$U\frac{\partial f}{\partial x} + V\frac{\partial f}{\partial y} = (U, V)\cdot\left(\frac{\partial f}{\partial x}, \frac{\partial f}{\partial y}\right) = \mathbf{S}\cdot\textbf{grad } f. \hspace{0.8cm} \text{(26.10)}$$

Now consider the directional-derivative formula (26.4), regarding it as representing the rate of change of $f(x, y)$ in the direction θ:

$$\frac{\mathrm{d}f}{\mathrm{d}s} = \frac{\partial f}{\partial x}\cos\theta + \frac{\partial f}{\partial y}\sin\theta.$$

To recast this in the form of (26.10), we require the vector

$$\mathbf{i}\cos\theta + \mathbf{j}\sin\theta.$$

This is a *unit vector* (i.e. it has length unity) because

$$(\cos^2 \theta + \sin^2 \theta)^{\frac{1}{2}} = 1;$$

so put

$$\boldsymbol{i} \cos \theta + \boldsymbol{j} \sin \theta = \hat{\boldsymbol{s}},$$

where $\hat{\boldsymbol{s}}$ is a unit vector pointing in the desired direction, and (26.10) becomes

---

Directional derivative

In the direction of a unit vector $\hat{\boldsymbol{s}}$, the rate of change of $f(x, y)$ is given by

$$\frac{\mathrm{d}f}{\mathrm{d}s} = \hat{\boldsymbol{s}} \cdot \mathbf{grad}\ f; \tag{26.11}$$

that is to say, $\mathrm{d}f/\mathrm{d}s$ is equal to the component of $\mathbf{grad}\ f$ in the direction of $\hat{\boldsymbol{s}}$.

Equation (26.11) can be written in a different way. If $\boldsymbol{a}$ and $\boldsymbol{b}$ are two vectors, then the angle between them, ϕ, can be obtained from the identity

$$\boldsymbol{a} \cdot \boldsymbol{b} = |\boldsymbol{a}|\,|\boldsymbol{b}| \cos \phi$$

(see (9.5)). If we put $\boldsymbol{a} = \hat{\boldsymbol{s}}$ and $\boldsymbol{b} = \mathbf{grad}\ f$, and use the fact that $|\hat{\boldsymbol{s}}| = 1$, we obtain an alternative form of (26.11).

Directional derivative (alternative form)

$$\frac{\mathrm{d}f}{\mathrm{d}s} = |\mathbf{grad}\ f| \cos \phi, \tag{26.12}$$

where ϕ is the angle between $\mathbf{grad}\ f$ and the required direction.

By using (26.12), the perpendicularity of the directions of steepest ascent and the contours of $f(x, y)$, proved in Section 26.3, can be recovered.

Problems, Chapter 26

26.1. Use the incremental approximation (26.1) or (26.2) to estimate the change δz due to changes δx and δy as specified, and check the percentage error by calculating the exact result.
(a) $z = x^2 + y^2$ at $(3, 1)$, $\delta x = 0.1$, $\delta y = 0.3$;
(b) $z = \sin xy$ at $(0.5, 1.2)$, $\delta x = 0.1$, $\delta y = -0.05$;
(c) $z = \mathrm{e}^{x^2 + 3y^2}$ at $(1, 1)$, $\delta x = 0.1$, $\delta y = 0.2$;
(d) $z = 1/(x^2 + y^2)^{\frac{1}{2}}$ at $(2, 1)$, $\delta x = -0.2$, $\delta y = 0.1$.

26.2. Given $z = x^2 - y^2$ and two points $P : (1.0, 2.1)$ and $Q : (1.1, 2.0)$, (a) estimate the change in z in going from P to Q; (b) estimate the change in going from Q to P; (c) explain in general terms why the second estimate is not precisely the negative of the first.

26.3. (See Example 26.2). Obtain the *exact algebraical form* of the error incurred in δf where

$$\delta f = f(x + \delta x, y + \delta y) - f(x, y),$$

by using the approximation $\delta f = \dfrac{\partial f}{\partial x} \delta x + \dfrac{\partial f}{\partial y} \delta y$

(a) for $f(x, y) = xy$ near the point $(2, 1)$;
(b) for $f(x, y) = x/y$ near the point $(2, 1)$.

26.4. The relation between the object distance u, the image distance v, and the focal length f of a thin lens is

$$\frac{1}{u} + \frac{1}{v} = \frac{1}{f}.$$

Suppose that the measured values of u and v are $u = 0.31(\pm 0.01)$, $v = 0.56(\pm 0.03)$; calculate the greatest possible error in estimating f, and the corresponding percentage error.

26.5. A viscous liquid is forced through a tube of diameter $d = 10(\pm 0.05 \times 10^{-3})$ and length $l = 0.1$ under a pressure $p = 10(\pm 5 \times 10^4)$, and is found to pass fluid at a rate $v = 0.625 \times 10^{-9}$ per unit time. The viscosity η is given by the formula

$$\eta = \frac{\pi}{128} \frac{pd^4}{vl}.$$

Find the maximum error in the viscosity estimate.

26.6. One root of the equation $x^2 + bx + c = 0$ is $x = \frac{1}{2}[-b + (b^2 - 4c)^{\frac{1}{2}}]$. Suppose that $b = 20.4$ and $c = 95.5$. Estimate the percentage error in the root which would arise if these were rounded to $b = 20$, $c = 96$.

26.7. The area S of a triangle with base b and base angles A and C is given by

$$S = \frac{\frac{1}{2}b^2 \tan A \tan C}{\tan A + \tan C}.$$

Suppose that nominally $b = 2$, $A = 30°$, $C = 60°$, but that C is found to be too large by 5%. By what amount should A be changed so that S would be restored to the correct area?

26.8. A certain type of experiment to measure surface tension S requires the formula $S = ahr^3/p^2$, where a is a constant and h, r, and p are measured quantities. Take the logarithm of the formula to find the fractional change in $\delta S/S$ in S, in terms of simultaneous fractional changes in h, r, and p.

26.9. Find the directional derivative df/ds of each of the following functions according to the data. Also,

for the given point, find the directions of the contour and the direction of steepest ascent.
(a) $f(x, y) = x^2 + y^2$ at $(1, 2)$, direction $\theta = 30°$;
(b) $f(x, y) = x^2y^2$ at $(2, 1)$, direction $\theta = -45°$;
(c) $f(x, y) = x^2y - xy^2 + 2$ at $(-1, 1)$, direction $\theta = 120°$;
(d) $f(x, y) = \sin xy$ at $(\frac{1}{2}, \pi)$, direction $\theta = -90°$;
(e) $f(x, y) = \cos(x^2 - y)$ at $(0, -\pi)$, direction $\theta = 0$;
(f) $f(x, y) = e^{x-y}$ at $(1, 1)$, direction $\theta = -45°$.

26.10. Find $\dfrac{dy}{dx}$ at the prescribed points on the curves given.
(a) $xy = 1$ at $(2, \frac{1}{2})$;   (b) $x^2 + y^2 = 25$ at $(3, 4)$;
(c) $1/x - 1/y = \frac{1}{2}$ at $(1, 2)$;
(d) $\frac{1}{10}x^2 + \frac{1}{15}y^2 = 1$ at $(2, 3)$;
(e) $x^3 + 2y^3 = 3$ at $(1, 1)$;
(f) $x^3y + 3x^2 - y^2 - 19 = 0$ at $(2, 1)$;
(g) $xy^2 - x^2y + 6 = 0$ at $(3, 2)$;
(h) $x^2 + y^2 = 4$ at $(2 \cos \theta, 2 \sin \theta)$;
(i) $x^2/a^2 + y^2/b^2 = 1$ at $(a \cos t, b \sin t)$;
(j) $x \cos y = y \sin x$ at (x, y);
(k) $y^2 - 4ax = 0$ at $(at^2, 2at)$.

26.11. The ideal-gas equation is $PV = RT$, where R is a constant. There are three variables: P is pressure, V is volume, and T is absolute temperature, for a fixed mass of gas. Show that

$$\left(\frac{\partial V}{\partial P}\right)_T = -\left(\frac{\partial T}{\partial P}\right)_V \bigg/ \left(\frac{\partial T}{\partial V}\right)_P.$$

(The notation $(\partial u/\partial v)_w$ means that the variable w is kept constant during differentiation when $u = g(v, w)$. Use (26.6).)

26.12. Find the cartesian equation of the tangent line at a point (x, y) on each of the following curves. (Find dy/dx first.)
(a) $x^2 + y^2 = a^2$;   (b) $x^2/a^2 + y^2/b^2 = 1$;
(c) $a^2x^2 - b^2y^2 = c$;
(d) $xy = 1$;   (e) $x^{\frac{2}{3}} + y^{\frac{2}{3}} = 1$;
(f) $ax^2 + 2hxy + by^2 + 2gx + 2fy + c = 0$.

26.13. Suppose that the curves $f(x, y) = \alpha$ and $g(x, y) = \beta$ intersect at right angles at a point (a, b). Find dy/dx at the point for each curve and deduce that, at (a, b),

$$\frac{\partial f}{\partial x} \frac{\partial g}{\partial x} + \frac{\partial f}{\partial y} \frac{\partial g}{\partial y} = 0.$$

Use this result to confirm that, in the following cases, the two systems of curves are orthogonal (i.e. they always intersect at right angles). Here α and β are the parameters for the two systems – by varying

them we obtain all the curves for the systems.
(a) $x^2 + y^2 = \alpha$, $y/x = \beta$;
(b) $x^2 - y^2 = \alpha$, $xy = \beta$;
(c) $y^3 - x^3 = \alpha$, $1/y + 1/x = \beta$;
(d) $(x^2 + y^2)/x = \alpha$, $(x^2 + y^2)/y = \beta$.

26.14. Let (x, y) be any point on the curve $y^3 - x^3 = 1$. Find an expression for dy/dx at the point. Since this expression holds good for every point on the curve, it is a differential equation, having the given curve as one of its solution curves. Verify this by solving it, and obtain the other solutions.

26.15. (**Numerical**). Form the differential equation for the following families of curves, in which c is the parameter; then use the numerical solution method of Section 20.2 to obtain a contour map of the functions concerned.
(a) $x^2 + 2y^2 = c$, $c > 0$;   (b) $x^2 + xy - y^3 = c$;
(c) $\dfrac{x^2 + y}{x + y^2} = c$;   (d) $xy\,e^{-x} = c$.

26.16. Form the differential equation for each system of curves, and deduce the differential equation for the orthogonal (perpendicular) system. Solve it to obtain the orthogonal system.
(a) $y^2 - x^2 = c$;   (b) $y^3 + x^3 = c$;
(c) $y^2 = cx$;   (d) $e^y - e^x = c$.

26.17. Find the curves of steepest ascent from an arbitrary point (a, b) for each of the following functions.
(a) $\frac{1}{2}x^2 + y^2$;   (b) x^3y^3;   (c) $\frac{1}{2}y^2 - y - x^2$.

26.18. Implicit differentiation of y with respect to x can be carried out as follows when $f(x, y)$ is given explicitly. Consider $f(x, y) = x^2 + 2xy + y^2 = c$. Then, by differentiating this equation and treating y as a function of x, we obtain

$$2x + 2x\frac{dy}{dx} + 2y + 2y\frac{dy}{dx} = 0,$$

from which dy/dx can be found. Check that (26.6) gives the same result.

26.19. Find normal vectors to the curves below, and find the angle between them at the intersection given.

(a) $xy = 2$, $x^2 - y^2 = -3$, at intersection $(1, 2)$.
(b) $y = x^3$, $x^2 + \frac{1}{4}y^2 = 36$, at intersection $(2, 8)$.
(c) $x^2 + xy + y^2 = 3$, $x + y = 2$, at intersection $(1, 1)$; interpret your result geometrically.
(d) $ax^2 + 2hxy + by^2 + c = 0$ and

$$ax_0x + h(x_0 + x)(y_0 + y) + by_0y + c = 0,$$

at any point (x_0, y_0) which lies on the first curve.

26.20. Find d^2y/dx^2 on the following curves.
(a) $x^4 - y^4 = 1$;   (b) $xy = 1$;   (c) $xy\,e^{xy} = 1$.

26.21. Obtain **grad** f, where $f(x, y)$ is given by the following. Give its components, its direction, and its magnitude at the points specified.
(a) $1/(x + y)$ at $(1, -2)$;   (b) y/x at $(2, 0)$;
(c) $y^2 - 3x^2 + 1$ at $(0, 0)$;   (d) $1/x - 1/y$ at $(2, 1)$;
(e) $1/r$, where r is the polar coordinate, $r = (x^2 + y^2)^{\frac{1}{2}}$; confirm that the gradient vector points in a radial direction.

26.22. Use the gradient vector to obtain a *unit* vector perpendicular to the following curves at the points given
(a) $2x - 3y + 1 = 0$ at any point;
(b) $x^2 + y^2 = 5$ at $(2, 1)$;
(c) $x^2 + y^2 = r^2$ at (x_0, y_0) on the circle;
(d) $x^2/a^2 + y^2/b^2 = 1$ at (x_0, y_0) on the ellipse;
(e) $y = 3x^2 - 2$ at $(2, 10)$.

26.23. Use the property (26.9) to find the angle of intersection of the following curves at the point of intersection given.
(a) $y^2 - x^2 = -3$ and $x^3 - y^3 = 7$ at $(2, 1)$;
(b) $x^2y - xy^2 = 0$ and $x/y - y/x = 0$ at $(2, 2)$;
(c) $x^2 + y^2 + 2x - 4y + 4 = 0$ and $y = x^2 + 2x + 2$ at $(-1, 1)$; explain the result geometrically.

26.24. Use (26.12) to prove the results given in Section 26.3 for a general $f(x, y)$: that (a) the directions of most rapid increase and decrease through a point (x, y) are perpendicular to the direction of the contour through the point; (b) the maximum rate of increase from the point is equal to $|$**grad** $f|$ at the point.

27 Chain rules, restricted maxima, coordinate systems

27.1 Chain rule for a single parameter

Suppose that x and y depend on, or are functions of, another variable t (say) which we call the **parameter**. It might represent time, for example. We shall write

$$x = x(t), \qquad y = y(t).$$

As t varies, the point (x, y) follows a curve of some sort which is said to be **defined parametrically**. The curve also has a characteristic **direction**, which is the direction the curve is described as t **is increasing**, and is indicated by an arrow. We then have a **directed path**.

Example 27.1. Show that both of the following parametrizations define a unit semicircle, centred at the origin, in the upper half-plane, traced anticlockwise: (a) $x = \cos t$, $y = \sin t$, where t increases from 0 to π; (b) $x = -u$, $y = (1 - u^2)^{\frac{1}{2}}$, where u increases from -1 to 1.

(a) The shape of the curve is obtainable by eliminating t:

$$x^2 + y^2 = \cos^2 t + \sin^2 t = 1;$$

so the points lie on the unit circle. Also, as t increases from 0 to π, y is positive and x decreases from 1 to -1. The path is the upper semicircle from $(1, 0)$ to $(-1, 0)$, described in a single direction, as shown in Fig. 27.1a.

(b) $x^2 + y^2 = (-u)^2 + (1 - u^2) = 1$. As u increases from -1 to 1, y remains positive while x decreases from 1 to -1. The path is as in (a): see Fig. 27.1b.

Given a function $f(x, y)$ which can take values all over the x, y plane, the function

$$g(t) = f(x(t), y(t))$$

picks out only the values on the path $(x(t), y(t))$. As we move along this path, the function value varies, and we might be concerned with the rate at which it changes with t. (This is generally different from

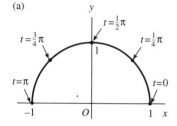

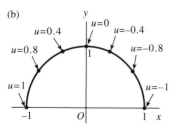

Fig. 27.1

the rate at which $f(x, y)$ changes with *distance along the path*, which is equal to the directional derivative (26.4), and corresponds to using arclength s as the parameter.)

To find df/dt, suppose that t increases from t to $t + \delta t$. Then, on the curve $(x(t), y(t))$, x changes from x to $x + \delta x$ and y to $y + \delta y$. Divide (26.2), the incremental approximation, by δt:

$$\frac{\delta f}{\delta t} \approx \frac{\partial f}{\partial x}\frac{\delta x}{\delta t} + \frac{\partial f}{\partial y}\frac{\delta y}{\delta t}.$$

Let $\delta t \to 0$. Then '$\approx$' becomes '$=$', $\dfrac{\delta x}{\delta t} \to \dfrac{dy}{dt}$, and $\dfrac{\delta y}{\delta t} \to \dfrac{dy}{dt}$, and we have the **chain rule** (or **total derivative**):

Chain rule for one parameter

Given $f(x, y)$, $x = x(t)$ and $y = y(t)$,

$$\frac{df}{dt} = \frac{\partial f}{\partial x}\frac{dx}{dt} + \frac{\partial f}{\partial y}\frac{dy}{dt} \qquad (27.1)$$

(and similarly with z in place of f, if we write $z = f(x, y)$).

This expression is like the chain rule (3.3) for functions of a single variable with an extra term in it for the variable y. Partial derivative rather than ordinary derivative signs are then written as necessary.

Example 27.2. Let $f(x, y) = xy - y^2$, $x = t^2$, $y = t^3$. (a) Find df/dt using the chain rule; (b) find df/dt by substitution.

(a) $\dfrac{\partial f}{\partial x} = y$, $\quad \dfrac{\partial f}{\partial y} = x - 2y$, $\quad \dfrac{dx}{dt} = 2t$, $\quad \dfrac{dy}{dt} = 3t$.

Therefore, by (23.1),

$$\frac{df}{dt} = \frac{\partial f}{\partial x}\frac{dx}{dt} + \frac{\partial f}{\partial y}\frac{dy}{dt} = y(2t) + (x - 2y)3t^2$$

$$= 2t^4 + (t^2 - 2t^3)3t^2 = 5t^4 - 6t^5.$$

(This expression can be written in various ways in terms of x and y, for example as $5x^2 - 6xy$, or $5yx^{\frac{1}{2}} - 6x^2y^{\frac{1}{3}}$. These all look very different, but they all take the same values since x and y are *connected* by the fact that (x, y) lies on the given curve.)

(b) By substitution,

$$f(x(t), y(t)) = xy - y^2 = t^2t^3 - (t^3)^2 = t^5 - t^6.$$

Therefore, as before

$$\frac{df}{dt} = 5t^4 - 6t^5.$$

Example 27.3. Prove the implicit-differentiation formula (26.6) by using the chain rule with x treated as the parameter.

If $f(x, y) = c$, then there is a solution $y = y(x)$ for which

$$f(x, y(x)) = c$$

is *automatically* true for every value of x involved (that is, it is an **identity**). Therefore

$$\frac{df(x, y(x))}{dx} = 0.$$

Comparing this with (27.1), the chain rule, we have x in place of t for the parameter. In terms of the chain rule, we therefore have

$$0 = \frac{\partial f}{\partial x}\frac{dx}{dx} + \frac{\partial f}{\partial y}\frac{dy}{dx} = \frac{\partial f}{\partial x} + \frac{\partial f}{\partial y}\frac{dy}{dx}.$$

From this we recover the implicit derivative formula

$$\frac{dy}{dx} = -\frac{\partial f}{\partial x}\bigg/\frac{\partial f}{\partial y}.$$

The chain rule is more useful for obtaining general results, as in Example 27.3, than in working out special instances such as Example 27.2.

27.2 Restricted maxima and minima: the Lagrange multiplier

Consider the simple function

$$f(x, y) = x + y.$$

This has no maxima, minima, or other stationary points, since $z = x + y$ represents an inclined plane. However, if we travel around the plane *on a particular path*, we are likely to encounter high points and low points, and points where we are momentarily travelling on the level. Suppose that we walk on the circular path $x^2 + y^2 = 1$, shown on a map in Fig. 27.2.

Then A is the highest point; this is where we were walking uphill but then turn downhill: this is a **local maximum point on the path**. If we plotted a graph of elevation against time, this point would show up as local maximum on the graph.

The clue which reveals A to be a maximum is that *one of the contours of $x + y$ is a tangent to the path at A*. Those nearby contours that the path crosses are all lower than the one through A. Similarly, at B, there is a **local minimum** for the path.

This is an example of a **restricted stationary-point problem**, the 'restriction' being the condition that the only points considered are those that lie on a particular curve. A general statement of the problem is as follows.

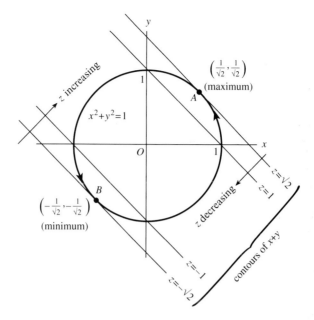

Fig. 27.2

Restricted stationary-point problem

Find the stationary points of $f(x, y)$ subject to the condition $g(x, y) = c$. **(27.2)**

Very simple problems of this type can be solved by an elementary method, as in the following example.

Example 27.4. Find the maximum possible area a rectangle may have if the perimeter is restricted to length 10 units.

Call the sides x and y. Then we require the maximum of

$$A = f(x, y) = xy \tag{i}$$

subject to the restriction

$$P = g(x, y) = 2x + 2y = 10. \tag{ii}$$

From the perimeter equation, we have $y = 5 - x$; so the area can be expressed in terms of x only:

$$A = x(5 - x).$$

This has a turning point where $\mathrm{d}A/\mathrm{d}x = 0$, or

$$5 - 2x = 0,$$

that is, at $x = \frac{5}{2}$. The perimeter equation (ii) gives correspondingly $y = \frac{5}{2}$, so the desired shape is a square of area $\frac{25}{4}$.

However, although the following problem looks very similar, there turns out to be a difficulty.

Example 27.5. Find the maxima/minima of $z = f(x, y) = x^2 - y^2$ on the circle $g(x, y) = x^2 + y^2 = 1$.

On the given curve,

$$y^2 = 1 - x^2. \tag{i}$$

The values taken by $z = x^2 - y^2$ on this curve are given in terms of x by

$$z = x^2 - (1 - x^2) = 2x^2 - 1.$$

The stationary points of this function are where

$$\frac{\mathrm{d}}{\mathrm{d}x}(2x^2 - 1) = 0;$$

which is at $x = 0$. At $x = 0$, the curve equation (i) gives $y = \pm 1$, so we have found the points $A : (0, 1)$ and $A' : (0, -1)$. These are in fact minima, and they are shown on the path in Fig. 27.3a.

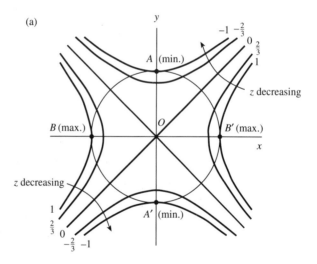

Fig. 27.3
(a) Contour map of $x^2 - y^2$, showing also the curve $x^2 + y^2 = 1$. Here A and A' are minima, and B and B' are maxima.

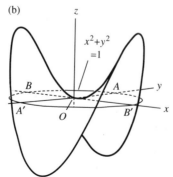

(b) The path $x^2 + y^2 = 1$ in the (x, y) plane, with the corresponding values of $z = x^2 - y^2$ shown.

However, there are plainly two *maxima* also, at B and B', which are completely missed by the process above. We could have found them (but lost A and A') if we had substituted for x instead of y by means of $x^2 = 1 - y^2$. You can see the reason for losing A and A' if you sketch the function $2x^2 - 1$ between $x = \pm 1$. The maxima are at the ends, but cannot be found by differentiating; see also Example 4.8.

We can get over this difficulty by parametrizing the curve $g(x, y) = c$ as in the following Example, which repeats Example 27.5.

Example 27.6. Find the stationary points of $x^2 - y^2$ on the curve $x^2 + y^2 = 1$.

Put

$$x = \cos t, \qquad y = \sin t, \quad 0 \leqslant t < 2\pi,$$

then the circle $C : x^2 + y^2 = 1$ is traced once, anticlockwise, starting and ending at $(1, 0)$. On C,

$$f(x, y) = \cos^2 t - \sin^2 t.$$

As we go along the path C, stationary points are encountered where

$$0 = \frac{\mathrm{d}f\big(x(t), y(t)\big)}{\mathrm{d}t} = 2 \cos t(-\sin t) - 2 \sin t \cos t$$

$$= -4 \sin t \cos t = -2 \sin 2t.$$

The solutions of this equation in the range $0 \leqslant t < 2\pi$ are $t = 0, \frac{1}{2}\pi, \pi, -\frac{1}{2}\pi$, which correspond to the points $(1, 0)$, $(0, 1)$, $(-1, 0)$, $(0, -1)$. Therefore this approach successfully found all the stationary points on the path, which the method of Example 27.5 failed to do.

We shall now describe the **Lagrange-multiplier method** for solving the restricted stationary-value problem (27.2). This uses the parametric idea, but all reference to a parameter is eliminated eventually so that we do not have to invent a parametrization and then go through the resulting algebra.

By thinking of time as a possible parameter t, and $P : (x(t), y(t))$ as a point moving along the curve with velocity $(\mathrm{d}x/\mathrm{d}t, \mathrm{d}y/\mathrm{d}t)$, it can be seen that $g(x, y) = c$ can be expressed parametrically so that (a) that path is traced exactly once as t moves through its range, and (b) that $\mathrm{d}x/\mathrm{d}t$ and $\mathrm{d}y/\mathrm{d}t$ are never both zero together (if t is time, this means that the moving point P never pauses).

Then as $P : (x(t), y(t))$ moves along $g(x, y) = c$, the points Q where $\mathrm{d}f/\mathrm{d}t = 0$ are the stationary points of $f(x(t), y(t))$. Therefore, by the chain rule (27.1),

$$\frac{\partial f}{\partial x}\frac{\mathrm{d}x}{\mathrm{d}t} + \frac{\partial f}{\partial y}\frac{\mathrm{d}y}{\mathrm{d}t} = 0.$$

To get rid of dx/dt and dy/dt, which are special to the particular parametrization chosen, we need another equation. On the curve, $g(x, y)$ has a constant value c, so $dg/dt = 0$ at every point including Q; so, by the chain rule,

$$\frac{\partial g}{\partial x}\frac{dx}{dt} + \frac{\partial g}{\partial y}\frac{dy}{dt} = 0.$$

These last two equations can be regarded as a pair of homogeneous algebraic equations of dx/dt and dy/dt. From Section 10.4, the equations have a nontrivial solution if and only if the determinant of the coefficients is zero, so at the (unknown) point Q

$$\frac{\partial f}{\partial x}\frac{\partial g}{\partial y} - \frac{\partial f}{\partial y}\frac{\partial g}{\partial x} = 0.$$

This can be written alternatively in the form

$$\frac{\partial f}{\partial x}\bigg/\frac{\partial g}{\partial x} = \lambda, \qquad \frac{\partial f}{\partial y}\bigg/\frac{\partial g}{\partial y} = \lambda, \qquad \text{(27.3a, b)}$$

where λ ('lambda') is a new unknown constant, called the **Lagrange multiplier** for the problem. We have lost some information here, because the condition $dg/dt = 0$ does not distinguish between one value of c and another, so we have to reassert the condition

$$g(x, y) = c. \qquad \text{(27.3c)}$$

Looking back, we have three unknowns: x and y (the coordinates of any stationary point) and λ, another constant. To determine these, there are three equations: (27.3a, b) and (27.3c). Finally, we summarize the method.

Lagrange-multiplier method for the restricted stationary-value problem

To find the statonary points of $f(x, y)$ subject to $g(x, y) = c$, solve the following equations for x, y, λ:

$$g(x, y) = c, \qquad \text{(i)}$$

$$\frac{\partial f}{\partial x} - \lambda\frac{\partial g}{\partial x} = 0, \qquad \text{(ii)} \qquad \text{(27.4)}$$

$$\frac{\partial f}{\partial y} - \lambda\frac{\partial g}{\partial y} = 0. \qquad \text{(iii)}$$

(The value of λ can usually be discarded.)

Notice that all reference to the parameter t has disappeared. There are many way of proving (27.4), but this is probably the simplest for two dimensions. The problem is treated for three dimensions in Section 28.8.

Example 27.7. Find the stationary points of $x^2 - y^2$ on the curve $x^2 + y^2 = 1$ (compare Examples 27.5 and 27.6).

In (27.4), $f(x, y) = x^2 - y^2$ and $g(x, y) = x^2 + y^2 = 1$. The equations to be solved, in the order of (27.4), become

$$x^2 + y^2 = 1, \tag{i}$$

$$2x - \lambda(2x) = 0 \quad \text{or} \quad (1 - \lambda)x = 0, \tag{ii}$$

$$-2y - \lambda(2y) = 0 \quad \text{or} \quad (1 + \lambda)y = 0. \tag{iii}$$

From (ii), either $\lambda = 1$ or $x = 0$. Taking these possibilities in order:

If $\lambda = 1$, then (iii) gives $y = 0$; consequently (i) gives $x = \pm 1$.

Therefore we have found the points $(1, 0)$ and $(-1, 0)$ which we called B' and B in Example 27.5.

If $x = 0$, then (i) gives $y = \pm 1$. We have therefore found the points $(0, 1)$ and $(0, -1)$ which we called A and A' in Example 27.5.

The equations obtained are often awkward to solve. It is best to be very systematic, not wandering aimlessly between the equations. Be careful not to overlook possibilities (such as that (ii) in Example 27.7 is solved by $\lambda = 1$); and check at the end that the solutions actually fit. The values found for λ do have a special significance in certain subjects but otherwise can be thrown away.

Example 27.8. Find the rectangle of maximum area which can be placed symmetrically in the ellipse $x^2 + 4y^2 = 1$ as shown in Fig. 27.4.

Suppose that one of the vertices, say A, is at (x, y). We shall require that x and y be positive, since this sufficient for the geometrical condition. The area is equal to $4xy = f(x, y)$, while x and y are subject to $g(x, y) = x^2 + 4y^2 = 1$.

The three equations, taken in the order of (27.4), become

$$x^2 + 4y^2 = 1, \tag{i}$$

$$2y - \lambda x = 0, \tag{ii}$$

$$x - 2\lambda y = 0. \tag{iii}$$

Suppose that neither x nor y is zero (that could not give a maximum). Then, from (ii) and (iii),

$$\lambda = 2y/x = x/2y, \tag{iv}$$

so $x = \pm 2y$. However, these must have the same sign for positive area, and we postulated that x and y should be positive. Therefore

$$x = 2y > 0; \tag{v}$$

so, from (iv) again,

$$\lambda = 1. \tag{vi}$$

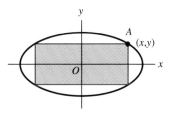

Fig. 27.4

Use (v) to substitute for x in (i): we get $8y^2 = 1$, or (rejecting negative values of y)

$$y = 1/2\sqrt{2},$$

and (v) gives correspondingly

$$x = 1/\sqrt{2}.$$

The sides have length $1/\sqrt{2}$ and $\sqrt{2}$, so the area is 1.

27.3 Curvilinear coordinates in two dimensions
Suppose that x and y are functions of **two parameters**, u and v. To indicate this, write

$$x = x(u, v), \qquad y = y(u, v).$$

This situation arises when we change coordinates from (x, y) to another system. For example, the equations

$$x = u \cos v, \qquad y = u \sin v,$$

represent polar coordinates, with u as the radial and v as the angular coordinate. Now hold v constant; put

$$v = \beta,$$

say, and let u vary. Then

$$x = u \cos \beta, \qquad y = u \sin \beta.$$

Here u is the only active parameter; as it varies, (x, y) traces a radial straight line. Suppose instead that u is held constant, say

$$u = \alpha;$$

then, as v varies, (x, y) follows the circle

$$x = \alpha \cos v, \qquad y = \alpha \sin v.$$

The point where the two curves intersect can be described either by

$$u = \alpha, \qquad v = \beta$$

in the new (polar) coordinates, or in the original coordinates by

$$x = \alpha \cos \beta, \qquad y = \alpha \sin \beta.$$

In general, if we have

$$x = x(u, v), \qquad y = y(u, v),$$

and vary u and v together in an arbitrary way, then the corresponding points (x, y) will completely cover some **area** in the x, y plane. If, however, we put $u = \alpha$ and vary v, then put $v = \beta$ and vary u, we obtain two curves in parametric form:

$$(x(\alpha, v), y(\alpha, v)) \quad \text{and} \quad (x(u, \beta), y(u, \beta)).$$

By choosing different values for α and β, we produce a net consisting of two independent systems of curves. This can serve as a new **coordinate system**.

(a)

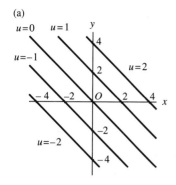

(b)

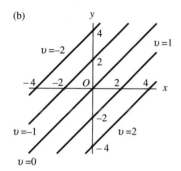

(c)

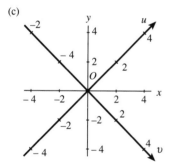

Fig. 27.5

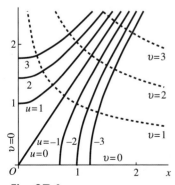

Fig. 27.6

Example 27.9. Sketch the coordinate system defined by

$$x = u + v, \qquad y = u - v.$$

Put $u = \alpha$ (constant) and vary v in the equations

$$x = \alpha + v, \qquad y = \alpha - v.$$

Eliminating the active parameter v between the two equations:

$$y = -x + 2\alpha,$$

which is a straight line. By taking different values of α, we obtain a system of parallel straight lines as in Fig. 23.5a.
Put $u = \beta$ and vary v:

$$x = u + \beta, \qquad y = u - \beta.$$

Therefore

$$y = x - 2\beta,$$

which gives another family of parallel straight lines, obtained by taking various values for the constant β, as in Fig. 27.5b.

The two families happen to be at right angles. Taken together, as in Fig. 27.5c, they form a left-handed system of cartesian coordinates (u, v) with origin at $x = 0$, $y = 0$.

New coordinates (u, v) can also be defined in the form

$$u = u(x, y), \qquad v = v(x, y).$$

For example, the system

$$u = (x^2 + y^2)^{\frac{1}{2}}, \qquad v = \arctan y/x$$

defines polar coordinates, u (radial) and v (angular). To sketch the curves corresponding to constant u or v, we put

$$\alpha = u(x, y) \quad \text{or} \quad \beta = v(x, y);$$

each of these gives the corresponding curve implicitly.

Example 27.10. Sketch the coordinate system (u, v) described by

$$u = y^2 - 2x^2, \qquad v = x^{\frac{1}{2}}y.$$

The curve $u = \alpha$ is obtained in terms of x and y by solving

$$\alpha = y^2 - 2x^2.$$

These curves (for various values of α) are in fact recognizable without solving, being a system of hyperbolas with asymptotes

$$y = \pm\sqrt{2}x.$$

The curves $v = \beta$ are given in $x^{\frac{1}{2}}y = \beta$, so $y = \beta/x^{\frac{1}{2}}$. The system is sketched in Fig. 27.6 for the first quadrant. Notice that $v = 0$ on both $x = 0$ and $y = 0$: the connection between (x, y) and (u, v) is not one-to-one over the whole (x, y) plane.

27.4 Orthogonal coordinates

Suppose that we have a (u, v) system of coordinates defined either by $x = x(u, v)$, $y = (u, v)$, or by $u = u(x, y)$, $v = v(x, y)$, and the curves $u = \alpha$ and $v = \beta$ always intersect at right angles for any constants α and β. Then the (u, v) system is said to be an **orthogonal system of coordinates**. For example polar coordinates are orthogonal. Coordinate systems which are not orthogonal are seldom used because of the complexity of the formulae connected with them. A test for orthogonality is the following.

Conditions for an orthogonal system of coordinates

The (u, v) system is orthogonal if

either (a) $u = u(x, y)$, $v = v(x, y)$, and

$$\frac{\partial u}{\partial x}\frac{\partial v}{\partial x} + \frac{\partial u}{\partial y}\frac{\partial v}{\partial y} = 0; \tag{27.5}$$

or   (b) $x = x(u, v)$, $y = y(u, v)$, and

$$\frac{\partial x}{\partial u}\frac{\partial x}{\partial v} + \frac{\partial y}{\partial u}\frac{\partial y}{\partial v} = 0.$$

We prove this result as follows.

(a) Consider the curve from each family which passes through (x, y). According to (26.7), normal vectors to the two curves at (x, y) are $\mathbf{n}_1 = (\partial u/\partial x, \partial u/\partial y)$ and $\mathbf{n}_2 = (\partial v/\partial x, \partial v/\partial y)$ respectively. The curves meet in a right angle if their normals do so, and the condition for this is $\mathbf{n}_1 \cdot \mathbf{n}_2 = 0$, which is equivalent to the condition given in (27.5a).

(b) Consider the curves $u = \alpha$ and $v = \beta$ which pass through a point P which has new coordinates (α, β). Their parametric equations are

$$x = x(\alpha, v), \quad y = y(\alpha, v), \quad \text{for the curve } u = \alpha,$$

$$x = x(u, \beta), \quad y = y(u, \beta), \quad \text{for the curve } v = \beta.$$

Their slopes at P are respectively given by

$$\left(\frac{\mathrm{d}y}{\mathrm{d}x}\right)_P = \left(\frac{\partial y}{\partial v}\bigg/\frac{\partial x}{\partial v}\right)_P \quad \text{and} \quad \left(\frac{\mathrm{d}y}{\mathrm{d}x}\right)_P = \left(\frac{\partial y}{\partial u}\bigg/\frac{\partial x}{\partial u}\right)_P.$$

The condition for the curves to be perpendicular is that the product should equal -1, and this is equivalent to the result in (27.5b).

Example 27.11. Confirm that the following coordinate systems (u, v) are orthogonal. (a) $u = y^2 - 2x^2$, $v = x^{\frac{1}{2}}y$; (b) $x = 2uv$, $y = u^2 - v^2$.

For (a), use (27.5a). We have

$$\frac{\partial u}{\partial x} = -4x, \quad \frac{\partial v}{\partial x} = \tfrac{1}{2}x^{-\frac{1}{2}}y, \quad \frac{\partial u}{\partial y} = 2y, \quad \frac{\partial v}{\partial y} = x^{\frac{1}{2}};$$

so

$$\frac{\partial u}{\partial x}\frac{\partial v}{\partial x} + \frac{\partial u}{\partial y}\frac{\partial v}{\partial y} = -4x(\tfrac{1}{2}x^{-\frac{1}{2}}y) + 2y(x^{\frac{1}{2}}) = 0.$$

For (b), use (27.5b); notice how this condition is differently structured from (27.5a). We have

$$\frac{\partial x}{\partial u} = 2v, \quad \frac{\partial x}{\partial v} = 2u, \quad \frac{\partial y}{\partial u} = 2u, \quad \frac{\partial y}{\partial v} = -2v;$$

so

$$\frac{\partial x}{\partial u}\frac{\partial x}{\partial v} + \frac{\partial y}{\partial u}\frac{\partial y}{\partial v} = 2v(2u) + 2u(-2v) = 0.$$

27.5 The chain rule for two parameters

Suppose that we have a new set of coordinates defined by

$$x = x(u, v), \qquad y = y(u, v),$$

and a function $f(x, y)$: an arbitrary **function of position**. The function $f(x, y)$ can be expressed in terms of the new coordinates; for example if

$$x = u^2 - v^2, \quad y = 2uv, \quad \text{and} \quad f(x, y) = x^2 + y^2,$$

then

$$f(x, y) = (u^2 - v^2)^2 + (2uv)^2 = (u^2 + v^2)^2$$

when evaluated at the same point.

If we put

$$z = f(x, y),$$

then the derivatives $\partial z/\partial u$ and $\partial z/\partial v$ indicate how z, or $f(x, y)$, changes as we move around in the new coordinates. Consider the derivative

$$\frac{\partial z}{\partial u}$$

in which v is held constant, at $v = \beta$ say. *Since only u varies, we are able to adopt the single-variable chain rule (27.1), with u instead of t.* However, we must write $\partial x/\partial u$ and $\partial y/\partial u$ instead of dx/du and dy/du in order to indicate that another variable v is present, although it is regarded as constant for the differentiation. We obtain the following.

Chain rule for two parameters

If $x = x(u, v)$, $y = y(u, v)$, $z = f(x, y)$, then

$$\frac{\partial z}{\partial u} = \frac{\partial z}{\partial x}\frac{\partial x}{\partial u} + \frac{\partial z}{\partial y}\frac{\partial y}{\partial u},$$

$$\frac{\partial z}{\partial v} = \frac{\partial z}{\partial x}\frac{\partial x}{\partial v} + \frac{\partial z}{\partial y}\frac{\partial y}{\partial v}.$$

(27.6)

(Or f may be written instead of z.)

Example 27.12. Use the chain rule (27.6) to obtain $\partial z/\partial v$ where $x = u^2 - v^2$, $y = 2uv$, and $z = xy$; check the result by substitution.

For the chain rule, we require

$$\frac{\partial z}{\partial x} = y, \quad \frac{\partial z}{\partial y} = x, \quad \frac{\partial x}{\partial v} = -2v, \quad \frac{\partial y}{\partial v} = 2u.$$

Then

$$\frac{\partial z}{\partial v} = \frac{\partial z}{\partial x}\frac{\partial x}{\partial v} + \frac{\partial z}{\partial y}\frac{\partial y}{\partial v} = -2yv + 2xu = 2u^3 - 6uv^2.$$

To check the result, write z in terms of u and v:

$$z = xy = (u^2 - v^2)2uv = 2u^3v - 2uv^3$$

Therefore $\dfrac{\partial z}{\partial v} = 2u^3 - 6uv^2$, as before.

There is clearly no advantage in using the chain rule for a simple explicit case such as this. The use of such rules is to obtain general results as in the following examples.

Example 27.13. Find expressions for $\partial z/\partial r$ and $\partial z/\partial \theta$ when $x = r\cos\theta$, $y = r\sin\theta$, and z is a function of position.

To use (27.6), put (r, θ) in place of (u, v):

$$\frac{\partial z}{\partial r} = \frac{\partial z}{\partial x}\frac{\partial x}{\partial r} + \frac{\partial z}{\partial y}\frac{\partial y}{\partial r} = \cos\theta\frac{\partial z}{\partial x} + \sin\theta\frac{\partial z}{\partial y},$$

$$\frac{\partial z}{\partial \theta} = \frac{\partial z}{\partial x}\frac{\partial x}{\partial \theta} + \frac{\partial z}{\partial y}\frac{\partial y}{\partial \theta} = -r\sin\theta\frac{\partial z}{\partial x} + r\cos\theta\frac{\partial z}{\partial y}.$$

Example 27.14. Find expressions for $\partial z/\partial x$ and $\partial z/\partial y$ in terms of $\partial z/\partial r$ and $\partial z/\partial \theta$, where $x = r \cos \theta$, $y = r \sin \theta$.

The appropriate form for chain rule (27.6) will be

$$\frac{\partial z}{\partial x} = \frac{\partial z}{\partial r}\frac{\partial r}{\partial x} + \frac{\partial z}{\partial \theta}\frac{\partial \theta}{\partial x}, \qquad \frac{\partial z}{\partial y} = \frac{\partial z}{\partial r}\frac{\partial r}{\partial y} + \frac{\partial z}{\partial \theta}\frac{\partial \theta}{\partial y}.$$

To find $\partial r/\partial x$ etc., use the alternative form for polar coordinates:

$$r = (x^2 + y^2)^{\frac{1}{2}}, \qquad \theta = \arctan y/x;$$

then

$$\frac{\partial r}{\partial x} = \frac{x}{(x^2 + y^2)^{\frac{1}{2}}} = \frac{r \cos \theta}{r} = \cos \theta;$$

$$\frac{\partial \theta}{\partial x} = \frac{1}{1 + (y/x)^2}\left(-\frac{y}{x^2}\right) = -\frac{y}{x^2 + y^2} = -\frac{r \sin \theta}{r^2} = -\frac{\sin \theta}{r}.$$

Therefore

$$\frac{\partial z}{\partial x} = \cos \theta \frac{\partial z}{\partial r} - \frac{\sin \theta}{r}\frac{\partial z}{\partial \theta}.$$

Similarly $\partial r/\partial y$ and $\partial \theta/\partial y$ can be calculated to give

$$\frac{\partial z}{\partial y} = \sin \theta \frac{\partial z}{\partial r} + \frac{\cos \theta}{r}\frac{\partial z}{\partial \theta}.$$

(These can also be obtained by treating the pair of expressions for $\partial z/\partial r$ and $\partial z/\partial \theta$ obtained in Example 27.13 as if they were a pair of simultaneous equations for $\partial z/\partial x$ and $\partial z/\partial y$, and solving them.)

Example 27.15. Supposing that no further information is provided, simplify the expression

$$\frac{\partial P}{\partial U}\frac{\partial U}{\partial M} + \frac{\partial P}{\partial V}\frac{\partial V}{\partial M}.$$

We may understand from the notation that

$$P = P(U, V).$$

The partial derivative notation $\partial U/\partial M$ and $\partial V/\partial M$ indicates that

$$U = U(M, \ldots) \quad \text{and} \quad V = V(M, \ldots),$$

at least one more variable being present: the expression does not tell us its name. The chain rule automatically simplifies the expression to

$$\frac{\partial P}{\partial U}\frac{\partial U}{\partial M} + \frac{\partial P}{\partial V}\frac{\partial V}{\partial M} = \frac{\partial P}{\partial M}.$$

Notice how the expressions in (27.6) are formed. Suppose that

$$P = P(U, V), \quad Q = Q(U, V), \quad U = (X, Y), \quad V = V(X, Y).$$

To form e.g. $\dfrac{\partial P}{\partial x}$, write

$$\frac{\partial P}{\partial x} = \frac{\partial P}{\partial x} \,\_\, + \frac{\partial P}{\partial x} \,\_\,,$$

then fill in the spaces in the first term with ∂U and the second with ∂V.

Example 27.16. Prove that if (x, y) and (u, v) are coordinates related by

$$x = x(u, v) \quad \text{and} \quad y = y(u, v), \tag{i}$$

or alternatively by

$$u = u(x, y) \quad \text{and} \quad v = v(x, y), \tag{ii}$$

then

$$\begin{bmatrix} \dfrac{\partial x}{\partial u} & \dfrac{\partial x}{\partial v} \\[2ex] \dfrac{\partial y}{\partial u} & \dfrac{\partial y}{\partial v} \end{bmatrix} \begin{bmatrix} \dfrac{\partial u}{\partial x} & \dfrac{\partial u}{\partial y} \\[2ex] \dfrac{\partial v}{\partial x} & \dfrac{\partial v}{\partial y} \end{bmatrix}$$

is equal to the unit matrix I_2.

In the first matrix, the relations (i) are implied, and in the second the relations (ii). By multiplying the matrices we obtain

$$\begin{bmatrix} \dfrac{\partial x}{\partial u}\dfrac{\partial u}{\partial x} + \dfrac{\partial x}{\partial v}\dfrac{\partial v}{\partial x} & \dfrac{\partial x}{\partial u}\dfrac{\partial u}{\partial y} + \dfrac{\partial x}{\partial v}\dfrac{\partial v}{\partial y} \\[3ex] \dfrac{\partial y}{\partial u}\dfrac{\partial u}{\partial x} + \dfrac{\partial y}{\partial v}\dfrac{\partial v}{\partial x} & \dfrac{\partial y}{\partial u}\dfrac{\partial u}{\partial y} + \dfrac{\partial y}{\partial v}\dfrac{\partial v}{\partial y} \end{bmatrix}.$$

Each of these elements has the right shape for the representation of a derivative by the chain rule (27.6), though the variable combinations occupying the various positions may seem unusual. The matrix becomes

$$\begin{bmatrix} \dfrac{\partial x}{\partial x} & \dfrac{\partial x}{\partial y} \\[2ex] \dfrac{\partial y}{\partial x} & \dfrac{\partial y}{\partial y} \end{bmatrix} = \begin{bmatrix} 1 & 0 \\ 0 & 1 \end{bmatrix}.$$

27.6 The use of differentials

Problems are sometimes made easier by working directly with the incremental approximation (26.2): if $z = f(x, y)$, then

$$\delta z \approx \frac{\partial z}{\partial x}\, \delta x + \frac{\partial z}{\partial y}\, \delta y;$$

this can be more fruitful than searching for a chain rule or other formula which will work. It is customary in certain applications, particularly in thermodynamics, to write this formula in the form

$$dz = \frac{\partial z}{\partial x}\, dx + \frac{\partial z}{\partial y}\, dy,$$

in which '$\approx$' becomes '$=$' and dx, dy, dz are put in place of δx, δy, δz. Such expressions can be manipulated in the same way as the differential forms described in Section 20.4 for functions of a single variable (the theory, however, is somewhat difficult). Here we shall adopt '$=$' for brevity, but retain δx etc.

Example 27.17. Find a vector normal to the curve $f(x, y) = c$ at a point (x, y) on the curve. (Compare Section 22.5.)

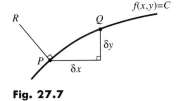

Fig. 27.7

Let P be (x, y) and Q a nearby point $(x + \delta x, y + \delta y)$ also on the curve (see Fig. 27.7). Put

$$z = f(x, y).$$

Then, since z is constant on the curve (it equals c),

$$\delta z = 0 = \frac{\partial z}{\partial x}\, \delta x + \frac{\partial z}{\partial y}\, \delta y,$$

where the derivatives are evaluated at P. This can be written

$$\left(\frac{\partial z}{\partial x}, \frac{\partial z}{\partial y} \right) \cdot (\delta x, \delta y) = 0.$$

But $(\delta x, \delta y) = \overline{PQ}$ is in the direction of the tangent at P (more and more nearly as PQ becomes smaller, of course), so $(\partial z/\partial x, \partial z/\partial y)$ is a vector in the direction of the normal, as we found in Section 26.7.

Example 27.18. Show that the coordinate system (u, v) defined by

$$x = 2uv \quad \text{and} \quad y = v^2 - u^2$$

is orthogonal.

We have to show that any two curves given respectively by $u = \alpha$ and $v = \beta$ intersect in a right angle, as in Fig. 27.8.

If u and v are allowed to vary arbitrarily, then

$$\delta x = 2v\, \delta u + 2u\, \delta v, \qquad \delta y = -2u\, \delta u + 2v\, \delta v. \tag{i}$$

But u does not vary on the curve $u = \alpha$, so $\delta u = 0$ and (i) becomes $\delta x = 2u\, \delta v$, $\delta y = 2v\, \delta v$.

The vector $\overline{PQ}$ points nearly in the direction of the tangent at P:

$$\overline{PQ} = (\delta x, \delta y) = (2u\, \delta v, 2v\, \delta v). \tag{ii}$$

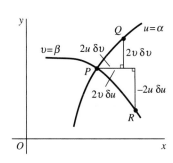

Fig. 27.8

Similarly, on the curve $v = \beta$, we have $\delta v = 0$; so
$$\delta x = 2v\,\delta u, \qquad \delta y = -2u\,\delta u. \tag{iii}$$
$\overline{PR}$ points in the direction of the tangent to $v = \beta$, and
$$\overline{PR} = (\delta x, \delta y) = (2v\,\delta u, -2u\,\delta u).$$
From (ii) and (iii), we have
$$\overline{PQ}\cdot\overline{PR} = (2u\,\delta v, 2v\,\delta v)\cdot(2v\,\delta u, -2u\,\delta u)$$
$$= 4uv\,\delta u\,\delta v - 4uv\,\delta u\,\delta v = 0,$$
so the curves intersect in a right angle.

Problems, Chapter 27

27.1. Find a parametrization $(x(t), y(t))$ suitable for the following curves, specifying the range of t required to traverse the curve exactly once, in the anticlockwise direction if the curve is closed.
(a) $x^2 + y^2 = 25$;   (b) $\frac{1}{4}x^2 + \frac{1}{9}y^2 = 1$;   (c) $xy = 4$;
(d) $x^2 - y^2 = 1$ (try using the identity
$1 + \tan^2 A = 1/\cos^2 A$);
(e) $\frac{1}{4}x^2 - \frac{1}{9}y^2 = 1$;   (f) $y^2 = 4ax$;
(g) $(x-1)^2 + (y-2)^2 = 9$;   (h) $2x - 5y + 2 = 0$.

27.2. For each of the following cases, obtain df/dt in terms of t by means of the chain rule (27.1).
(a) $f(x, y) = x^2 + y^2$, $x(t) = t$, $y(t) = 1/t$;
(b) $f(x, y) = x^2 - y^2$, $x(t) = \cos t$, $y(t) = \sin t$;
(c) $f(x, y) = xy$, $x(t) = 2\cos t$, $y(t) = \sin t$;
(d) $f(x, y) = x\sin y$, $x(t) = 2t$, $y(t) = t^2$;
(e) $f(x, y) = 4x^2 + 9y^2$, $x(t) = \frac{1}{2}\cos t$, $y(t) = \frac{1}{3}\sin t$.

27.3. Two athletes run around concentric circular tracks of radius r and R with speeds v and V respectively. They start on the same radial line. By using time as a parameter, find the rate of change with time of the distance between them.

27.4. Use the Lagrange-multiplier method to solve the following problems.
(a) Find the maximum area of a rectangle having perimeter of length 10.
(b) Find the rectangle with area 9 which has shortest perimeter.
(c) Find the stationary points of $x^2 + 2y^2$ subject to $x^2 + y^2 = 1$.
(d) Find the largest rectangle in the first quadrant of the (x, y) plane which has two of its sides along $x = 0$ and $y = 0$ respectively, and a vertex on the line $2x + y = 1$.

(e) Find the minimum distance of the straight line $x + 2y = 1$ from the point $(1, 1)$. (It is easier to consider the *square* of the distance.)
(f) Find the shortest distance from the origin to the curve $x^2 + 8xy + 7y^2 = 225$.
(g) With reference to Fig. 27.4, find the rectangle in the ellipse which has the minimum perimeter.
(h) Find the stationary points of $(x - y + 1)^2$ on $y = x^2$.
(i) Show that in general there are three normals to a parabola from any given point inside it.

27.5. Find the stationary points of $f(x, y)$ on $g(x, y) = c$ (i) by parametrizing the given path as in Example 27.6, (ii) by using the Lagrange-multiplier technique, in each of the following cases.
(a) $f(x, y) = x^2 + y^2$ on $g(x, y) = xy = 1$;
(b) $f(x, y) = x^2 + y^2$ on $(x-1)^2 + y^2 = 1$;
(c) $f(x, y) = x^2 + 4y^2$ on $x^2 + y^2 = 1$;
(d) $f(x, y) = 3x - 2y$ on $x^2 - y^2 = 4$;
(e) $f(x, y) = xy$ on $g(x, y) = x^2 + y^2 = 1$ (compare this with (a).)

27.6. Show by means of sketches that, for the restricted stationary-value problem, a stationary point can be expected at any point where the curve $g(x, y) = c$ is tangential to a contour of $f(x, y)$.
Use this observation to derive the Lagrange-multiplier principle. (Hint: consider the normals at the point of tangency; or use implicit differentiation to get expressions for the directions of the curves there.)
There are cases when a stationary point can occur although the curves are not tangential there. Try to identify these cases by sketching various possibilities. (Hint: they correspond to $\lambda = 0$.)

27.7. A change of coordinates from (x, y) to (u, v) is specified by each of the following. Show that the new coordinate system is orthogonal.

(a) $u = 2x + 3y,\ v = -3x + 2y$;
(b) $u = xy,\ v = x^2 - y^2$;
(c) $u = x^2 + 2y^2,\ v = y/x^2$;
(d) $u = xy^2,\ v = y^2 - 2x^2$;
(e) $u = x + 1/x + y^2/x,\ v = y - 1/y + x^2/y$;
(f) $x = 2u - v,\ y = u + 2v$;
(g) $x = u^2 - v^2,\ y = 2uv$;
(h) $x = u/(u^2 + v^2),\ y = v/(u^2 + v^2)$;
(i) $x = u^2 - v^2,\ y = -2uv$.

27.8. Let $r(t)$ and $\theta(t)$ be polar coordinates which are functions of a parameter t.

(a) Express dx/dt and dy/dt in terms of dr/dt, $d\theta/dt$, r, and θ.

(b) Use (a) to obtain expressions for d^2x/dt^2 and d^2y/dt^2.

(c) Prove that

$$\cos\theta\,\frac{d^2x}{dt^2} + \sin\theta\,\frac{d^2y}{dt^2} = \frac{d^2r}{dt^2} - r\left(\frac{d\theta}{dt}\right)^2,$$

$$\cos\theta\,\frac{d^2y}{dt^2} - \sin\theta\,\frac{d^2x}{dt^2} = \frac{1}{r}\frac{d}{dt}\left(r^2\,\frac{d\theta}{dt}\right).$$

(These two equations express the radial and tangential components of acceleration, given on the left, in terms of polar coordinates.)

27.9. Use the chain rule (27.6) to find $\partial f/\partial u$ and $\partial f/\partial v$ in terms of u and v in each of the following cases.

(a) $f(x, y) = 2x - y,\ x = uv,\ y = u^2 - v^2$;
(b) $f(x, y) = y/x,\ x = u + v,\ y = u - v$;
(c) $f(x, y) = y^2,\ x = u^2 + v^2,\ y = v/u$;
(d) $f(x, y) = (x - y)/(x + y),\ x = v,\ y = u - v$.

27.10. By using the chain rule (27.6) twice, obtain $\partial^2 f/\partial u^2$, $\partial^2 f/\partial v^2$, and $\partial^2 f/\partial u\,\partial v$ in each of the following cases.

(a) $f(x, y) = y/x,\ x = u + v,\ y = u - v$;
(b) $f(x, y) = x^2 + y^2,\ x = uv,\ y = u^2 - v^2$;
(c) $f(x, y) = y^2,\ x = uv,\ y = v$.

27.11. Find expressions for $\partial f/\partial u$, $\partial f/\partial v$, $\partial^2 f/\partial u^2$, $\partial^2 f/\partial v^2$, and $\partial^2 f/\partial u\,\partial v$ if

$$f(x, y) = g(x^2 - y^2),\ x = u + v,\ y = u - v.$$

(The expressions will involve the functions $g'(x^2 - y^2)$ etc.)

27.12. Let $w = w(u, v)$, $u = u(x, y)$, $v = v(x, y)$, where u and v are related in such a way that

$$\frac{\partial u}{\partial x} = \frac{\partial v}{\partial y}, \qquad \frac{\partial u}{\partial y} = -\frac{\partial v}{\partial x}.$$

Prove that

$$\frac{\partial^2 u}{\partial x^2} + \frac{\partial^2 u}{\partial y^2} = 0, \qquad \frac{\partial^2 v}{\partial x^2} + \frac{\partial^2 v}{\partial y^2} = 0.$$

Use the chain rule (27.6) to prove that

$$\frac{\partial^2 w}{\partial x^2} + \frac{\partial^2 w}{\partial y^2} = \left[\left(\frac{\partial u}{\partial x}\right)^2 + \left(\frac{\partial v}{\partial y}\right)^2\right]\left[\frac{\partial^2 w}{\partial x^2} + \frac{\partial^2 w}{\partial y^2}\right].$$

27.13. Let r and θ be the usual polar coordinates, and $z = f(x, y)$; show that:

(a) $\left(\dfrac{\partial z}{\partial x}\right)^2 + \left(\dfrac{\partial z}{\partial y}\right)^2 = \left(\dfrac{\partial z}{\partial r}\right)^2 + \dfrac{1}{r^2}\left(\dfrac{\partial z}{\partial \theta}\right)^2$;

(b) $\dfrac{\partial^2 z}{\partial x^2} + \dfrac{\partial^2 z}{\partial y^2} = \dfrac{\partial^2 z}{\partial r^2} + \dfrac{1}{r}\dfrac{\partial z}{\partial r} + \dfrac{1}{r^2}\dfrac{\partial^2 z}{\partial \theta^2}$.

28 Functions of any number of variables

28.1 The incremental approximation; errors

For functions of three and more variables, simple pictorial representations are not available. Nevertheless many of the important formulae follow the pattern of the two-variable case, simply containing more terms of the same type. This follows from the incremental approximation (26.1), extended to three and more variables.

Suppose that $f(x, y, z, \ldots)$ is any function of N ($\geqslant 3$) variables. Some examples were mentioned at the beginning of Chapter 25. The **partial derivatives** $\partial f/\partial x$, $\partial f/\partial y$, $\partial f/\partial z, \ldots$, have the same meaning as they did there: during differentiation, all the variables except the named one are treated as constants.

Higher derivatives are defined as with functions of two variables; for example

$$\frac{\partial^3 f}{\partial x\, \partial y\, \partial z} = \frac{\partial}{\partial x} \frac{\partial}{\partial y} \frac{\partial f}{\partial z}.$$

It follows from the result for second derivatives (Section 25.4) that

$$\frac{\partial^3 f}{\partial x\, \partial y\, \partial z} = \frac{\partial^3 f}{\partial y\, \partial x\, \partial z} = \frac{\partial^3 f}{\partial z\, \partial y\, \partial x}$$

and so on: the derivatives may be taken in any order.

The **incremental approximation** has the same form as (26.1) and (26.2), simply containing further terms corresponding to the extra variables:

Incremental approximation for $f(x, y, z, \ldots)$

For small enough increments δx, δy, $\delta z, \ldots$:

$$\delta f = f(x + \delta x, y + \delta y, z + \delta z, \ldots) - f(x, y, z, \ldots)$$

$$\approx \frac{\partial f}{\partial x} \delta x + \frac{\partial f}{\partial y} \delta y + \frac{\partial f}{\partial z} \delta z + \cdots. \qquad (28.1)$$

If we put $w = f(x, y, z, \ldots)$, this can be written

$$\delta w \approx \frac{\partial w}{\partial x} \delta x + \frac{\partial w}{\partial y} \delta y + \frac{\partial w}{\partial z} \delta z + \cdots.$$

To prove (28.1), the idea of a tangent plane is not available, so we must go directly for the linear approximation to the function. Put $w = f(x, y, z, \ldots)$, and consider a fixed 'point' $P : (x, y, z, \ldots)$ and another nearby point $Q : (x + \delta x, y + \delta y, z + \delta z, \ldots)$. Then w changes to $w + \delta w$. We assume that the relation between δw and $\delta x, \delta y, \delta z, \ldots$ is **close to linear** for small $\delta x, \delta y, \delta z, \ldots$; that is to say,

$$\delta w = A\,\delta x + B\,\delta y + C\,\delta z + \cdots + \varepsilon, \tag{28.2}$$

where $A, B, \ldots$ are certain constants and the **error ε is of a lower order of magnitude than the δ-quantities** (compare Example 26.2 for two variables).

In order firstly to find A, vary only x, so that

$$\delta x \neq 0, \qquad \delta y = \delta z = \cdots = 0.$$

Put these into (28.2) and divide by δx, giving

$$\frac{\delta w}{\delta x} = A + \frac{\varepsilon}{\delta x}.$$

Now let $\delta x \to 0$. Then $\delta w / \delta x \to \partial w / \partial x$, and (since ε is of lower order of magnitude than δx) $\varepsilon / \delta x \to 0$. Therefore

$$\frac{\partial w}{\partial x} = A.$$

Similarly $\partial w / \partial y = B$, and so on, which gives the result (28.1).

The incremental approximation (28.1) can be used to estimate **errors** as in Section 26.2.

Small-error formula

If $w = f(x, y, z, \ldots)$, then (approximately)

$$\Delta w = \frac{\partial w}{\partial x}\Delta x + \frac{\partial w}{\partial y}\Delta y + \frac{\partial w}{\partial z}\Delta z + \cdots, \tag{28.3}$$

where $x, y, z, \ldots$ stand for the measured values and Δ stands for

$$\text{error} = (\text{measured value}) - (\text{exact value}).$$

Example 28.1. In a triangle ABC, $\cos C = (c^2 - a^2 - b^2)/2ab$. In a particular case, the measured side lengths are $a = 3$, $b = 4$, and $c = 5.5$ units. Possible errors of measurement lie between ± 0.1 units. Find the error in estimating C in the worst case.

For ease of differentiation put

$$w = \cos C = \frac{c^2}{2ab} - \frac{a}{2b} - \frac{b}{2a}.$$

Then

$$\Delta(\cos C) = \Delta w \approx \frac{\partial w}{\partial a} \Delta a + \frac{\partial w}{\partial b} \Delta b + \frac{\partial w}{\partial c} \Delta c,$$

where

$$\frac{\partial w}{\partial a} = -\frac{c^2}{2a^2 b} - \frac{1}{2b} + \frac{b}{2a^2},$$

and similarly for the other two derivatives. From the measurements,

$$\frac{\partial w}{\partial a} = -0.323, \qquad \frac{\partial w}{\partial b} = -0.388, \qquad \frac{\partial w}{\partial c} = 0.458.$$

Therefore $\Delta(\cos C) \approx -0.323\,\Delta a - 0.388\,\Delta b + 0.458\,\Delta c$.

To obtain ΔC from $\Delta(\cos C)$:

$$\Delta(\cos C) \approx \left(\frac{d}{dC} \cos C\right) \Delta C = (-\sin C)\,\Delta C,$$

where C is in *radians*. The value of C estimated from the cosine rule using the measured values is 1.350 radians, and $\sin 1.350 = 0.976$. Therefore

$$\Delta C \approx -0.334\,\Delta a - 0.398\,\Delta b + 0.469\,\Delta c.$$

The magnitude of ΔC is a maximum if by chance the errors are

$$\Delta a = \mp 0.1, \qquad \Delta b = \mp 0.1, \qquad \Delta c = \pm 0.1,$$

and then

$$\Delta C \approx \pm 0.120.$$

which is about a 9% error.

28.2 Implicit differentiation

There is an analogy with the implicit-differentiation formula (26.6). Suppose that

$$f(x, y, z, \ldots) = 0. \tag{28.4}$$

This condition implies that any one of the variables depends on, or is a function of, all the others. For example, if the variables are x, y, z, and r, and

$$x^2 + y^2 + z^2 - r^2 = 0,$$

then

$$y = \pm(r^2 - x^2 - z^2)^{\frac{1}{2}}.$$

Subject to (28.4) we can therefore talk about partial derivatives such as $\partial y/\partial x$: we think of y as being a function of the other variables, but with all the variables except x and y held constant.

Suppose that $(x, y, z, \ldots)$ and $(x + \delta x, y + \delta y, z + \delta z, \ldots)$ both satisfy condition (28.4). Then $\delta f = 0$ and the incremental approximation gives

$$\frac{\partial f}{\partial x} \delta x + \frac{\partial f}{\partial y} \delta y + \frac{\partial f}{\partial z} \delta z + \cdots \approx 0. \tag{28.5}$$

Suppose next that all the variables except x and y are kept constant, so that $\delta x \neq 0$ and $\delta y \neq 0$, but $\delta z = \cdots = 0$, Equation (28.5) becomes $(\partial f / \partial x)\,\delta x + (\partial f / \partial y)\,\delta y \approx 0$, so that

$$\frac{\delta y}{\delta x} \approx -\frac{\partial f}{\partial x} \Big/ \frac{\partial f}{\partial y}.$$

Now let $\delta x \to 0$ and the equation becomes

$$\frac{\partial y}{\partial x} = -\frac{\partial f}{\partial x} \Big/ \frac{\partial f}{\partial y}:$$

Implicit differentiation

If $f(x, y, z, \ldots) = 0$, then

$$\frac{\partial y}{\partial x} = -\frac{\partial f}{\partial x} \Big/ \frac{\partial f}{\partial y}. \tag{28.6}$$

Any other two variables may be substituted for x and y.

Example 28.2. For a fixed mass of gas, an equation of the form $f(P, V, T) = 0$ holds (the 'equation of state'), where P, V, and T represent the pressure, volume, and temperature respectively. Show that

(a) $\dfrac{\partial P}{\partial T}\dfrac{\partial T}{\partial V} = -\dfrac{\partial P}{\partial V}$, (b) $\dfrac{\partial P}{\partial T}\dfrac{\partial T}{\partial V}\dfrac{\partial V}{\partial P} = -1.$

The relation $f(P, V, T) = 0$ implies that any of P, V, or T is a function of the other two variables: $P = P(V, T)$, $V = V(T, P)$, and $T = T(P, V)$. If we put, say $P = P(V, T)$ constant, then implicit differentiation, by (28.6), gives $\partial V / \partial T$ or $\partial T / \partial V$ in terms of $\partial f / \partial V$ and $\partial f / \partial T$ (where we are reminded of the 'constant P' condition by the partial derivative signs instead of $\mathrm{d}V/\mathrm{d}T$ and $\mathrm{d}T/\mathrm{d}V$). Similarly we obtain $\partial P / \partial T$, $\partial T / \partial P$, $\partial P / \partial V$, and $\partial V / \partial T$.

(a) $\dfrac{\partial P}{\partial T} = -\dfrac{\partial f}{\partial T} \Big/ \dfrac{\partial f}{\partial P}$ and $\dfrac{\partial T}{\partial V} = -\dfrac{\partial f}{\partial V} \Big/ \dfrac{\partial f}{\partial T}$

(from (28.6)). Therefore

$$\frac{\partial P}{\partial T}\frac{\partial T}{\partial V} = \frac{\partial f}{\partial V} \Big/ \frac{\partial f}{\partial P} = -\frac{\partial P}{\partial V} \quad \text{(using (28.6) again)}.$$

(b) By repeating the process (a) with different variables,

$$\frac{\partial P}{\partial T}\frac{\partial T}{\partial V}\frac{\partial V}{\partial P} = \left(-\frac{\partial f}{\partial T} \Big/ \frac{\partial f}{\partial P} \right)\left(-\frac{\partial f}{\partial V} \Big/ \frac{\partial f}{\partial T} \right)\left(-\frac{\partial f}{\partial P} \Big/ \frac{\partial f}{\partial V} \right) = -1.$$

There are many more similar formulae obtainable by permuting P, T, V: these identities are important in the theory of thermodynamics.

28.3 Chain rules

The **chain rule for a single parameter** t is obtained exactly as in the case of a single variable: divide (28.1) by δt and take the limit; to give the following formula.

Chain rule for one parameter

Given $f(x, y, z, \ldots)$, where $x = x(t)$,

$$y = y(t), z = z(t), \ldots,$$

$$\frac{\mathrm{d}f}{\mathrm{d}t} = \frac{\partial f}{\partial x}\frac{\mathrm{d}x}{\mathrm{d}t} + \frac{\partial f}{\partial y}\frac{\mathrm{d}y}{\mathrm{d}t} + \frac{\partial f}{\partial z}\frac{\mathrm{d}z}{\mathrm{d}t} + \cdots$$

(or with w in place of f if $w = f(x, y, z, \ldots)$).

(28.7)

Notice that $(x(t), y(t), z(t))$ defines a directed path in three dimensions.

In the case of more than one parameter, the results of Section 26.5 may be extended as follows.

Chain rule for more than one parameter

For a function $f(x, y, z, \ldots,)$, where x, y, $z, \ldots$ are functions of parameters $u, v, \ldots$, we have

$$\frac{\partial f}{\partial u} = \frac{\partial f}{\partial x}\frac{\partial x}{\partial u} + \frac{\partial f}{\partial y}\frac{\partial y}{\partial u} + \frac{\partial f}{\partial z}\frac{\partial z}{\partial u} + \cdots,$$

$$\frac{\partial f}{\partial v} = \frac{\partial f}{\partial x}\frac{\partial x}{\partial v} + \frac{\partial f}{\partial y}\frac{\partial y}{\partial v} + \frac{\partial f}{\partial z}\frac{\partial z}{\partial v} + \cdots,$$

(28.8)

and so for any other parameters. (If $w = f(x, y, z, \ldots)$ then w may be written in place of f.)

28.4 The gradient vector in three dimensions

The **gradient vector function**, introduced for two dimensions in Section 26.6, extends to any number of dimensions, though we shall restrict consideration to three variables in this section.

In the equations we have obtained, such as (28.1) and (28.8), there repeatedy occurs the triplet of elements $\partial f/\partial x$, $\partial f/\partial y$, $\partial f/\partial z$; added, together with various multipliers. We can manipulate this group as a unit by regarding

$$\left(\frac{\partial f}{\partial x}, \frac{\partial f}{\partial y}, \frac{\partial f}{\partial z}\right), \quad \text{or} \quad \frac{\partial f}{\partial x}\hat{\boldsymbol{\imath}} + \frac{\partial f}{\partial y}\hat{\boldsymbol{\jmath}} + \frac{\partial f}{\partial z}\hat{\boldsymbol{k}},$$

as a vector function—**the gradient of** f, now in three dimensions—and denote it by

$$\mathbf{grad}\ f \quad \text{or} \quad \nabla f,$$

as before. As in Section 26.6, we can also think of **grad** or ∇ standing alone as an operator: an instruction to carry out the process

$$\hat{\boldsymbol{\imath}}\frac{\partial}{\partial x} + \hat{\boldsymbol{\jmath}}\frac{\partial}{\partial y} + \hat{\boldsymbol{k}}\frac{\partial}{\partial z}, \quad \text{or} \quad \left(\frac{\partial}{\partial x}, \frac{\partial}{\partial y}, \frac{\partial}{\partial z}\right),$$

on some scalar function $f(x, y, z)$. The definition is stated for reference as follows.

Gradient vector function (three dimensions)

For a scalar function $f(x, y, z)$:

$$\mathbf{grad}\ f \quad \text{or} \quad \nabla f$$

$$= \left(\frac{\partial f}{\partial x}, \frac{\partial f}{\partial y}, \frac{\partial f}{\partial z}\right) = \hat{\boldsymbol{\imath}}\frac{\partial f}{\partial x} + \hat{\boldsymbol{\jmath}}\frac{\partial f}{\partial y} + \hat{\boldsymbol{k}}\frac{\partial f}{\partial z}; \qquad (28.9)$$

Alternatively, **grad** or ∇ stands for the **operator**

$$\left(\hat{\boldsymbol{\imath}}\frac{\partial}{\partial x} + \hat{\boldsymbol{\jmath}}\frac{\partial}{\partial y} + \hat{\boldsymbol{k}}\frac{\partial}{\partial z}\right), \quad \text{or} \quad \left(\frac{\partial}{\partial x}, \frac{\partial}{\partial y}, \frac{\partial}{\partial z}\right).$$

Example 28.3. Let $f(x, y, z) = x^2 + y^2 + z^2$. Obtain (a) the vector function **grad** $f(x, y, z)$; (b) the value of **grad** $f(x, y, z)$ at the point $(1, 2, 3)$; (c) an expression for the magnitude (or length) of **grad** $f(x, y, z)$.

(a) $\mathbf{grad}\ f(x, y, z) = \left(\dfrac{\partial f}{\partial x}, \dfrac{\partial f}{\partial y}, \dfrac{\partial f}{\partial z}\right) = (2x, 2y, 2z);$

or one can use the 'operator' idea and the other way of writing a vector:

$$\mathbf{grad}\ f = \left(\hat{\boldsymbol{\imath}}\frac{\partial}{\partial x} + \hat{\boldsymbol{\jmath}}\frac{\partial}{\partial y} + \hat{\boldsymbol{k}}\frac{\partial}{\partial z}\right)(x^2 + y^2 + z^2)$$

$$= \hat{\boldsymbol{\imath}}(2x) + \hat{\boldsymbol{\jmath}}(2y) + \hat{\boldsymbol{k}}(2z).$$

(b) At $x = 1$, $y = 2$, $z = 3$,

$$\mathbf{grad}\ f = (2, 4, 6).$$

(c) The magnitude or length $|\boldsymbol{v}|$ of a vector $\boldsymbol{v} = (a, b, c)$ is $|\boldsymbol{v}| = (a^2 + b^2 + c^2)^{\frac{1}{2}}$; so

$$|\mathbf{grad}\ f| = [(2x)^2 + (2y)^2 + (2z)^2]^{\frac{1}{2}} = 2(x^2 + y^2 + z^2)^{\frac{1}{2}}.$$

Expressions which occur in the theory frequently take the form

$$U\frac{\partial f}{\partial x} + V\frac{\partial f}{\partial y} + W\frac{\partial f}{\partial z}, \qquad (28.10)$$

where U, V, and W may be constants, or various functions. If we put

$$\hat{\imath}U + \hat{\jmath}V + \hat{k}W = \mathbf{S},$$

where $\mathbf{S}$ is another vector (compare Section 26.10), then we can write (28.10) in the form

$$U\frac{\partial f}{\partial x} + V\frac{\partial f}{\partial y} + W\frac{\partial f}{\partial z} = (U, V, W) \cdot \left(\frac{\partial f}{\partial x}, \frac{\partial f}{\partial y}, \frac{\partial f}{\partial z}\right) = \mathbf{S} \cdot \mathbf{grad}\, f,$$

as in the following example.

Example 28.4. Suppose that the concentration of plankton in the sea is $C(x, y, z, t)$. A whale travels on the path $x = x(t)$, $y = y(t)$, $z = z(t)$, where t is time. Show that, on the path of the whale,

$$\frac{dC}{dt} = \frac{\partial C}{\partial t} + \mathbf{v} \cdot \mathbf{grad}\, C,$$

where $\mathbf{v}$ is its velocity.

  By the chain rule (28.7),

$$\frac{dC}{dt} = \frac{\partial C}{\partial x}\frac{dx}{dt} + \frac{\partial C}{\partial y}\frac{dy}{dt} + \frac{\partial C}{\partial z}\frac{dz}{dt} + \frac{\partial C}{\partial t},$$

after putting $dt/dt = 1$ into the final term. The whale's velocity is

$$\mathbf{v} = \left(\frac{dx}{dt}, \frac{dy}{dt}, \frac{dz}{dt}\right),$$

so that

$$\frac{dC}{dt} = \frac{\partial C}{\partial t} + \mathbf{v} \cdot \mathbf{grad}\, C.$$

(If the whale drifted with the motion of the sea, $\mathbf{v}$ would represent the velocity of the current. This case is related to the concept of **material derivative** in fluid mechanics. Instead of C there is a quantity such as the density or momentum of a particular piece of fluid, whose variation we follow as the fluid moves around.)

28.5 Normal to a surface

An equation of the form

$$g(x, y, z) = k$$

represents a **surface** in three dimensions, because we can imagine 'solving' the equation for z in order to obtain equivalent equation(s):

$$z = f(x, y).$$

Thus, if $x^2 + y^2 + z^2 = 1$, then $z = \pm(1 - x^2 - y^2)^{\frac{1}{2}}$. The normal to the surface can be expressed neatly as follows.

Normal (perpendicular) to a surface

Let P be any point on a **surface** $g(x, y, z) = k$. Then (28.11)
grad g, evaluated at P, is normal to the surface at P.

(Compare (26.9), for the normal to a curve in two dimensions.) The proof is as follows. In Fig. 28.1, $P : (x, y, z)$ is the given point on the surface and $Q : (x + \delta x, y + \delta y, z + \delta z)$ is *any* nearby point on the surface. Then

$$g(x + \delta x, y + \delta y, z + \delta z) - g(x, y, z) = 0,$$

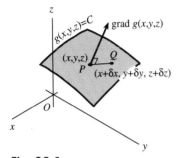

Fig. 28.1

or

$$\delta g = 0.$$

Therefore, by the incremental formula (28.1),

$$0 = \frac{\partial g}{\partial x} \delta x + \frac{\partial g}{\partial y} \delta y + \frac{\partial g}{\partial x} \delta z$$

$$= (\mathbf{grad}\ g) \cdot (\delta x, \delta y, \delta z).$$

This shows that **grad** g is perpendicular to the vector $(\delta x, \delta y, \delta z)$. But $\overline{PQ}$ can be chosen to point in any direction from P in the surface, so the only possibility is that **grad** g is perpendicular to the surface itself at P.

We already know (from (25.7)) that a vector normal to a surface described in the form $z = f(x, y)$ is

$$\left(\frac{\partial f}{\partial x}, \frac{\partial f}{\partial y}, -1 \right).$$

This is reconciled with (28.11) if we write its equation in the form

$$g(x, y, z) = f(x, y) - z = 0.$$

28.6 Equation of the tangent plane

Suppose that a surface is specified in the form $g(x, y, z) = k$, and that the point $P : (x, y, z)$ is on the surface. The conditions that the tangent plane must satisfy are (a) it contains the point P; and (b) it is perpendicular to the normal vector **grad** $g(x, y, z)$ evaluated at P, as in (28.11). These conditions are satisfied by the equation

$$\left(\frac{\partial g}{\partial x} \right)_P (x - a) + \left(\frac{\partial g}{\partial y} \right)_P (y - b) + \left(\frac{\partial g}{\partial z} \right)_P (z - c) = 0. \quad (28.12)$$

It can be seen that the expression is zero when $x = a, y = b, z = c$. Also the coefficient vector

$$\left(\left(\frac{\partial g}{\partial x} \right)_P, \left(\frac{\partial g}{\partial y} \right)_P, \left(\frac{\partial g}{\partial z} \right)_P \right) = [\mathbf{grad}\ g]_P$$

is perpendicular to the plane. Therefore we may state the equation of the plane as follows.

> **Tangent plane to the surface** $g(x, y, z) = k$ **at**
> $P : (a, b, c)$
>
> (28.13)
>
> $$\left(\frac{\partial g}{\partial x}\right)_P (x - a) + \left(\frac{\partial g}{\partial y}\right)_P (y - b) + \left(\frac{\partial g}{\partial z}\right)_P (z - c) = 0.$$

28.7 Directional derivative in terms of gradient

The vector **grad** f contains the necessary information to calculate the rate of change of $f(x, y, z)$ in any direction. In Fig. 28.2, let $P : (a, b, c)$ be any point. Suppose that we require the rate of change with distance of $f(x, y, z)$ in the direction PR.

Choose a nearby point $Q : (x + \delta x, y + \delta y, z + \delta z)$ on PR, and put

$$PQ = \delta s = (\delta x^2 + \delta y^2 + \delta z^2)^{\frac{1}{2}}.$$

(where δs is a standard symbol for a small element of distance). Then

$$\frac{\delta x}{\delta s} = \cos \alpha, \qquad \frac{\delta y}{\delta s} = \cos \beta, \qquad \frac{\delta z}{\delta s} = \cos \gamma,$$

where $\cos \alpha$, $\cos \beta$, $\cos \gamma$ are the **direction cosines** of PQ. Now divide the incremental approximation (28.1) through by δs and take the limit as $\delta s \to 0$. We obtain an expression for the rate of change of $f(x, y, z)$ with distance in any direction:

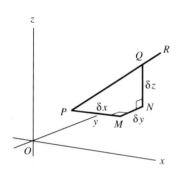

Fig. 28.2

> **Directional derivative in three dimensions**
> In the direction having direction cosines ($\cos \alpha$, $\cos \beta$,
> $\cos \gamma$):
>
> (28.14)
>
> $$\frac{\partial f}{\partial s} = \frac{\partial f}{\partial x} \cos \alpha + \frac{\partial f}{\partial y} \cos \beta + \frac{\partial f}{\partial z} \cos \gamma.$$

In the two-dimensional version (26.4), the coefficients $\cos \theta$ and $\sin \theta$ are equal to the two-dimensional direction cosines, $\cos \theta$ and $\cos(\frac{1}{2}\pi - \theta)$, so (26.4) is compatible with (28.14).

The **direction cosines** $\cos \alpha$, $\cos \beta$, $\cos \gamma$ have the property $\cos^2 \alpha + \cos^2 \beta + \cos^2 \gamma = 1$ (see Section 9.10), so they are the components of a **unit vector** $\hat{s}$ which points in the desired direction. Therefore (28.14) can be written differently:

> **Directional derivative in three dimensions in terms of the gradient**
> In the direction of the unit vector $\hat{s}$,
> $$\frac{\mathrm{d}f}{\mathrm{d}s} = \hat{s} \cdot \mathbf{grad}\, f, \qquad (28.15)$$
> which is the component of $\mathbf{grad}\, f$ in direction $\hat{s}$.

As in Section 26.6, the result (28.15) can be expressed in a third way. If $\mathbf{a}$ and $\mathbf{b}$ are two vectors, and ϕ is the angle between them, then $\mathbf{a} \cdot \mathbf{b} = |\mathbf{a}||\mathbf{b}| \cos \phi$. Putting $\hat{s}$ for $\mathbf{a}$ and $\mathbf{grad}\, f$ for $\mathbf{b}$ in (24.15), and using the fact that $\hat{s} = 1$, we obtain the next result.

> **Directional derivative in 3 dimensions**
> $$\frac{\mathrm{d}f}{\mathrm{d}s} = |\mathbf{grad}\, f| \cos \phi. \qquad (28.16)$$
> where ϕ is the angle between $\mathbf{grad}\, f$ and the unit direction vector $\hat{s}$.

Now take a function $f(x, y, z)$, and a point $P : (x_1, y_1, z_1)$ as in Fig. 28.3. By means of (28.16), we can explore the rate of variation of $f(x, y, z)$ in all directions, by pointing $\hat{s}$ in the required directions. The only thing that changes when we do this is the angle ϕ. It can be seen from (28.16) that (i) If $\phi = \frac{1}{2}\pi$, then $\mathrm{d}f/\mathrm{d}s = 0$, which is consistent with $\hat{s}$ pointing tangentially to the surface $f(x, y, z) = f_P$, where f_P is the value of f at P. (ii) $\mathrm{d}f/\mathrm{d}s$ takes its maximum value $|\mathbf{grad}\, f|$, when $\phi = 0$. That is to say, $\mathbf{grad}\, f$ points in the direction of most rapid increase of f; i.e. it is normal to the surface $f(x, y, z) = f_P$.

It is worth noticing that, for a fixed angle ϕ, the unit vector $\hat{s}$ may point anywhere along the generators of a cone having axis $\mathbf{grad}\, f$, as shown in Fig. 28.3. The directional derivative $\mathrm{d}f/\mathrm{d}s$ is the same in all these directions.

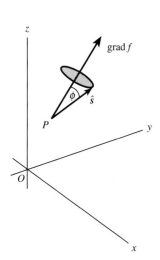

Fig. 28.3

Example 28.5. Let $f(x, y, z) = 4 - x^2 - \frac{1}{2}y^2 - \frac{1}{2}z^2$ represent the atmospheric concentration of a chemical which attracts insects. (a) Write down $\mathbf{grad}\, f$ at (x, y, z). (b) Find a unit vector $\hat{s}$ which points in the direction of most rapid rate of increase in $f(x, y, z)$ at the point $(1, 1, 1)$. (c) An insect sets off from $(1, 1, 1)$ and flies a short distance δs in the direction given by (b). Find its new coordinates (approximately).

(a) $\mathbf{grad}\, f(x, y, z) = \left(\dfrac{\partial f}{\partial x}, \dfrac{\partial f}{\partial y}, \dfrac{\partial f}{\partial z} \right)$

$\qquad = (-2x, -y, -z)$, or $-2x\hat{\imath} - y\hat{\jmath} - z\hat{k}$.

(b) By (ii) above, **grad** f always points in the required direction; at the point $(1, 1, 1)$, its components are $(-2, -1, -1)$. To obtain the corresponding *unit* vector $\hat{s}$, divide by the length $[(-2)^2 + (-1)^2 + (-1)^2]^{\frac{1}{2}} = \sqrt{6}$, obtaining $\hat{s} = (-2/\sqrt{6}, -1/\sqrt{6}, -1/\sqrt{6})$.

(c) The insect moves a distance δs from the point $P : (1, 1, 1)$ along $\hat{s}$ (see Fig. 28.4), so its *vector* displacement $\overline{PQ}$ is

$$\hat{s}\,\delta s = \left(-\frac{2}{\sqrt{6}}\,\delta s, \; -\frac{1}{\sqrt{6}}\,\delta s, \; -\frac{1}{\sqrt{6}}\,\delta s \right).$$

The components of this vector are the x, y, z displacements

$$\delta x = -\frac{2}{\sqrt{6}}\,\delta s, \qquad \delta y = -\frac{1}{\sqrt{6}}\,\delta s, \qquad \delta z = -\frac{1}{\sqrt{6}}\,\delta s.$$

The new coordinates are therefore

$$x = 1 - \frac{2}{\sqrt{6}}\,\delta s, \qquad y = 1 - \frac{1}{\sqrt{6}}\,\delta s, \qquad z = 1 - \frac{1}{\sqrt{6}}\,\delta s.$$

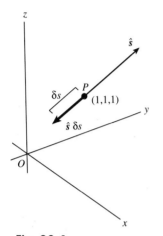

Fig. 28.4

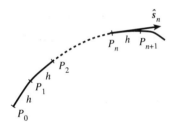

Fig. 28.5

Example 28.6. Using Example 28.5(c) as a model, give a systematic method, suitable for computation, for approximating to the path of an insect which always flies in the direction of most rapidly increasing concentration.

Figure 28.5 shows notionally the path of such an insect. It starts at $P_0 : (x_0, y_0, z_0)$, and the path consists of short steps of equal length h (instead of δs, for the purpose of programming). The progression is $P_0, P_1, P_2, \ldots, P_n, P_{n+1} \ldots$, with coordinates numbered (x_n, y_n, z_n) for $n = 0, 1, 2, \ldots$. At each P_n, the insect moves in the unit-vector direction $\hat{s}_n$, where $s_n = \mathbf{grad}\, f(x, y, z)$ and, as in Example 28.5c,

$$\begin{aligned}
\hat{s}_n &= [(\mathbf{grad}\, f)/|\mathbf{grad}\, f|]_{P_n} \\
&= (-2x_n, -y_n, -z_n)/(4x_n^2 + y_n^2 + z_n^2)^{\frac{1}{2}} \\
&= (a_n, b_n, c_n) \quad \text{(say)}. \tag{28.17}
\end{aligned}$$

For the general step from P_n to P_{n+1}, we obtain the small displacement components $\delta x_n, \delta y_n, \delta z_n$ in the $x, y,$ and z directions (in general these will differ from step to step):

$$(\delta x_n, \delta y_n, \delta z_n) = \hat{s}_n h = (a_n h, b_n h, c_n h),$$

from (28.17). Therefore

$$\begin{aligned}
(x_{n+1}, y_{n+1}, z_{n+1}) &= (x_n + \delta x_n, y_n + \delta y_n, z_n + \delta z_n) \\
&= (x_n + a_n h, y_n + b_n h, z_n + c_n h). \tag{28.18}
\end{aligned}$$

Equations (28.17)–(28.18), with the starting point (x_0, y_0, z_0) given, form a step-by-step process which is easy to computerize. The following table of the early stages was calculated with $h = 0.5$; the starting point in this case is the point $(1, 1, 1)$, where $f(x, y, z) = 4 - x^2 - \frac{1}{2}y^2 - \frac{1}{2}z^2$ as in Example 28.5.

n	x_n	y_n	z_n
0	1	1	1
1	0.959	0.980	0.979
2	0.919	0.959	0.959
3	0.878	0.938	0.938
4	0.839	0.917	0.917
5	0.799	0.895	0.895

If a surface is defined by the equation

$$f(x, y, z) = c,$$

where c is a constant, it is called a **level surface of the function** f (it is the analogy of a contour in the theory for functions of two variables). According to (28.11), therefore, we can say in different language:

> **Normal to a level surface of** $f(x, y, z)$
>
> **grad** f, evaluated at a point P, is perpendicular to the   **(28.19)** level surface of $f(x, y, z)$ through P.

It follows that the insect in Example 28.6 crosses perpendicularly all the level surfaces that it meets.

28.8 Stationary points

Stationary points (which include **maxima and minima**) are more difficult to discuss in more than two dimensions, since we no longer have the horizontal tangent plane to refer to. We should expect a stationary point of $f(x, y, z, \ldots)$ to occur at any point Q where

$$\frac{\partial f}{\partial x} = \frac{\partial f}{\partial y} = \frac{\partial f}{\partial z} = \cdots = 0, \tag{28.20}$$

since all our previous formulae have been merely extended versions of the two-dimensional case. To show that this criterion is the right one, choose *any* path through Q, and suppose that we describe it parametrically by

$$x = x(t), \quad y = y(t), \quad z = z(t), \quad \ldots \quad .$$

Then, if (28.20) holds at Q, the chain rule (28.7) together with (28.20) gives

$$\frac{df}{dt} = \frac{\partial f}{\partial x}\frac{dx}{dt} + \frac{\partial f}{\partial y}\frac{dy}{dt} + \cdots = 0.$$

Therefore a turning point of $f\big(x(t), y(t), z(t), \ldots\big)$ is encountered at Q on *every* path passing through Q, and this is what we should wish to happen for the point Q to be described as stationary.

Stationary points of $f(x, y, z, \ldots)$

The stationary points are the solutions $(x, y, z, \ldots)$ of
the equations   (28.21)

$$\frac{\partial f}{\partial x} = \frac{\partial f}{\partial y} = \frac{\partial f}{\partial z} = \cdots = 0.$$

Example 28.7. Find the stationary points of the function

$$f(x, y, z) = x^2 + y^2 + z^2 - xy - 2yz - zx - z.$$

The conditions (28.21) become

$$\partial f/\partial x = 2x - y - z = 0,$$
$$\partial f/\partial y = -x + 2y - 2z = 0,$$
$$\partial f/\partial z = -x - 2y + 2z - 1 = 0.$$

These are solved by $x = -\frac{1}{2}$, $y = -\frac{5}{8}$, $z = -\frac{3}{8}$.

Restricted stationary-value problems (see Section 27.2) may
occur in any number of dimensions. In three dimensions, the
restriction may either be to **values of** $f(x, y, z)$ **on some given
curve**, or to **values on some given surface**.

To help visualize a three-dimensional situation; suppose that a
fish swims through a field of pollution of density $P = f(x, y, z)$. At
some point in the sea the pollution is at an overall maximum, but
this is of no concern to the fish if it does not swim through it.
However, it will notice highs and lows along its own path even if
there is nothing special about such points from an overall viewpoint.
These are restricted maxima and minima on the fish's path. Suppose
that the path of the fish is expressed parametrically:

$$x = x(t), \quad y = y(t), \quad z = z(t).$$

Then the stationary points peculiar to the path are where $\mathrm{d}f/\mathrm{d}t = 0$.
By the chain rule (28.7), these are the points where

$$\frac{\partial f}{\partial x}\frac{\mathrm{d}x}{\mathrm{d}t} + \frac{\partial f}{\partial y}\frac{\mathrm{d}y}{\mathrm{d}t} + \frac{\partial f}{\partial z}\frac{\mathrm{d}z}{\mathrm{d}t} = 0.$$

When written in terms of t, this is an equation giving the critical
values of t. (We must be careful to avoid a parametrization such
that $\mathrm{d}x/\mathrm{d}t = \mathrm{d}y/\mathrm{d}t = \mathrm{d}z/\mathrm{d}t = 0$ at some point on the path: at such
a point, a nonexistent stationary point would be predicted.)

> **Stationary points of** $f(x, y, z)$ **on the path**
> $\big(x(t), y(t), z(t)\big)$
> The stationary points are the solutions of   (28.22)
> $$\frac{\partial f}{\partial x}\frac{dx}{dt} + \frac{\partial f}{\partial y}\frac{dy}{dt} + \frac{\partial f}{\partial z}\frac{dz}{dt} = 0.$$

(In particular cases it might be easier to substitute $x(t)$, $y(t)$, $z(t)$ directly into $f(x, y, z)$ for the turning points of f with respect to t.)

It is more usual for restricted stationary-value problems to be formulated in a way that avoids parametric considerations. The **restriction to a surface** is the easier case. Instead of a fish in the body of the sea, consider a crab which confines itself to the undulating seabed described by an equation of the form

$$g(x, y, z) = c,$$

encountering there the local pollution, given throughout the sea by $f(x, y, z)$. The crab does not know about the rest of the sea, but as it moves around it will meet highs and lows (and other stationary points) unconnected with possibly more extreme pollution in the body of the sea. A stationary point will be found at a point Q on the surface $g(x, y, z) = c$ if

$$\frac{df}{ds} = 0 \quad \text{at } Q$$

in all directions $\hat{s}$ **from** Q which do not point into the body of the sea, but are **tangential to the surface** $g(x, y, z) = c$.

Figure 28.6 shows such a point Q, and various tangential directions denoted by unit vectors $\hat{s}$ pointing away from Q. From (28.15), the required condition is

$$\frac{df}{ds} = 0 = \hat{s} \cdot \mathbf{grad}\, f \quad \text{at } Q, \text{ for all such } \hat{s}. \qquad (28.23)$$

In other words, **grad** f *must be perpendicular to the surface at Q* (ignoring for the moment the chance that **grad** f might be zero at Q). But, by (28.11), **grad** g is *always perpendicular to the surface* $g(x, y, z) = c$—in particular at Q. Therefore **grad** f and **grad** g, evaluated at Q, are parallel vectors; so

$$\mathbf{grad}\, f = \lambda\, \mathbf{grad}\, g \quad \text{at } Q,$$

where λ is an (unknown) constant, called a **Lagrange multiplier** for the problem. By writing **grad** f and **grad** g in their components, we obtain

$$\frac{\partial f}{\partial x} - \lambda\frac{\partial g}{\partial x} = 0, \qquad \frac{\partial f}{\partial y} - \lambda\frac{\partial g}{\partial y} = 0, \qquad \frac{\partial f}{\partial z} - \lambda\frac{\partial g}{\partial z} = 0.$$

$$(28.24\text{a,b,c})$$

We now have three equations for the four unknowns: (x, y, z) (the

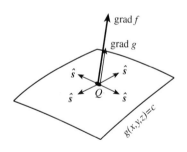

Fig. 28.6

position of Q) and λ. To find another equation, notice that (28.24a,b,c) would be unaffected if we had $g(x, y, z)$ equal to some constant other than c, so it is necessary to reassert the particular surface:

$$g(x, y, z) = c. \tag{28.24d}$$

The very special case mentioned above, that (28.23) is satisfied alternatively if (by chance) **grad** $f = 0$ at Q, is still governed by the same equations. When they are solved, we should merely find that $\lambda = 0$. (We should not usually realize this in advance.) The case corresponds to the unrestricted stationary-point problem (see (28.20)), where the point found happens to lie in the specified surface.

Restricted stationary-point problem:
stationary points of $f(x, y, z)$ **subject to**
$g(x, y, z) = c$

Solve for x, y, z, λ the equations

$$g = c, \tag{i}$$

$$\frac{\partial f}{\partial x} - \lambda \frac{\partial g}{\partial x} = 0, \tag{ii}$$

$$\frac{\partial f}{\partial y} - \lambda \frac{\partial g}{\partial y} = 0, \tag{iii}$$

$$\frac{\partial f}{\partial z} - \lambda \frac{\partial g}{\partial z} = 0. \tag{iv}$$

(28.25)

Example 28.8. Find the stationary points of $x^2 + y^2 + yz + zx$ on the hyperboloid $x^2 + y^2 - z^2 = 1$.

The four equations (28.25) are

$$x^2 + y^2 - z^2 = 1, \tag{i}$$

$$2x + z - 2\lambda x = 0, \quad \text{or} \quad (2 - 2\lambda)x + z = 0; \tag{ii}$$

$$2y + z - 2\lambda y = 0, \quad \text{or} \quad (2 - 2\lambda)y + z = 0; \tag{iii}$$

$$y + x + 2\lambda z = 0, \quad \text{or} \quad x + y + 2\lambda z = 0. \tag{iv}$$

Equations (ii), (iii), and (iv) constitute a set of homogeneous linear algebraic equations for x, y, z. The only possibilities are *either* that $x = y = z = 0$, which is excluded since these values do not satisfy (i), *or* that the determinant of the coefficients is zero:

$$\det \begin{bmatrix} 2 - 2\lambda & 0 & 1 \\ 0 & 2 - 2\lambda & 1 \\ 1 & 1 & 2\lambda \end{bmatrix} = 0,$$

so that $(1 - \lambda)(2\lambda^2 - 2\lambda + 1) = 0$. The only real solution is
$$\lambda = 1.$$
The equations then become
$$z = 0, \quad z = 0, \quad x + y + 2z = 0,$$
or
$$z = 0, \quad y = -x. \tag{v}$$
Substitute for y in terms of x into (i). Then
$$2x^2 = 1, \quad \text{or} \quad x = \pm 1/\sqrt{2}.$$
Therefore, using (v) again gives the stationary points
$$(\pm 1/\sqrt{2}, \quad \mp 1/\sqrt{2}, \quad 0).$$

For the corresponding problem of finding the **stationary points of** $f(x, y, z)$ **on a specified curve**, as for the fish problem discussed at the beginning of the section, the curve will be assumed to be specified by the intersection of two surfaces:
$$g(x, y, z) = c_1, \qquad h(x, y, z) = c_2.$$
The method of solution involves two Lagrange multipliers:

Restricted stationary-point problem:
Stationary points of $f(x, y, z)$ **subject to**
$g(x, y, z) = c_1$ **and** $h(x, y, z) = c_2$.

Solve for x, y, z, λ, μ the equations

$$g = c_1 \qquad h = c_2, \qquad \text{(i), (ii)}$$

$$\frac{\partial f}{\partial x} - \lambda \frac{\partial g}{\partial x} - \mu \frac{\partial h}{\partial x} = 0, \qquad \text{(iii)}$$

$$\frac{\partial f}{\partial y} - \lambda \frac{\partial g}{\partial y} - \mu \frac{\partial h}{\partial y} = 0, \qquad \text{(iv)}$$

$$\frac{\partial f}{\partial z} - \lambda \frac{\partial g}{\partial z} - \mu \frac{\partial h}{\partial z} = 0. \qquad \text{(v)}$$

(28.26)

We shall not give the proof in full. Briefly, the situation is shown in Fig. 28.7. Q is a stationary point on the curve of intersection and $\hat{s}$ a unit vector tangential to it at Q. Since
$$\frac{\mathrm{d}f}{\mathrm{d}s} = \hat{s} \cdot \mathbf{grad}\, f = 0 \quad \text{at } Q,$$
$\mathbf{grad}\, f$ is perpendicular to $\hat{s}$ at Q. For the same reason as in the earlier case, $\mathbf{grad}\, g$ and $\mathbf{grad}\, h$ are also perpendicular to $\hat{s}$ at Q. Therefore the three vectors $\mathbf{grad}\, f$, $\mathbf{grad}\, g$, $\mathbf{grad}\, h$ all lie in the same plane (which is perpendicular to $\hat{s}$), so $\mathbf{grad}\, f$ can be expressed in terms of the other two vectors:
$$\mathbf{grad}\, f = \lambda\, \mathbf{grad}\, g + \mu\, \mathbf{grad}\, h,$$

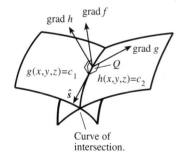

Fig. 28.7

grad h grad f

grad g

$g(x,y,z)=c_1$ Q

$h(x,y,z)=c_2$

$\hat{s}$

Curve of intersection.

where λ and μ are certain constants, the **Lagrange multipliers** for this problem. Then split this equation into its components to obtain (28.26).

Example 28.9. Find the stationary points of $x^2 + y^2 + z^2$ on the curve of intersection of the vertical cylinder $x^2 + y^2 = 1$ with the plane $x + y + z = 1$. (This is an inclined ellipse.)

Here $f(x, y, z) = x^2 + y^2 + z^2$, $g(x, y, z) = x^2 + y^2$, and $h(x, y, z) = x + y + z$. The equations to be solved become

$$x^2 + y^2 = 1, \tag{i}$$
$$x + y + z = 1, \tag{ii}$$
$$2x - \lambda 2x - \mu = 0, \quad \text{or} \quad 2x(1 - \lambda) = \mu, \tag{iii}$$
$$2y - \lambda 2y - \mu = 0, \quad \text{or} \quad 2y(1 - \lambda) = \mu, \tag{iv}$$
$$z - \mu = 0. \tag{v}$$

From (iii) and (iv), *either* (a) $\lambda = 1$, so that $\mu = 0$, *or* (b) $\lambda \neq 1$, so that $x = y$. We consider these possibilities in order.

(a) *The case* $\lambda = 1$, $\mu = 0$. (We cannot deduce anything about x and y from (iii) and (iv) if this is true.) From (v) we obtain $z = 0$, so (i) and (ii) become

$$x^2 + y^2 = 1, \qquad x + y = 1.$$

The solutions are $x = 0$, $y = 1$, and $x = 1$, $y = 0$. Then we have found two solutions:

$$(0, 1, 0) \quad \text{and} \quad (1, 0, 0).$$

(b) *The case* $\lambda \neq 1$, $x = y$. From (i), $x = \pm 1/\sqrt{2}$, $y = \pm 1/\sqrt{2}$. Equation (ii) then gives $z = 1 - x - y = 1 \mp \sqrt{2}$. Thus we have two more solutions:

$$(1/\sqrt{2}, 1/\sqrt{2}, 1 - \sqrt{2}) \quad \text{and} \quad (-1/\sqrt{2}, -1/\sqrt{2}, 1 + \sqrt{2}).$$

For a **restricted stationary-point problem in N variables**, there may be up **to $N - 1$ restricting equations, or constraints, with the corresponding number of Lagrange multipliers**. The equations to be solved then follow the pattern of (28.25) and (28.26).

The identification a **maximum** or **minimum** is usually of most interest. The general question is difficult, but sometimes it is fairly obvious. For instance, in the last example, the values of f are restricted to a closed curve, so the corresponding values of f make it clear that the points (a) give minima of f, and points (b) give maxima.

28.9 The envelope of a family of curves
Figure 28.8a shows the straight lines

$$y = \alpha - \alpha^2 x,$$

for several values of α, which we call the **parameter of the family**

(a)

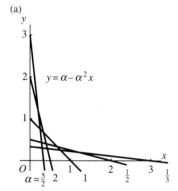

$$y = \alpha - \alpha^2 x$$

$$\alpha = \tfrac{5}{2} \quad 2 \quad 1 \quad \tfrac{1}{2} \quad \tfrac{1}{3}$$

(b)

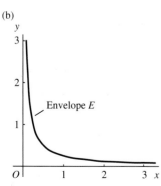

Envelope E

Fig. 28.8

of straight lines. The 'boundary' of the family is starting to form itself into a curve E, which is sketched in Fig. 28.8b. The reason why the curve E is sharply defined is because all the straight lines are tangential to it, and therefore reinforce it along its length. The curve is called the **envelope of the family** $y = \alpha x - \alpha^2 x$, where α is the **parameter of the family**.

The family does not have to consist of straight lines. Suppose that the family is described by

$$f(x, y, \alpha) = 0.$$

To find the envelope (Fig. 28.9) consider two close values of the parameter, α and $\alpha + \delta\alpha$, the corresponding curves of the family being

$$f(x, y, \alpha) = 0 \quad \text{and} \quad f(x, y, \alpha + \delta\alpha) = 0.$$

The intersection point R is the point where

$$f(x, y, \alpha) = f(x, y, \alpha + \delta\alpha) \quad (=0).$$

Therefore, at the point R,

$$\frac{f(x, y, \alpha + \delta\alpha) - f(x, y, \alpha)}{\delta\alpha} = 0,$$

Now let $\delta\alpha \to 0$. Then R and Q come together at P on the envelope, and this equation becomes

$$\frac{\partial f(x, y, \alpha)}{\partial \alpha} = 0, \tag{28.27}$$

at P. Also P lies on the curve

$$f(x, y, \alpha) = 0. \tag{28.28}$$

(We had not yet used the fact that f is zero rather than some other constant.) If we eliminate α between (28.27) and (28.28), we obtain an equation in x and y which describes the envelope.

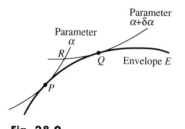

Parameter $\alpha + \delta\alpha$

Parameter α

R

Q Envelope E

P

Fig. 28.9

Envelope of the family of curves $f(x, y, \alpha) = 0$, **where α is a parameter.**

The result of eliminating α between the equations **(28.29)** $f(x, y, \alpha) = 0$ and $\dfrac{\partial f}{\partial \alpha} = 0$ contains the envelope.

(The solution might also include the track of other peculiarities.)

Example 28.10. Find the envelope of the family of straight lines $y = \alpha - \alpha^2 x$, where α is the parameter. (See Fig. 28.8.)

Let $f(x, y, \alpha) = y - \alpha + \alpha^2 x$. Then

$$\frac{\partial f}{\partial \alpha} = -1 + 2\alpha x = 0$$

Therefore
$$\alpha = 1/2x. \qquad\qquad\text{(i)}$$
On the envelope, also
$$y - \alpha + \alpha^2 x = 0; \qquad\qquad\text{(ii)}$$
so, from (i), $y - 1/2x + 1/4x = 0$, or
$$y = 1/4x,$$
which is a rectangular hyperbola (see Fig. 28.7b).

Problems, Chapter 28

28.1. Write down the incremental approximation for δf in the following cases.
(a) $f(x, y, z) = 2x + 3y^2 + 4z^2 - 3$;
(b) $f(x, y, t) = (x^2 + y^2)^{-\frac{1}{2}} e^{-t}$;
(c) $f(r, \theta, t) = e^{-t} r \cos \theta$;
(d) $f(x, y, z, t) = x^2 + y^2 + z^2 - t^2$;
(e) $f(x_1, y_1, x_2, y_2) = (x_1 - x_2)^2 + (y_1 - y_2)^2$;
(f) $f(x, y, z, t) = (1/r) e^{-(x^2+y^2)/t}$. Compare with the expression for δg when $g(r, t) = (1/r) e^{-r^2/t}$ in polar coordinates.

28.2. The distance d between two points (x_1, y_1, z_1) and (x_2, y_2, z_2) in a plane is given by $d^2 = (x_1 - x_2)^2 + (y_1 - y_2)^2 + (z_1 - z_2)^2$. Find approximately the change from $(1, 1, 2), (1, 2, 1)$ to $(1.1, 0.9, 1.8), (0.9, 2.1, 1.1)$.

28.3. R_1, R_2, R_3, R_4 are resistances in a circuit whose overall resistance is R, arranged so that
$$1/R = 1/R_4 + (R_1 + R_2)/(R_1 R_2 + R_2 R_3 + R_3 R_1).$$
Find an expression for δR in terms of $\delta R_1, \delta R_2, \delta R_3$, and δR_4.
Suppose that initially $R_1 = 3, R_2 = 10, R_3 = 5$, and $R_4 = 10$, and that R_1 becomes 3.2 and R_2 becomes 9.8. Estimate the change in R_3 necessary if R is to remain unaltered.

28.4. The equation $2x^3 - 3x - 45 = 0$ has a solution $x = 3$. Find an approximate solution to the equation $2.1x^3 - 2.9x - 47 = 0$.

28.5. Estimate the maximum possible error and the corresponding percentage error in w for the following cases.
(a) $w = yz + zx + xz$, $x = 2$ (± 0.1), $y = 3$ (± 0.2), $z = 1$ (± 0.1).
(b) $w = (x - y)(y - z)(z - x)$, $x = 1(\pm 0.1)$, $y = 2(\pm 0.1)$, $z = 3(\pm 0.1)$.
(c) $w = (x + y + z - t)^{-1}$, where it is known only that $x = 1.2, y = 2.9, z = 1.9$, and $t = 2.1$ after round-

ing to one decimal place. Compare with the exact maximum and percentage errors.

28.6. Estimate the maximum error and the maximum percentage error for the following.
(a) (For c) $c^2 = a^2 + b^2 - 2ab \cos A$ (the 'cosine rule' for a triangle ABC). Here $a = 2$ (± 0.1), $b = 4$ (± 0.1), $A = 135°$ $(\pm 2°)$. [Note $\Delta(c^2) \approx 2c \, \Delta c$.]
(b) (For d)
$$d^2 = (x_1 - x_2)^2 + (y_1 - y_2)^2 + (z_1 - z_2)^2,$$
where the measured values $(x_1, y_1, z_1) = (1, 2, 1)$ and $(x_2, y_2, z_2) = (2, 1, 1)$ have been rounded to one significant figure. [Note $\Delta(d^2) \approx 2d \, \Delta d$.]
(c) (For A) The area of a triangle with sides a, b, c is given by
$$A = [s(s - a)(s - b)(s - c)]^{\frac{1}{2}},$$
where $s = \frac{1}{2}(a + b + c)$. Consider the case when $a = 2$, $b = 4, c = 3$, all with possible errors as large as ± 0.1. [You can substitute s directly into the formula for A or A^2, but it is easier algebraically to obtain two simultaneous equations, with numerical coefficients, involving $\Delta A, \Delta s, \Delta a, \Delta b, \Delta c$.]

28.7. (Section 24.2). If $f(x, y, z, w) = c$ (a constant), then any of the four variables is a function of the other three. Use (28.6) to show that
(a) $\dfrac{\partial x}{\partial y} \dfrac{\partial y}{\partial x} = 1$; (b) $\dfrac{\partial x}{\partial y} \dfrac{\partial y}{\partial z} = -\dfrac{\partial x}{\partial z}$.
(c) Simplify $\dfrac{\partial x}{\partial y} \dfrac{\partial y}{\partial z} \dfrac{\partial z}{\partial w} \dfrac{\partial w}{\partial x}$.
Test the truth of the results in the cases:
(i) $x + 2y + 3z + 4w = 5$; (ii) $xy^2 z^3 w = 1$.

28.8. Assume that the following relations define z *implicitly* as a function of x and y. Write down the relation between $\delta x, \delta y, \delta z$ at the points prescribed. Without solving for z, deduce $\partial z/\partial x$ and $\partial z/\partial y$ at the points.

(a) $2x - 3y + 4z = 1$ at points satisfying the condition;
(b) $x^2 + y^2 + z^2 = 14$ at $(1, 2, -3)$;
(c) $4x^3 + y^4 + 9z^3 - xyz^2 = 13$ at $(1, 1, 1)$;
(d) $x^2 - z^2 = 9$ at $x = 5$, $y = y_0$, $z = 4$. (this is a hyperbolic cylinder).

28.9. (a) Compare the result of using the chain rule (28.7) with that of direct substitution in order to find df/dt when $f(x, y, z) = xy/z$ and $x = t, y = 4t, z = 2t$.
(b) The same parametrization as in (a), but with the function $f(x, y, z) = \sin(xy/z)$.
(c) Obtain an expression for df/dt on the path in (a) when $f(x, y, z) = g(xy/z)$, g being any function, and confirm that it works with case (b). (Hint: express the result in terms of g'.)

28.10. Cylindrical coordinates r, θ, z are shown in Fig. 28.10. They are related to x, y, z by $x = r\cos\theta$, $y = r\sin\theta$, $z = z$.

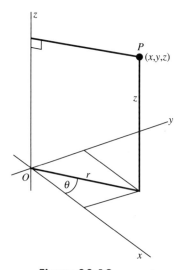

Figure 28.10

(a) Given $f(x, y, z)$, use the chain rule (28.8) with r, θ, z as the parameters to express $\partial f/\partial r, \partial f/\partial\theta, \partial f/\partial z$ in terms of $\partial f/\partial x, \partial f/\partial y, \partial f/\partial z$.
(b) Regarding (a) as a pair of equations for $\partial f/\partial x$ and $\partial f/\partial y$, show that

$$\frac{\partial f}{\partial x} = \cos\theta\,\frac{\partial f}{\partial r} - \frac{\sin\theta}{r}\frac{\partial f}{\partial\theta} \quad \text{and}$$

$$\frac{\partial f}{\partial y} = \sin\theta\,\frac{\partial f}{\partial r} + \frac{\cos\theta}{r}\frac{\partial f}{\partial\theta}.$$

(c) The results (b) show that the differentiation operations $\partial/\partial x$ and $\partial/\partial r$ are equivalent respectively

to the polar forms

$$\cos\theta\,\frac{\partial}{\partial r} - \frac{\sin\theta}{r}\frac{\partial}{\partial\theta} \quad \text{and} \quad \sin\theta\,\frac{\partial}{\partial r} + \frac{\cos\theta}{r}\frac{\partial}{\partial\theta}.$$

Use this fact to confirm that

$$\frac{\partial^2 f}{\partial x^2} + \frac{\partial^2 f}{\partial y^2} = \frac{\partial^2 f}{\partial r^2} + \frac{1}{r}\frac{\partial f}{\partial r} + \frac{1}{r^2}\frac{\partial^2 f}{\partial\theta^2}.$$

28.11. Obtain the vector function **grad** f for each of the following.
(a) $x + y + z$; (b) $2x - 3y + 5z - 6$;
(c) $x^2 + y^2 + z^2$; (d) $x^3 + 3z^3 - 1$ (in three dimensions);
(e) $x^2 - \frac{1}{4}y^2 + \frac{1}{9}z^2$;
(f) $1/r$, where $r = (x^2 + y^2 + z^2)^{\frac{1}{2}}$; confirm that the gradient vector points in the direction of the position vector (x, y, z).

28.12. Obtain a vector which is normal to the following surfaces at the points specified, and construct a unit vector from it.
(a) $x - 2y + z = 0$ at any point;
(b) $y^2 + z^2 = 2$ at any point;
(c) $x^2 + y^2 + z^2 = 9$ at $(2, 1, -2)$;
(d) $\frac{1}{4}x^2 + \frac{1}{9}y^2 + \frac{1}{16}z^2 = 3$ at $(2, 3, 4)$;
(e) $x^3y + zx^3 = 5$ at $(1, 2, 3)$;
(f) $\dfrac{1}{x} + \dfrac{1}{y} + \dfrac{1}{z} = 1$ at $(2, 3, 6)$;
(g) $(x^2 + 4y^2 - z^2)^{-1} = \frac{1}{16}$ at $(4, 1, 2)$.

28.13. By finding the gradient vectors, obtain the angle between the following surfaces at the point of intersection given.
(a) $x^2 + y^2 + z^2 = 9$, $x^2 - z^2 = 0$ at $(2, 1, 2)$;
(b) $x^2 - y^2 + z^2 = 1$, $2x - 3y + z + 1 = 0$ at $(2, 2, 1)$;
(c) $x^2 + y^2 - z^2 = 0$, $3x + 4y - 5z = 50$ at $(3, 4, 5)$; explain the result.

28.14. (a) Find **grad** f for

$$f(x, y, z) = A\,e^{\alpha(2x^2 + 4y^2 + z^2)^{\frac{1}{2}}},$$

where A and α are constants. Deduce that the vector $(2x, 4y, z)$ points in the direction of **grad** f.
(b) Let $f(x, y, z) = g[u(x, y, z)]$, where g and u are two other functions. Show that

$$\textbf{grad}\, f = \left(g'(u)\,\frac{\partial u}{\partial x}, g'(u)\,\frac{\partial u}{\partial y}, g'(u)\,\frac{\partial u}{\partial z}\right),$$

and deduce that **grad** u points either in the same or in the opposite direction to **grad** f.

28.15. Write down expressions for the directional derivative of the following at the point (x, y, z), in terms of a unit direction vector $\hat{s}$.

(a) $x + 2y + 3z$; (b) $x^2 - y^2 - 3z$;
(c) $(x - 1)^3 + y^3 + z^3$.

28.16. Find df/ds for the following functions f, taken at the point $(2, 3, 2)$ in the direction $\hat{s} = (\frac{1}{4}\sqrt{2}, \frac{1}{4}\sqrt{2}, \frac{1}{2}\sqrt{3})$.
(a) $x - y + 2z$; (b) $xy + yz + zx$;
(c) $(xy + yz + zx)^2$;
(d) $x^2 - y^2 + 5$ (in three dimensions: this represents a vertical cylinder).

28.17. The equations for two surfaces, $f(x, y, z) = a$, $g(x, y, z) = b$, where a and b are constants, together represent their curve of intersection, C. Show that the vector product $\mathbf{grad}\, f \times \mathbf{grad}\, g$, evaluated at a point on C, points in the direction of C. Use this to find a unit vector $\hat{s}$ in the direction of C in the following cases.
(a) $2x + 3y - z = 1$, $x - y - z = 0$, at any common point.
(b) $x + y = 0$, $x - z = 0$, at any common point.
(c) $x^2 + y^2 + z^2 = 6$, $x - y + z = 0$, at $(1, 2, 1)$
(d) $x^2 + (y - 1)^2 = 1$, $x^2 + (y - 2)^2 = 4$, at $x = 0$, $y = 0$, and any value of z. Explain what is happening here.
(e) $xy + yz + zx = 3$, $x + y + z = 3$, at $(1, 1, 1)$.

28.18. Find the stationary points of the following functions with respect to all the variables named in f.
(a) $f(x, y, z) = x^2 + y^2 + z^2$;
(b) $f(x, y, z) = x^3 - 3x + y^3 - 3yz + 2z^2$;
(c) $f(x, y, z) = xy + yz + zx + y - z$;
(d) $f(x, y, z) = x/z + y/x + z/y$;
(e) $f(x, y, z, \lambda) = (x + y + z) - \lambda(x^2 + y^2 + z^2 - 1)$;
(f) $f(x, y, z) = x^4 + y^4 + z^4 - 2(x - y + z)^2$.

28.19. Find the stationary points of $x^2 + y^2 + z^2$ on the path

$$x = \cos t, \qquad y = \sin t, \qquad z = \sin \tfrac{1}{2}t,$$

where $0 < t < 4\pi$.

28.20. At the point (x, y, z) in the air, an insecticide maintains a concentration

$$s = C \exp\{-\alpha[2(x - 1)^2 + 4y^2 + z^2]\},$$

where C is a constant. An insect is trying to escape by following the path of most rapid decrease in concentration.
(a) Show that, when it is at (x, y, z), its direction is that of the vector $(2(x - 1), 4y, z)$.
(b) In a short interval of time δt, it moves to $(x + \delta x, y + \delta y, z + \delta z)$. Show that, approximately,

$$\frac{\delta x}{2(x - 1)} = \frac{\delta y}{4y} = \frac{\delta z}{z}.$$

(c) By letting $\delta t \to 0$, show that its path is described by the two differential equations

$$\frac{dz}{dx} = \frac{z}{2(x - 1)}, \qquad \frac{dz}{dy} = \frac{z}{4y}.$$

(Such simultaneous equations would often be written $dx/2(x - 1) = dy/4y = dz/z$.)
(d) Show that the general solution of these equations, which expresses a path in space, can be written as

$$z = Ay^{\frac{1}{4}} = B(x - 1)^{\frac{1}{2}},$$

where A and B are arbitrary constants.
(e) Assuming that the insect starts at $(0, 1, 1)$, find its path.

28.21. Use the Lagrange-multiplier technique of (28.25)–(28.26) to solve the following restricted stationary-point (SP) problems.
(a) SPs of $x + y + z$ subject to $1/x + 1/y + 1/z = 1$.
(b) SPs of xyz subject to $1/x + 1/y + 1/z = 1$.
(c) SPs of $x^2 + y^2 + z^2$ subject to $ax + by + cz = 1$.
(d) SPs of $xy + yz + zx$ subject to $xyz = 1$. (This corresponds to finding the rectangular block of given volume which has the smallest surface area.)
(e) The problem of the rectangular block of greatest volume which can be fitted into an ellipsoid leads to the problem: find the SPs of xyz subject to $x^2/a^2 + y^2/b^2 + z^2/c^2 = 1$.
(f) SPs of $x^2 + 4y^2 + z^2$ on the intersection of the two planes $x - y - 2z = 0$ and $z = 1$.
(g) SPs of $x^2 - y^2 - z^2$ on the straight line

$$\frac{x - 1}{2} = \frac{y - 2}{-1} = \frac{z - 2}{3}.$$

(h) SPs of xyz subject to $xy + yz + zx = 1$. (Compare (d).)
(i) SPs of $x - y - 2z$ on the intersection of $z = 1$ with $x^2 + 4y^2 + z^2 = 6$. (Compare (f).)

28.22. (Numerical). (a) Write a program to carry out the numerical scheme suggested in Example 28.6 *in two dimensions* in order to obtain the curves along which a function $f(x, y)$ increases most rapidly. It may also be possible for you to display the curves on the screen.
(b) Use (a) to obtain a numerical solution of the following problems. Try a succession of decreasing step lengths h in order to ensure plotting accuracy.
(c) The altitude H of a part of a hill is given in km by

$$H = 0.5 - x^2 - 4y^2.$$

Start e.g. with $(x, y) = (2, 2)$, and go to the summit. (For comparison, the exact solution to the problem is $y = \frac{1}{8}x^4$; the summit is at the origin.)

(d) Plot the track of most rapid *descent* from the point (3, 2) in the case where $H = \frac{1}{2} + x^2 - y^2$. (To descend, use negative h. The shape is a saddle: viewed from the origin, H increases east and west and decreases north and south.)

(e) In certain types of fluid flow in two dimensions, the velocity vector $v(x, y)$ is equal to the gradient of a single scalar **potential** function $\phi(x, y, z)$: $v = \textbf{grad }\phi$. A **streamline** through any point is in the direction of v. Plot some streamlines on and outside of the circle $x^2 + y^2 = 1$ when

$$\phi(x, y) = x\left(1 + \frac{1}{x^2 + y^2}\right).$$

28.23. Often a function $f(x, y, z)$ takes the form of 'a function of a function': $w = f(x, y, z) = g\big(u(x, y, z)\big)$. (An example is

$$w = f(x, y, z) = \sin xyz : u = xyz, w = \sin u.)$$

(a) Write down several examples of functions which can be regarded in this way.

(b) Show that

$$\frac{\partial f}{\partial x} = g'(u)\frac{\partial u}{\partial x}, \qquad \frac{\partial f}{\partial y} = g'(u)\frac{\partial u}{\partial y}, \qquad \frac{\partial f}{\partial z} = g'(u)\frac{\partial u}{\partial z}.$$

(You only need the one-variable chain rule (3.3).)

(c) Check the correctness of the formulae (b) in the cases when (i) $w = e^{x^2 - y^2 + z^2}$, (ii) $w = \sin(xy/z)$.

(d) Using the results (b), rewrite the chain rule (28.7) in the form appropriate to functions of the form $g\big(u(x, y, z)\big)$.

(e) The path $x = \cos t$, $y = \sin t$, $z = t$ represents a helix whose axis is the z axis. Find an expression in terms of t for df/dt on the path when $f(x, y, z) = g(xy/z)$, where g is any function. Confirm the result for any simple case.

28.24. Often a function takes the form

$$w = f(u, v, w),$$

where u, v, and w are themselves functions of x, y, and z. Write down a version of the chain rule (24.8) which enables $\partial f/\partial x$, $\partial f/\partial y$, $\partial f/\partial z$ to be found. (In this case, x, y, and z function like parameters and u, v, and w like the principal variables.) Use this result to prove the following results:

(a) If $w = f(x - y, y - z, z - x)$, where f is any function, then

$$\frac{\partial w}{\partial x} + \frac{\partial w}{\partial y} + \frac{\partial w}{\partial z} = 0.$$

Check your result with the function

$$(x - y)(y - z)(z - x).$$

(b) If $w = f(y/x, z/x)$, where f is any function of *two* variables, then

$$x\frac{\partial w}{\partial x} + y\frac{\partial w}{\partial y} + z\frac{\partial w}{\partial z} = 0.$$

Try it out with the function $x/y + y/z + z/x$, noting that

$$\frac{y}{x}\bigg/\frac{z}{x} = \frac{y}{z}.$$

28.25. Let $f(x, y, z, t) = e^{j(k_1 x + k_2 y + k_3 z - \omega t)}$, where j is the complex element ($j^2 = -1$), and k_1, k_2, k_3, and ω are constants. Show that

$$\frac{\partial^2 f}{\partial x^2} + \frac{\partial^2 f}{\partial y^2} + \frac{\partial^2 f}{\partial z^2} = \frac{1}{c^2}\frac{\partial^2 f}{\partial t^2},$$

where $c = \omega/(k_1 + k_2 + k_3)$. (This is called the **wave equation** in three dimensions, and $f(x, y, z, t)$ is one of its solutions.)

Prove that $g(k_1 x + k_2 y + k_3 z - \omega t)$, where g is any function of a single variable, is also a solution.

28.26. Find the envelopes of the following families.
(a) $y = \alpha x + \alpha^2 x$ (parameter α);
(b) $y + \alpha^2 x = \alpha$ (parameter α);
(c) $\dfrac{x}{\alpha} + \dfrac{y}{1 - \alpha} = 1$ (parameter α);
(d) $x\cos\theta + y\sin\theta = 1$ (parameter θ).

28.27. The cross-sectional profile of a long cylindrical mirror is the semicircle $x^2 + y^2 = 1$ in the right-hand half plane. Rays from the left, parallel to the x axis, fall on the mirror.
(a) Show that the equation of the ray reflected from the point $(\cos\theta, \sin\theta)$ on the mirror is $x\sin 2\theta - y\cos 2\theta = \sin\theta$.
(b) By regarding θ as the parameter, show that the envelope of these reflected rays is given by $x^2 + y^2 = \frac{1}{4}(3y^{\frac{2}{3}} + 1)$. (In optics, this envelope is called the **caustic** of the reflected rays.)

29 Double integration

29.1 Repeated integrals with constant limits

Before explaining how they arise, we show first how a **repeated integral** is written and evaluated. The following is an example of a repeated integral with **constant limits**:

$$I = \int_0^1 \int_0^2 (xy + y^2 - 1)\, dx\, dy.$$

There are two stages of integration; first with respect to x, then with respect to y, this being determined by the order in which dx and dy appear under the integral signs.

The reader is recommended to copy the following procedure step-by-step at first.

(i) Put brackets round the **inner integral**, which is the first to be evaluated:

$$I = \int_0^1 \left(\int_0^2 (xy + y^2 - 1)\, dx \right) dy.$$

(ii) Make it clear which variable connects with which limits of integration, by explicitly labelling them as shown:

$$I = \int_{y=0}^1 \left(\int_{x=0}^2 (xy + y^2 - 1)\, dx \right) dy.$$

(iii) Evaluate the inner integral with respect to the first variable (here x), **treating the other variable (y) as a constant**:

$$\int_{x=0}^2 (xy + y^2 - 1)\, dx = \left[\tfrac{1}{2}x^2 y + y^2 x - x \right]_{x=0}^2$$

$$= 2y + 2y^2 - 2.$$

This process eliminates the variable x.

(iv) Use the result of (iii) as the integrand of the **outer integral**:

$$I = \int_{y=0}^1 (2y + 2y^2 - 2)\, dy = \left[y^2 + \tfrac{2}{3}y^3 - 2y \right]_{y=0}^1 = -\tfrac{1}{3}.$$

This eliminates the variable y, so that the final **result is a definite**

number. If you find you are left with an x or y in the result, then you have not followed the process correctly.

Example 29.1. Evaluate the repeated integrals

(a) $I = \displaystyle\int_0^1 \int_2^4 (xy + 1)\,dx\,dy$, $\qquad$ (b) $J = \displaystyle\int_2^4 \int_0^1 (xy + 1)\,dy\,dx$.

(a) $I = \displaystyle\int_{y=0}^1 \left(\int_{x=2}^4 (xy + 1)\,dx \right) dy$.

The inner integral becomes

$$\int_{x=2}^4 (xy + 1)\,dx = \left[\tfrac{1}{2}x^2 y + x \right]_{x=2}^4 = 8y + 4 - (2y + 2)$$

$$= 6y + 2.$$

This forms the integrand of the outer integral:

$$I = \int_{y=0}^1 (6y + 2)\,dy = \left[3y^2 + 2y \right]_{y=0}^1 = 5.$$

(b) Here the order of the symbols $\int_{x=2}^4$ and $\int_{y=0}^1$ has been reversed, and also the order of dx and dy. In other words, the same processes are to be carried out, but in the reverse order. The details, however, look different.

We have

$$J = \int_{x=2}^4 \left(\int_{y=0}^1 (xy + 1)\,dy \right) dx.$$

The inner integral is with respect to y, and we treat x as a constant:

$$\int_{y=0}^1 (xy + 1)\,dy \quad (x \text{ constant}).$$

This is equal to

$$\left[x(\tfrac{1}{2}y^2) + y \right]_{y=0}^1 = \tfrac{1}{2}x + 1.$$

The outer integral becomes

$$J = \int_{x=2}^4 (\tfrac{1}{2}x + 1)\,dx = 5,$$

which is the same as I in (a).

In this example, it makes no difference whether we integrate with respect to x or y first. Later we show that this is always true when the repeated integral has constant limits.

29.2 Examples leading to repeated integrals with constant limits

Figure 29.1a represents a heap of grain in a rectangular silo of length 8 m and breadth 4 m. The top surface of the grain is curved, with

(a)

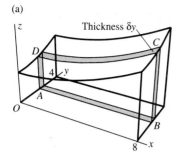

(b)

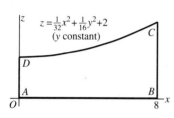

Fig. 29.1

the equation

$$z = \tfrac{1}{32}x^2 + \tfrac{1}{16}y^2 + 2 \quad (0 \leqslant x \leqslant 4;\ 0 \leqslant y \leqslant 8),$$

and problem is to find the volume, V say, of the grain.

Imagine the grain divided into thin vertical plane slices, parallel to the (x, z) plane, the thickness of a slice being δy. A typical slice is shown in Fig. 29.1a, and the value of y is constant on its faces. It is lifted out and displayed in elevation separately in Fig. 29.1b. The face area is given by

$$\text{Area ABCD} = \int_{x=0}^{8} (\tfrac{1}{32}x^2 + \tfrac{1}{16}y^2 + 2)\, dx,$$

in which y takes the current constant value. Therefore its volume, δV say, is given by

$$\delta V \approx \left(\int_{x=0}^{8} (\tfrac{1}{32}x^2 + \tfrac{1}{16}y^2 + 2)\, dx \right) \delta y.$$

When we take the sum of all the elements δV and let $\delta y \to 0$, we obtain in the usual way

$$V = \int_{y=0}^{4} \left(\int_{x=0}^{8} (\tfrac{1}{32}x^2 + \tfrac{1}{16}y^2 + 2)\, dx \right) dy.$$

The result has therefore taken the form of a repeated integral of the kind described in Section 29.1. In evaluating it, the inner integral gives the cross-sectional area of a slice on which y is held constant:

$$\int_{x=0}^{8} (\tfrac{1}{32}x^2 + \tfrac{1}{16}y^2 + 2)\, dx = \left[\tfrac{1}{96}x^3 + \tfrac{1}{16}y^2 x + 2x \right]_{x=0}^{8}$$

$$= \tfrac{64}{3} + \tfrac{1}{2}y^2.$$

Finally

$$V = \int_{y=0}^{4} (\tfrac{64}{3} + \tfrac{1}{2}y^2)\, dy = \left[\tfrac{64}{3}y + \tfrac{1}{6}y^3 \right]_{y=0}^{4} = 96.$$

It can be seen that, if we had taken the slices parallel to the (y, z) plane, the process would have led to the integral

$$V = \int_{x=0}^{8} \left(\int_{y=0}^{4} (\tfrac{1}{32}x^2 + \tfrac{1}{16}y^2 + 2)\, dy \right) dx.$$

The integrand is the same, and the result must be the same, when the integrations over x and y are carried out in the opposite order.

In general, any repeated integral with *constant limits*,

$$\int_{c}^{d} \int_{a}^{b} f(x, d)\, dx\, dy,$$

can be interpreted when $f(x, y)$ is positive as the volume of material in a box standing on the rectangle specified by $a \leqslant x \leqslant b, c \leqslant y \leqslant d$, when the material in it has depth $f(x, y)$. If $f(x, y)$ **is negative** over any part of this area, then it will obviously make a **negative**

contribution to the integral. This **signed-volume analogy** is closely similar to the signed-area analogy (13.13). We can use the idea to say that the order of integration for repeated integrals having constant limits is immaterial:

> **Changing order of integration in a repeated integral with constant limits**
>
> $$\int_c^d \int_a^b f(x, y)\, dx\, dy = \int_a^b \int_c^d f(x, y)\, dy\, dx.$$
>
> (29.1)

There is frequently an advantage to be had from changing the order of integration in this way.

Example 29.2. Evaluate $I = \displaystyle\int_0^1 \int_0^2 x\, e^{xy}\, dx\, dy$.

The inner integral is

$$\int_{x=0} x\, e^{xy}\, dx,$$

which, though not very difficult, does involve integration by parts. To avoid this, try the alternative order of integration:

$$I = \int_{x=0}^2 \left(\int_{y=0}^1 x\, e^{xy}\, dy \right) dx.$$

The inner integral is

$$\int_{y=0}^1 x\, e^{xy}\, dy = \left[x\, \frac{1}{x}\, e^{xy} \right]_{y=0}^1 = e^x - 1.$$

Then

$$I = \int_0^2 (e^x - 1)\, dx = \left[e^x - x \right]_0^2 = e^2 - 3,$$

which is much simpler.

29.3 Repeated integrals over other regions

Suppose now that the silo, loaded with grain, has the triangular shape OPQ shown in Fig. 29.2; to take a definite instance, we could consider its depth $f(x, y)$ to be the same as before: $f(x, y) = \frac{1}{32}x^2 + \frac{1}{16}y^2 + 2$; but our discussion will hold for a general function f. Again we measure the volume V by summing the volumes of slices parallel to the x axis, having thickness δy. Figure 29.2b shows a typical slice lifted out and viewed in (x, z) axes in order to obtain its face area, and Fig. 29.2c shows the base of the silo in plan view; the sloping side has the equation $y = \frac{1}{2}x$; and the slice chosen is marked AB.

(a)
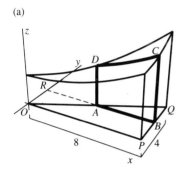

(b)
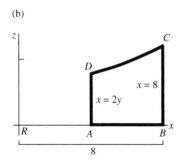

(c)
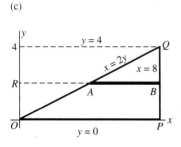

Fig. 29.2

The slices all have different x values at their starting points, and these values depend on y, so **the limits of integration are not constant** in this case. In order to determine the range of integration of the slice at level y, it is necessary to refer to the triangular area in Fig. 29.2c, called the **region of integration** for this problem. The equation of the side OQ is

$$y = \tfrac{1}{2}x$$

for $0 \leqslant x \leqslant 8$. Since we need x in terms of y, we express this as

$$x = 2y,$$

and it is helpful to write it on OQ, as shown, together with the simpler information required for the other limits of integration.

The face area of the slice at level y is therefore given by

$$\text{area } ABCD = \int_{x=2y}^{8} f(x, y)\,\mathrm{d}x.$$

Its volume δV is equal to (Area $ABCD \times \delta y$), so

$$\delta V \approx \left(\int_{x=2y}^{8} f(x, y)\,\mathrm{d}x \right) \delta y,$$

and finally the whole volume V is

$$V = \int_{0}^{4} \int_{2y}^{8} f(x, y)\,\mathrm{d}x\,\mathrm{d}y.$$

Notice that **the limits of integration have nothing to do with the integrand** $f(x, y)$, **but depend only on the shape of the region of integration** in the (x, y) plane (in this case the triangle in Fig. 29.2c). The limits of integration are the same no matter what the integrand.

Example 29.3. Evaluate the integral $I = \displaystyle\int_{0}^{1} \int_{2y}^{2} (x + y)\,\mathrm{d}x\,\mathrm{d}y$.

Write the integral

$$I = \int_{y=0}^{1} \left(\int_{x=2y}^{2} (x + y)\,\mathrm{d}x \right) \mathrm{d}y.$$

The inner integral is

$$\int_{x=2y}^{2} (x + y)\,\mathrm{d}x = \left[\tfrac{1}{2}x^2 + yx \right]_{x=2y}^{2}$$

$$= (2 + 2y) - (2y^2 + 2y^2) = 2 + 2y - 4y^2.$$

(Follow the calculation carefully.) Then

$$I = \int_{y=0}^{1} (2 + 2y - 4y^2)\,\mathrm{d}y = \left[2y + y - \tfrac{4}{3}y^3 \right]_0^1 = \tfrac{5}{3}.$$

(a)

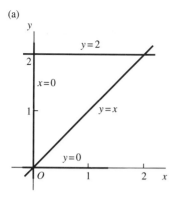

(b)

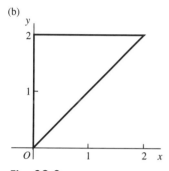

Fig. 29.3

Example 29.4. Evaluate the integral $I = \int_0^2 \int_0^y xy\,dx\,dy$, and sketch the region of integration.

$$I = \int_{y=0}^2 \left(\int_{x=0}^y xy\,dx \right) dy.$$

The inner integral is

$$\int_{x=0}^y xy\,dx = \left[\tfrac{1}{2}x^2 y \right]_{x=0}^y = \tfrac{1}{2}y^3.$$

Therefore

$$I = \int_0^2 \tfrac{1}{2}y^3\,dy = \tfrac{1}{8}\left[y^4 \right]_0^2 = 2.$$

A sketch of the region of integration can be constructed in the following way. The region of integration consists of the points (x, y) which simultaneously satisfy

(i) $0 \leqslant y \leqslant 2$ and   (ii) $0 \leqslant x \leqslant y$.

One way of finding these is to sketch the *boundaries* of the required region. These are the lines

$$y = 0, \quad y = 2, \quad \text{and} \quad x = 0, \quad x = y,$$

and they are shown on Fig. 29.3a. The region consists of any points which lie between both pairs of boundaries (i) and (ii). This is the triangle shown in Fig. 29.3b.

29.4 Changing the order of integration for nonrectangular regions

The rule (29.1) shows that, if the region of integration is a rectangle with sides parallel to the axes, then changing the order of integration simply involves performing the same operations in the opposite order. If we do the same thing with the previous example, we get

$$\int_0^2 \int_0^y xy\,dx\,dy = \int_0^y \int_0^2 xy\,dy\,dx.$$

This is obviously nonsense: the answer contains y, whereas we ought to get the answer 2 again. In fact the new form means nothing at all.

The integral

$$I = \int_0^2 \int_0^y f(x, y)\,dx\,dy$$

was constructed and evaluated by first considering the volume over a typical 'strip' parallel to the x axis. To reverse the order of integration we have to begin with strips parallel to the y axis, as in Fig. 29.4. The boundaries of the triangle are indicated, the diagonal being labelled $y = x$ rather than $x = y$, because this is the way around to specify the corresponding limit of integration for y. Then

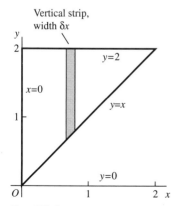

Fig. 29.4

the inner integral is

$$\int_{y=x}^{2} f(x, y)\, dy.$$

The outer integral involves all x between 0 and 2, so finally we have

$$\int_{0}^{2} \int_{0}^{y} f(x, y)\, dx\, dy = \int_{0}^{2} \int_{x}^{2} f(x, y)\, dy\, dx.$$

Each case has to be considered individually in this way.

Example 29.5. Change the order of integration in the repeated integral

$$\int_{0}^{\frac{1}{2}} \int_{-\sqrt{(1-4y^2)}}^{\sqrt{(1-4y^2)}} y\, dx\, dy$$

and so evaluate the integral.

Write the integral in the form

$$I = \int_{y=0}^{\frac{1}{2}} \left(\int_{x=-\sqrt{(1-4y^2)}}^{\sqrt{(1-4y^2)}} y\, dx \right) dy.$$

The limits of integration express the boundaries of the region of integrations:

$$x = (1 - 4y^2)^{\frac{1}{2}}, \quad x = -(1 - 4y^2)^{\frac{1}{2}}, \qquad y = 0, \quad y = \tfrac{1}{2}.$$

These are shown in Fig. 29.5 (the curved part can be written $x^2 + 4y^2 = 1$: an ellipse with semi-axes equal to 1 and $\tfrac{1}{2}$). Figure 29.5a shows how the form given is obtained, by starting with **horizontal strips** which end at $x = \pm(1 - 4y^2)^{\frac{1}{2}}$.

In Fig. 29.5b, the position with regard to **vertical strips** is shown,

(a)

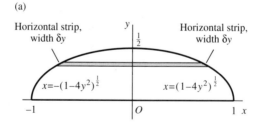

(b)

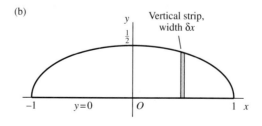

Fig. 29.5

for which the inner integral will be over y. When the order of integration is changed by this means, we obtain

$$I = \int_{x=-1}^{1} \left(\int_{y=0}^{\frac{1}{2}\sqrt{(1-x^2)}} y \, dy \right) dx.$$

The inner integral is now

$$\int_{0}^{\frac{1}{2}\sqrt{(1-x^2)}} y \, dy = \left[\tfrac{1}{2}y^2\right]_{0}^{\frac{1}{2}\sqrt{(1-x^2)}} = \tfrac{1}{8}(1-x^2).$$

Therefore

$$I = \tfrac{1}{8} \int_{-1}^{1} (1-x^2) \, dx = \tfrac{1}{8}\left[x - \tfrac{1}{3}x^3\right]_{-1}^{1} = \tfrac{1}{6}.$$

It is left to the reader to try it in the original form; it is perfectly possible, but more complicated.

29.5 Double integrals

The repeated-integral notation is very informative and self-contained: we need no other information in order to work one out, apart from that provided by the data given in the integral. It even suggests the coordinates to be used, and gives explicitly the boundary of the region of integration. However, problems do not always fall easily into this form.

Suppose we have a lake of any shape, as shown in Fig. 29.6a, whose depth is $f(x, y)$: notional contours are suggested. We want to find the volume of water in the lake.

Call the area covered by the lake the **region of integration** $\mathcal{R}$ for the problem. Construct a **mesh** on $\mathcal{R}$ consisting of small area elements δA as in Fig. 29.6b: the mesh may be quite arbitrary for the present purpose. A typical area element δA is at P. Below δA is a depth of water we shall denote by $f(P)$ (we shall not use $f(x, y)$ because cartesian coordinates might not be the ones we eventually want to use). The volume δV in the vertical column of water below δA at the point P is approximated by

$$\delta V \approx f(P) \, \delta A.$$

If we add up all the volume elements in the usual way, we obtain the total volume V. Denote this operation by

$$V = \sum_{\mathcal{R}} \delta V,$$

thus indicating a certain region of integration $\mathcal{R}$, without specifying it exactly. $\mathcal{R}$ can be obtained by reference to the diagram, or might be specified separately in words or in some other way. Now let all the area elements δA tend to zero, while becoming more numerous in order to cover $\mathcal{R}$. We obtain

$$V = \sum_{\mathcal{R}} \delta V = \lim_{\delta A \to 0} \sum_{\mathcal{R}} f(P) \, \delta A.$$

(a)

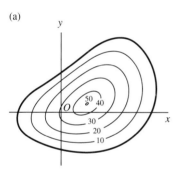

(b)

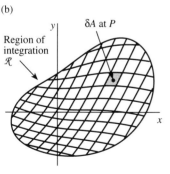

Region of integration $\mathcal{R}$

δA at P

Fig. 29.6

It is natural to write this as some kind of integral as we did in one dimension (see Section 13.1). There are several notations; we shall write

$$V = \iint_{\mathcal{R}} f(P)\, dA,$$

which is to be read: the **double integral of f over the region** $\mathcal{R}$. Unlike a repeated integral it does not give any clue as to how to evaluate it.

As a rule, the argument that gives rise to a double integral by way of a certain summation will not have anything to do with volume; but, as a result of the summation that it represents, the **signed-volume analogy** referred to in Section 29.2 will always hold good:

Double integral $I = \iint_{\mathcal{R}} f(P)\, dA$ and the signed-volume analogy.

(i) I stands for $\lim\limits_{\delta a \to 0} \sum_{\mathcal{R}} f(P)\, \delta A$; $\mathcal{R}$ represents a given region in a plane; and δA is a typical area element of $\mathcal{R}$, taken at the point P. The summation is over all the elements δA of $\mathcal{R}$. **(29.2)**
(ii) (Signed-volume analogy) Whatever its origin, the integral is numerically equal to the signed volume between a surface $z = f(x, y)$ and the plane $z = 0$, taken over the region $\mathcal{R}$. (Where z is negative the contribution counts as negative.)

Example 29.6. A flat plate occupying a region $\mathcal{R}$ is acted on at every point P on the plate by a variable normal stress $\sigma(P)$ per unit area. Express the total (resultant) force F on the plate as a double integral.

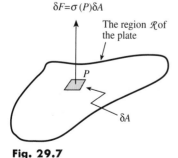

$\delta F = \sigma(P)\delta A$

The region $\mathcal{R}$ of the plate

P

δA

Fig. 29.7

The position is as in Fig. 29.7: $\mathcal{R}$ is the plate. The force δF acting on a typical area element δA at P is given by

$$\delta F \approx \sigma(P)\, \delta A.$$

Add up the contributions of all the elements covering $\mathcal{R}$, and take the limit as the mesh becomes finer and finer. We obtain

$$F = \lim_{\delta A \to 0} \sum_{\mathcal{R}} \sigma(P)\, \delta A = \iint_{\mathcal{R}} \sigma(P)\, dA.$$

Confronted with a double integral, one has to decide on what coordinate system to use (say cartesian or polar coordinates) so as to turn it into a repeated integral. In the following example, cartesian coordinates (x, y) are appropriate.

Example 29.7. We have shown in Example 29.6 that the resultant force F on any flat plate R subject to a normal distribution of stress $\sigma(P)$ is given by $F = \iint_R \sigma(P)\, \mathrm{d}A$. Find the force on a rectangular plate of sides 2 and 3 units when $\sigma = 3(r^2 - 2)$, where r is the distance from one of the corners.

This double-integral expression is perfectly general, applying to any plate, any distribution of force and any coordinates. We have to reformulate the problem for this case. Place the rectangle as shown, with the corner to which the data refers at the origin. A suitable mesh is the rectangular mesh, with δA having sides δx and δy: the area element is $\delta A = \delta x\, \delta y$. Also $\sigma = 3(x^2 + y^2 - 2)$.

We can add (which ultimately means integrate) the contributions
$$\delta F \approx 3(x^2 + y^2 - 2)\, \delta x\, \delta y$$
in any order that is convenient. Suppose we decide to add the contributions along each horizontal strip at level y, and then to add the results from the strips. Then from the strip at level y we obtain the contribution

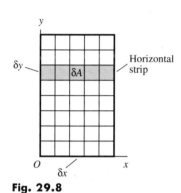

Fig. 29.8

$$\left(\int_{x=0}^{2} 3(x^2 + y^2 - 2)\, \mathrm{d}x \right) \delta y,$$

after letting $\delta x \to 0$. When we add the contributions from all the strips and let $\delta y \to 0$, we have repeated integral

$$F \int_0^3 \int_0^2 3(x^2 + y^2 - 2)\, \mathrm{d}x\, \mathrm{d}y = 3 \int_0^3 (\tfrac{2}{3} + 2y)\, \mathrm{d}y = 60.$$

The following examples show the adaptability of the notation. In each case, R represents the region in question, with P a representative point of R and $\mathrm{d}A$ the corresponding element.

(i) *Area.* Area of R: $\iint_R \mathrm{d}A$.

(ii) *Variable surface density.* Total mass of a thin flat plate, of variable mass $\sigma(P)$ per unit area: $\iint_R \sigma(P)\, \mathrm{d}A$.

(iii) *Moments.* Moment of (ii) about the x axis: $\iint_R y\sigma(P)\, \mathrm{d}A$.

(iv) *Moments of inertia.* Moment of inertia of (ii) about the y axis: $\iint_R x^2\sigma(P)\, \mathrm{d}A$.

(v) *Probability.* A function $f(x, y)$ is eligible to be a probability density function for random variables X and Y over a region $\mathcal{R}$ if

$$f(x, y) \geqslant 0 \text{ and } \iint_{\mathcal{R}} f(x, y)\, dA = 1.$$

The probability that (X, Y) lies in a subregion S is then $\iint_{S} f(x, y)\, dA$. (Here it is helpful to retain x and y: we are not obliged to use P if it does not have the right associations.)

(vi) *Vector resultant.* A force per unit area, $\boldsymbol{f}(P)$ (stress), variable in direction and magnitude, is applied to the surface of a flat plate $\mathcal{R}$. The resultant force $\boldsymbol{F}$ is given by $\boldsymbol{F} = \iint_{\mathcal{R}} \boldsymbol{f}(P)\, dA.$

In order to interpret or evaluate the integral, we write $\boldsymbol{f}$ in its components $\boldsymbol{f} = \hat{\boldsymbol{\imath}} f_1 + \hat{\boldsymbol{\jmath}} f_2 + \hat{\boldsymbol{k}} f_3$: the original double integral with a vector function as integrand is really three double integrals in one.

The resultant force $\boldsymbol{F} \cdot \hat{\boldsymbol{s}}$ in a fixed direction $\hat{\boldsymbol{s}}$ is $\iint_{\mathcal{R}} \boldsymbol{f}(P) \cdot \hat{\boldsymbol{s}}\, dA.$

This integrand $\boldsymbol{f} \cdot \hat{\boldsymbol{s}}$ is not a vector. It can be rewritten in any convenient way: for example as $|\boldsymbol{f}| \cos \theta$, where θ is the angle between $\boldsymbol{f}$ and the perpendicular.

29.6 Polar coordinates

If the boundary of $\mathcal{R}$ in a double integral is circular, or if the boundary is a circular sector, it might be easiest to work it out using polar coordinates. However, when we do this, the integrand changes in an important way.

Figure 29.9a shows an annular sector $\mathcal{R}$ whose boundaries are specified by $r = a$, $r = b$, $\theta = \alpha$, $\theta = \beta$, where r and θ are polar coordinates. We want to evaluate

$$\iint_{\mathcal{R}} f(P)\, dA = \lim_{\delta A \to 0} \sum_{\mathcal{R}} f(P)\, \delta A, \qquad (29.3)$$

where P is a representative point of $\mathcal{R}$, and δA for the moment permits any kind of division of $\mathcal{R}$ into small area elements. We want to put everything in terms of polar coordinates. This process must include a suitable choice of elements δA, so that the summation, or integration, (25.3) can be carried out in an orderly way over the δA elements—the equivalent of 'strips' in (x, y) coordinates.

The mesh suitable for this purpose is also shown in Fig. 29.9a, and one of the area elements δA is shown in Fig. 29.9b. It is nearly a rectangle, with sides δr and $r\, \delta \theta$, so

$$\delta A \approx r\, \delta r\, \delta \theta.$$

The sum in (29.3) therefore becomes, in polar coordinates,

$$\lim_{\substack{\delta r \to 0 \\ \delta \theta \to 0}} \sum_{\mathcal{R}} f(r, \theta)(r\, \delta r\, \delta \theta).$$

(a)

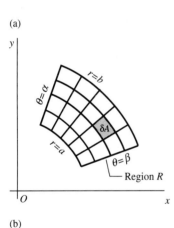

(b)

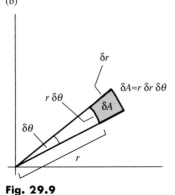

Fig. 29.9

The sum of the elements along the radial line θ is

$$\left(\int_{r=a}^{b} f(r, \theta) r \, dr \right) \delta\theta$$

after letting $\delta r \to 0$. Now add up the contributions from all these narrow sectors, ranging from $\theta = \alpha$ to $\theta = \beta$, and let $\delta\theta \to 0$. We obtain the repeated integral

$$\int_{\theta=\alpha}^{\beta} \left(\int_{r=a}^{b} f(r, \theta) r \, dr \right) d\theta.$$

Notice that this contains **an extra element r in the integrand**.

Double integral in polar coordinates, when the region of integration $\mathcal{R}$ is an annular sector

If the sector $\mathcal{R}$ is the region $a \leqslant r \leqslant b, a \leqslant \theta \leqslant \beta$, then   **(29.4)**

$$\iint_{\mathcal{R}} f(P) \, dA = \int_{\alpha}^{\beta} \int_{a}^{b} f(r, \theta) r \, dr \, d\theta.$$

Example 29.8. Find the volume V between the two planes $x + y + z = 4$ and $z = 0$, over the quadrant $0 \leqslant r \leqslant 1, 0 \leqslant \theta \leqslant \frac{1}{2}\pi$.

Here $\mathcal{R}$ is the region $0 \leqslant r \leqslant 1, 0 \leqslant \theta \leqslant \frac{1}{2}\pi$ in the plane $z = 0$. Expressed as a double integral, the required volume V is

$$V = \iint_{\mathcal{R}} f(P) \, dA = \iint_{\mathcal{R}} z \, dA.$$

Here z is given by

$$z = 4 - x - y = 4 - r \cos \theta - r \sin \theta = f(r, \theta).$$

Then by (29.4),

$$V = \int_{0}^{\frac{1}{2}\pi} \int_{0}^{1} (4 - r \cos \theta - r \sin \theta) r \, dr \, d\theta$$

$$= \int_{\theta=0}^{\frac{1}{2}\pi} \left(\int_{r=0}^{1} (4r - r^2 \cos \theta - r^2 \sin \theta) \, dr \right) d\theta$$

$$= \int_{\theta=0}^{\frac{1}{2}\pi} \left[2r^2 - \tfrac{1}{3}r^3 \cos \theta - \tfrac{1}{3}r^3 \sin \theta \right]_{r=0}^{1} d\theta$$

$$= \int_{0}^{\frac{1}{2}\pi} (2 - \tfrac{1}{3} \cos \theta - \tfrac{1}{3} \sin \theta) \, d\theta = \pi - \tfrac{2}{3}.$$

Example 29.9. A circular disc of radius 0.1 m has a surface charge density $\sigma = 10^{-6} (1 + 10^3 r^3 \sin \frac{1}{2}\theta)$ coulomb m^{-2}. Find the total charge.

The total charge Q is given by $\displaystyle\iint_{\mathcal{R}} \sigma(P)\,dA$, where the region $\mathcal{R}$ is the disc $0 \leqslant r \leqslant 0.1$, $0 \leqslant \theta \leqslant 2\pi$ (if in doubt, sketch it). Remembering that, in polar coordinates, $\delta A = r\,dr\,d\theta$ (or reading straight from (29.4)), we have

$$
\begin{aligned}
Q &= \int_0^{2\pi} \int_0^{0.1} \sigma(r,\theta)\,r\,dr\,d\theta \\
&= \int_0^{2\pi} \int_0^{0.1} 10^{-6}(1 + 10^3 r^3 \sin\tfrac{1}{2}\theta)\,r\,dr\,d\theta \\
&= 10^{-6} \int_{\theta=0}^{2\pi} \left(\int_{r=0}^{0.1} (r + 10^3 r^4 \sin\tfrac{1}{2}\theta)\,dr \right) d\theta \\
&= 10^{-6} \int_{\theta=0}^{2\pi} \left[\tfrac{1}{2} r^2 + \tfrac{1}{5} 10^3 r^5 \sin\tfrac{1}{2}\theta \right]_{r=0}^{0.1} d\theta \\
&= 10^{-8} \int_0^{2\pi} (\tfrac{1}{2} + \tfrac{1}{5} \sin\tfrac{1}{2}\theta)\,d\theta = 10^{-8} \left[\tfrac{1}{2}\theta - \tfrac{2}{5} \cos\tfrac{1}{2}\theta \right]_0^{2\pi} \\
&= 3.94 \times 10^{-8}.
\end{aligned}
$$

Since the repeated integral has constant limits, the same result would be obtained by integrating in the reverse order (see (29.1)).

Example 29.10. The curve $r = \cos\theta$ ($0 \leqslant \theta \leqslant \tfrac{1}{4}\pi$), together with the radii from the origin to its ends, forms the boundary of the region $\mathcal{R}$, and is shown in Fig. 29.10. Obtain (a) the area of $\mathcal{R}$, and (b) its moment about the y axis.

(a) In general, the area of a region $\mathcal{R}$ is $\displaystyle\iint_{\mathcal{R}} dA$. In this case, we shall add up the contributions δA along radial sectors inclined at angle θ, one of which is shown, and then sum the results for all these sectors to obtain the total area A. We can indicate this, together with the range for r and θ, by writing

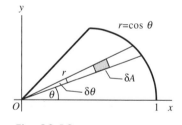

Fig. 29.10

$$
A = \sum_{\theta=0}^{\theta=\frac{1}{4}\pi} \sum_{r=0}^{r=\cos\theta} \delta A \approx \sum_{\theta=0}^{\theta=\frac{1}{4}\pi} \sum_{r=0}^{r=\cos\theta} (r\,\delta r\,\delta\theta).
$$

When we let δr and $\delta\theta$ tend to zero, we have a repeated integral with a variable limit:

$$
A = \int_0^{\frac{1}{4}\pi} \int_0^{\cos\theta} r\,dr\,d\theta = \int_{\theta=0}^{\frac{1}{4}\pi} \left[\tfrac{1}{2} r^2 \right]_{r=0}^{\cos\theta} d\theta = \tfrac{1}{2} \int_0^{\frac{1}{4}\pi} \cos^2\theta\,d\theta
$$

$$
= \tfrac{1}{16}\pi + \tfrac{1}{8}.
$$

(b) The moment δM, about the y axis, of an area element δA is

$$
\delta M \approx x\,\delta A;
$$

so, as a double integral, the total moment M is given by

In the same way as in (a), but with an extra factor

$$x = r \cos \theta,$$

we have

$$M = \int_0^{\frac{1}{4}\pi} \int_0^{\cos \theta} r^2 \cos \theta \, dr \, d\theta = \tfrac{1}{32}\pi + \tfrac{1}{12}.$$

29.7 Separable integrals

Suppose that we have a repeated integral I **with constant limits**, whose integrand $f(x, y)$ is the product of a function of x only with a function of y only:

$$I = \int_c^d \int_a^b f(x, y) \, dx \, dy = \int_c^d \int_a^b g(x)h(y) \, dx \, dy.$$

It is called a **separable integral** because of the following property.
   The inner integral is

$$\int_a^b g(x)h(y) \, dx = h(y) \int_a^b g(x) \, dx,$$

since y is held constant. Therefore

$$I = \int_c^d h(y) \left(\int_a^b g(x) \, dx \right) dy$$

The integral $\int_a^b g(x) \, dx$, once worked out, is just a constant, so we can take it out from under the y integral, obtaining

$$I = \int_a^b g(x) \, dx \int_c^d h(y) \, dy,$$

which is simply the product of two ordinary integrals. We have the following result:

Separable integrals

$$\int_c^d \int_a^b g(x)h(y) \, dx \, dy = \int_a^b g(x) \, dx \int_c^d h(y) \, dy. \qquad (29.5)$$

This can sometimes speed up the working when evaluating integrals, but the following example proves an important result by applying (29.5) the other way round.

Example 29.11. Prove that $\displaystyle\int_0^\infty e^{-x^2} \, dx = \tfrac{1}{2}\pi^{\frac{1}{2}}.$

Put $\displaystyle I = \int_0^\infty e^{-x^2} \, dx.$

The name given to the variable of integration in a definite integral is a matter of indifference, so we can equally well put

$$I = \int_0^\infty e^{-y^2}\, dy.$$

The product is I^2:

$$I^2 = \int_0^\infty e^{-x^2}\, dx \int_0^\infty e^{-y^2}\, dy.$$

By (29.5), this can be written as a repeated integral

$$I^2 = \int_0^\infty \int_0^\infty e^{-x^2} e^{-y^2}\, dx\, dy, \qquad (29.6)$$

because this repeated integral is separable.

Regard x and y as cartesian coordinates. The region of integration $\mathcal{R}$ is the whole of the first quadrant $(0 \leqslant x < \infty; 0 \leqslant y < \infty)$ in Fig. 29.11. Now change to polar coordinates, putting

$$e^{-x^2} e^{-y^2} = e^{-(x^2+y^2)} \quad \text{and} \quad x^2 + y^2 = r^2.$$

The area element is $dA = r\, dr\, d\theta$, and the same region $\mathcal{R}$ is described in polars by

$$0 \leqslant r < \infty, \qquad 0 \leqslant \theta \leqslant \tfrac{1}{2}\pi.$$

Then

$$I^2 = \int_0^{\frac{1}{2}\pi} \int_0^\infty e^{-r^2} r\, dr\, d\theta$$

$$= \int_0^{\frac{1}{2}\pi} d\theta \int_0^\infty e^{-r^2} r\, dr \ \text{(by (29.5) again)}$$

$$= \tfrac{1}{2}\pi \tfrac{1}{2} \big[e^{-r^2} \big]_{r=0}^\infty = \tfrac{1}{4}\pi.$$

Therefore $I = \tfrac{1}{2}\pi^{\frac{1}{2}}$.

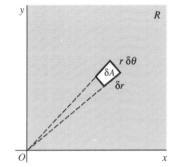

Fig. 29.11

Example 29.12. Prove the convolution theorem (see (23.11), Laplace transforms): that is, if $F(s)$ and $G(s)$ are the Laplace transforms of $f(t)$ and $g(t)$ respectively, then $F(s)G(s)$ is the Laplace transform of $\int_0^t f(\tau)g(t-\tau)\, d\tau$.

Consider the Laplace transform $L(s)$ of $\int_0^t f(\tau)g(t-\tau)\, d\tau$:

$$L(s) = \int_0^\infty e^{-st} \int_0^t f(\tau)g(t-\tau)\, d\tau\, dt$$

$$= \int_0^\infty \int_0^t e^{-st} f(\tau)g(t-\tau)\, d\tau\, dt.$$

The region of integration is the triangle in the (τ, t) plane shown in Figure 29.12. Change the order of integration by summing vertical

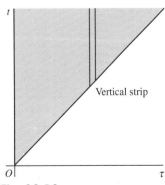

Vertical strip

Fig. 29.12

strips: we find that

$$L(s) = \int_0^\infty \int_\tau^\infty e^{-st} f(\tau)g(t - \tau) \, dt \, d\tau.$$

Now change the variable in the inner integral from t to u, where

$$u = t - \tau,$$

remembering that τ is constant in the inner integral. We obtain

$$L(s) = \int_0^\infty \int_0^\infty g(u)f(\tau) \, e^{-s(u + \tau)} \, du \, d\tau$$

$$= \int_0^\infty \int_0^\infty e^{-su} \, e^{-s\tau} \, g(u)f(\tau) \, du \, d\tau$$

$$= \int_0^\infty e^{-su} \, g(u) \, du \int_0^\infty e^{-s\tau} \, f(\tau) \, d\tau$$

(since the integral is separable)

$$= F(s)\,G(s).$$

Problems, Chapter 29

29.1. Evaluate the following repeated integrals with constant limits.

(a) $\displaystyle\int_0^1 \int_1^2 xy^2 \, dx \, dy;$ (b) $\displaystyle\int_0^1 \int_0^1 y \, e^{xy} \, dx \, dy;$

(c) $\displaystyle\int_c^d \int_a^b dx \, dy;$ (d) $\displaystyle\int_a^b \int_c^d dx \, dy;$

(e) $\displaystyle\int_c^d \int_a^b dy \, dx;$ (f) $\displaystyle\int_0^{\frac{1}{2}} \int_0^{\frac{1}{2}\pi} y \sin xy \, dx \, dy;$

(g) $\displaystyle\int_{-1}^1 \int_{-1}^1 x^2 \, dx \, dy;$ (h) $\displaystyle\int_1^2 \int_0^1 x^2 \, dy \, dx;$

(i) $\displaystyle\int_0^1 \int_{-1}^1 (xy^2 - x^2 y) \, dx \, dy;$

(j) $\displaystyle\int_{-1}^1 \int_0^1 (xy^2 - x^2 y) \, dy \, dx;$

(k) $\displaystyle\int_0^1 \int_0^1 (x + y^2 + 1)^2 \, dx \, dy;$

(l) $\displaystyle\int_0^{\frac{1}{2}\pi} \int_0^{\frac{1}{2}\pi} \cos(x + y) \, dy \, dx;$

(m) $\displaystyle\int_1^2 \int_0^1 \frac{x}{y} \, dx \, dy.$

29.2. Find the signed volume between the given surfaces and the plane $z = 0$ over the specified rectangular regions.

(a) $z = xy, \quad 0 \leqslant x \leqslant 1, \ 0 \leqslant y \leqslant 1;$

(b) $z = xy, \quad -1 \leqslant x \leqslant 1, \ 0 \leqslant y \leqslant 1$ (explain the result);

(c) $z = x + y, \quad -1 \leqslant x \leqslant 2, \ -2 \leqslant y \leqslant 1;$

(d) $z = -1, \quad a \leqslant x \leqslant b, \ c \leqslant y \leqslant d;$

(e) $z = 2x - y + 3, \quad 0 \leqslant x \leqslant 1, \ 0 \leqslant y \leqslant 1;$

(f) $z = 1/(x + y), \quad 1 \leqslant x \leqslant 2, \ 0 \leqslant y \leqslant 1;$

(g) $z = (x + 2y - 1)^2, \quad -2 \leqslant x \leqslant 1, \ -1 \leqslant y \leqslant 1;$

29.3. In the following problems, the region of integration is not rectangular. In each case, sketch the region of integration and indicate a typical strip for the inner integral.

(a) $\displaystyle\int_0^1 \int_0^y dx \, dy;$ (b) $\displaystyle\int_0^1 \int_y^1 x^2 y \, dx \, dy;$

(c) $\displaystyle\int_0^1 \int_0^x x^2 y \, dy \, dx$ (compare (b));

(d) $\displaystyle\int_0^1 \int_0^y (x + y^2)^2 \, dx \, dy;$ (e) $\displaystyle\int_0^1 \int_{-y}^y y \, dx \, dy;$

(f) $\displaystyle\int_0^2 \int_0^y y^2 \sin xy \, dx \, dy;$

(g) $\displaystyle\int_0^2 \int_0^{1 - \frac{1}{2}x} x^2 \, dy \, dx;$ (h) $\displaystyle\int_0^1 \int_0^{\sqrt{(1 - y^2)}} x \, dx \, dy;$

(i) $\displaystyle\int_0^1 \int_0^{\sqrt{(1 - x^2)}} x \, dy \, dx.$

29.4. Find the volume of the wedge-shaped object having one curved surface which is part of the cylinder

$x^2 + y^2 = 1$, and whose flat surfaces are $z = 0$ and the plane $z = 2y$. (Consider the simple wedge in $z \geqslant 0$ only.)

29.5. Reverse the order of integration in each of the following cases. It is *necessary* to sketch the region of integration and to indicate a typical strip corresponding to the new order of integration, as in Section 29.4.

(a) $\displaystyle\int_0^1 \int_0^y f(x, y)\, dx\, dy$;   (b) $\displaystyle\int_0^1 \int_y^1 f(x, y)\, dx\, dy$;

(c) $\displaystyle\int_1^2 \int_0^{y+1} f(x, y)\, dx\, dy$;

(d) $\displaystyle\int_0^1 \int_{-\sqrt{(1-y^2)}}^{\sqrt{(1-y^2)}} f(x, y)\, dx\, dy$;

(e) $\displaystyle\int_2^4 \int_0^{\frac{1}{2}y} f(x, y)\, dx\, dy$;   (f) $\displaystyle\int_0^1 \int_{x^3}^{x^2} f(x, y)\, dy\, dx$;

(g) $\displaystyle\int_0^1 \int_{-1+x}^{1-x} f(x, y)\, dy\, dx$ (it becomes the sum of two integrals);

(h) $\displaystyle\int_{-1}^1 \int_{1-\sqrt{(x^2-1)}}^{1+\sqrt{(x^2-1)}} f(x, y)\, dy\, dx$.

29.6. Change the order of integration in the following, and hence evaluate them. It is *necessary* to sketch the region of integration R and to indicate the strip corresponding to the new inner integral.

(a) $\displaystyle\int_0^{\frac{1}{2}} \int_0^{\frac{1}{2}\pi} x \sin xy\, dx\, dy$;

(b) $\displaystyle\int_0^2 \int_0^{2(y-1)} x^2\, dx\, dy$;

(c) $\displaystyle\int_0^1 \int_0^y x^2 e^{xy}\, dx\, dy$;

(d) $\displaystyle\int_0^\infty \int_{\frac{1}{4}y-2}^{y-2} x^2 y\, e^{-x^2 y^2}\, dx\, dy$;

(e) $\displaystyle\int_{-1}^1 \int_{-2}^2 y(1 - x^2 - y^2)^2\, dx\, dy$;

(f) $\displaystyle\int_1^2 \int_0^1 \frac{y}{x^2 + y^2}\, dx\, dy$;

(g) $\displaystyle\int_1^\infty \int_0^\infty \frac{x}{(x+y)^3}\, dx\, dy$;

(h) $\displaystyle\int_0^1 \int_y^1 y(x^2 - y^2)^{\frac{1}{4}}\, dx\, dy$;

(i) $\displaystyle\int_0^2 \int_{-y-1}^{y-1} x^2\, dx\, dy$ (the integral must be split into two parts);

(j) $\displaystyle\int_0^1 \int_y^1 \frac{y\, dx\, dy}{(x^2 - y^2)^{\frac{1}{2}}}$.

29.7. The symbol $\displaystyle\iint_R f(P)\, dA$ represents a double integral taken over the region R, and dA is the area element at the point P in R (see Section 29.5). In the following cases, the region R is described, and $f(P)$ given in cartesian coordinates. Evaluate the integrals.

(a) R is the rectangle with corners at $(1, 1)$, $(2, 1)$, $(2, 4)$, and $(1, 4)$, and $f(P) = x^2 + y^2$.

(b) R is the equilateral triangle with vertices at $(0, -1)$, $(3^{\frac{1}{2}}, 0)$, and $(0, 1)$, and $f(P) = x$.

(c) R is the circle of radius 2, centred at the origin, and $f(P) = y^2$.

29.8. As in Problem 29.7, but polar coordinates are to be used for the evaluation (see Section 29.6). Remember the change in the area element; see (29.4).

(a) R is the disc $x^2 + y^2 \leqslant 1$, and $f(P) = x^2 + y^2$.

(b) R is the disc $x^2 + y^2 \leqslant 1$, and $f(P) = y^2$.

(c) R is the area whose boundary consists of the x axis between $x = 0$ and 2, the y axis between $y = 0$ and 2, and a quarter of the circle $x^2 + y^2 = 4$. Also $f(x) = y/x$.

(d) R is the sector $1 \leqslant r \leqslant 2$, $0 \leqslant \theta \leqslant \frac{1}{2}\pi$, and $f(P) = xy$.

(e) R is the disc $x^2 + y^2 \leqslant 4$, and $f(x, y) = \arctan y/x$.

(f) R is the first quadrant of the plane, and $f(x, y) = e^{-4(x^2 + y^2)}$.

(g) Show that the volume of a sphere of radius a is $\frac{4}{3}\pi a^3$. (Consider the hemisphere
$$0 \leqslant z \leqslant (a^2 - x^2 - y^2)^{\frac{1}{2}}.)$$

(h) R is the half plane $y \geqslant 0$, and $f(P) = y\, e^{-(x^2 + y^2)}$. [Hint: separate the integral: see (29.5).]

Line integrals

30

30.1 Illustrating a line integral

Consider the following scenario. The success of a museum is measured by two variables. These are the monthly income x from visitors, and the monthly income y from grants and donations. The variation is smoothed out so that they form a continuous record. The exhibitions director receives a bonus increase I which rewards success and penalises failure in promoting attendance: when attendance changes by a small amount δx, positive or negative, there is a change in bonus (up or down) of δI, where

$$\delta I \simeq f(x, y)\, \delta x. \tag{30.1}$$

Figure 30.1 charts the fortunes, or the **state** (x, y) of the museum over a period, starting at state A and arriving at state B, in the form of a curve joining A and B. Time does not register on this diagram, except that direction of development as time increases is indicated by the arrow. The directed curve is called the **path from A to** B, denoted by (AB) (there may be more letters in the brackets). Suppose that the bonus at the starting state A is I_A, and at the state B it is I_B. Then the problem is to find the change in bonus over the period, $I_{(AB)}$:

$$I_B - I_A = I_{(AB)}. \tag{30.2}$$

Divide up the path into many short segments such as PP' (Fig. 30.1). The increment δI over a typical segment is given by

$$(\delta I)_{\text{along } PP'} \approx f(x, y)\, \delta x.$$

Add the contributions of all the segments to obtain $I_{(AB)}$:

$$I_{(AB)} \approx \sum_{(AB)} f(x, y)\, \delta x \tag{30.3}$$

Given a specific function $f(x, y)$ and a *specific path* (AB), $I_{(AB)}$ could in principle be computed by carrying out the summation (30.3) numerically, taking δx very small, and allowing for the fact that δx is *sometimes positive and sometimes negative*. We have to split up the path for this purpose: in Fig. 30.1, δx is negative along (AC) and

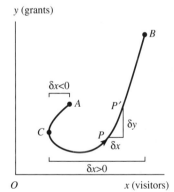

Fig. 30.1

positive along (CB). If the path is vertical along a section, then δx will be *zero*, and there will be a zero change in the bonus along this part despite the fact that y is changing.

In imagination, let $\delta x \to 0$. Then '$\approx$' becomes '$=$'. It is natural to write the result as a kind of integral:

$$I_{(AB)} = \lim_{\delta x \to 0} \sum_{(AB)} f(x, y)\, \delta x = \int_{(AB)} f(x, y)\, \mathrm{d}x. \tag{30.4}$$

where the notation reminds us that we take values of (x, y) which lie on (AB) and take account of the sign of δx at each point on the path.

The integral in (30.4) is called a **line integral**. It is not straight-forwardly an ordinary integral because **the direction, left to right or right to left, at every point must be taken into account**. The director is losing money along (AC). In order to arrive at B from A many paths are possible; we shall see later that, in general, $I_{(AB)}$ will depend on the total history, on what path has been followed between A and B, and we say that the integral is **path-dependent**.

To show how to intepret (30.4) in terms of ordinary integrals, suppose that the bonus function $f(x, y)$ is given by

$$f(x, y) = x + \tfrac{1}{2}y \quad \text{so that} \quad \delta I = (x + \tfrac{1}{2}y)\, \delta x$$

(in suitable units). Suppose that the museum starts off at state A in Figs 30.2a,b; it thrives in Fig. 30.2a and declines in Fig. 30.2b.

Consider the case (a). The graph AB can be expressed in principle as a function of x, and the curve chosen for illustration is

$$y = x^2 - 7x + 15.$$

Then

$$I_{(AB)} = \int_{(AB)} (x + \tfrac{1}{2}y)\, \mathrm{d}x = \int_{(AB)} [x + \tfrac{1}{2}(x^2 - 7x + 15)]\, \mathrm{d}x.$$

Now δx is **positive**; so, regarded as the limit of the sum in (30.4), this is just an **ordinary integral**. After simplifying the integrand, we have

$$I = \int_{x=3}^{5} \tfrac{1}{2}(x^2 - 5x + 15)\, \mathrm{d}x = 11.33.$$

For the case (b); the equation of the curve joining C and A is

$$y = x^2 - 3x + 3,$$

so that $I_{(AC)}$ becomes

$$I_{(AC)} = \int_{(AC)} \tfrac{1}{2}(x^2 - x + 3)\, \mathrm{d}x.$$

In this case, however, the δx in the sum (30.4) are all *negative*: x is *decreasing*. To turn this into an ordinary integral, we have therefore

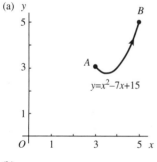

(a) y

$y = x^2 - 7x + 15$

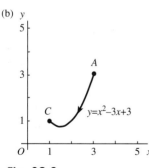

(b) y

$y = x^2 - 3x + 3$

Fig. 30.2

to reverse the sign:

$$I_{(AC)} = -\int_{x=1}^{3} \tfrac{1}{2}(x^2 - x + 3)\,dx = -5.33. \tag{30.5}$$

There is a reduction of bonus for bringing the museum to the edge of ruin.

While we are still observing things, notice first that, in connection with the sign change for negative δx on (AC) in (30.5), we can write

$$I = -\int_{1}^{3} \tfrac{1}{2}(x^2 + x + 3)\,dx = (+)\int_{3}^{1} \tfrac{1}{2}(x^2 + x + 3)\,dx.$$

In other words, we obtain the correct result by setting the x coordinate of the **starting point** as the **lower limit**, and that of the **end point** as the **upper limit**, whether x is constantly decreasing or constantly increasing along the path, and this is a general result.

Lastly we compare the result for the parabolic path (AC) just considered with a straight path from A to C whose equation is

$$y = x.$$

Then

$$\int_{(AC)} (x + \tfrac{1}{2}y)\,dx = \int_{3}^{1} \tfrac{3}{2}x\,dx = \tfrac{3}{2}\big[\tfrac{1}{2}x^2\big]_3^1 = -6.$$

This is different from (30.5), so we must in general expect that line integrals will be path-dependent.

The following summary generalizes the special case we have discussed.

The line integral $I_{(AB)} = \displaystyle\int_{(AB)} f(x, y)\,dx$

(a) Definition: $I_{(AB)} = \displaystyle\lim_{\delta x \to 0} \sum_{(AB)} f(x, y)\,\delta x$, where (x, y) takes values on the path (AB) from A to B.

(b) $I_{(AB)}$ is path-dependent, and $I_{(BA)} = -I_{(AB)}$.

(c) If δx has constant sign on the path $y = y(x)$ from $C : (x_C, y_C)$ to $D : (x_D, y_D)$, then

$$\int_{(CD)} f(x, y) = \int_{x_C}^{x_D} f\big(x, y(x)\big)\,dx.$$

(30.6)

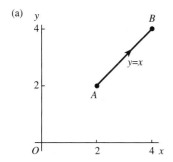

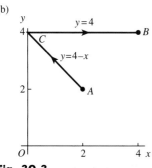

Fig. 30.3

Example 30.1. Evaluate the two line integrals

(a) $\displaystyle\int_{(AB)} xy\,dx$, (b) $\displaystyle\int_{(ACB)} xy\,dx$,

shown in Figs 30.3a,b respectively.

(a) On (AB), $y = x$ and $\delta x > 0$. Here $A = (2, 2)$ and $B = (4, 4)$; so, by (30.6),

$$I_{(AB)} = \int_{(AB)} xy \, dx = \int_2^4 x^2 \, dx = [\tfrac{1}{3}x^3]_2^4 = \tfrac{56}{3}.$$

(b) The path (ACB) has to be broken into two parts: (AC), on which $\delta x < 0$, and (CB), on which $\delta x > 0$, where $C = (0, 4)$. Then

$$I_{(ACB)} = \int_{(ACB)} xy \, dx = \int_{(AC)} xy \, dx + \int_{(CB)} xy \, dx.$$

On (AC), $y = 4 - x$; on (CD), $y = 4$. Therefore

$$I_{(ACB)} = \int_{(AC)} x(4 - x) \, dx + \int_{(CB)} 4x \, dx$$

$$= \int_2^0 (4x - x^2) \, dx + \int_0^4 4x \, dx$$

$$= [2x^2 - \tfrac{1}{3}x^3]_2^0 + 2[x^2]_0^4 = \tfrac{80}{3}.$$

Despite (30.6c), reduction to ordinary integrals over x is not usually the best way to evaluate line integrals, as will be seen later on.

30.2 General line integrals in two and three dimensions

Line integrals of the type

$$\int_{(AB)} g(x, y) \, dy$$

are to be understood in a similar way:

$$\int_{(AB)} g(x, y) \, dy \quad \text{means} \quad \lim_{\delta y \to 0} \sum_{(AB)} g(x, y) \, dy,$$

in which the sign of δy is **positive** on a segment along which y is **increasing** and **negative** on a segment where y is **decreasing**. We can similarly consider paths and functions in three dimensions:

$$\int_{(AB)} f(x, y, z) \, dx, \quad \int_{(AB)} g(x, y, z) \, dy, \quad \int_{(AB)} h(x, y, z) \, dz,$$

and string these types together to obtain a general integral

$$\int_{(AB)} (f \, dx + g \, dy + h \, dz).$$

In the following definition, (AB) is a directed path in three dimensions with representative point $P : (x, y, z)$, and f, g, h are any three functions, their values at P being denoted by f_P, g_P, h_P:

General line integrals

(a) $\displaystyle\int_{(AB)} (f\,dx + g\,dy + h\,dz)$

$$= \lim_{\delta x, \delta y, \delta z \to 0} \sum_{(AB)} (f_P\,\delta x + g_P\,\delta y + h_P\,\delta z).$$

(δx, δy, δz are positive (negative) where x, y, z are increasing (decreasing)). **(30.7)**

(b) $\displaystyle\int_{(AB)} (f\,dx + g\,dy + h\,dz)$

$$= -\int_{(BA)} (f\,dx + g\,dy + h\,dz).$$

(c) For two dimensions, suppress the variable z.
(d) The above integrals generally depend on the path between A and B.

To organize such an integral in order to take account of the signs of δx, δy, δz is often difficult. For example, if the path (AB) consists of an ellipse inclined to the three axes, each term must be broken into two sections, making six sections in all, to ensure constancy of sign of δx, δy, or δz along each. However, if a **parametric representation of the path** is adopted, the correct interpretation is obtained **automatically**.

Consider the integral with respect to x;

$$\int_{(AB)} f(x, y, z)\,dx,$$

where (AB) is parametrized as

$$x = x(t),\ y = y(t),\ z = z(t),$$

so that, as the parameter t increases or decreases from t_A at A to t_B at B, the path is traced exactly once in the right direction. Then, in the short interval from t to $t + \delta t$. the change in δx is approximated by

$$\delta x = \frac{dx}{dt}\,\delta t,$$

and δx automatically has the right sign. Now put $(dx/dt)\,\delta t$ into the defining sum in place of δx; correspondingly $(dx/dt)\,dt$ will go into the original integral in place of dx. After doing the same thing with the y and z integrals, we have the following result.

Parametric evaluation of line integrals

(a) If $x = x(t)$, $y = y(t)$, $z = z(t)$, and $P : (x, y, z)$ covers (AB) exactly once in the correct direction as t increases or decreases from t_A to t_B, then

$$\int_{(AB)} (f\,dx + g\,dy + h\,dz) \tag{30.8}$$

$$= \int_{t_A}^{t_B} \left(f\frac{dx}{dt} + g\frac{dy}{dt} + h\frac{dz}{dt} \right) dt.$$

(b) For the two-dimensional case, omit z.

Example 30.2. Evaluate $I = \displaystyle\int_{(AB)} (x^2 + y)\,dy$, where (AB) is the path shown in Fig. 30.4.

On (AB), $y = -x$, so we can use $x = t$, $y = -t$, with t running from $t = 1$ to $t = -1$. This covers (AB) once in the right direction (it is like using x as the parameter). Then

$$x^2 + y = t^2 - t \quad \text{and} \quad \frac{dy}{dt} = -1,$$

so

$$I = \int_{1}^{-1} (t^2 - t)\,dt = -\left[\tfrac{1}{3}t^3 - \tfrac{1}{2}t\right]_{1}^{-1} = \tfrac{2}{3}.$$

It is immaterial what parametrization is used, so long as it satisfies the conditions in (30.8). The following example compares two parametrizations.

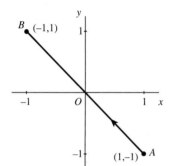

Fig. 30.4

Example 30.3. Evaluate $I = \displaystyle\int_{(AB)} (x\,dy - y\,dx)$, where (AB) is the semicircle shown in Fig. 30.5, by means of two parametrizations:
(a) $x = \cos t$, $y = \sin t$ for $-\tfrac{1}{2}\pi \leqslant t \leqslant \tfrac{1}{2}\pi$; (b) $x = (1 - t^2)^{\frac{1}{2}}$, $y = t$, for $-1 \leqslant t \leqslant 1$.
  (a) $x = \cos t$, $y = \sin t$; so

$$\frac{dx}{dt} = -\sin t \qquad \frac{dy}{dt} = \cos t.$$

Then

$$I = \int_{-\frac{1}{2}\pi}^{\frac{1}{2}\pi} [\cos t \cos t - \sin t(-\sin t)]\,dt$$

$$= \int_{-\frac{1}{2}\pi}^{\frac{1}{2}\pi} [\cos^2 t + \sin^2 t]\,dt = \int_{-\frac{1}{2}\pi}^{\frac{1}{2}\pi} dt = \pi.$$

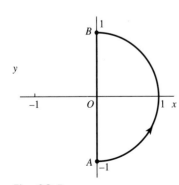

Fig. 30.5

(b) $x = (1 - t^2)^{\frac{1}{2}}$, $y = t$; so

$$\frac{dx}{dt} = -t(1 - t^2)^{-\frac{1}{2}}, \qquad \frac{dy}{dt} = 1.$$

Then

$$I = \int_{-1}^{1} \{(1 - t^2)^{\frac{1}{2}} - t[-t(1 - t^2)^{-\frac{1}{2}}]\} \, dt$$

$$= \int_{-1}^{1} \frac{dt}{(1 - t^2)^{\frac{1}{2}}} = [\arcsin t]_{-1}^{1} = \pi.$$

Example 30.4. Evaluate $I = \int_{(AB)} (x \, dx + y \, dy + z \, dz)$, where (AB) is the path $x = a \cos t$, $y = a \sin t$, $z = bt$ between $t = 0$ and 4π. $((AB)$ is a **helix** along the z axis.)

$$\frac{dx}{dt} = -a \sin t, \qquad \frac{dy}{dt} = a \cos t, \qquad \frac{dz}{dt} = b, \quad \text{so the integral}$$

becomes

$$I = \int_{0}^{4\pi} (-a^2 \cos t \sin t + a^2 \sin t \cos t + b^2 t) \, dt$$

$$= b^2 \int_{0}^{4\pi} t \, dt = 8b^2 \pi^2.$$

30.3 Paths parallel to the axis

Sometimes it is necessary to evaluate line integrals along line segments which are parallel to the axes. In such cases the easiest approach is a direct one, as shown in the following example.

Example 30.5. (See Fig. 30.6.) Evaluate the line integral $\int_{(AB)} x \, dy$
(a) over the path (AOB);   (b) over the path (AQB).

In this method, we refer to the sums in the definition (30.7).

(a) On (AO), $\delta y = 0$ since y is constant; so $\int_{(AO)} x \, dy = 0$. On (OB), $x = 0$; so $\int_{(OB)} x \, dy = 0$. Therefore

$$\int_{(AOB)} x \, dy = \int_{(AO)} x \, dy + \int_{(OB)} x \, dy = 0.$$

(b) On (AQ), $x = 1$; so $\int_{(AQ)} x \, dy = \int_{(AQ)} 1 \, dy = 1.$

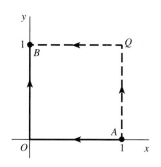

Fig. 30.6

On (QB), y is constant; so $\delta y = 0$. Therefore $\int_{(QB)} x \, dy = 0$. Hence

$$\int_{(AQB)} x \, dy = \int_{(AQ)} x \, dy + \int_{(QB)} x \, dy = 1.$$

30.4 Path independence and perfect differentials

Despite the fact that the value of a line integral taken between two given points usually depends on the path chosen, there are many cases of physical importance for which the value is **independent of the path**: all paths between A and B lead to the same value. To show that such cases can exist, the following two examples have integrands for which the integral is independent of the path.

Example 30.6. (In two dimensions). Show that $I = \int_{(AB)} (y \, dx + x \, dy)$ is independent of the path chosen from A to B.

We can write the integrand in terms of a **perfect differential** (see Section 20.4):

$$y \, dx + y \, dy = d(xy).$$

Now express the integral in the form

$$I = \int_{(AB)} (y \, dx + x \, dy) = \int_{(AB)} d(xy). \tag{i}$$

From (30.7), the meaning of the integral is the limit of a certain sum, which can be recast in the form

$$\lim_{\delta x, \delta y \to 0} \sum_{(AB)} (y \, dx + x \, dy) = \lim_{\delta(xy) \to 0} \sum_{(AB)} \delta(xy). \tag{ii}$$

As we travel along (AB), the value of (xy) starts at $x_A y_A$, where the values are taken at A, then goes by steps $\delta(xy)$ until it attains the value $x_B y_B$. In other words,

$$\sum_{(AB)} \delta(xy) = x_B y_B - x_A y_A.$$

which is independent of what path connects A to B.

Example 30.7. Prove that

$$I = \int_{(AB)} [(y + z) \, dx + (z + x) \, dy + (x + y) \, dz]$$

is independent of the path from A to B.

We can use the idea of differentials in exactly the same way: it is suggested by the incremental approximation (28.1). If we put

$$f(x, y, z) = yz + zx + xy,$$

the incremental approximation gives

$$\delta f = (y + z)\,\delta x + (z + x)\,\delta y + (x + y)\,\delta z,$$

which parallels the corresponding statement involving differentials:

$$d(yz + zx + xy) = (y + z)\,dx + (z + x)\,dy + (x + y)\,dz.$$

In the same way as in Example 30.6, we have

$$I = \int_{(AB)} [(y + z)\,dx + (z + x)\,dy + (x + y)\,dz]$$

$$= \int_{(AB)} d(yz + zx + xy)$$

$$= (yz + zx + xy)_B - (yz + zx + xy)_A,$$

the suffices A and B meaning the values of the brackets at the end points A and B. This is independent of the path.

In general, suppose that we can recognize the functions f, g, and h in the differential form

$$f(x, y, z)\,dx + g(x, y, z)\,dy + h(x, y, z)\,dz$$

as being expressible in terms of a single function $S(x, y, z)$ in the following way:

$$f = \frac{\partial S}{\partial x}, \qquad g = \frac{\partial S}{\partial y}, \qquad h = \frac{\partial S}{\partial z}.$$

Then we can write $f\,dx + g\,dy + h\,dz$ as a perfect differential

$$f\,dx + g\,dy + h\,dz = \frac{\partial S}{\partial x}\,dx + \frac{\partial S}{\partial y}\,dy + \frac{\partial S}{\partial z}\,dz = dS.$$

This can be substituted into our integrals—or the corresponding expression with δ instead of d substituted into the sum which defines the integral—to yield

$$\int_{(AB)} (f\,dx + g\,dy + h\,dz) = \int_{(AB)} dS = S_B - S_A.$$

Provided that there is no ambiguity in the values to be assigned to S_B and S_A (this possibility is discussed in Section (30.10) below), this provides a way of evaluating the integral and possibly demonstrating path-independence.

Evaluation of integrals over perfect differentials

If $f\,dx + g\,dy + h\,dz = dS$, then

$$\int_{(AB)} (f\,dx + g\,dy + h\,dz) = S_B - S_A$$

(30.9)

(a)

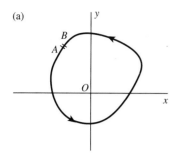

(b)

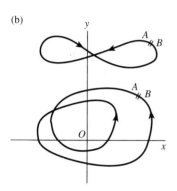

Fig. 30.7
(a) A simple closed path.
(b) Closed paths which are
not simple.

(a)

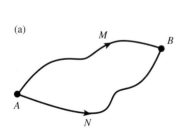

(b)

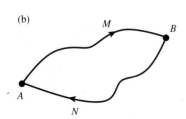

Fig. 30.8

30.5 Closed paths

A closed path is one that returns to its starting point, so that B has the same coordinates as A, as in Figs 30.7a,b. We shall discuss only **simple closed paths**. These do not cross over themselves, as do the curves in Fig. 30.7b.

It is clear from the definition (30.7) that, when A and B are the same point, and the path is closed, their position on the curve will not affect the value of the integral. Consequently its coordinates are not usually stated; a closed path is indicated by a symbol such as C, and the integral is written

$$\int_C (f\,dx + g\,dy + h\,dz).$$

In three dimensions, the **direction** along C is specified by extra information, such as by an arrow on a sketch of the curve. However, in **two** dimensions, a convention operates: if it is not otherwise indicated, the **standard direction is anticlockwise**.

Example 30.8. Evaluate $I = \int_C (x\,dy - y\,dx)$, where C is the ellipse $x^2/a^2 + y^2/b^2 = 1$, described in the standard direction.

The ellipse can be parametrized for the anticlockwise direction by

$$x = a\cos t \quad \text{and} \quad y = b\sin t \quad (0 \leqslant t \leqslant 2\pi),$$

where we assume a and b to be positive. We had a choice for the range of t, because we can start at any point on the ellipse. For the choice 0 to 2π, the path starts and ends at $(a, 0)$. Then

$$\frac{dx}{dt} = -a\sin t, \quad \frac{dy}{dt} = b\cos t;$$

so

$$I = \int_0^{2\pi} [a\cos t(b\cos t) - b\sin t(-a\sin t)]\,dt$$

$$= ab\int_0^{2\pi} dt = 2\pi ab.$$

The following result is sometimes useful:

A criterion for general path independence

If $\int_C (f\,dx + g\,dy + h\,dz) = 0$ for every closed curve C, then $\int_{(AB)} (f\,dx + g\,dy + h\,dz)$ is path-independent for every A and B. (30.10)

To prove this, see Fig. 30.8a. Here A and B are *any* two points.

while (AMB) and (ANB) are *any* two paths from A to B. If we reverse the direction of the path (ANB) we have Fig. 30.8b, which is a closed curve C. Suppose that we know the integral around every closed curve to be zero. Then

$$0 = \int_{(AMBNA)} (f\,dx + g\,dy + h\,dz)$$

$$= \int_{(AMB)} (f\,dx + g\,dy + h\,dz) + \int_{(BNA)} (f\,dx + g\,dy + h\,dz)$$

$$= \int_{(AMB)} (f\,dx + g\,dy + h\,dz) - \int_{(ANB)} (f\,dx + g\,dy + h\,dz).$$

Therefore the integrals along (AMB) and (ANC) are equal.

Path independence for *any two particular points* is sufficient to ensure path independence between *all pairs of points*:

> **Path independence between given points.**
>
> Let A and B be any *given* points. Then, if $I_{(AB)}$ is path-independent, $I_{(PQ)}$ is path-independent for every pair of points P and Q.   **(30.11)**

The fixed points A and B are shown on Figure 30.9, and (P, Q) is any other pair of points. $(AKB), (AP), (QB),$ and (PRQ) are arbitrary paths joining the points specified by their brackets. Since $I_{(AB)}$ is independent of the path joint A and B,

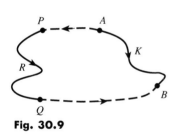

$$I_{(APRQB)} = I_{(AKB)},$$

so

$$I_{(AP)} + I_{(PRQ)} + I_{(QB)} = I_{(AKB)},$$

or

$$I_{(PRQ)} = I_{(AKB)} - I_{(AP)} - I_{(QB)}.$$

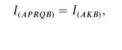

Fig. 30.9

But the right-hand side does not depend on which path was chosen for (PQ). Therefore $I_{(PQ)}$ is independent of the path joining P and Q, which proves the result.

30.6 Green's theorem

The following theorem connects a **two-dimensional line integral around a closed curve** C with a certain **double integral over the region** $\mathcal{A}$ **enclosed by** C, as shown in Fig. 30.10. The functions $P(x, y)$ and $Q(x, y)$ which occur are assumed to be 'smooth' in the region considered. 'Smoothness' has a technical meaning, that all the first derivatives of P and Q are continuous, but it will be enough for us to say that P and Q must have no jumps or infinities on C and its interior $\mathcal{A}$ taken together. The theorem is:

> **Green's theorem in a plane**
>
> C is a simple closed path containing a region $\mathcal{A}$; $P(x, y)$ and $Q(x, y)$ are smooth functions. Then
>
> $$\int_C (P \, dx + Q \, dy) = \iint_{\mathcal{A}} \left(\frac{\partial Q}{\partial x} - \frac{\partial P}{\partial y} \right) d\mathcal{A}, \qquad \textbf{(30.12)}$$
>
> where $d\mathcal{A}$ is the area element, and the line integral direction is anticlockwise.

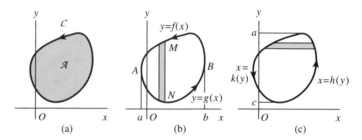

Fig. 30.10
(a) The diagram for Green's theorem. (b) For the integration of $\partial P/\partial y$. (c) For the integration of $\partial Q/\partial x$.

Although the result is true in general, we shall prove it only for a curve like that in Fig. 30.10a, for which lines parallel to the axes cut the curve in at most two points.

Consider $\iint_{\mathcal{A}} \dfrac{\partial P}{\partial y} \, d\mathcal{A}$. We shall integrate it by vertical strips as in Fig. 30.10b. Suppose that the top part AMB of C, between $x = a$ and b, has the equation $y = f(x)$, and the lower part ANB is $y = g(x)$. Then

$$\iint_{\mathcal{A}} \frac{\partial P}{\partial y} \, d\mathcal{A} = \int_a^b \int_{g(x)}^{f(x)} \frac{\partial P}{\partial y} \, dy \, dx$$

$$= \int_a^b \left[P\big(x, f(x)\big) - P\big(x, g(x)\big) \right] dx$$

$$= \int_{(BMA)} P(x, y) \, dx - \int_{(ANB)} P(x, y) \, dx$$

$$= -\int_{(AMB)} P(x, y) \, dx - \int_{(ANB)} P(x, y) \, dx$$

$$= -\int_C P(x, y) \, dx. \qquad \textbf{(i)}$$

Similarly, but by using horizontal strips as in Fig. 30.10c,

$$\iint_{\mathcal{A}} \frac{\partial Q}{\partial x} \, d\mathcal{A} = \int_c^d \int_{k(y)}^{h(y)} \frac{\partial Q}{\partial x} \, dx \, dy = \int_C Q \, dy. \qquad \textbf{(ii)}$$

By subtracting (i) and (ii) we obtain the result required.

Example 30.9. Show that if C is a simple closed curve, the geometrical area it encloses is equal to $\frac{1}{2} \displaystyle\int_C (x\,dy - y\,dx)$.

Put $P = -y$ and $Q = x$ in Green's theorem:

$$\frac{1}{2} \int_C (-y\,dx + x\,dy) = \frac{1}{2} \iint_{\mathcal{A}} \left(\frac{\partial}{\partial x}(x) - \frac{\partial}{\partial y}(-y) \right) d\mathcal{A}$$

$$= \iint_{\mathcal{A}} d\mathcal{A},$$

which is the geometrical area enclosed.

Green's theorem (30.12) enables us to produce a criterion by which path independence can be recognized:

> **A condition for path-independence**
>
> If $\partial P/\partial y = \partial Q/\partial x$, then for any points A and B, **(30.13)** $\displaystyle\int_{(AB)} (P\,dx + Q\,dy)$ is independent of the path (AB).

To prove this, let C be any closed curve. Its interior is denoted by $\mathcal{A}$. Then, by Green's theorem,

$$\int_C (P\,dx + Q\,dy) = \iint_{\mathcal{A}} \left(\frac{\partial Q}{\partial x} - \frac{\partial P}{\partial y} \right) d\mathcal{A} = 0.$$

Therefore, by (30.10), we have path independence.

30.7 Line integrals and work

A particle follows a certain path (AB) in three-dimensional space under the action of various forces and its own inertia. Consider *one* of the forces, $\boldsymbol{F}(x, y, z)$, which might be the contribution of a force field such as gravity, or a point force such as friction or the tension in a string. $\boldsymbol{F}$ is a **vector**, and it does not necessarily point along the path of the particle.

The path can be parametrized by using t, the time, as the parameter, and its position specified briefly by its position vector $\boldsymbol{r}(t)$:

$$\big(x(t),\, y(t),\, z(t)\big) = \boldsymbol{r}(t)$$

as in Fig. 30.11a. In a time interval δt, the particle moves from P to Q, and $\boldsymbol{r}(t)$ changes to $\boldsymbol{r}(t + \delta t)$, the change being denoted by $\delta \boldsymbol{r}$. Figure 30.11a also shows the force $\boldsymbol{F}$ acting on the particle when it is at P. During the interval δt, the work δW done by $\boldsymbol{F}$ *alone* on the

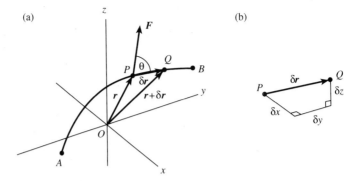

Fig. 30.11

particle is approximated by

$$\delta W = (\text{component of } \boldsymbol{F} \text{ in direction } PQ) \times (\text{distance } PQ)$$
$$= (|\boldsymbol{F}| \cos \theta)|\delta \boldsymbol{r}| = \boldsymbol{F} \cdot \delta \boldsymbol{r}.$$

The total work $W_{(AB)}$ done on the particle by $\boldsymbol{F}$ along the path (AB) is given by

$$W_{(AB)} = \sum_{(AB)} \delta W \approx \sum_{(AB)} \boldsymbol{F} \cdot \delta \boldsymbol{r}.$$

When the step length goes to zero the sum can be written as an integral:

$$W_{(AB)} = \int_{(AB)} \boldsymbol{F} \cdot \mathrm{d}\boldsymbol{r}.$$

This integral has an ordinary meaning when it is written in $(\mathrm{d}x, \mathrm{d}y, \mathrm{d}z)$ form by splitting $\boldsymbol{F}$ and $\delta \boldsymbol{r}$ into their components (see Fig. 30.11b):

$$\boldsymbol{F} = F_1 \hat{\boldsymbol{\imath}} + F_2 \hat{\boldsymbol{\jmath}} + F_3 \hat{\boldsymbol{k}} \quad \text{and} \quad \delta \boldsymbol{r} = \delta x \, \hat{\boldsymbol{\imath}} + \delta y \, \hat{\boldsymbol{\jmath}} + \delta z \, \hat{\boldsymbol{k}}.$$

Then

$$\boldsymbol{F} \cdot \delta \boldsymbol{r} = F_1 \, \delta x + F_2 \, \delta y + F_3 \, \delta z,$$

and

$$W_{(AB)} \approx \sum_{(AB)} (F_1 \, \delta x + F_2 \, \delta y + F_3 \, \delta z).$$

Finally, taking the limit as $\delta x, \delta y, \delta z$ approach zero, we obtain the exact result.

Work done by a force along a path (AB)

$$W_{(AB)} = \int_{(AB)} \boldsymbol{F} \cdot \mathrm{d}\boldsymbol{r} = \int_{(AB)} (F_1 \, \mathrm{d}x + F_2 \, \mathrm{d}y + F_3 \, \mathrm{d}z). \qquad \textbf{(30.14)}$$

(For two dimensions, suppress z.)

Example 30.10. A field of force F is constant everywhere. Show that the work done by F alone on a particle which moves from a fixed point A to a fixed point B is independent of the path followed.

Put $F = a\hat{\imath} + b\hat{\jmath} + c\hat{k}$ where a, b, c are constants. Then, by (30.14), $W_{(AB)}$ is given by

$$W_{(AB)} = \int_{(AB)} (a\,dx + b\,dy + c\,dz) = \int_{(AB)} d(ax + by + cz)$$
$$= (ax + by + cz)_B - (ax + by + cz)_A,$$

which is a quantity independent of the path followed.

Example 30.11. (a) r is a position vector in axes x, y, z, and $r = |r|$. Show that, in terms of differentials,

$$d(r^{-1}) = -(x/r^3)\,dx - (y/r^3)\,dy - (z/r^3)\,dz.$$

(b) The gravitational force F of the earth acting on a particle of mass m at a distance r from the earth's centre O is equal to $-m\gamma r/r^3$ (γ constant; F is directed towards O and has magnitude $m\gamma/r^2$). Show that the work done by F when the particle moves between any two points A and B is equal to $m\gamma(r_B^{-1} - r_A^{-1})$ (it is path-independent).

(a) $r = (x^2 + y^2 + z^2)^{\frac{1}{2}}$, so by (28.1)

$$\delta(r^{-1}) \approx \frac{\partial}{\partial x}(r^{-1})\,\delta x + \frac{\partial}{\partial y}(r^{-1})\,\delta y + \frac{\partial}{\partial z}(r^{-1})\,\delta z$$

$$= -\frac{x}{(x^2 + y^2 + z^2)^{\frac{3}{2}}}\,\delta x - \frac{y}{(x^2 + y^2 + z^2)^{\frac{3}{2}}}\,\delta y$$

$$\quad -\frac{z}{(x^2 + y^2 + z^2)^{\frac{3}{2}}}\,\delta z$$

$$= -(x/r^3)\,\delta x - (y/r^3)\,\delta y - (z/r^3)\,\delta z.$$

The corresponding differential relation is

$$d(r^{-1}) = -(x/r^3)\,dx - (y/r^3)\,dy - (z/r^3)\,dz.$$

(b) Refer to Fig. 30.12 and (30.14). The components of F, are $(F_1, F_2, F_3) = (-m\gamma x/r^3,\ -m\gamma y/r^3,\ -m\gamma z/r^3)$, obtained by splitting r into its components. Therefore the work $W_{(AB)}$ done is

$$W_{(AB)} = \int_{(AB)} (F_1\,dx + F_2\,dy + F_3\,dz)$$

$$= m\gamma \int_{(AB)} \left(-\frac{x}{r^3}\,dx - \frac{y}{r^3}\,dy - \frac{z}{r^3}\,dz\right)$$

$$= m\gamma \int_{(AB)} d(r^{-1}) \quad \text{(from (a))}$$

$$= m\gamma(r_B^{-1} - r_A^{-1}).$$

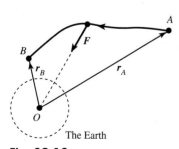

Fig. 30.12

Examples 30.10 and 30.11 illustrate cases where the work done by a force between two fixed points is independent of the path between them, but this is not a universal state of affairs: for example, it is not the case for the force field $(y, -x, 0)$.

30.8 Conservative fields

We have spoken in a general way of a field of force and its action on a particle. By a **particle** we mean an object small enough for its exact shape, physical constitution, state of rotation, and so on to be unimportant on the scale of the problem being considered; it behaves in the way we imagine a point should behave.

However, the magnitude of the force exerted upon it by gravity, electrostatic influence, etc. will still depend on the mass or charge assigned to the particle. We need a way to specify the strength of the force field itself, a **field intensity**, which is independent of what particle we put into it. When the field strength is specified, we should be able to deduce its effect on any particle.

This is not always quite straightforward, because the introduction of a new particle into (say) an electrostatic field might change the distribution of charge that constitutes the **source of the field**, so that in effect we would be putting the particle into a modified situation. The case is similar with gravity: if an asteroid enters the moon's gravitational field, the moon will respond by moving, and the field entered will change, if only by a little. For the purpose of defining field intensity, we imagine that somehow such an effect is prevented from taking place. Subject to this, we have the following definition.

Field intensity f_P at P

f_P is equal to the vector force that would act on a particle of unit mass (charge, etc.) at P if the sources are assumed to be unaffected by the particle.   (30.15)

Therefore, if the gravitational field intensity is G_F at P, the force with which the field acts on a particle of mass m at P is mG_P. One can alternatively imagine a particle of extremely small mass μ to be introduced as a test particle. Then f_P will be equal to μ^{-1} times the force exerted on such a particle.

Consider the action of a field of intensity $f(x, y, z)$ on a **unit particle** which is travelling on a path (AB) (Fig. 30.13). We shall consider not the work done *by f* on the particle, but the **work done against the field by the particle**, which has the opposite sign. Denote this quantity generally by v. The work done against f in a step PQ is given by

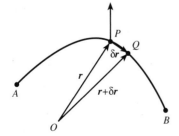

Fig. 30.13

$$\delta v \approx -f \cdot \delta r. \tag{30.16}$$

The total work along the path is the limit of the sum of the δv, which can be expressed as a line integral as before:

$$v_{(AB)} = -\int_{(AB)} \boldsymbol{f} \cdot dr = -\int_{(AB)} (f_1\, dx + f_2\, dy + f_3\, dz),$$

where $\boldsymbol{f} = (f_1, f_2, f_3)$.

The important case is when $v_{(AB)}$ is **independent of the path**, in which case the **field is said to be a conservative field**. In practice our 'field' will not usually consist of the whole of space; some space will be occupied by impenetrable bodies, or we might be interested only in the region R inside a metal cage. According to (30.11) it is only necessary to check path-independence for any single pair of points (A, B) within this region. Therefore:

Conservative field in a region $\mathcal{R}$

Let A and B be two given points in $\mathcal{R}$. Then $\boldsymbol{f}(x, y, z)$ is conservative in $\mathcal{R}$ if $v_{(AB)} = -\displaystyle\int_{(AB)} \boldsymbol{f} \cdot dr$ is independent of the path from A to B. (Or equivalently if $\displaystyle\int_C \boldsymbol{f} \cdot dr$ is zero for every closed path C in $\mathcal{R}$: see (30.10).) $\qquad$ **(30.17)**

The constant field of Example 30.10 and the gravitation field of Example 30.11 are conservative.

30.9 Potential for a conservative field

Suppose that $\boldsymbol{f}(x, y, z)$ is *conservative* in a region $\mathcal{R}$, and that A is a fixed point in $\mathcal{R}$. Since $\boldsymbol{f}$ is conservative, the integral

$$v_{(AP)} = -\int_{(AP)} \boldsymbol{f} \cdot dr,$$

where P is another point in $\mathcal{R}$, is independent of the path (AP), and so its value depends only on the location (x, y, z) of P. Therefore we shall write

$$v_{(AP)} = V(x, y, z) \quad \text{or} \quad V_P, \qquad\qquad \textbf{(30.18)}$$

in which we have suppressed the coordinates of A since they are constant.

In Fig. 30.14, suppose that (AP) is a fixed reference path from A to $P : (x, y, z)$. Let $Q : (x + \delta x, y, z)$ be a point close to P, displaced from it a distance δx in the x direction only. Choose a path (AQ) consisting of two parts: the selected path (AP) and a *straight-line extension* (PQ) from P to Q. The choice of these paths rather than any others does not affect the values of V_P and V_Q since the field is

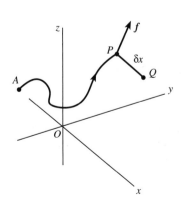

Fig. 30.14

conservative. Then, from (30.18),

$$v_{(AQ)} - v_{(AP)} = V_Q - V_P = V(x + \delta x, y, z) - V(x, y, z).$$

But also, from (30.16), $\delta v = -\boldsymbol{f} \cdot \delta \boldsymbol{r}$ with $\delta y = \delta z = 0$; so

$$v_{(AQ)} - v_{(AP)} \approx -f_1 \, \delta x,$$

where f_1 is the x component of $\boldsymbol{f}$. Equating the last two results and dividing by δx; we obtain

$$f_1 = -[V(x + \delta x, y, z) - V(x, y, z)]/\delta x.$$

When $\delta x \to 0$, this becomes

$$f_1 = -\frac{\partial V}{\partial x},$$

and similarly

$$f_2 = -\frac{\partial V}{\partial y} \quad \text{and} \quad f_3 = -\frac{\partial V}{\partial z}.$$

Therefore

$$\boldsymbol{f} = f_1 \hat{\boldsymbol{i}} + f_2 \hat{\boldsymbol{j}} + f_3 \hat{\boldsymbol{k}}$$
$$= -\left(\frac{\partial V}{\partial x} \hat{\boldsymbol{i}} + \frac{\partial V}{\partial y} \hat{\boldsymbol{j}} + \frac{\partial V}{\partial z} \hat{\boldsymbol{k}} \right),$$

or

$$\boldsymbol{f} = -\mathbf{grad} \, V. \tag{30.19}$$

We call V a **potential function for the field** $\boldsymbol{f}$, or simply a **potential**. The single scalar function $V(x, y, z)$ contains all the information necessary to define the *three* scalar components of $\boldsymbol{f}$: $f_1(x, y, z)$, $f_2(x, y, z)$, $f_3(x, y, z)$. The point A is commonly taken to be at infinity: the reader might recognise the idea of 'the work required to bring a particle in from infinity' in mechanics. However, if we choose a different reference point A, it only changes V by an additive constant, and does not, therefore, affect the truth of (30.19); we get the same $\boldsymbol{f}$ whatever location A has. We sum up this result as follows.

Potential V of a conservative field $\boldsymbol{f}$

If $\boldsymbol{f}(x, y, z)$ is conservative in a region $\mathcal{R}$, then

$$\boldsymbol{f} = -\mathbf{grad} \, V$$

in $\mathcal{R}$, where V is a scalar potential function for $\boldsymbol{f}$. V is defined in the region $\mathcal{R}$ by

$$V_P = -\int_{(AP)} \boldsymbol{f} \cdot d\boldsymbol{r},$$

where A is a fixed point.

(30.20)

As an example of a potential, the gravitational field from a particle of mass M, namely $\boldsymbol{f} = -M\gamma \boldsymbol{r}/r^3$, has the potential $V = -M\gamma/r$.

This can be checked from the working of Example 30.10. The potential function V is equal to the work done against the field in moving a unit particle form a fixed point A to the current point P in cases when the field is conservative. Therefore the **potential energy** of a particle of mass m relative to the reference point A is equal to mV. Alternatively, V can be regarded as energy stored by the field, like energy stored in a spring.

30.10 Single-valuedness of potentials

There is a connection between the question of single-valuedness in a perfect differential and the conservative property of a force field. There exist fields $f(x, y, z)$ which have a potential, but are not conservative, because they do not satisfy the condition (30.17), that

$$\int_{(AB)} f \cdot dr \text{ should be independent of path.}$$

Potential field

If there is a scalar function V such that

$$f = -\mathbf{grad}\ V,$$

then $f(x, y, z)$ is called a **potential field**.

(30.21)

We can test whether such a field is conservative or not:

Condition for a potential field to be conservative

If $f = -\mathbf{grad}\ V$, and V is single-valued, then f is conservative. In this case, the work $v_{(AB)}$ in moving a unit particle from A to B against the field is equal to $V_B - V_A$.

(30.22)

This is proved as follows:

$$v_{(AB)} = -\int_{(AB)} f \cdot dr = \int_{(AB)} (\mathrm{grad}\ V) \cdot dr$$

$$= \int_{(AB)} \left(\hat{\imath} \frac{\partial V}{\partial x} + \hat{\jmath} \frac{\partial V}{\partial y} + \hat{k} \frac{\partial V}{\partial z} \right) \cdot (\hat{\imath}\, dx + \hat{\jmath}\, dy + \hat{k}\, dz)$$

$$= \int_{(AB)} \left(\frac{\partial V}{\partial x} dx + \frac{\partial V}{\partial y} dy + \frac{\partial V}{dz} dz \right) = \int_{(AB)} dV \text{ (see (28.1))}$$

$$= V_B - V_A.$$

Provided that $\int_{(AB)} dV$ is independent of the path from A to B, the value to be assigned to $V_B - V_A$ is unambiguous and we say that V is single-valued. However, the values of V may depend not only on the position, *but also on the way in which the position was reached* (analogously to the time spent reaching a point on the other side of a road being dependent on whether you cross directly or via the underpass). For example, in the plane, let

$$V = \theta,$$

where θ is the polar angle traversed in reaching the current position, measured continuously from a given starting point. What do we mean by

$$\int_{(AB)} d\theta?$$

Fig. 30.15 shows two paths from A to B: (ACB) goes from A to B more or less directly, and (ADB) circles the origin completely first. The definition of the integral is that

$$\int_{(AB)} d\theta = \lim_{\delta\theta \to 0} \sum_{(AB)} \delta\theta,$$

where the summation is carried out by taking small steps along the path. On (ACB), θ passes smoothly from $\theta = 0$ to $\theta = \frac{1}{2}\pi$, so

$$\int_{(ACB)} dV = \int_{(ACB)} d\theta = \theta_B - \theta_A = \frac{1}{2}\pi.$$

On (ADB), θ starts at $\theta_A = 0$ and increases smoothly through values $\frac{1}{2}\pi$, π, $\frac{3}{2}\pi$, 2π, to $\theta_B = \frac{5}{2}\pi$. Therefore

$$\int_{(ADB)} dV = \int_{(ADB)} d\theta = \theta_B - \theta_A = \frac{5}{2}\pi.$$

Therefore V in this case is path-dependent; if the potential of a force field is given by

$$V = \theta,$$

where θ is the *traversed* polar angle, then the field is strictly not conservative: various paths from A to B involve different amounts of work by a unit particle moving in the field.

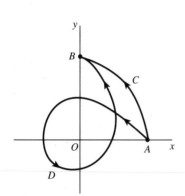

Fig. 30.15.

Example 30.12. Show that the two-dimensional field

$$f(x, y, z) = \frac{-y}{x^2 + y^2}\,\hat{\imath} + \frac{x}{x^2 + y^2}\,\hat{\jmath}$$

is not a conservative field.

Apart from physical constants this represents the circumferential magnetic field around a straight wire carrying a current, or the velocity field of a vortex. Put $\mathbf{r} = x\hat{\imath} + y\hat{\jmath}$; then

$$f\cdot r = x\,\frac{-y}{x^2 + y^2} + y\,\frac{x}{x^2 + y^2} = 0.$$

Therefore the field is perpendicular to the radius vector at every

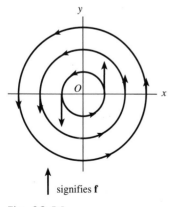

signifies **f**

Fig. 30.16

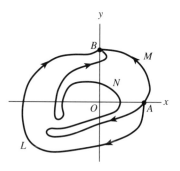

Fig. 30.17

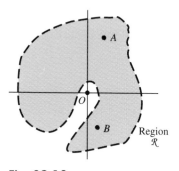

Fig. 30.18

point, as in Fig. 30.16. It is easy to confirm that

$$f = -\text{grad } V,$$

where V is a (path-dependent) *continuous* function such that

$$\tan V = y/x.$$

Thus we may take $V = \theta$ as described in the case we just discussed. (We cannot write $V = \arctan y/x$, because this function is discontinuous across the y axis: it would have an infinite gradient there.) The figure makes it obvious that the field is not conservative: more work is done if you take a unit magnetic pole against the field fifty times around the origin in order to travel between two points than if you go directly.

The field in Example 30.12 is not conservative, but whole classes of paths *are* equivalent. Suppose that, as in Fig. 30.17, we have two paths, (AMB) and (ANB), which can be steadily deformed into each other (as if A and B were conected by a piece of elastic) *without passing over the origin*. Then these two paths are equivalent. In this case, θ starts at $\theta_A = 0$; although the value of θ wanders about on (ANB), increasing and decreasing, it still ends at the value $\theta_B = \frac{1}{2}\pi$, as on the path (AMB).

However, (AMB) *cannot* be deformed into the third path (ALB) *without passing over the origin*; by following it around, it can be seen that $\theta = -\frac{3}{2}\pi$ for this path.

Suppose that we confine consideration to a 'patch', or region $\mathcal{R}$ as in Fig. 30.18, which neither contains nor surrounds the origin O. Then, *within this region*, the field behaves as if it were conservative, because any path from A to B inside the region can be deformed into any other without crossing the origin. We could not tell, from experiments confined to $\mathcal{R}$, that the field is not conservative over the whole plane.

Problems, Chapter 30

30.1 (Section 30.1.) Evaluate the following line integrals where (AOB) is shown in Fig. 30.19a.

(a) $\displaystyle\int_{(AOB)} x \, dx$; (b) $\displaystyle\int_{(AOB)} y \, dx$; (c) $\displaystyle\int_{(AOB)} x^2 \, dx$.

30.2. Evaluate the following integrals, $\mathcal{P}$ represents the parabolic path (AOB) on $y^2 = x$, shown in Fig. 30.19b.

(a) $\displaystyle\int_{\mathcal{P}} x \, dx$; (b) $\displaystyle\int_{\mathcal{P}} y \, dx$; (c) $\displaystyle\int_{\mathcal{P}} x^2 \, dx$;

(d) $\displaystyle\int_{\mathcal{P}} (x + y) \, dy$; (e) $\displaystyle\int_{\mathcal{P}} xy^2 \, dy$;

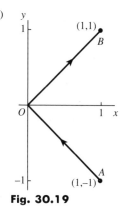

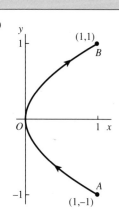

Fig. 30.19

(f) $\displaystyle\int_{P} (x\,dx + y\,dy);$ (g) $\displaystyle\int_{P} (\tfrac{1}{2}\,dx - y\,dy);$

(h) $\displaystyle\int_{P} (y\,dx - x\,dy).$

30.3. (Section 30.2). Evaluate the following line integrals over the various paths P, which are specified parametrically.

(a) $\displaystyle\int_{P} xy^2\,dx;$ P is $x = t^2$, $y = t$; $0 \leqslant t \leqslant 1$.

(b) $\displaystyle\int_{P} (x\,dy - y\,dx);$ P is $x = \cos t$, $y = \sin t$; $0 \leqslant t \leqslant \pi$.

(c) $\displaystyle\int_{P} (z\,dx - x\,dy + y\,dz);$ P is $x = t + 1$, $y = t$, $z = 2t$; $0 \leqslant t \leqslant 1$.

(d) $\displaystyle\int_{P} (x^2\,dx + y^2\,dy + z^2\,dz);$ P is $x = \cos t$, $y = \sin t$, $z = t$; $0 \leqslant t \leqslant 2\pi$.

(e) Compare (c) when P joins the same two points, $(1, 0, 0)$ to $(2, 1, 2)$, but $x = t^2 + 1$, $y = 2t - t^2$, $z = 2t^2$; $0 \leqslant t \leqslant 1$.

30.4. (Section 30.2). The line integral $\displaystyle\int_{(AB)} f(x, y)\,dy$, where the path (AB) is described by the curve $y = k(x)$, can be written formally as

$$\int_{(AB)} f(x, k(x))\,\frac{dk}{dx}\,dx.$$

Apply this formula to $\displaystyle\int_{(AB)} (x + y)\,dy$, taken over the parabolic path in Fig. 30.19b. Express it as the sum of two ordinary integrals over x. (This is like using x as the parameter in Section 30.1.)

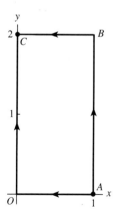

Fig. 30.20

30.5. (Section 3.) The references are to Fig. 30.20.)

(a) $\displaystyle\int_{(ABC)} dx;$ (b) $\displaystyle\int_{(AOC)} dy;$

(c) $\displaystyle\int_{(ABC)} (x\,dy - y\,dx);$ (d) $\displaystyle\int_{(AOC)} (x\,dy - y\,dx);$

(e) $\displaystyle\int_{(ABC)} y\,dy;$ (f) $\displaystyle\int_{(AOC)} y\,dy;$

(g) $\displaystyle\int_{(ABC)} (y\,dx + x\,dy);$ (h) $\displaystyle\int_{(AOC)} (y\,dx + x\,dy).$

30.6. (Section 30.4.) The integrands given are perfect differentials; P represents any path having the right direction which joins the two given points.

(a) $\displaystyle\int_{P} (x\,dx + y\,dy + z\,dz);$ P is $(-1, 1, -1)$ to $(1, -1, 1)$.

(b) $\displaystyle\int_{P} (yz\,dx + zx\,dy + xy\,dz);$ P is $(0, 0, 0)$ to $(1, 1, 1)$.

(c) $\displaystyle\int_{P} e^{x^2 + y^2 + z^2}\,(x\,dx + y\,dy + z\,dz);$ P is $(0, 0, 0)$ to $(1, 1, 1)$.

(d) $\displaystyle\int_{P} [(y + z)\,dx + (z + x)\,dy + (x + y)\,dx];$ P is $(1, 1, 1)$ to $(0, 1, 0)$.

(e) $\displaystyle\int_{P} [\cos(xy + yz + zx)]$
$\times [(y + z)\,dx + (z + x)\,dy + (x + y)\,dz];$ P is $(1, 0, \pi)$ to $(0, \pi, 1)$.

(f) $\displaystyle\int_{P} (xy^2\,dx - x^2y\,dy);$ P is $(1, 1)$ to $(2, 2)$.

30.7. (Section 30.5). Evaluate the following two-dimensional line integrals over the closed paths C given, the direction being anticlockwise.

(a) $\displaystyle\int_{C} (x^2\,dy - y^2\,dx);$ C is the circle $x^2 + y^2 = 4$.

(b) $\displaystyle\int_{C} \left(\frac{x}{y}\,dx + \frac{y}{x}\,dy \right);$ C is the ellipse $\tfrac{1}{4}x^2 + \tfrac{1}{9}y^2 = 1$; use the parametrization $x = 2\cos\theta$, $y = 3\sin\theta$.

30.8. Evaluate the following (all the paths C are closed).

(a) $\displaystyle\int_{C} (y\,dx + z\,dy + x\,dz);$ C is $x = \sin t$, $y = \cos t$, $z = \sin t$; $0 \leqslant t \leqslant 2\pi$.

(b) $\displaystyle\int_{(ABC)} (y\,dx + z\,dy + x\,dz);$ (ABC) is the triangle $A : (1, 0, 0)$, $B : (0, 1, 0)$, $C : (0, 0, 1)$.

(c) $\int_C (yz\,dx + zx\,dy + xy\,dz)$; C is any closed path.

30.9. Show that $\int_{(AB)} (yx^2\,dx + \frac{1}{3}x^3\,dy)$ is path-independent between any two points A and B. Use this fact to evaluate the integral along the spiral path given in polar coordinates (r, θ) by $r = e^{\theta/2\pi}$ for $0 \leqslant \theta \leqslant 2\pi$.

30.10 Show that if $\int_{(AB)} (f\,dx + g\,dy)$ is independent of the path (AB) for every two points A and B, then the integral around every closed path is zero. [Hint: A and B may coincide.]

30.11. Show that if the variables are changed in a perfect differential form, it remains a perfect differential. Illustrate this by transforming the identity $y\,dx + x\,dy = d(xy)$ into polar coordinates.

30.12. (Green's theorem, Section 30.6.) Confirm the truth of Green's theorem (30.12) for some very simple cases for which you know you can work out both the line integral and the double integral involved.

30.13. (Green's theorem, Section 30.6.) Check the correctness of the area formula, Example 30.9, by evaluating the line integral $\frac{1}{2}\int_C (x\,dy - y\,dx)$ taken around the following closed paths.
(a) The circle $x^2 + y^2 = 4$.
(b) The ellipse $\frac{1}{4}x^2 + \frac{1}{9}y^2 = 1$.
(c) The triangle with vertices $(-1, 0)$, $(2, 0)$, $(0, 4)$.

30.14. Find the area of the star-shaped region bounded by the curve $x^{\frac{2}{3}} + y^{\frac{2}{3}} = 1$, by parametrizing its equation as in Example 30.9.

30.15. The gravitation force F arising from a particle of mass M at the origin upon a particle of mass m at a point with position vector r is given by $f = -\gamma Mmr/r^3$. Find the work done by F on a particle which travels in from infinity to r.

30.16. Use Green's theorem with (30.10) to decide whether the following represent conservative fields (in two dimensions) or not in the stated regions.
(a) $(x^2 - y^2, 2xy)$; all x, y.
(b) $(\frac{1}{2}\ln(x^2 + y^2), \arctan(y/x))$; $x > 0$.

30.17. A force field has field intensity $f(x, y, z) = y\hat{\imath} + z\hat{\jmath} + x\hat{k}$. Is f conservative? Find the work done against the field by a unit particle moving in a straight line from $(0, 0, 0)$ to $(1, 1, 1)$.

30.18. A force f is given by $f(x, y, z) = yz\hat{\imath} + xz\hat{\jmath} + xy\hat{k}$. Show that it is conservative. Find the work done against f along the path $x = \cos t$, $y = \sin t$, $z = \sin t \cos t$; $-\frac{1}{2} \leqslant t \leqslant \frac{1}{2}\pi$. Are you doing this the easiest way?

30.19. Prove that a force field f having the form $f = r^\alpha \hat{r}$, where α is any constant, r is distance from the origin, $\hat{r}$ is the *unit* position vector and $\hat{r} = r/r$, is a conservative field. [Hint: start by putting $r = (x^2 + y^2 + z^2)^{\frac{1}{2}}$, and guess something that f might be the gradient of. If you cannot guess, then use the fact that grad $F(r) = \hat{r}(\partial F/\partial r)$.]

30.20. Generalize Problem 30.19 to a field $f = \hat{r}f(r)$. What is the potential of such a field?

30.21. Confirm that Green's theorem still holds for boundary C of the annular region $\mathcal{A}$ between the circles $x^2 + y^2 = 1$ and $x^2 + y^2 = 4$ for the line integral

$$\int_C [(2x - y^3)\,dx - xy\,dy].$$

What are the directions on C?

30.22. Show that $\int_C (5x^4y\,dx + x^5\,dy) = 0$ holds for any closed curve C for which Green's theorem is true.

30.23. Sketch the curve given parametrically by

$$x = \cos t - \frac{1}{2}\sin 2t, \quad y = \sin t; \quad 0 \leqslant t < 2\pi.$$

Using Green's theorem, find the area enclosed by the curve.

31

Sets

31.1 Notation

We are often interested in grouping together objects that have common characteristics or features. We might be interested in the integers 1, 2, 3, 4, or in all the integers. The set of all points in a plane would consist of pairs of numbers of the form (x, y), where x and y are coordinates which can take any real values. These examples all involve numbers, but the elements of sets can be other objects such as functions, or matrices, or Fourier series, or Laplace transforms, etc.

A **set** is a collection of objects or **elements**. The elements in the set can be defined by a rule or in any descriptive manner. Sets are usually denoted by capital letters such as S, A, B, X, etc, and their elements by lowercase letters such as s, a, b, x, etc. The elements in a set can be listed between braces $\{\ldots\}$. If the set A consists of just two numbers 0 and 1, then we write

$$A = \{0, 1\}, \quad \text{or} \quad A = \{1, 0\}, \tag{31.1}$$

the order being a matter of indifference. We say that 0 and 1 are **elements** or **members** of the set A, or **belong to** A. We write

$$0 \in A, \qquad 1 \in A,$$

read as '0 belongs to the set A', etc. The number 2 does not belong to A, and we write

$$2 \notin A,$$

that is, '2 does not belong to the set A'.

The set defined by (31.1) is the **binary** set which could represent the *on* and *off* states of a system. This could be the state of a light switch, for example.

Sets can be either **finite**, having a finite number of elements, or **infinite**, in which case the set contains an infinite number of elements. Thus the set given by (31.1) defines a finite set A, while

$$B = \{1, 2, 3, \ldots\},$$

the list of positive integers, defines an infinite set.

Some of the more common sets have their own special symbols:

> $\mathbb{R}$, the set of all real numbers
> $\mathbb{R}^+$, the set of positive real numbers (excludes zero)
> $\mathbb{Z}$, the set of all integers (positive, negative and zero)
> $\mathbb{N}^+$, the set of positive integers   (31.2)
> $\mathbb{N}^-$, the set of negative integers
> $\mathbb{Q}$, the set of rational numbers (that is, numbers of
> the form p/q where $q \neq 0$ and p are integers)

Often the elements are defined by a rule rather than by a list or formula. We write the set as

$$S = \{x \mid x \text{ satisfies specified rules}\},$$

which can be translated as 'S is the set of values of x which satisfy the stated rules'. The restrictions occur after the vertical |. Thus

$$S = \{x \mid x \in \mathbb{N}^+ \text{ and } 2 \leqslant x \leqslant 8\}$$

is an alternative way of writing $S = \{2, 3, 4, 5, 6, 7, 8\}$. As another example,

$$S = \{x \mid x \in \mathbb{R} \text{ and } 0 \leqslant x \leqslant 1\}$$

is the closed interval $[0, 1]$.

31.2 Equality, union, and intersection

Two sets A and B are said to be **equal** if they contain exactly the same elements. If this is the case, we write

$$A = B.$$

For example,

$$A = \{1, 2, 3\}, \qquad B = \{3, 2, 1\}, \qquad C = \{3, 1, 2, 1\}$$

are all equal, that is, $A = B = C$. The order of the elements is immaterial, and repeated elements are discounted.

In a given context, the set of all elements of interest in a given context is known as the **universal set**. It could be the set $\mathbb{R}$ (the set of real numbers), or the set of all complex numbers, but it will vary from application to application.

We now define how sets can be combined to create new sets. The **union** of two sets A and B is the set of all elements that belong to A, or to B, or to both. It is written as

$$A \cup B = \{x \mid x \in A \text{ or } x \in B \text{ or both}\},$$

and read as 'A union B'.

Example 31.1. Find the union of

$$A = \{x \mid x \in \mathbb{R} \text{ and } 0 \leqslant x \leqslant 2\} \quad \text{and}$$

$$B = \{x \mid x \in \mathbb{R} \text{ and } 1 \leqslant x \leqslant 3\}.$$

The elements in the union have to belong to one or other of the intervals $0 \leqslant x \leqslant 2$, or $1 \leqslant x \leqslant 3$, or to both. The interval $0 \leqslant x \leqslant 3$ contains all these numbers. Hence

$$A \cup B = \{x \mid \mathbb{R} \text{ and } 0 \leqslant x \leqslant 3\}.$$

The **intersection** of two sets A and B is the set $A \cap B$ that contains all elements common to both A and B. It is written and defined by

$$A \cap B = \{x \mid x \in A \text{ and } x \in B\}.$$

Example 31.2. Find the intersection of the sets A and B in Example 31.1.

The elements in the intersection have to belong simultaneously to both intervals, that is, to the overlapping part of the intervals $[0, 2]$ and $[1, 3]$, which is $[1, 2]$. Thus

$$A \cap B = \{x \mid x \in \mathbb{R} \text{ and } 1 \leqslant x \leqslant 2\}.$$

In the definitions of $A \cup B$ and $A \cap B$ above, we can see that the logical operation 'or' is associated with union, while 'and' is associated with intersection.

If $A \cap B$ has no elements, than A and B are said to be **disjoint**. The set with no elements is called the **empty set** and denoted by $\varnothing$. Thus, if A and B are disjoint, then $A \cap B = \varnothing$.

The **complement** of a set A is the set of all those elements which belong to the universal set U but do not lie in A. We denote this set by $\bar{A}$ (the notations A^c and A' are also frequently used). Hence, the complement of A is

$$\bar{A} = \{x \mid x \in U, x \notin A\}.$$

We say that A is a **subset** of B, expressed as $A \subseteq B$, if every element of A also belongs to the set B. It follows that $A \subseteq U$ if $B \subseteq U$. If there are elements of B which are not in A, then A is called a **proper subset** of B and written $A \subset B$. The statement $A \subseteq B$ includes the possibility that $A = B$, while $A \subset B$ does not. If $A \subseteq B$ and $B \subseteq A$, then all elements in A are contained in B, and *vice versa*; in other words, $A = B$.

The sets of integers $\mathbb{Z}$ and rational numbers $\mathbb{Q}$ are proper subsets of the real numbers $\mathbb{R}$, that is

$$\mathbb{Z} \subset \mathbb{R}, \quad \text{and} \quad \mathbb{Q} \subset \mathbb{R}.$$

We can summarize the results as follows.

(a) **Union**: $A \cup B = \{x \mid x \in A \text{ or } x \in B \text{ or both}\}$.
(b) **Intersection**: $A \cap B = \{x \mid x \in A \text{ and } x \in B\}$.
(c) **Complement**: $\bar{A} = \{x \mid x \notin A\}$.
(d) **Empty set**: $\varnothing$, the set with no elements   (31.3)
(e) **Subset**: $A \subseteq B$ means that A is a **subset** of B.
(f) **Proper subset**: means that $A \subset B$.

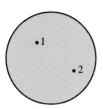

Fig. 31.1

31.3 Venn diagrams

Useful graphical views and interpretations of sets and operations on them can be provided by **Venn diagrams**. We represent sets by regions in the plane, with the interpretation that the region stands for those elements belonging to the given set. The diagrams are symbolic: the set $A = \{1, 2\}$, for example, could be represented by two numbers in the circle as shown in Fig. 31.1. The set consists of just the two marked points, not all the coordinate points enclosed within the curve. Usually, sets are represented by the interiors of circles, but any closed curves can be used. In a given context, all the sets are subsets of a certain **universal set** U, whose nature will differ according to the context.

If the universal set is represented by a rectangle, then a subset A of U could be represented by a circle within the rectangle shown in Fig. 31.2. This is a Venn diagram for U and A. Remember that A

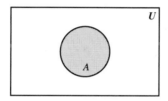

Fig. 31.2
Venn diagram for the universal set U and a set A.

could represent an infinite number of elements, or include just one element, or be the empty set $\varnothing$. The union, intersection, complement, and proper subset can be represented by the Venn diagrams shown in Fig. 31.3. The shaded regions indicate the elements defined by the operations.

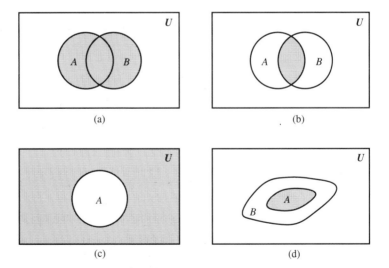

Fig. 31.3
(a) Union $A \cup B$.
(b) Intersection $A \cap B$.
(c) Complement $\bar{A}$.
(d) Proper subset $A \subset B$.

From the definitions of union, intersection, and complement, or from Venn diagrams, the following laws of the algebra of sets can be deduced:

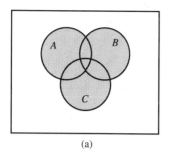

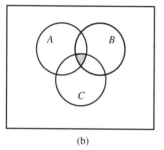

Fig. 31.4
(a) $(A \cup B) \cup C$ or
$A \cup (B \cup C)$. (b) $(A \cap B) \cap C$
or $A \cap (B \cap C)$.

$$A \cup A = A, \qquad A \cap A = A.$$
Commutative laws:
$$A \cup B = B \cup A, \qquad A \cap B = B \cap A.$$
Associative laws:
$$(A \cup B) \cup C = A \cup (B \cup C),$$
$$(A \cap B) \cap C = A \cap (B \cap C).$$
Distributive laws:
$$A \cap (B \cup C) = (A \cap B) \cup (A \cap C),$$
$$A \cup (B \cap C) = (A \cup B) \cap (A \cup C).$$

(31.4)

Sets also satisfy the following identity and complementary laws:

Identity laws: $A \cup \varnothing = A, \qquad A \cap U = A.$

Complementary laws:
$$A \cup \bar{A} = U, \qquad A \cap \bar{A} = \varnothing, \qquad \bar{\bar{A}} = A.$$

(31.5)

For example, $\bar{A}$ consists of all elements that do not belong to A, and none that do; so there are no elements common to A and $\bar{A}$. Therefore $A \cap \bar{A} = \varnothing$.

The **difference** of the sets A and B, written as $A \backslash B$, consists of the set of those elements that belong to A but do not belong to B. Thus
$$A \backslash B = \{ x \mid x \in A \text{ and } x \notin B \}.$$
(The notation $A - B$ is also used for $A \backslash B$). Figure 31.5 shows a Venn diagram for $A \backslash B$.

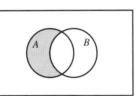

Fig. 31.5
Venn diagram for the
difference $A \backslash B$ (shaded).

Example 31.3. Using Fig. 31.6 as the Venn diagram of two sets A and B, mark by shading the following sets:
(a) $A \cup \bar{B}$, (b) $A \cap \bar{B}$, (c) $\bar{A} \cap \bar{B}$, (d) $\bar{A} \cup \bar{B}$, (e) $\overline{A \cup B}$, (f) $\overline{A \cap B}$.
Venn diagrams of the sets are shown in Fig. 31.7.

The previous example confirms **de Morgan's laws**, which are

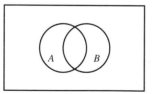

Fig. 31.6

$$\overline{A \cup B} = \bar{A} \cap \bar{B}, \qquad \overline{A \cap B} = \bar{A} \cup \bar{B}.$$

(31.6)

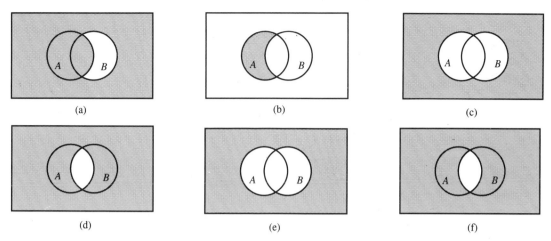

(a)   (b)   (c)

(d)   (e)   (f)

Fig. 31.7

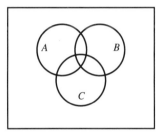

Fig. 31.8

Example 31.4. Using Fig. 31.8 as the Venn diagram of three sets A, B, and C, shade the following sets:

(a) $(A \cap B) \cup C$,   (b) $(A \cap B) \cap C$,   (c) $(A \cap B) \cap (A \cup C)$,
(d) $(A \cup B) \cup (A \cap C)$.

The required sets are shown in Fig. 31.9.

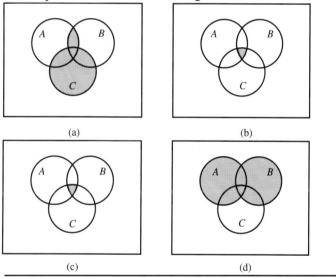

(a)   (b)

(c)   (d)

Fig. 31.9

Example 31.5. Show that $(A \cap B) \cup (A \cap \bar{B}) = A$.

By the distributive law,

$$(A \cap B) \cup (A \cap \bar{B}) = A \cap (B \cup \bar{B})$$
$$= A \cap U \quad \text{(by the complementary law)}$$
$$= A \quad \text{(by the identity law)}.$$

Example 31.6. Show that $(A \cup B) \cup (A \backslash B) = A \cup B$.

From Fig. 31.5, we can observe that $A \backslash B = A \cap \bar{B}$. Hence

$$
\begin{aligned}
A \cup B) \cup (A \backslash B) &= (A \cup B) \cup (A \cap \bar{B}) \\
&= A \cup (B \cup (A \cap \bar{B})) \quad \text{(associative law)} \\
&= A \cup ((B \cup A) \cap (B \cup \bar{B})) \\
&\qquad\qquad\qquad\qquad \text{(distributive law)} \\
&= A \cup ((B \cup A) \cap U) \\
&= A \cup (B \cup A) \quad \text{(identity law)} \\
&= (B \cup A) \cup A \quad \text{(commutative law)} \\
&= B \cup (A \cup A) = B \cup A = A \cup B.
\end{aligned}
$$

Alternatively, and more intuitively, we may notice that, since $A \backslash B$ is a subset of A, it is therefore also a subset of $A \cup B$, and so adds nothing to $A \cup B$ when united with it.

31.4 Sets and events

Suppose that two *distinguishable* coins a and b are spun simultaneously. We might be interested in the outcome that both coins show heads (H_a and H_b). This could be denoted by the set $A = \{(H_a, H_b)\}$. Similarly, outcomes of head and tail could be represented by the set $B = \{(H_a, T_b), (T_a, H_b)\}$, and two tails by $C = \{(T_a, T_b)\}$. The universal set for this problem is

$$
U = \{(H_a, H_b), (H_a, T_b), (T_a, H_b), (T_a, T_b)\}.
$$

Obviously, A and B are proper subsets of U. (We could dispense with the subscripts a and b if the understanding is that the first term refers to a and the second to b.) In a series of **trials** (that is, the spinning of the two coins), what is the set of outcomes in which either two heads or two tails occur? The answer is the union $A \cup C$. The two sets A and C in this experiment have no common elements. Thus $A \cap C = \varnothing$. The universal set in this type of application, which is the preamble to an investigation of the probability of an event occurring, is also known as the **sample space**. Generally, in probability theory, sets of outcomes are known as **events**: they can represent any collection of outcomes from the sample space.

In the problem with two *indistinguishable* coins, we might be only concerned with the total outcome, not with which coins are heads or tails. In this case the sample space would have just the three elements

$$
U = \{(H, H), (H, T), (T, T)\},
$$

where, for example, (H, T) means the outcome in which either coin is heads and the other tails.

Example 31.7. Suppose that A, B, C are three events in the sample space. Write down the sets which represent the event that: (a) A occurs, but B and C do not; (b) A, B, and C all occur.

(a) The event that B or C occurs will be $B \cup C$. The event that neither B nor C occurs will be the complement $\overline{B \cup C}$. The required set will be the intersection of this set and A, namely

$$A \cap \overline{(B \cup C)}.$$

By de Morgan's first law (31.6), this is equivalent to $A \cap (\bar{B} \cap \bar{C})$ or

$$A \cap \bar{B} \cap \bar{C}$$

(which is unambiguous, by the associative law of intersection in (31.4)).

(b) Events B and C occur in the set $B \cap C$. Events A and $B \cap C$ occur in the event $A \cap (B \cap C)$, which may unambiguously be written

$$A \cap B \cap C.$$

Two events are said to be **mutually exclusive** if they cannot occur together in a single trial, which in set terms is equivalent to two sets being disjoint. Consider the following illustrative application of a single die, which is rolled and the score noted. The outcome of a random experiment can be observed in many ways. A player could be interested in even or odd scores, the score 2 or not, or scores which are factors of 6 or not. In each case the sample space is divided into disjoint sets or mutually exclusive events, each of them providing an **exhaustive** list of outcomes. For example, if A stands for the event of an even score, than $\bar{A}$ must represent an odd score, and $U = A \cup \bar{A}$. Similarly, if A_i denotes a score of i where $i = 1, 2, \ldots, 6$, then

$$U = A_1 \cup A_2 \cup \cdots \cup A_6.$$

Any set can be divided into mutually exclusive events. Consider the set or event $A \cup B$. The sets $A \cap \bar{B}$, $A \cap B$, and $\bar{A} \cap B$ **partition** $A \cup B$ into mutually exclusive events (see Fig. 31.3a). Hence

$$A \cup B = (A \cap B) \cup (\bar{A} \cap B) \cup (A \cap \bar{B}).$$

An event A in the sample space, which also contains B, can be divided as

$$A = (A \cap B) \cup (A \cap \bar{B});$$

we can interpret this as saying that A can occur either with B or without B.

If the sample space is partitioned into the n mutually exclusive events $A_1, A_2, \ldots, A_n$, and, if A is an event associated with a trial, then

$$A = (A \cap A_1) \cup (A \cap A_2) \cup \cdots \cup (A \cap A_n).$$

This means that, if A occurs, then it must occur as one, and only one, of the events $A_1, A_2, \ldots, A_n$. It might happen that $A \cap A_1 = \emptyset$ for some intersections, but this does not matter.

Example 31.8. Suppose that two distinguishable dice are rolled. How many elements does the sample space contain? Let A denote the event {the sum of the outcomes is 7}, and B the event {at least one die shows 3}. List all the elements of A, B, $A \cup B$, and $A \cap B$.

The set U is given by

$$S = \{(i, j) \mid i, j = 1, 2, 3, 4, 5, 6\},$$

which has $6 \times 6 = 36$ elements, since the dice are distinguishable and (i, j) is distinct from (j, i). The events or sets are given by

$$A = \{(i, j) \mid i + j = 7\}, \quad B = \{(i, j) \mid \text{either } i = 3 \text{ or } j = 3 \text{ or both}\}$$

$$A \cup B = \{(1, 3), (1, 6), (2, 3), (2, 5), (3, 1), (3, 2), (3, 3), (3, 4),$$
$$(3, 5), (3, 6), (4, 3), (5, 2), (5, 3), (6, 1), (6, 3)\},$$

$$A \cap B = \{(3, 4), (4, 3)\}.$$

Example 31.9. In a manufacturing process, a product passes through three production stages and is given a quality check at all three stages, which it either passes or fails. Let P_i represent the set of products passing the quality check at stage i. Draw a Venn diagram of the process. Interpret the quality failures of the products in the sets given by $\bar{P}_1$, $P_2 \backslash (P_1 \cup P_3)$ and $(P_1 \cup P_2) \cap P_3$. What set represents the completely satisfactory products?

A production run of 1000 occurs, of which 98 fail all stages, 20 pass only stage P_1, 31 only stage P_2, and 17 only stage P_3; 814 pass stages P_1 and P_2, 902 stages P_2 and P_3, and 800 stages P_3 and P_1. Determine the final number which pass all quality checks.

$\bar{P}_1$ represents all products which fail the P_1 quality check.

$P_2 \backslash (P_1 \cup P_3)$ represents those products which pass only P_2 stage.

$(P_1 \cup P_2) \cap P_3$ represents those products which are satisfactory at stages P_3 and P_1 or P_2. The set $P_1 \cap P_2 \cap P_3$ represents those products which are satisfactory at all stages.

The numbers associated with each subset of the universal set U are shown in Fig. 31.10. Since 98 fail all quality checks, then the number of elements in $P_1 \cup (P_2 \cup P_3)$ is 902. In the figure, k represents the number of products which pass all the quality checks. Hence $800 - k$, for example, represents those products which are satisfactory in stages P_1 and P_2, but fail in P_3. Thus $P_1 \cup P_2 \cup P_3$ contains

$$902 = 20 + 31 + 17 + (814 - k) + (902 - k) + (800 - k) + k$$

products. Hence $902 = 2584 - 2k$, and so

$$k = 841.$$

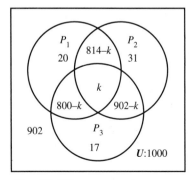

Fig. 31.10

Of the 1000 products manufactured, 841 passed all the quality checks.

In the previous example, we are really interested in the *numbers* of elements in each of the sets. For example, the number of elements in U is 1000 and the number of elements in $P_2 \backslash (P_1 \cup P_3)$, those products which pass only stage 2, is 31. We write

$$n(U) = 1000, \qquad n[P_2 \backslash (P_1 \cup P_3)] = 31.$$

The number of elements in the set S is $n(S)$: this number is known as the **cardinality** of S. Many sets can have **infinite cardinality**. For example, $n(\mathbb{Q})$, where $\mathbb{Q}$ is the set of rational numbers, is an infinite number. We write $n(\mathbb{Q}) = \infty$. The empty set $\varnothing$ has no elements: hence $n(\varnothing) = 0$.

The following results apply to *finite sets*. If two finite sets A and B are disjoint, then they have no elements in common. It follows that

$$n(A \cup B) = n(A) + n(B).$$

This result applies to any number of disjoint sets. It is clear that they must be disjoint, since otherwise elements would be counted more than once.

This last result is also a useful method of counting elements when combined with a Venn diagram. Consider just two sets A and B as shown in Fig. 31.11. The sets representing each of the subsets in the Venn diagram $A \backslash B$, $A \cap B$, and $B \backslash A$ are shown in Fig. 31.11. Since these sets are disjoint, then we can obtain a formula for the number of elements in the union of A and B, namely

$$n(A \cup B) = n(A \backslash B) + n(A \cap B) + n(B \backslash A). \tag{31.7}$$

For sets A and B separately,

$$n(A) = n(A \backslash B) + n(A \cap B), \qquad n(B) = n(B \backslash A) + n(A \cap B). \tag{31.8}$$

Elimination of $n(A \backslash B)$ and $n(B \backslash A)$ between (31.7) and (31.8) leads to the alternative result

$$n(A \cup B) = n(A) + n(B) \backslash n(A \cap B).$$

For three finite sets A, B, and C the corresponding result is

$$n(A \cup B \cup C) = n(A) + n(B) + n(C) + n(A \cap B \cap C)$$
$$- n(B \cap C) - n(C \cap A) - n(A \cap B).$$

This result can be constructed from the Venn diagram.

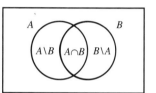

Fig. 31.11

Counting elements in the union of two sets.

Problems

31.1. (Section 31.1). List the elements in the following sets:

(a) $S = \{x \mid x \in \mathbb{N}^+ \text{ and } 3 \leqslant x \leqslant 10\}$;

(b) $S = \{x \mid x \in \mathbb{N}^+ \text{ and } -2 \leqslant x \leqslant 4\}$;

(c) $S = \{x \mid x \in \mathbb{Z} \text{ and } -2 \leqslant x \leqslant 4\}$;

(d) $S = \{x \mid x \in \mathbb{N}^+, \mathbb{N}^- \text{ and } -2 \leqslant x \leqslant 4\}$;

(e) $S = \{1/x \mid x \in \mathbb{N}^+ \text{ and } 3 \leqslant x \leqslant 8\}$;

(f) $S = \{x^2 \mid x \in \mathbb{N}^+ \text{ and } |x| \leqslant 3\}$;

(g) $S = \{x + jy \mid x \in \mathbb{N}^+, y \in \mathbb{N}^+, 1 \leqslant x \leqslant 4, 2 \leqslant y \leqslant 5\}$.

31.2. (Section 31.3). Show on Venn diagrams the following sets:
(a) $A \cup \bar{B}$;   (b) $\bar{A} \cap \bar{B}$;   (c) $A \cap (B \cup C)$;
(d) $(A \cap B) \cup (B \cap C)$;   (e) $\overline{A \cap B}$;
(f) $(A \backslash B) \cap C$;   (g) $A \backslash (B \cap C)$;
(h) $\overline{(A \backslash B) \cup (B \backslash C)}$.

31.3. (Section 31.2). Determine the union $A \cup B$ of each of the following pairs of sets A and B:
(a) $A = \{x \mid x \in \mathbb{R} \text{ and } -1 \leqslant x \leqslant 2\}$,
    $B = \{x \mid x \in \mathbb{R} \text{ and } -1 \leqslant x \leqslant 4\}$;
(b) $A = \{x \mid x \in \mathbb{R} \text{ and } -1 \leqslant x < 0\}$, $B = \{x \mid x \in \mathbb{R}$ and $0 < x < 1\}$;
(c) $A = \{1, 2, 3, 4\}$, $B\{-4, -3, -2, -1\}$;
(d) $A = \{y \mid y = \cos x, x \in \mathbb{R}, \text{ and } 0 \leqslant x \leqslant \frac{1}{2}\pi\}$,
    $B = \{y \mid y = \sin x, x \in \mathbb{R}, \text{ and } -\frac{1}{2}\pi \leqslant x \leqslant \frac{1}{2}\pi\}$.

31.4. (Section 31.2). Determine the intersections $A \cap B$ of the following sets:
(a) $A = \{x \mid x \in \mathbb{R}, \text{ and } -2 \leqslant x \leqslant 1\}$,
    $B = \{x \mid x \in \mathbb{R}, \text{ and } -1 \leqslant x \leqslant 2\}$;
(b) $A = \{x \mid x \in \mathbb{N}^+ \text{ and } -5 \leqslant x \leqslant 2, B = \{x \mid x \in \mathbb{R}$, and $-5 \leqslant x \leqslant 2\}$;
(c) $A = \{n \mid n = 1/m \text{ and } m \in \mathbb{N}^+\}$, $B = \{n \mid n = 1/m^2$ and $m \in \mathbb{N}^+\}$;
(d) $A = \{x \mid x \in \mathbb{R} \text{ and } x^2 - 3x + 2 = 0\}$,
    $B = \{x \mid x \in \mathbb{R} \text{ and } 2x^2 + x - 3 = 0\}$;
(e) $A = \{x \mid x \in \mathbb{R} \text{ and } |x| \leqslant 2\}$,
    $B = \{x \mid x \in \mathbb{R} \text{ and } |x - 1| \leqslant 1\}$.

31.5. (Section 31.3). Construct a set formula for the shaded sets of Fig. 31.12:

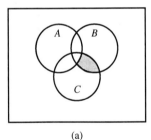

(a)

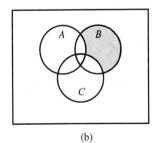

(b)

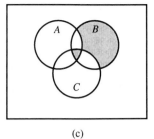

(c)

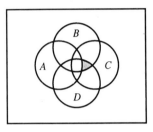

(d)

Fig. 31.12

31.6. The set S consists of products, each of which is given n pass/fail tests, numbered 1 to n. The set S_r consists of those products that pass test r. What is the set of products that
(a) fails all tests,
(b) fails only test 1,
(c) fails some tests?

31.7. At Keele University, all first-year students must take three subjects of which at least one must be a science subject, and at least one must be a humanities or social science subject. Let A be the set of all first-year students in a given year, A_1 the set of students who take exactly one science subject, B_1 the set of students who take just one humanities subject, and B_2 the set of those who take two social science subjects. Draw a Venn diagram to represent the different sets of students classified by groups of subjects. Give set formulae for students who take
(a) just one social science subject
(b) no humanities subject,
(c) one subject from each group.

31.8. (Section 31.3). The rules listed in (31.4) illustrate the **duality principle** which states that every statement involving sets which is true *for all sets* has a dual in which $\cup$ and $\cap$ are interchanged, and $\varnothing$ and U are interchanged everywhere.
    Use Venn diagrams to establish the following:
(a) $(A \backslash B) \cap C = (A \cap C) \backslash B$;

(b) $A \cap (B \cup C) = (A \cap B) \cup (A \cap C)$.
What are their dual identities?

31.9. Three sets A, B, and C satisfy

$$A \cap B \cap C = (A \cap C) \cup (B \cap C).$$

Explain why the dual result of Problem 31.8 is not necessarily true. What condition of the duality principle is violated?

31.10. The **cartesian product** of two sets A and B is the set of all **ordered pairs** $\{(a, b)\}$, where $a \in A$ and $b \in B$. It is written as

$$A \times B = \{(a, b) \mid a \in A \text{ and } b \in B\}.$$

If $A = B$, then we write $A \times A = A^2$. Let $A = \{1, 2\}$ and $B = \{1, 2, 3\}$; write down all the elements in the sets $A \times B$, $B \times A$, A^2, and B^2.

31.11. (Section 31.4). Consider the sample space of events in which two coins are spun. In the first case the coins are indistinguishable, and in the second distinguishable, as described in Section 31.4. In which case is the sample space a cartesian product?

31.12. The cartesian product extends to the products of three or more sets. Thus

$$A \times B \times C = \{(a, b, c) \mid a \in A \text{ and } b \in B \text{ and } c \in C\}.$$

Let $A = \{1, 2, 3\}$, $B = \{0, 1\}$, and $C = \{1, 2\}$. Write down all the elements in

$$A \times B \times C, \ A^2 \times C, \ (A \cup B) \times C, \ (A \cap B) \times C.$$

31.13. At the end of a production process, 500 electrical components pass through three quality checks P, Q, and R. It is found that 38 components fail check P, 29 fail Q, 30 fail R, 7 fail P and Q, 5 fail Q and R, 8 fail R and P and 3 fail all checks. Determine how many components:
(a) pass all checks, (b) fail just one check,
(c) fail just two checks.

31.14. (Section 31.4). Two distinguishable dice are rolled and the scores noted. Formulate the set for the sample space. How many elements does the set have? Let
(a) A denote the event {the sum of the outcomes is 5},
(b) B denote the event {at least one die shows 4}.
   Express the sets of these events in formula terms. List all the elements in A, B, $A \cup B$, and $A \cap B$.

31.15. (Section 31.4). Suppose that A, B, and C are three events (subsets) of the sample space (set) S.

Write down the set formulae for the events:
(a) only B occurs,
(b) exactly one of A, B, C occurs,
(c) none of A, B or C occurs.

31.16. (Section 31.4). Suppose that a sample space U includes the events A and B. Show that the number of elements in $A \cup B$ can be expressed as

$$n(A \cup B) = n(A \cap B) + n(\bar{A} \cap B) + n(A \cap \bar{B}),$$

(this is an alternative version of (31.7)).
   Suppose that two distinguishable dice are rolled. Let A denote the event {the sum of the outcomes is 6} and B the event {both dice show the same number}. Find $A \cap B$, $\bar{A} \cap B$, and $A \cap \bar{B}$, and find $n(A \cup B)$ using the formula above.

31.17. (Section 31.4). For three finite sets A, B, and C, show that the number of elements in the union of the sets is given by

$$n(A \cup B \cup C) = n(A) + n(B) + n(C)$$
$$+ n(A \cap B \cap C) - n(B \cap C)$$
$$- n(C \cap A) - n(A \cap B).$$

31.18. If A and B are two finite sets, explain why, for the cartesian product (defined in Problem 31.10 above),

$$n(A \times B) = n(A)n(B).$$

31.19 The menu in a restaurant contains three courses: 4 starters (set A), 5 main courses (set B), and 3 sweets (set C). Customers can choose either the full menu or, alternatively, a main course and a sweet. In terms of cartesian products what is the set of all possible meals (the answer is really a *set of pairs and triples*). For how many different orders can customers ask?

31.20. Given $A = \{1, 2, 3\}$, $B = \{3, 4\}$, and $C = \{2, 3, 4, 5\}$, find the elements in the sets $B \cup C$, $B \cap C$, and the cartesian products $A \times B$ and $A \times C$. Verify that

$$A \times (B \cup C) = (A \times B) \cup (A \times C),$$
$$A \times (B \cap C) = (A \times B) \cap (A \times C).$$

(This example suggests general results which are true for all sets.)

32 Boolean algebra: logic gates and switching functions

32.1 Laws of Boolean algebra

We are now going to present some new operations between special entities. They have some analogies with ordinary addition and multiplication, and the symbols for them will be similar—but not the same, since we need to emphasize that these are Boolean operations. The algebra involved is named after George Boole (1815–64) who first developed the modern ideas of symbolic logic. Boolean algebra has applications in logic and switching circuits.

Consider a set B which consists of just two elements 0 and 1, that is, $B = \{0, 1\}$. Suppose that we denote the **sum** of two elements a and b of B by $a \oplus b$ (the notations $\vee$, $\cup$, and $+$, and the alternative term **join** are also used); we denote the **product** of the two elements by $a \cdot b$ (the notations $\wedge$, $\cap$, $\times$, and $*$, or simply **juxtaposition**, and the alternative term **meet** are also in use) and the **complement** of a by $\bar{a}$ ($\sim a$ and $\neg a$ are used in logic). These **binary operations** applied to the members of B are defined to give the elements shown in Table 32.1.

Table 32.1 *Binary operations*

Sum			Product			Complement	
a	b	$a \oplus b$	a	b	$a \cdot b$	a	$\bar{a}$
0	0	0	0	0	0	0	1
0	1	1	0	1	0	1	0
1	0	1	1	0	0		
1	1	1	1	1	1		

Thus, for example

$$0 \oplus 1 = 1, \quad 1 \oplus 1 = 1, \qquad 0 \cdot 0 = 0, \quad 1 \cdot 1 = 1, \qquad \bar{0} = 1, \quad \bar{1} = 0.$$

The elements of B are known as **Boolean variables**. We have restricted our set B to one with just two elements or binary digits, because this is the main application in circuits and computer design, but definitions can be interpreted for more general sets. A **Boolean algebra** is a set with the operations $\oplus$, $\cdot$, and $^-$ defined on it, together with the following laws on any elements a, b, c which belong to B:

Commutative laws:
$$a \oplus b = b \oplus a, \qquad a \cdot b = b \cdot a;$$

Associative laws:
$$a \oplus (b \oplus c) = (a \oplus b) \oplus c, \qquad a \cdot (b \cdot c) = (a \cdot b) \cdot c; \qquad (32.1)$$

Distributive laws:
$$a \cdot (b \oplus c) = (a \cdot b) \oplus (a \cdot c),$$
$$a \oplus (b \cdot c) = (a \oplus b) \cdot (a \oplus c).$$

In addition, the set must contain distinct identity elements 0 and 1 for the operations $\oplus$ and $\cdot$ respectively. For these identities we must have the **identity laws**

$$a \oplus 0 = 1, \qquad a \cdot 1 = a.$$

Finally, the **complement laws** must hold:

$$a \oplus \bar{a} = 1, \qquad a \cdot \bar{a} = 0.$$

To summarize, we can say that a Boolean algebra consists of the collection

$$(B, \oplus, \cdot, {}^-, 0, 1),$$

in other words, a set, the binary operations $\oplus$ and $\cdot$, the complement $^-$, and the identity elements 0 and 1.

The binary set $B = \{0, 1\}$, which particularly concerns us in this chapter, consists simply of the two identity elements. We can check that the definitions in Table 32.1 satisfy the laws in (32.1). They look familiar if Chapter 31 has already been covered: they are essentially the laws of set operations with sum $\oplus$ and product $\cdot$ replacing union $\cup$ and intersection $\cap$, and with 1 replacing the universal set U and 0 the empty set $\varnothing$.

Just as with sets, we can deduce further laws, some of which are listed in (32.2):

Absorption laws: $a \oplus a \cdot b = a, \qquad a \cdot (a \oplus b) = a;$

de Morgan's laws: $\overline{a \oplus b} = \bar{a} \cdot \bar{b}, \qquad \overline{a \cdot b} = \bar{a} \oplus \bar{b};$

Identity laws:
$$1 \oplus a = a \oplus 1 = 1, \qquad 0 \cdot a = a \cdot 0 = 0; \qquad (32.2)$$

Reflexive law: $\bar{\bar{a}} = a.$

Note that $\cdot$ takes precedence over $\oplus$. Thus, in the first absorption law, $a \oplus a \cdot b$ means $a \oplus (a \cdot b)$; in the second absorption law, the parentheses are essential.

We will prove one of the absorption laws to illustrate how proofs are approached in Boolean algebra.

Example 32.1. Prove that $a \oplus a \cdot b = a$.

For all $a,b \in B$

$$a \oplus a \cdot b = a \cdot 1 \oplus a \cdot b \quad \text{(identity law)}$$
$$= a \cdot (1 \oplus b) \quad \text{(distributive law)}.$$

Now

$$1 \oplus b = (1 \oplus b) \cdot 1 \quad \text{(identity law)}$$
$$= 1 \cdot (b \oplus 1) \quad \text{(associative law)}$$
$$= (b \oplus \bar{b}) \cdot (b \oplus 1) \quad \text{(complement law)}$$
$$= b \oplus \bar{b} \cdot 1 \quad \text{(distributive law)}$$
$$= b \oplus \bar{b} \quad \text{(identity law)}$$
$$= 1 \quad \text{(complement law)}.$$

Finally

$$a \oplus a \cdot b = a \cdot 1 = a.$$

32.2 Logic gates and truth tables

Any element made up from the elements of B and the operations $\oplus$, $\cdot$, and $^-$ is known as a **Boolean expression**. For example,

$$a \oplus b, \quad a \oplus \bar{b}, \quad a \oplus \bar{a} \cdot b,$$

are examples of Boolean expressions. For the binary set, the elements 1 and 0 can represent 'on' or 'off' states in digital circuits. The basic components in a computer are **logic gates** which can produce an ouptut from inputs. All the outputs and inputs can be in one of two states, usually either low voltage (0) or high voltage (1).

The fundamental Boolean operations of $\oplus$, $\cdot$, and $^-$ correspond to devices known respectively as the OR gate, AND gate and NOT gate. As with circuit components such as resistance and inductance, each has its own symbol.

The OR gate has two inputs and a single output represented by the symbol in Fig. 32.1. The output is $f = a \oplus b$. The inputs a and b can each take either of the values 0 or 1. Hence there are four possible inputs into the device as listed in Table 32.2. The final column f can be completed using sum rule in Table 32.1. Then, if a is 'on' (1) and b is 'off' (0), the output f is 'on' (1). Table 32.2 is known as the **truth table** of the OR gate.

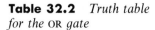

Table 32.2 *Truth table for the OR gate*

a	b	$f = a \oplus b$
0	0	0
0	1	1
1	0	1
1	1	1

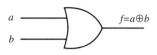

Fig. 32.1
The OR gate.

Table 32.3 *Truth table for the AND gate*

a	b	$f = a \cdot b$
0	0	0
0	1	0
1	0	0
1	1	1

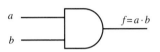

Fig. 32.2
The AND gate.

The symbol and truth table for the AND gate are shown in Fig. 32.2 and Table 32.3. Again the device has two inputs and the single output $f = a \cdot b$, the product of a and b.

Table 32.4 *Truth table for the NOT-gate*

a	$f = \bar{a}$
0	1
1	0

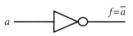

Fig. 32.3
The NOT gate.

Finally the NOT gate is shown in Fig. 32.3 with its truth table given as Table 32.4. The NOT gate has a single input and a single output which is the complement of its input.

There is further jargon associated with these gates. The output $a \oplus b$ is known as the **disjunction** of a and b, while $a \cdot b$ is known as the **conjunction** of a and b, and $\bar{a}$ is called the **negation** of a.

These devices can be connected in series and parallel to create new logic devices, each of which will have its own truth table.

A series connection between a NOT gate and an AND gate is shown in Fig. 32.4a. The output $a \cdot b$ of the AND gate becomes the input of the NOT gate which results in the output $\overline{a \cdot b}$. This device is known as the NAND gate, and it has its own symbolic representation shown in Fig. 32.4b. Its truth table is given in Table 32.5.

A series connection between a NOT gate and an OR gate produces the NOR gate as shown in Fig. 32.5a. The output f is the complement of the sum of a and b. The NOR gate also has its own symbol shown in Fig. 32.5b. It has the truth table shown in Table 32.6.

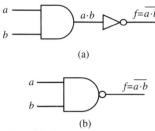

Fig. 32.4
The NAND gate.

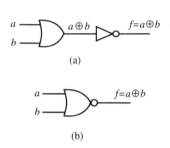

Fig. 32.5
The NOR gate.

Table 32.5 *Truth table for the NAND gate*

a	b	$f = \overline{a \times b}$
0	0	1
0	1	1
1	0	1
1	1	0

Table 32.6 *Truth table for the NOR gate*

a	b	$f = \overline{a \oplus b}$
0	0	1
0	1	0
1	0	0
1	1	0

32.3 Logic networks

The five gates introduced in the previous section can be linked in series and parallel combinations to create further **logic networks**. Some examples are presented here.

Example 32.2. Construct the Boolean expression for the output f of the device shown in Fig. 32.6.

Starting from the left in Fig. 32.6, the upper AND gate produces an output $a \cdot b$ and the lower OR gate has an output $c \oplus d$. These become the inputs into the OR gate on the right. Hence the final output is

$$f = a \cdot b \oplus c \oplus d.$$

Since there are four inputs, the output can be determined for each of the $2^4 = 16$ possible inputs.

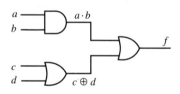

Fig. 32.6

Example 32.3. Figure 32.7 shows a logical network with three inputs a, b, c, and four devices. Find a Boolean expression for the output f. Write down the truth table for the system.

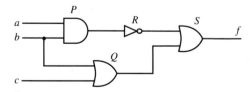

Fig. 32.7

The input b is the same in both devices P and Q. The output from the AND gate P is $a \cdot b$, and the output from R is $\overline{a \cdot b}$. The output from Q is $b \oplus c$. Hence the inputs $\overline{a \cdot b}$ and $b \oplus c$ into S produce an output

$$f = \overline{a \cdot b} \oplus b \oplus c.$$

The truth table for this network is given in Table 32.7.

Table 32.7

a	b	c	$\overline{a \cdot b}$	$b \oplus c$	$\overline{a \cdot b} \cdot (b \oplus c)$
0	0	0	0	0	0
0	0	1	0	1	1
0	1	0	0	1	1
0	1	1	0	1	1
1	0	0	0	0	0
1	0	1	0	1	1
1	1	0	1	1	1
1	1	1	1	1	1

Example 32.4. Show that, using just the NOR gate, it is possible to build a logic network to model any Boolean expression.

Given inputs a and b, we have to show that devices can be constructed using just NOR gates with outputs of $a \oplus b$, $a \cdot b$, and $\bar{a}$. For inputs of a and b, the single NOR gate generates an output of $\overline{1 \oplus b}$. Figure 32.8 shows three devices which simulate the required outputs.

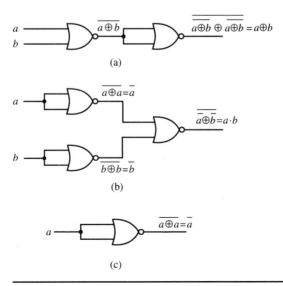

Fig. 32.8

Example 32.5. Design a logic network using OR, AND, and NOT gates to reproduce the Boolean expression $f = a \cdot \bar{b} \oplus a$ for inputs a and b.

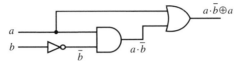

Fig. 32.9

From input b we obtain $\bar{b}$ by a NOT gate. The inputs a and $\bar{b}$ are then fed into an AND gate to produce $a \cdot \bar{b}$. Finally a spur from the a input and the $a \cdot \bar{b}$ output are fed into an OR gate as shown in Fig. 32.9.

32.4 The inverse truth-table problem
In this problem we attempt to re-create a Boolean expression for a given truth table. For example, Table 32.8 is a truth table for two

Table 32.8

a	b	f
0	0	0
0	1	1
1	0	1
1	1	0

inputs a and b. The question posed is: can we provide a general construction of a Boolean expression which will generate this truth table? In the final column, note which outputs are 1. For each case in which $f = 1$, consider the product $a \cdot b$ if $a = b = 1$, $\bar{a} \cdot b$ if $a = 0$ and $b = 1$, etc. using, in the product, the element or its complement according as the table entry is 1 or 0. Now form the sum of these elements. Put

$$f = a \cdot \bar{b} \oplus \bar{a} \cdot b, \qquad (32.4)$$

where it can be checked that, if $a = b$, then $f = 0$, and if a and b are not the same, then $f = 1$.

This particular gate is known as the **exclusive-OR** gate, or EXOR gate, and has its own symbol shown in Fig. 32.10. This form of f obtained by the construction just described is known as the **disjunctive normal form**.

The method can be applied to more complex truth tables. Table 32.9 shows an output for three inputs. The output 1 appears in rows 2, 4, 5, 7, 8. The disjunctive normal form for a corresponding Boolean expression is, following the rules for products of elements and their complements,

$$f = \bar{a} \cdot \bar{b} \cdot c \oplus \bar{a} \cdot b \cdot c \oplus a \cdot \bar{b} \cdot \bar{c} \oplus a \cdot b \cdot \bar{c} \oplus a \cdot b \cdot c.$$

Fig. 32.10
The exclusive-OR gate.

Table 32.9

a	b	c	f
0	0	0	0
0	0	1	1
0	1	0	0
0	1	1	1
1	0	0	1
1	0	1	0
1	1	0	1
1	1	1	1

Thus in row 2 write $a \cdot \bar{b} \cdot c$, since $a = b = 0$ but c is 1, and so on. Check that f does give the required output. The disjunctive normal form always guarantees an answer, but it is not necessarily the simplest or most efficient in circuit architecture.

32.5 Switching circuits

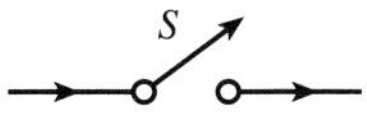

A circuit of **on–off** switches can also be represented by Boolean expressions. For example, Fig. 32.11 shows a simple on–off switch in part of a circuit. Current flows if the switch S is in the *on* or closed position ($a = 1$), and does not flow if the switch is in the *off* or open position ($a = 0$). The variable a represents the state of the switch.

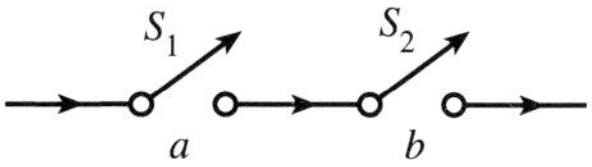

Consider two switches S_1 and S_2 in series (Fig. 32.12). Current only flows if both switches are closed, that is, when $a_1 = 1$ and $a_2 = 1$, where a_1 and a_2 represent the states of the switches. Hence the truth table for the series switches is as shown in Table 32.10. Thus the state of current flow is given by $f = a \cdot b$, the product of a and b.

Table 32.10   *Truth table for two switches in series*

a	b	f
0	0	0
0	1	0
1	0	1
1	1	1

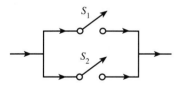

Fig. 32.11
On–off switch.

Fig. 32.12
Two switches in series.

Fig. 32.13
Two switches in parallel.

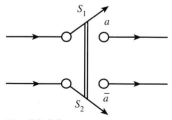

Fig. 32.14
Complement of a switch using a rigid tie.

Similarly two switches in parallel (Fig. 32.13) correspond to the sum of a and b. The truth table is given in Table 32.11. The final column indicates that $f = a \oplus b$.

The complement of a, the state of switch S_1, is another switch S_2 in the circuit which is always in the complementary state to S_1, off when S_1 is on and vice versa. It can be represented symbolically by Fig. 32.14, in which the switches S_1 and S_2 are joined by a rigid tie.

These devices are analogous to the gates of Section 32.3. For switching circuits, the Boolean expressions are often referred to as **switching functions**.

Table 32.11   *Truth table for two switches in parallel*

a	b	f
0	0	0
0	1	1
1	0	1
1	1	1

Example 32.6. Find a switching function f for the system shown in Fig. 32.15.

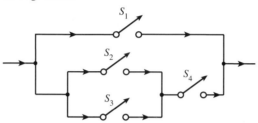

Fig. 32.15

Let a_1, a_2, a_3, a_4 represent respectively the states of each switch S_1, S_2, S_3, S_4. Since S_2 and S_3 are in parallel, their output will be $a_2 \oplus a_3$. This combined in series with a_4 will give an outut of $(a_2 \oplus a_3) \cdot a_4$. In turn, this is in parallel with S_1. Hence, the final output is

$$(a_2 \oplus a_3) \cdot a_4 \oplus a_1.$$

Example 32.7. A light on a staircase is controlled by two switches S_1 and S_2, one at the bottom of the stairs and one at the top. Switches can be separately 'up' or 'down'. If both switches are up, the light is off. Either switch changed to down switches the light on, and any subsequent change to a switch alters the state of the light. Design a truth table for the circuit.

The truth table is shown in Table 32.12, where the state of S_i ($i = 1, 2$) is $a_i = 0$ when the switch is up (off) and $a_i = 1$ when the switch is down (on). The light on is $f = 1$, and the light off is $f = 0$. This truth table is the same as that for the exclusive-OR gate in Section 32.4. Hence, from (32.4), the circuit can be represented by the switching function

$$f = a_1 \cdot \bar{b}_2 \oplus \bar{a}_1 \cdot a_2.$$

The actual circuit is shown in Fig. 32.16, where S_1 and S_2 are one-pole two-way switches. At S_1, the state a_1 represents the switch 'up' and its complement $\bar{a}_1$ is the switch down. A similar state operates at S_2.

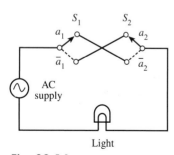

Fig. 32.16
Two-switch light control.

Table 32.12

Switch S_1	Switch S_2	Light	a_1	a_2	f
up	up	off	0	0	0
down	up	on	1	0	1
down	down	off	1	1	0
up	down	on	0	1	1

Problems, Chapter 32

32.1. Read through Example 32.1. Now prove the other absorption law:

$$a \cdot a \oplus b = a.$$

(Example 32.1 and this result illustrates the **duality principle**, which states that any theorem which can be proved in Boolean algebra implies another theorem with $\cdot$ and $\oplus$ interchanged for the same elements.)

32.2. (Section 32.1). Prove the de Morgan result

$$\overline{a \oplus b} = \bar{a} \cdot \bar{b},$$

by showing that $(a \oplus b) \oplus (\bar{a} \cdot \bar{b}) = 1$. Explain how the duality result (Problem 32.1) gives the other de Morgan theorem.

32.3. (Section 32.1). Let B be the Boolean algebra with the two elements 0 and 1. For arbitrary $a,b \in B$, prove the following:
(a) $a \cdot (\bar{a} \oplus b) = a \cdot b$;   (b) $(a \oplus b) \cdot (a \oplus \bar{b}) = a$;
(c) $(a \oplus b) \cdot \bar{a} \cdot \bar{b} = 0$.

32.4. (Section 32.1). Using the laws of Boolean algebra for the set with two elements 0 and 1, show that:
(a) $a \cdot b \oplus a \cdot \bar{b} = a$;   (b) $a \oplus \bar{a} \cdot \bar{b} \cdot c = a \oplus \bar{b} \cdot c$.
Use the result to obtain the truth tables in each case.

32.5. (Section 32.4). In Problem 32.4b, it is shown that

$$a \oplus \bar{a} \cdot \bar{b} \cdot c = a \oplus \bar{b} \cdot c.$$

Design two sequences of gates which give the same output for the inputs a, b, and c. The resultant gates are said to be **logically equivalent**.

32.6. (Section 32.4). Design a circuit of gates to produce the output

$$(a \oplus \bar{b}) \cdot (a \oplus \bar{c}).$$

Construct the truth table for this Boolean expression.

32.7. (Section 32.1). Show that the Boolean expressions $(a \oplus b) \cdot (\bar{a} \oplus b) \oplus a$ and $a \oplus b$ are equivalent.

32.8. (Section 32.1). Show that the following Boolean expressions are equivalent:
(a) $a \oplus b$;   (b) $a \oplus b \cdot b$.

32.9. (Section 32.3). Find a Boolean expression f which corresponds to the truth table shown in Table 32.13.

Table 32.13

a	b	c	f
0	0	0	1
0	0	1	1
0	1	0	0
0	1	1	0
1	0	0	1
1	0	1	1
1	1	0	0
1	1	1	0

32.10. (Section 32.3). Construct Boolean expressions for the output f in the devices shown in Fig. 32.17a–**d**. Construct the truth tables in each case.

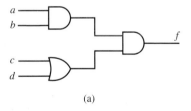

(a)

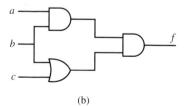

(b)

Fig. 32.17 (*Continued*)

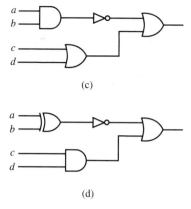

(c)

(d)

Fig. 32.17 (*Continued*)

32.11. Find the outputs f and g in the logic circuits shown in Fig. 32.18. This device can represent **binary addition** in which g is the 'carry' in the binary table shown in Table 32.14. The output g gives the '1' in the '10' in the binary sum $1 + 1 = 10$.

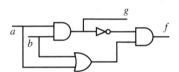

Fig. 32.18

Table 32.14

x	y	$x + y$
0	0	0
0	1	1
1	0	1
1	1	10

32.12. (Section 32.3). Reproduce the logic gate in Fig. 32.6 using just the NOR gate.

32.13. (Section 32.4). Using the disjunctive normal form, construct a Boolean expression f for the truth tables given in Tables 32.15 and 32.16.

32.14. (Section 32.3). Show that any Boolean expression can be modelled using just a NAND gate. (Hint: use a method similar to that explained in Example 32.4.)

32.15. (Section 32.4). Find switching functions for the switching circuits shown in Figs 32.19a,b.

Table 32.15

a	b	f
0	0	0
0	1	1
1	0	1
1	1	1

Table 32.16

a	b	c	f
0	0	0	1
0	0	1	0
0	1	0	0
0	1	1	1
1	0	0	1
1	0	1	0
1	1	0	1
1	1	1	0

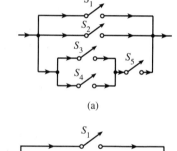

(a)

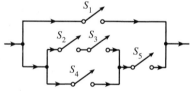

(b)

Fig. 32.19

32.16. A lecture theatre has three entrances and the lighting can be controlled from each entrance; that is, it can be switched on or off independently. The light is 'on' if the output f equals 1 and 'off' if $f = 0$. Let $a_i = 1$ ($i = 1, 2, 3$) when switch i is up, and let $a_i = 0$ ($i = 1, 2, 3$) when it is down. Construct a truth table for the state of the lighting for all states of the switches. Also specify a Boolean expression which will control the lighting.

Graph theory: circuit analysis and signal flow

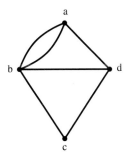

Fig. 33.1

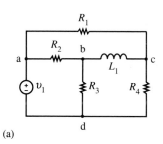

(a)

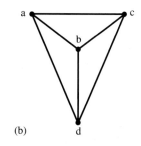

(b)

Fig. 33.2

33.1 Examples of graphs

A graph is a network or diagram composed of **points**, or **nodes** or **vertices**, joined together by lines or **edges**, each of which has a vertex at each end. Figure 33.1 shows a graph which has four vertices {a, b, c, d} and six edges {ab, ab, ad, bd, bc, cd}. Two vertices are not joined in this graph, namely a and c, while a and b are joined by two edges. Generally, it is not the shape of the graph which is important; it is usually the presence and number of edges which is significant.

Here are some practical examples of situations and objects which can be usefully represented by graphs.

(i) *Electrical circuits.* Figure 33.2a shows an electrical circuit with three resistors R_1, R_2, and R_3, an inductor L, and a voltage source V_1. Each edge has just one component, and the joins between components are the vertices (the term node is frequently used in circuit theory) in the graph. Care has to be taken with the definition of nodes (see Section 33.6): they are not necessarily where three or more wires meet. This circuit has four vertices a, b, c, d, and it can be represented by the graph in Fig. 33.2b. The presence of a line or edge between two nodes in the graph indicates that there is a component between the nodes.

Figure 33.3 shows another circuit with six vertices in which the boxes indicate electrical components. The wires joining c to f and b to e cross over each other. In the design of printed circuits, it is useful to know whether the circuit can be redrawn so that no wires cross. Such a graph, with no edges crossing, is known as a **planar** graph. The graph in Fig. 33.2 is planar, but the graph of the circuit in Fig. 33.3 has no planar drawing: at least two edges will cross in any plane diagram of it.

(ii) *Chemical molecules.* The molecule of ethanol can be represented by Fig. 33.4a. In its graph representation in Fig. 33.4b, the vertices represent **atoms** and the edges **bonds**. The number of bonds which meet at an atom is the **valency** of the atom. Thus carbon (C) has

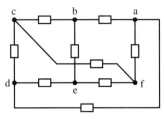

Fig. 33.3

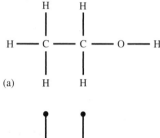

(a)

(b)

Fig. 33.4
Ethanol molecule.

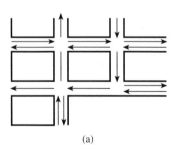

(a)

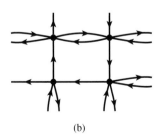

(b)

Fig. 33.5

valency 4, oxygen (O) valency 2, and hydrogen (H) valency 1. Generally in graphs, the number of edges that meet at a vertex is known as the **degree** of the vertex.

(ii) *Road maps*. Road maps and street plans are graphs with roads as edges and junctions as vertices. However, most road networks include one-way streets. Hence graphs need to be modified to indicate directions in which movement or flow is permitted. Figure 33.5a shows a typical section of a street plan with some one-way streets. We have to associate directions with the edges as shown in the graph of the plan in Fig. 33.5b. Note that two-way streets now have two directed edges associated with them. This is an example of a **directed graph**, which is also known by the shortened term **digraph**.

(iv) *Shortest paths*. Figure 33.6 shows a digraph with **weights** associated with each edge. The graph could represent routes between towns S and F which pass through intermediate towns A, B, . . . , the weights associated with each directed edge could stand for distances or times. This graph is shown as a digraph, but weights could be present without directions in some cases. We might be interested in this example in the shortest distance between the start (S) and the finish (F).

33.2 Definitions and properties of graphs
As we have seen, a **graph** is an object composed of vertices and edges with one vertex at each end of every edge. An edge which joins a vertex to itself is known as a **loop**. If two or more edges join the same two vertices then they are known as **multiple edges**. A graph with no loops or multiple edges is known as a **simple graph**. A graph with loops and/or multiple edges is known as a **multigraph**.

A graph in which every vertex can be reached from every other vertex along a succession of edges is said to be **connected**. Otherwise the graph is said to be **disconnected**. A connected graph is in one piece: a disconnected graph is in two or more pieces.

The **degree** of a vertex x is the number of edges that meet there, denoted by $\deg(x)$. If, in a graph G, all the vertices have the same degree r, then G is said to be **regular of degree** r.

Example 33.1. Find the degree of the vertices in the graph in Fig. 33.1.

Three edges meet at the vertex a. Hence $\deg(a) = 3$. Four edges meet at b. Hence $\deg(b) = 4$. Similarly, $\deg(c) = 2$ and $\deg(d) = 3$.

A simple graph in which every vertex is joined to every other vertex by just one edge is called a **complete graph**.

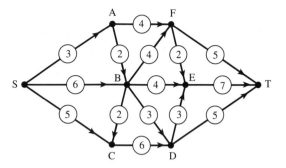

Fig. 33.6

Figure 33.7 shows some examples of the various graphs described above.

Since every edge has a vertex at each end, it follows that the sum of all the vertex degrees equals twice the number of edges. This is known as the **handshaking lemma**. For example, from Example 33.1,

$$\deg(a) + \deg(b) + \deg(c) + \deg(d) = 3 + 4 + 2 + 3 = 12,$$

which is twice the number of edges in the graph shown in Fig. 33.1.

There two immediate consequences of the handshaking lemma:

(i) the sum of all the vertex degrees in a graph is an even number;

(ii) the number of vertices of odd degree is even.

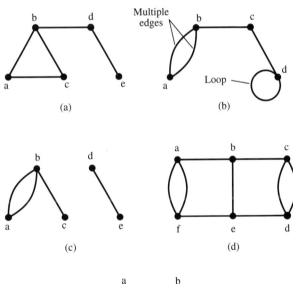

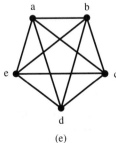

Fig. 33.7

(a) Connected simple graph.
(b) Connected multigraph.
(c) Disconnected multigraph.
(d) Regular graph of degree 3.
(e) Complete graph with five vertices: $\deg(a) = 4$.

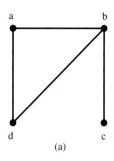

a   b

d   c

(a)

(b)

Fig. 33.8

33.3 How many simple graphs are there?

Graphs can be described as **labelled**, in which case the vertices are distinguishable as in Fig. 33.8a or **unlabelled** as in Fig. 33.8b. If we look at graphs with just three vertices, there are eight labelled simple graphs as shown in Fig. 33.9, but there are just four distinct unlabelled graphs as shown in Fig. 33.10. In Fig. 33.9, the labelled graphs (2), (3), and (4) will correspond to the same unlabelled graph.

The number of labelled simple graphs with n vertices is fairly easy to calculate. Between any two vertices, there is the possibility of an edge. Any vertex can be joined to $n - 1$ other vertices. Since this will duplicate edges, there will be $\frac{1}{2}n(n - 1)$ possible edges. Each edge may be either present or not. Hence the number of possible combinations of present and absent edges will be $2^{\frac{1}{2}n(n-1)}$, which is number of labelled graphs. Thus there must be $2^{\frac{1}{2}4(4-1)} = 2^6 = 64$ labelled graphs with four vertices; of these, 11 can be identified as unlabelled graphs. The latter graphs are shown in Fig. 33.11. Of the 11 unlabelled graphs it can be seen that six are connected and four are regular.

For applications involving circuits, the main interest is in connected graphs. The numbers of the various categories of graphs up to $n = 7$ vertices are given in Table 33.1 . It can be seen from the table that the number of unlabelled graphs is a considerable reduction on the labelled set, and that regular graphs are comparatively rare. The counting of unlabelled graphs does not follow from a simple formula.

Table 33.1

n	1	2	3	4	5	6	7
Labelled graphs	1	2	8	64	1024	32 768	2 097 152
Unlabelled graphs	1	2	4	11	34	156	1044
Connected graphs	1	1	2	6	21	112	853
Regular graphs	1	2	2	4	3	8	6

33.4 Paths and cycles

Consider a graph G. Suppose we follow a succession of connected edges between two vertices a and z, along which there may be repeated edges and vertices. This is known as a **walk** between a and z. If all the edges are different (that is, no edge is covered more than once), then the walk defines what is known as a **path**. A path is said to be **closed** if the first and last vertices are the same. If all the vertices on a path are different, except possibly the end pair, then the succession defines a **simple path**. If a simple path starts and ends at a vertex, then it is known as a **cycle** or **circuit**. For example, in Fig. 33.12, a–f–b–c–d is a simple path between a and d, but a–b–f–e–b–c–d is only a path since vertex b is passed through twice. Also, a–b–c–d–e–f–a is an example of a cycle.

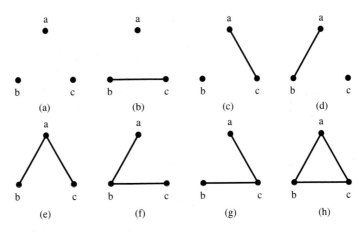

Fig. 33.9
Labelled graphs with three vertices.

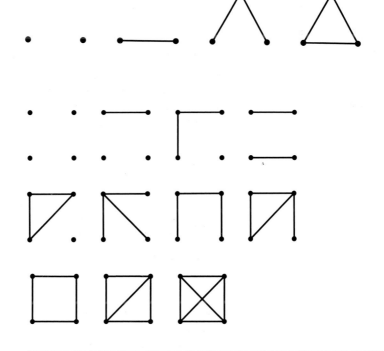

Fig. 33.10
Unlabelled graphs with three vertices.

Fig. 33.11
All unlabelled graphs with four vertices.

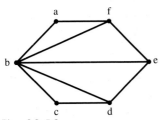

Fig. 33.12

Example 33.2. Electrical circuits are usually such that every edge of its representative graph is part of a cycle. List all the distinct cycles in the circuit in Fig. 33.2(a).

The edge of the circuit is repeated in Fig. 33.13. The complete list of cycles is:

3-edge cycles: a–b–c–a,   a–b–d–a,   a–d–c–a,   b–d–c–b;
4-edge cycles: a–b–d–c–a,   a–d–b–c–a,   a–b–c–d–a.

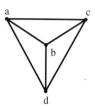

Fig. 33.13

Some graphs have special closed-path and cycle properties. A connected graph G is said to be **eulerian** if there exists a closed path that includes every edge in G. A connected graph G is said to be **hamiltonian** if there exists a cycle that includes every vertex in G. The graph in Fig. 33.13 is hamiltonian but not eulerian. One hamiltonian cycle in its graph is a–b–d–c–a. Note that this cycle does not have to cover *every* edge in the graph.

The graph in Fig. 33.14 is both eulerian and hamiltonian. An eulerian path is

a–b–c–d–e–f–g–e–c–g–b–f–a,

and a Hamiltonian cycle is

a–b–c–d–e–g–f–a.

It can be shown that a graph is eulerian if and only if every vertex has even degree. This provides an easy test for the eulerian property of a graph.

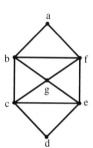

Fig. 33.14

33.5 Trees

A connected graph which has no cycles is known as a **tree**. An example of a tree is shown in Fig. 33.15. The edges in a tree are called **branches**.

Suppose that a graph G consists of the set $V(G)$ of vertices and the set $E(G)$ of edges. Then any graph whose vertices and edges are subsets of $V(G)$ and $E(G)$ respectively is called a **subgraph**. It is important to note that the subgraph must be a graph whose vertices and edges come from G; and only edges that join two vertices of the subgraph are permitted in the subset of $E(G)$.

Suppose that G is a connected graph.

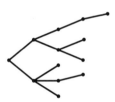

Fig. 33.15
An example of a tree.

> A **spanning tree** of G is a subgraph of G which is a tree and includes all vertices of G.

Figure 33.16a shows a connected graph G and Figure 33.16b shows a spanning tree of G. Graphs can have many different spanning trees. The set of edges that are not part of the spanning tree (the dashed edges in Figure 33.16b) is known as the **cotree** and its edges are called **links**.

Consider a tree with n vertices. Construct the tree from a chosen vertex by adding edges. Each edge added must introduce a new vertex, since otehwise a cycle would be created and the graph would no longer be a tree. Hence a tree with n vertices must have just $n - 1$ branches. It follows that a graph with n vertices must have a cotree with $e - n + 1$ links, where e is the number of edges of the graph.

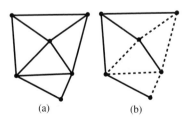

     (a)   (b)

Fig. 33.16
(a) Connected graph. (b) The some graph with a spanning tree.

We now introduce a method by which we can disconnect a graph into two subgraphs which together contain all the original vertices, by removing a minimum set of edges in the graph.

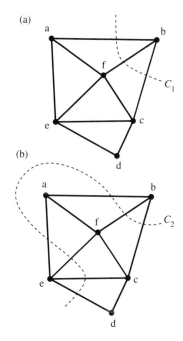

Fig. 33.17
A cutset of a graph.

In a connected graph, a **cutset** is a minimal set of edges whose removal disconnects the graph into two subgraphs so that any proper subset of the cutset does not disconnect the graph.

In other words, there must be no redundancy in the cutset. Thus, for example in Fig. 33.17a, the cut C_1, which removes the edges ab, bf, and bc, defines a cutset; but C_2 in Fig. 33.17b does not define a cutset, since the subset {ab, bf, bc} of edges disconnects the graph.

33.6 Electrical circuits: cutset and nodal analysis

In this section we give a brief description of the representation of circuits by graphs, and show how Kirchhoff's laws can be applied to cutsets of the resulting graphs. Figure 33.18a shows a plan of a circuit with seven resistors, a voltage supply, and two capacitors. This particular circuit has ten components and ten edges. Note that A will be a vertex or **node** (a preferred term in circuits) but that the joins B, C, and D are not separate nodes but can be coalesced into a single node. The equivalent graph is shown in Fig. 33.18b: it has five nodes and ten edges. Note that it is a multigraph with two nodes joined by two edges and two nodes joined by four edges.

A **circuit loop** in the circuit is a cycle in the graph.

Kirchhoff's laws have already been stated in Section 19.4, but for convenience they are given again here in graph terms. They state (i) that *the algebraic sum of the voltages around any loop is zero*, and (ii) that *the algebraic sum of the currents entering a node is zero*.

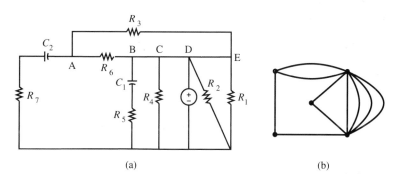

Fig. 33.18
A circuit and its graph.

Figure 33.19a shows a circuit with two voltage sources v_1 and v_{10}. The remaining voltage differences along the edges of the circuit are shown in Fig. 33.19b. The notation $+v_6-$, for example, on edge ae means that the voltage difference between e and a, $v_e - v_a$, is v_6. Thus, Kirchhoff's first law applied to the loop a–b–f–e–a implies that

$$v_1 - v_4 - v_7 - v_6 = 0.$$

The currents in each edge are defined in Fig. 33.19c. Kirchhoff's second law applied to the node f, for example, implies the result

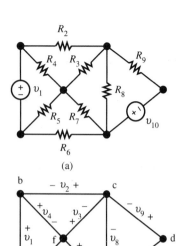

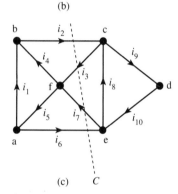

Fig. 33.19

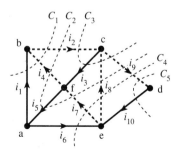

Fig. 33.20
Fundamental cutsets.

$$i_4 - i_3 - i_7 + i_5 = 0.$$

Consider the cutset C shown in Fig. 33.19c. By repeated use of Kirchhoff's second law to each node on one side of the cut, we can derive the **cutset equation** which states that the algebraic sum of the currents crossing C is zero. In this example,

$$i_6 - i_7 - i_3 + i_2 = 0.$$

Figure 33.16b shows a spanning tree for the circuit, which has five branches. Any cutset of the original graph that contains one and only one branch of a spanning tree (the rest of the cutset consisting of links) is known as a **fundamental cutset** of the circuit. We can associate five fundamental cutsets with the spanning tree of Fig. 33.16b. Five such cutsets $C_1, \ldots, C_5$ are shown in Fig. 33.20.

With the cuts we can associate five cutset equations:

$$C_1 \quad i_1 = i_2 - i_4, \tag{33.1}$$

$$C_2 \quad i_2 = i_4 + i_5 - i_7 - i_8 + i_9, \tag{33.2}$$

$$C_3 \quad i_3 = i_2 + i_8 - i_9, \tag{33.3}$$

$$C_4 \quad i_6 = i_7 + i_8 - i_9, \tag{33.4}$$

$$C_5 \quad i_{10} = i_9. \tag{33.5}$$

These equations must be independent, since each one contains a current from a branch of the spanning tree that does not appear in any other equation. Further, any non-fundamental cutset equation will be a linear combination of the five fundamental cutset equations. The number of branches in the spanning tree defines the number of independent current equations. Kirchhoff's second law applied to every node of it has inbuilt redundancy.

We require ten equations in all to find the currents in all the edges. The remaining equations can be obtained from a cycle or loop analysis associated with the links—of which there will be $e - n + 1$ in general, and five in our working example.

Start with the spanning tree as shown in Fig. 33.21, and progressively re-introduce the links. Suppose that we start by putting back link bf. This creates a cycle a–b–f–a, and Kirchhoff's first law for this loop gives

$$v_4 = v_1 + v_5. \tag{33.6}$$

Each time a link is introduced, a cycle or loop is created, and Kirchhoff's law can be applied to that loop. Hence if the links bc, ef, ec, and cd follow, then the voltage equations associated with the loops are

$$v_2 = -v_1 - v_3 - v_5, \tag{33.7}$$

$$v_7 = -v_5 + v_6, \tag{33.8}$$

$$v_8 = -v_7 + v_5 - v_6 - v_{10}, \tag{33.9}$$

$$v_9 = -v_3 + v_5 - v_6 - v_{10}. \tag{33.10}$$

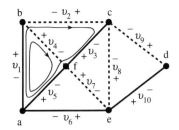

Fig. 33.21

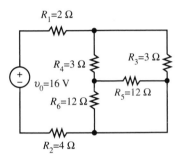

Fig. 33.22
Circuit with one voltage source.

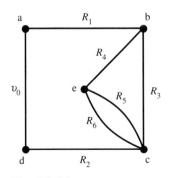

Fig. 33.23
Graph representation of the circuit shown in Fig. 33.22.

The voltages $v_2, v_3, \ldots, v_9$ can be obtained from Ohm's law:

$$v_n = R_n i_n \quad (n = 2, \ldots, 9). \tag{33.11}$$

Equations (33.1) to (33.11) contain 18 equations in total which can be used to determine all the 10 currents and 8 voltages in terms of the given voltages v_1 and v_{10}.

Example 33.3. Using the cutset method, find all currents and voltages across the resistors in the circuit shown in Fig. 33.22.

The graph corresponding to this circuit is shown in Fig. 33.23. A tree associated with this graph together with its links is shown in Fig. 33.24, together with the cuts C_1, C_2, C_3, and C_4. Currents $i_0, i_1, \ldots, i_6$ have now been added to the figure. The cutset equations associated with each of these cuts are:

$$i_0 = i_1, \tag{33.12}$$

$$i_2 = -i_1, \tag{33.13}$$

$$i_6 = i_3 - i_5 - i_1, \tag{33.14}$$

$$i_4 = -i_3 + i_1. \tag{33.15}$$

Now successively replace the links ec, bc, and ab in Fig. 33.25 to produce the voltage equations:

$$v_5 = v_6, \tag{33.16}$$

$$v_3 = v_4 - v_6, \tag{33.17}$$

$$v_1 = -v_4 + v_6 + v_2 - v_0. \tag{33.18}$$

Finally, by Ohm's law,

$$v_1 = 2i_1, \tag{33.19}$$

$$v_2 = 4i_2, \tag{33.20}$$

$$v_3 = 3i_3, \tag{33.21}$$

$$v_4 = 3i_4, \tag{33.22}$$

$$v_5 = 12i_5, \tag{33.23}$$

$$v_6 = 4i_6, \tag{33.24}$$

Equations (33.12) to (33.24) together with the given voltage

$$v_0 = 16 \tag{33.25}$$

are the equations that determine the currents and voltages. Using (33.16) to (33.24), eliminate the voltages in (33.16) to (33.18):

$$12i_5 = 4i_6, \tag{33.26}$$

$$3i_3 = 3i_4 - 4i_6, \tag{33.27}$$

$$2i_1 = -3i_4 + 4i_6 + 4i_2 - 16. \tag{33.28}$$

Eliminate i_2 and i_6, using (33.13) and (33.26), from the remaining

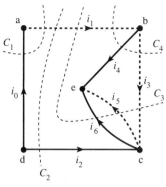

Fig. 33.24
A spanning tree and cutsets associated with the graph in Fig. 33.23.

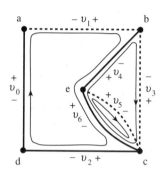

Fig. 33.25

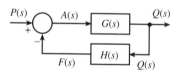

Fig. 33.26
Negative-feedback control system.

current equations:

$$3i_5 = i_3 - i_5 - i_1, \tag{33.29}$$

$$i_4 = -i_3 + i_1, \tag{33.30}$$

$$3i_3 = 3i_4 - 12i_5, \tag{33.31}$$

$$2i_1 = -3i_4 + 12i_5 - 4i_1 - 16. \tag{33.32}$$

Addition of (33.29) and (33.30) results in

$$i_4 = -4i_5, \tag{33.33}$$

and addition of (33.31) and (33.32) yields

$$3i_3 = -6i_1 - 16. \tag{33.34}$$

Eliminate i_4 and i_3 between (33.30) and (33.31), using (33.33) and (33.34):

$$-4i_5 = 3i_1 + \tfrac{16}{3},$$

$$-6i_1 - 16 = -24i_5.$$

Finally, elimination of i_5 leads to $-6i_1 - 16 = 18i_1 + 32$, or

$$i_1 = -2A.$$

The remaining currents can now be calculated:

$$i_2 = 2A, \qquad i_3 = -\tfrac{4}{3}A, \qquad i_4 = -\tfrac{2}{3}A, \qquad i_5 = \tfrac{1}{6}A, \qquad i_6 = \tfrac{1}{2}A.$$

33.7 Signal-flow graphs

Figure 33.26 shows a **block diagram** of a **negative-feedback control system**. The input into the system is $P(s)$ and the output $Q(s)$. All operations are defined by their transfer functions (see Section 23.4). The boxes represent devices or controllers. The circle represents a sum operator, and the return sign on $F(s)$ indicates positive or negative feedback. The output signal $Q(s)$ is fed back into the input through $H(s)$, and it is a **negative feedback** which will reduce the output. In a later problem, we shall consider a device with a positive feedback. Thus the input into $G(s)$ is

$$A(s) = P(s) - F(s), \tag{33.35}$$

The boxes each produce outputs given by the transfer functions

$$Q(s) = G(s)A(s), \tag{33.36}$$

$$F(s) = H(s)Q(s), \tag{33.37}$$

We wish to find $Q(s)$ in terms of $P(s)$, $G(s)$, and $H(s)$, from the equations (33.35) to (33.37). Thus, from (33.36)

$$Q(s) = G(s)A(s) = G(s)[P(s) - F(s)],$$
$$= G(s)[P(s) - H(s)Q(s)].$$

Hence the output transfer function is

$$Q(s) = \frac{G(s)}{1 + G(s)H(s)} P(s).$$

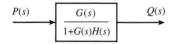

Fig. 33.27
Block-reduced diagram for Fig. 33.26.

This is the closed-loop transfer function. The actual signal can be obtained by finding the inverse Laplace transform for $Q(s)$. Hence the system is equivalent to that shown in Fig. 33.27.

If the feedback reinforces the input signal it is called **positive feedback**. Figure 33.28 shows a multiple-feedback control system with a positive and a negative feedback. The output signal is given by

$$Q(s) = \frac{G_1(s)G_2(s)G_3(s)}{1 - G_2(s)H_1(s) + G_1(s)G_2(s)G_3(s)H_2(s)} P(s), \quad (33.38)$$

which can be obtained by the method of **block-diagram reduction**. For example, the feedback through H_1 makes the system equivalent to that shown in Fig. 33.29. We can now combine the series devices which reduce the system to the negative-feedback control system considered at the beginning of this section. The details are omitted here.

This block-reduction method can get quite complicated for a complex feedback system. Instead of using block reduction in this way, represent the system by a **weighted digraph** as shown in Fig. 33.30, where the weights are the transfer functions—except that the edges representing the input and output are assigned weight 1 since they carry no devices. Also the negative feedback is replaced by $-H_2(s)$, to make sure that it reduces the input into $G_1(s)$. This is the **signal-flow graph** of the system. Let the inputs into the nodes be x_1, x_2, x_3, and x_4 as shown; then, for the positive feedback cycle,

$$x_3 = G_2 x_2, \qquad x_2 = G_1 x_1 + H_1 x_3.$$

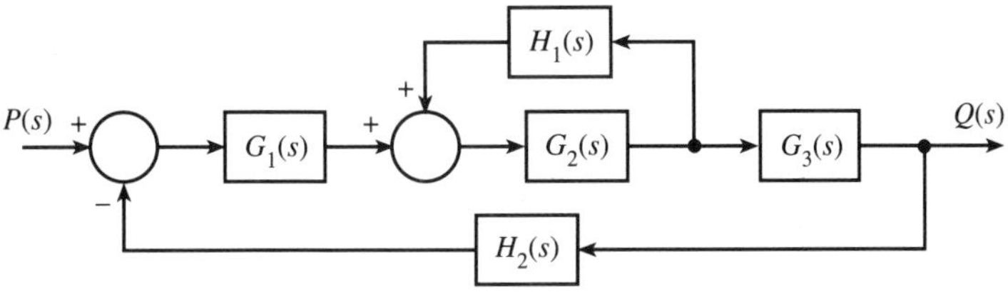

Fig. 33.28
A multiple-feedback control system.

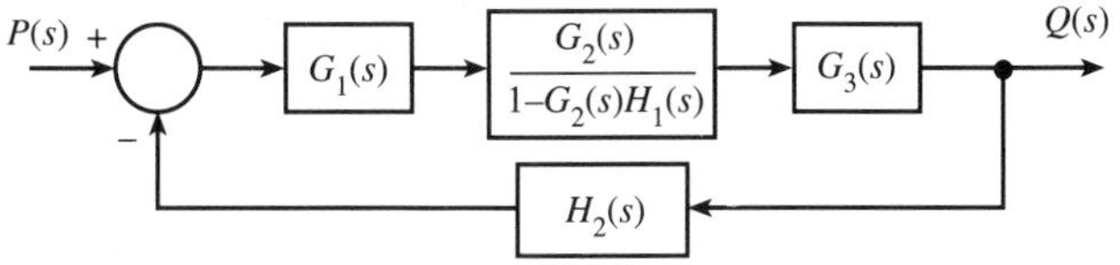

Fig. 33.29
First stage in the block reduction of the multiple-feedback control system.

Fig. 33.30
Signal-flow graph for the multiple-feedback control system shown in Fig. 33.28.

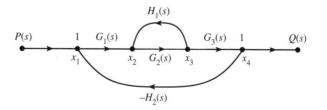

(The argument (s) has now been dropped from the working.) Hence

$$x_3 = \frac{G_1 G_2 x_1}{1 - G_2 H_1}.$$

In other words, we can replace in Fig. 33.31(a) by (b).

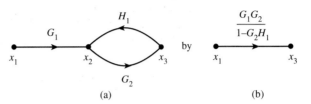

Fig. 33.31

(a)   (b)

There are other rules, and a complete list now follows for the replacements for subgraphs in the graph.

(a) *Multiple edges.* See Fig. 33.32. This follows since

$$x_2 = Gx_1 + Hx_1 = (G + H)x_1.$$

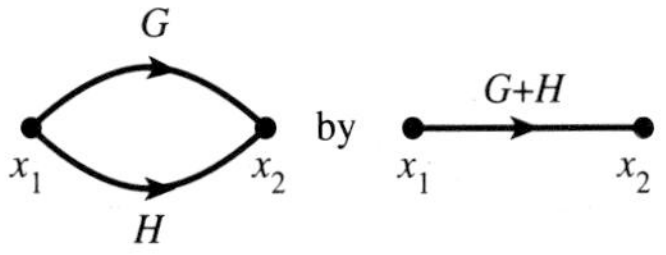

(b) *Edges in series.* See Fig. 33.33. This follows since

$$x_3 = Hx_2 = H(Gx_1) = HGx_1.$$

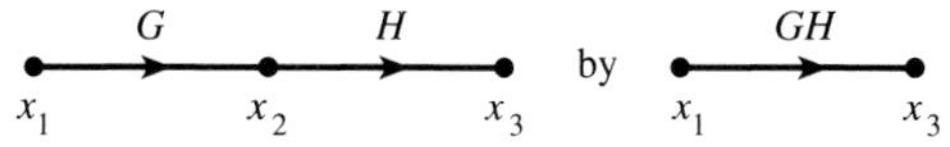

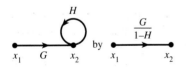

Fig. 33.32
Multiple edges.

Fig. 33.33
Edges in series.

(c) *Cycles.* See Fig. 33.34. This follows since

$$x_3 = Hx_2 \quad \text{and} \quad x_2 = Gx_1 + Jx_3.$$

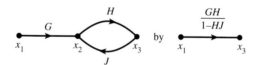

Fig. 33.34
Cycle.

Assume that $HJ \neq 1$; otherwise there is infinite gain.

(d) *Loops.* See Fig. 33.35. This follows since

$$x_2 = Gx_1 + Hx_2$$

with $H \neq 1$.

(e) *Stems.* See Fig. 33.36. This follows since

$$x_2 = Gx_1, \qquad x_3 = Hx_2 = HGx_1, \qquad x_4 = Jx_2 = JGx_1.$$

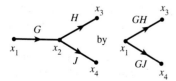

Fig. 33.35
Loop.

Apply these rules to the successive reduction of the feedback system in Fig. 33.30. The sequence of steps in the reduction of the signal-control graph to a single-edge graph is shown in Fig. 33.37. The weight of the final edge agrees with the output in equation (33.38).

Essentially the operations in a signal-flow graph are those applied to a *weighted digraph* as illustrated in the following example.

Fig. 33.36
Stem.

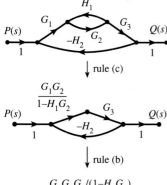

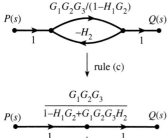

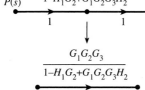

Fig. 33.37
Successive steps in the
reduction of the signal-flow
graph of the control system
shown in Fig. 33.28.

Example 33.4. Find the output–input relation in the signal-flow
graph shown in Fig. 33.38.

Applying rule (a) to the multiple edge, and rule (c) to the cycle, the
graph is reduced to Fig. 33.39. Apply the series rule to the divided
edges to give Fig. 33.40. Finally the multiple edge and series rules
give Fig. 33.41. Thus the output is given by

$$q = \frac{abd}{1 - bc} + he(g + f).$$

In the actual control system a, b, c, ... will be transfer functions.

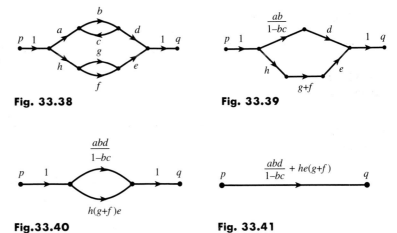

Fig. 33.38

Fig. 33.39

Fig. 33.40

Fig. 33.41

33.8 Planar graphs

As we remarked in Section 33.1, planar graphs are important in
circuit design since planar circuits can be manufactured as a single
board. A planar graph is a graph that can be drawn with no edges
crossing or meeting except at vertices. The standard example of a
simple application which cannot be represented by a planar graph
is the delivery of three services, water (W), gas (G), and electricity
(E), to three houses A, B, C (Fig. 33.42). This graph has no plane
drawing. The re-organization of the graph in Fig. 33.43 shows the
impossibility of this; if W and C are connected last then this edge
must cross either AE or BG.

The graph in Fig. 33.42 is an example of a **bipartite graph** in
which one set of vertices may be connected to another set of vertices,
but not to vertices in the same set. If every vertex in one set is
connected by one edge to every vertex in the other set then it is

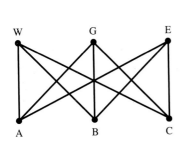

Fig. 33.42
Bipartite graph $K_{3,3}$.

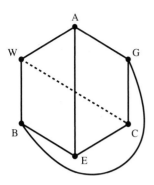

Fig. 33.43

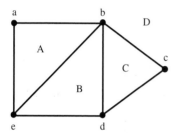

Fig. 33.44
A planar graph with five vertices,
seven edges, and four faces.

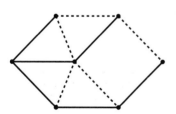

Fig. 33.45
A graph with a spanning tree.

called a **complete bipartite graph**. If the sets have m and n vertices respectively, then the notation $K_{m,n}$ denotes the complete bipartite graph. Figure 33.42 shows the graph $K_{3,3}$ and this graph is not planar. Check that the graphs $K_{2,2}$ and $K_{2,3}$ are planar.

In planar graphs there is a relation between the numbers of vertices, edges, and faces. In a plane drawing of a graph, the plane is divided into regions called **faces**. One face is the region external to the graph. Figure 33.44 shows a planar graph with five vertices and seven edges, and with four faces: A, B, C, and the external face D.

A remarkable formula, due to Euler, links the numbers of vertices, edges, and faces of a graph.

Theorem (Euler). Suppose that the graph G has a planar drawing, and let v be the number of vertices, e the number of edges, and f the number of faces of G. Then

$$v - e + f = 2.$$

Proof. For the graph G, define a spanning tree (see, for example, Fig. 33.45). The spanning tree must have n vertices and $n - 1$ edges (see Section 33.5). It must also have just one face. Since

$$n - (n - 1) + 1 = 2,$$

Euler's formula holds for the spanning tree. Successively replace the other edges in the graph. Each time an extra edge is added, a face is divided and one extra face is added. However, algebraically, this cancels the additional edge in the accumulation to Euler's formula for the spanning tree. Hence

$$v - e + f = 2$$

for the reconstructed graph G.

A complete graph with n vertices is a simple graph in which every vertex is joined to every other vertex. For the n-vertex graph, it is denoted by K_n. The graphs of K_2, K_3, K_4, and K_5 are shown in Fig. 33.46. Of these graphs, K_2, K_3, and K_4 are planar, but K_5 and all succeeding complete graphs are not.

The graphs $K_{3,3}$ and K_5 are the keys to tests for planarity of graphs, and whether it is possible to design, for example, a plane printed-circuit board to make the required connections between electronic components. It was proved by Kuratowski in 1930 that every non-planar graph contains subgraphs which are either $K_{3,3}$ or K_5, or $K_{3,3}$ or K_5 with additional vertices on their edges.

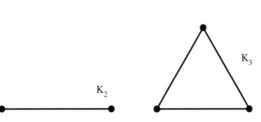 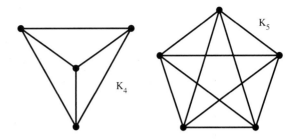

Fig. 33.46
The complete graphs K_n for $n = 2, 3, 4, 5$.

Problems, Chapter 33

33.1. (Section 33.2). Write down the degree of each vertex in the graph in Fig. 33.47.

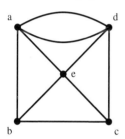

Fig. 33.47

33.2. (Section 33.2). Draw the complete graph with six vertices. How many edges does it have?

33.3. (Section 33.2). Sketch the 21 connected unlabelled graphs with five vertices. How many of them are planar?

33.4. Sketch the eight regular graphs with six vertices. How many of them are connected?

33.4 The **adjacency matrix** of a graph G with no loops is a vertex–vertex matrix, in which the element in the ith row and jth column is 0 if vertices i and j are not joined by an edge, and r if i and j are joined by r edges. Thus, if we number the vertices a, b, c, d as 1, 2, 3, 4 respectively, then the adjacency matrix of the graph in Fig. 33.1 is

$$A = \begin{bmatrix} 0 & 2 & 0 & 1 \\ 2 & 0 & 1 & 1 \\ 0 & 1 & 0 & 1 \\ 1 & 1 & 1 & 0 \end{bmatrix}.$$

Note that the leading diagonal has zeros if there are no loops. The adjacency matrix is a formula for the graph.

Evaluate A^2. What is the interpretation of the matrix in terms of the edges of G?

33.6. Draw the graphs defined by the following adjacency matrices:

(a) $\quad A = \begin{bmatrix} 0 & 1 & 1 & 1 & 1 \\ 1 & 0 & 1 & 1 & 1 \\ 1 & 1 & 0 & 1 & 1 \\ 1 & 1 & 1 & 0 & 1 \\ 1 & 1 & 1 & 1 & 0 \end{bmatrix},$

(b) $\quad A = \begin{bmatrix} 0 & 2 & 0 & 0 \\ 2 & 0 & 1 & 1 \\ 0 & 1 & 0 & 1 \\ 0 & 1 & 1 & 0 \end{bmatrix}.$

33.7. Write down the adjacency matrices of the graphs in Fig. 33.7. Note that a single loop introduces an element 1 into the appropriate position on the leading diagonal. What characterizes the matrix of a disconnected graph?

33.8. (Section 33.4). How many different cycles pass through a single vertex in a complete graph with four vertices?

33.9. (Section 33.4). List all paths between vertices a and f in the graph shown in Fig. 33.48. Identify which paths in the list are simple.

33.10. (Section 33.4). Is the graph in Fig. 33.48 eulerian? If it is find an eulerian closed path. Is it hamiltonian?

33.11. (Section 33.5). Construct a spanning tree for the graph shown in Fig. 33.48. Draw its cotree. Show

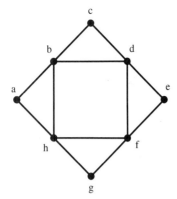

Fig. 33.48

that there is a spanning tree in which no vertex has degree more than two.

33.12. Figure 33.49 shows a graph with seven vertices.

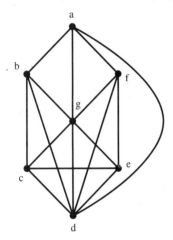

Fig. 33.49

(a) Decide whether the graph is eulerian.
(b) Construct a spanning tree for the graph. How many branches does the tree have?
(c) Draw a cutset which disconnects the vertices a, b, g, f from the vertices c, d, e.

33.13. Figure 33.50 shows a digraph. How many paths are there between a and e? Which of them are simple? Can you find a four-edge cycle?

33.14. (Section 33.6). Figure 33.51 shows a circuit with an independent current source i_0. Represent the circuit by a graph. How many vertices does the graph have?

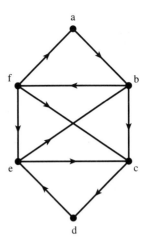

Fig. 33.50

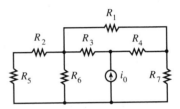

Fig. 33.51

33.15. (Section 33.6). A circuit is represented by the graph shown in Fig. 33.52. The current i_0 is from an independent source, and all other edges contain a resistor in which the current i_1 passes through a resistor R_1 and so on. Define a spanning tree for the graph. How many fundamental cutsets are required? Write down the current equations associated with each of the cutsets. Do a loop analysis of the circuit, and find all currents and voltages if $R_k = 1$ for $k = 1, \ldots, 7$.

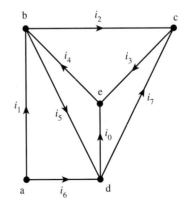

Fig. 33.52

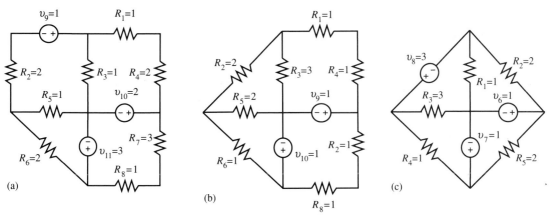

Fig. 33.53

33.16. (Section 33.6). Figures 33.53a,b,c show some circuits with applied voltages and resistors with resistances in ohms. Represent each by a graph, and analyse each of them using the cutset and loop analysis methods. Find the unknown currents and voltages.

33.17. Complete the block-reduction method for the multi-feedback control system shown in Fig. 33.28.

33.18. (Section 33.5). Figure 33.54 shows a positive feedback control system. If $P(s)$ is the system input,

find its output $Q(s)$, and the transfer function of a single equivalent device.

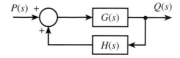

Fig. 33.54

33.19. (Section 33.7). Find the outputs in the systems shown in Fig. 33.55a,b by progressively replacing parts of the system by equivalent devices until just

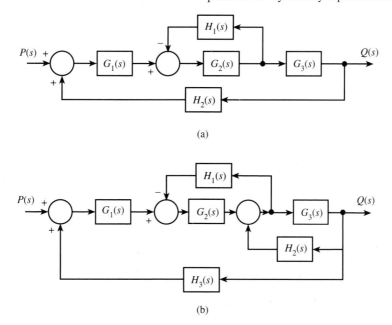

Fig. 33.55

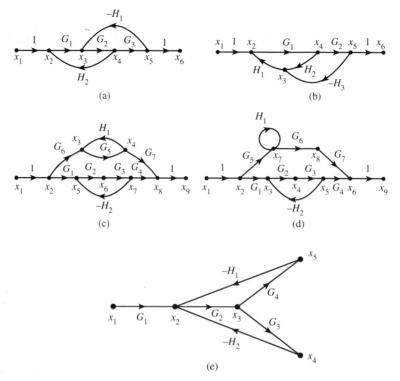

(a) (b) (c) (d)

(e)

Fig. 33.56
(a) (b) (c) (d) (e).

one device remains. Find the transfer function of the resulting equivalent single device.

33.20. (Section 33.7). Reduce each of the signal flow graphs in Figs 33.56a,b,c,d to an equivalent single edge, and (e) to a stem, and find the transfer function in each case.

33.21. (Section 33.8). Label the edges, vertices, and faces of the graphs shown in Figs 33.57a,b and verify Euler's formula.

33.22. (Section 33.8). Show that the bipartite graph $K_{2,3}$ has a planar representation.

33.23. (Section 33.8). The complete graph K_5 does not have a plane drawing. What is the minimum number of edge crossings in a plane representation of the graph?

33.24. (Section 33.1). List all the paths between S and F in the network given in Fig. 33.6, and hence find the shortest and longest paths. (This method of simply listing all paths can become very extensive for larger networks: efficient algorithms are really required to reduce the number of calculations.)

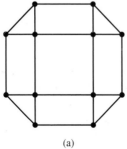

(a)

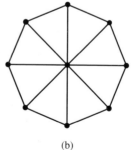

(b)

Fig. 33.57

34 Difference equations

34.1 Discrete variables

In many applications, functions can only take discrete values—that is, they cannot (for various reasons) take a continuous spectrum of values. It is reasonable to model the temperature in a room by a function which varies continuously with time—most of the calculus in this book is concerned with such functions. On the other hand, the population size of a country can only take integer values. As births and deaths occur, the population size is *discontinuous* in time, and the graph of population size against time will. be a **step function**. Between births and deaths the population number will be constant so that we are only concerned with changes which take place at these events. In this problem jumps occur at variable time intervals.

We can obtain discrete data from a continuous signal or function by sampling the signal at regular time steps rather than keeping a continuous record. This is often the situation in microprocessor-driven operations.

Let us start by considering a simple financial application which generates discrete values. In compound interest the sum of £P_0 is invested in an account to which interest accrues annually at a compound rate of $100I\%$. If £P_1 is the amount in the account at the end of the first year, then

$$P_1 = (1 + I)P_0. \tag{34.1}$$

Let £P_n be the sum after n years. Then, similarly

$$P_n = (1 + I)P_{n-1}. \tag{34.2}$$

This is an example of a **difference equation** or **recurrence relation**. It gives the values of P_n at the integer values $0, 1, 2, \ldots$ in terms of the immediately preceding value. Treating the variable as n, the **difference** in this case is 1. The notation $P(n)$ instead of P_n is often used to emphasize the function aspect of P but we have chosen the more economical subscript form P_n.

It is fairly easy to solve (34.2) by repeated application of the formula starting with (34.1). Thus

$$P_2 = (1 + I)P_1 = (1 + I)^2 P_0,$$
$$P_3 = (1 + I)P_2 = (1 + I)^3 P_0,$$

and so the formula

$$P_n = (1 + I)^n P_0 \qquad\qquad (34.3)$$

holds at least for values of n up to 3. Suppose that (34.3) holds for $n = k$. Then (34.2) implies that

$$P_{k+1} = (1 + I)P_k = (1 + I)^{k+1} P_0.$$

So the same formula holds for P_{k+1}. Hence, if the result is true for k then it is also true for $k + 1$. Equation (34.1) confirms that it is true for $k = 1$. It follows sequentially that it is true for $n = 2$, $n = 3$, and so on. (This method of proof is known as **induction**.)

Example 34.1. £1000 is invested for 5 years at the following rates: (a) 5% annually; (b) $\frac{5}{12}$% calendar monthly; (b) $\frac{5}{365}$% daily (ignoring leap years). Calculate the final amount in the account in each case.

In each case the formula is

$$P_n = (1 + I)^n P_0,$$

with $P_0 = 1000$, but the I and n differ.

(a) This is the original problem with $n = 5$ and $I = 0.05$. Hence

$$P_5 = (1 + 0.05)^5 \times 1000 = 1.05^5 \times 1000 = 1276.28$$

(in £, to the nearest penny).

(b) This account has 12 *compounding periods* each year, giving a total of 60 over the 5 years. Hence we require

$$P_{60} = \left(1 + \frac{0.05}{12}\right)^{60} \times 1000 = 1283.36.$$

(c) For the daily rate, there are $365 \times 5 = 1825$ compounding periods. Thus we require

$$P_{1825} = \left(1 + \frac{0.05}{365}\right)^{1825} \times 1000 = 1284.00.$$

There is a slight gain with increasing number of compounding periods.

Example 34.2. The sum of £50 000 is borrowed over 25 years and repaid in equal annual instalments, the interest on the outstanding balance in any year being 8%. Find the annual repayments. This scheme equalizes the repayments over the term of the loan.

Let us solve the general mortgage-repayment problem. Suppose that £P is borrowed over N years at an interest rate of $100I$% on the outstanding loan. Let the amount outstanding after m years be £Q_m. Thus $Q_0 = P$ and $Q_N = 0$. Let £A be the annual repayment. Then this must include the interest on the debt still owed and capital repayment. Thus

$$A = IQ_{m-1} + (Q_{m-1} - Q_m),$$

or

$$Q_m - (1 + I)Q_{m-1} = -A \quad (m = 1, 2, \ldots, N). \tag{34.4}$$

It can be done, but it is not quite so obvious now how to iterate Q_m from Q_0. It is sometimes helpful to look for any *constant* solutions of the difference equation. In this case, let $Q_m = C$ for $m = 1, 2, \ldots, N$. Then

$$C - (1 + I)C = -A, \quad \text{or} \quad C = A/I.$$

This is known as a **fixed point** or an **equilibrium value** of the difference equation.

Put

$$Q_m = C + U_m = A/I + U_m$$

into (34.4). Then

$$(A/I + U_m) - (1 + I)(A/I + U_{m-1}) = -A.$$

Hence U_m satisfies

$$U_m - (1 + I)U_{m-1} = 0,$$

which is the difference equation (34.1) again. Thus

$$U_m = (1 + I)^m U_0 = (1 + I)^m(Q_0 - A/I),$$

and so

$$Q_m = A/I + U_m = A/I + (1 + I)^m(Q_0 - A/I).$$

Finally, the **boundary conditions** $Q_0 = P$ and $Q_N = 0$ imply

$$0 = A/I + (1 + I)^N(P - A/I).$$

Hence

$$A = \frac{IP(1 + I)^N}{(1 + I)^N - 1}.$$

For the data given, $P = 50\,000$, $I = 0.08$, $N = 25$, and consequently

$$A = \frac{0.08 \times 50\,000 \times 1.08^{25}}{1.08^{25} - 1} = 4683.94 \quad \text{(to 2 d.p.)},$$

which represents the annual payments in £ to the nearest penny. The total repayment over the 25 years is

$$AN = A \times 25 = 117\,098.47 \quad \text{(to 2 d.p.)},$$

In the first year, in £.

$$Q_1 = A/I + (1 + I)(P - A/I) = P(1 + I) - A = 49\,316.06$$

which indicates, as one might expect, that interest payments dominate in the early years.

34.2 Difference equations: general properties

Any equation of the form

$$u_n = f(u_{n-1}, u_{n-2}, \ldots, u_{n-m}) \qquad (34.5)$$

(where m is an integer $\geqslant 1$) for any successive sequence of integers n, which may terminate or not, is known as a **difference equation**. The term **discrete dynamical system** is also frequently used. Thus

$$u_n = 2u_{n-1} + 2, \qquad (34.6)$$

$$u_n = 3u_{n-1} + 2u_{n-2} + n^2, \qquad (34.7)$$

$$u_{n+1} = ku_n(1 - u_n), \qquad (34.8)$$

are examples of difference equations.

The number m in (34.5) is known as the **order** of the difference equation: it is the difference between the largest and smallest subscripts attached to u, namely

$$n - (n - m) = m.$$

Thus (34.6) and (34.8) are first-order difference equations, while (34.7) is second-order. The sequence of integers attached to u can be translated (that is, any integer can be added to the index) without affecting the difference equation. The difference equation

$$u_{n+2} = 3u_{n+1} + 2u_n + (n + 2)^2,$$

is the same as (34.7): n has been replaced by $n + 2$ throughout.

Given **initial conditions**, the successive terms are very easy to compute. For a first-order difference equation, we can assume that u_0 is given, but it could be any term, say u_r, which is taken as the initial condition. Generally, our aim is to find a **sequence** $\{u_n\}$ and a formula for u_n for $n \geqslant r$ which satisfies the difference equation.

The difference equation (34.8) (which is known as the **logistic equation**) with $k = 2$ is

$$u_{n+1} = 2u_n(1 - u_n). \qquad (34.9)$$

Suppose that we put $u_0 = \tfrac{1}{4}$; then the sequence

$$u_1 = \frac{3}{8}, \quad u_2 = \frac{15}{32}, \quad u_3 = \frac{255}{512}, \quad u_4 = \frac{65\,535}{131\,072}, \ldots,$$

follows by successive substitution. This sequence of numbers is actually approaching the value $\tfrac{1}{2}$ as n increases. We can sketch the sequence by discrete values at integer values of x in the usual cartesian axes. The series of dots in Fig. 34.1 is a graphical representation of the sequence.

The implied limiting value of u_n as $n \to \infty$ for this sequence suggests that $u_n = \tfrac{1}{2}$ is a solution of the difference equation (34.9), and this can be confirmed. We can find all constant solutions by simply putting $u_n = u$ for all n. From (34.9), the constant solutions are given by

$$u = 2u(1 - u), \qquad \text{or } 2u^2 - u = 0,$$

which implies that $u = 0$ and $u = \tfrac{1}{2}$ are solutions. These are also

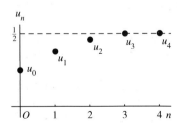

Fig. 34.1
Iterations of the sequence
$u_{n+1} = 2u_n(1 - u_n)$ with $u_0 = \tfrac{1}{4}$

known as the **fixed points** or **equilibrium values** of the difference equation.

Fixed points or equilibrium values

For any first-order difference equation $u_{n+1} = f(u_n)$, **(34.10)**
its fixed points are given by solutions of

$$u = f(u).$$

You might notice that the solutions of (34.9) vary qualitatively with the initial conditions. If $0 < u_0 < 1$, then $u_n \to \frac{1}{2}$ as $n \to \infty$; but, if $u_0 > 1$ or $u_0 < 0$, then u_n becomes unbounded for large n. We shall discuss the logistic equation further in Section 34.5.

For second-order difference equations, the same process gives equilibrium values. For example, if

$$u_{n+2} - 2u_{n+1} + 4u_n = 6,$$

then the equilibrium value is given by

$$u - 2u + 4u = 6, \quad \text{or} \quad u = 2.$$

On the other hand, the second-order difference equation (34.7) has no fixed points since

$$u - 3u - 2u - n^2 = -4u - n^2$$

can never be constant for constant u.

34.3 First-order difference equations and the cobweb
An alternative method of representing solutions of difference equations graphically is the **cobweb**. Consider the first-order difference equation

$$u_{n+1} = f(u_n) = \tfrac{1}{2}u_n + 1.$$

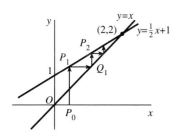

Fig. 34.2

Plot the lines $y = x$ and $y = \frac{1}{2}x + 1$ (Fig. 34.2). Select an initial value, say, $u_0 = \frac{1}{2}$, which corresponds to P_0 on the x axis. Then

$$u_1 = \tfrac{1}{2} \cdot \tfrac{1}{2} + 1 = \tfrac{5}{4},$$

which we can represent by $P_1 : (u_0, u_1)$ on the line $y = \frac{1}{2}x + 1$. Locate the point $Q_1 : (u_1, u_1)$ on $y = x$, which can be achieved by drawing P_1Q_1 parallel to the x axis. Next P_2 on $y = \frac{1}{2}x + 1$ can be found by drawing Q_1P_2 parallel to the y axis. Its ordinate must be u_2. Repeat the process by drawing lines between $y = x$ and $y = \frac{1}{2}x + 1$ using the same rules.

The usefulness of the method is that a graphical representation and interpretation of the solutions can be achieved by a simple line drawing. It is particularly helpful for finding fixed points and assessing their **stability**. We can see that this difference equation has a fixed point, which is **stable** since *all* cobwebs approach the fixed point.

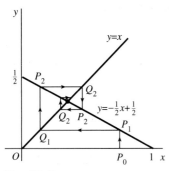

Fig. 34.3
Cobweb for $u_{n+1} = -\frac{1}{2}u_n + \frac{1}{2}$
with $u_0 = \frac{3}{4}$.

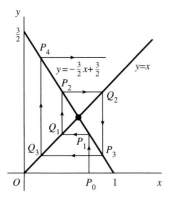

Fig. 34.4
Cobweb for $u_{n+1} = -2u_n + 2$
with $u_0 = \frac{3}{4}$.

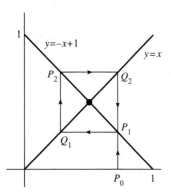

Fig. 34.5
Cobweb for $u_{n+1} = -u_n + 1$
with $u_0 = \frac{3}{4}$.

Example 34.3. sketch a cobweb solution for
$$u_{n+1} = -ku_n + k,$$
for (a) $k = \frac{1}{2}$, (b) $k = \frac{3}{2}$, (c) $k = 1$, using the initial value $u_0 = \frac{3}{4}$ in each case.

(a) Plot the lines $y = x$ and $y = -\frac{1}{2}x + \frac{1}{2}$. They intersect at the fixed point $(\frac{1}{3}, \frac{1}{3})$. Starting from $P_0 : (\frac{3}{4}, 0)$, the cobweb traces $P_0P_1Q_1P_2Q_2P_3 \ldots$ in Fig. 34.3. Evidently it approaches the fixed point as $n \to \infty$, indicating stability.

(b) The lines are $y = x$ and $y = -\frac{3}{2}x + \frac{3}{2}$. The fixed point is at $(\frac{3}{5}, \frac{3}{5})$, and the cobweb path is $P_0P_1Q_1P_2Q_2 \ldots$ in Fig. 34.4. The path moves away from the fixed point implying its instability.

(c) The lines are $y = x$ and $y = -x + 1$ with fixed point $(\frac{1}{2}, \frac{1}{2})$. The path starting at $P_0 : (\frac{3}{4}, 0)$ follows the rectangle $P_1Q_1P_2Q_2$, indicating **periodicity** (Fig. 34.5). This is true for any starting value except that of the fixed point itself.

Graphs of the sequences u_n versus n are shown in Fig. 34.6.
The stability of the fixed point of the first-order linear difference equation can be summarized as follows.

Stability
The first-order difference equation $u_{n+1} = -ku_n + a$ has
(a) a stable fixed point if $-1 < k < 1$,   **(34.11)**
(b) a fixed point that is not stable if $|k| > 1$,
(c) a periodic fixed point if $k = 1$,

34.4 Constant-coefficient linear difference equations
Any difference equation of the form
$$u_n + a_{n-1}u_{n-1} + \cdots + a_{n-m}u_{n-m} = f(n),$$
where the a_i ($i = n - m, \ldots, n - 1$) are constants, is a constant-coefficient linear difference equation. We shall look in detail at the second-order case
$$u_{n+2} + 2au_{n+1} + bu_n = f(n),$$ **(34.12)**
where a and b are constants and $f(n)$ is a given function. The methods generalize in a fairly obvious way to higher-order systems.

There are many parallels between the difference equation (34.12) and second-order constant-coefficient equations (Chapters 16–17). The equation is said to be **homogeneous** if $f(n) = 0$, and **inhomogeneous** otherwise, just as in the case of second-order differential equations. However, this section is self-contained and reference back is not necessary. The general solution of the inhomogeneous case requires that of the homogeneous case: hence we start with the latter.

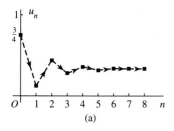

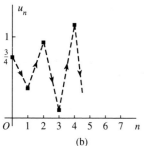

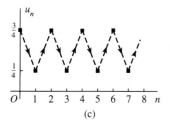

Fig. 34.6
Solutions of
$u_{n+1} = -k(u_n + 1)$ for
(a) $k = \frac{1}{2}$, (b) $k = 2$, (c) $k = 1$.

Homogeneous equations. We can see how to proceed by looking at the first-order constant-coefficient equation

$$u_{n+1} + cu_n = 0. \tag{34.13}$$

As can be seen from (34.2) or verified directly, the general solution of this equation is

$$u_n = Ac^n, \tag{34.14}$$

where A is any constant. Notice that we could equally well write

$$u_n = Ac^{n-1}, \quad \text{or} \quad u_n = Ac^{n+1};$$

the result would be equally correct, although A would take different values for the same initial condition. The significant property of (34.13) and its solution (34.14) is that u_{n+1} is a constant multiple of u_n.

With this in view, we attempt to find solutions of

$$u_{n+2} + 2au_{n+1} + bu_n = 0 \tag{34.15}$$

in the form

$$u_n = p^n,$$

where p is a constant. Thus

$$\begin{aligned} u_{n+2} + 2au_{n+1} + bu_n &= p^{n+2} + 2ap^{n+1} + bp^n, \\ &= (p^2 + 2ap + b)p^n, \\ &= 0, \end{aligned}$$

for all n, if $p = 0$ or

$$p^2 + 2ap + b = 0. \tag{34.16}$$

The case $p = 0$ leads to the self-evident *trivial* solution $u_n = 0$. We are interested in solutions of (34.16), which is known as the **characteristic equation** of (34.15).

There are various cases to consider. Suppose that the roots of (34.16) are the distinct numbers p_1 and p_2. Hence $u_n = p_1^n$ and $u_n = p_2^n$ are solutions of (34.15). Since this equation is homogeneous and *linear*, it follows that any linear combination of p_1^n and p_2^n is also a solution. We state this as follows.

Distinct roots

The general solution of $u_{n+2} + 2au_{n+1} + bu_n = 0$ for (34.17) distinct roots p_1 and p_2 of $p^2 + 2ap + b = 0$ is

$$u_n = Ap_1^n + Bp_2^n, \quad \text{for any constants } A \text{ and } B.$$

Example 34.4. Find the general solution of

$$u_{n+2} - u_{n+1} - 6u_n = 0. \tag{34.18}$$

The characteristic equation of (34.18) is

$$p^2 - p + 6 = 0, \quad \text{or} \quad (p - 3)(p + 2) = 0.$$

The roots are $p_1 = 3$, $p_2 = -2$. Hence the general solution is

$$u_n = A \cdot 3^n + B(-2)^n.$$

Example 34.5. Find the solution of

$$u_{n+2} + 2u_{n+1} - 3u_n = 0$$

that satisfies $u_0 = 1$, $u_1 = 2$.

The characteristic equation is

$$p^2 + 2p - 3 = 0, \quad \text{or} \quad (p + 3)(p - 1) = 0.$$

The roots are $p_1 = -3$, $p_2 = 1$. Hence the general solution is

$$u_n = A(-3)^n + B \cdot 1^n = A(-3)^n + B.$$

From the initial conditions,

$$u_0 = 1 = A + B, \qquad u_1 = 2 = -3A + B.$$

Hence $A = -\frac{1}{4}$ and $B = \frac{5}{4}$. The required solution is

$$u_n = -\frac{1}{4} \cdot (-3)^n + \frac{5}{4}.$$

The characteristic equation can have equal roots, which is a special case. Consider the difference equation

$$u_{n+2} - 2au_{n+1} + a^2 u_n = 0,$$

where $a \neq 0$. Its characteristic equation is

$$p^2 - 2ap + a^2 = 0, \quad \text{or} \quad (p - a)^2 = 0,$$

which has the repeated root $p = a$. One solution is Aa^n; but we require a second independent solution. Let $u_n = v_n a^n$. Then

$$0 = u_{n+2} - 2au_{n+1} + a^2 u_n = a^{n+2} v_{n+2} - 2a^{n+2} v_{n+1} + a^{n+2} v_n,$$
$$= a^{n+2}(v_{n+2} - 2v_{n+1} + v_n),$$

so that

$$v_{n+2} - 2v_{n+1} + v_n = 0.$$

since $a \neq 0$. It can be verified that this equation has a solution $v_n = n$. Hence a further independent solution is $u_n = Bna^n$.

Equal roots

The general solution of $u_{n+2} - 2au_{n+1} + a^2 u_n = 0$ is   **(34.19)**

$$u_n = (A + Bn)a^n$$

Roots can also be complex. Consider the difference equation

$$u_{n+2} + 2u_{n+1} + 2u_n = 0.$$

Its characteristic equation is

$$p^2 + 2p + 2 = 0$$

with roots $p_1 = -1 + \mathrm{j}$, $p_2 = -1 - \mathrm{j}$. The method still works.

However, the general solution becomes

$$u_n = A(-1 + j)^n + B(-1 - j)^n.$$

For a real-valued problem, the constants A and B will be complex conjugates which ensure that u_n is real. The solution can be cast in real form by using the polar forms (Section 6.3) of the complex numbers. In this case

$$-1 \pm j = \sqrt{2}\, e^{\pm \frac{3}{4}\pi j}.$$

Hence

$$u_n = A2^{\frac{1}{2}n}\, e^{\frac{3}{4}\pi j n} + B2^{\frac{1}{2}n}\, e^{-\frac{3}{4}\pi j n},$$
$$= 2^{\frac{1}{2}n}[A(\cos \tfrac{3}{4}\pi n + j \sin \tfrac{3}{4}\pi n) + B(\cos \tfrac{3}{4}\pi n - j \sin \tfrac{3}{4}\pi n)],$$
$$= 2^{\frac{1}{2}n}(C \cos \tfrac{3}{4}\pi n + D \sin \tfrac{3}{4}\pi n),$$

where $C = A + B$ and $D = (A - B)j$.

Complex roots, $\alpha \pm j\beta = r\, e^{\pm \theta j}$

The general complex solution of $u_{n+2} + 2au_{n+1} + bu_n = 0$, where $a^2 < b$, is

$$u_n = A(\alpha + j\beta)^n + B(\alpha - j\beta)^n. \tag{34.20}$$

The general real solution is

$$u_n = r^n(C \cos \theta n + D \sin \theta n).$$

Example 34.6. Obtain the general solution of

$$u_{n+2} + u_n = 0.$$

The characteristic equation is

$$p^2 + 1 = 0,$$

giving roots $p_1 = j$, $p_2 = -j$. Hence

$$u_n = Aj^n + B(-j)^n.$$

In polar form, $j = e^{\frac{1}{2}\pi j}$, $-j = e^{-\frac{1}{2}\pi j}$. Hence the real form of the solution is

$$u_n = C \cos \tfrac{1}{2}\pi n + D \sin \tfrac{1}{2}\pi n.$$

Inhomogeneous equations. The general inhomogeneous equation is

$$u_{n+2} + 2au_{n+1} + bu_n = f(n) \tag{34.21}$$

(see (34.12)). Let $u_n = v_n + q_n$, where v_n is the *general solution* of the corresponding *homogeneous equation*. Substitute this form of u_n into (34.21):

$$(v_{n+2} + q_{n+2}) + 2a(v_{n+1} + q_{n+1}) + b(v_n + q_n) = f(n),$$

or

$$(v_{n+2} + 2av_{n+1} + bv_n) + (q_{n+2} + 2aq_{n+1} + bq_n) = f(n).$$

y

Since v_n satisfies the homogeneous equation, it follows that

$$q_{n+2} + 2aq_{n+1} + bq_n = f(n),$$

which means that q_n must be a **particular solution** of the inhomogeneous equation. As in differential equations, v_n is known as the **complementary function**.

We construct particular solutions by appropriate choices of functions usually containing adjustable parameters which are suggested by the form of the function $f(n)$. If a particular choice fails, then we reject it and try something else. For example, if $f(n) = k$, a constant, then we might try $q_n = C$. This will work provided that the homogeneous equation does not itself have a constant solution, a point which the next two examples illustrate.

Example 34.7. Obtain the general solution of

$$u_{n+2} - u_{n+1} - 6u_n = 4.$$

From Example 34.4, the complementary function is

$$v_n = 3^n A + (-2)^n B.$$

For the particular solution, we try $q_n = C$, since $f_n = 4$. Then

$$q_{n+2} - q_{n+1} - 6q_n - 4 = C - C - 6C - 4,$$
$$= -6C - 4 = 0,$$

if $C = -\frac{2}{3}$. Hence $q_n = -\frac{2}{3}$, and the general solution is

$$u_n = 3^n A + (-2)^n B - \frac{2}{3}.$$

Note that the two unknown constants in the general solution occur in the complementary function.

Example 34.8. Obtain the general solution of

$$u_{n+2} + 2u_{n+1} - 3u_n = 4.$$

From Example 34.5, the complementary function is

$$v_n = (-3)^n A + B.$$

In this case we expect the choice $q_n = C$ to fail, since **it must make the left-hand side of the difference equation vanish**. When this happens, we try

$$q_n = Cn.$$

Then

$$q_{n+2} + 2q_{n+1} - 3q_n - 4 = C(n + 2) + 2C(n + 1) - 3Cn - 4$$
$$= 2C + 2C - 4 = 4C - 4 = 0,$$

If $C = 1$. Hence the general solution is

$$u_n = (-3)^n + B + n.$$

Table 34.1 lists some simple forcing terms $f(n)$ with suggested

Table 34.1

$f(n)$	Trial solution q_n
k (a constant)	C; or Cn, if C fails; or Cn^2, if C and Cn fail; etc.
k^n	Ck^n; or Cnk^n, if Ck^n fails; etc.
n n^p (p an integer)	$C_0 + C_1 n$ $C_0 + C_1 n + \cdots + C_p n^p$ (may need higher powers of n in special cases)
$\sin kn$ or $\cos kn$	$C_1 \cos kn + C_2 \sin kn$

forms of particular solution and alternatives containing parameters to be determined by direct substitution.

Example 34.9. Find the general solution of

$$u_{n+2} - 4u_n = n.$$

The characteristic equation is

$$p^2 - 4 = 0, \quad \text{or} \quad (p - 2)(p + 2) = 0.$$

The roots are $p_1 = 2$, $p_2 = -2$. Hence the complementary function is

$$v_n = 2^n A + (-2)^n B.$$

For the particular solution, try (choosing from Table 34.1)

$$q_n = C_0 + C_1 n.$$

Then

$$q_{n+2} - 4q_{n+1} - n = C_0 + C_1(n + 2) - 4C_0 - 4C_1 n - n,$$
$$= (-3C_0 + 2C_1) + n(-3C_1 - 1).$$

The right-hand side vanishes for all n if

$$-3C_0 + 2C_1 = 0, \qquad -3C_1 - 1 = 0.$$

Hence $C_1 - \frac{1}{3}$, $C_0 = 2C_1/3 = -\frac{2}{9}$, and the general solution is

$$u_n = 2^n A + (-2)^n B - \frac{2}{9} - \frac{1}{3} n.$$

34.5 The logistic difference equation

Consider again the logistic difference equation

$$u_{n+1} = \alpha u_n (1 - u_n), \tag{34.22}$$

where α is a constant. This nonlinear equation can model population growth of generations. If u_n represents the population size of generation n and α is the birth-rate, then we might expect the population size of the next generation to be αu_n in the absence of any inhibiting factors such as lack of resources or overcrowding. If $\alpha > 1$, then the population model given by the first-order difference

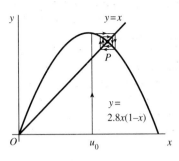

Fig. 34.7
Cobweb solution for
$u_{n+1} = 2.8u_n(1 - u_n)$ showing
a solution starting from
$x = u_0$ approaching the fixed
point at P.

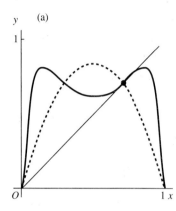

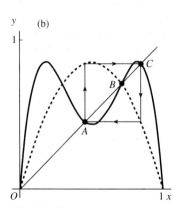

Fig. 34.8
(a) Graph of $y = f(f(x))$ for
the critical case $\alpha = 3$.
(b) Graph of $y = f(f(x))$ for
$\alpha = 3.4$ showing fixed points
O, A, B, C.

equation $u_{n+1} = \alpha u_n$ would imply that the population would grow to infinity, since the equation has the solution $u_n = \alpha^n u_0$. To counter this possibility, we can introduce a feedback term $-\alpha u_n^2$ which will tend to reduce population growth when the population is large.

Fixed points of the equation (34.22) occur where

$$u = \alpha u(1 - u)$$

at $u = 0$ and $u = 1 - 1/\alpha$. We can adapt the cobweb method of Section 34.3 to this nonlinear difference equation by plotting graphs of the parabola $y = f(x) = \alpha x(1 - x)$ and the straight line $y = x$. Fixed points of the difference equation will occur where the line and the parabola intersect. The values of x at these points will be given by the solutions of

$$\alpha x(1 - x) = x, \quad \text{or} \quad x(\alpha x - 1 - \alpha) = 0.$$

The points are $x = 0$ and $P : x = 1 - 1/\alpha$. We shall only look at the case $\alpha > 1$, so that one fixed point is in the first quadrant, $x > 0$, $y > 0$. A cobweb solution for the case $\alpha = 2.8$ is shown in Fig. 34.7.

Notice that, for this choice of α, the fixed point P is *stable*; that is, the cobweb solution approaches P. The slope of the graph of $y = \alpha x(1 - x)$ at P determines the stability or instability of the solutions. The slope at P is $m = \alpha - 2\alpha x$, where $x = 1 - 1/\alpha$. Hence

$$m = \alpha - 2\alpha(1 - 1/\alpha) = -\alpha + 2.$$

As with the cobweb for two intersecting lines for the linear difference equation in Section 34.3, the fixed point P is **locally stable** if $m > -1$, in that all cobweb paths starting close to $1 - 1/\alpha$ approach the fixed point P as $n \to \infty$. This corresponds to $\alpha < 3$. Notice also that, if $1 < \alpha < 2$, then the $y = x$ intersects the parabola $y = \alpha x(1 - x)$ between the origin and its maximum value.

For $\alpha \geqslant 3$ the solutions become more complicated. The fixed point at the origin is unstable: hence there is no *stable* fixed point to which solutions can approach. We can obtain a clue as to what happens if we look at the function of a function given by

$$y = f(f(x)) = \alpha[\alpha x(1 - x)][1 - \alpha x(1 - x)],$$
$$= \alpha^2 x(1 - x) - \alpha^3 x^2(1 - x)^2.$$

when $\alpha = 3$, this curve intersects $y = x$ at $x = 0$ and at P only. This can be checked by noting that

$$x = \alpha^2 x(1 - x) - \alpha^3 x^2(1 - x)^2$$

can be written as

$$x(27x^3 - 54x^2 + 36x - 8) = 0, \quad \text{or} \quad x(3x - 2)^3 = 0,$$

when $\alpha = 3$. Graphs of the curves $y = f(x)$ and $y = f(f(x))$ for $\alpha = 3$ are shown in Fig. 34.8a. The fixed point P is at $(\frac{2}{3}, \frac{2}{3})$. As α increases, three fixed points develop on the line $y = x$. Further graphs of the two functions $y = f(x)$ and $y = f(f(x))$ for $\alpha = 3.4$ are shown in Fig. 34.8(b), together with the line $y = x$. There are fixed points at O, A, B, C, of which A and C are stable. While A and C

are fixed points of this equation, they can be associated with 2-**cycles** or **period**-2 solutions in the original function, as shown by the cobweb on $y = \alpha x(1 - x)$ superimposed on Fig. 34.8b. This phenomenon is known as **period doubling**. It first appears when the fixed point P on $y = f(x)$ ceases to be stable at $\alpha = 3$. The 2-cycle then grows in 'amplitude' as α increases. The solution is said to **bifurcate** at $\alpha = 3$. This type of bifurcation, where the stable solution becomes suddenly unstable and throws off two stable solutions on either side, is an example of a **pitchfork bifurcation**, called this because of its fork-like appearance.

For general α, the fixed points of $y = f(f(x))$ occur where

$$x = \alpha^2 x(1 - x) - \alpha^3 x^2(1 - x)^2.$$

To solve this equation, put $1 - x = u$. Then u satisfies

$$1 - u = \alpha^2(1 - u)u - \alpha^3(1 - u)^2 u^2,$$

which can be written as $G(u) = 0$, where

$$G(u) = (u - 1)(\alpha^3 u^3 - \alpha^3 u^2 + \alpha^2 u - 1). \tag{34.23}$$

One obvious solution of $G(u) = 0$ is $u = 1$, while the cubic factor has the solution $u = 1/\alpha$ as can be verified. Hence

$$G(u) = (u = 1)(\alpha u - 1)[\alpha^2 u^2 + \alpha(1 - \alpha)u + 1].$$

At A and C, u satisfies

$$\alpha^2 u^2 + \alpha(1 - \alpha)u + 1 = 0. \tag{34.24}$$

Hence, at the two fixed points A and C,

$$\left.\begin{array}{r}x_1\\x_2\end{array}\right\} = \frac{1 + \alpha \mp \sqrt{[(\alpha + 1)(\alpha - 3)]}}{2\alpha} \quad (\alpha > 3),$$

respectively, while $x = 1/\alpha$ at B. Period doubling occurs since $f(x_1) = x_2$ and $f(x_2) = x_1$, and transfers between x_1 and x_2 can take place around the square cobweb in Fig. 34.8b.

The fixed point A will remain stable if the absolute value of its slope is less than 1. The same condition will also apply at C. In fact the critical slope is -1, and we will find the value of α at which this occurs. We have

$$\frac{\mathrm{d}}{\mathrm{d}x} f(f(x)) = \alpha^2 - 2\alpha^2 x - \alpha^3(2x - 6x^2 + 4x^3)$$

$$= \alpha^2 - 2\alpha^2(1 + \alpha)x + 6\alpha^3 x^2 - 4\alpha^3 x^3. \tag{34.25}$$

We require the value of α given by

$$\frac{\mathrm{d}}{\mathrm{d}x} f(f(x)) = -1,$$

$$\text{or} \quad 4\alpha^3 x^3 - 6\alpha^3 x^2 + 2\alpha^2(1 + \alpha)x - \alpha^2 = 1, \tag{34.26}$$

when x satisfies (34.24) with $u = 1 - x$, which is when

$$\alpha^2 x^2 - \alpha(1 + \alpha)x + \alpha + 1 = 0. \tag{34.27}$$

Remove the x^3 term from (34.26) by multiplying (34.27) by $4\alpha x$, and subtracting it from (34.26). Then

$$-2\alpha^2(\alpha - 2)x^2 + 2\alpha(\alpha - 2)(\alpha + 1)x - (1 + \alpha^2) = 0. \quad \textbf{(34.28)}$$

Equations (34.27) and (34.28) must have the *same* roots in x. In each case, make the coefficient of x^2 equal to 1. The results for comparison are

$$x^2 - \frac{(\alpha + 1)}{\alpha}x + \frac{(\alpha + 1)}{\alpha^2} = 0,$$

$$x^2 - \frac{(\alpha + 1)}{\alpha}x + \frac{(\alpha^2 + 1)}{2\alpha^2(\alpha - 2)} = 0.$$

These equations have the same roots if

$$\frac{\alpha + 1}{\alpha^2} = \frac{(\alpha^2 + 1)}{2\alpha^2(\alpha - 2)}, \quad \textbf{(34.29)}$$

or

$$\alpha^2 - 2\alpha - 5 = 0.$$

We are interested in values of $\alpha > 3$, so that the required root of (34.29) is $\alpha = 1 + \sqrt{6} = 3.449\ldots$. In fact the slopes at both A and C both become -1 for this value of α. Thus, for

$$3 < \alpha < 1 + \sqrt{6},$$

the 2-cycle solution is stable.

At $\alpha = 1 + \sqrt{6}$, the system bifurcates again into a 4-cycle or period-4 solution, which corresponds to the set of stable fixed points of $y = f(f(f(f(x))))$. A graph of this function for $\alpha = 3.54$ is shown in Fig. 34.9 together with the 8 fixed points. The cycle doubles again at about $\alpha = 3.544, \ldots$ and so on. The intervals between the bifurcations of the period doubling rapidly decrease, until a limit is reached at about $\alpha = 3.570, \ldots$ beyond which **chaos** occurs. The iterations are no longer periodic for most values of α beyond this point, although there are some brief intervals of periodicity.

The sequence of period doubling bifurcations is known as the **Feigenbaum sequence**, and it has certain universal aspects in that it is not just a consequence of the logistic equation, but has common features with other difference equations which generate period doubling.

The simplest way to view the progressively complex behaviour is through a computer-drawn picture of the iterations of

$$u_{n+1} = \alpha u_n(1 - u_n)$$

for stepped increases in α starting at $\alpha = 2.8$ up to $\alpha = 3.8$, which covers the main area of interest. The result is shown in Fig. 34.10. The series of single dots for each α in $2.8 \leqslant \alpha \leqslant 3$ indicates the fixed point, which then bifurcates into a stable 2-cycle **attractor** for $3 < \alpha \leqslant 1 + \sqrt{6}$. This in turn bifurcates into a stable 4-cycle attractor at $\alpha = 1 + \sqrt{6}$ and so on. The effect of infinite period doubling is that the solution is ultimately nonperiodic. The generally chaotic and noisy behaviour of the difference equation can clearly be seen

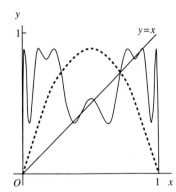

Fig. 34.9
Fixed points of
$y = f(f(f(f(x))))$ for
$\alpha = 3.54$, given by the
intersection of the curve and
the line $y = x$.

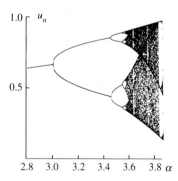

Fig. 34.10
Period doubling for the
logistic equation for increasing
α, followed by chaotic
iterations beyond about
$\alpha = 3.57$.

in the large number of dots for larger values of α. These nonperiodic sets are known as **strange attractors**. The successive iterates of the logistic equation wander about in a seemingly random but bounded manner, and never settle into a periodic solution. However, within the chaotic band of α values, there appear windows of periodic cycles. Problem 34.26, for example, confirms that there is a 3-cycle around $\alpha = 3.83$.

The logistic equation can be thought of as a relatively simple model example. Many similar nonlinear difference equations also exhibit similar period-doubling bifurcations and strange attractors.

Problems, Chapter 34

34.1. £1000 is invested over 10 years at an interest rate of 6% annually. Find the final total investment. What should the *monthly* investment rate be to achieve the same final total?

34.2. The sum of £50 000 is borrowed over 25 years and the money is repaid in equal annual instalments. The interest rate on the outstanding balance in any year is 10%. Find what the annual repayments would be. After 5 years, the interest rate is reduced to 9%.
(a) Find the required adjustment to the annual repayments for the loan to be repaid over the original term.
(b) If the repayments are not changed, by how much will the mortgage term be reduced?

34.3. Find the fixed points of the following difference equations:
(a) $u_{n+1} = u_n(2 - u_n)$;
(b) $u_{n+1} = u_n(1 + u_n)(2 - 3u_n)$;   (c) $u_{n+1} = \sin u_n$;
(d) $u_{n+1} = \frac{1}{2}\sin u_n$;   (e) $u_{n+1} = e^{u_n} - 1$.

34.4. Given the initial value u_0 in each case, calculate the sequence of terms up to u_5 for each of the following first-order difference equations:
(a) $u_{n+1} = 2u_n(3 - u_n)$, $u_0 = 1$;
(b) $u_{n+1} = 2u_n(1 - u_n)$, $u_0 = \frac{1}{2}$;
(c) $u_{n+1} = 3.2u_n(1 - u_n)$, $u_0 = \frac{1}{2}$;
(d) $u_{n+1} = 4u_n(1 - u_n)$, $u_0 = \frac{1}{2}$.

34.5. (Section 34.3). Sketch the cobweb solutions for the following first-order equations with the stated initial conditions, and discuss the stability of the fixed point:
(a) $u_{n+1} = \frac{1}{2}u_n + \frac{1}{2}$, $u_0 = \frac{1}{2}$ and $u_0 = \frac{3}{2}$;
(b) $u_{n+1} = 2u_n - 2$, $u_0 = \frac{1}{2}$ and $u_0 = \frac{3}{2}$;
(c) $u_{n+1} = -u_n + 2$, $u_0 = \frac{1}{2}$ and $u_0 = \frac{3}{4}$;

(d) $u_{n+1} = -\frac{1}{2}u_n + \frac{3}{2}$, $u_0 = \frac{1}{2}$ and $u_0 = \frac{3}{2}$;
(e) $u_{n+1} = -2u_n + 3$, $u_0 = \frac{1}{2}$ and $u_0 = \frac{3}{2}$.

34.6. The function $f(n)$ satisfies
$$f(n) = f(\tfrac{1}{2}n) + 1.$$
Put $n = 2^m$ and $g(m) = f(2^m)$, and show that
$$g(m) = g(m - 1) + 1.$$
Hence find $f(n)$ given that $f(1) = 0$.

34.7. Use the method suggested in the previous problem to solve
$$f(n) = f(\tfrac{1}{3}n) + \tfrac{5}{8},$$
given the initial condition $f(1) = 0$.

34.8. (Section 34.3). Find the general solutions of the following difference equations:
(a) $u_{n+2} + 2u_{n+1} - 3u_n = 0$;   (b) $u_{n+2} - 9u_n = 0$;
(c) $u_{n+2} + 9u_n = 0$;   (d) $u_n - 4u_{n-1} + 5u_{n-2} = 0$;
(e) $u_{n+2} - 4u_{n+1} + 4u_n = 0$;
(f) $u_{n+3} - u_{n+2} + u_{n+1} - u_n = 0$;
(g) $u_{n+3} - u_n = 0$;
(h) $u_{n+3} - 3u_{n+2} + 3u_{n+1} - u_n = 0$;
(i) $u_{n+2} - u_{n+1} - u_n + u_{n-1} = 0$.

34.9. Express the solution of the initial-value problem
$$u_{n+2} - 6u_{n+1} + 13u_n = 0, \qquad u_0 = 0, \quad u_1 = 1,$$
in real form.

34.10 Find the difference equation satisfied by
$$u_n = A \cdot 2^n + B \cdot (-5)^n,$$
for all A and B.

34.11. Obtain particular solutions of the following inhomogeneous difference equations:
(a) $u_{n+2} + 2u_{n+1} - 3u_n = f(n)$, where
    (i) $f(n) = 2^n$; (ii) $f(n) = n$; (iii) $f(n) = 2$,
    (iv) $f(n) = (-3)^n$.

(b) $u_{n+2} + 2u_{n+1} + 2u_n = f(n)$, where
(i) $f(n) = 1$; (ii) $f(n) = n + 3$;
(iii) $f(n) = \cos\frac{3}{4}\pi n$.

(c) $u_{n+3} - 3u_{n+2} + 3u_{n+1} + u_n = f(n)$, where
(i) $f(n) = 1$, (ii) $f(n) = n$, (iii) $f(n) = n^2$.

(d) $u_{n+2} - 6u_{n+1} + 9u_n = f(n)$, where (i) $f(n) = 2^n$,
(iii) $f(n) = 3$; (iii) $f(n) = 3^n$; (iv) $f(n) = n3^n$.

34.12. A ball bearing is dropped from a height $z = h_0$ on to a metal plate, and the coefficient of restitution between the ball and the plate is ε, where $0 < \varepsilon < 1$. Set up a difference equation for the maximum height reached after n impacts. Solve the equation. (Assume that a ball dropped from a height h hits the plate with speed $v = \sqrt{(2gh)}$, where g is the acceleration due to gravity. The rebound speed of the ball is εv.) Instead of being stationary, the plate now oscillates so that it is moving upwards at a speed u (a constant) at the moment of each impact with the ball. Find the difference equation for h_n. Show that the difference equation has a fixed point and interpret its meaning.

34.13. $D_n(x)$ is the $n \times n$ determinant defined by

$$D_n(x) = \begin{vmatrix} 2x & 1 & 0 & \cdots & 0 \\ 1 & 2x & 1 & \cdots & 0 \\ \cdots & \cdots & \cdots & \cdots & \cdots \\ 0 & 0 & 0 & \cdots & 2x \end{vmatrix} \quad (n > 2),$$

$$D_2(x) = \begin{vmatrix} 2x & 1 \\ 1 & 2x \end{vmatrix}, \quad D_1(x) = 2x,$$

Show that

$$D_n(x) = 2xD_{n-1}(x) - D_{n-2}(x).$$

Solve the difference equation for $x \neq 1$ and $x = 1$.

34.14. Let $\{u_n\}$ $(n = 0, 1, \ldots)$ be a sequence. The power series

$$f(u_n, x) = \sum_{n=0}^{\infty} u_n x^n$$

is known as the **generating function** of the sequence. Thus, for example, if $u_n = (-1)^n/n!$, then

$$f(u_n, x) = \sum_{n=0}^{\infty} \frac{(-1)^n}{n!} x^n = e^{-x},$$

which means that e^{-x} is the generating function of $\{u_n\}$.

The generating function of $\{u_{n+1}\}$ is

$$f(u_{n+1}, x) = \sum_{n=0}^{\infty} u_{n+1} x^n = \frac{1}{x} \sum_{n=0}^{\infty} \frac{(-1)^{n+1}}{(n+1)!} x^{n+1}$$

$$= \frac{1}{x}\left(\sum_{n=0}^{\infty} \frac{(-1)^n}{n!} x^n - 1\right) = \frac{1}{x}[f(u_n, x) - 1].$$

Consider the difference equation

$$u_{n+2} + u_{n+1} - 2u_n = 0, \qquad u_0 = 1, \quad u_1 = -2.$$

By taking the generating function of the equation, show that

$$f(u_n, x) = \frac{1}{1 + 2x}.$$

Using the binomial theorem find u_n.

34.15. A Fibonacci sequence is defined as a sequence in which any term is the sum of the two preceding terms. For the Fibonacci sequence starting with $u_1 = 1$, $u_2 = 2$, find and solve the difference equation for u_n.

34.16. Solve the initial-value difference equation

$$3u_{n+2} - 2u_{n+1} - u_n = 0, \qquad u_1 = 2, \quad u_2 = 1,$$

and show that $u_n \to \frac{5}{4}$ as $n \to \infty$.

34.17. A symmetric **random walk** takes place on the integer steps on the line between $x = 0$ and $x = N$. At any position $x = r$ $(1 \leqslant r \leqslant N - 1)$, the probability that the walker moves to either $x = r + 1$ or $x = r - 1$ at any stage is $\frac{1}{2}$. The probability u_k that the walker reaches $x = 0$ first, given an initial position $x = k$, satisfies the difference equation

$$u_k = \tfrac{1}{2}u_{k-1} + \tfrac{1}{2}u_{k+1}, \qquad u_0 = 1, \quad u_N = 0,$$

for $1 \leqslant k \leqslant N - 1$. Find u_k. What is the probability that the walker reaches $x = N$ first?

If d_k is the expected number of steps in the walk before it reaches 0 or N, then d_k satisfies

$$d_k = \tfrac{1}{2}(1 + d_{k+1}) + \tfrac{1}{2}(1 + d_{k-1}), \qquad d_0 = d_N = 0$$

for $1 \leqslant k \leqslant N - 1$. Find the expected duration of the walk.

34.18. Show that $u_n = n!$ is a solution of the second-order difference equation

$$u_{n+2} = (n + 2)(n + 1)u_n.$$

By using the substitution $u_n = v_n n!$, find a second independent solution.

34.19. Given that

$$s_n = \sum_{k=1}^{n} k^3,$$

find a first-order difference equation for s_n. Solve the equation to find a formula for the sum s_n.

34.20. Show that the difference equation

$$u_{n+2} + 2au_{n+1} + bu_n = 0$$

can be expressed as

$$z_{n+1} = Az_n,$$

where

$$z_n = \begin{bmatrix} u_n \\ v_n \end{bmatrix}, \qquad A = \begin{bmatrix} -2a & b \\ -1 & 0 \end{bmatrix}.$$

Deduce that

$$z_n = A^n z_0.$$

Consider the case with $a = 1$ and $b = -8$. Find the eigenvalues of A and use the methods of Section 11.6 to find a formula for A^n. Hence solve the difference equation for u_n in terms of u_0 and u_1.

34.21. (Section 34.5). Consider the logistic equation

$$u_{n+1} = \alpha u_n(1 - u_n).$$

Draw cobweb solutions starting at $u_0 = \frac{1}{2}$ for the cases $\alpha = 2.7$, $\alpha = 2.9$, and $\alpha = 3.3$. What do you infer about the stability of the fixed point in the first quadrant?

34.22. (Section 34.5). In the logistic equation $u_{n+1} = \alpha u_n(1 - u_n)$, for what positive values of α is the origin a stable fixed point?

34.23. (Section 34.5). Find the two stable values between which u_n ultimately oscillates in the logistic equation $u_{n+1} = 3.25u_n(1 - u_n)$.

34.24. Consider the difference equation

$$u_{n+1} = \alpha(\tfrac{1}{2} - |u_n - \tfrac{1}{2}|).$$

Sketch the function $y = f(x) = \alpha(\tfrac{1}{2} - |x - \tfrac{1}{2}|)$ for $\alpha = \frac{3}{2}$. Where are the equilibrium points of the difference equation for $\alpha > 1$? Show that the origin is stable if $\alpha < 1$, and unstable if $\alpha > 1$. What happens if $\alpha = 1$?

Sketch the graph of $y = f(f(x))$ for $\alpha = 2$. Show that there exists a 2-cycle and locate the periodic values of u_n.

34.25. Find the fixed points of

$$u_{n+1} = \alpha u_n(1 - u_n^3),$$

for all α. Determine the slope of $y = f(x) = \alpha x(1 - x^3)$ at the nonzero fixed point. Confirm that this fixed point is stable if $\alpha < \frac{5}{3}$ and unstable if $\alpha > \frac{5}{3}$. Sketch cobweb solutions for $\alpha = 1.2$, 1.4, 1.8.

34.26. By starting from $u_0 = 0.957417$, compute u_1, $u_2, \ldots, u_5$ for the difference equation

$$u_{n+1} = \alpha u_n(1 - u_n), \quad \alpha = 3.83,$$

and confirm that the logistic equation appears to have a 3-cycle for this value of α.

34.27. Find the fixed points of the difference equation

$$u_{n+1} = \alpha u_n(1 - u_n)^2,$$

in the three cases (a) $\alpha = 9$, (b) $\alpha = 4$, (c) $\alpha = \frac{9}{4}$. Discuss the stability of the fixed points in each case.

35 Applications projects using symbolic computing

35.1 Symbolic computation

There have been a number of significant advances in symbolic computation and computer algebra manipulation in recent years. These are systems which bring together symbolic, numerical, and graphical operations in one software package. The mathematical methods introduced in this book are particularly appropriate contexts in which to have a first look at such systems.

The software *Mathematica*† has been used extensively in the production of the drawings of curves and surfaces, and in the checking of examples and problems in this text. At an elementary level, *Mathematica* is particularly helpful, for example, with operations such as differentiation (including partial derivatives), the construction of Taylor series, elementary algebraic operations involving matrices and linear equations, elementary integration (including repeated integrals), and difference equations; but most topics in this book can be approached to some extent using *Mathematica*. It is also useful in curve sketching in that a quick view of the general feature of a curve can be obtained, which can then be revised and edited to produce detailed graphs as required.

It is not the purpose of this book to provide an introduction to *Mathematica*. There are a number of texts which do, including the handbook that comes with the system.

A few useful titles are listed below:

Wolfram, S. (1991). *Mathematica: a system for doing mathematics by computer* (2nd edn). Addison-Wesley, Redwood City, California.
Abell, M. L., and Braselton, J. P. (1992). *Mathematica by example.* Academic Press, San Diego, California.
Blackman, N. (1992). *Mathematica: a practical approach.* Prentice-Hall, Englewood Cliffs, New Jersey.
Skeel, R. D., and Keeper, J. B. (1993). *Elementary numerical computing with Mathematica* McGraw-Hill, New York.

† *Mathematica* is a registered trade mark of Wolfram Research Inc.

35.2 Projects

The following projects are listed by chapter. They are selective samples of problems and do not cover every topic in the book. The intention is that they can be approached using mainly built-in *Mathematica* commands: very few problems require programming in *Mathematica*. It is generally inadvisable to attempt these problems by hand, since many could involve a great deal of manipulation, although some projects are prompted by examples and problems in the relevant chapters. Some commands which might be used for the projects are listed in the answers in Appendix G, together with sample programs and outputs for selected projects.

It is worth emphasizing that computer algebra systems usually generate outputs or answers without explanation of how the outputs were arrived at, unless the programming within them is investigated. Outputs can go wrong for many mathematical reasons. For example, a curve can oscillate too frequently for the built-in point spacing to detect, which can result in a false graph. This can be corrected by increasing the number of plot points, but the potential difficulty has to be recognized at the formulation stage. Symbolic computation is not a substitute for understanding mathematical techniques.

Mathematica notebooks for each project, with comments, are available on disk in PC Microsoft Windows format. *Mathematica* version 2.2 is required. Enquiries about the disk (3.5 in) should be directed to

Department of Mathematics, Keele University, Keele, Staffordshire
ST5 5BG, England.

Chapter 1

1. Draw the graphs of $y = x^3$, $y = (x - 1)^3$, $y - 1 = x^3$, $y - 1 = (x - 1)^3$ for $-1.5 \leqslant x \leqslant 2.5$. How do they differ?

2. (a) Plot the points $(n, n^2 + 1)$ for $n = 1, 2, 3, 4, 5$.
(b) Plot the points in (a) but with successive points joined by straight lines.
(c) Plot $y = x^2$ between $x = 0$ and $x = 5$.
(d) Show the curves from (b) and (c) on the same graph.

3. Plot curves defined by the following relations between x and y.
(a) $x^2 + 3y^2 = 4$; $-2 \leqslant x \leqslant 2$;
(b) $x^2 + 2y^2 - xy + 2y = 4$; $-3 \leqslant x \leqslant 3$;
(c) $x^4 + 2y^2 - xy - 2x^2y = 4$; $-2 \leqslant x \leqslant 3$.

4. Define the function $f(x) = x(1 - x^2)$. Plot the graphs
(a) $y = f(x)$;   (b) $y = f(1 - x)$;   (c) $y = f(-x)$;   (d) $y = f(|x|)$,
all for $-2 \leqslant x \leqslant 2$.

5. Define the Heaviside function $H(t)$ and the signum function sgn t. Plot graphs of the following functions on $-4 \leqslant t \leqslant 4$:
(a) $H(t)$;   (b) sgn t;   (c) $H(t) + H(-t)$;   (d) sgn(sin t).

6. Plot the graphs of the curves defined by the following polar equations:

(a) $r = \frac{1}{2}(1 - \cos \theta)$ for $0 \leqslant \theta \leqslant \pi$ (cardioid).

(b) $r = (4 \sin^2 \theta - 1) \cos \theta$ for $0 \leqslant \theta < 2\pi$ (folium).

7. Express

$$\frac{1}{(x-1)(x-2)(x-3)(x-4)(x-5)}$$

in partial fractions.

Chapter 2

1. Define the function

$$f(x) = \frac{x \sin x - 1 + \cos x}{\sin 2x + 2 - 2 e^x}.$$

Find $\lim_{x \to 0} f(x)$. Plot the function for $-0.5 \leqslant x \leqslant -0.001$ and for $0.001 \leqslant x \leqslant 0.5$, and check graphically that this agrees with the limit.

2. Find the derivative of

$$f(x) = 7x^2 + 8x^3 + 9x^4 + 10x^5 + 11x^6 + 12x^7$$

and its values $f'(0.2)$ and $f'(0.4)$.

3. Find the derivative of

$$f(x) = x^4 + 2x^3 - 3x^2 - 2x + 4.$$

Find the approximate values of x where $f'(x) = 0$, using a numerical solution routine. Plot graphs of $y = f(x)$ and $y = f'(x)$ on the same axes and compare the zeros of $f'(x)$ with the zero slopes on $y = f(x)$.

4. Find the equation of the tangent to the curve

$$y = x \sin 2x$$

at $x = 0.7$. Plot the graphs of the curve and its tangent.

5. Find the first three derivatives of

$$f(x) = x \sin^2 x + x^2 \sin(x^2),$$

and confirm that the first nonzero higher derivative at $x = 0$ is $f^{(3)}(0) = 6$.

6. Plot the graphs of $y = f(x)$, $y = f'(x)$, and $y = f''(x)$ for

$$f(x) = x^2(x^2 - 3)$$

in the interval $-2 \leqslant x \leqslant 2.5$. (This should confirm the results from Problem 2.19.)

Chapter 3

1. Display rules for the derivatives of the following general forms:
(a) $f(x)g(x)$;   (b) $f(x)/g(x)$;   (c) $f(g(x)))$;   (d) $f(x)g(x)h(x)$;
(e) $f(x)g(x)/h(x)$;   (f) $f(h(x))/h(x)$.

2. Find the first derivatives of
$$f(x) = e^{\sin x \cos^2 x} \sin x.$$
The function is periodic. What is its minimum period? Plot its graph and the graph of $f'(x)$ over one cycle. Estimate where $f(x)$ is stationary and then find each of the roots of $f'(x) = 0$ to 5 decimal places using a root-finding routine.

3. If
$$x^2 + 2y^2 - xy - 2yx^2 = 4,$$
find dy/dx as a function of x and y.

Chapter 4

1. Display rules for the first and second derivatives with respect to x of the following general forms:
(a) $f(x^2)$;   (b) $f(\sin x)$;   (c) $f(\sin(x^2))$.

2. Find the first and second derivatives of
$$f(x) = 0.1x^5 - 0.5x^4 + 0.2x^3 + x^2 - 0.7x + 2.2.$$
Estimate the roots of $f'(x) = 0$ from a graph of $y = f(x)$. Then find the roots to five decimal places by a root-finding routine. Calculate $f''(x)$ at each stationary point, and confirm the second-derivative test for stationary points. Points of inflection are given by $f''(x) = 0$. Find their locations on the original graph of $y = f(x)$.

3. Plot the graph of
$$y = \frac{x^2 - 1}{2x + 1},$$
and its asymptotes $y = \frac{1}{2}x - \frac{1}{4}$ and $x = -\frac{1}{2}$ (see Fig. 4.13).

4. Plot the graph of $y = f(x) = x^5 - 2x^3 + x^2 - 3x + 1$ in the interval $-1 \leqslant x \leqslant 3$, and estimate the roots of $f(x) = 0$ in this interval. Set up a Newton routine
$$x_{n+1} = x_n - \frac{f(x_n)}{f'(x_n)},$$
for calculating the roots of $f(x) = 0$, and find, starting at $x = 0.5$ and 1.6, the roots to ten significant figures. What is the smallest number of iterations required in each case to calculate the roots to ten significant figures?

5. Plot the graph of $y = x + \sin 5x$ in the interval $0 \leqslant x \leqslant 25$ using
(a) the default plotting routine,
(b) plotting with 20 plot points,
(c) plotting with 50 plot points.
Explain why the graphs are different for this type of function.

Chapter 5

1. Obtain formulas for the Taylor polynomials for the following functions centred at $x = a$ as far as $(x - a)^3$:
(a) $f(x)$;   (b) $[f(x)]^2$;   (c) $f(x)g(x)$;   (c) $e^{f(x)}$. State the coefficient of $(x - a)^2$ in each case.

2. Find Taylor expansions about $x = 0$ up to and including x^5 for each of the following functions:
(a) e^x;   (b) $(x + 1)\cos x$;   (c) $\ln(1 + \sin x)$;   (d) $\exp(\sin(e^x - 1))$.

3. Find the Taylor polynomials for $(\sin^2 x)/x^2$ up to and including x^N for $N = 2, 4, 6$. Plot the graphs of the function and its Taylor polynomials for $0.001 \leqslant x \leqslant 2$, and compare them. At approximately what values of x do the Taylor polynomials visibly part company from the exact function?

4. Find the Taylor polynomials for $\ln x$ about $x = 1$ for $N = 6$. Construct an error function which is the difference of $\ln x$ and its Taylor polynomial. Show that, at 2.159 approximately, this error starts to exceeds 0.2 as x increases. Plot this error function against x for $1 \leqslant x \leqslant 2.2$.

Chapter 6

1. Solve, for the complex number a, the equation $z = 0$ where
$$z = \frac{(2 + 3j)^4}{(1 - 5j)^3} + \frac{(a - 2j)}{(1 + 5j)^4}.$$

2. If $z = x + jy$, find the real and imaginary parts of $z\, e^z \cos z$.

3. Find the 13 roots of $z^{13} = 1 + j$, and plot the roots on the Argand diagram.

4. Let $z_1 = 1 - 2j$, $z_2 = 3 + j$. Plot the following points on the Argand diagram:
$$z_1 + z_2, \quad \bar{z}_1 + \bar{z}_2, \quad z_1 - z_2, \quad \bar{z}_1 + z_2, \quad z_1 z_2, \quad z_1/z_2.$$

5. Find $|z|$ and $\operatorname{Arg} z$, where
$$z = \frac{(1 + 2j)^4}{(1 + 3j)} - \frac{2(3 - 4j)^3}{1 + 4j}.$$

Chapter 7

1. Let
$$A = \begin{bmatrix} 1 & 2 & 3 & 4 \\ -2 & 3 & -4 & 1 \\ 3 & 4 & 1 & 2 \\ 4 & -1 & 2 & 3 \end{bmatrix}, \quad B = \begin{bmatrix} 1 & 0 & -1 & 0 \\ 1 & -2 & 1 & 2 \\ -3 & 1 & -3 & 1 \\ 2 & 1 & 2 & 1 \end{bmatrix},$$

$$C = \begin{bmatrix} 3 & 1 & 2 & 1 \\ p & p & 1 & 2 \\ 1 & -2 & -3 & 2 \\ 2 & 1 & 0 & -1 \end{bmatrix}.$$

Find and compare
(a) AB and BA; (b) $A(BC)$ and $(AB)C$; (c) $(A + B)^T$ and $A^T + B^T$; (d) $(AB)^T$ and $B^T A^T$.

2. Find the inverse of

$$\begin{bmatrix} 1 & x_1 & x_1^2 \\ 1 & x_2 & x_2^2 \\ 1 & x_3 & x_3^2 \end{bmatrix}$$

(see Problem 7.18). Find the equation of the parabola of the form $y = a + bx + cx^2$ through the points $(-1, -2)$, $(\frac{1}{2}, -1)$, and $(\frac{5}{2}, 2)$.

3. Let

$$A = \begin{bmatrix} \frac{1}{3} & \frac{1}{3} & \frac{1}{6} & \frac{1}{6} \\ \frac{1}{4} & \frac{1}{2} & \frac{1}{8} & \frac{1}{8} \\ \frac{1}{8} & \frac{1}{4} & \frac{3}{8} & \frac{1}{4} \\ \frac{1}{2} & \frac{1}{6} & \frac{1}{6} & \frac{1}{6} \end{bmatrix}.$$

Find A^2, A^4, A^8, A^{16}. How do you expect A^n to behave as $n \to \infty$?

Chapter 8
1. Let

$$A = \begin{bmatrix} 1 & -1 & 2 & 3 \\ 3 & 1 & 0 & -3 \\ 2 & -1 & 3 & -1 \\ 2 & -1 & 2 & 4 \end{bmatrix}, \quad B = \begin{bmatrix} 2 & 4 & -3 & 1 \\ 0 & -1 & 4 & 3 \\ -2 & -2 & 3 & 1 \\ -2 & 5 & 6 & -5 \end{bmatrix}.$$

Find $\det A$, $\det B$, $\det A^{-1}$, and $\det AB$. Confirm that

$$\det A^{-1} = 1/\det A, \qquad \det A \det B = \det AB.$$

2. Factorize the following determinants:

(a) $\begin{vmatrix} 1 & 1 & 1 \\ a & b & c \\ a^2 & b^2 & c^2 \end{vmatrix}$; (b) $\begin{vmatrix} 1 & 1 & 1 & 1 \\ a & b & c & d \\ a^2 & b^2 & c^2 & d^2 \\ a^3 & b^3 & c^3 & d^3 \end{vmatrix}$; (c) $\begin{vmatrix} 1 & 1 & 1 & 1 \\ a & b & c & d \\ a^2 & b^2 & c^2 & d^2 \\ a^4 & b^4 & c^4 & d^4 \end{vmatrix}$.

3. Find the values of a for which

$$\begin{vmatrix} 5 & a & -1 & 1 \\ 2 & 1 & a & 2 \\ 3 & a & 1 & 4 \\ -1 & 0 & a & 2 \end{vmatrix}$$

is zero.

Chapter 9

1. The area of a triangle whose vertices are the points with position vectors a, b, and c is given by the formula

$$\tfrac{1}{2}|b \times c + c \times a + a \times b|.$$

Devise a program based on this formula to determine the area for general vertices. What is the area if $a = (1, 0, 1)$, $b = (2, -1, 1)$, and $c = (1, 1, 2)$? Plot a diagram showing the triangle.

2. A tetrahedron has vertices with position vectors

$$a = (1, -1, 2), \quad b = (-1, 2, 3), \quad c = (2, -1, 3), \quad d = (1, 3, -2).$$

Find its surface area. Draw a three-dimensional plot showing the tetrahedron viewed from the point with position vector $(2.1, -2.4, 1.5)$.

3. Plot the curve which has the position vector

$$r = (2 \cos t)\hat{\imath} + (2 \sin t)\hat{\jmath} + 0.3t\hat{k}$$

from $t = 0$ to $t = 20$. What is the curve called? The position vector represents a particle moving along the curve. Find the velocity vector $\dot{r}$ and the acceleration vector $\ddot{r}$ of the particle. Show that $\dot{r} \cdot \ddot{r} = 0$.

Chapter 10

1. Use a row-reduction routine to solve the linear equations

$$x + 2y - 3z = q,$$
$$2x + py + z = -1,$$
$$x - 2y - z = 4,$$

where p and q are two parameters. Determine for what values for p and q, (a) the equations have a unique solution, (b) no solution, (c) an infinite set of solutions.

2. Use a row-reduction method to solve the linear equations

$$x + 2y + pz = 5,$$
$$3x + 2y + z = q,$$
$$2x - y + 4z = 7,$$

where p and q are two parameters. Confirm that

$$z = \frac{63 - 5q}{11 + 7p} \quad (p \neq -\tfrac{11}{7}),$$

and discuss the nature of solutions for all values of p and q.

3. Using a row-reduction instruction, show that

$$
\begin{aligned}
x_1 + \qquad\quad 3x_3 \qquad\quad &= 5, \\
-x_1 + x_2 - \quad x_3 + x_4 &= -1, \\
x_1 + 2x_2 + 11x_3 \qquad &= 4, \\
-x_1 + 2x_2 + \quad 3x_3 + x_4 &= 3,
\end{aligned}
$$

is an inconsistent set of equations.

Chapter 11

1. Find the eigenvalues and eigenvectors of

$$
A = \begin{bmatrix}
-6 & 1 & 2 & 0 \\
1 & 0 & -3 & -1 \\
2 & 1 & -6 & 0 \\
-2 & 2 & 0 & -3
\end{bmatrix}.
$$

How many linearly independent eigenvectors does A have?
  Find the eigenvalues of the following matrices:
(a) A^{-1}; (b) A^2; (c) $A + kI$.

2. Find the eigenvalues and eigenvectors of

$$
A = \begin{pmatrix}
1 & 2 & 1 \\
2 & 1 & 1 \\
1 & 1 & 2
\end{pmatrix}.
$$

Construct a matrix C of eigenvalues and confirm that

$$A = CDC^{-1},$$

where D is a diagonal matrix of eigenvectors. Obtain the general formula for

$$A^n = CD^nC^{-1}.$$

3. Find the inverse and transpose of

$$
A = \tfrac{1}{3}\begin{bmatrix}
1 & 2 & 2 \\
2 & 1 & -2 \\
2 & -2 & 1
\end{bmatrix},
$$

and verify that A is an orthogonal matrix. Find the eigenvalues of A. What expected property do they have?

4. Find the eigenvalues of

$$A = \begin{bmatrix} 5 & 5 & -6 & 2 \\ -3 & 13 & -6 & 2 \\ -3 & 7 & 0 & 2 \\ 3 & -15 & 12 & 2 \end{bmatrix}.$$

Find the expression $\det(A - \lambda I_4)$, and demonstrate the Cayley–Hamilton theorem of Problem 11.23.

Chapter 12

1. Plot the graphs of the derivative $dy/dx = \sin 2x$ and the equation of the curve through $(\pi, -1)$ of which this is the derivative (see Example 12.7).

2. Plot the graph of

$$\frac{dy}{dx} = x e^{-x} + \sin x - x^2 \cos 2x,$$

for $0 \leqslant x \leqslant 10$. Show that an antiderivative which is zero when $x = 0$ is

$$y = 2 + \tfrac{1}{4}[-4(1 + x)e^{-x} - 4 \cos x - 2x \cos 2x$$
$$+ \sin 2x - 2x^2 \sin 2x].$$

Plot the graph of the signed area between $x = 0$ and $x = 10$.

Chapter 13

1. Set up a program to compute the area under the curve $y = f(x)$ between $x = a$ and $x = b$ using the approximation

$$h \sum_{n=0}^{N-1} f(x_n),$$

where $h = (b - a)/N$ and $x_n = a + nh$. Apply the method to the following functions, limits and subdivision numbers:
(a) $f(x) = x^2$, $1 \leqslant x \leqslant 3$, $N = 200$;
(b) $f(x) = x e^{-x}$, $0 \leqslant x \leqslant 3$, $N = 20$;
(c) $f(x) = x^3 \sin x$, $0 \leqslant x \leqslant \pi$, $N = 30$;
(d) $f(x) = \cos(e^{-x})$, $0 \leqslant x \leqslant 1$, $N = 25$.
In cases (a), (b), and (c), compare the numerical result with the areas obtained by integration. In these cases, how many subdivisions are required to obtain a numerical result correct to three decimal places? In (a), show that over 10,000 steps are required. Why is this?

2. Use a symbolic integration program to obtain the following indefinite integrals:

(a) $\displaystyle\int (\ln x)^3 \, dx$;　(b) $\displaystyle\int \sin^5 x \cos^3 x \, dx$;　(c) $\displaystyle\int x^2 e^x \sin x \, dx$;

(d) $\int \sqrt{(1-x^2)}\,dx$; (e) $\int \dfrac{dx}{x(x+1)(x+2)(x+3)}$; (f) $\int \dfrac{dx}{(1-x^3)}$.

Check each answer by recovering the integrands by differentiation.

3. Evaluate the following definite integrals:

(a) $\displaystyle\int_1^2 x(\ln x)^3\,dx$; (b) $\displaystyle\int_0^1 \dfrac{x\,dx}{\sqrt{(5+4x-4x^2)}}$; (c) $\displaystyle\int_0^{\frac12} \dfrac{x^3\,dx}{(1-x^2)^{\frac52}}$;

(d) $\displaystyle\int_0^1 \sum_{n=0}^{100} \dfrac{x^n}{n!}\,dx$.

4. Find
$$I(a) = \int_1^a (\ln x)^3\,dx.$$

Find the limit
$$\lim_{b\to 0} \dfrac{bI(1/b)}{(-\ln b)^3}.$$

How does $I(a)$ behave as $a \to \infty$? Does
$$\int_1^\infty (\ln x)^3\,dx$$
exist?

5. A cylindrical hole of circular cross-section and radius b is drilled through a sphere of radius $a > b$, the axis of the hole passing through the centre of the sphere. Find the volume of the remaining object. Display a diagram of the object for some values of a and b.

Chapter 14
1. Plot the graph of the polar equation $r = \sin 5\theta$ for $0 \leqslant \theta \leqslant 2\pi$. Find the area enclosed by the five 'petals' of the curve.
   Show that the area of the $2n + 1$ petals of $r = \sin(2n + 1)\theta$ $(n \geqslant 1)$ in dependent of n.

2. Devise a program to generate the trapezium rule:
$$\int_a^b f(x)\,dx \approx \dfrac{b-a}{N}\left[\tfrac12 f(a) + (f(x_1) + f(x_2)\right.$$
$$\left. + \cdots + f(x_{N-1})) + \tfrac12 f(b)\right].$$

Apply the program to the integral
$$\int_0^2 e^{-2x}\sin^2 x\,dx,$$

and compare the result with the exact value of the integral. Investigate how many steps are required to obtain a result accurate to three decimal places.
   Apply the program also to Problem 14.20.

3. A thin plane metal plate consists of an isosceles triangle of height h and base length $2a$ with a semicircle of radius a attached symmetrically by its diameter to the base of the triangle. Find the location of its centroid on its axis of symmetry.

4. Set up a program to generate Simpson's rule

$$\int_a^b f(x)\, dx \approx$$

$$\frac{b-a}{3N}\left(f(a) + f(b) + 4\sum_{k=1}^{\frac{1}{2}N} f(x_{2k-1}) + 2\sum_{k=1}^{\frac{1}{2}N-1} f(x_{2k})\right),$$

where N is an even number. Apply the method to $f(x) = e^{-x^2}$, with $b = 1$, $a = 0$. Compare results with the trapezium rule above.

Chapter 15
1. Illustrate the substitution method in integration by writing a program to integrate

$$\int \frac{x-2}{\sqrt{(5 + 4x - x^2)}}\, dx,$$

using the substitutions $x = u + 2$, $u = 3 \sin t$. Integrate directly and through the substitutions.

2. Integrate the following, and compare your answers with computer-integrated ones:

(a) $\displaystyle\int \frac{x\, dx}{4x^2 + 1}$;   (b) $\displaystyle\int \tan x\, dx$;   (c) $\displaystyle\int \cos^4 x\, dx$;

(d) $\displaystyle\int \frac{x\, dx}{\sqrt{(x-1)}}$;   (e) $\displaystyle\int \frac{\sin^3 x}{\cos x}\, dx$.

3. Computer-integrate the infinite integrals

$$I_{10} = \int_0^\infty t^{10}\, e^{-t}\, dt, \quad I_{11} = \int_0^\infty t^{11}\, e^{-t}\, dt,$$

and confirm that $I_{11}/I_{10} = 11$.

4. Computer-integrate the following infinite integrals:

(a) $\displaystyle\int_0^\infty e^{-x} \sin x\, dx$;   (b) $\displaystyle\int_1^\infty \frac{\ln x}{x^{10}}\, dx$;   (c) $\displaystyle\int_1^\infty x^3 e^{-ax^2}\, dx$.

5. Evaluate the integral

$$f(a) = \int_1^a \frac{(\ln x)^6}{x^2}\, dx$$

for $a > 1$. Find $f(10)$, $f(20)$, and $f(\infty)$. The results indicate that $f(a)$ tends to a limit very slowly as $a \to \infty$. Find where

$$g(x) = \frac{(\ln x)^6}{x^2}$$

has a maximum value, and plot the graph $y = g(x)$ for $1 \leqslant x \leqslant 100$.

Chapter 16

1. Solve the differential equation $\dot{x} + x = 0$, for the initial conditions (a) $x(0) = 0$, (b) $x(0) = 1$, (c) $x(0) = 2$ and plot the solutions on the same axes for $0 \leqslant t \leqslant 2$.

2. Solve the differential equations
(a) $2\ddot{x} + 3\dot{x} + x = 0$,   (b) $\ddot{x} + 2\dot{x} + 2x = 0$,   (c) $\ddot{x} + 2\dot{x} + x = 0$,

each for the six sets of initial conditions:
 (i) $x(0) = 0$, $\dot{x}(0) = 1$;   (ii) $x(0) = 0$, $\dot{x}(0) = 2$;
(iii) $x(0) = 0$, $\dot{x}(0) = 3$;   (iv) $\dot{x}(0) = 0$, $x(0) = 1$;
 (v) $\dot{x}(0) = 0$, $x(0) = 2$;   (vi) $\dot{x}(0) = 0$, $x(0) = 3$.
Plot all solutions on the same axes for each differential equation, for $0 \leqslant t \leqslant 5$.

Chapter 17

1. Solve the differential equation $2\ddot{x} + 3\dot{x} + x = \cos t$ subject to $\dot{x}(0) = 0$, $x(0) = 1$. Plot the solution for $0 \leqslant t \leqslant 50$.

2. Solve the differential equation $\ddot{x} + x = \cos t$ subject to $x(0) = 0$, $\dot{x}(0) = 0$. Plot the solution for $0 \leqslant t \leqslant 20$.

Chapter 18

1. Solve the differential equation $\ddot{x} + x = 0$ subject to the initial conditions $x(0) = 1$, $\dot{x}(0) = 0$. Also solve $\ddot{x} + \sin x = 0$, by a built-in numerical solution method for $0 \leqslant t \leqslant 10$ subject to the same initial conditions. Plot both solutions for $0 \leqslant t \leqslant 10$. Comparison of the plotted solutions will indicate by how much the period decreases when the linear approximation is used. Re-run the programs for different amplitudes $x(0)$.

Chapter 19

1. Draw the phasor diagram of the sum of the three phasors of
$$u(t) = 2 \cos 10t, \quad v(t) = \cos(10t - \tfrac{1}{2}\pi), \quad w(t) = 3 \cos(10t + \tfrac{1}{4}\pi)$$
(see Example 19.6).

Chapter 20

1. Draw the lineal element diagram of $dy/dx = xy$, produced by a standard package in the square $\{0 \leqslant x \leqslant 1, 0 \leqslant y \leqslant 1\}$ (see Section 20.1). Compare this with the exact solution (see Section 20.1) drawn through the points $(0, 0.2)$, $(0, 0.4)$, and $(0, 0.6)$.

2. Repeat the above process for the differential equation $dy/dx = x - y$ of Example 20.1.

3. Design a program for Euler's method (Section 20.2) for the initial-value problem
$$\frac{dy}{dx} = xy^2, \qquad y(0) = 1,$$

(see Example 20.4) with step length $h = 0.2$ and 5 steps. Run the program for the cases $h = 0.1$ and $h = 0.01$ and compare the results.

4. Plot numerical solutions for

$$\frac{dy}{dx} = \frac{3y - x}{3x - y}$$

(Example 20.14 and Fig. 20.11) using built-in routines. As with many equations of this type it is often easier to solve the equivalent simultaneous equations

$$\frac{dx}{dt} = 3x - y, \qquad \frac{dy}{dt} = 3y - x,$$

numerically for various initial values of $x(0)$ and $y(0)$.

Chapter 21
1. By splitting the differential equation $\ddot{x} + 2x^3 = 0$ into the system

$$\dot{x} = y, \qquad \dot{y} = -2x^3,$$

and plotting four phase paths respectively through the four points

$$(x(0), y(0)) = (0.3, 0), (0.6, 0), (0.9, 0), (1.2, 0)$$

over the interval $-1.5 \leqslant x \leqslant 1.5$, show that the solutions appear to be periodic.

2. Plot phase paths for the van der Pol equation

$$\ddot{x} + 10(x^2 - 1)\dot{x} + x = 0$$

showing the limit cycle. Also show the corresponding (t, x) graph of the periodic solution (the periodic solution has an initial value close to $x(0) = 2$, $\dot{x}(0) = 0$).

Chapter 22
1. Computer algebra systems are quite efficient at finding Laplace transforms of complicated expressions involving standard functions. Test the system with the following transforms:

(a) $L\{t^8 e^{-t}\}$;   (b) $L\{t^2 e^{-t} \cos t\}$;   (c) $L\left\{\dfrac{d^3 x}{dt^3}\right\}$;

(d) $L\{f(t)\}$ where $f(t) = \begin{cases} 1 & \text{if } 0 \leqslant t \leqslant a \\ 0 & \text{if } t > a \end{cases}$;

(e) $L\{e^t / t^{\frac{1}{2}}\}$;   (f) $L\{\cosh at\}$.

2. Solve

$$\dot{x} + 2x = e^{-t}, \qquad x(0) = 3,$$

using a Laplace-transform package, and compare the answer with that of Example 22.12. Plot the input e^{-t} and the output against t for $0 \leqslant t \leqslant 3$.

3. Using a Laplace-transform package, solve the system

$$\ddot{x} + 2\dot{x} + x = a \cos \omega t, \qquad x(0) = 0, \ \dot{x}(0) = 0.$$

Plot the input and output functions for $a = 1, \omega = 1$ and $0 \leqslant t \leqslant 30$. Estimate the eventual amplitude of the periodic output.

4. Find the functions whose Laplace transforms are:

(a) $\dfrac{1}{s(s + 1)(s + 2)(s + 3)}$; (b) $\dfrac{e^{-s}}{(s^2 + 4)(s + 1)}$.

Plot the functions in each case.

5. Consider the function $f(t) = \ln t$. Show the Laplace-transform package produces the transform

$$\frac{1}{s}(\gamma + \ln s),$$

where γ is *Euler's constant* given by

$$\gamma = \lim_{m \to \infty} \left(\sum_{k=1}^{m} \frac{1}{k} - \ln m \right).$$

Derive a program to calculate Euler's constant. It should give $\gamma = 0.577215 \ldots$.

Chapter 23

1. Find the Laplace transform of the solution of

$$\ddot{x} + \omega^2 x = a\delta(t - 1), \qquad x(0) = \dot{x}(0) = 0,$$

which has impulse input applied at time $t = 1$. Invert the transform and plot the output for $\omega = 4, a = 1$ (see Example 23.3).

2. Following the previous project, solve more complicated problem with two impulses:

$$2\ddot{x} + 3\dot{x} + 2x = a\delta(t - \pi) \cos t + b\delta(t - 2\pi),$$

$x(0) = \dot{x}(0) = 0$.

Plot the output for $a = b = 1$.

3. Let $f(t) = t^3$, $g(t) = \cos t$. Find the convolution

$$\int_0^t f(t - u)g(u) \, du.$$

Then verify that

$$L\{f(t)\}L\{g(t)\} = L\left\{ \int_0^t f(t - u)g(u) \, du \right\}.$$

Chapter 24

1. Consider the period 2 sawtooth function defined over its fundamental interval $-1 < t \leqslant 1$ by $f(t) = t$. Find its general Fourier coefficient and output its first four terms. Plot and compare the graphs of this truncated series and the sawtooth for $-3 < t \leqslant 3$.

2. Repeat the previous problem but with the function

$$f(t) = \begin{cases} 1 & (0 \leqslant t < 1), \\ -1 & (-1 \leqslant t < 0). \end{cases}$$

Plot the graphs of $f(t)$ and the first 12 terms of its Fourier series. The graph should show Gibbs' phenomenon, in which the Fourier series approximation overshoots the function at discontinuities. You can try it with (say) 20 terms or more, but you should include more interpolating points in these cases.

3. Find the Fourier coefficients of the 2π-periodic function defined by

$$f(x) = x^6 - 5\pi^2 x^4 + 7\pi^4 x^2$$

on the interval $-\pi \leqslant x < \pi$. What is the sum of the series

$$\sum_{n=1}^{\infty} \frac{(-1)^{n+1}}{n^6} ?$$

Chapter 25

1. Plot the saddle surface $z = x^2 - y^2$ in the cylinder $x^2 + y^2 \leqslant 1$, using a three-dimensional parametric plot routine with parameters r and u where

$$(x, y, z) = (r \cos u, \quad r \sin u, \quad r^2 \cos 2u).$$

Also draw a contour plot of the surface in the (x, y) plane on the square $-1 \leqslant x \leqslant 1, -1 \leqslant y \leqslant 1$.

2. Plot the surface $z = xy(x^2 - y^2)$ in the cylinder $x^2 + y^2 \leqslant 1$ using the same routine as in Project 25.1 above, but with the parametric equations

$$(x, y, z) = (r \cos u, r \sin u, \tfrac{1}{4} r^3 \sin 4u).$$

How would you describe this saddle? Draw its contour plot in the square $-1 \leqslant x \leqslant 1, -1 \leqslant y \leqslant 1$.

3. For the function

$$f(x, y) = e^{x^2 y} \sin(xy) + x \ln(x^2 + y^3),$$

verify that

$$\frac{\partial^2 f}{\partial y \, \partial x} = \frac{\partial^2 f}{\partial x \, \partial y}.$$

4. Plot the surface given by $z = \cos xy$ over $-\pi \leqslant x \leqslant \pi, -\tfrac{1}{2}\pi \leqslant y \leqslant \tfrac{1}{2}\pi$. Find the partial derivatives at $(\tfrac{1}{4}\pi, 1)$ and construct the equation of the tangent plane there. Finally plot the surface and its tangent plane.

5. Find the stationary points of

$$f(x, y) = 0.3x^3 + 0.2y^2 - x^2 y - xy + 2y$$

numerically by solving

$$\frac{\partial f}{\partial x} = \frac{\partial f}{\partial y} = 0.$$

Plot the contours on the (x, y) plane for $-3 \leqslant x \leqslant 3$, $-9 \leqslant y \leqslant 3$.

Find the values of the second derivatives at each stationary point and check the second derivative tests (25.9) at each point.

6. Find the least-squares straight-line fit to the points

$$(0, 1.1), \quad (1, 2), \quad (2, 2.9), \quad (3, 3.9), \quad (4, 4.5), \quad (5, 5.1),$$

in the (x, y) plane. Plot the data and the least-squares straight-line fit. If you are using a built-in routine, check your results against that given by (25.10).

Chapter 26
1. Find the family of curves orthogonal to that of

$$\frac{dy}{dx} = y\,e^{-x}.$$

Plot both families of curves for $|x| \leqslant 2$, $|y| \leqslant 2$.

Chapter 27
1. Find where the function

$$f(x, y) = x^3 - 2xy - x + 3y^2$$

is stationary subject to the condition $x^2 + 2y^2 = 1$. Devise a program which uses the Lagrange-multiplier method (27.4): here is a suggested line of approach. First plot the contours of $z = f(x, y)$ and the curve $x^2 + 2y^2 = 1$. Locate the approximate coordinates of any point of tangency. Then use a built-in root-finding scheme to locate the stationary values. There should be four.

Chapter 28
1. Find the equation of the tangent plane to the surface

$$x^3 y + zx + xy^2 z = -3$$

at $(1, 2, -1)$.

2. Show graphically the intersection of the cylinder $x^2 + y^2 = 1$ and the plane $x + y + z = 1$ (Example 28.9).

3. Find the envelope of the family of curves

$$y(a^2 - 1 + ax) = x$$

with parameter a. Plot the envelope and a sample of touching curves in $-3 \leqslant x \leqslant 3$.

Chapter 29

1. By repeated integration, evaluate the integral

$$\int_{-1}^{1} \int_{0}^{1} (x + y\,e^{-xy} + xy)\,dx\,dy,$$

using a symbolic routine. Plot the surface

$$z = x + y\,e^{-xy} + xy$$

over $0 \leqslant x \leqslant 1$, $-1 \leqslant y \leqslant 1$. Interpret the integral as the volume under the surface. Does the integral contain 'negative' volumes under the surface? Plot the positive part of the surface over the same rectangle.

2. Evaluate the repeated integral

$$\int_{0}^{a} \int_{-\sqrt{(a^2 - y^2)/a}}^{\sqrt{(a^2 - y^2)/a}} x^2 y\,dx\,dy.$$

Plot the region of integration in the (x, y) plane, and then check that the integral has the same value with the order of the integration reversed.

Chapter 30

1. Let

$$f(x, y, z) = xy\hat{\imath} + yz\hat{\jmath} + (z - y)x\hat{k}.$$

Find f as a function of t on the line $x = t$, $y = t$, $z = t$. Evaluate the line integral

$$\int f \cdot dr$$

on this line between $(0, 0, 0)$ and $(1, 1, 1)$.

   Repeat the process with the curve $x = t^2$, $y = t^3$, $z = t^4$ and the same end points. Plot both paths of integration.

Chapter 31

1. A and B are the sets of integers defined by

$$A = \{2n + 5(-1)^n \,|\, n \in \mathbb{N}^+, 1 \leqslant n \leqslant 100\},$$
$$B = \{n^2 - n + 1 \,|\, n \in \mathbb{N}^+, 1 \leqslant n \leqslant 10\}.$$

Produce lists of the elements in $A \cup B$ and $A \cap B$. How many elements do each of these sets have?

2. Let A, B, and C be the following sets:

$$A = \{n(n - 1) \,|\, n \in \mathbb{N}^+, 2 \leqslant n \leqslant 100\},$$
$$B = \{|n^2 - 100n| \,|\, n \in \mathbb{N}^+, 1 \leqslant n \leqslant 160\},$$
$$C = \{4n \,|\, n \in \mathbb{N}^+, 1 \leqslant n \leqslant 2200\}.$$

Verify the first distributive law

$$A \cap (B \cup C) = (A \cap B) \cup (A \cap C).$$

How many elements are there in the set $A \cap (B \cup A)$?

Chapter 33

1. Draw the labelled drawings of the bipartite graphs $K_{5,6}$ and $K_{6,6}$. Answer the following for each graph by the built-in diagnostic test.
(a) How many edges has each graph?
(b) Is the graph eulerian? If it is, list an eulerian cycle.
(c) Is it hamiltonian? If it is, list a hamiltonian cycle.

2. Check the complete graphs K_n, $2 \leqslant n \leqslant 7$ and the bipartite graphs $K_{i,j}$ $(2 \leqslant i \leqslant 5;\ i \leqslant j \leqslant 6)$ for planarity, using a built-in diagnostic test.

Chapter 34

1. Rework Example 34.2 using a symbolic package for solving difference equations. Solve the mortgage difference equation
$$Q_m - (1 + I)Q_{m-1} = -A,$$
with $I = 0.08$ and $Q_0 = P = 50{,}000$ (in £). Given that $Q_{25} = 0$, find A. List the outstanding debt Q_m each year m to the nearest £. Plot (a) the outstanding debt against years and (b) the annual interest repayments $A - IQ_m$ against years.

2. Solve the following homogeneous difference equations:
(a) $u_{n+2} - u_{n+1} - 12u_n = 0$;   (b) $u_{n+2} + 2u_{n+1} + 2u_n = 0$;
(c) $u_{n+2} + 4u_{n+1} + 4u_n = 0$;
(d) $u_{n+3} + 3u_{n+2} + 3u_{n+1} + u_n = 0$, $u_0 = 0$, $u_1 = 1$, $u_2 = -1$.

3. Solve the following inhomogeneous difference equations
(a) $u_{n+2} - u_{n+1} - 12u_n = 2 + n + n^2$;   (b) $u_{n+2} - u_{n+1} + 4u_n = 2^n$;
(c) $u_{n+3} + 3u_{n+2} + 3u_{n+1} + u_n = n^2$, $u_0 = 0$, $u_1 = 1$, $u_2 = -1$.

4. Devise a program to generate cobweb plots for the first-order difference equation
$$u_{n+1} = -ku_n + k$$
for (a) $k = \frac{1}{2}$, (b) $k = \frac{3}{2}$, (c) $k = 1$, with initial value $u_0 = \frac{3}{4}$ in each case (see Example 34.3).

5. Display cobweb plots for the logistic difference equation
$$u_{n+1} = \alpha u_n(1 - u_n)$$
for selected values of α. Some suggested values are:
(a) $\alpha = 2.8$ to show a stable fixed point;
(b) $\alpha = 3.4$: find the period 2 solution;
(c) $\alpha = 3.5$: find the period 4 solution;
(d) $\alpha = 3.7$: chaotic output;
(e) $\alpha = 3.83$: should be stable to locate a stable period 3 solution.

Appendix

Appendix A: Some algebraical rules
(a) Index laws for real numbers
(i) $a^0 = 1$.

(ii) $a^p a^q = a^{p+q}$.

(iii) $a^{-p} = 1/a^p$.

(iv) $(a^p)^q$ or $(a^q)^p = a^{pq}$ (so $a^{p/q} = (a^p)^{1/q}$ or $(a^{1/q})^p$).

For example, $a^{\frac{1}{2}} = \sqrt{a}$ because $(a^{\frac{1}{2}})^2 = a$ and $(\sqrt{a})^2 = a$. Conventionally, $a^{\frac{1}{2}}$ and $\sqrt{a}$ represent the **positive root** when we are talking about real numbers (for complex numbers, see Chapter 6). It may be necessary to restrict a to be positive so that $a^{p/q}$ is a real number. For example, $(-8)^{\frac{1}{2}}$ or $\sqrt{(-8)}$ is not real: there is no real number whose square is equal to -8. But $(-8)^{\frac{1}{3}}$ or $\sqrt[3]{(-8)} = -2$.

(b) Quadratic equations
(i) $ax^2 + bx + c = 0$ has the solutions
$$x_1, x_2 = [-b \pm \sqrt{(b^2 - 4ac)}]/2a.$$

(ii) In terms of x_1 and x_2, the factors are
$$ax^2 + bx + c = a(x - x_1)(x - x_2).$$

(iii) The coefficients are given in terms of x_1 and x_2 by
$$b = -a(x_1 + x_2), \, c = ax_1 x_2.$$

(c) Binomial theorem
(i) If n is a positive integer (or whole number)
$$(a + b)^n = a^n + na^{n-1}b + \frac{n(n-1)}{2!} a^{n-2}b^2$$
$$+ \frac{n(n-1)(n-2)}{3!} a^{n-3}b^3 + \cdots + b^n.$$

There are $(n + 1)$ terms in this sum, and it is symmetrical in a and b.

An important special case is
$$(1 + x)^n = 1 + nx + \frac{n(n-1)}{2!} x^2 + \frac{n(n-1)(n-2)}{3!} x^3 + \cdots + x^n.$$

(ii) *Pascal's triangle*. Each entry (apart from the 1s) is the sum of two previous entries – the one above, and the one above and to the left – as illustrated by the underlined group:

$$
\begin{array}{llllll}
n = 1 & & 1 & 1 & & \\
n = 2 & & 1 & 2 & 1 & \\
n = 3 & & 1 & \underline{3} & \underline{3} & 1 \\
n = 4 & & 1 & 4 & \underline{6} & 4 & 1
\end{array}
$$

and so on. Thus

$$(1 + x)^4 = 1 + 4x + 6x^2 + 4x^3 + x^4.$$

(d) Factorization
$a^2 - b^2 = (a + b)(a - b),$
$a^3 - b^3 = (a - b)(a^2 + ab + b^2),$
$a^3 + b^3 = (a + b)(a^2 - ab + b^2).$

(e) Constants
$e = 2.71828182\ldots,$ $\qquad \pi = 3.14159265\ldots,$
$1 \text{ radian} = 57.29578\ldots^\circ,$ $\quad 1^\circ = 0.01745\ldots \text{ radian},$
$360^\circ = 2\pi \text{ radian}.$

Appendix B: Trigonometric formulae
(a) Relation between trigonometric functions
$\sin^2 A + \cos^2 A = 1,$
$\tan A = \sin A/\cos A; \quad \sec A = 1/\cos A; \quad \operatorname{cosec} A = 1/\sin A.$

(b) Addition formulae
$\sin(A \pm B) = \sin A \cos B \pm \cos A \sin B,$
$\cos(A \pm B) = \cos A \cos B \mp \sin A \sin B,$
$\tan(A \pm B) = (\tan A \pm \tan B)/(1 \mp \tan A \tan B).$

(c) Addition formulae: special cases
$\sin 2A = 2 \sin A \cos A,$
$\cos 2A = \cos^2 A - \sin^2 A$
$\qquad = 2 \cos^2 A - 1 = 1 - 2 \sin^2 A,$
$\tan 2A = 2 \tan A/(1 - \tan^2 A),$
$\sin 3A = 3 \sin A - 4 \sin^3 A,$
$\cos 3A = 4 \cos^3 A - 3 \cos A.$

(d) Product formulae
$\sin A \sin B = \frac{1}{2}[\cos(A - B) - \cos(A + B)],$
$\cos A \cos B = \frac{1}{2}[\cos(A - B) + \cos(A + B)],$
$\sin A \cos B = \frac{1}{2}[\sin(A - B) + \sin(A + B)].$
$\sin C + \sin D = 2 \sin \frac{1}{2}(C + D) \cos \frac{1}{2}(C - D),$
$\sin C - \sin D = 2 \sin \frac{1}{2}(C - D) \cos \frac{1}{2}(C + D),$
$\cos C + \cos D = 2 \cos \frac{1}{2}(C + D) \cos \frac{1}{2}(C - D),$
$\cos C - \cos D = -2 \sin \frac{1}{2}(C + D) \sin \frac{1}{2}(C - D).$

(e) *Product formulae: special cases*

$\sin^2 A = \frac{1}{2}(1 - \cos 2A)$,

$\cos^2 A = \frac{1}{2}(1 + \cos 2A)$,

$\sin^3 A = \frac{1}{4}(3 \sin A - \sin 3A)$,

$\cos^3 A = \frac{1}{4}(3 \cos A + \cos 3A)$.

(f) *Cosine rule*

For a triangle ABC, $c^2 = a^2 + b^2 - 2ab \cos C$.

(g) *Trigonometric equations*

In the following, n represents any integer (i.e. any whole number, positive or negative); x is in radians.

(i) $\sin x = 0$ and $\tan x = 0$ when $x = n\pi$; $\cos x = 0$ when $x = \frac{1}{2}\pi + n\pi$.

(ii) The following formulae show how to obtain all the solutions of certain equations when one solution has been obtained (for example, a hand calculator or a computer gives only one solution of $\sin x = -\frac{1}{2}$, namely $x = \arcsin(-\frac{1}{2}) = -0.5236\ldots$).

If $\sin \alpha = c$, then all the solutions of $\sin x = c$ are

$$x = n\pi + (-1)^n \alpha.$$

If $\cos \beta = c$, then all the solutions of $\cos x = c$ are

$$x = 2n\pi \pm \beta$$

If $\tan \gamma = c$, then all the solutions of $\tan x = c$ are

$$x = n\pi + \gamma.$$

The equation $\tan x = c$ occurs rather frequently: notice particularly that, if γ is a solution, then so are $\gamma \pm \pi$.

Appendix C: Areas and volumes

(a) The area of a triangle is $\frac{1}{2}bh$, where b is the length of one side and h its height from that side.

(b) The circumference of a circle is $2\pi r$, where r is its radius.

(c) The area of a circle is πr^2, where r is its radius.

(d) The area of a circle sector is $\frac{1}{2}r^2\theta$, where r is its radius and θ the angle of the sector in radians.

(e) The volume of a sphere is $\frac{4}{3}\pi r^3$, where r is its radius.

(f) The surface area of a sphere is $4\pi r^2$, where r is its radius.

(g) The volume of a cone is $\frac{1}{3}Ah$, where h is its height and A the cross-sectional area of its base.

(h) The area of an ellipse is πab, where a and b are the lengths of its semi-axes.

Appendix D: A table of derivatives

y	$\dfrac{dy}{dx}$
c (constant)	0
x^n (n any constant)	nx^{n-1}
e^{ax}	$a\,e^{ax}$
k^x $(k > 0)$	$k^x \ln k$
$\ln x$ $(x > 0)$	x^{-1}
$\sin ax$	$a \cos ax$
$\cos ax$	$-a \sin ax$
$\tan ax$	$a/\cos^2 ax$
$\cot ax$	$-a/\sin^2 x$
$\sec ax$	$(a \sin ax)/\cos^2 ax$
$\operatorname{cosec} ax$	$-(a \cos ax)/\sin^2 ax$
$\arcsin ax$	$a/(1 - a^2x^2)^{\frac{1}{2}}$
$\arccos ax$	$-a/(1 - a^2x^2)^{\frac{1}{2}}$
$\arctan ax$	$a/(1 + a^2x^2)$
$\sinh ax$	$a \cosh ax$
$\cosh ax$	$a \sinh ax$
$\tanh ax$	$a/\cosh^2 ax$
$\sinh^{-1} ax$	$a/(1 + a^2x^2)^{\frac{1}{2}}$
$\cosh^{-1} ax$	$a/(a^2x^2 - 1)^{\frac{1}{2}}$
$\tanh^{-1} ax$	$a/(1 - a^2x^2)$

Appendix E: A table of integrals

$f(x)$	$\displaystyle\int f(x)\,dx$ (C is an arbitrary constant.)
x^m $(m \neq -1)$	$\dfrac{1}{m + 1}\,x^{m+1} + C$
x^{-1}	$\ln|x| + C$, or $\ln|Cx|$
e^{ax}	$(1/a)\,e^{ax} + C$
k^x $(k > 0)$	$k^x/\ln k + C$
$\ln x$ $(x > 0)$	$x \ln x - x + C$
$\sin ax$	$-(1/a) \cos ax + C$
$\cos ax$	$(1/a) \sin ax + C$
$\tan ax$	$-(1/a) \ln|\cos ax| + C$ or $-(1/a) \ln|C \cos ax|$
$\cot ax$	$(1/a) \ln|\sin ax| + C$ or $(1/a) \ln|C \sin ax|$
$\sec ax$	$-(1/2a) \ln[(1 - \sin ax)/(1 + \sin ax)] + C$
$\operatorname{cosec} ax$	$(1/2a) \ln[(1 - \cos ax)/(1 + \cos ax)] + C$
$1/(x^2 + a^2)$	$(1/a) \arctan(x/a) + C$
$1/(x^2 - a^2)$	$(1/2a) \ln|(x - a)/(x + a)| + C$ or $(1/a) \tanh^{-1}(x/a) + C$
$1/(a^2 - x^2)^{\frac{1}{2}}$	$\arcsin(x/a) + C$ (or $-\arccos(x/a) + C$)
$1/(a^2 + x^2)^{\frac{1}{2}}$	$(1/a) \sinh^{-1}(x/a) + C$ or $\ln[x + (x^2 + a^2)^{\frac{1}{2}}] + C$
$1/(x^2 - a^2)^{\frac{1}{2}}$	$\ln[x + (x^2 - a^2)^{\frac{1}{2}}] + C$

Appendix F: A table of Laplace transforms and inverses

In the following table, n and m represent a positive integer or zero. The constants k and c are arbitrary unless otherwise indicated.

Transforms		Inverses	
$f(t)$	$F(s) = \displaystyle\int_0^\infty e^{-st} f(t)\, ds$	$F(s)$	$f(t)$
t^n	$\dfrac{n!}{s^{n+1}}$	$\dfrac{1}{s^m}$	$\dfrac{1}{(m-1)!} t^{m-1}$
e^{kt}	$\dfrac{1}{s-k}$	$\dfrac{1}{s-k}$	e^{kt}
$t^n e^{kt}$	$\dfrac{n!}{(s-k)^{n+1}}$	$\dfrac{1}{(s-k)^m}$	$\dfrac{1}{(m-1)!} t^{m-1} e^{kt}$
$\cos kt$	$\dfrac{s}{s^2+k^2}$	$\dfrac{s}{s^2+k^2}$	$\cos kt$
$\sin kt$	$\dfrac{k}{s^2+k^2}$	$\dfrac{1}{s^2+k^2}$	$\dfrac{1}{k}\sin kt$
$t\cos kt$	$\dfrac{s^2-k^2}{(s^2+k^2)^2}$	$\dfrac{s^2-k^2}{(s^2+k^2)^2}$	$t\cos kt$
$t\sin kt$	$\dfrac{2ks}{(s^2+k^2)^2}$	$\dfrac{s}{(s^2+k^2)^2}$	$\dfrac{1}{2k} t\sin kt$
$H(t-c)\ (c>0)$	e^{-cs}/s	e^{-cs}/s	$H(t-c)$
$\delta(t-c)\ (c>0)$	e^{-cs}	e^{-cs}	$\delta(t-c)$

Summary of rules

In the following rules, $F(s) \leftrightarrow f(t)$.

Scale rule (22.5). $f(kt) \leftrightarrow \dfrac{1}{k} F\left(\dfrac{s}{k}\right)$ and $F(ks) \leftrightarrow \dfrac{1}{k} f\left(\dfrac{t}{k}\right)$ $(k>0)$

Shift rule, or multiplication by e^{kt} (22.7).   If k is any constant,
$$e^{kt} f(t) \leftrightarrow F(s-k)$$

Powers of t (22.8).   If n is a positive integer, then
$$t^n f(t) \leftrightarrow (-1)^n \frac{d^n F(s)}{ds^n}.$$

Derivatives (22.12).
$$\frac{df(t)}{dt} \leftrightarrow sF(s) - f(0), \qquad \frac{d^2 f(t)}{dt^2} \leftrightarrow s^2 F(s) - sf(0) - f'(0).$$

Delay rule (22.15).   If $c > 0$, then $\mathrm{e}^{-cs}F(s) \leftrightarrow f(t - c)\mathrm{H}(t - c)$ (where H is the Heaviside unit function).

$1/s$ *as an integration operator* (23.1).   If $F(t) \leftrightarrow f(s)$, then

$$\frac{1}{s}F(s) \leftrightarrow \int_0^t f(\tau)\, \mathrm{d}\tau$$

Convolution theorem (23.11).   If $g(t) \leftrightarrow G(s)$ and $f(t) \leftrightarrow F(s)$, then

$$F(s)G(s) \leftrightarrow \int_0^t g(t - \tau)f(\tau)\, \mathrm{d}\tau \left(= \int_0^t g(\tau)f(t - \tau)\, \mathrm{d}\tau \right).$$

Appendix G: Answers to selected problems
Chapter 1

1.2. (a) $y = -2x + 3$;   (b) $y = 1$;   (c) $y = \frac{2}{3}x - \frac{1}{3}$.
Intersections are A: $(2, 1)$, B: $(\frac{5}{4}, \frac{1}{2})$, C: $(1, 1)$.

$$AB = \sqrt{\tfrac{5}{2}}, \quad AC = 1, \ BC = \sqrt{\tfrac{15}{16}}.$$

1.3. (b) Slope $= \frac{1}{3}$. Intersection with axes at $(2, 0)$, $(0, -\frac{2}{3})$.

1.4. (b) $(y + 2)/(x + 1) = -2$, so $y = -2x - 4$.
(d) $(y - 2)/(x - 1) = 3$, so $y = 3x - 1$.

1.6. Hint: choose α suitably.

1.7. (b) Centre $(1, 0)$, radius 2.
(d) Centre $(\frac{1}{2}, -\frac{1}{2})$, radius $\sqrt{7}$.

1.9. (b) $x = -\frac{3}{5} \pm \frac{1}{10}\sqrt{14}$, $y = -\frac{1}{5} \pm \frac{1}{5}\sqrt{14}$.

1.14. (b) 1.
(d) $-1/\sqrt{2}$.   (f) $-2/\sqrt{3}$.

1.16. (b) $\cos x$;   (d) $-\cos x$.

1.17. (b) $2 \cos \frac{1}{2}(x + y) \sin \frac{1}{2}(x - y)$.

1.18. In the following, n represents any integer:
(b) $\frac{1}{2}\pi + n\pi$;   (d) $\frac{1}{6} + \frac{1}{3}n$;   (f) $2n + 1$.

1.19. (b) amp. $= 1.5$; ang. freq. $= 0.2$;
period $= 31.41$; phase $= 0.48$.

1.20. (b) $\frac{1}{2}x - \frac{3}{2}$;   (d) $\arcsin \frac{1}{2}x$, $0 \leqslant x \leqslant 2$.
(f) $\arccos(\arcsin x)$, $0 \leqslant x \leqslant \sin 1$.
(h) $-\frac{1}{2} + (1 + 4x)^{\frac{1}{2}}$, $x \geqslant -\frac{1}{4}$.

1.22. (b) $\frac{1}{3} \ln 2$;   (d) $\frac{1}{3} \ln \frac{1}{3}$, or $-\frac{1}{3} \ln 3$.
(f) 2;   (h) $(e^e - 1)^{\frac{1}{2}}$;
(l) Hint: write $\sinh 2x = \frac{1}{2}(e^{2x} - e^{-2x})$ and obtain a
quadratic equation for e^{2x}. $x = \frac{1}{2} \ln(2 + \sqrt{17})$.

1.26. Hint: $x = \sinh y = \frac{1}{2}(e^y - e^{-y})$. Form an equation for e^y and solve it.

1.28. $5 \cos(2t + 0.927)$.

1.29. 1.386.

1.30. Tidal period $= 12.56$ hr. It floats for 9.20 hr.
Hint: It floats when $\sin 0.5t \geqslant -0.666$. Sketch $y = \sin 0.5t$ and $y = 0.666$ and find the intersections.

1.33. The vertex is $(-4, 7)$.

1.35. (b) $2/(x + 2) - 1/(x + 1)$.
(d) $1/2x - 1/(x + 1) + 1/2(x + 2)$.
(f) $1/4x - 1/4(x + 2) - 1/2(x + 2)^2$.
(h) $1/2(x - 3) + 1/2(x + 1)$.

1.37. (b) $1/2(x - 1) + 1/2(x^2 + 1) - x/2(x^2 + 1)$.

1.38. (b) $x - 3 - 1/(x + 1) + 8/(x + 2)$.

1.39. (b) $1 + 1/2 + 1/5 + 1/10 + 1/17$.

1.40. (b) $\displaystyle\sum_{n=2}^{6} (\tfrac{1}{3})^n = (\tfrac{1}{3})^2 + (\tfrac{1}{3})^3 + \cdots + (\tfrac{1}{3})^6$

$$= (\tfrac{1}{3})^2\{1 + \tfrac{1}{3} + \cdots + (\tfrac{1}{3})^4\}.$$

Now (1.31) gives the sum in the brackets. Finally we obtain 121/729.
(e) $-341/1024$.

Chapter 2

2.1. (b) 0.5;   (e) 2;   (g) 1.

2.2. (c) 6;   (e) -0.25;   (g) -4.

2.3. (c) $-1/x^2$;   (f) $4x$.

2.4. (c) -8.

2.5. (c) 32, -32.

2.8. (c) $dE/dT = 4kT^3$.

2.9. (b) $7x^6 - 18x^5 + 1$.

2.10. $dy/dx = \tan \alpha$, where α is the inclination angle.

2.11. Use the formula for $\tan(A - B)$ in Appendix B(b). (c) $40.2°$.

2.12. (b) $\frac{1}{2}$;   (d) 1;   (g) 2;   (i) $\pi/180 = 0.0175$.

2.15. (a) $2 \cos x + 3 \sin x$.

2.16. (b) $y = 24x - 39$;   (d) $y = e^{-1}x$.

2.17. (b) $6x - 2, 6, 0$.

Chapter 3

3.1. (b) $x \cos x + \sin x$;   (f) $2x \ln x + x$.

3.2. (b) $1/(1 + x)^2$;   (f) $(x^2 - 2x \sin x \cos x)/x^4 \cos x$.
(m) nx^{n-1}.

3.3. (d) $f\dfrac{dg}{dx} + g\dfrac{df}{dx}$, $g\dfrac{d^2f}{dx^2} + 2\dfrac{df}{dx}\dfrac{dg}{dx} + f\dfrac{d^2g}{dx^2}$,

$$g\dfrac{d^3f}{dx^3} + 3\dfrac{d^2f}{dx^2}\dfrac{dg}{dx} + 3\dfrac{df}{dx}\dfrac{d^2g}{dx^2} + f\dfrac{d^3g}{dx^3}.$$

3.4. (b) $-2 \cos x \sin x$;   (e) $2 \sin x/\cos^3 x$.
(j) $12x^2(x^3 + 1)^3$;   (n) $-3 e^{-3x}$.
(s) $a^x = e^{x \ln a}$, so $d(a^x)/dx = (\ln a)a^x$.

3.5. (f) $\frac{1}{2}x^{-\frac{1}{2}}$; (i) $-\frac{1}{2}x^{-\frac{3}{2}}$.

3.6. (f) $e^{-t}(\cos t - \sin t)$; (k) $2\sin x(\cos x - \sin x)/x^3$.

3.9. (c) $(-2x\sin x^2)/\cos x^2$. The original function only has a meaning when $\cos x^2 > 0$.

3.10. (b) $e^t(\cos t + t\cos t - t\sin t)$.

3.11. (b) $dy/dx = -y^{\frac{1}{2}}/x^{\frac{1}{2}}$. This can be written in other ways; for example put $y^{\frac{1}{2}} = 1 - x^{\frac{1}{2}}$ from the equation of the curve.

3.15. (b) -5.

3.16. (b) $dy/dx = \pm x/\sqrt{(1-x^2)}$.

Chapter 4

4.1. (b) $2t^2$; (c) $4t^3$.

4.2. (c) $x = e^{-1}$ (max.); (g) $x = 0$ (min.). (i) $x = -1/\sqrt{3}$ (min.), $x = 1/\sqrt{3}$ (max.). (o) Point of inflection at $x = 0$; (t) Points of inflection at $x = n\pi$; maxima at $x = (2n + \frac{1}{2})\pi$; minima at $(2n - \frac{1}{2})\pi$.

4.5. If base $= x$ and rectangle height $= y$, then $A = xy + \frac{1}{8}\pi x^2$ (constant), and $P = (1 + \frac{1}{2}\pi)x + 2y$. Substitute for y from the formula for A to express P in terms of x only. The minimum of P is reached when $x = [2A/(1 + \frac{1}{4}\pi)]^{\frac{1}{2}}$.

4.10. (b) $\delta y \approx -0.2$ (exact value $-0.227\ldots$). (d) $\delta y \approx -0.4$ (exact value -0.5).

4.11. (a) $\delta v \approx -0.11$; (d) $\delta A \approx -0.16$.

Chapter 5

5.1. (b) $(1 + x)^{\frac{1}{2}} \approx 1 + \frac{1}{2}x - \frac{1}{8}x^2 + \frac{1}{16}x^3$. For two decimal places, we need $|\frac{1}{16}x^3| < 0.005$, or $-0.43 < x < 0.43$; (d) To four terms,

$$\sin 2x \approx 2x - 1.333x^3 + 0.267x^5 + 0.025x^7,$$

where (for this context) the coefficients are rounded to three decimal places. For two-decimal accuracy, we need $-0.79 < x < 0.79$.

5.3. (b) The terms in the expansion of $\sin x$ are of size $|x|^{2n-1}/(2n-1)!$ with $n = 1, 2, \ldots$. We need to choose n so that this is less than 0.00005 when $x = \pm 2$. The first value within the limits is $n = 7$. The polynomial is

$$x - \frac{1}{3!}x^3 + \frac{1}{5!}x^5 - \frac{1}{7!}x^7 + \frac{1}{9!}x^9 - \frac{1}{11!}x^{11} + \frac{1}{13!}x^{13}.$$

5.4. (b) $\frac{1}{2}\pi - x$.

5.5. (b) $\frac{1}{2} - \frac{1}{4}x + \frac{1}{8}x^2 + \cdots$, $-2 < x < 2$.

(h) $1 - \frac{1}{2!}x + \frac{1}{4!}x^2 - \cdots$, valid for all x.

5.6. (b) $1 + \frac{1}{2}x - \frac{1}{8}x^2$.

5.7. (b) $\tan x \approx (x - \frac{1}{6}x^3 + \frac{1}{120}x^5)(1 - \frac{1}{2}x^2 + \frac{1}{24}x^4)$
$$\approx x - \frac{5}{6}x^3 + \frac{2}{15}x^5.$$

5.8. (d) $\ln(1 + x + x^2) = \ln[x^2(1 + 1/x + 1/x^2)]$
$$= 2\ln x + \ln(1 + 1/x + 1/x^2).$$

Then treat $1/x + 1/x^2$ as the small variable.

5.11. (b) Suppose that the first nonzero derivative is the Nth: $f^{(N)}(c) \neq 0$. Consider whether N is even or odd, and whether $f^{(N)}(c)$ is positive or negative.

Chapter 6

6.1. (b) $3 \pm j$.

6.3. (b) $3 - 5j$; (d) $9 + 3j$; (f) $1 + 6j$.

6.5. (d) $-\frac{13}{25} + \frac{9}{25}j$.

6.6. (a) $-4j$; (c) $-\frac{1}{5} + \frac{8}{5}j$.

6.7. (a) $-j$; (c) $-2j$.

6.8. (b) $16.233 - 0.167j$; (d) 88.669.

6.9. (b) $|z_2| = 8$; $\text{Arg } z_2 = \frac{3}{4}\pi$. (d) $|z_4| = 3$; $\text{Arg } z_4 = \pi$.

6.10. (b) $y = 2$; (d) The parabola, $y^2 = 4x$; (f) $y = x$.

6.11. (a) $\sqrt{2}e^{\frac{3}{4}\pi j}$; (d) $14e^{-\frac{1}{3}\pi j}$; (g) $e^2 e^j$; (j) $\sqrt{2}e^{-\frac{3}{4}\pi j}$.

6.16. (a) $2n\pi j$ $(n = 0, \pm 1, \pm 2, \ldots)$; (c) $(2n + 1)\pi j$.

6.18. (a) $e^{\frac{1}{2}\pi}$.

6.23. (a) $x^2 + y^2 + 2xy$. (d) $\cos x \cosh y - j\sin x \sinh y$.

6.28. $2 - j$, $-1 - j$, $-1 + j$.

6.29. (b) $e^{2\cos\theta}\cos(2\sin\theta)$.

Chapter 7

7.2. $x = -2$, $y = 1$.

7.6. $A = \begin{bmatrix} -10 & -5 \\ 20 & 10 \end{bmatrix}$.

7.7. $A^2 + C^2 = \begin{bmatrix} -5 & 6 & 16 \\ -8 & 11 & 2 \\ -6 & -6 & -7 \end{bmatrix}$.

7.11. $(A^{2n-1})^{-1}$.

7.16. $x = -17$, $y = -2$, $z = 8$.

Chapter 8

8.1. (c) 1;   (e) -1.

8.4. (b) 1728;   (d) -8132.

8.6. $(b - c)(c - a)(a - b)(a + b + c)$.

8.14. $x = a, b, c, -a - b - c$.

8.16. $\det(AB) = -36$, $A^{-1} = \frac{1}{2}\begin{bmatrix} 7 & 1 & -5 \\ -2 & 0 & 2 \\ 1 & 1 & -1 \end{bmatrix}$.

Chapter 9

9.15. (a) 6;   (e) $-7\hat{\jmath} - 12\hat{k}$.

9.17. $\lambda = -\frac{1}{2}$.

9.19. Area of the triangle is 5 units.

9.20. (b) $-3\hat{\imath} + 2\hat{\jmath} - \hat{k}$;   (f) $-4\hat{\imath} - 2\hat{\jmath} - 6\hat{k}$.

9.25. (b) Vector moment of force about $(-2, 1, 2)$ is $\frac{1}{3}(16\hat{\imath} - 24\hat{\jmath} - 16\hat{k})$.

9.26. Minimum speed at time $-1 + \frac{2}{3}\sqrt{3}$.

9.28. Minimum distance apart 13.4 m.

9.34. Point of intersection is $(\frac{5}{2}, 4, -\frac{1}{2})$.

Chapter 10

10.1. (c) $x_1 = 1$, $x_2 = -\frac{21}{5}$, $x_3 = -5$.
(e) $x_1 = 2$, $x_2 = -1$, $x_3 = 2$.

10.7. $x_1 = 40$, $x_2 = 88$, $x_3 = -68$, $x_4 = -59$.

10.9. (b)
$$\frac{1}{25}\begin{bmatrix} 5 & 0 & -5 \\ -6 & 10 & 1 \\ 7 & 5 & 3 \end{bmatrix}.$$

(e)
$$\begin{bmatrix} 1 & 0 & 0 & 0 & 0 \\ -1 & 1 & 0 & 0 & 0 \\ 0 & -1 & 1 & 0 & 0 \\ 0 & 0 & -1 & 1 & 0 \\ 0 & 0 & 0 & -1 & 1 \end{bmatrix}.$$

10.12. The shadow on the z plane has vertices at the points $(0, -1, 0)$, $(0, -3, 0)$, $(2, -1, 0)$.

10.16. Nontrivial solutions if $k = 1, -1, 4$.

10.18. Nontrivial solutions if $k = -6, -1, 3, 4$.

10.22. $x_1 = 1.3758$, $x_2 = 0.9732$, $x_3 = -0.1007$, $x_4 = -0.0134$.

Chapter 11

11.1. (b) Eigenvalues 4, 9; Eigenvectors $\begin{bmatrix} -3 \\ 2 \end{bmatrix}, \begin{bmatrix} 1 \\ 1 \end{bmatrix}$.

(e) Eigenvalues $3 - 4\sqrt{2}$, $3 + 4\sqrt{2}$; Eigenvectors $\begin{bmatrix} -1 - 2\sqrt{2} \\ 7 \end{bmatrix}, \begin{bmatrix} -1 + 2\sqrt{2} \\ 7 \end{bmatrix}$.

11.4. (c) Eigenvalues $(-2, 2, 3)$; Eigenvectors
$$\begin{bmatrix} 0 \\ -1 \\ 2 \end{bmatrix}, \begin{bmatrix} 1 \\ 0 \\ 0 \end{bmatrix}, \begin{bmatrix} 0 \\ 2 \\ 1 \end{bmatrix}.$$

11.7. $a = -2$.

11.14. The matrix C is given by
$$C = \begin{bmatrix} 7 & -1 & -1 \\ -1 & -1 & 1 \\ -1 & 0 & 2 \end{bmatrix}.$$

11.18. $\lim_{n \to \infty} A^n = \frac{1}{3}\begin{bmatrix} 1 & 1 & 1 \\ 1 & 1 & 1 \\ 1 & 1 & 1 \end{bmatrix}$.

11.24. Eigenvalues are 0, 4, 4, 12.

11.28. $A^{3n} = I_3$, $A^{3n+1} = A$, $A^{3n+2} = A^2$.

Chapter 12

12.1. (a) $\frac{1}{6}x^6 + C$; $\frac{3}{5}x^5 + C$; $\frac{1}{2}x^4 + C$; $\frac{4}{9}x^3 + C$; $3x^2 + C$; $3x + C$; C.
(g) $e^x + C$; $-e^{-x} + C$; $\frac{5}{2}e^{2x} + C$; $-2e^{-\frac{1}{2}x} + C$; $-\frac{3}{2}e^{-2x} + C$.
(k) $x + \ln x + C$ $((x + 1)/x = 1 + x^{-1})$; $2x - 2x^{\frac{1}{2}} + C$; $\ln|x| - 2x^{-1} - \frac{1}{2}x^{-2} + C$.

12.2. (b) $-\frac{1}{5}(1 - x)^5 + C$; $-2(1 - x)^{\frac{1}{2}} + C$; $-\frac{3}{4}(1 - x)^{\frac{4}{3}} + C$.

12.3. (b) $-\ln|1 - x| + C$; $-\frac{1}{5}\ln|4 - 5x| + C$.

12.4. (c) $\frac{3}{8}x + \frac{1}{4}\sin 2x + \frac{1}{32}\sin 4x + C$.

12.5. $x^2e^x - 2xe^x + 2e^x + C$.

12.6. (a) -2π;   (h) $-\ln 2$.

12.7. (c) $4 - x^2 \geqslant 0$ if $-1 \leqslant x \leqslant 2$, and $4 - x^2 \leqslant 0$ if $2 \leqslant x \leqslant 3$. The geometrical area is

$$[F(x)]^2_{-1} - [F(x)]^3_2,$$

where $F(x) = 4x - \frac{1}{3}x^3$.

12.8. (a) $At + B$; (b) $\frac{1}{6}t^3 + At + B$.

Chapter 13

13.1. (b)

$$\lim_{\delta x \to 0} \sum_{x=-1}^{x=1} x^5 \, \delta x = \int_{-1}^{1} x^5 \, dx = [\tfrac{1}{6}x^6]^1_{-1} = 0.$$

13.2. (b) $\displaystyle\int (x+1)^{\frac{1}{2}} \, dx = \tfrac{2}{3}(x+1)^{\frac{3}{2}} + C.$

13.3. (c) $\displaystyle\int_0^2 dx = [x]^2_0 = 2$; (i) $\tfrac{2}{3}(2^{\frac{3}{2}} - 1)$.

13.4. (b) $\displaystyle\int_{-1}^{1} (x^2 - 1) \, dx = [\tfrac{1}{3}x^3 - x]^1_{-1} = -\tfrac{4}{3}.$

13.5. (b) $\displaystyle\int_0^\infty e^{-\frac{1}{2}v} \, dv = -2[e^{-\frac{1}{2}v}]^\infty_0 = -2(0-1) = 2.$

13.6. (b) $2/\pi$; (h) $\displaystyle\int_0^T (1 - e^{-t}) \, dt = T + e^{-T} - 1.$

$$\frac{1}{T}(T + e^{-T} - 1) = 1 + T^{-1} e^{-T} - T^{-1} \to 1$$

as $T \to \infty$.

13.7. The integrands are (a) even; (b) odd; (c) even; (d) odd.

13.9. (b) The exact result is $(\pi/2)^{\frac{1}{2}}$.

13.10. (e) $\tfrac{1}{2}(x+1)^{\frac{1}{2}} \sin(x+1) - \tfrac{1}{2}x^{-\frac{1}{2}} \sin x.$

13.11. $x \leqslant -1$: $\tfrac{1}{2}$ (constant); $-1 \leqslant x \leqslant 1$: $\tfrac{1}{2}x^2$; $x \geqslant 1$: $\tfrac{1}{2}$ (constant).

Chapter 14

14.1. $5.\dot{3} \times 10^{-3}$.

14.2. $\displaystyle\int_2^4 (20 - 10t) \, dt = -20,$ $x(4) = -17.$

14.3. (b) $\tfrac{1}{2}\pi$; (g) π.

14.5. (a) $\tfrac{4}{3}\pi ab^2$.

14.6. $v = \displaystyle\int_{\frac{1}{2}}^1 \pi x^2 \, dy = \int_{\frac{1}{2}}^1 \pi(2y)^2 \, dy = \tfrac{7}{6}\pi.$

14.7. Put $x = 0$ at A; moment $= \displaystyle\int_0^L mx \, dx = \tfrac{1}{2}L^2.$

14.8. 1.27.

14.9. 0.015 gm.

14.12. A sketch shows that $x(x - 1) \geqslant -x$ if $0 \leqslant x \leqslant 2$. Therefore the area is

$$\lim_{\delta x \to 0} \sum_{x=0}^{x=2} [x(x-1) - (-x)]\delta x = \int_0^2 x^2 \, dx = \tfrac{8}{3}.$$

14.13. $\tfrac{3}{2}$.

14.14. π.

14.15. In a plane perpendicular to the end, y is downward and x is horizontal; the origin is at the top. Area elements are horizontal strips of width δy in the end face. Force $= \tfrac{1}{2}\rho g L H^2$. Moment $= \tfrac{1}{3}\rho g L H^3$.

14.16. Distance of centre of mass from vertex is $\tfrac{3}{4}H$.

14.17. $\tfrac{1}{12}\sigma a^3 b$ (σ = mass per unit area).

14.18. (a) $\tfrac{1}{4}\sigma BH^3$; (b) $\tfrac{13}{64}\sigma HB^3$, where σ is mass per unit area.

Chapter 15

15.1. (c) $-\tfrac{1}{3}e^{-3x} + C$; (f) $-\tfrac{1}{12}(3 - 2x)^6 + C$. (j) $(2x - 3)^{\frac{3}{2}} + C$; (n) $\tfrac{1}{2}\ln|2x + 3| + C$. (o) $-\ln|1 - x| + 1/(1 - x) + C$.

15.2. (b) $-\tfrac{2}{3}\cos\tfrac{1}{2}(3t - 1) + C$; (e) $-\tfrac{2}{3}(-t)^{\frac{3}{2}} + C$.

15.3. (d) $\tfrac{1}{2}\sin(x^2 + 3) + C$. (i) $\tfrac{1}{4}(1 - x^2)^{-2} - \tfrac{1}{2}(1 - x^2)^{-1} + C$.

15.4. (c) $\tfrac{1}{6}\sin^3 2x + C$. (g) Put $\cot 2x = \cos 2x/\sin 2x$, then $u = \sin 2x$, giving $\tfrac{1}{2}\ln|\sin 2x| + C$; (j) $\tfrac{1}{3}\cos^3 x - \cos x + C$.

15.5. (b) $205/32$; (e) $-\ln 2$; (h) $\tfrac{1}{2}\ln 2$. (k) Zero; (n) $(2/\omega)\cos\phi$.

15.6. (b) $\tfrac{1}{2}\pi$; (d) $\tfrac{1}{4}\pi + \tfrac{1}{2}$; (f) $\tfrac{3}{8}\pi$.

15.7. (e) $\tan x - x + C$; (f) $-x^{-1} + \arctan(x^{-1}) + C$. (k) $\tfrac{1}{2}[\arcsin x + x(1 - x^2)^{\frac{1}{2}}] + C$.

15.8. (b) $\tfrac{1}{2}\ln|x/(x + 2)| + C$. (d) $\ln|x + 1| - \tfrac{1}{2}\ln|2x + 1| + C$. (f) $\ln|x| - \tfrac{1}{2}\ln(x^2 + 1) + C$. (i) $\tfrac{1}{2}\ln[(1 + \sin x)/(1 - \sin x)] + C$.

15.9. (b) $\tfrac{1}{2}\ln(x^2 - 2x + 3) + C$; (e) $\ln(e^x + e^{-x}) + C$. (f) $2\ln(x^{\frac{1}{2}} + 1) + C$.

15.10. (b) $\frac{1}{3}x\,e^{3x} - \frac{1}{9}e^{3x} + C$.
(f) $2x\sin\frac{1}{2}x + 4\cos\frac{1}{2}x + C$.
(i) $\frac{1}{2}x^2\ln x - \frac{1}{4}x^2 + C$.
(j) $x^{n+1}[\ln x - 1/(n+1)]/(n+1)$.
(k) Hint: bring together the two terms $\int(\ln x/x)\,dx$.

15.11. (a) Hint: there are two stages required; see Example 15.20.

15.12. Hint: the same integral occurs on both sides but with a different factor.

15.13. (b) Zero;   (d) $\frac{1}{2}$;   (h) $\frac{1}{2}\pi$.

15.15. $F(0) = \frac{1}{2}\pi$, $F(1) = 1$, $F(4) = \frac{3}{16}\pi$, $F(5) = \frac{8}{15}$.

15.16. (a) $2\ln^3 2 - 6\ln^2 2 + 12\ln 2 - 6$.
(b) $F(0) = 2$, $F(1) = \pi$, $F(4) = -\pi^4 + 12\pi^2 + 48$, $F(4) = -\pi^5 + 2\pi^3 + 120\pi$.

Chapter 16

16.2. (b) $x = A\,e^{\frac{1}{2}t}$;   (e) $x = A\,e^{-\frac{4}{3}t}$;   (i) $x = A\,e^t$.

16.3. (b) $x = e^{\frac{1}{3}(t-1)}$;   (d) $x = 10\,e^{-(t+1)}$.

16.4. $I(t) = I_0\,e^{-Rt/L}$. I reduces to a fraction $1/n$ of itself in any interval of length $L\ln n/R$.

16.5. (a) $A(t) = C\,e^{-kt}$ (C arbitrary);   (b) The half-life $T = \dfrac{1}{k}\ln 2$ years. The information implies that $e^{-20k} = 1 - 0.175 = 0.825$, so $k = 0.0096$. Therefore $T = 72$ years.

16.6. If $N(t)$ is the number, then $\delta N \approx 20(\frac{1}{2}N)\delta t$ so the equation is $dN/dt = 10N$. In the second experiment there is an average death-rate of 1 per rabbit per year, so $dN/dt = 9N$.

16.7. (b) $A\,e^t + B\,e^{-2t}$;   (e) $A\,e^{t/2\sqrt 3} + B\,e^{-t/2\sqrt 3}$.
(l) $A\,e^{-3t} + Bt\,e^{-3t}$.
(n) $A + Bt$ (this is an exception to (16.10)).

16.9. (b) $\frac{2}{3}(e^t - e^{-2t})$;   (d) (The general solution is $A\,e^{-x} + Bx\,e^{-x}$), $y = e(x-1)e^{-x}$.

16.10. (b) $A\cos 3t + B\sin 3t$.
(d) $A\cos\omega_0 t + B\sin\omega_0 t$.
(f) $e^t(A\cos t + B\sin t)$.
(i) $e^{-\frac{2}{3}t}(A\cos\frac{1}{3}\sqrt 2t + B\sin\frac{1}{3}\sqrt 2t)$.

16.11. (c) $a\cos\omega_0 t + (b/\omega_0)\sin\omega_0 t$.

16.12. $\theta = \alpha\cos(g/l)^{\frac{1}{2}}t$.

16.13. The initial angular velocity $d\theta/dt$ is v/l;

$$\theta = \frac{v}{(lg)^{\frac{1}{2}}}\sin\left(\frac{g}{l}\right)^{\frac{1}{2}}t.$$

16.14. $\theta = 0.035\,e^{-0.033t}\sin 1.436t$.

16.16. $A = (Mg/P)\,e^{-\rho g(y-H)}$.

Chapter 17

17.1. (b) $-\frac{1}{3}t^3 - \frac{1}{3}t^2 - \frac{2}{9}t - \frac{11}{27}$.
(d) $\frac{3}{5}e^{-2t}$;   (i) $-\frac{2}{15}\sin 3t$.
(k) $-\frac{3}{25}\cos 2t + \frac{4}{25}\sin 2t$.

17.2. (d) $\frac{1}{5}(6\cos t - 3\sin t)$.
(f) $-\frac{2}{137}(4\cos 2t + 11\sin 2t)$.
(h) $-\frac{3}{65}e^t(7\cos 2t - 4\sin 2t)$.

17.3. (b) $-\frac{3}{4}t\cos 2t$.

17.4. (b) $\frac{1}{2}t^2\,e^t$;   (e) $\frac{1}{2}t\,e^t\sin t$.

17.5. (c) $A\,e^{\frac{1}{2}t} + B\,e^{-\frac{1}{2}t} - 1 - \frac{3}{17}\cos 2t$.
(i) $A\cos x + B\sin x + x^2 - 1 + \frac{1}{5}e^{3x}$.

17.6. (c) $-\frac{1}{2} + A\,e^{t^2}$.
(g) $(\sin x - \cos x - x\cos x + A)/(x+1)$.
(l) $(x+1)\ln|x+1| + 1 + A(x+1)$.

17.9. $11\frac{1}{2}$ minutes.

Chapter 18

18.1. (b) $3\cos(\omega t + \pi)$;   (e) $3\cos(2t + \frac{1}{2}\pi)$.
(h) $5\cos(2t + \phi)$, $\phi = -\arctan\frac{4}{3}$.

18.2. x leads y by π.

18.3. (b) (i) 0.318 cycles/sec. (ii) 0.316 cycles/sec. (iii) About 3 cycles.

18.4. (b) $C = \sqrt{(2 - \sqrt 2)}$, $\phi = \arctan[1/(\sqrt 2 - 1)]$.

18.7. The solutions are of exponential type.

18.8. $x = 2\,e^{-4t} - 8\,e^{-6t}$.

18.9. $A\,e^{-kt} + Bt\,e^{-kt}$.

18.10. (a) Period $= 0.9516$.
(b) Amplitude $= 10/[(36 - \omega^2)^2 + \omega^2]^{\frac{1}{2}}$, phase $= -\arctan[\omega/(36 - \omega^2)]$.
(c) Resonance: $\omega = 5.958$.

Chapter 19

19.1. (b) $-2\,e^{\frac{1}{2}\pi j}$ ($2\,e^{-\frac{1}{2}\pi j}$ in standard form).

19.2. (d) $2\,e^{-\frac{3}{4}\pi j}$; $2\cos(\omega t - \frac{3}{4}\pi)$.
(i) $e^{1.97j}$; $\cos(\omega t + 1.97)$.

19.3. (b) $1 - e^{-\frac{1}{2}\pi j} = 1 + j = \sqrt 2\,e^{\frac{1}{4}\pi j}$.

19.4. (b) $1 - 3\,e^{-\frac{1}{2}\pi j} + e^{\frac{1}{2}\pi j} = 1 + 4j = \sqrt{17}\,e^{j\phi}$, where $\phi = \arctan 4 = 1.33$.

19.6. (b) $R + \omega L\mathbf{j}$; (d) $R/(1 + \omega RC\mathbf{j})$.
(i) $R + j\omega L/(1 - \omega^2 LC)$.
(k) $j\omega RL/[R(1 - \omega^2 LC) + j\omega L]$.

19.7. $V = ZI$ and $V = 2$.
(d) $I = 2(1 + j\omega RC)/R$; $|I| = 2(1 + \omega^2 RC)^{\frac{1}{2}}/R$;
$\arg I = \arctan(\omega RC)$.

19.8. (b) $V_1/V_0 = \frac{2}{13}(3 - 2\mathbf{j})$; $V_0/I_1 = \frac{1}{2}(5 - \mathbf{j})$.

Chapter 20

20.4. (b) $2x^2 - y^2 = C$; (g) $y = x/(1 + Cx)$.
(k) $x = \pm 2^{-\frac{1}{2}}(C - t^3)^{-\frac{1}{2}}$ for $t^3 < C$.
(n) $\arctan y + \arctan x = \frac{1}{4}\pi$. Take the tangent of this expression and use the formula for $\tan(A + B)$; we find that $y = (1 - x)/(1 + x)$.

20.6. (b) $y = \frac{1}{16}(x^2 + C)^2$ for $x^2 + C > 0$. $y = 0$ is also a solution; (d) Those parts of the curves $y = \sin(\ln|x| + C)$ for which x and dy/dx have the same sign. Also $y = \pm 1$ are solutions.

20.7. (b) $y^3 - 3xy = C$, $y = 0$, $y = \pm\sqrt{2x}$.
(d) $xy - y^2 - x^2 = C$.
(f) $y^3 + y - x^3 = C$, $y = 0.889x$.
(h) $y + \cos y\, y + \sin x = C$; (j) $e^{x+y} + y - x = C$.

20.8. (b) $xy + y/x = C$; (d) $x/y + y - x = C$.
(f) $y/x - x/y - 1/x = C$; (g) $x^2/2y^2 + 1/xy = C$.

20.12. (b) $x^{-1}(1 + 2y^2/x^2)^{\frac{1}{4}} = C$.
(d) $(2y + x)x^2/(2y - 1)y^3 = C$.

Chapter 21

21.2. (b) $y = Cx$ (this is not covered by (21.22)).
(d) $xy = C$ (a saddle).

21.4. (b) Saddle (i.e. unstable). $m = \frac{1}{2}(-3 \pm \sqrt{13})$.
(f) Stable spiral; directions are clockwise round origin.

21.5. (b) Equilibrium points at $(1, 1)$ and $(-1, -1)$. $(-1, -1)$ is a saddle with separatrices in directions $m = 0$ and 3 from $(-1, -1)$. $(1, 1)$ is a stable spiral, anticlockwise about $(1, 1)$; (d) Equilibrium points at $(-1, 0)$, $(0, 0)$, $(0, 1)$; all centres, anticlockwise about the points.

Chapter 22

22.1. (b) $4/(s + 1)$; (d) $18/s^3 - 1/s$.
(g) $(3 - s)/(s^2 + 1)$.

22.2. (b) $1/s - 2/(s + 2)$; (e) $(3s - 4)/(s^2 + 4)$.
(g) $\frac{1}{2}[1/s - s/(s^2 + 4)]$.

22.3. (b) $1/(s + 2)^2$; (d) $(s - 2)/(s^2 - 4s + 8)$.
(i) $(s^2 - 9)/(s^2 + 9)^2$; (l) $24/(s + 1)^5$.

22.5. (b) 1; (d) $\frac{1}{8}t^4$; (g) $\frac{3}{2}e^{\frac{1}{2}t}$.
(j) $\frac{1}{2}e^t + \frac{3}{2}e^{-t}$; (m) $2\cos 2t - \frac{1}{2}\sin 2t$.
(q) $\frac{1}{2}e^t t^2$; (s) $\frac{1}{3}(\cos t - \cos 2t)$.

22.6. (e) $(2s^2 + 3s - 2)X(s) - 10s - 11$.

22.7. (b) $2e^t + e^{-2t}$; (e) $3e^{-t}\cos 2t$.
(f) $y = \frac{1}{4}e^x + \frac{1}{4}e^{-x} - \frac{1}{2}\cos x$.

22.8. (b) $3 - 3\cos t + \sin t$.
(e) $-\frac{1}{8}e^{-t} + \frac{9}{8}e^t - \frac{1}{4}t\,e^t + \frac{1}{2}t^2\,e^t$.
(i) $-\frac{7}{6}e^t + \frac{1}{2}e^{-t} + \frac{3}{4}e^{2t} - \frac{1}{12}e^{-2t}$.

22.9. (b) $x = \frac{3}{8} + \frac{5}{8}e^{4t} + \frac{1}{2}t\,e^{4t}$,
$$y = -\frac{3}{16} + \frac{3}{16}e^{4t} + \frac{1}{4}t\,e^{4t}.$$

22.10. (b) $e^t(\frac{1}{2}A + \frac{1}{2}B + \frac{3}{2}) + e^{-t}(\frac{1}{2}A - \frac{1}{2}B + \frac{3}{2}) - 3$,
where A and B are arbitrary. This is the same as $C\,e^t + D\,e^{-t} - 3$, where C and D are arbitrary.

22.13. $e^{-2}\,e^{-2s}[(s + 1)^2 - 1]/[(s + 1)^2 + 1]^2$
$$= e^{-2}\,e^{-2s}s(s + 2)/(s^2 + 2s + 2).$$

22.14. (b) $H(t)\sin t - H(t - 1)\cos(t - 1)$.

22.15. (b) $(\frac{1}{8}e^{2t} + \frac{1}{8}e^{-2t} - \frac{1}{4})H(t)$,
$$-(\frac{1}{8}e^{2(t-1)} + \frac{1}{8}e^{-2(t-1)} - \frac{1}{4})H(t - 1).$$
(d) $\frac{1}{2}H(t)t\sin t - \frac{1}{2}H(t - \pi)(t - \pi)\sin(t - \pi)$.

Chapter 23

23.3. Hint for working: $s^2 + 2ks + \omega^2$ has real factors when $k^2 > \omega^2$; so put $s^2 + 2ks + \omega^2 = (s - \alpha)(s - \beta)$, where $\alpha, \beta = -k \pm (k^2 - \omega^2)^{\frac{1}{2}}$. Then $x(t)$ is given by
$$(\alpha - \beta)^{-1}[(\alpha + \kappa)\,e^{\alpha t} - (\beta + \kappa)\,e^{\beta t}]H(t)$$
$$+ I(\alpha - \beta)^{-1}[\,e^{\alpha(t - t_0)} - e^{\beta(t - t_0)}]H(t - t_0),$$
where $\kappa = 1 + 2k$.

23.4. By proceeding as suggested, we obtain
$$u(x) = Ax + \frac{1}{6}Bx^3 + (Mg/6K)(x - \frac{1}{2}l)^3 H(x - \frac{1}{2}l).$$
The conditions at $x = l$ give $A = -\frac{1}{12}l^2$, $B = -\frac{1}{2}$. This problem could be solved by integrating the equation four times, and linking the solutions over $[0, \frac{1}{2}l]$ and $[\frac{1}{2}l, l]$ by the condition that $u(x)$, $u'(x)$, $u''(x)$ are continuous at $x = \frac{1}{2}l$, but this is automatically secured in the Laplace-transform method.

23.5. (b) $2s/(6s^2 + s + 1)$.

23.6. (b) $V_2/V_1 = 3/(20s^2 + 12s + 5)$;
$$V_2/I = 3/(4s^2 + 1).$$

23.7. (b) t; (f) $1 - \cos t$; (h) $\frac{1}{2}(-t\cos t + \sin t)$.
(j) $n!m!t^{n+m+1}/(n + m + 1)!$.

23.8. (b) $\dfrac{1}{2\omega}\displaystyle\int_0^t f(\tau)(e^{\omega(t-\tau)} - e^{-\omega(t-\tau)})\,d\tau.$

23.9. (b) $e^t.$

Chapter 24

24.1. (b) $a_n = 0,\quad b_n = -(-1)^n/n.$

(e) $a_n = 0,\quad b_n = \dfrac{2}{\pi n}[1 + (-1)^n - 2\cos(\tfrac{1}{2}n\pi)].$

24.2. (b) $b_n = 0,\quad a_0 = \dfrac{2\pi^2}{3},$

$a_n = \dfrac{4}{n^2}(-1)^n \ (n = 1, 2, \ldots).$

(c) $b_n = 0,\quad a_n = \dfrac{4(-1)^n}{\pi(4n^2 - 1)}.$

24.3. (a) $a_0 = \tfrac{1}{2}\pi,$

$a_{2n} = 0,\quad a_{2n-1} = -\dfrac{2}{\pi n^2},$

$b_n = -\dfrac{(-1)^n}{n}\quad (n = 1, 2, \ldots).$

24.5. Series sum is $\tfrac{1}{4}\pi.$

24.8. $F = 2.$

24.10. $a_0 = 0,\qquad a_n = 0,\quad b_n = \dfrac{4\beta}{\pi n^3}[1 - (-1)^n]$

$(n = 1, 2, \ldots).$

24.16. (a) $\displaystyle\sum_{n=1}^{\infty}\dfrac{4}{(2n-1)\pi}\sin(2n-1)\pi t.$

24.18. $\dfrac{2}{\pi} - 4\displaystyle\sum_{n=1}^{\infty}\pi(4n^2 - 1)\cos 2n\omega t.$

24.23. (b) $R(t) = \dfrac{1}{2} + \dfrac{4}{\pi^2}\left(\cos t + \dfrac{1}{3^2}\cos 3t\right.$

$\left. + \dfrac{1}{5^2}\cos 5t + \cdots\right).$

24.26. (b) $\displaystyle\sum_{n=0}^{\infty}\dfrac{1}{(2n+1)^2} = \dfrac{\pi^2}{8}.$

Chapter 25

25.3. (c) $4x - 2y - 1;\ -6y - 2x - 1.$
(f) $y - 2;\ x - 1;$ (i) $2y/(x+y)^2;\ -2x/(x+y)^2.$
(k) $x(x^2 + y^2)^{-\frac{1}{2}};\ y(x^2 + y^2)^{-\frac{1}{2}}.$

25.4. (c) $\dfrac{\partial V}{\partial x} = g'(r)\cos\theta;\ \dfrac{\partial V}{\partial y} = g'(r)\sin\theta.$

25.8. $\partial^2 f/\partial x^2,\ \partial^2 f/\partial y^2,$ and $\partial^2 f/\partial x\,\partial y = \partial^2 f/\partial y\,\partial x$
are given in order: (b) 2, 4, 3. (d) $2y/x^3,\ 0,\ -1/x^2.$
(h) $108(3x - 4y)^2,\ 192(3x - 4y)^2,\ 144(3x - 4y)^2.$
(k) $-r^{-3} + 3x^2 r^{-5},\ -r^{-3} + 3y^2 r^{-5},\ 3xyr^{-5},$ where
$r = (x^2 + y^2)^{\frac{1}{2}}.$

25.10. (b) $2x + 2y - z = 4;$ one normal is $(2, 2, -1).$
(d) $3x + 4y + 8z = 41;$ one normal is $(-\tfrac{3}{8}, -\tfrac{1}{2}, -1).$

25.11. $78.9°$ or $101.1°.$

25.12. (b) $(1, -1)$, min; (d) $(n\pi, m\pi)$; max. if n and
m odd, min if n and m even, otherwise neither.
(h) $(0, 0)$ neither; $(1, 1)$ minimum; (k) $(0, 0)$, neither.

25.14. (a) $a = b = c = 7;$ (b) $a = b = c = 4.$

25.15. The maximum is 9, attained at $(2, \pm 1).$

25.16. Minimum distance $= \sqrt{2}.$

Chapter 26

26.1. (b) $\delta z = 0.0718\ldots$ (exactly). The incremental
approximation gives $\delta z \approx 0.0784.$ Percentage error $=$
$9.1\%.$

26.3. (b) $(x\,\delta y^2 - y^2\,\delta x\,\delta y)/y^2(y + \delta y).$

26.6. $-5\%.$

26.7. $1.67°$ reduction, approximately.

26.9. (b) $-2\sqrt{2};$ (d) Zero (it is the same in all
directions).

26.10. (b) $-\tfrac{3}{4};$ (e) $-\tfrac{1}{2};$ (j) 1.

26.12. (b) $x_1 x/a^2 + y_1 y/b^2 = x_1^2/a^2 + y_1^2/b^2.$
(f) $ax_1 x + h(y_1 x + x_1 y) + byy_1 + g(x + x_1) +$
$f(y + y_1) + c = 0.$

26.16. (b) $x^{-1} - y^{-1} = $ const; (d) $e^x + e^y = $ const.

26.17. (b) $y^2 - x^2 = b^2 - a^2.$

26.18. (b) Depth $= 2^{-\frac{2}{3}}V^{\frac{1}{3}};$ square base, side $2^{\frac{1}{3}}V^{\frac{1}{3}}.$

26.19. (b) $49.8°$ or $130.2°.$
(d) Hint: compare Problem 26.12f.

26.20. (d) $-(x^2 + y^2)/y^3.$

26.21. (b) $(0, \tfrac{1}{2});$ (d) $(-\tfrac{1}{4}, 1).$

26.22. (b) $(4, 2);$ (d) $(2x_0/a^2, 2y_0/b^2).$

26.23. (b) $\phi = 0.$

Chapter 27

27.2. (b) $-4\sin t\cos t;$ (d) $2\sin(t^2) + 4t^2\cos(t^2).$

27.3. It is easiest to start by expressing the distance

D in terms of polar coordinates (r, θ), (R, ϕ) by using the cosine rule (Appendix B(f)). Then

$$\frac{dD}{dt} = \frac{(Rv - rV)\sin(\phi - \theta)}{[R^2 + r^2 - 2Rr\cos(\phi - \theta)]^{\frac{1}{2}}},$$

where $\theta = vt/r$, $\phi = Vt/R$.

27.4. (b) $x = y = 3$;   (e) The coordinates of the nearest point on the given line are $(\frac{3}{5}, \frac{1}{5})$. Distance $= \sqrt{13}/5$.

27.5. (b) $(0, 0)$, $(2, 0)$. (A suitable parametrization is $x = 1 + \cos t$, $y = \sin t$.)
(d) $(\pm 6/\sqrt{5}, \pm 4/\sqrt{5})$. (A suitable parametrization would be $x = 2/\cos t$, $y = 2\tan t$.)

27.8. (b)
$$\ddot{x} = -2\dot{r}\dot{\theta}\sin\theta + \ddot{r}\cos\theta - \dot{\theta}^2 r\cos\theta - \ddot{\theta}r\sin\theta,$$
$$\ddot{y} = 2\dot{r}\dot{\theta}\cos\theta + \ddot{r}\sin\theta - \dot{\theta}^2 r\sin\theta + \ddot{\theta}r\cos\theta.$$

27.9. (c) $\partial f/\partial u = 2v^2/u^3$, $\partial f/\partial v = 2v/u$.

27.10. (b) $\partial^2 f/\partial u^2 = 12u^2 - 2v^2$, $\partial^2 f/\partial u\,\partial v = -4uv$, $\partial^2 f/\partial v^2 = -2u^2 + 12v^2$.

27.11. It is easiest to put $x^2 - y^2$ in terms of uv. Finally,
$$\partial^2 f/\partial u^2 = 16v^2 g''(4uv), \quad \partial^2 f/\partial v^2 = 16u^2 g''(4uv),$$
$$\partial^2 f/\partial u\,\partial v = 4g'(4uv) + 16uvg''(4uv).$$

Chapter 28
28.1. (b) $\delta f \approx -x(x^2 + y^2)^{-\frac{3}{2}}e^{-t}\,\delta x$
$$- y(x^2 + y^2)^{-\frac{3}{2}}e^{-t}\,\delta y - (x^2 + y^2)^{-\frac{1}{2}}e^{-t}\,\delta t.$$
(e) $\delta f \approx 2(x_1 - x_2)\,\delta x_1 - 2(x_1 - x_2)\,\delta x_2 +$
$$2(y_1 - y_2)\,\delta y_1 - 2(y_1 - y_2)\,\delta y_2.$$

28.2. -0.07.

28.3. It is easiest to write $\delta(1/R) \approx -\delta R/R^2$. We obtain $\delta R \approx 0.198\,\delta R_1 + 0.018\,\delta R_2 + 0.334\,\delta R_3$. The required δR_3 is -0.108.

28.4. Put $ax^3 - 3x - 45 = f(a, b, c, x)$ and use (28.1).

28.5. (a) Hint: use logarithmic differentiation: $\delta w \approx -\frac{3}{2}\delta x + \frac{3}{2}\delta z$ and $\delta w \approx 2(\pm 0.3)$. What is the significance of the absence of a term in δy?

28.6. (b) Maximum $|\delta w| \approx 0.14$; max. percentage error 10%.

28.8. (b) $2\delta x + 4\delta y - 6\delta z = 0$. For $\partial z/\partial x$, put $\delta y = 0$: $\partial z/\partial t = \frac{1}{3}$. Similarly $\partial z/\partial y = \frac{2}{3}$.

28.11. (b) $(2, -3, 5)$;   (d) $(3x^2, 0, 9z^2)$.
(f) $(-x/r^3, -y/r^3, -z/r^3)$, where $r = (x^2 + y^2 + z^2)^{\frac{1}{2}}$.

28.12. (b) $(0, 2y, 2z)$. Unit vector $= (0, y/(y^2 + z^2)^{\frac{1}{2}}, z/(y^2 + z^2)^{\frac{1}{2}})$.

28.13. (b) $\cos\phi = 11/3\sqrt{14}$, so $\phi = 11.5°$ (i.e. the angle of intersection of smallest magnitude).

28.15. (b) $\hat{s}\cdot(2x, -2y, -3)$.

28.16. (b) (Check that $\hat{s}$ as given is a unit vector.)
$$\frac{df}{ds} = 7.51.$$

28.17. (b) $-2\hat{\imath} - 2\hat{k}$.

28.18. (b) $(\pm 1, 0, 0)$ and $(\pm 1, \frac{1}{4}, \frac{3}{16})$.
(d) $x = y = z$ is a line of stationary points (excluding the origin).
(e) $x = y = z = \pm 1/\sqrt{3}$, $\lambda = \pm\frac{1}{2}\sqrt{3}$.

28.12. $t = \frac{1}{2}\pi + n\pi$. These are alternately max. and min. Since $x^2 + y^2 + z^2 = 1 + \sin^2\frac{1}{2}t$ has period 4π, $n = 0, 1, 2, 3$ supplies the only four distinct points.

28.21. (b) $(3, 3, 3)$;   (e) $(a/\sqrt{3}, b/\sqrt{3}, c/\sqrt{3})$.
(g) $(\frac{4}{3}, \frac{7}{3}, 1)$.

28.26. (b) $y = 1/2x$;   (d) $x^2 + y^2 = 1$.

Chapter 29
29.1. (b) $e - 2$;   (g) $(d - c)(b - a)$;   (i) $-\frac{1}{3}$.
(l) $\frac{1}{2}\ln 2$.

29.2. (b) Zero. Refer to the signed volume analogy (29.2b);   (f) $\ln\frac{27}{4} - 1$.

29.4. $\frac{4}{3}$.

29.5. (b) $\displaystyle\int_0^1 \int_0^x f(x, y)\,dy\,dx$.

(d) $\displaystyle\int_{-1}^1 \int_0^{\sqrt{(1-x^2)}} f(x, y)\,dy\,dx$.

(g) $\displaystyle\int_{-1}^0 \int_0^{1+y} f(x, y)\,dx\,dy + \int_0^1 \int_0^{1-y} f(x, y)\,dx\,dy$.

29.6. (b) $\frac{2}{3}$;   (d) $\frac{3}{2}$;   (h) $\frac{1}{12}$.

29.7. (b) 1.

29.8. (b) $\frac{1}{4}\pi$;   (d) $\frac{15}{8}$;   (f) $-\frac{1}{4}\pi^2$.

Chapter 30
30.1. (b) 1.

30.2. (b) $\frac{4}{3}$;   (d) $\frac{2}{3}$;   (f) 0.

30.3. (b) π; (d) $\frac{8}{3}\pi^3$.

30.5. (b) 2; (d) 0; (g) 0.

30.6. (b) 1; (d) 3; (f) $\frac{15}{2}$.

30.7. (b) 0.

30.8. (b) $-\frac{3}{2}$.

30.9. Zero.

30.10. Put $x = x(u, v)$ and $y = y(u, v)$, where u and v are the new coordinates. Then put $dx = \dfrac{\partial x}{\partial u}\,du + \dfrac{\partial x}{\partial v}\,dv$ etc.

30.14. $\frac{3}{8}\pi$.

30.16. (b) nonconservative.

Chapter 31

31.1. (c) $-2, -1, 1, 2, 3, 4$; (f) $1, 4, 9$.

31.3. (c) $A \cup B = \{-4, -3, -2, -1, 1, 2, 3, 4\}$.

31.4. (b) $A \cap B = \{x \mid x \in \mathbb{N}^+ \text{ and } -5 \leqslant x \leqslant 2\}$.
(d) $A \cap B = \{1\}$.

31.5. (b) $B \setminus (A \cup C)$; (d) $(B \cup C) \setminus A$.

31.6. (b) $A_1 \setminus (A \cup A_2 \cup \cdots \cup A_r)$.

31.7. (b) $[(A \setminus A_1) \setminus B_1] \cup B_2$.

31.10. $A^2 = \{(1, 1), (1, 2), (2, 1), (2, 2)\}$.

31.13. (b) 66.

31.14. $A \cap B = \{(1, 4), (4, 1)\}$.

31.15. (a) $B \setminus (A \cup C)$.

Chapter 32

32.6. See table below.

32.10. (b) $(a \cdot b) \cdot (b \oplus c)$; (d) $\overline{(\bar{a} \cdot b \oplus a \cdot b)}(c \cdot d)$.

32.15. (a) If a_1 represents the state of switch S_1, etc., then the switching function is

$$(a_1 \oplus a_2) \oplus [(a_3 \oplus a_4) \cdot a_5].$$

32.16. See the table.

Solution for 32.6

a	b	c	$(a \oplus \bar{b}) \cdot (a \oplus \bar{c})$
0	0	0	1
0	0	1	0
0	1	0	0
0	1	1	0
1	0	0	1
1	0	1	1
1	1	0	1
1	1	1	1

Solution for 32.16

a_1	a_2	a_3	f
0	0	0	0
0	0	1	1
0	1	0	1
0	1	1	0
1	0	0	1
1	0	1	0
1	1	0	0
1	1	1	1

Chapter 33

33.3. Twenty are planar.

33.4. Five are connected.

33.8. Four not including reversed order.

33.13. There are three different paths between a and e.

33.14. Five vertices.

33.15. $i_1 = -\frac{1}{3}i_0$, $i_2 = -\frac{1}{12}i_0$, $i_3 = -\frac{5}{6}i_0$, $i_4 = \frac{11}{12}i_0$, $i_5 = \frac{2}{3}i_0$, $i_6 = \frac{1}{3}i_0$, $i_7 = -\frac{3}{4}i_0$.

33.16. (b) $i_1 = -\frac{11}{24}$, $i_2 = \frac{1}{8}$, $i_3 = \frac{1}{3}$, $i_4 = \frac{11}{24}$, $i_5 = \frac{3}{8}$, $i_6 = -\frac{1}{4}$, $i_7 = 0$, $i_8 = 0$.

33.17. The transfer function is

$$Q = \frac{PG_1G_2G_3}{1 - G_2H_1 + G_1G_2G_3H_2}.$$

33.19. (a) The transfer function is

$$Q = \frac{PG_1G_2G_3}{1 + G_2H_1 + G_1G_2G_3H_2}.$$

33.20. (a) $\dfrac{G_1G_3}{(1 - G_1G_2H_2)(1 + G_3H_1)}$.

(d) $\dfrac{G_1G_2G_3G_4}{1 + G_2G_3H_2} + \dfrac{G_5G_6G_7}{1 - H_1}$.

33.23. $SAFT$, length 12.

Chapter 34

34.1. £1790.85, 4.87%.

34.2. (b) 16.9 years.

34.3. (b) $0, \frac{1}{6}(-1 \pm \sqrt{13})/6$.

34.6. $f(n) = (\ln n)/\ln 2$.

34.8. (b) $u_n = A3^n + B(-3)^n$.
(c) $u_n = 3^n(A \cos \frac{1}{2}n\pi + B \sin \frac{1}{2}n\pi)$.

34.8. (i) $u_n = A + Bn + C(-1)^n$.

34.11. (a)(ii) $u_n = \frac{12}{7}n - \frac{8}{7}n^2$.; (b)(ii) $u_n = \frac{1}{5}n + \frac{11}{25}$.
(c)(iii) $u_n = \frac{1}{2}n^2$.
(d)(iii) $u_n = -\frac{1}{9}n^2 3^n$.

34.13. $D_n(1) = 2n - 1$.

34.16. $u_n = \frac{5}{4} - \frac{4}{9}(-\frac{1}{3})^n$.

34.17. $d_k = k(N - k)$.

34.19. $s_n = \frac{1}{4}n^2(1 + n)^2$.

34.22. $0 < \alpha < 1$.

34.23. Oscillates between 0.4953 and 0.8124.

34.24. The periodic values of the 2-cycle are 0.4 and 0.8.

Chapter 35

The answers contain suggested *Mathematica* commands for solutions to the projects. Some full programs and outputs are also included for selected problems. Note that there are frequently several ways of presenting answers using *Mathematica* programs. Inevitably these answers and comments are brief: a working knowledge of *Mathematica* using mainly built-in commands is required.

35-1.1. Use the **Plot**[ ] command.

35-1.2. Use **Table**[ ] to define the set of points and **ListPoint**[ ] to plot the points. The curve can be plotted using **Plot**[ ] and all displayed with **Show**[ ].

35-1.3. Here is a program and a selected output.

In[1]:=

```
<<Graphics`ImplicitPlot`
```

In[3]:=

```
ImplicitPlot[x^4+2y^2-x y-2 y x^2==4,{x,-4,4},
AxesLabel->{x,y},PlotRange->{{-2,2.2},{-1.5,3.2}}]
```

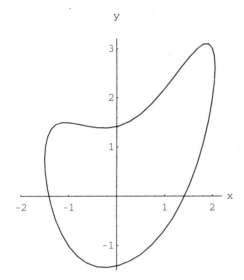

Out[3]=

```
-Graphics-
```

35-1.4. Use **Plot**[ ] commands.

35-1.5. Use **Plot**[ ] commands.

35-1.6. Here is a program which uses the command **ParametricPlot**[ ].

The polar variables are (r,t): equations are defined by r1[t] and r2[t].

```
r1[t_]:=.5 (1+Cos[t])
```

This the graph of the cardioid.

```
ParametricPlot[{r1[t] Cos[t],r1[t] Sin[t]},{t,0,2 Pi},
AspectRatio->1]
```

This is the graph of the folium.

```
r2[t_]:=Cos[t]*(4*(Sin[t])^2-1)
```

```
ParametricPlot[{r2[t] Cos[t],r2[t] Sin[t]},{t,0,2 Pi},
AspectRatio->1]
```

35-1.7. Partial fractions can be generated by **Apart**[ ].

35-2.1. Limits can be obtained by the **Limit**[ ] command.

35-2.2. Use either **D**[ ] or $f'[x]$.

35-2.3. Numerical solutions can be found for the equation $g(x) = 0$ by using **NSolve**$[g[x] = = 0]$, and curves plotted by **Plot**[ ].

35-2.4. Define the function $f[x] = x \sin[2 * x]$ and a tangent function.

tangent$[x, a] = f'[a] * (x - a) + f[a]$.

35-2.5. Use **D**[ ].

35-2.6. Use **Plot**[ ].

35-3.1. Product rule given by **D**$[f[x] * g[x], x]$.

35-3.2. Stationary values can be found by using **FindRoot**[ ].

35-3.3. Implicit differentiation can be implemented using **Dt**[ ] and the explicit derivative obtained by **Solve**[. . . ,**Dt**[ ]].

35-4.1. Use **D**[ ] command.

35-4.2. A possible program is included here.

This line defines the function

```
f[x_]=0.1*x^5-0.5*x^4+0.2*x^3+x^2-0.7*x+2.2
```

This draws its graph between x=-2 and x=4.

```
Plot[f[x],{x,-2,4},AspectRatio->Automatic,PlotRange->
{-2.8,3.5},AxesLabel->{x,y}]
```

D[ ] determines its derivatives

```
D[f[x],x]
D[f[x],{x,2}]
```

The following commands find the values of x at the four stationary values. Initial estimates for the FindRoot[ ] commands are estimated from the graph above.

```
FindRoot[f'[x]==0,{x,-1}]
FindRoot[f'[x]==0,{x,.5}]
FindRoot[f'[x]==0,{x,1.5}]
FindRoot[f'[x]==0,{x,3}]

x1=-0.93972;x2=0.352685;x3=1.27565;x4=3.31138;
```

The following values of the second derivatives check the second derivative test for each stationary point.

```
f''[x1]
f''[x2]
f''[x3]
f''[x4]

Plot[f''[x],{x,-2,4},AspectRatio->Automatic,
PlotRange->{-2.8,3.5}]
```

The following commands determine the points where f"(x) is zero, the points of inflection on the curve.

```
FindRoot[f''[x]==0,{x,-0.5}]
FindRoot[f''[x]==0,{x,1}]
FindRoot[f''[x]==0,{x,2.5}]
```

35-4.3. Use **Plot[ ]**

35-4.4. Estimate the location of the root using **Plot[ ]**, define a function

$$h[x] = x - f[x]/f'[x],$$

and then use **NestList[ ]**.

35:4.5. A suggested program is included here.

The graphs of y=x+sin(5x) drawn for 0<x<25 in (a) the default Plot[ ] form, (b) Plot[ ] with 20 plot points, (c) Plot [ ] with 50 plot points. Notice the differences.

```
curve1=Plot[x+Sin[5*x],{x,0,25},AxesLabel->{x,y}]
```

```
curve2=Plot[x+Sin[5*x],{x,0,25},PlotPoints->20,
PlotStyle->RGBColor[0,0,1],AxesLabel->{x,y}]
```

```
curve3=Plot[x+Sin[5*x],{x,0,25},PlotPoints->50,
PlotStyle->RGBColor[1,0,0],AxesLabel->{x,y}]
```

```
Show[curve1,curve2,curve3]
```

The default plot misses a beat around x=21, whilst a scheme with 20 plot points misses the oscillations completely. Plot[ ] with 50 plot points is sufficient to detect all the oscillations of the functions. The Plot[ ] algorithm can be deceived.

35-5.1. Series[ ] defines the Taylor series up to any required number of terms. **Normal[ ]** defines the corresponding Taylor polynomial.

35-5.2. Use **Series[ ]** and **Normal[ ]**.

35-5.3. A possible program and output is included here.

The function is ((sin x)/x)^2.

In[1]:=

```
Series[((Sin[x])/x)^2,{x,0,2}]
```

Out[1]=

$$1 - \frac{x^2}{3} + O[x]^3$$

In[2]:=

```
f1[x_]=Normal[%]
```

Out[2]=

$$1 - \frac{x^2}{3}$$

In[3]:=

```
Series[((Sin[x])/x)^2,{x,0,4}]
```

Out[3]=

$$1 - \frac{x^2}{3} + \frac{2 x^4}{45} + O[x]^5$$

In[4]:=

```
f2[x_]=Normal[%]
```

Out[4]=

$$1 - \frac{x^2}{3} + \frac{2 x^4}{45}$$

In[5]:=

```
Series[((Sin[x])/x)^2,{x,0,6}]
```

Out[5]=

$$1 - \frac{x^2}{3} + \frac{2 x^4}{45} - \frac{x^6}{315} + O[x]^7$$

In[6]:=

```
f3[x_]=Normal[%]
```

Out[6]=

$$1 - \frac{x^2}{3} + \frac{2 x^4}{45} - \frac{x^6}{315}$$

The actual curve is in black and the approximations in various levels of gray.

In[10]:=

```
Plot[{f1[x],f2[x],f3[x],(Sin[x]/x)^2},{x,0.001,3},
PlotRange->{0,1.1},PlotStyle->{{GrayLevel[.2]},
{GrayLevel[.4]},{GrayLevel[.6]},{GrayLevel[0]}},
AxesLabel->{x,""}]
```

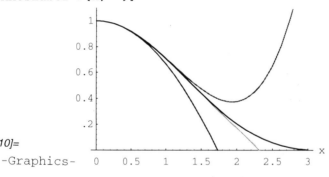

Out[10]= -Graphics-

35-5.4. Use **Series[ ]** followed by **Normal[ ]**, **FindRoot[ ]** and **Plot[ ]**.

35-6.1. Use **Solve[ ]**.

35-6.2. Requires package **Algebra`Relm`** and command **ComplexExpand[ ]**.

35-6.3. List the roots using **Table[ ]** and plot them using **ListPlot[ ]**.

35-6.4. Use **ListPlot[ ]**.

35-6.5. **Abs[ ]** and **Arg[ ]** give the modulus and principal argument of a complex number.

35-7.1. Matrices are entered as lists by rows. Transpose and Inverse are given by the commands **Transpose[ ]** and **Inverse[ ]**.

35-7.2. Requires the **Inverse[ ]**.

35-7.3. Can use either products or **MatrixPower[ ]** to calculate powers of matrices.

35-8.1. Determinant of a matrix is given by **Det**[ ].

35-8.2. Use **Det**[ ] followed by **Factor**[ ].

35-8.3. Use commands **Det**[ ] and **Solve**[ ].

35-9.1. Requires package **LinearAlgebra`CrossProduct`**. The Use **Cross**[ ] for vector products and **Polygon**[ ] to show the triangle.

35-9.2. A program is given below.

Requires package LinearAlgebra`CrossProduct`.

```
<<LinearAlgebra`CrossProduct`
a={1,-1,2};b={-1,2,3};c={2,-1,3};d={1,3,-2};
p1=Cross[b,c]+Cross[c,d]+Cross[d,b];
p2=Cross[c,d]+Cross[d,a]+Cross[a,c];
p3=Cross[d,a]+Cross[a,b]+Cross[b,d];
p4=Cross[a,b]+Cross[b,c]+Cross[c,d];
```

Uses the area formula from the previous program.

```
area=0.5*(Sqrt[p1.p1]+Sqrt[p2.p2]+Sqrt[p3.p3]+
Sqrt[p4.p4])
N[%]
```

The following steps define the triangular faces of the tetrahedron

```
poly1=Polygon[{b,c,d}];
poly2=Polygon[{a,c,d}];
poly3=Polygon[{a,b,d}];
poly4=Polygon[{a,b,c}];
```

The next instruction displays the tetrahedron. Try different viewpoints.

```
Show[Graphics3D[{poly1,poly2,poly3,poly4}],
Axes->True,AxesLabel->{x,y,z}
,ViewPoint->{2.100,-2.400,1.500},
AspectRatio->.8]
```
INP9-2

35-9.3. Plot the path using **ParametricPlot3D**[ ].

35-10.1. Can use **LinearSolve**[ ] to solve linear equations.

35-10.2. **RowReduce**[ ] reduces the matrix of coefficients to echelon form.

35-10.3. Use **RowReduce**[ ].

35-11.1. Use commands **Eigenvalues**[ ] and **Eigenvectors**[ ].

35-11.2. Program is listed below.

```
a={{1,2,1},{2,1,1},{1,1,2}};
Eigenvalues[a]
f=Eigenvectors[a]
```

The transpose command creates the matrix C.

```
c=Transpose[f]
```

And its inverse.

```
Inverse[c]
```

```
d=DiagonalMatrix[{-1,1,4}]
c.MatrixPower[d,n].Inverse[c]
```

The next line works out the n-th power of A directly

```
MatrixPower[a,n]
MatrixPower[a,n]-c.MatrixPower[d,n].Inverse[c]
```

The last steps check that the two are the same.

35-11.3. Check **Inverse**[ ]−**Transpose**[ ].

35-11.4. Use **Eigenvalues**[ ], **Det**[ ], **IdentityMatrix**[ ], and **MatrixPower**[ ].

35-12.1. Use **Plot**[ ].

35-12.2. Command **Integrate**[ ] will perform the necessary integration.

35-13.1. A sample program is given below.

This method simply divides the interval into equal steps and the approximation is the sum the products of the step lengths and the function values at the step points.

```
f[x_]=x^2;
a=1;b=3;h=(b-a)/100;
```

The next step works out the sums

```
N[Sum[h*f[a+i*h],{i,0,99}]]
```

The next instruction evalautes the integral. These steps are repeated for each function except for (d) which does not have an elementary integral.

```
N[Integrate[f[x],{x,1,3}]]
g[x_]=x*Exp[-x];
b=3;a=0;h=(b-a)/20;
N[Sum[h*g[a+i*h],{i,0,19}]]
N[Integrate[x*Exp[-x],{x,0,3}]]
p[x_]=x^3*Sin[x];
b=Pi;a=0;h=(b-a)/30;
N[Sum[h*p[a+i*h],{i,0,29}]]
N[Integrate[x^3*Sin[x],{x,0,Pi}]]
q[x_]=Cos[Exp[-x]];
b=1;a=0;h=(b-a)/25;
N[Sum[h*q[a+i*h],{i,0,24}]]
```

Try the sums with different values for h.

35-13.2. Essentially use **Integrate**[ ], although simplification is possible with commands **Simplify**[ ], **Together**[ ], and **ExpandAll**[ ].

35-13.3. Use **Integrate**[ ].

35-13.4. Try **Integrate**[ ] for a suitable definite integral followed by **Limit**[ ].

35-13.5. Program and output shown below.

The following diagrams show sphere and the cylindical hole.

In[14]:=

```
s1=ParametricPlot3D[{Cos[t]*Cos[u],Sin[t]*Cos[u],
Sin[u]},{t,0,2*Pi},{u,-.35*Pi,.35*Pi},Boxed->False,
Axes->False,DisplayFunction->Identity]
s2=ParametricPlot3D[{0.454*Sin[t],0.454*Cos[t],u},
{t,0,2*Pi},
{u,-.891,.891},Boxed->False,Axes->False,
DisplayFunction->Identity]
Show[s1,s2,Shading->False,DisplayFunction->
$DisplayFunction]
```

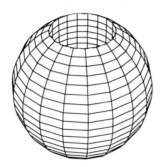

Explain why the volume is given by the following integral.

In[17]:=

```
Integrate[Pi*(a^2-x^2-b^2),
{x,-Sqrt[a^2-b^2],Sqrt[a^2-b^2]}]
```

Out[17]=

$$\frac{4\,(a^2 - b^2)^{3/2}\,Pi}{3}$$

35-14.1. Requires package **Graphics`Graphics`** and then **PolarPlot**[ ] and **Integrate**[ ].

35-14.2. Possible program show below.

```
f[x_]=Sin[x]^2*Exp[-2*x];
b=2;a=0;n=10;
```

The following steps define the trapezium rule given by TrapI.

```
h=(b-a)/n;
TrapI=N[h*(0.5*f[a]+Sum[f[a+i*h],{i,1,n-1}]+0.5*f[b])]
Int=Integrate[Sin[x]^2*Exp[-2*x],{x,0,2}]//N
TrapI-Int
```

The last step checks the difference between the result form the trapezium rule and the exact integral,which is possible in this example. Try different values for the number of steps n.

35-14.3. Uses **Plot**[ ] and **Integrate**[ ].

35-14.4. Program given below.

A program for Simpson's rule.
```
f[x_]=Exp[x^2];
b=1;a=0;n=4;
h=(b-a)/n;
Simp=N[h*(f[a]+Sum[(3+(-1)^(i+1))*f[a+i*h],
{i,1,n-1}]+f[b])/3]
```

35-15.1. Program given below.

The substitutions are x=g(u) and u=j(v).
```
f[x_]=(x-2)/Sqrt[5+4*x-x^2];
g[u_]=u+2;
x=g[u];
j[v_]=3*Sin[v];
u=j[v];k[v_]=f[g[j[v]]]D[g[j[v]],v];Simplify[%]
m[v_]=Integrate[%,v]
```
The following is by direct integration using the built-in Integrate[ ].
```
Clear[x,u,v]
ClearAll[f]
f[x_]=(x-2)/Sqrt[5+4*x-x^2];
Integrate[f[x],x]
```

35-15.2. Use **Integrate**[ ].

35-15.3. Use **Integrate**[ ].

35-15.4. Use **Integrate**[ ].

35-15.5. Program given below.

```
f[x_]=Log[x]^6/x^2;
```
This tests the limit of f(x) as x-> infinity. It is necessary that f(x)-> 0 but not sufficient for the integral to exist.
```
Limit[f[x],x->Infinity]
g[a_]=Integrate[Log[x]^6/x^2,{x,1,a}]
N[g[10]]
N[g[20]]
```
The following graph indicates that f(x) tends to 0 as x-> infinity.
```
Plot[f[x],{x,1,100}]
D[f[x],x]
NSolve[f'[x]==0,x]
```

35-16.1. Program and output are given below.

The solution of the equation is given by DSolve[ ].

In[18]:=

```
sol=DSolve[x'[t]+x[t]==0,x[t],t]
```

Out[18]=

$$\{\{x[t] \rightarrow \frac{C[1]}{E^t}\}\}$$

The solutions for the different initial values can be put in one Plot[ ] command.

In[19]:=

```
Plot[{sol[[1,1,2]]/.C[1]->0,sol[[1,1,2]]/.C[1]->1,
sol[[1,1,2]]/.C[1]->2},{t,0,2},AxesLabel->{t,x}];
```

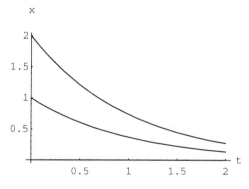

35-16.2. Use **DSolve[ ]** to solve the differential equation. Plotting can be achieved by adapting the program in the previous project.

35-17.1. Program and output shown below.

```
In[24]:=
sol=DSolve[{2*x''[t]+3*x'[t]+x[t]==Cos[t],
x'[0]==0,x[0]==1},x[t],t]
```

```
Out[24]=
```

$$\{\{x[t] \rightarrow \frac{-1}{2\,E^t} + \frac{8}{5\,E^{t/2}} + \frac{-Cos[t] + 3\,Sin[t]}{10}\}\}$$

```
In[25]:=
graph=Plot[sol[[1,1,2]],
{t,0,50},AxesLabel->{t,x},
PlotStyle->{Thickness[.006]}];
```

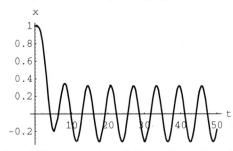

The forced periodic output is quickly achieved as the transient dies out.

35-17.2. Use **DSolve**[ ].

35-18.1. Program given below.

This is the solution of the linearised pendulum equation for simple harmonic motion.

```
sol=DSolve[{x''[t]+x[t]==0,x[0]==1,
x'[0]==0},x[t],t]

graph1=Plot[sol[[1,1,2]],{t,0,10},PlotStyle->
RGBColor[1,0,0]]
```

The following equation is the nonlinear pendulum equation solved numerically for comparison. Both solutions are shown finally together.

```
appsol=NDSolve[{x''[t]+Sin[x[t]]==0,x[0]==1,
x'[0]==0},x,{t,0,10}]

graph2=Plot[Evaluate[x[t]/.appsol],{t,0,10}]

Show[graph1,graph2]
```

35-19.1. Program given below.

```
x1=3/Sqrt[2];y1=3/Sqrt[2];x2=2;y2=0;x3=0;y3=-1;

ArgandPhasor={Thickness[.01],Line[{{0,0},{x1,y1},{x1+x2,
y1+y2},{x1+x2+x3,y1+y2+y3}}]};

Show[Graphics[ArgandPhasor],Axes->True,AspectRatio->
Automatic]

Print[{x1+x2+x3,y1+y2+y3}//N]
```

35-20.1. Requires package **Graphics`PlotField`** and then **PlotVectorField**[ ] and **DSolve**[ ] for the solutions of the differential equation.

35-20.2. As in 20.1.

35-20.3. Program listing given below.

```
f[x_,y_]=x*y^2;
h=.2;
y[0]=1;
x[n_]=n*h;
y[n_]:=y[n]=y[n-1]+h*f[x[n-1],y[n-1]];

yvalue=Table[y[i],{i,1,6}]

points=Table[{x[i-1],yvalue[[i]]},{i,1,6}]
Euler=ListPlot[points]
```

Euler gives a list of points computed by Euler's method.

```
Clear[f,x,y]

DSolve[{y'[x]==x*y[x]^2,y[0]==1},y[x],x]
```

The solution of the differentail equation is given by Exact.

```
Exact=Plot[1/(1-0.5*x^2),{x,0,1},PlotStyle->
RGBColor[1,0,0]]

Show[Euler,Exact]
```

35-20.4. Numerical solution of the differential equation is given by **NDSolve**[ ] and the plotting by **ParametricPlot**[ ].

35-21.1. Sample program shown.

See Example 21.2. The second-order equation is replaced by two first-order differential equations.

```
sol1=NDSolve[{x'[t]==y[t],
y'[t]==-2 x[t]^3,x[0]==0.3,y[0]==0},{x,y},{t,0,25}]
d1=ParametricPlot[Evaluate[{x[t],y[t]}/.sol1],{t,0,25},
AspectRatio->Automatic,PlotRange->All,DisplayFunction->
Identity]
sol2=NDSolve[{x'[t]==y[t],
y'[t]==-2 x[t]^3,x[0]==.6,y[0]==0},{x,y},{t,0,20}]
d2=ParametricPlot[Evaluate[{x[t],y[t]}/.sol2],{t,0,20},
AspectRatio->Automatic,PlotRange->All,DisplayFunction->
Identity]
sol3=NDSolve[{x'[t]==y[t],
y'[t]==-2 x[t]^3,x[0]==.9,y[0]==0},{x,y},{t,0,16}]
d3=ParametricPlot[Evaluate[{x[t],y[t]}/.sol3],{t,0,16},
AspectRatio->Automatic,PlotRange->All,DisplayFunction->
Identity]
sol4=NDSolve[{x'[t]==y[t],
y'[t]==-2 x[t]^3,x[0]==1.2,y[0]==0},{x,y},{t,0,12}]
d4=ParametricPlot[Evaluate[{x[t],y[t]}/.sol4],{t,0,12},
AspectRatio->Automatic,PlotRange->All,DisplayFunction->
Identity]
```

The final Show[ ] displays 4 phase paths.

```
Show[d1,d2,d3,d4,AxesLabel->{x,y},
DisplayFunction->$DisplayFunction]
```

35-21.2. See the previous program.
The next eight projects require the package **Calculus`LaplaceTransform`**.

35-22.1. The transforms are given by **LaplaceTransform**[ ], but use **Integrate**[ ] instead for the finite transform.

35-22.2. Inverse transforms are given by **InverseLaplaceTransform**[ ].

35-22.3. Sample program below.

Requires package Calculus`LaplaceTransform`.

```
<<Calculus`LaplaceTransform`
Solution of second-order equation.
Clear[a,w,s,t,x]
lap=LaplaceTransform[x''[t]+2*x'[t]+x[t]-
a*Cos[w*t],t,s]
lap1=lap /.{LaplaceTransform[x[t],t,s]->lapx,
x[0]->0,x'[0]->0}
lap3=Solve[lap1==0,lapx]
solution=InverseLaplaceTransform[lap3[[1,1,2]],s,t]
a=1;w=1;
```

Inputs and outputs are shown in the diagram.

```
Plot[{solution,Cos[t]},{t,0,30},AxesLabel->{x,t},
PlotRange->{-1.5,1.5},PlotPoints->50,
PlotStyle->{RGBColor[1,0,0],RGBColor[0,0,1]}]
Clear[a,w]
```

35-22.4. Use **InverseLaplaceTransform**[ ].

35-22.5. Use **LaplaceTransform**[ ].

35-23.1. Additionally requires package **Calculus`DiracDelta`**. See 22.3.

35-23.2. Additionally requires package **Calculus`DiracDelta`**. See 22.3.

35-23.3. Define convolution as **Integrate**$[f[t-u]*g[u], \{u,0,t\}]$, and then apply **LaplaceTransform** of the convolution.

35-24.1. Uses **Plot**[ ], **Integrate**[ ], **Sum**[ ], and **Show**[ ].

35-24.2. See the previous project.

35-24.3. See the following program.

Requires package Calculus`FourierTransform`.

```
<<Calculus`FourierTransform`
```

The function is even. Hence only Fourier cosine coefficients are present.

```
Remove[fs]
fs=FourierCosSeriesCoefficient[x^6-5*Pi^2*x^4+
7*Pi^4*x^2,{x,-Pi,Pi},n]
rules1={Cos[Pi n_]->(-1)^n,Sin[Pi n_]->0}
fs=Simplify[fs/.rules1]
a0=FourierCosSeriesCoefficient[x^6-5*Pi^2*x^4+
7*Pi^4*x^2,{x,-Pi,Pi},0]
```

The sum of the series can be obtained by putting x=0 in the Fourier series: the answer is 31(Pi)^6/30240.

35-25.1. Program and partial output shown below.

Requires package Graphics`ParametricPlot`

```
<<Graphics`ParametricPlot3D`
```

This plots the surface.

In[29]:=
```
ParametricPlot3D[{r*Cos[u],r*Sin[u],r^2*Cos[2*u]},
{r,0,1},{u,0,2*Pi},AxesLabel->{x,y,z},
PlotPoints->{15,25},Shading->False]
```

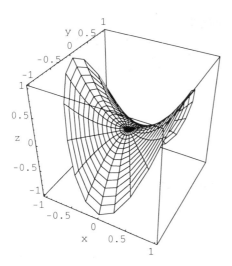

Out[29]=

```
-Graphics3D-
```

This plots the contour curves on the (x,y) plane.

```
ContourPlot[x^2-y^2,{x,-1,1},{y,-1,1},
ContourSmoothing->Automatic,ColorFunction->Hue,
FrameLabel->{x,y}]
```

35-25.2. Requires package **Graphics`ParametricPlot3D`** and then the commands **ParametricPlot3D**[ ] for the surface, and **ContourPlot**[ ] for the contours of the surface.

35-25.3. Use **D**[ ] for partial derivatives.

35-25.4. Program given below.

```
ClearAll[f,fx,fy,x,y]
```
The following defines a function of two variables.
```
f[x_,y_]:=Sin[x*y]
surface=Plot3D[f[x,y],{x,-Pi,Pi},{y,-Pi/2,Pi/2}]
```
The partial derivartives follow.
```
D[f[x,y],x]
fx[x_,y_]:=y*Cos[x*y]
D[f[x,y],y]
fy[x_,y_]:=x*Sin[x*y]
tangentfunction=f[Pi/4,1]+fx[Pi/4,1]*(x-Pi/4)+
fy[Pi/4,1]*(y-1)
```
z=tangent is the equation of the tangent plane.
```
tangent=Plot3D[tangentfunction,{x,-Pi,Pi},
{y,-Pi/2,Pi/2}]
contact={Line[{{Pi/4,1,0},{Pi/4,1,2}}],
Thickness[.02]};
```

Vertical line indicates the point of tangency. The surface and the tangent plane
are put on the same figure. Try other tangent planes and viewpoints.

```
Show[surface,tangent,Graphics3D[contact],
AxesLabel->{x,y,z},BoxRatios->{1,1,.8},
ViewPoint->{1.990,-2.400,1.750}]
```

35-25.5. Program listing given below.

```
ClearAll[f]
f[x_,y_]:=.3x^3+.2*y^2-x^2*y-x*y+x+2*y

fx=D[f[x,y],x]

fy=D[f[x,y],y]
```

The stationary values are found numerically by NSolve[ ].

```
NSolve[{fx==0,fy==0},{x,y}]
```

The following gives a contour plot of the surface showing all the stationary
points.

```
ContourPlot[f[x,y],{x,-3,3},{y,-9,3},Contours->30,
ContourSmoothing->Automatic,
FrameLabel->{x,y},PlotPoints->20]
```

The next cell checks formula (25.9) for the first stationary point.

```
(D[f[x,y],x,x]*D[f[x,y],y,y]-D[f[x,y],x,y]^2)/.
{x->-0.620464,y->-5.58872}
D[f[x,y],x,x]/.{x->-0.620464,y->-5.58872}
D[f[x,y],y,y]/.{x->-0.620464,y->-5.58872}
```

(0.620464,-5.58872) is a minimum.

Apply the formula to the other two points.

35-25.6. Program listing given below.

The command Fit[ ] does the least squares for you.

```
Fit[{{0,1.1},{1,2},{2,2.9},{3,3.9},{4,5},{5,5.1}},
{1,x},x]
```

The following plots the straight line.

```
leastsquares=Plot[%,{x,-0.5,5.5}]

Data={{0,1.1},{1,2},{2,2.9},{3,3.9},{4,5},{5,5.1}}

dataplot=ListPlot[Data,AxesOrigin->{0,0},
PlotRange->{-.5,7.5},PlotStyle->PointSize[0.01]]
```

Show[ ] displays both the data and the straight line fit to the data.

```
Show[dataplot,leastsquares]
```

35-26.1. Program listing given below.

Curves given by f(x,y)=c.

```
f[x_,y_]:=y*Exp[-x]

D[y[x]*Exp[-x],x]

diff=Solve[%==0,y'[x]]

diff[[1,1,2]]

orthogonal=1/diff[[1,1,2]]
```

Differential equation of the orthogonal trajectories.

```
DSolve[y'[x]==-orthogonal,y[x],x]
```

'Curves' are solutions of the differential equation, and 'trajectories' are their orthogonal trajectories.

```
curves=ContourPlot[f[x,y],{x,-2,2},{y,-2,2},
ContourShading->False,
ContourSmoothing->Automatic,Contours->11,
PlotPoints->25]
```

```
trajectories=ContourPlot[y^2+2*x,{x,-2,2},{y,-2,2},
ContourShading->False,ContourSmoothing->Automatic,
PlotPoints->25]
```

```
Show[curves,trajectories]
```

35-27.1. Program listing given below.

Requires package Graphics`ImplicitPlot`.

The problem is to find where f(x,y) is stationary subject to the condition g(x,y)=0.

```
<<Graphics`ImplicitPlot`
```

```
ClearAll[f,g]
```

```
f[x_,y_]:=x^3-2*x*y-x+3*y^2
g[x_,y_]:=x^2+2*y^2-1
```

```
lagrange=ContourPlot[f[x,y],{x,-1.5,1.5},
{y,-1.5,1.5},
ContourShading->False,Contours->15,
ContourSmoothing->
Automatic,PlotPoints->25,AxesLabel->{x,y}]
```

```
constraint=ImplicitPlot[g[x,y]==0,{x,-1.5,1.5},
{y,-1.5,1.5},PlotStyle->RGBColor[1,0,0],
AxesLabel->{x,y}]
```

The following figure shows the intersection of the contours of f(x,y) and the curve g(x,y)=0. Tangencies between them indicate approximate stationary values which are then used to find numerical locations using a root-finding routine. There are four stationary values.

```
Show[lagrange,constraint]
```

```
D[f[x,y]-p*g[x,y],x]
```

```
D[f[x,y]-p*g[x,y],y]
```

The approximations for x and y are taken from the figure above.

```
FindRoot[{-1-2*p*x+3*x*x-2*y==0,
-2*x+6*y-4*p*y==0,x^2+2*y^2==1},{x,.2},{y,-.6},
{p,1}]
```

```
FindRoot[{-1-2*p*x+3*x*x-2*y==0,
-2*x+6*y-4*p*y==0,x^2+2*y^2==1},{x,1},{y,0},{p,1},
MaxIterations->400]
```

```
FindRoot[{-1-2*p*x+3*x*x-2*y==0,
-2*x+6*y-4*p*y==0,x^2+2*y^2==1},{x,-1.3},{y,-.3},
{p,1}]
```

```
FindRoot[{-1-2*p*x+3*x*x-2*y==0,
-2*x+6*y-4*p*y==0,x^2+2*y^2==1},{x,0.5},{y,-0.8},
{p,1.4}]
```

35-28.1. Requires package **Calculus`VectorAnalysis`** and then **Grad[ ]**.

35-28.2. Program and output displayed.

> This diagram shows the intersection of the plane and the cylinder in Example 28.9.

In[1]:=

```
cylinder=ParametricPlot3D[{Sin[t],Cos[t],u},
{t,0,2*Pi},{u,-2,3},AxesLabel->{x,y,z},
DisplayFunction->Identity]
```

In[2]:=

```
plane=ParametricPlot3D[{u*Sin[t],u*Cos[t],
1-u*Sin[t]-u*Cos[t]},{t,0,2*Pi},{u,-2,2},
AxesLabel->{x,y,z},DisplayFunction->Identity]
```

In[3]:=

```
Show[cylinder,plane,ViewPoint->{1.929,-1.306,1.553},
DisplayFunction->$DisplayFunction,Shading->False]
```

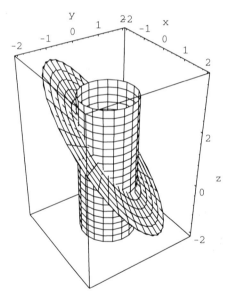

35-28.3. Requires package **Graphics`ImplicitPlot`**.

35-29.1. Use **Integrate[ ]** and **Plot3D[ ]**.

35-30.1. Program and output displayed.

In[17]:=

```
ClearAll[x,y,z,r,t,f]
```

In[18]:=

```
x[t_]:=t;y[t_]=t;z[t_]=t;
f[x_,y_,z_]:={x*y,y*z,x*(z-y)}
f[x[t],y[t],z[t]]
r[x_,y_,z_]:={x,y,z}
```

Out[20]=

$$\{t^2, t^2, 0\}$$

First line integral

```
Integrate[f[x[t],y[t],z[t]].D[r[x[t],y[t],z[t]],t],
{t,0,1}]
```

Out[22]=

$$\frac{2}{3}$$

In[23]:=

```
ClearAll[x,y,z,f,r]
```

In[24]:=

```
x[t_]:=t^2;y[t_]=t^3;z[t_]=t^4;
f[x_,y_,z_]:={x*y,y*z,x*(z-y)}
f[x[t],y[t],z[t]]
r[x_,y_,z_]:={x,y,z}
```

Out[26]=

$$\{t^5, t^7, t^2(-t^3 + t^4)\}$$

Second line integral

```
Integrate[f[x[t],y[t],z[t]].D[r[x[t],y[t],z[t]],t],
{t,0,1}]
```

Out[28]=

$$\frac{341}{630}$$

The two paths of integration are shown below

In[29]:=

```
ParametricPlot3D[{{t,t,t},{t^2,t^3,t^4}},{t,0,1},
AxesLabel->{x,y,z}]
```

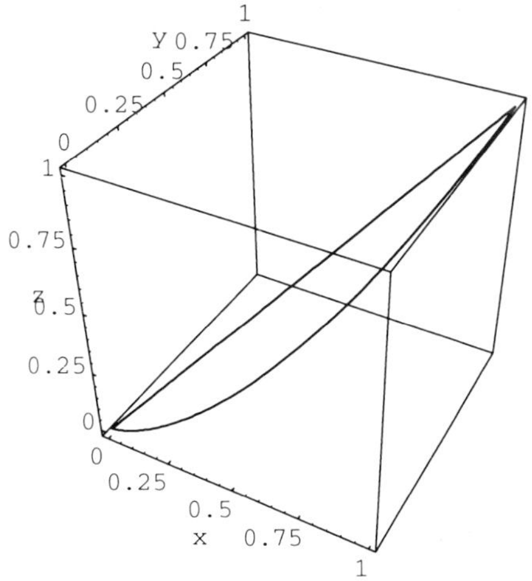

35-31.1. Use **Intersection[ ]** and **Union[ ]**. The command **Length[ ]** gives the number of elements in a set.

35-31.2. Use **Table[ ]** to generate the sets, and then the distributive law on **Intersection[ ]** and **Union[ ]**.

35-33.1. Requires the package **DiscreteMath`Combinatorica`**. The tests are **EulerianQ[ ]** and **HamiltonianQ[ ]**.

35-33.2. Requires the package **DiscreteMath`Combinatorica`**. The test for planarity is **PlanarQ[ ]**.

  The remaining projects require the package **DiscreteMath`RSolve`**.

35-34.1. Use **RSolve[ ]** to solve the difference equation.

35-34.2. Use **RSolve[ ]**.

35-34.3. Use **RSolve[ ]**.

35-34.4. A sample program and output is shown below.

        The difference equation is u(n+1)=-ku(n)+k with k=1/2.

In[30]:=

```
k:=1/2
```

In[31]:=

```
f[x_]:=-k*x+k
```

In[32]:=

```
x:=3/4
```

In[36]:=

```
cobweb1=Table[Line[{{Nest[f,x,n],Nest[f,x,n]},
{Nest[f,x,n],Nest[f,x,n+1]},
{Nest[f,x,n+1],Nest[f,x,n+1]}}],{n,4}];
```

In[34]:=

```
diff1=Plot[-k*x+k,{x,0,1},DisplayFunction->Identity]

Show[diff1,
Graphics[Line[{{x,f[x]},{f[x],f[x]}}]],
Graphics[Line[{{0,0},{0.75,0.75}}]],
Graphics[cobweb1],
AspectRatio->Automatic,
DisplayFunction->$DisplayFunction]
```

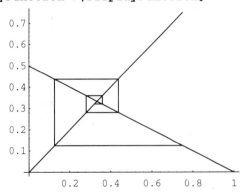

This solution is stable. Find other cobwebs for k=3/2 and k=1.

35-34.4 See 34.4 for the programming method.

Index

Trigonometric formulae

(a) Relation between trigonometric functions

$\sin^2 A + \cos^2 A = 1$,

$\tan A = \sin A/\cos A$; $\quad \sec A = 1/\cos A$; $\quad \operatorname{cosec} A = 1/\sin A$.

(b) Addition formulae

$\sin(A \pm B) = \sin A \cos B \pm \cos A \sin B$,

$\cos(A \pm B) = \cos A \cos B \mp \sin A \sin B$,

$\tan(A \pm B) = (\tan A \pm \tan B)/(1 \mp \tan A \tan B)$.

(c) Addition formulae: special cases

$\sin 2A = 2 \sin A \cos A$,

$\cos 2A = \cos^2 A - \sin^2 A$

$\qquad = 2 \cos^2 A - 1 = 1 - 2 \sin^2 A$,

$\tan 2A = 2 \tan A/(1 - \tan^2 A)$,

$\sin 3A = 3 \sin A - 4 \sin^3 A$,

$\cos 3A = 4 \cos^3 A - 3 \cos A$.

(d) Product formulae

$\sin A \sin B = \frac{1}{2}[\cos(A - B) - \cos(A + B)]$,

$\cos A \cos B = \frac{1}{2}[\cos(A - B) + \cos(A + B)]$,

$\sin A \cos B = \frac{1}{2}[\sin(A - B) + \sin(A + B)]$.

$\sin C + \sin D = 2 \sin \frac{1}{2}(C + D) \cos \frac{1}{2}(C - D)$,

$\sin C - \sin D = 2 \sin \frac{1}{2}(C - D) \cos \frac{1}{2}(C + D)$,

$\cos C + \cos D = 2 \cos \frac{1}{2}(C + D) \cos \frac{1}{2}(C - D)$,

$\cos C - \cos D = -2 \sin \frac{1}{2}(C + D) \sin \frac{1}{2}(C - D)$.

(e) Product formulae: special cases

$\sin^2 A = \frac{1}{2}(1 - \cos 2A)$,

$\cos^2 A = \frac{1}{2}(1 + \cos 2A)$,

$\sin^3 A = \frac{1}{4}(3 \sin A - \sin 3A)$,

$\cos^3 A = \frac{1}{4}(3 \cos A + \cos 3A)$.

(f) Cosine rule

For a triangle ABC, $c^2 = a^2 + b^2 - 2ab \cos C$.

Areas and volumes

(a) The area of a triangle is $\frac{1}{2}bh$, where b is the length of one side and h its height from that side.

(b) The circumference of a circle is $2\pi r$, where r is its radius.

(c) The area of a circle is πr^2, where r is its radius.

(d) The area of a circle sector is $\frac{1}{2}r^2\theta$, where r is its radius and θ the angle of the sector in radians.

(e) The volume of a sphere is $\frac{4}{3}\pi r^3$, where r is its radius.

(f) The surface area of a sphere is $4\pi r^2$, where r is its radius.

(g) The volume of a cone is $\frac{1}{3}Ah$, where h is its height and A the cross-sectional area of its base.

(h) The area of an ellipse is πab, where a and b are the lengths of its semi-axes.

A table of derivatives

y	$\dfrac{dy}{dx}$
c (constant)	0
x^n (n any constant)	nx^{n-1}
e^{ax}	$a\,e^{ax}$
k^x ($k > 0$)	$k^x \ln k$
$\ln x$ ($x > 0$)	x^{-1}
$\sin ax$	$a \cos ax$
$\cos ax$	$-a \sin ax$
$\tan ax$	$a/\cos^2 ax$
$\cot ax$	$-a/\sin^2 x$
$\sec ax$	$(a \sin ax)/\cos^2 ax$
$\operatorname{cosec} ax$	$-(a \cos ax)/\sin^2 ax$
$\arcsin ax$	$a/(1 - a^2x^2)^{\frac{1}{2}}$
$\arccos ax$	$-a/(1 - a^2x^2)^{\frac{1}{2}}$
$\arctan ax$	$a/(1 + a^2x^2)$
$\sinh ax$	$a \cosh ax$
$\cosh ax$	$a \sinh ax$
$\tanh ax$	$a/\cosh^2 ax$
$\sinh^{-1} ax$	$a/(1 + a^2x^2)^{\frac{1}{2}}$
$\cosh^{-1} ax$	$a/(a^2x^2 - 1)^{\frac{1}{2}}$
$\tanh^{-1} ax$	$a/(1 - a^2x^2)$

A table of integrals

$f(x)$	$\displaystyle\int f(x)\,dx$ (C is an arbitrary constant.)
x^m ($m \neq -1$)	$\dfrac{1}{m+1} x^{m+1} + C$
x^{-1}	$\ln\lvert x\rvert + C$, or $\ln\lvert Cx\rvert$
e^{ax}	$(1/a)\,e^{ax} + C$
k^x ($k > 0$)	$k^x/\ln k + C$
$\ln x$ ($x > 0$)	$x \ln x - x + C$
$\sin ax$	$-(1/a) \cos ax + C$
$\cos ax$	$(1/a) \sin ax + C$
$\tan ax$	$-(1/a) \ln\lvert\cos ax\rvert + C$ or $-(1/a) \ln\lvert C \cos ax\rvert$
$\cot ax$	$(1/a) \ln\lvert\sin ax\rvert + C$ or $(1/a) \ln\lvert C \sin ax\rvert$
$\sec ax$	$-(1/2a) \ln[(1 - \sin ax)/(1 + \sin ax)] + C$
$\operatorname{cosec} ax$	$(1/2a) \ln[(1 - \cos ax)/(1 + \cos ax)] + C$
$1/(x^2 + a^2)$	$(1/a) \arctan(x/a) + C$
$1/(x^2 - a^2)$	$(1/2a) \ln\lvert(x - a)/(x + a)\rvert + C$ or $(1/a) \tanh^{-1}(x/a) + C$
$1/(a^2 - x^2)^{\frac{1}{2}}$	$\arcsin(x/a) + C$ (or $-\arccos(x/a) + C$)
$1/(a^2 + x^2)^{\frac{1}{2}}$	$(1/a) \sinh^{-1}(x/a) + C$ or $\ln[x + (x^2 + a^2)^{\frac{1}{2}}] + C$
$1/(x^2 - a^2)^{\frac{1}{2}}$	$\ln[x + (x^2 - a^2)^{\frac{1}{2}}] + C$